ISOLATION AND PURIFICATION OF PROTEINS

BIOTECHNOLOGY AND BIOPROCESSING SERIES

1. Membrane Separations in Biotechnology, *edited by W. Courtney McGregor*
2. Commercial Production of Monoclonal Antibodies: A Guide for Scale-Up, *edited by Sally S. Seaver*
3. Handbook on Anaerobic Fermentations, *edited by Larry E. Erickson and Daniel Yee-Chak Fung*
4. Fermentation Process Development of Industrial Organisms, *edited by Justin O. Neway*
5. Yeast: Biotechnology and Biocatalysis, *edited by Hubert Verachtert and René De Mot*
6. Sensors in Bioprocess Control, *edited by John V. Twork and Alexander M. Yacynych*
7. Fundamentals of Protein Biotechnology, *edited by Stanley Stein*
8. Yeast Strain Selection, *edited by Chandra J. Panchal*
9. Separation Processes in Biotechnology, *edited by Juan A. Asenjo*
10. Large-Scale Mammalian Cell Culture Technology, *edited by Anthony S. Lubiniecki*
11. Extractive Bioconversions, *edited by Bo Mattiasson and Olle Holst*
12. Purification and Analysis of Recombinant Proteins, *edited by Ramnath Seetharam and Satish K. Sharma*
13. Drug Biotechnology Regulation: Scientific Basis and Practices, *edited by Yuan-yuan H. Chiu and John L. Gueriguian*
14. Protein Immobilization: Fundamentals and Applications, *edited by Richard F. Taylor*
15. Biosensor Principles and Applications, *edited by Loïc J. Blum and Pierre R. Coulet*
16. Industrial Application of Immobilized Biocatalysts, *edited by Atsuo Tanaka, Tetsuya Tosa, and Takeshi Kobayashi*
17. Insect Cell Culture Engineering, *edited by Mattheus F. A. Goosen, Andrew J. Daugulis, and Peter Faulkner*
18. Protein Purification Process Engineering, *edited by Roger G. Harrison*
19. Recombinant Microbes for Industrial and Agricultural Applications, *edited by Yoshikatsu Murooka and Tadayuki Imanaka*
20. Cell Adhesion: Fundamentals and Biotechnological Applications, *edited by Martin A. Hjortso and Joseph W. Roos*
21. Bioreactor System Design, *edited by Juan A. Asenjo and José C. Merchuk*
22. Gene Expression in Recombinant Microorganisms, *edited by Alan Smith*
23. Interfacial Phenomena and Bioproducts, *edited by John L. Brash and Peter W. Wojciechowski*
24. Metabolic Engineering, *edited by Sang Yup Lee and Eleftherios T. Papoutsakis*

25. Biopharmaceutical Process Validation, *edited by Gail Sofer and Dane W. Zabriskie*
26. Membrane Separations in Biotechnology: Second Edition, Revised and Expanded, *edited by William K. Wang*
27. Isolation and Purification of Proteins, *edited by Rajni Hatti-Kaul and Bo Mattiasson*

ADDITIONAL VOLUMES IN PREPARATION

ISOLATION AND PURIFICATION OF PROTEINS

edited by
Rajni Hatti-Kaul
Bo Mattiasson

Lund University
Lund, Sweden

CRC Press is an imprint of the
Taylor & Francis Group, an **informa** business

CRC Press
Taylor & Francis Group
6000 Broken Sound Parkway NW, Suite 300
Boca Raton, FL 33487-2742

First issued in paperback 2019

ISBN-13: 978-0-8247-0726-2 (hbk)
ISBN-13: 978-0-367-39549-0 (pbk)

Library of Congress Cataloging-in-Publication Data
A catalog record for this book is available from the Library of Congress.

Visit the Taylor & Francis Web site at
http://www.taylorandfrancis.com

and the CRC Press Web site at
http://www.crcpress.com

Preface

The whole area of protein purification has come a long way from being just a laboratory practice. Since earlier times, purification has remained an essential practice to enable one to characterize the proteins by their structural and functional properties, which may be necessary, for determining their role in cells and tissues or their potential for a certain application, etc. Today, it constitutes a key segment of protein production, which is a major biotechnology industry. Hence, increasing demands for process efficiency, predictability, and economy have encouraged constant improvements in the existing separation technology, and also development of some innovations.

Although the various separation techniques used in protein purification have been dealt with in great detail in a number of books, this new book can be seen as a supplement to already existing volumes, providing an overall approach to how to tackle the task of protein purification. The reader is given an idea of where the area of separation technology started from and where it stands today. The influence of other fields such as genetic technology, biochemical engineering, polymer chemistry, computer modeling, and automation on downstream processing is also presented.

This book provides a picture of the state of the art in separation technology as applied to protein purification. The conventional techniques used for protein separation, although still being used, are treated in brief as sufficient documentation about them already exists. With those techniques as the starting point, we focus on the developments that have taken place around them to make them more efficient and cost-effective. An important issue is the trend toward designing compact downstream processing. The possibility of doing this by introduction of a separation technique that fulfils the functions of two or more separate techniques or by integrating the different techniques into one unit operation is shown. Recombinant DNA technology has had a great impact on bringing forward the limitations of existent downstream processing. The ways in which genetic techniques can be used to simplify the purification are discussed. Information is also given on techniques available for process monitoring and possibilities for process control, which becomes a necessity for large-scale processes and for integration of different stages of a purification scheme. Finally, essential information for commercial protein production, e.g., regulatory and economic considerations, and protein formulation, is included.

The underlying theme of the book is that it may not always be necessary to follow a long and difficult path to obtain a pure protein; proper planning and a smart choice and integration of separation techniques can be used to fulfil the need for an efficient, clean, and cost-effective process.

Leading researchers from both universities and industries have contributed to the chapters forming the book. They have provided first-hand knowledge of their area of expertise.

The book will arouse the interest of scientists in both universities and industries dealing with protein production. It can also be used as a reference or textbook for teachers and students in the areas of biochemistry, microbiology, and biotechnology.

Rajni Hatti-Kaul
Bo Mattiasson

Contents

Contributors

M. R. Aires-Barros *Instituto Superior Técnico, Lisbon, Portugal*
Gene Burton *Bayer Corporation, Berkeley, California, U.S.A.*
J. M. S. Cabral *Instituto Superior Técnico, Lisbon, Portugal*
Richard Carrillo *Bayer Corporation, Berkeley, California, U.S.A.*
Howard A. Chase *University of Cambridge, Cambridge, England*
Robert H. Clemmitt *BioProducts Laboratory, Elstree, Hertfordshire, England*
M. T. Cunha *Instituto Superior Técnico, Lisbon, Portugal*
Ruth Freitag *Swiss Federal Institute of Technology, Lausanne, Switzerland*
Igor Yu. Galaev *Lund University, Lund, Sweden*
Siddhartha Ghose* *Aston University, Birmingham, England*
Munishwar Nath Gupta *Indian Institute of Technology, New Delhi, India*
Per-Erik Gustavsson *Lund University, Lund, Sweden*
Rajni Hatti-Kaul *Lund University, Lund, Sweden*
Sophia Hober *Royal Institute of Technology, Stockholm, Sweden*
Drew N. Kelner *Bayer Corporation, Berkeley, California, U.S.A.*
Woo-Sik Kim *Kyunghee University, Suwon, Korea*
Ashok Kumar *Lund University, Lund, Sweden*
Peter Kumpalume* *Aston University, Birmingham, England*

**Current affiliation*: University of Cambridge, Cambridge, England

Per-Olof Larsson *Lund University, Lund, Sweden*
E. K. Lee *Hanyang University, Ansan, Korea*
Bo Mattiasson *Lund University, Lund, Sweden*
Ricardo A. Medronho *Federal University of Rio de Janeiro, Rio de Janeiro, Brazil*
Dieter Melzner *Sartorius AG, Göttingen, Germany*
Joakim Nilsson *Royal Institute of Technology, Stockholm, Sweden*
Per-Åke Nygren *Royal Institute of Technology, Stockholm, Sweden*
Kerstin Plate *University of Hannover, Hannover, Germany*
Oskar-Werner Reif *Sartorius AG, Göttingen, Germany*
Ipsita Roy *Indian Institute of Technology, New Delhi, India*
Gail Sofer *Bioreliance Corporation, Rockville, Maryland, U.S.A.*
Stefan Ståhl *Royal Institute of Technology, Stockholm, Sweden*
Mathias Uhlén *Royal Institute of Technology, Stockholm, Sweden*
Roland Ulber *University of Hannover, Hannover, Germany*
D. Q. Wang *Bayer Corporation, Berkeley, California, U.S.A.*
Wei Wang *Bayer Corporation, Berkeley, California, U.S.A.*

Introduction

Isolation and purification comprise the downstream processing (DSP) stage of the manufacture of bioproducts from raw materials such as fermentation media, cell homogenates, or other biological materials. Isolation involves primary recovery operations, such as broth conditioning, clarification, cell rupture, debris and nucleic acid removal, refolding, and concentration, with the objective of obtaining the product in solution from the production system. The resulting process stream contains the target protein mixed with other proteins and also other contaminants that differ in certain physicochemical properties. These differences are exploited in the purification process, in which a sequence of high-resolution operations, invariably based on chomatography, brings the product to a specified level of purity. The degree of purification required is dictated by the ultimate application of the product. The proteins meant for the therapetic use require extremely high levels (greater than 99%) of purity while industrial enzymes are relatively crude, concentrated solutions formulated to meet actual customer requirements for activity and stability. The product purity in turn determines the cost of purification, higher purity demanding higher costs due to the larger number of processing steps required and resultant lower yields.

Downstream processing of proteins started to attract attention toward the end of the 20th century when advances in recombinant DNA technology made it possible to produce almost any protein in a host of choice. Several protein-based therapeutics including a variety of monclonal antibodies have already reached the market, while many more are in different stages of development in biotechnology companies around the world. With the increase in industrial production of high-value proteins, it was realized that the isolation and purification accounted for a significant fraction—in some cases up to 80%—of the total costs of the production process (Bonnerjea et al. 1986; Wheelwright 1987; Curling 2001). Thus, there has been a general consensus that reducing the number of separation steps and maximizing the yield at each step are necessary for an economical process. At the same time, with growing requirements of product purity and safety, increasing attention is being given to the importance of elimination from the process stream of contaminants arising from the production host such as residual host cell proteins, viruses, endotoxins and nucleic acids; culture media; leachates from the separation media, etc.; and also of the various isoforms of the product arising from posttranslational modification, denaturation, aggregation, etc. (Kalyanpur 2002). Downstream processing has thus been challenged with demands of high yields, resolving power, and cost efficiency. This has triggered remarkable developments in improvising process tools and innovative strategies for protein separation.

This book provides a comprehensive description of the various unit operations used for protein isolation and purification. Chapter 1 details the means for disruption of cells or tissue for making the target protein accessible to purification. Particular attention is given to chromatography as it forms the backbone for high-resolving protein separations (Chapters 2, 3, 10, 11, and 12). The genetic strategies used to modify recombinant proteins so as to facilitate selective capture are presented in Chapter 4. The techniques commonly employed for removal of solids (Chapters 5 and 6) and volume reduction (Chapters 6 and 7), as well as the techniques used for direct capture of the target protein from unclarified feedstock (Chapters 9 and 10), are described. A relatively recent use of crystallization as a powerful means of purification as an alternative to chromatography is discussed (Chapter 8.) Also described are the analytical tools available for process monitoring of downstream processing (Chapter 14). Finally, even aspects related to industrial production such as scale up (Chapter 13), product formulation (Chapter 15), and regulatory considerations (Chapter 16) are examined.

1. RATIONAL DESIGN OF MULTISTAGE DOWNSTREAM PROCESSING

The complexity of the raw materials and also that of the proteins, and the latter's limited stability under various environmental conditions, make the design of downstream processing an art. Being a multistage operation, it is worthwhile to conceive a proper downstream scheme at an early stage if large-scale purification is to be developed. Among the various separation techniques, a larger variety is available for laboratory-scale operations in which aspects of cost or recyclability of reagents are less relevant while the options for process-scale operations can be limited. A process that has been designed for small quantities may not be optimal for manufacturing large quantities. Moreoever, the procedure is difficult to change once the purification procedure is set and regulatory approval of the product is in progress. Hence, one should use in the laboratory only such techniques that can realistically be used on a large scale.

There has been a great deal of interest in the development of systematic methods for synthesis of purification. According to Wheelwright (1987), downstream process design starts at the end (i.e., by defining the product: its desired purity, application, stability, etc.). The other prerequisite is the characterization of raw material. Information on the type of material (e.g., bacterial fermentation, mammalian cell culture, transgenics), major contaminants (e.g., host cell proteins, albumin, product variants), and the presence of solid bodies (e.g., cells, membrane debris, inclusion bodies) as well as on physical properties of the product (charge and titration curves, isoelectric point, molecular weight, surface hydrophobicity, specific binding properties, thermal stability, etc.) is helpful in exploring all the separation options available, including the constraints on conditions during processing such as range of pH and temperature.

A set of heuristics has been suggested as a guide to select the individual separation steps (Wheelwright 1987). Some of the recommendations are to choose: (1) separation processes based on different physical properties rather than multiple steps based on the same property, (2) processes that exploit the greatest differences in the physical properties of the product and the impurities, (3) the step reducing the process volume significantly as an early-stage operation, thus leading to a lower cost of processing, and (4) the most expensive step toward the end, when the amount of material to be treated in significantly reduced and most of the impurities have been removed.

Attempts have been made to apply mathematical models and computer-aided design for optimizing individual operations as well as to facilitate synthesis of a whole process (Vasquez-Alvarez et al. 2001). Another significant approach has been the use of artificial intelligence tools (e.g., expert systems that operate through the manipulation of heuristic rules, algebraic equations describing behavior of proteins, and databases that contain the characteristics of protein molecules) (Asenjo and Maugeri 1992). Some examples of expert systems such as Protein Purification Advisor,

Reactivate Planning P8, FPLC Assistant (Pharmacia), and Prot_Ex has been reported (Eriksson et al. 1991a; 1991b; Leser and Asenjo 1992). For implementing an expert system, recovery and purification sections of downstream processing have been treated separately because the information and available heuristic knowledge are different in each part (Leser and Asenjo 1992). For recovery operations, Prot_Ex could use heuristic rules from literature and human experts, while an important amount of quantitative data on physiochemical properties of both the protein and the main contaminating ones was needed to make a good selection of unit operations for purification. Selection of the purification sequence could be based on the separation selection coefficient (SSC) criterion that characterizes the ability of the purification step to separate two proteins or the purity criterion that compares the final purity level obtained after a particular chromatography step (Lienqueo et al. 1999; Lienqueo and Asenjo 2000).

The suitability of each separation step—both individually and in conjunction with the other steps—must be determined, so as to take into account the whole process rather than focusing on refining an isolated step. The possibility of process integration without affecting product purity and safety may also be evaluated. In this direction, one more recent approach has been to introduce certain selectivity (e.g. pseudoaffinity) early in the separation train, in order to reduce volume and remove a large fraction of impurities, while more high resolving techniques can then be applied to the smaller volumes left for processing (Kaul and Mattiasson 1992; Hatti-Kaul and Mattiasson 2001).

2. PRE-DOWNSTREAM FACTORS INFLUENCING PROTEIN ISOLATION

Choice of the raw material for production of a particular protein is determined by many factors, but is currently influenced greatly by recombinant DNA technology. Mammalian cells, *Escherichia coli*, and yeast are the main heterologous hosts for recombinant proteins, while transgenic animals and plants present attractive alternatives for reducing the upstream costs of protein production (Curling 2001; Mison and Curling 2000; Kalyanpur 2002).

Upstream processing and bioreactor conditions influence the DSP significantly and must be taken into consideration during process design. The host organism used often determines the choice of solid–liquid separation method (Chapter 5) and eventual cell lysis method (Chapter 1) to be used. Extra- /intracellular or periplasmic location of the target product, quantity of the product formed, and formation of inclusion bodies are also considerations dependent on the host. Excretion of proteins into extracellular medium or even periplasm is often preferred so as to obviate the harsh cell disruption step, and also due to fewer contaminants that need to be dealt with during purification. On the other hand, expression of recombinant proteins as inclusion bodies in *E. coli* cells is rather commonplace, with the advantage that the protein is quite pure. The recovery of the product in active form, however, is often complicated, requiring solubilization of the aggregated protein under denaturing conditions followed by refolding of the solubilized protein that usually results in low yields of the active protein. Alternative solubilization methods that do not rely on high denaturant concentrations, and development of systems offering efficient and scalable refolding, would benefit industrial implementation of these processes (Hart et al. 1994; De Bernardez Clark 2001; Lee et al. 2001).

The nature of the medium used for fermentation has an effect on the growth form of many microorganisms and influences biomass removal, protein recovery, and even the effluent treatment. High viscosity of the medium will pose problems during biomass removal and so does a rich medium that promotes the formation of colloids.

In view of the above considerations, a strategy adopted by many industries is to express the protein in a standard production host for which cultivation conditions and the nature of contaminants are well characterized. Furthermore, a genetically fused purification tag is often used to distinguish the target protein from the contaminants and thus facilitate selective purification. A number of companies are in the business of providing affinity tag expression/purification systems (Constans 2002b).

3. FROM DISCRETE UNIT OPERATIONS TO INTEGRATED PROCESSING

Downstream processing of a protein has traditionally been planned with the aim of chromatography being the ultimate purification stage. Since chromatography in a packed bed has been the norm, a prerequisite is to process the crude feedstock during the isolation stage to yield a protein preparation in a clear and concentrated form. Developing the isolation stage of downstream processing normally involves choosing between alternative operations available (e.g., between high-pressure homogenization and bead milling for cell disruption, centrifugation and filtration for cell separation, and precipitation and ultrafiltration for concentration). On the other hand, design of the purification stage involves setting up an optimal chromatographic sequence with maximum yield. While size exclusion chromatography is used mainly as a polishing step for removal of aggregates, desalting, and buffer exhange, process scale purification relies on adsorption chromatography in which the separation principles commonly used are ion exchange, hydrophobic interaction, and affinity. Screening of the chromatograhic media is often complicated by the numerous products commercially available (Rathore 2001). Besides selectivity for a particular separation, issues such as flow performance and physical and chemical stability of the matrix need to be considered during selection (Levison et al. 1997).

All the steps in a traditional downstream processing scheme are often performed as discrete packages. To avoid the losses accompanying each step and to achieve better process efficiency, the recommendation has been to introduce integrated and selective processes to fish out the target product from crude feedstock in a minimal number of downstream steps. The integrated processes could involve coupling of a capture step with the bioreactor and/or that of clarification, concentration, and preliminary purification into a single step. Selective approaches for product harvest are achieved by integrating affinity interactions with unit operations used for early stage processing. Several innovative approaches have been developed to meet the new needs (Kaul and Mattiasson 1992; Hatti-Kaul and Mattiasson 2001).

The techniques providing an integrated approach include extraction in aqueous two-phase systems (Hatti-Kaul 2000; Chapter 9) and expanded-bed adsorption chromatography (Mattiasson 1999; Chapter 10). They allow processing of whole-cell broth or homogenates to obtain a sufficiently/partially pure product without the need for individual clarification, concentration, and purification steps. Two-phase extraction has the advantage of having a capacity for high particulate loads and has also been found to be suitable for recovery of proteins from inclusion bodies and membranes (Hatti-Kaul 2000; 2001). However, expanded-bed adsorption seems to have gained favor in the biotechnology industry and its applications are facilitated by the availability of various matrices with different functionalities. The problem that still remains is the fouling of the matrix that can affect the performance for repeated use and is discouraging for process validation.

Extraction and expanded-bed adsorption are rendered more selective by further integration with affinity interactions. Similarily, selectivity has also be incorporated in other high-throughput processes, such as membrane filtration and precipitation to yield membrane affinity filtration (Chapter 6) and affinity precipitation (Chapter 7) processes, respectively. When using an affinity-based approach for "fishing" out the product from a crude material, one is of course limited to the use of affinity ligands that may lack absolute selectivity but are biologically and chemically stable.

Simultaneously, yet another development has been the application of crystallization (Chapter 8), a technique normally applied to ultrapure protein solutions to enable structural determination by X-ray crystallography, as a large-scale purification technique supported by ultrafiltration.

4. IMPROVEMENTS IN PROCESS TOOLS

Bioseparation advances have typically relied on the development of new techniques and materials. Although many of the separation principles have remained unchanged, developments have occurred in improving instrumentation for allowing rapid processing and contained operation, and also in the analytical tools with potential for on-line process monitoring.

Of significance are the developments in membrane processes that are currently used throughout downstream processing for broth clarification, product concentration, buffer exchange, desalting, and sterile filtration. New membrane materials, modules, and process designs have been developed with improvements in selectivity of membranes while maintaining their inherent high-throughput characteristics. High-performance tangential flow filtration and membrane chromatography are emerging new techniques for protein purification (van Reis and Zydney 2001; Kalyanpur 2002).

Developments have been seen in new commercial chromatography matrices based on different organic and inorganic materials. The matrix format is no longer limited to bead-shaped gels; adsorbents are being manufactured as monoliths, disks, and membranes for continous chromatography (Chapters 11 and 12). Matrices with pores in which the binding is not limited by slow process of diffusion have been made [e.g., the POROS and hybrid HyperD sorbents (Chapter 12; Schwartz et al. 2001)]. Some matrices with supermacropores capable of even separating cells have been presented (Arvidsson et al. 2002). Furthermore, different modes of operation have been worked out, such as displacement chromatography and radial flow chromatography. The latter has also led to new column designs, as increase in bed height is not necessary for scale-up.

Other chromatographic separation mechanisms have been investigated in recent years, especially with application for purification of monclonal antibodies that constitute the largest therapeutic products. Improvements have been made in hydroxyapatite chromatography, which can potentially be used for separation of two proteins with similar isoelectric point, molecular weight, and hydrophobicity (Constans 2002a). Another technique is hydrophobic charge induction chromatography, based on the use of dual-mode ligands that are designed to combine a molecular interaction supported by a mild hydrophobic association effect in the absence of salts, and when environmental pH is changed, the ligand becomes ionically charged, resulting in desorption of the protein (Guerrier et al. 2000; 2001; Boschetti 2002).

Much attention has been given to ligand development for affinity chromatography with respect to robustness and selectivity, and also aspects related to validation. Much of the effort is being put into probing the combinatorial libraries based on a variety of chemical motifs for suitable affinity ligands (Amatschek et al. 1999; Romig et al. 1999; Teng et al. 1999; Lowe 2001; Sato et al. 2002). Combinatorial libraries containing random linear or constrained peptides, gene fragments, cDNA, and antibody libraries, presented on biological vehicles—especially bacteriophages—are widely used. From molecular models of the proteins, de novo synthesis of new mimetic chemical entities can be achieved from the molecules selected from chemical combinatorial libraries (Curling 2001). Synthetic chemical ligands (e.g., those based on triazinyl dyes) have been successfully developed for a number of proteins. Combinatorial chemical techniques are now beginning to have an impact on the discovery and design of ligands for glycoproteins, metal binding ligands for fusion peptides, etc. (Lowe 2001). A combination of the structural knowledge obtained by X-ray crystallography, NMR, or homology structures with defined or combinatorial chemical synthesis and advanced computational tools has provided a powerful route to the rational design of affinity ligands for simple protein purification.

5. PROTEIN PURIFICATION IN POST-GENOMIC ERA

As researchers attempt to understand the vast amount of genetic information, proteomics has become a major focus in the biotechnology industry since gene function is derived from the protein product it encodes. The challenge of studying proteins holistically is driving the development of analytical and preparative tools that allow the resolution and characterization of complex sets of protein mixtures in a high-throughput mode and the subsequent purification of target therapeutic protein. High-throughput processing typically involves automated instrumentation. Robotic systems for nucleic acid purification have recently been adapted for protein purification and are available from several companies, including Qiagen, Gilson, and Packard (Lesley 2001). Unlike DNA, however, protein complexity and diversity make the task of parallel processing rather difficult (Lesley 2001). On a small scale, parallel processing usually involves use of a 96-well plate format. Specialized 96-well plates clear cell debris via vacuum filtration and are also used to retain chromatography resin. Cell lysis is typically achieved using a combination of lysozyme and freeze-thaw cycles or non-ionic detergents. Addition of nucleases helps to reduce viscosity and facilitate removal of cell debris at the low g forces commonly used with microtiter plates.

Parallel expression and purification of the gene products are often simplified by utilization of purification tags. At the same time, however, instrumentation in the form of a minaturized high-throughput alternative for evaluation of chromatographic functionalities and binding and wash conditions has been developed for reducing chromatography development time and costs. This involves a family of protein biochips that carry functional groups typical of those used for chromatography adsorbents and that have been designed to bind proteins and peptides from complex mixtures (Santambien et al. 2002). After washing the individual spots under selective conditions, the retained target protein and/or impurity components bound to the array are analyzed by time-of-flight mass spectrometry. Besides allowing extremely rapid determination of the most effective combination of chromatographic modes in a protein purification scheme, the technology can be used for analysis of chromatography fractions to track target and impurity protein during process-scale purification.

It is apparent that advances in materials science will acompany the transformation of genetics and biology into protein products. Nanoscale materials are predicted to have a direct impact on genomics, proteomics, and high-throughput processes. Eventual production of an identified therapeutic product will, in all likelihood, depend on the sophistication of affinity purification.

REFERENCES

Amatschek K, Necina R, Hahn R, Schallau E, Schwinn H, Josic D, Jungbauer A (2000) Affinity chromatography of human blood coagulation factor VIII on monoliths with peptides from a combinatorial library. J High Resol. Chromatogr. 23, 47–58.

Arvidsson P, Plieva F, Svina IN, Lozinsky VI, Fexby S, Bülow L, Galaev IYu, Mattiasson B (2002) Chromatography of microbial cells using continuous macroporous affinity and ion exchange columns. J. Chromatogr. (in press).

Bonnerjea J, Oh S, Hoare M, Dunnill P (1986) Protein purification: the right step at the right time. Bio/Technol. 4, 954–958.

Boschetti E (2002) Antibody separation by hydrophobic charge induction chromatography. Trends Biotechnol. 20, 333–337.

Constans A (2002a) Protein purification I: liquid chromatography. Scientist 40–42.

Constans A (2002b) Protein purification II: affinity tags. Scientist 37–40.

Curling J (2001) Biospecific affinity chromatography: intelligent combinatorial chemistry. Gen. Eng. News 21 (20) 54–54.

De Bernardez Clark E (2001) Protein refolding for industrial processes. Curr. Opin. Biotechnol. 12, 202–207.

Eriksson H, Sandahl K, Forslund G and Österlund B (1991a) Knowledge-based planning for protein purification. Chemom. Intell. Lab Syst. 13, 173–184.

Eriksson H, Sandahl K, Brewer J, Österlund B (1991b) Reactive planning for chromatography. Chemom. Intell. Lab Syst. 13, 185–194.

Guerrier L, Girot P, Schwartz W, Boschetti E (2000) New method for the selective capture of antibodies under physiological conditions. Bioseparation 9, 211–221.

Guerrier L, Flayeux I, Boschetti E (2001) A dual-mode approach to the selective separation of antibodies and their fragments. J. Chromatogr. B 755, 37–46.

Hart RA, Lester PM, Reifsnyder DH, Ogez JR, Builder SE (1994) Large scale in situ isolation of periplasmic IGF-I from *E. coli.* Bio/Technol. 12, 1113–1117.

Hatti-Kaul R, ed (2000) Aqueous Two-Phase Systems: Methods and Protocols. Humana Press, Totowa, NJ.

Hatti-Kaul R (2001) Aqueous two-phase systems: a general overview. Mol. Biotechnol. 19, 269–277.

Hatti-Kaul R, Mattiasson B (2001) Downstream processing in biotechnology. In: Basic Biotechnology, Ratledge C, Kristiansen B, eds, Cambridge University Press, Cambridge, pp. 187–211.

Kalyanpur M (2002) Downstream processing in the biotechnology industry. Mol. Biotechnol. 22, 87–98.

Kaul R, Mattiasson B (1992) Secondary purification. Bioseparation 3, 1–26.

Lee CT, Mackley MR, Stonestreet, Middelberg APJ (2001) protein refolding in an oscillatory flow reactor. Biotechnol. Lett. 23, 1899–1901.

Leser EK, Asenjo JA (1992) Rational design of purification processes for recombinant proteins. J. Chromatogr. 584, 43–57.

Lesley SA (2001) High-throughput proteomics: protein expression and purification in the postgenomic world. Prot. Expr. Purif. 22, 159–164.

Levison PR, Mumford C, Streater M, Brandt-Nielsen A, Pathirana ND, Badger SE (1997) Performance comparison of low pressure ion-exchange chromatography media for protein separation. J. Chromatogr. A 760, 151–158.

Lienqueo ME, Salgado JC and Asenjo JA (1999) An expert system for selection of protein purfication processes: experimental validation. J. Chem. Technol. Biotechnol. 74, 293–299.

Lienqueo ME and Asenjo JA (2000) Use of expert systems for the synthesis of downstream protein processes. Comp. Chem. Eng. 24, 2339–2350.

Lowe CR (2001) Combinatorial approaches to affinity chromatography. Curr Opin Cem Biol 5, 248–256.

Mattiasson B, ed (1999) Expanded Bed Chromatography, Kluwer Academic, Dordrecht.

Mison D, Curling J (2000) The industrial production costs of recombinant therapeutic proteins in transgenic corn. BioPharm. May, pp. 48–54.

Rathore A.S. (2001) Resin screening to optimize chromatographic separations. LCGC 19, 616–622.

Romig TS, Bell C, Drolet DW (1999) Aptamer affinity chromatography: combinatorial chemistry applied to protein purification. J. Chromatogr. B 731, 275–284.

Santambien P, Brenac V, Schwartz WE, Boschetti E, Spencer J (2002) Rapid "on-chip" protein analysis and purification. Gen. Eng. News 22, 44–46.

Sato AK, Sexto DJ, Morganelli LA, Cohe EH, Wu QL, Coley GP, Streltsova Z, Lee SW, Devlin M, DeOliveira DB, Enright J, Kent RB, Wescott CR, Rensohoff TC, Ley AC, Ladner RC (2002) Development of mammalian serum albumin affinity purification media by peptide phage display. Biotechnol. Prog. 18, 182–192.

Schwartz WE, Santambien P, Boschetti E, Tunon P (2001) Enhanced-diffusion chromatographic sorbents: improved binding capacity and capture efficiency for ion-exchange industrial chromatography. Gen. Eng. News 21, 38.

Teng SF, Sproule K, Hussain A, Lowe CR (1999) A strategy for the generation of biomimetic ligands for affinity chromatrography. Combinatorial synthesis and biological evaluation of an IgC binding ligand. J. Mol. Recogn. 12, 67–75.

Van Reis R, Zydney A (2001) Membrane separations in biotechnology. Curr. Opin. Biotechnol. 12, 208–211.

Vasquez-Alvarez E, Lienqueo ME and Pinto JM (2001) Optimal synthesis of protein purification processes. Biotechnol. Prog. 17, 685–696.

Wheelwright SM (1987) Designing downstream processes for large-scale protein purification. Biol/Technol. 5, 789–793.

1
Release of Protein from Biological Host

Rajni Hatti-Kaul and Bo Mattiasson
Lund University, Lund, Sweden

1. BIOLOGICAL HOSTS FOR PROTEIN PRODUCTION

The means to gain access to the product from a biological source is the primary consideration during downstream processing of proteins. Traditionally, the choice of a protein source was restricted to the natural biological material producing it in sufficient amounts. But recombinant DNA technology has completely changed the scenario by allowing protein production in a biological host of choice. Microorganisms clearly constitute the most common production systems for industrial enzymes and other proteins. Mammalian cell hosts are preferred only when correct posttranslational modification is essential for the function of eukaryotic proteins. Insect cells are also emerging as production systems; however, the posttranslational processing differs from the mammalian cell systems.

Despite the advantages of microorganisms as protein source, several protein products are still economically produced from animal and plant materials possibly because of sufficiently high amounts of these products in such sources. Animal tissues and organs provide excellent sources for some lipases, esterases, proteases, and other proteins. Hen egg white continues to be a good source for lysozyme. Cultivated plants provide adequate source for proteases like bromelain and ficin. Moreover, transgenic animals and plants are also appearing as commercial production hosts for heterologous proteins, the former being preferred for therapeutic proteins and the latter for industrial enzymes.

Access to the product is simple and inexpensive when the protein is produced extracellularly. Separation of biomass or any other particulate matter is the only requirement for obtaining the protein in a clarified form that is processed further for concentration and purification. Bulk enzymes are invariably produced extracellularly by *Bacillus* species and fungi, as are the proteins produced by mammalian cell culture. Protein production in transgenic animals is often directed to milk; here the clarification step involves the removal of lipids and casein.

Proteins associated with the cells, on the other hand, need to be released prior to commencement of downstream processing. The method of protein release is often determined by the kind of host cells/tissue harboring it and its exact location. Most cell-associated proteins are located in the cytoplasm, the soluble fraction of the cells, whereas some are present in the plasma membrane enveloping the cytoplasm and others in the periplasmic space separating the membrane from the cell wall. In *Escherichia coli*, which is often the microbial host of choice for production of recombinant proteins, the foreign protein normally accumulates in the cytoplasm either in the soluble form or, in the case of overproduction, as insoluble inclusion bodies. *Saccharomyces cerevisiae* is another industrial host particularly when a GRAS (generally regarded as safe) organism is required for a particular process, and its secretory mechanism does not function for all proteins that are usually produced as soluble intracellular products.

Different types of cells vary basically in the wall structure that provides elasticity and mechanical strength to the cells, and hence in their resistance to cell breakage. Cell wall is lacking in animal cells, which are thus easy to break. In gram-positive bacteria, peptidoglycan constitutes the major cell wall component associated with teichoic acids and polysaccharides, whereas in gram-negative bacteria, the peptidoglycan structure is covered by another wall layer composed of proteins, phospholipids, lipopoteins, and lipopolysaccharides (Fig. 1). There are also substantial variations in the structure among different bacterial species, and also in the presence of capsules and slime layers, that further influence the resistance of cells to disruption. The cell wall in yeasts is made of glucans, mannans, and proteins, which form a highly cross-linked structure (Fig. 1), whereas in a typical mature plant cell it is composed of cellulose and other polysaccharides.

2. GENETIC APPROACHES FOR EXPORT OF RECOMBINANT PROTEINS

In order to avoid the cell disruption stage and the problems related to subsequent processing, various strategies have been attempted at the genetic

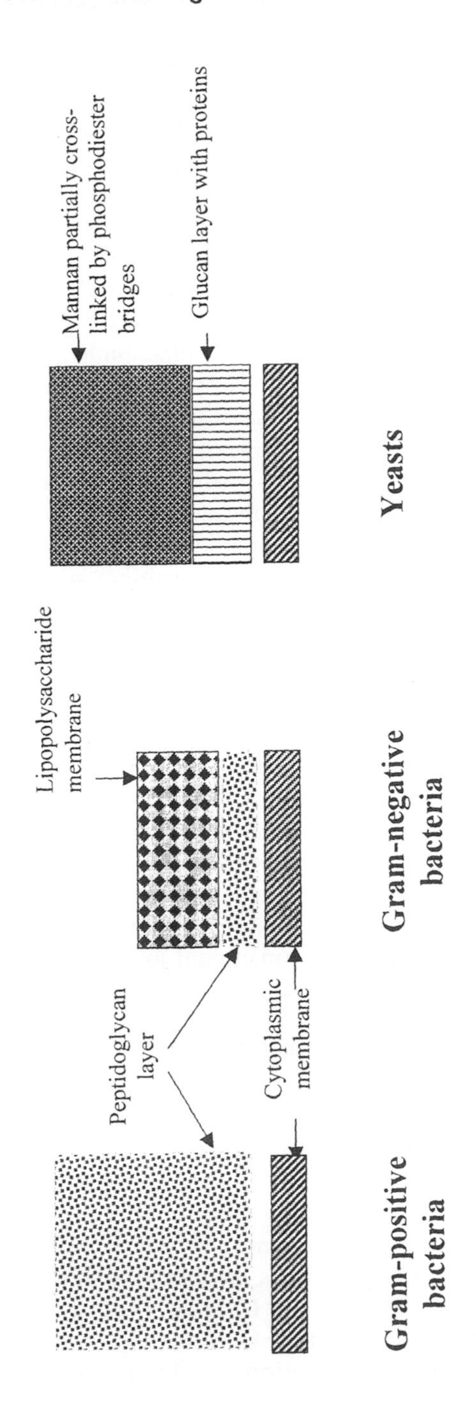

Figure 1 Schematic presentation of cell envelopes of bacteria and yeast.

level to make the cells secrete the protein. Expression of proteins in periplasm is also preferred since the amount of contaminating proteins is significantly lower than in the cytoplasm. Furthermore, because of its oxidizing environment the periplasm constitutes an attractive destination for the production of eukaryotic proteins (Baneyx 1999). Polypeptides destined for export are synthesized as preproteins containing a signal sequence 18–30 amino acids long at the amino terminus, which is cleaved during the translocation by inner-membrane-associated signal peptidases. Many signal sequences derived from naturally occurring secretory proteins (e.g., OmpA, OmpT, PelB, β-lactamase, and alkaline phosphatase) allow the translocation of heterologous proteins across the inner membrane when fused to their amino-terminal ends (Baneyx 1999; Pines and Inouye 1999).

Excretion of human growth hormone to the culture medium of *E. coli* cells was reported using excretion vector (pEAP8) (Kato et al. 1987). The passage through the bacterial inner membrane was made possible using the penicillinase signal sequence of an alkalophilic *Bacillus* sp. whereas the outer membrane was rendered permeable by the insertional activation of the *kil* gene (derived from plasmid pMB9) (Kobayashi et al. 1986). In another study, induction of β-lactamase synthesis from a strong *tac* promoter resulted in the release of more than 90% of the total soluble enzyme activity into the medium (Geourgiou et al. 1988). Excretion was not significant when the promoter was not fully induced or when the enzyme synthesis was directed by a weaker promoter. Permeability of the outer membrane of the excreting cells appeared to be affected as seen by higher sensitivity to detergents and also by release of periplasmic enzymes into the medium.

Abrahamsen et al. (1986) reported the extracellular secretion in *E. coli* of insulin-like growth factor I as a fusion protein with a protein A fragment composed of two IgG-binding fragments (EE). The presence in the protein A gene of a stress-induced promoter functional in *E. coli* was suggested.

3. RELEASE OF CELL-BOUND PROTEIN

The different strategies that may be used for releasing the cell-bound proteins include breaking the cells structure by mechanical forces (e.g., homogenisation or grinding with abrasives), damaging preferentially the cell wall (e.g., by drying or enzymatic lysis), or lysing primarily the membranes (e.g., by treatment with chemicals). Different mechanical and nonmechanical methods of disruption of microbial cells have been reviewed earlier in several papers (Harrison 1991; Schütte and Kula 1993; Chisti and Moo-Young 1986; Middelberg 1995). For most of the methods used, a number of factors, including type of organism, cultivation conditions, storage conditions, growth phase, viscosity and rheological properties of the cell suspension, pH, and so forth, influence the outcome of the disruption. Hence, expressing proteins into standard organisms like *E. coli* and yeasts that have established disruption procedures minimizes the extent of optimization required for cell breakage.

Development of a cell disruption process cannot be isolated from further downstream processing because the operation affects the physical properties of the suspension, such as viscosity, density, particle size, and settleability, which influence the subsequent solid-liquid separation step (Chisti and Moo-Young 1986). The product release can also be accompanied by its denaturation, due to the stress forces generated by the procedure used to rupture the cells, digestion by the proteases liberated from the cells, or interaction with other nonprotein constituents of the cells. A practical consideration of importance when evaluating the current methods available for industrial scale breakage of microbial cells is to ensure the availability both of small-scale equipment for development and optimization of the process, and of equipment of suitable capacity for the proposed large-scale operations.

3.1 Mechanical Disruption of Microbial Cells

Mechanical disruption is the most common mode for protein release from cells despite the higher capital and operating costs, stress to the proteins, and generation of viscous homogenate with fine cell debris as compared with the nonmechanical methods. The disruption is based primarily on liquid or solid shear forces.

The liquid shear cell disruption is often associated with the cavitation phenomenon that involves formation of vapor cavities in liquid due to local reduction in pressure that could be affected by ultrasonic vibrations, local increase in velocity, and so forth. Collapse and rebound of the cavities will occur until an increase in pressure causes their destruction. On the collapse of the cavitation bubble, a large amount of energy is released as mechanical energy in the form of elastic waves that disintegrate into eddies. According to Doulah (1977), the eddies larger than the dimension of the cell will move it from place to place whereas the smaller eddies will impart motions of different intensities to the cell, creating a pressure difference across the cell. When the kinetic energy content of the cell exceeds the cell wall strength, the cell disintegrates. Cavitation also produces free radicals that lead to protein denaturation (Save et al. 1997).

An established technique for cell disruption in small samples that involves cavitation is ultrasonication, i.e., using ultrasonic vibrations covering a frequency range extending upward from 20 kHz (Chisti and Moo-

Young 1986; Harrison 1991). This technique is not used at industrial scale primarily because the ultrasonication energy absorbed into suspensions ultimately appears as heat, and good temperature control is necessary. On the other hand, hydrodynamic cavitation results from pumping of a cell suspension through a downstream constriction, which causes a local increase in velocity (Save et al. 1994, 1997; Harrison 1991). Disruption of yeast (Save et al. 1994, 1997) and *Alcaligenes eutrophus* (Harrison 1991) by such a process was shown to be more energy efficient than the ultrasonic cavitation. However, in contrast to ultrasonication, the disruption was influenced significantly by the concentration of cells.

The established large-scale cell disruption operations based on liquid and solid shear are high-pressure homogenization and bead milling, respectively, with the former being the more common of the two. These will be described in some detail here. Much of the industrial disruption equipment has been inherited from chemical and food processing industries. Although no radical changes have occurred in cell disruption technologies over the years, major efforts have been put into introducing sterilization and containment as secondary features for biotechnology applications in order to ensure safe operation with recombinant organisms (Foster 1995).

High-Pressure Homogenization

A high-pressure homogenizer comprises a positive displacement pump that delivers fluid at a relatively constant flow to a homogenizing valve (Engler 1990; Pandolfe 1993; Pandolfe and Kinney 1998). The fluid exits through an adjustable gap between the valve and the valve seat and ultimately impinges on an impact ring as seen in Fig. 2. The flow restriction in the valve assembly drives up pressure (in the range of 50 and 120 MPa), which is then drastically reduced as the high-velocity jet of fluid emerges from the opening. This process applied to a microbial cell suspension results in the disruption of cells. Disruption follows first-order process at a given pressure in a high-pressure homogenizer. The extent of protein release is represented by (Hetherington et al. 1971; Schütte and Kula 1993):

$$\ln\left(\frac{R_m}{R_m - R}\right) = kNP^a \qquad (1)$$

where R_m and R are the maximal amount of protein available for release and the protein amount released at a certain time, respectively (kg protein/kg cells), k is the first-order rate constant, N the number of passages, P the operating pressure, and a the pressure exponent.

The values of the constants are contingent on the nature of the organism and its culture conditions. However, the cell concentration over a rela-

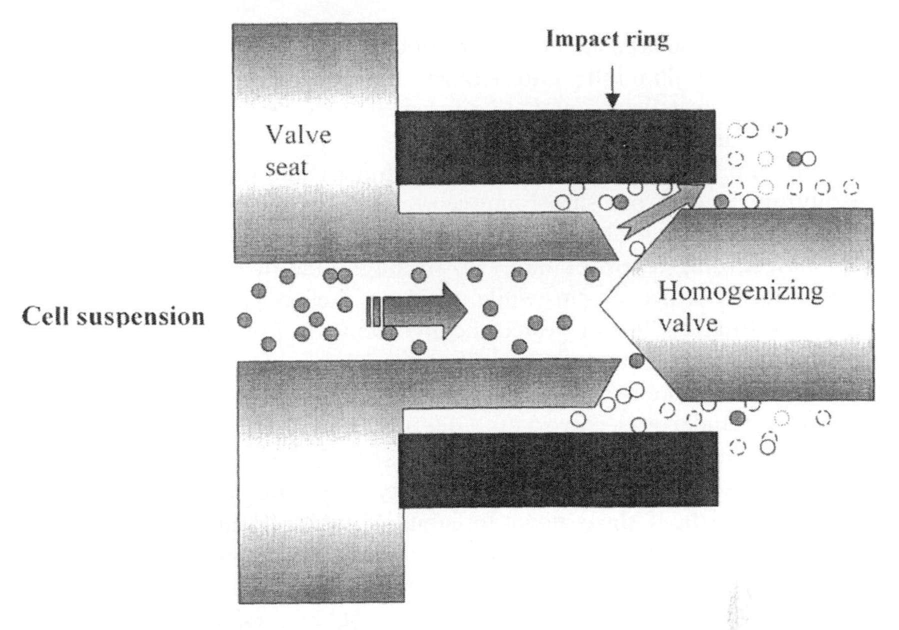

Figure 2 Flow path of a cell suspension through a homogenizing valve in a high-pressure homogenizer.

tively broad range is seen to have little influence on the disruption (Hetherington et al. 1971). Studies have shown that specific enzymes are released at different rates depending on their location in the cell; the periplasmic enzymes at a rate faster than the total protein, all cytoplasmic enzymes at the same rate as total protein, whereas membrane-bound enzyme slower (Follows et al. 1971). High-pressure homogenization is preferred for disruption of nonfilamentous organisms since the mycelial organisms result in clogging of the homogenizing valve (Zetelaki 1969; Keshavarz et al. 1990a).

Optimization of the cell disruption process is achieved by variation in the operating pressure and valve design (Engler 1990, Schütte and Kula 1993). Increasing the operating pressure reduces the number of passes required for disruption and hence the operating costs (Engler 1990; Pandolfe 1993). It also indirectly reduces the severe downstream clarification problems arising due to the fine cell debris resulting from further disintegration of already disrupted cells on repeated passage through the machine. However, the rise in pressure is also accompanied by a linear increase in temperature (2.5°C per 10 MPa) that necessitates rapid cooling of the homogenized suspension. While maintaining the pressure, alteration in homogenizing valve geometry can make significant changes in the protein yield (Pandolfe 1993; Pandolfe and Kinney 1998), e.g., disruption of baker's yeast was more efficient with "knife edge" valve seat than the flat type unit (Hetherington et al. 1971).

Several investigators have sought to understand the mechanism of cell disruption by high-pressure homogenizer (Brookman 1974; Engler 1990; Keshavarz Moore et al. 1990b; Middelberg 1995; Kleinig and Middelberg 1998). According to Brookman (1974), magnitude of pressure drop is the major causative mechanism, while other researchers (Keshavarz Moore et al. 1990b; Engler 1990) suggested the involvement of two mechanisms—one relating to homogenization zone within the valve unit and the other with the impingement in the exit zone. The latter theory was confirmed by recent studies of Lander and coworkers (2000) with polysaccharides as a model shear-sensitive compound, where a similarity in the breakage pattern in terms of molecular size and polydispersity was observed with the results obtained from fluid shear flows in capillary tubes. Their results indicated that breakage occurs primarily by fluid shear in the valve unit, with a contribution by cavitation occurring in the impingement section where the jet of fluid strikes the impingement plate and the bubbles collapse as a result of repressurization of this impact. The occurrence of cavitation was indicated by the detection of free radicals in a pressure range from 11 to 35 MPa (Shirgaonkar et al. 1998; Lander et al. 2000).

The traditional form of high-pressure homogenizer is the Manton-Gaulin APV type, for which special designs for biotechnology applications have been developed (Pandolfe 1993). A few other types of equipment are now available. Microfluidizer (Microfluidics, Newton, MA) is one in which the cell suspension is driven at a flow rate of 250–600 ml/min by an air-powered intensifier pump at constant pressure (20–159 MPa) into an interaction chamber where the stream is split into two, which pass through precisely defined fixed-geometry microchannels (100 μm), and are later directed head-on at each other before emerging at atmospheric pressure (Schütte and Kula 1993) (Fig. 3). The equipment can be operated in a batch or continuous mode. The extent of disruption and release of the soluble proteins in a microfluidizer, besides being dependent on the pressure, number of passes through the interaction chamber, and growth conditions of the organism, may also be influenced by the initial cell concentration (Sauer et al. 1989; Schütte and Kula 1993; Choi et al. 1997). This led to the introduction of an additional exponent b in Eq. (1), which varied linearly with cell concentration and dilution rate.

$$\ln\left(\frac{R_{\mathrm{m}}}{R_{\mathrm{m}} - R}\right) = kN^{b}P^{a} \tag{2}$$

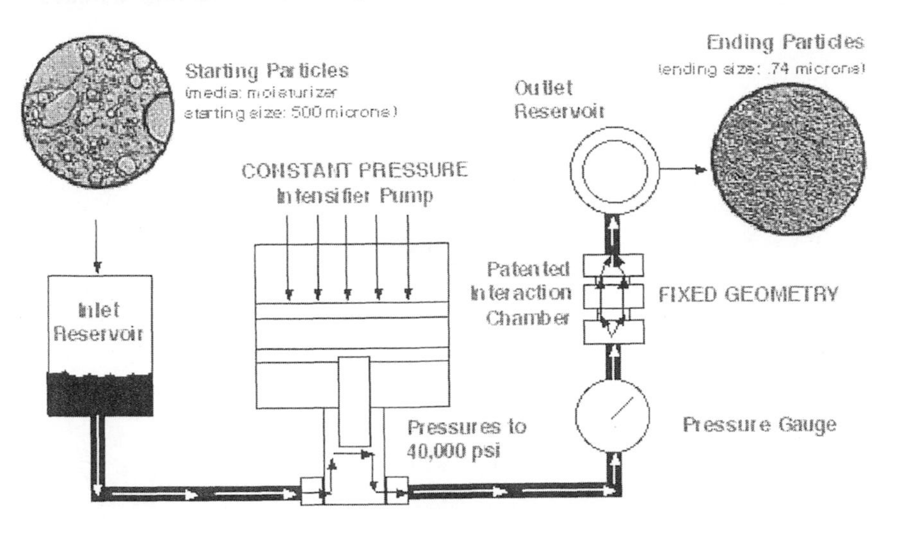

Figure 3 Schematic flow of a cell suspension through a microfluidizer processor (Microfluidics Corporation).

A different cell disrupter manufactured by Constant Systems Ltd (Warwick, UK) has been designed specifically for biotechnology work to allow high-level containment and cleaning in place (Foster 1992, 1995; Lovitt et al. 2000). The apparatus utilizes a hydraulically driven pump to drive the process liquid at very high pressure (up to 275 MPa) through a small jet (0.1–0.2 mm diameter) into a disruption chamber maintained at atmospheric pressure (Lovitt et al. 2000) (Fig. 4). The jet of liquid strikes a target surface in the disruption chamber, spreading radially and then axially down over the inner surface of the chamber, which is cooled by a cooling jacket. This construction minimizes temperature increase; during disruption of baker's yeast, a temperature increase of only 9°C was observed in the disrupted sample at an operating pressure of 240 MPa (Lovitt et al. 2000). Operating pressure, velocity of the liquid flow through the jet, and size of the jet are the critical parameters influencing the profile of the soluble protein.

Bead Mill Disruption

Stirring a cell suspension with glass beads is an effective method of disruption of organisms (Song and Jacques 1997). The process is normally preformed in a bead mill, such as Dyno-Mill (W.A. Bachofen, Basel, Switzerland), which is a continuously operating system with horizontal or

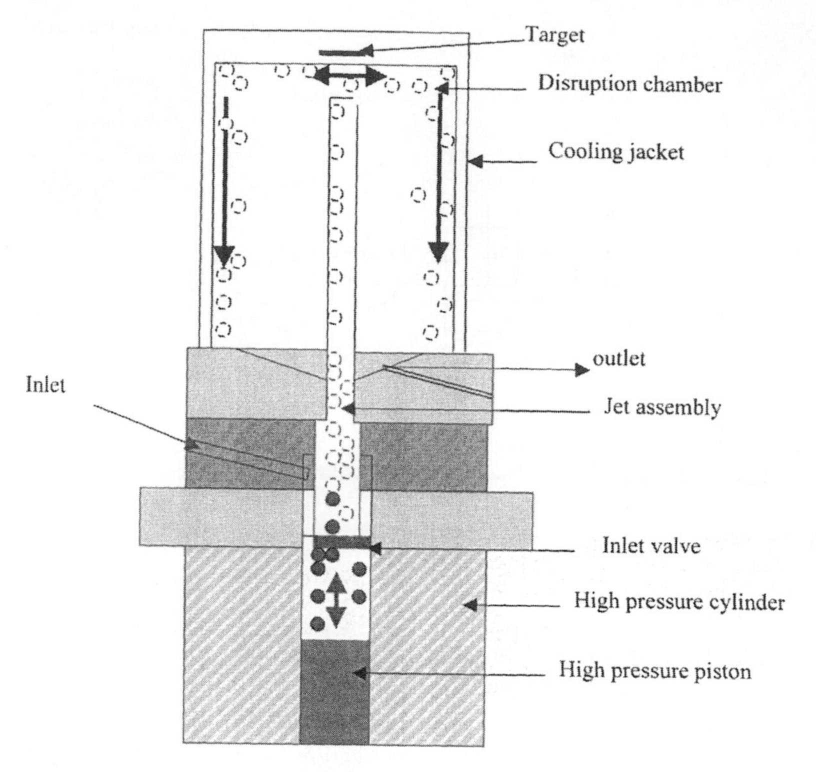

Figure 4 Illustration of a Constant Systems Series B disrupter head (Lovitt et al. 2000).

vertical grinding container (the former being favored in biotechnology) for dispersion and wet grinding in a completely enclosed system (Fig. 5) (Schütte and Kula 1990, 1993; Middelberg 1995).

Dyno-Mills are available in different designs and sizes, from small laboratory models (0.15–0.6 L) to large mills for production plants (600 L). The mill chamber is made of stainless steel, but is glass at the laboratory scale. Specially designed agitator disks mounted symmetrically on a shaft transfer kinetic energy to the spherical beads, while an external pump feeds the cell suspension into the mill (Fig. 6). The beads are made of wear-resistant materials such as glass, zirconium oxide, zirconium silicate, titanium carbide, etc. They are retained in the mill by means of a dynamic gap separator or a special slot screen. The shape and arrangement of the disks are important for optimal energy input and even distribution of the grinding beads. The beads are accelerated in a radial direction, forming stream layers

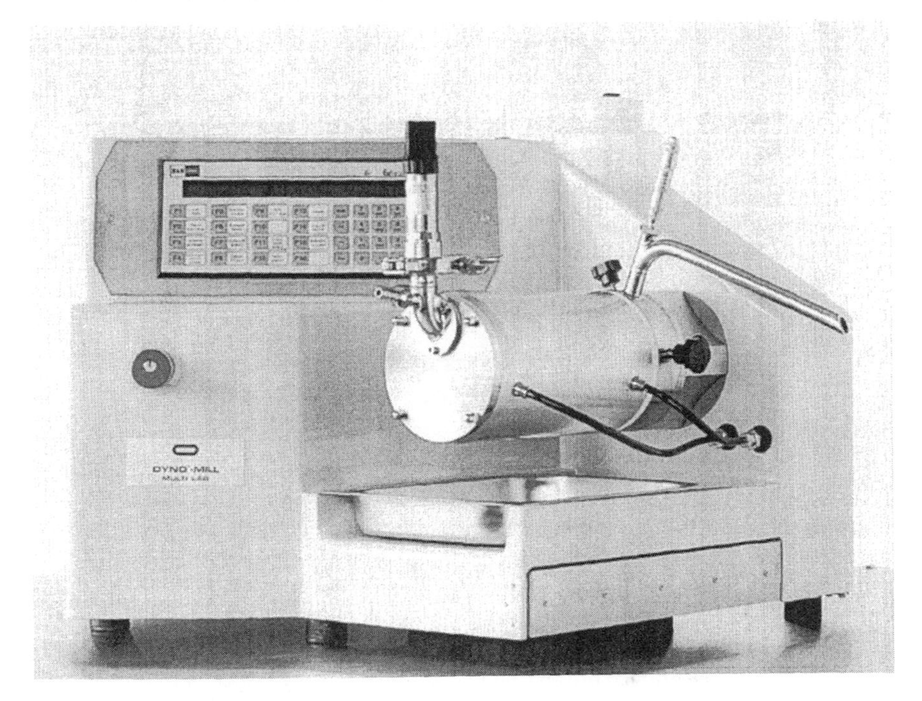

Figure 5 Dyno-Mill Multi-Lab (W.A. Bachofen AG). The available grinding container size: 0.3 to 1.4 L, grinding beads: 0.2–2.5 mm. The production capacity is in the range of 5–40 kg/h depending on the product being processed, size of the grinding container, and operating mode (batch or continuous).

of different velocities and thus creating high shear forces. The differential velocity between the layers is influenced by the size and density of the beads and distance to fixed or rotating parts in the grinding chamber. The optimal size of glass beads is related to the size of the microorganisms: for yeasts diameters of > 0.5 mm and for bacteria < 0.5 mm are considered optimal (Schütte and Kula 1990, 1993). It also depends on agitator design and mill geometry. The cylinder is filled with beads to about 80–90%. Higher loading leads to excess power consumption and heat generation. The energy of the grinding elements is maximal at the periphery of the agitator disk, and minimal at the chamber wall and near the axis. As with the high-pressure homogenizer, removal of the energy dissipated in the broth is an important factor in the scale-up of the bead mill, which is achieved by efficient cooling by a spiral flow of cooling water in the outer jacket.

Cell disruption in a bead mill is also a first-order process, the soluble protein release being represented as

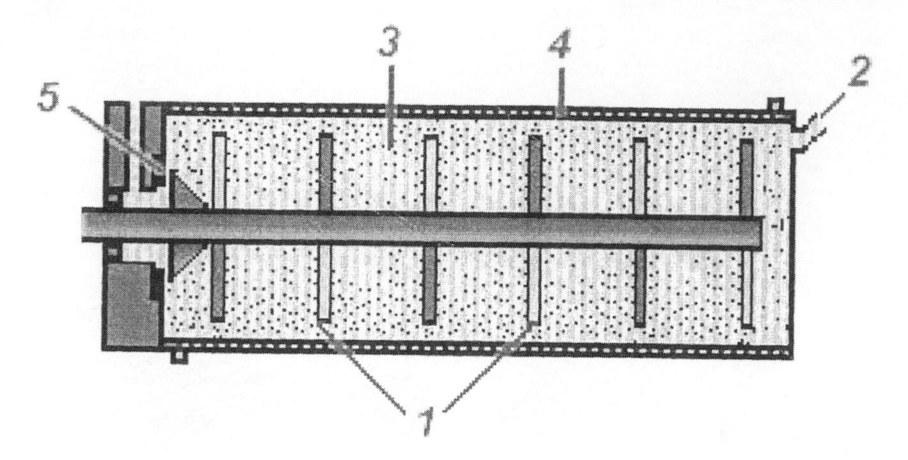

Figure 6 Principle of the Dyno-Mills (W.A. Bachofen AG). The agitator disks (1) transmit the power of the electric motor to the grinding beads (2), the motion of which imparts stress on the cells to be disrupted. The cell suspension is pumped into the inlet (3), moved through the grinding chamber (4), and exposed for a certain amount of time to the stress of the moving grinding beads. Most of the energy introduced by the agitator disks is converted to heat. This heat (~ 70% of the power consumption of the motor) has to be carried off by means of the cooling liquid, which circulates in the cooling jacket (5) of the grinding chamber. Before leaving the grinding chamber, the grinding beads are separated from the product by means of a separation system (6).

$$\ln\left(\frac{R_{\mathrm{m}}}{R_{\mathrm{m}} - R}\right) = \ln D = kt \tag{3}$$

where D is the reciprocal fraction of the unreleased protein, t is the time of treatment, and k is a first-order rate constant that depends on the history of the organism and a number of operational parameters such as design and the speed of the agitator, bead loading, bead size, and initial cell concentration (Kula et al. 1990; Schütte and Kula 1990; Heim and Solecki 1999; Ricci-Silva et al. 2000). As noted during high-pressure homogenization, the rate of protein release during bead milling is also indicative of its location in the cell (Hetherington et al. 1971; Schütte and Kula 1990; Torner and Asenjo 1991). Cell rupture by the beads is caused by a combination of high shear and impact with the cells, and is also influenced by frequency and strength of collision events between the beads. Bead milling is effective for breaking tough cells, including yeast, other fungi, cyanobacteria, microalgae, and spores.

Optimization of the operational parameters in a bead mill is done in a batch mode (Schütte and Kula 1990). The information can then be used to determine the feed rate for a continuous disruption process. In practice, the mean residence time is set at three times the necessary time in batch processing. Continuous breakage was seen to be more efficient than the batch process, which was attributed to the flow of liquid past the agitators eliminating dead spots in the chamber (Woodrow and Quirk 1982).

CoBall-Mill (Fryma-Maschinen, Rheinfelden, Switzerland) is an alternative form of bead mill (van Gaver and Huyghebaert 1990) in which the disruption chamber is composed of a conical rotor fitting into a conical stator (Fig. 7). The narrow gap between rotor and stator, which is about four times larger than the diameter of the disruption beads, is partially filled with the beads and the cell suspension is pumped continuously through it. The volume of the disruption chamber is only 25% of the disruption volume required for a conventional ball mill with the same throughput, but the surface area to volume ratio is much higher providing a good temperature control. Additional cooling capacity is provided in the mill's stator and lid, and also in rotors. The rotary movement of the rotor in the disruption gap forces the disruption media radially outward, with the kinetic energy of the disruption media increasing progressively on their path through the gap. Finally, a separator system stops the beads leaving through the mill outlet and returns them through a duct to the disruption chamber inlet. The CoBall-Mills available have capacities ranging from 10–30 L/h (MS-12) to 800–2000 L/h (MS-100) (van Gaver and Huyghebaert 1990).

3.2 Physical Rupture of Microbial Cells

Among the traditional physical rupture techniques are desiccation (by slow drying in air, drum drying, etc.) followed by extraction of the microbial powder, repeated freezing and thawing of cell suspension, and osmotic shock (by, e.g., suspending a cell suspension in a solution with high salt concentration) (Harrison 1991; Middelberg 1995). These methods have not gained popularity due to low protein yields and the length of time required in comparison with mechanical methods.

Exposure to high temperature can be an effective approach to cell disruption but is limited to heat-stable products. Heating to 50–55°C disrupts the outer membrane of gram-negative bacteria like *E. coli* and releases proteins from the periplasmic space (Tsuchido et al. 1985). Cytoplasmic proteins were released by heating at 90°C for 10 min from *E. coli*, but *Bacillus megaterium* gave only low levels of released proteins at high temperatures (Watson et al. 1987). The protein release was higher from the stationary cells than from exponentially growing cells. If applicable, ther-

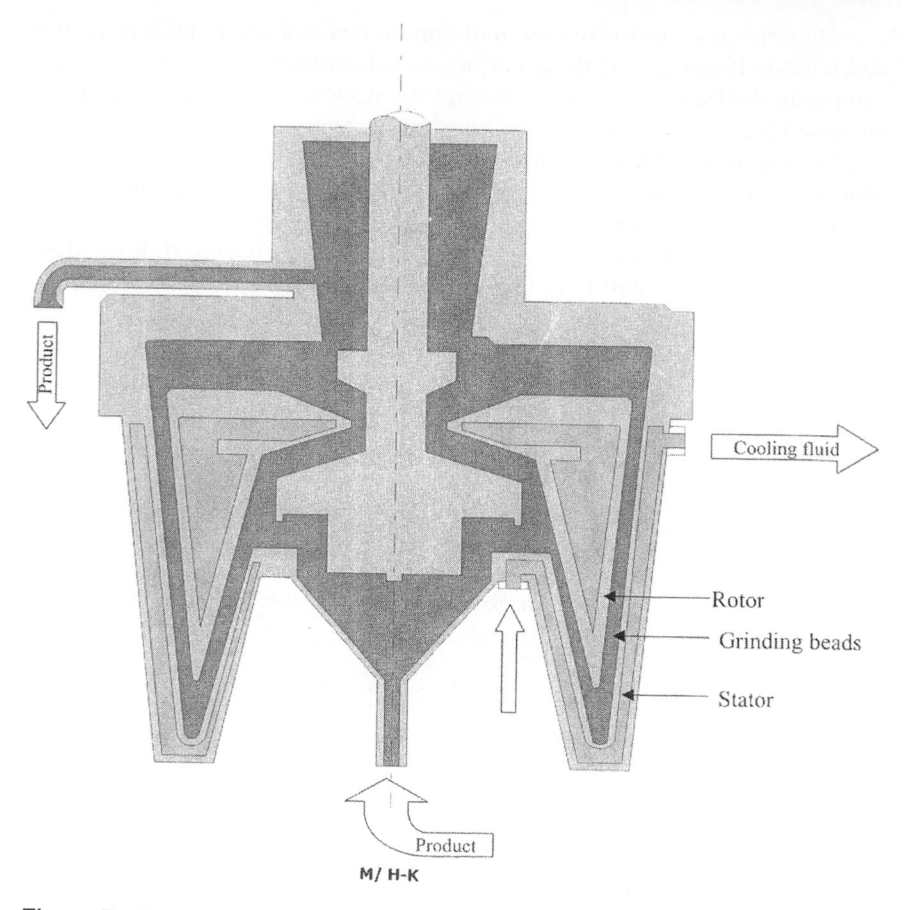

Figure 7 Schematic diagram of a disruption chamber of a CoBall Mill (van Gaver and Huyghebaert 1990).

molysis has the advantage of killing the microbial host, thus minimizing the possibility of its release into the environment.

Additional physical disruption methods based on other principles have now been developed. These include nebulization and decompression.

Nebulization

A cell disruption apparatus, BioNeb (Glas-Col Apparatus Co., Terre Haute, Indiana) functions on the principle that liquid flow in capillaries takes place in layers, i.e., the layers in the center flow faster than the layers toward the side of the tube (Lewis 1993). A cell placed in the area of the differential speed will stretch and break due to its two ends experiencing different flows. Nebulization occurs when a gas is blown over a surface of liquid. The gas flow causes droplets to form, which are connected to the liquid for a short time with the neck size about half the diameter of the droplet. The "neck" between the liquid surface and the emerging droplet resembles a tiny capillary tube for a millisecond. In BioNeb, the cells are sheared within that neck because of the differential flow. The method has been shown to work well on a variety of cell types.

Decompression

When pressurized, the microbial cells are gradually penetrated and filled with gas. After saturation by the gas, the applied pressure is suddenly released when the absorbed gas rapidly expands within the cells leading to their mechanical rupture. This concept of sequential pressurization and explosive decompression was reported first in 1951 (Fraser 1951) and developed later by other researchers as a potential unit operation to recover intracellular enzyme and recombinant DNA products from microbial cells (Lin et al. 1991, 1992). While developing a sterilization method for heat-sensitive materials, Nakamura et al. (1994) showed that the wet cells of baker's yeast were completely destroyed when the microorganisms saturated with CO_2 gas at 40°C and 40 atm for more than 3 h were subjected to a sudden discharge of pressure. The initial rate of gas desorption is considered to be related to the magnitude of cell disruption. A commercial instrument, Cell Disruption Bomb from Parr Instruments Co. (Moline, Illinois), working on a similar principle, employs nitrogen decompression and targets delicate mammalian cells (Lewis 1993). The technique can be applied on large sample volumes without the problem of heat generation.

3.3 Disruption of Microbial Cells Using Lytic Agents

Disruption processes utilizing chemicals or enzymes as lytic agents have also been extensively investigated. In comparison with other disruption methods, these tend to be expensive and also require subsequent removal of the lytic agents from the process stream. On scale-up, disruption by lytic agents can be effected within a sealed vessel and contained to the same degree as a fermentor.

Chemicals as Lytic Agents

Different classes of chemicals can permeabilize membrane layers of the microbial cells by different mechanisms (Naglak et al. 1990), but only a few of them find application in protein release.

Treatment with ethylenediaminetetraacetate (EDTA) is used to release periplasmic proteins from gram-negative bacteria (Neu and Heppel 1964; Ryan and Parulekar 1991) as it disrupts the outer membrane of the bacteria by binding Mg^{2+} and Ca^{2+} ions that cross-link the lipopolysaccharide (LPS) molecules. The release of LPS occurs rapidly and is independent of temperature; however, the concentration of EDTA required for permeabilization is dependent on the microbial strain (Marvin et al. 1989). The space created by the release of LPS is apparently filled by phospholipid molecules from the inner membrane, creating areas of enhanced permeability (Nikaido and Nakae 1979).

Antibiotics cause lysis of gram-negative bacteria by different mechanisms. The common class of β-lactam antibiotics inhibits peptidoglycan synthesis in growing cells, which are not able to maintain their osmotic pressure and hence disrupt (Spratt 1980; Kohlrausch and Höltje 1991). Because of high costs and dependence of their effect on growth state of the cells, the use of antibiotics for large-scale disruption is not known. The assembly of the peptidoglycan layer is also inhibited by chaotropic agents, such as guanidine hydrochloride and urea, that disrupt the water structure and result in weakening of hydrophobic associations (Naglak et al. 1990). These chemicals have been found to cause lysis of growing *E. coli* cells but not of the cells in the stationary phase (Ingram 1981). Lysis of *E. coli* cells on treatment with ethanol was also ascribed to a chaotropic-like effect (Ingram 1981). The use of chaotropic agents is also quite expensive at large scale; moreover, there is an inherent danger of protein denaturation.

Organic solvents and detergents mainly cause dissolution of lipids in the periplasmic membrane and the outer membrane (de Smet et al. 1978; Jones 1992). Detergents are invariably used for solubilization of membrane proteins. The nonionic detergent Triton X-100 is most commonly used, but other detergents, including cholate, sodium dodecyl sulfate (SDS), or Brij are also able to act on the cell membranes (Schütte and Kula 1993). The large-scale use is, however, limited; one of the very few procedures using Triton X-100 involves recovery of cholesterol oxidase from *Nocardia* sp. (Buckland et al. 1974). Detergents having low cloud point (i.e., temperature at which a detergent solution separates into a detergent-enriched phase and aqueous phase), such as Triton X-114, present attractive alternatives to the other detergents for recovery and significant purification of integral membrane proteins (Bordier 1981; Pyrde 1986; Sanchez-Ferrer et al. 1989; Ramelmeier et al. 1991; Quina and Hinze 1999). The proteins can be sufficiently concentrated by temperature-induced phase separation of the detergent. However, subsequent removal of the detergent from the protein requires additional processing.

Release of intracellular proteins by treatment of cells with detergent or organic solvent has also been investigated. In general, low protein yields and no release of high molecular weight proteins have been observed (Schütte and Kula 1993). Deutscher (1974) applied the toluene (0.5–10%, 5–15 min, 37°C) treatment for selective release of low molecular weight (~ 50,000) transfer ribonucleic acid (tRNA) nucleotidyltransferase from *E. coli* cells. Initial plasmolysis by solvents followed by autolysis by addition of water allows the release of larger enzymes like carboxypeptidase (Breddam and Beenfeldt 1991). The rate of this process is highly dependent on pH; essentially complete autolysis is achieved at pH 8.0 within 20 h using the best solvents. Straight chain alcohols of medium chain length, i.e., C_6-C_9 (1.2 mL/100 g yeast) and solvents like trichloroethane, chloroform, and ether (2.5–10 mL/100 g yeast) were found to be efficient in accelerating autolysis in yeast cells (Breddam and Beenfeldt 1991). A variety of nonionic and ionic detergents (~ 2.5%) were also shown to cause autolysis of the yeast cells, with Triton X-100 and *N*-lauroylsarcosine being the most efficient (Breddam and Beenfeldt 1991). This autolysis process was most efficient when the compressed yeast had been initially plasmolysed by treatment with sodium chloride followed by addition of detergent and then water. Fenton (1982) demonstrated that treatment of yeast cells with concentrated solvent (80%) for about 90 min and subsequent dilution with phosphate buffer and shaking for an additional 15 h resulted in release of high amounts of β-galactosidase, a high molecular weight intracellular enzyme.

Release of the intracellular protein by solvent or detergent is further enhanced in the presence of EDTA (Naglak et al. 1990). An effective permeabilization method was achieved by combination of chaotropic agent (0.1 M guanidine hydrochloride) with a detergent (0.5–2% Triton X-100), which can interact synergistically on the microbial cell envelope (Hettwer and Wang 1989; Naglak and Wang 1990). The combination of the detergent with dipotassium hydrogen phosphate has been suggested as a cheaper and non-denaturing alternative to guanidine hydrochloride for permeabilization of *E. coli* cells (Zhao and Yu 2001).

Alkaline lysis is an effective but harsh method of cell rupture. Alkali added to a cell suspension reacts with the cell walls in a number of ways, including the saponification of lipids in the cell walls. Although there are some reports on protein release (Harrison 1991; Schütte and Kula 1993), the method is not very useful for protein isolation in general because of the sensitivity of proteins to alkaline conditions.

Enzymatic Lysis of Cells

The advantage of using enzymes for cell lysis, besides selectivity during the product release, is operation under mild conditions (Asenjo and Andrews 1990). Enzymes hydrolyse the walls of microbial cells, and when sufficient wall has been removed, the internal osmotic pressure bursts the periplasmic membrane allowing the intracellular components to be released (Asenjo and Andrews 1990). A variety of microorganisms have been found to produce microbial lytic enzyme systems, most of which are relatively small (10–30 kDa) and thus relatively easily separable from protein products after cell lysis (Andrews and Asenjo 1987). Availability, costs, and the need for removal in subsequent purification steps presently limit application of the lytic enzymes.

The effect of lytic enzymes is specific for particular groups of microorganisms, attributed to the differences in the cell wall composition. While enzyme treatment alone is sufficient to damage the wall of gram-positive bacteria, the lysis of gram-negative cells requires passage of enzyme through the outer wall layer, which is aided by the addition of EDTA or a detergent. The best known lytic enzyme for bacteria is lysozyme from hen egg white, which catalyzes the hydrolysis of β-1,4-glycosidic bonds in the peptidoglycan layer of bacterial cell walls. Lysozyme has been used for several large-scale processes of enzyme extraction (Schütte and Kula 1993).

Complete breakage of yeast cell wall is achieved by synergistic action of a specific wall-lytic protease to degrade the outer layer of protein-mannan and a lytic β-(1-3)-glucanase to degrade the inner wall layer of glucan. Selective release of recombinant 60-nm protein particles (virus-like particles) from yeast cells has been achieved by treatment with pure glucanase component of the lytic enzyme complex of *Oerskovia* and *Cytophaga* sp. (Asenjo et al. 1993). Treatment of the enzymatically lysed pellet with Triton X-100 further increased the amount of virus-like particles released.

3.4 Combined Mechanical and Nonmechanical Disruption of Microbial Cells

A combination of enzymatic/chemical lysis with mechanical disintegration has been suggested for enhancing the efficiencies of the respective methods, with savings in time and energy and facilitating subsequent processing by reducing the amount of cell debris. Pretreatment of cells with enzymes, e.g., *Bacillus cereus* with cellosyl, and *Candida utilis* and *Saccharomyces cerevisiae* with zymolyase, has shown dramatically increased release of intracellular enzymes with fewer passes during mechanical disruption (Kula et al. 1990; Vogels and Kula 1992; Baldwin and Robinson 1990, 1994). Heat pretreatment could be another alternative for heat-stable products (Vogels and Kula 1992). Weakening of the cell wall of the gram-negative bacterium *Alcaligenes eutrophus* by various forms of chemical and thermal pretreatment, e.g., by high pH (10.5) shock, addition of SDS at 70°C, increasing monovalent cations concentration at 60°C, addition of EDTA, and lysis by lysozyme (Harrison et al. 1991), and addition of chaotropic agents (Becker et al. 1983), also led to easier mechanical disruption of the cells.

Pretreatment of recombinant *E. coli*, expressing human growth hormone inclusion bodies, with guanidine hydrochloride and Triton X-100 reduced the number of passes required for cell disruption and the number of steps required for the recovery of protein from inclusion bodies (Bailey et al. 1995). The exponentially growing cells treated with 1.5 M guanidine HCl and 1.5% Triton X-100 gave adequate disruption after one pass at 41 MPa, with a particle size distribution similar to that for untreated cells disrupted after one pass at 62 MPa. This combination of reagents was also used to wash the inclusion bodies free of some contaminating proteins without solubilization of the growth hormone.

3.5 Disruption of Animal and Plant Tissue

Absence of cell walls makes the disintegration of mammalian tissue rather easy. The tissue is cut into small pieces using a domestic or an industrial meat grinder. Finer grinding is done by homogenization of the tissue suspension in buffer in a blender-type homogenizer at laboratory scale or milling through a colloid mill at pilot or industrial scale.

The plant cell wall, on the other hand, is very rigid. Preparation of an extract from a fleshy or nonfibrous tissue is done by rapid homogenization of the tissue suspension (in a suitable ice-cold buffer) in a precooled Waring blender. The more fibrous material, which is difficult to macerate, is frozen and ground to a dry powder before suspending in the extraction buffer for homogenization (Gegenheimer 1990). Most problems during disruption of plant material are caused by phenolic compounds, including tannins and pigments present largely in the vacuoles but that mix with the proteins and cause inactivation once the intact plant cell is destroyed (Loomis and Battaile 1966; Loomis 1974). The use of various phenol scavengers has been suggested, the most satisfactory agent being an insoluble form of polyvinylpyrrolidone (PVP) (trade name Polyclar AT) that has the advantages of easy separation from the enzymes and noninterference with protein assays (Hatti-Kaul and Mattiasson 1996). The phenols not binding to PVP can be removed by adsorption to Amberlite XAD-4, a hydrophobic adsorbent composed of porous polystyrene resin (Loomis et al. 1979). The use of Amberlite XAD-4 and Polyclar AT together is even more effective than use of the respective materials individually. It is also common practice to add reducing agents such as ascorbate, thiols, and so forth, to prevent the accumulation of quinones (Anderson 1968).

Selective release of proteins (and not the vacuole contents) from plant cells is of course possible and extremely attractive. It is achieved by enzymatic digestion of the cell wall by cellulases and pectinases.

4. ANALYSIS OF CELL DISRUPTION

Monitoring of the progress of microbial cell breakage often relies on off-line microscopic examinations or spectrophotometric assays. While microscopic observation allows one to distinguish intact and broken cells, quantitation is tedious and gives no information on the quality of the product. Phase contrast microscopy is superior to ordinary light microscopy as it is capable of distinguishing between an intact cell and a partially disrupted cell. Partial disruption may indicate release of the product, and as the host cells remain as large particles they can be easily separated. Optical density (OD) of cell suspension for assessing disruption efficiency is not always satisfactory unless a relationship between OD and product release has been proven.

Indirect methods, such as determination of the amount of released activity and protein, are often used but are limited by the lack of knowledge of a 100% value. It may be noted that the activity per unit weight of cells may vary with the cultivation conditions, time and conditions of storage, and disruption conditions employed. Automated flow injection analysis for on-line monitoring of enzyme activity released during disintegration of cell suspension has been demonstrated (Recktenwald et al. 1985). The dielectric measurement of cell suspension has also been proposed as a method for on-line monitoring of the cell condition directly in the disruption process (Morita et al. 1999). The advantage of this technique lies in its noninvasive assay and applicability to opaque and turbid suspensions.

The size of the insoluble fragments generated during cell disruption is an important parameter for the subsequent solid/liquid separation step, but is normally difficult to measure. The size range may vary from the original cell size (1–10 μm) down to the colloidal range. The dynamic range of different particle size analyzers does not span the entire range of interest. Counting devices such as a Coulter counter have intrinsic limits below 1 μm, and particles below 0.3–0.5 μm cannot be detected. Photon correlation spectroscopy and other light scattering techniques have been found to be applicable for monitoring the extent of cell breakage (Clark et al. 1987; Agerkvist and Enfors 1990).

Sedimentation analysis suffers from the unknown density of cell fragments and the small density difference between fragments and aqueous media. Irreproducibility in results may arise due to time-dependent changes in the sample, such as agglomeration. Particle size distribution during cell disruption using a centrifugal photosedimentometer (Shimadzu-CP 2-10) has been reported. The instrument consists of a disk rotating at a certain speed onto which two glass cells may be attached (van Gaver and Huyghebaert 1990). The changing optical density of the sample in one glass cell during centrifugation is registered with the aid of a photometric cell. Another method applied for monitoring disruption of *E. coli* cells is based on the use of analytical disk centrifuges, which consists of a pair of spinning disks between which a fluid (e.g., 10% glycerol-water) is added to give a rotating annulus of "spin fluid" (Middelberg et al. 1991). A less dense buffer or water is introduced into the inner radius of this annulus to establish a density gradient. The sample suspension is then introduced into the inner radius and the particles settle radially outward due to the centrifugal force, with each particle having a characteristic settling rate dependent on its Stokes diameter. The size distribution is determined by measuring the optical density of the spin fluid near the outer radius as a function of time.

5. CONCLUDING REMARKS

The interactions of the cell disruption step with the upstream and downstream process cannot be ignored. The disruption is dependent on the properties of the cells and hence the fermentation conditions. It has also been shown that efficiency of disruption may be affected if the cells have been stored frozen (Milburn and Dunnill 1994). Besides aiming at liberation of maximal amount of the product without undue damage, the cell disruption method has to be chosen or optimized with consideration of the subsequent downstream processing. Mechanical methods of disruption are preferred because of short residence times, lower operating costs, and contained operation. Comparison of *E. coli* cell disruption by bead mill, Manton-Gaulin homogenizer, and microfluidizer showed no significant difference in protein and enzyme release but considerably different physical properties of the cell homogenates, which influences centrifugation and filtration (Agerkvist and Enfors 1990). A higher degree of disruption (i.e., higher product release) would normally mean that the degree of separation is drastically reduced and hence may not lead to higher product recovery. It was shown that increasing the degree of disruption in the bead mill only slightly affects the separation degree at centrifugation but had a strong effect in the homogenizers (Agerkvist and Enfors 1990). A combination of chemical/enzymatic and mechanical disruption appears to be the most suitable for obtaining the product and homogenate with desirable characteristics. Integrating product release and recovery, say, by extraction in an aqueous two-phase system, can provide additional benefits by overcoming the problems with debris removal (Ariga et al. 1994; Hart et al. 1994).

ACKNOWLEDGMENTS

The authors are grateful to the Swedish Competence Center for Bioseparation for financial support. Microfluidics Corporation and W.A. Bachofen AG are acknowledged for their interest and for providing the figures for their equipment. The authors also thank Jing Liu for his artistic help with a figure.

REFERENCES

Abrahamsen L., Moks, T., Nilsson B., Uhlén M. (1986) Secretion of heterologous gene products to the culture medium of *Escherichia coli*. Nucl. Acids Res. 14, 7487–7500.

Agerkvist I., Enfors S-O. (1990) Characterization of *E. coli* cell disintegrates from a bead mill and high pressure homogenizers. Biotechnol. Bioeng. 36, 1083–1089.

Anderson J.W. (1968) Extraction of enzymes and subcellular organelles from plant tissues. Phytochem. 7, 1973–1988.

Andrews B.A., Asenjo J.A. (1987) Enzymatic lysis and disruption of microbial cells. Trends Biotechnol. 5, 273–277.

Ariga O., Miyakawa I., Aota T., Sano Y. (1994) Simultaneous release and purification of gene product in an aqueous two-phase system. J. Ferment. Bioeng. 77, 71–74.

Asenjo J.A., Andrews B.A. (1990) Enzymatic cell lysis for product release. In: Separation Processes in Biotechnology (Asenjo J.A., ed.), Marcel Dekker Inc., New York, pp. 143–175.

Asenjo J.A., Ventom A.M., Huang R-B., Andrews B.A. (1993) Selective release of recombinant protein particles (VLPs) from yeast using a pure lytic glucanase enzyme. Bio/Technol 11, 214–217.

Bailey S.M., Blum P.H., Meagher M.M. (1995) Improved homogenization of recombinant *Escherichia coli* following pretreatment with guanidine hydrochloride. Biotechnol. Prog. 11, 533–539.

Baldwin C., Robinson C.W. (1990) Disruption of *Saccharomyces cerevisiae* using enzymatic lysis combined with high pressure homogenization. Biotechnol. Tech. 4, 329–334.

Baldwin C., Robinson C.W. (1994) Enhanced disruption of *Candida utilis* using enzymatic pretreatment and high pressure homogenization. Biotechnol. Bioeng. 43, 46–56.

Baneyx F. (1999) Recombinant protein expression in *Escherichia coli*. Curr. Opin. Biotechnol. 10, 411–421.

Becker T., Ogez J.R., Builder S.E. (1983) Downstream processing of proteins. Biotechnol. Adv. 1, 247–261.

Bordier C. (1981) Phase separation of integral membrane proteins in Triton X-114 solution. J. Biol. Chem 256, 1604–1607.

Breddam K., Beenfeldt T. (1991) Acceleration of yeast autolysis by chemical methods for production of intracellular enzymes. Appl. Microbiol. Biotechnol. 35, 323–329.

Brookman J.S.G. (1974) Mechanism of cell distintegration in a high pressure homogenizer. Biotechnol. Bioeng. 16, 371–383.

Buckland B.C., Richmond W., Dunnill P., Lilly M.D. (1974) Large scale isolation of intracellular microbial enzymes. Cholesterol oxidase from *Nocardia*. In: Industrial Aspects of Biochemistry, Vol. 30, Part I (Spencer B., ed.), North-Holland-American Elsevier, Amsterdam, pp. 65–79.

Chisti Y., Moo-Young M. (1986) Disruption of microbial cells for intracellular products. Enzyme Microb. Technol. 8, 194–204.

Choi H., Laleye L., Amantea G.F., Simard R.E. (1997) Release of aminopeptidase from *Lactobacillus casei* sp. *casei* by cell disruption in a microfluidizer. Biotechnol. Tech. 11, 451–453.

Clark D.J., Stansfield A.G., Jepras R.I., Collinge T.A., Holding F.P., Atkinson T. (1987) Some approaches to downstream processing monitoring and the development of new separation techniques. In: Separations for Biotechnology (Verrall M.S. and Hudson M.J., eds.), Ellis Horwood, pp. 419–429.

De Smet M.J., Kingma J., Witholt B. (1978) The effect of toluene on the structure and permeability of the outer and cytoplasmic membranes of *Escherichia coli*. Biochim. Biophys. Acta 506, 64–80.

Deutscher M.P. (1974) Preparation of cells permeable to macromolecules by treatment with toluene: studies of transfer ribonucleic acid nucleotidyltransferase. J. Bacteriol. 118, 633–639.

Doulah M.S. (1977) Mechanism of disintegration of biological cells in ultrasonic cavitation. Biotechnol. Bioeng. 19, 649–660.

Engler C.R. (1990) Cell disruption by homogenizer. In: Separation Processes in Biotechnology (Asenjo J.A., ed.), Marcel Dekker, New York, 95–105.

Fenton D.M. (1982) Solvent treatment for β-D-galactosidase release from yeast cells. Enzyme Microb. Technol. 4, 229–232.

Follows M., Hetherington P.J., Dunnill P., Lilly M.D. (1971) Release of enzymes from baker's yeast by disruption in an industrial homogenizer. Biotechnol. Bioeng. 13, 549–560.

Foster D. (1992) Cell disruption: breaking up is hard to do. Bio/Technol. 10, 1539–1541.

Foster D. (1995) Optimizing recombinant product recovery through improvements in cell-disruption technologies. Curr. Opin. Biotechnol. 6, 523–526.

Fraser D. (1951) Bursting bacteria by release of gas pressure. Nature 167, 33–34.

Gegenheimer P. (1990) Preparation of extracts from plants. Meth. Enzymol. 182, 174–193.

Geourgiou G., Shuler M.L., Wilson D.B. (1988) Release of periplasmic enzymes and other physiological effects of β-lactamase overproduction in *Escherichia coli*. Biotechnol. Bioeng. 32, 741–748.

Harrison S.T.L. (1991) Bacterial cell disruption: a key unit operation in the recovery of intracellular products. Biotechnol. Adv. 9, 217–240.

Harrison S.T.L., Dennis J.S., Chase H.A. (1991) Combined chemical and mechanical processes for the disruption of bacteria. Bioseparation 2, 95–105.

Hart R.A., Lester P.M., Relfsnyder D.H., Ogez J.R., Builder S.E. (1994) Large scale, in situ isolation of periplasmic IGF-I from *E. coli*. Bio/Technol. 12, 1113–1117.

Hatti-Kaul R., Mattiasson B (1996) Downstream processing of proteins from transgenic plants. In: Transgenic Plants: A Production System for Industrial and Pharmaceutical Proteins (Owen M.R.L. and Pen J., eds), John Wiley, Chichester, pp. 115–147.

Heim A., Solecki M. (1999) Disintegration of microorganisms in a bead mill with a multi-disk impeller. Powder Technol. 105, 389–395.

Hetherington P.J., Follows M., Dunnill P., Lilly M.D. (1971) Release of protein from baker's yeast by disruption in an industrial homogenizer. Trans. Inst. Chem. Eng. 49, 142.

Hettwer D., Wang H. (1989) Protein release from *E. coli* cells permeabilized with guanidine-HCl and Triton X-100. Biotechnol. Bioeng. 33, 886–895.

Ingram L.O. (1981) Mechanism of lysis of *Escherichia coli* by ethanol and other chaotropic agents. J. Bacteriol. 145, 331–336.

Jones M.N. (1992) Surfactant interactions with biomembranes and proteins. Chem. Soc. Rev. 21, 127–136.

Kato C., Kobayashi T., Kudo T., Furusato T., Murakami Y., Tanaka T., Baba H, Oishi T., Ohtsuka E., Ikehara M., Yanagida T., Kato H., Moriyama S., Horikoshi K. (1987) Construction of an excretion vector and extracellular production of human growth hormone from *Escherichia coli*. Gene 54, 197–202.

Keshavarz E., Bonnerjea J., Hoare M., Dunnill P. (1990a) Disruption of a fungal organism, *Rhizopus nigricans*, in a high-pressure homogenizer. Enzyme Microb. Technol. 12, 494–498.

Keshavaraz Moore E., Hoare M., Dunnill P. (1990b) Disruption of baker's yeast in a high-pressure homogenizer: new evidence on mechanism. Enzyme Microb. Technol. 12, 764–770.

Kleinig A.R., Middelberg A.P.J. (1998) On the mechanism of microbial cell disruption in high-pressure homogenisation. Chem. Eng. Sci. 53, 891–898.

Kobayashi T, Kato C, Kudo T., Horikoshi K. (1986) Excretion of the penicillinase of an alkalophilic *Bacillus* sp. through the *Escherichia coli* outer membrane is caused by insertional activation of the *kil* gene in plasmid pMB9. J. Bacteriol. 166, 728–732.

Kohlrausch U., Höltje J.-V. (1991) Analysis of murein and murein precursors during antibiotic-induced lysis of *Escherichia coli*. J. Bacteriol. 173, 3425–3431.

Kula M-R., Schütte H., Vogels G., Frank A. (1990) Cell disintegration for the purification of intracellular proteins. Food Biotechnol. 4, 169–183.

Lander R., Manger W., Scouloudis M., Ku A., Davis C., Lee A. (2000) Gaulin homogenization; a mechanistic study. Biotechnol. Prog. 16, 80–85.

Lewis R. (1993) New disrupters help cell biologists retrieve products. The Scientist 7(24), 18–19.

Lin H.M., Chan E.C., Chen C., Chen L.F. (1991) Disintegration of yeast cells by pressurized carbon-dioxide. Biotechnol. Prog. 7, 201–204.

Lin H.M., Yang Z.Y., Chen L.F. (1992) An improved method for disruption of microbial cells with pressurized carbon-dioxide. Biotechnol. Prog. 8, 165–166.

Loomis W.D. (1974) Overcoming problems of phenolics and quinones in the isolation of plant enzymes and organelles. Meth. Enzymol. 31, 528–544.

Loomis W.D., Battaile J. (1966) Plant phenolic compounds and isolation of plant enzymes. Phytochem. 5, 423–438.

Loomis W.D., Lile J.D., Sanstrom R.P., Burbott A.J. (1979) Adsorbent polystyrene as an aid in plant enzyme isolation. Phytochem. 18, 1049–1054.

Lovitt R.W., Jones M., Collins S.E., Coss G.M., Yau C.P., Attouch C. (2000) Disruption of bakers' yeast using a disrupter of simple and novel geometry. Process Biochem. 36, 415–421.

Marvin H.J., ter Beest M.B.A., Witholt B. (1989) Release of outer membrane fragments from wild-type *Escherichia coli* and from several *E. coli* lipopolysaccharide mutants by EDTA and heat shock treatments. J. Bacteriol. 171, 5262–5267.

Middelberg A.P.J. (1995) Process-scale disruption of microorganisms. Biotech. Adv. 13, 491–551.

Middelberg A.P.J., O'Neill B.K., Bogle D.L. (1991) A novel technique for the measurement of disruption in high-pressure homogenization: studies on *E. coli* containing recombinant inclusion bodies. Biotechnol. Bioeng. 38, 363–370.

Milburn P.T., Dunnill P. (1994) The release of virus-like particles from recombinant *Saccharomyces cerevisiae*: effect of freezing and thawing on homogenization and bead milling. Biotechnol. Bioeng. 44, 736–744.

Morita S., Umakoshi H., Kuboi R. (1999) Characterization and on-line monitoring of cell disruption and lysis using dielectric measurement. J. Biosci. Bioeng. 88, 78–84.

Naglak T.J., Wang H.Y. (1990) Protein release from the yeast *Pichia pastoris* by chemical permeabilization: comparison to mechanical disruption and enzymatic lysis. In: Separations for Biotechnology, Vol. 2 (Pyle D., ed.), Elsevier, London, pp. 55–64.

Naglak T.J., Hettwer D.J., Wang H.Y. (1990) Chemical permeabilization of cells for intracellular product release. In: Separation Processes in Biotechnology (Asenjo J.A., ed.), Marcel Dekker, NY, pp. 177–205.

Nakamura K., Enomoto A., Fukushima H., Nagai K., Hakoda M. (1994) Disruption of microbial cells by the flash discharge of high-pressure carbon dioxide. Biosci. Biotech. Biochem. 58, 1297–1301.

Neu H.C., Heppel L.A. (1964) On the surface localization of enzymes in *E. coli*. Biochem. Biophys. Res. Commun. 17, 215–219.

Nikaido H., Nakae T. (1979) The outer membrane of gram-negative bacteria. In: Advances in Microbial Physiology (Rose A.H. and Gareth Morris J., eds.), Academic Press, London, pp. 163–250.
Pandolfe W.D. (1993) A cell disruption homogenizer. Amer. Biotech. Lab. 11, pp. 16,18,20.
Pandolfe W.D., Kinney R.R. (1998) High-pressure homogenization. Chem. Process. 61, 39–43.
Pines O., Inouye M. (1999) Expression and secretion of proteins in *E. coli.* Mol. Biotechnol. 12, 25–34.
Pyrde J.G. (1986) Triton X-114: a detergent that has come in from the cold. Trends Biosci. 11, 160–163.
Quina F.H., Hinze W.L. (1999) Surfactant-mediated cloud-point extractions: an environmentally benign alternative separation approach. Ind. Eng. Chem. Res. 38, 4150–4168.
Ramelmeier R.A., Terstappen G.C., Kula M-R. (1991) The partitioning of cholesterol oxidase in Triton X-114 based aqueous two-phase systems. Bioseparation 2, 315–324.
Recktenwald A., Kroner K-H., Kula M-R. (1985) On-line monitoring of enzymes in downstream processing by flow injection analysis. Enzyme Microb. Technol. 7, 607–612.
Ricci-Silva M.E., Vitolo M., Abrahao-Neto J. (2000) Protein and glucose 6-phosphate dehydrogenase releasing from baker's yeast cells disrupted by a vertical bead mill. Process Biochem 35, 831–835.
Ryan W., Parulekar S.J. (1991) Recombinant protein excretion in *Escherichia coli* JM 103 [pUC8]: Effects of plasmid content, ethylenediaminetetraacetate, and phenethyl alcohol on cell membrane permeability. Biotechnol. Bioeng. 37, 430–444.
Sanchez-Ferrer A., Bru R., Garcia-Carmona F. (1989) Novel procedure for extraction of a latent grape polyphenoloxidase using temperature-induced phase separation in Triton X-114. Plant Physiol. 91, 1481–1487.
Sauer T., Robinson C.W., Glick B.R. (1989) Disruption of native and recombinant *Escherichia coli* in a high-pressure homogenizer. Biotechnol. Bioeng. 33, 1330–1342.
Save S.S., Pandit A.B., Joshi J.B. (1994) Microbial cell disruption: role of cavitation. Chem. Eng. J. 55, B67–B72.
Save S.S., Pandit A.B., Joshi J.B. (1997) Use of hydrodynamic cavitation for large scale microbial cell disruption. Trans I. Chem. E. 75, Part C, 41–49.
Schütte H., Kula M-R. (1990) Bead mill disruption. In: Separation processes in Biotechnology (Asenjo J.A., ed.), Marcel Dekker, New York, pp. 107–141.
Schütte H., Kula M-R. (1993) Cell disruption and isolation of non-secreted products. In: Biotechnology (Stephanopoulos G., ed.), VCH, New York, pp. 505–526.
Shirgaonkar I.Z., Lothe R.R., Pandit A.B. (1998) Comments on the mechanism of microbial cell disruption in high-pressure and high-speed devices. Biotechnol. Prog. 14, 657–660.

Song D.D., Jacques N.A. (1997) Cell disruption of *Escherichia coli* by glass bead stirring for the recovery of recombinant proteins. Anal. Biochem. 248, 300–301.

Spratt B.G. (1980) Biochemical and genetical approaches to the mechanism of action of penicillin. Phil. Trans. R. Soc. Lond. B 289, 273–283.

Torner M.J., Asenjo J.A. (1991) Kinetics of enzyme release from breadmaking yeast cells in a bead mill. Biotechnol. Tech. 5, 101–106.

Tsuchido T., Katsui N., Takeuchi A., Takano M., Shibasaki I. (1985) Destruction of the outer membrane permeability barrier of *Escherichia coli* by heat treatment. Appl. Environ. Microbiol. 50, 298–303.

van Gaver D., Huyghebaert A. (1990) Optimization of yeast cell disruption with a newly designed bead mill. Enzyme Microb. Technol. 13, 665–671.

Vogels G., Kula M-R. (1992) Combination of enzymatic and/or thermal pretreatment with mechanical cell disruption. Chem. Eng. Sci. 47, 123–131.

Watson J.S., Cumming R.H., Street G., Tuffnell J.M. (1987) Release of intracellular proteins by thermolysis. In: Separations for Biotechnology (Verrall M.S. and Hudson M.I., eds.), Ellis Horwood, London, pp. 105–109.

Woodrow J.R., Quirk A.V. (1982) Evaluation of the potential of a bead mill for the release of intracellular bacterial enzymes. Enzyme Microb. Technol. 4, 385–389.

Zetelaki K. (1969) Disruption of mycelia for enzymes. Process Biochem. 4, 19–22, 27.

Zhao F., Yu J. (2001) L-asparaginase release from *Escherichia coli* cells with K_2HPO_4 and Triton X100. Biotechnol. Prog. 17, 490–494.

2

Chromatography: The High-Resolution Technique for Protein Separation

Peter Kumpalume* and Siddhartha Ghose*
Aston University, Birmingham, England

1. INTRODUCTION

Chromatography is a technique in which the components of a mixture are separated depending on the rates at which they are carried through a stationary phase by a mobile phase, which can be a liquid or a gas. It therefore follows from this definition that same components will travel at the same rate under the same experimental conditions. Thus, by comparing the chromatogram of an unknown sample with that of known samples, chromatography can be used to identify and determine the components of mixtures. But perhaps it is in the separation and subsequent purification of substances that chromatography has found widespread use. In fact, it has been argued that no other separation method is as powerful and as generally applicable as is chromatography.

While many separation techniques are limited to certain types of substances, chromatography can be applied to a wide spectrum of compounds. Polar and nonpolar compounds can be separated on bonded phase partition chromatography, and those compounds containing charged groups can be separated by ion exchange chromatography. Molecules that have affinity for

**Current affiliation*: University of Cambridge, Cambridge, England other molecules or metals (Porath 1975) are easily purified by affinity chromatography (see Chap. 3). The substances that differ in size can be separated by molecular size exclusion chromatography. If the sample is a gas, gas chromatography can be used to separate the components. If two components elute at the same time, another mobile phase or stationary phase can be used until separation is achieved. It is due to this versatility that to date the power of chromatography is unrivaled.

For the purification of proteins, liquid column chromatography, using an insoluble matrix and an aqueous buffer as mobile phase, is commonly used. With the development of expanded bed chromatography (see Chap. 10), the power of chromatography has even been strengthened as it is now possible to reduce the cost of production by eliminating initial clarification steps during protein isolation from crude feeds.

2. HISTORY OF CHROMATOGRAPHY

The development of chromatography is often accredited to the Russian botanist Mikhail Tsvet, who reported his method in 1903 but gave a full account of it in Germany in 1906 (Heftmann 1973, 1983). However prior to his work there are reports of some researchers like Way (Rieman and Walton 1970) and Reed in England, and Day in the United States who conducted some chromatography experiments (Heftmann 1983). But it was only in the early 1930s that the potential of chromatography was recognized. Today it has become an indispensable tool in the separation and purification of biological substances, such as enzymes and other proteins, as well as inorganic ones.

As time passed, various forms of chromatography came into development. Paper chromatography (PC) was developed by Consden et al. (1944). Their work stemmed from earlier work conducted by Martin and Synge (1941) who used a liquid as a stationary phase (partition chromatography). Consden and coworkers replaced the column with sheets of filter paper. However, in their experiments the solvent flowed downward and it was William and Kirby (1948) who modified the technique so that the solvent flowed up the strip as is done today.

The development of thin layer chromatography (TLC) in the late 1930s and the improvement of this technique by Egon Stahl (Heftmann 1983) and Kirchner and coworkers (1951) complemented the use of PC. This meant that natural fats (James and Martin 1952) could now be separated. Tiselius and his group (1937) carried out extensive work on electrophoresis. His work was recognized by the award of the Nobel Prize in 1948.

It was probably with the purification of carotenoids (Zechmeister and von Cholnoky 1937) and steroids (Reichstein and Shoppee 1949) that column chromatography gained ground. The introduction of new techniques, such as carrier displacement (Tiselius and Hagdahl 1950) and gradient elution (Alm et al. 1952), shed new light in improving and understanding of chromatographic techniques.

The application of chromatography for the purification of proteins began in the 1960s. The 70-μm beads (Regnier and Gooding 1992) that were being used at the time for small molecules (at flow velocities of 60 cm/h) presented severe mass transfer problems because proteins have a lower diffusion rate. Hence, resolution was poor and longer process times had to be used. From the theoretical work of Martin and Synge (1941), the solution was to use small particles. However, because of their low permeability, special instrumentation (now called high-performance liquid chromatography, HPLC) was developed and was first applied in 1967 (Horvath et al. 1967). The ease with which column chromatography could be used with various adsorbents, e.g., ion exchange resins developed in 1935 (Adams and Holmes 1935) but first applied to proteins in 1951 (Moore and Stein 1951), molecular sieve gels (Regnier and Gooding 1992) and affinity adsorbents in the 1960s (Porath and Flodin 1959), and hydrophobic resins in the 1970s (Shaltiel 1974), revolutionized protein purification.

The early 1990s saw the development of yet another form of chromatography—expanded bed adsorption (EBA) chromatography pioneered by Chase and coworkers (Chase and Draeger 1992; McCreath et al. 1992) (see Chap. 10). By applying a feed in the upward direction, the bed can be expanded to a desired height. In such a system, the void fraction is larger than in packed beds; hence, unclarified feedstock can be applied to a column without performing the preliminary solid/liquid separation processes. This reduces both process times and product loss, and therefore economically EBA is very attractive.

Many people have contributed to the development and understanding of chromatography, but due to the space limitations, it is impossible to list them all. Interested readers should consult Ettre and Zlatkis (1975) and Heftmann (1973).

3. CLASSIFICATION

Chromatography is classified in several ways:

1. **Planar and column**. The classification is based on whether the process is conducted in a column or on a plane sheet of paper or glass. *Planar chromatography* includes thin-layer chromatography (TLC), PC, and electrophoresis. *Column chromatography*

includes liquid chromatography (LC), HPLC, and gas chromatography (TC).

2. **Mobile phase based**. The mobile phase in chromatography can be gas, liquid, or a supercritical fluid. Accordingly, the technique is classified as *gas*, *liquid*, or *supercritical fluid chromatography*.
3. **Interaction based**. Initially chromatography was classified as *partition* if the stationary phase is a liquid or *adsorption* if the stationary phase is a solid. With the development of new resins, chromatography is also classified as ion exchange if adsorption is based on *ion exchange* as opposed to physical adsorption, *hydrophobic interaction chromatography* (HIC) if adsorption is based on hydrophobic interaction, and *affinity chromatography* if the resin has specific affinity for the target molecule.

   Partition chromatography is classified into two groups: *liquid-liquid* if retention is by physical adsorption and *bonded phase* if covalent bonds are involved. The latter is known as *normal bonded phase* if the stationary phase is polar or *reverse bonded phase* if the stationary phase is nonpolar.
4. **Operational modes**. Until the late 1980s chromatography was performed in a *packed bed mode*, whereas the 1990s saw the development of *expanded bed adsorption* (EBA) chromatography wherein bed is allowed to expand to a desired height by pumping the mobile phase upward.

4. THE THEORY OF CHROMATOGRAPHY

There are two factors that govern chromatographic processes:

1. Distribution of the analyte between the stationary and the mobile phases. This governs retention and selectivity.
2. Axial dispersion. This determines column efficiency.

We shall discuss item 1 by using elution chromatography as an example.

4.1 Distribution, Retention, and Selectivity in Chromatographic Processes

In elution chromatography, a sample applied at the top of the column is continuously washed down the column by the mobile phase (mp) until it is eluted. The time it takes for it to be eluted is called the retention time, t_R. Since the analyte can only move down the column if it is in the mp, it therefore follows that t_R will depend on the extent to which it partitions itself between the stationary phase (sp) and the mp.

For analyte A, the equilibrium involved is

$$A_{mp} \rightleftharpoons A_{sp}$$

$$K = \frac{C_{sp}}{C_{mp}} \tag{1}$$

where C_{sp} and C_{mp} are the concentrations of A in the sp and mp, respectively. K is called the *partition ratio* or the *partition coefficient*. Equation (1) holds if the concentration of the analyte is not too·high in which case there is deviation from linearity. Fortunately, most chromatographic separations are performed in conditions where Eq. (1) holds.

When a solute is injected, it is retained by the sp due to solute-sp interactions. These interactions can be physical as well as electrostatic. If the column has length L, the average linear rate of analyte migration, v, is defined as

$$v = \frac{L}{t_R} \tag{2}$$

t_R can be obtained from the first moment (M_1) of the Gaussian distribution as in Eq. (3):

$$M_1 = \frac{1}{A}\int_0^\infty Ct\,dt \tag{3}$$

where A is the area of the peak and can be obtained from the zeroeth moment (M_0) of Gaussian distribution as shown in Eq. (4) and C is concentration.

$$M_0 = A = \int_0^\infty C\,dt \tag{4}$$

Usually t_R is just read from the peak on a chart recorder.

Similarly, the average linear rate of mp or the unretained species u is defined as

$$u = \frac{L}{t_M} \tag{5}$$

where t_M is the retention time of the mp or the unretained species. The unretained species should not interact with the stationary phase, and acetone meets this property.

Since the analyte (A) can only be eluted if it is in the mp, then

$$v = u \times \text{the time for the analyte (A) in mp}$$
$$= u \times \text{fraction of A in the mp}$$
$$= u \times \frac{C_{mp}V_{mp}}{C_{mp}V_{mp} + C_{sp}V_{sp}} \tag{6}$$
$$= \frac{u}{1 + K\frac{V_{sp}}{V_{mp}}}$$

where V_{sp} and V_{mp} are the volumes of the sp and the mp, respectively.

The retention time is related to the capacity factor k, also called the mass distribution ratio. It is defined as:

$$k = \frac{\text{amont of analyte in sp}}{\text{amount of analyte in mp}} \tag{7}$$

If we substitute this equation into Eq. (6) we obtain

$$v = \frac{u}{1 + k}$$

Substituting the expressions for v and u from Eqs. (2) and (5), respectively, and rearranging the equation we obtain

$$k = \frac{t_R - t_M}{t_M} \tag{8}$$

We can see from Eq. (8) that if k is very small, $t_R \approx t_M$, elution occurs very rapidly and accurate measurements of t_R are very difficult. On the other hand, if $t_R \gg t_M$, ($k > 20$), we shall have to wait too long to elute the analyte. Ideally $1 < k > 5$ is desired, and fortunately most chromatographic separations are performed under these conditions. Accurate measurements of k depend on accurate measurements of t_M. In practice this has always been a problem and because of this, relative retention has found greater use.

It is worth pointing out that in preparative chromatographic separations like ion exchange, where the sample application is followed by washing and elution, some of these considerations need to be modified. For example, during loading it is desirable that k be very large (more selective adsorption). On the other hand, during elution k should be small.

As was pointed out above, the capacity factor is related to the selectivity because if the selectivity is high, the sample will be more retained, i.e., high k. The selectivity factor, α, also called the relative retention, is defined as

$$\alpha = \frac{K_B}{K_A} \tag{9}$$

where B is the more retained species. Hence, by convention $\alpha > 1$. Since K_B and K_A are thermodynamic quantities, it follows that α is also a thermodynamic quantity, which at constant temperature will depend only on factors that affect adsorption, i.e., interactions between sp-solute, mp-solute, mp-sp, and solute-solute.

If we substitute the expressions for K from Eq. (7), we obtain

$$\alpha = \frac{k_B}{k_A} \tag{10}$$

From Eq. (10) we can see that α is a measure of the selectivity of a system for two species. It is obvious, therefore, that α will affect the resolution of a column.

Resolution, R_S, can be defined as the measure of the column to separate two species and is given as:

$$\begin{aligned} R_s &= \frac{2[t_{RB} - t_{RA}]}{W_A + W_B} \\ &= \frac{2\Delta t_R}{W_A + W_B} \end{aligned} \tag{11}$$

where W_A and W_B are bandwidth determined as shown in Fig. 1. R_S of 1.5 is required for complete (99.7%) separation. However, R_S of 1 (96% separation) is satisfactory for most chromatographic separations. It is a trivial matter to see that to improve R_S, a longer column should be used.

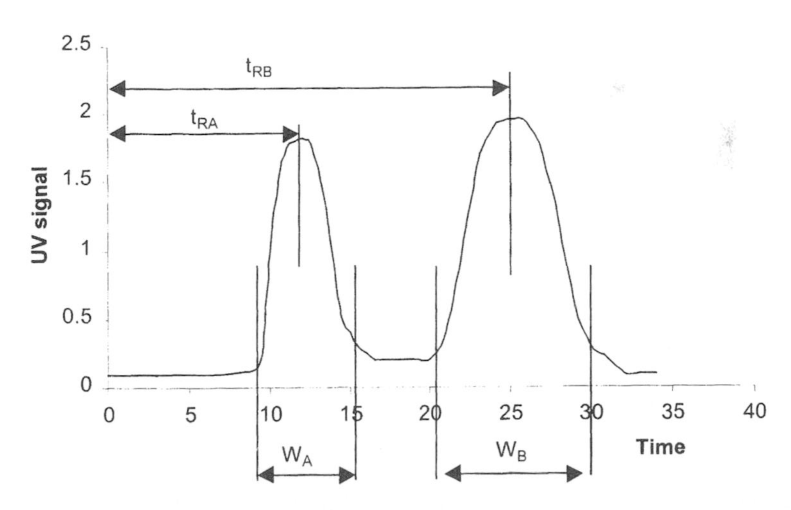

Figure 1 Resolution of peaks in chromatography.

Unfortunately, t_R becomes agonizingly long. The relationship of R_S to k and α can be manipulated here. High selectivity for species B implies high k; hence, t_R is large and resolution better. In fact, to improve R_S it is α and k that are altered rather than the column length because, as we shall see later, a longer column will lead to band broadening and thus lower column efficiency.

4.2 Axial Dispersion

Axial dispersion is longitudinal mixing (or back mixing) in a column. If axial dispersion is high, the efficiency of the column is low. Hence, in chromatographic separations, axial dispersion should be at a minimum.

According to Fick, molecular diffusion is defined as:

$$\frac{\partial C}{\partial t} = D\frac{\partial^2 C}{\partial x^2} \tag{12a}$$

where D is the molecular diffusion coefficient, x is the direction of diffusion.

Similarly, in a column where molecular diffusion is outweighed by the turbulent flow,

$$\frac{\partial C}{\partial t} = D_a\frac{\partial^2 C}{\partial t^2} \tag{12b}$$

where D_a is the axial dispersion coefficient, C is concentration, and t is time.

Since column chromatographic separations can be considered as closed systems, axial dispersion can be described by the axial dispersion model, which simplifies to

$$\frac{\partial C}{\partial \theta} = \left(\frac{D_a}{UL}\right)\frac{\partial^2 C}{\partial z^2} \tag{13}$$

where U is the superficial velocity $\theta = t/\bar{t}$ and L is column length. D_a/UL $(= Dv)$ is the axial dispersion number and the measure of the extent of axial dispersion. If the flow approaches plug flow (small dispersion), $Dv = D_a/UL \rightarrow 0$. For large dispersion, $Dv \rightarrow \infty$.

The inverse of DV (i.e., UL/D_a) is called the Peclet number (Pe). In chromatographic separations we want DV to be as small as possible (high Pe) for it is under these conditions that the efficiency of the column is high, i.e., there is less back mixing in the column. Being such an important parameter, we find it fitting to show how it can be obtained from an experiment.

The variance σ^2 of a continuous distribution is defined as

$$\sigma^2 = \frac{\int_0^\infty (t-\bar{t})^2 C dt}{\int_0^\infty C dt} \tag{14a}$$

$$= \frac{\int_0^\infty t^2 C dt}{\int_0^\infty C dt} - \bar{t}^2 \tag{14b}$$

If the measurements are made at finite number of equidistant locations (like from a tracer input), the variance simplifies to

$$\sigma^2 = \frac{\sum t_i^2 C_i}{\sum C_i} - \bar{t}^2 = \frac{\sum t_i^2 C_i}{C_i} - \left[\frac{\sum t_i C_i}{\sum C_i}\right]^2 \tag{15}$$

For closed systems,

$$\sigma_\theta^2 = \frac{\sigma^2}{\bar{t}^2} = 2\frac{D_a}{UL} - 2\left(\frac{D_a}{UL}\right)^2 [1 - \exp(-UL/D_a)] \tag{16}$$

Hence, the dispersion number D_a/UL can be calculated. For small dispersion,

$$\sigma_\theta^2 \approx 2\frac{D_a}{UL} \tag{17}$$

But perhaps the commonly used parameters for column efficiency are the number of theoretical plates, N, and the plate height, H (height equivalent to a theoretical plate, HETP). We now show the relationship of these parameters.

The peak variance is found to be directly proportional to column length:

$$\sigma^2 \alpha L$$

The proportionality constant is the plate height. Hence,

$$\sigma^2 = HL \tag{18}$$

Therefore, the plate height can be worked out easily. We want the band spread to be minimal (close to plug flow), i.e., σ should be small. Hence, high-efficiency columns have a low HETP.

The number of theoretical plates, N, is also related to σ as follows:

$$N = \frac{t_R^2}{\sigma^2} = \frac{L^2}{\sigma^2} \tag{19}$$

It is important that both σ and t_R or σ and L are in the same units. From Eqs. (18) and (19) it follows that

$$N = \frac{L}{\sigma^2} \times L$$
$$= \frac{L}{H}$$

From Eq. (17) we know that

$$Dv = \frac{1}{2}\frac{\sigma^2}{t^2}$$

hence,

$$Dv = \frac{1}{2N} \tag{20a}$$
$$= \frac{H}{2L} \tag{21a}$$

In Eq. (16), U is the superficial velocity. In Eq. (19), t_R is measured from interstitial velocity, U_i. Taking this into consideration, Eqs. (20a) and (21a), respectively, become

$$Dv = \frac{1}{2\epsilon N} \tag{20b}$$
$$= \frac{H}{2\epsilon L} \tag{21b}$$

where ϵ is the void fraction.

Accurate and reproducible measurements of σ, N, H, and Dv can also be obtained from the negative input signal as shown in Fig. 2.

4.3 Variables That Affect Column Efficiency

In Sec. 4.1 it was stated that adsorption is governed by the equilibrium $K = C_{sp}/C_{mp}$ [Eq. (1)]. According to this relationship, for adsorption to occur, the analyte must diffuse from the bulk of the sample solution to the surface of the adsorbent. Hence, column efficiency will be affected by:

Concentration gradient. This leads to longitudinal diffusion and is insignificant in LC. On the other hand, longitudinal diffusion is high in GC because diffusivities are high in the gaseous state.

Diffusion coefficient of the analyte in the mp. This depends on temperature and viscosity of the mp.

Thickness of the liquid coating. This is especially true in liquid–liquid chromatography.

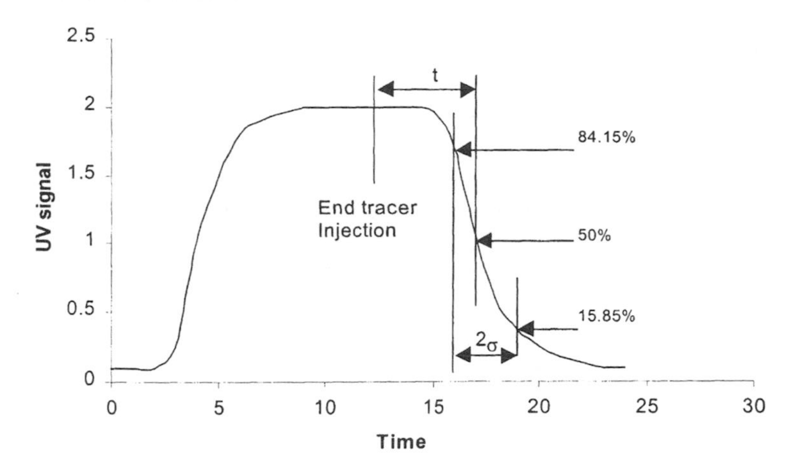

Figure 2 Determination of σ and t from a negative input signal.

Capacity factor. It was indicated that the capacity factor gives the migration rate of analytes on columns. The smaller it is, the lower the diffusion and hence the more efficient the column.

Diameter of the packing material. Adsorption depends on the surface area of the adsorbent, which in turn depends on the diameter of the particles. The smaller the diameter, the larger the surface area, the better the adsorption, and hence higher is the efficiency.

Velocity of the mp. The effect of mp can best be described by looking at its effect on HETP. Although it is easy to calculate HETP, understanding all the processes that contribute to HETP has been a problem yet to be solved. We shall consider one such equation here (the van Deemter equation), which has been used with some success.

$$H = A + \frac{B}{u} + C_s u + C_m u \tag{22}$$

where A and B are constants, C_s and C_m are the mass transfer coefficient in the sp and mp, respectively, and u is the flow velocity. The constant A takes into account column packing and the diameter of the packing material.

B/u is just the longitudinal diffusion and it decreases as flow velocity increases. This makes sense as at high flow velocity the analyte has less time to diffuse.

For adsorption to occur, time is needed for molecules to diffuse from the center of the mp to the surface of the sp. Thus, as flow velocity increases, there is less time for this equilibrium to be reached (in fact, it is never reached) hence, $H \alpha u$.

Another contribution of mp is known as eddy diffusion. As a molecules travels through the column, it can sample any of the multipaths available, consequently moving from one stream to the next. At low flow rates, diffusion averaging tends to occur, which reduces band spreading. At high flow rates there is less time for diffusion averaging to occur and the band spreads more. The overall effect of mp on HETP depends on which of the terms is dominant. At low flow rates B/u is dominant and hence HETP decreases (high column efficiency). It reaches a minimum as $C_s u$ and $C_m u$ become dominant and eventually increases with flow velocity. This is summarized in Fig. 3.

5. CHROMATOGRAPHY OF PROTEINS

Chromatography was initially developed for the separation of coloured compounds. With time, it became apparent that chromatography was the best tool for the large-scale purification of enzymes and other proteins. The major problem in the early days was the complex nature of the raw materials containing the proteins. A yeast homogenate sample contains hundreds of proteins that are chemically very similar and seem to behave in a similar way in a variety of solvents. The initial stationary phase depended on physical adsorption forces that are not selective at all. Separating a single protein from such a junk with such stationary phases was a demanding task. The successful application of chromatography for the purification of proteins is

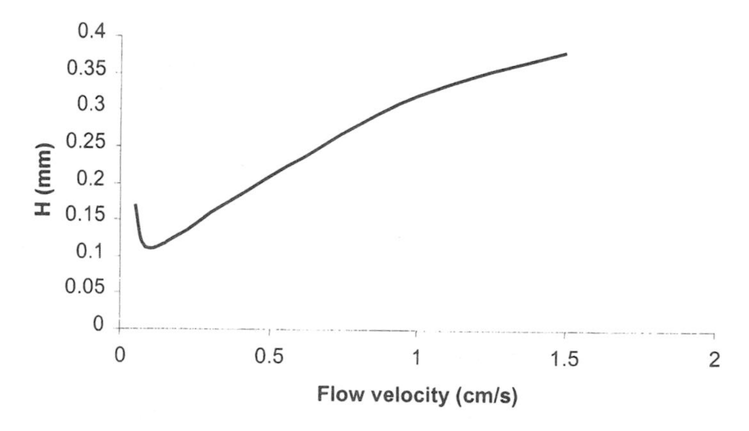

Figure 3 The effect of flow velocity on plate height, H.

greatly dependent on a thorough understanding of the nature and chemistry of the proteins to be separated.

5.1 The Proteins

Proteins are basically polymers composed of amino acids. The side chain of amino acids can be hydrophobic, e.g., phenyalanine, tyrosine, tryptophan, isoleucine, methionine, valine, etc. Seven amino acids are readily ionizable, i.e., histidine, aspartic and glutamic acids, cysteine, tyrosine, lysine, and arginine. A protein can be a dimer or tetramer, etc., and has a three-dimensional structure. Hence, the chemistry of the protein is a resultant chemistry of the amino aids and the three-dimensional nature of the protein. Ion exchange chromatography takes advantage of the ionizable groups on the proteins and hydrophobic interaction chromatography makes use of the hydrophobic side chains of the protein, whereas affinity chromatography makes use of the specific sequence of amino acids in the chain and the conformation of the protein.

Proteins are labile molecules and easily denatured by high temperature, extreme pH, organic solvents, and oxidative atmosphere. It is therefore of paramount importance that during any chromatographic run these factors be taken into account. If the protein is an enzyme, simple activity assays can be conducted to determine optimal conditions for the stability of the protein. It is often a good idea to add reducing agents such as β-mercaptoethanol or dithiothreitol to prevent oxidation. Speed is also essential in chromatographic runs. The longer the process the more the protein is exposed to the oxidative atmosphere.

5.2 The Chromatographic Matrix

A chromatographic matrix should (1) be insoluble in the buffer, (2) be hydrophilic, (3) be easily activated and coupled to a ligand (for affinity chromatography), (4) have large pores accessible to the protein, (5) have a large surface area, and (6) be physically and chemically stable to withstand the conditions during derivatization, sterilization, etc. (Janson and Rydén 1998). A variety of materials have been used as matrices. These include inorganic materials such as glass, silica, and hydroxyapatite; synthetic organic polymers like polyacrylamide, polystyrene, etc.; and polysaccharides. The latter are the most commonly used matrices, e.g., agarose based (e.g., Sepharose and Superose), dextran based (e.g., Sephadex) and cellulose based (e.g., Whatman and Sephacel). Agarose is particularly useful for adsorption chromatography because of its reasonable rigidity, stability, and the ease with which functional groups can be coupled to it.

In theory, a matrix should be used over and over again. However, this depends on cleaning procedures. There are some matrices that, even with the same cleaning procedures, lose their virgin properties after one use. They may either swell considerably (a typical problem with dextran-based exchangers) or lose their binding capacity to a large degree.

Selectivity of the Matrix

The sample containing the target protein has other proteins that can bind to the matrix. It is important, therefore, that the matrix show some selectivity for the target protein. This information can be obtained from a breakthrough curve, as is shown in Fig. 4. If the matrix has no selectivity, the target protein and the other proteins break through at the time (P_1). If the selectivity is good, the target protein breaks through well after the general proteins (P_2). Usually a higher selectivity entails a higher binding capacity and a higher purification fold. In a typical experiment conducted by Chang and Chase, they observed that Streamline SP had no selectivity for glucose-6-phosphate dehydrogenase, whereas Streamline DEAE showed very good selectivity (Chang and Chase 1996). Let us emphasize that the selectivity of the matrix will depend on experimental conditions, as discussed in Section 4.3, and on the nature of the target protein.

5.3 Chromatography Modes

There are basically two mechanisms used for chromatographic separation of proteins: adsorption and molecular sieve chromatography (Janson and

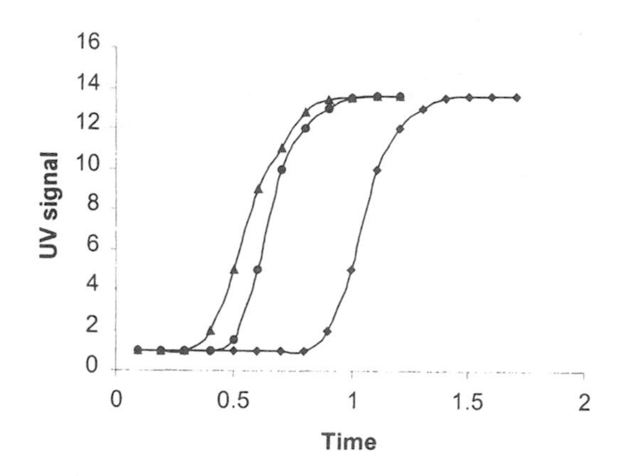

Figure 4 Typical breakthrough curves for proteins. General proteins (▲), P_1 (●) and P_2 (◆). The matrix has better selectivity for P_2 than for P_1.

Rydén 1998). The former includes adsorption to matrices derivatized with ion exchange, hydrophobic, and affinity ligands. Covalent chromatography, which involves formation of covalent bonds between the protein and the matrix, has been introduced as yet another mode of protein purification.

Amersham Pharmacia Biotech provides literature on all modes of chromatography, which may be useful to beginners in the area.

Ion Exchange Chromatography

The net charge of a protein depends on its amino acid sequence and on the pH of the buffer in which it has been solubilized. At its isoelectric point (pI), the net charge of a protein is zero. At pH values above this pI, a protein is negatively charged hence an anion. Therefore, if the matrix has positively charged groups (an anion exchanger), it can be used to purify the protein if the buffer pH is above the pI of the protein. Below the pI, the protein is positively charged; hence, a cation exchange matrix can be used to adsorb the protein. However, sometimes an ion exchanger may bind a protein having complementary charges on the surface even if the net charge is not suitable.

Operation Principle. There are four main stages in ion exchange purification of proteins: loading, washing, elution, and regeneration of the matrix.

The amount of sample to be loaded depends on the purpose of the process. For analytical purposes a small sample is injected at the top of the column and washed with a suitable elution buffer. The sample moves down the column until it is eluted. Today open columns are rarely used for analytical purposes. This is done efficiently with HPLC. In this section, we will discuss ion exchange as a preparative tool. Suffice to say that the basic components of today's open columns are similar to the basic components of HPLC, i.e., reservoir, pump, sample loop, column, detector, and fraction collector.

For preparative purposes, it is desirable to make good use of the capacity of the column. The sample is therefore applied until a set breakthrough percentage is reached (usually 5–10%). Sometimes loading to total breakthrough may be used. The mode used depends on the cost of the feed. The concentration of the feed should not be too high as it affects the pH of the flowthrough. For example, let's consider a column that has been equilibrated with phosphate buffer and a matrix that is an anion exchanger. The protein will have H^+ as its counterion and the column will have PO_4^{3-} as its counterion. When the sample is applied, the protein will displace the PO_4^{3-} ions from the matrix and HPO_4^{2-} is discharged. Hence, the pH of the flowthrough can be shown to be higher than that of the feed. If this pH change is high enough it can affect adsorption and stability of the adsorbed protein. Another problem is that the ionic strength of the flowthrough also increases at the rate of $1\,mM\,mg^{-1}\,ml^{-1}$ of protein adsorbed (Scopes 1993). It is recommended that the concentration of the protein to be adsorbed not be greater than 5 mg/ml (Scopes 1993). However, with proper buffering, this concentration can be as high as 15 mg/ml without any observable problem. From the work carried out in our laboratory (unpublished data), it may be advantageous to dilute the sample prior to loading as the purification fold was observed to be higher if the sample concentration was low.

During the passage of the sample through the column, some contaminating proteins get weakly adsorbed to the matrix whereas some just get trapped in the pores of the matrix. Washing gets rid of these proteins and other molecules. If the effluent passes through a UV detector (280 nm), washing is stopped when the signal returns to zero. Washing is usually done with the buffer used in the equilibration of the column.

In ion exchange chromatography, elution can be achieved in two ways: by changing the pH of the elution solution or by using a solution with a higher ionic strength than the protein. As stated at the beginning of this section, pH of the solution determines the overall charge on the protein. As the pH is decreased to below the pI, the protein will become positively charged (for anion exchange chromatography) and eventually desorb from the matrix. Resolution (purity of the targeted protein) is usually low. A word of caution should be given here. Much as the process is as easy as it appears, it is worth noting that the ionic strength of the loading buffer is usually low to maximize adsorption. This implies that the buffering capacity is low. Thus, a decrease in pH (or increase in pH for cation exchangers) is not as efficient as salt elution because the protein and to some extent the matrix will act as buffers (next section).

Salt elution is the most common means of eluting adsorbed proteins. Gradient or stepwise elution can be used. Gradient elution offers better resolution (hence higher purification) and sharper peaks (less band broadening) than stepwise elution. During elution with salt, proteins that are strongly adsorbed to the matrix move more slowly than those that are weakly bound. As such there is always a trailing edge. When elution is done by gradient elution, this trailing edge is continuously experiencing an increasing concentration of salt, and k decreases substantially thus decreasing this tailing. Figure 5 shows elution curves from gradient and stepwise elution. A simple gradient elution arrangement can be set up as in Fig. 6.

Ion exchange adsorbents are designed for repetitive use. However, the length of time that they maintain good binding characteristics depends on the cleaning procedures. Feed to columns contains some precipitated and

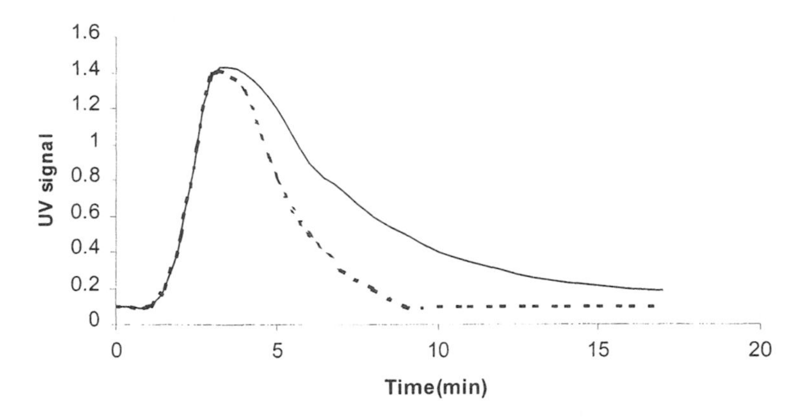

Figure 5 Elution profile of proteins by gradient (...) and one step elution (—).

denatured proteins and other molecules. These can affect the binding capacity and the selectivity of the matrix. Depending on the matrix, clumping can also occur. This affects the axial dispersion and can even cause severe channeling. It is therefore important that after each run the adsorbent be cleaned thoroughly free of these molecules.

The cleaning procedure should be optimized for each type of feed. For ion exchange adsorbents, 1 M NaOH has been used with success. It is cheap and does not contain harmful traces that can contaminate the product. It also sanitizes the system and destroys pyrogens. Amersham Pharmacia Biotech recommends running a high-salt concentrated solution, e.g., 1–2 M NaCl, through the ion exchange column before running the NaOH. Organic solvents, such as 70% ethanol, can be used as complements to NaOH. We found that ethanol also helps to break the lumps and hence reduces axial dispersion brought about by the clumping together of the particles.

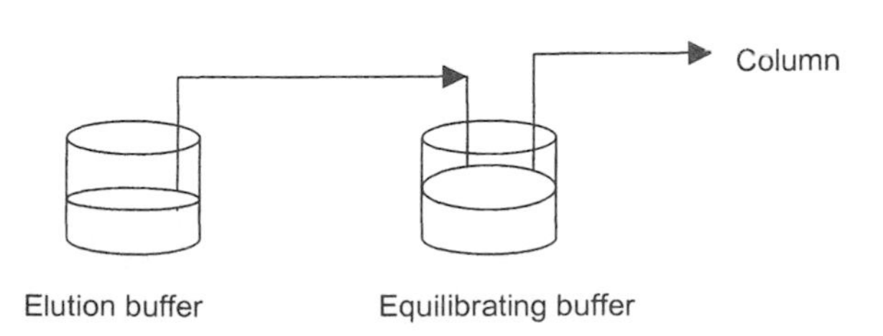

Figure 6 A simple arrangement for gradient elution.

Storage of the matrix is also crucial for its life span. Recommended buffers and other conditions for storage can be obtained from the manufacturers of the matrix.

Chromatofocusing

Chromatofocusing is a chromatographic technique in which proteins are eluted from an ion exchange column by a pH gradient. The proteins are separated according to their isoelectric points, pI. The protein with the highest pI elutes first. This technique was developed by Sluyterman and co-workers in late 1970s (Sluyterman and Elgersma 1978a,b; Sluyterman and Wijdenes 1981). As mentioned in the above section, protein elution may be achieved by changing the pH. When ordinary buffers are used, sudden pH changes occur in the column (Scopes 1993). As a result, proteins with different isoelectric points elute at the same time; hence, poor resolution is observed. In chromatofocusing, amphoteric buffers (which are polymeric) are used. They contain a large number of buffering species (both positive and negative) and have a strong constant buffering capacity over a wider range than ordinary buffers. The pH gradient in chromatofocusing is formed by the polybuffer (ampholyte) and the buffering action of the ion exchanger itself. Therefore, it is also important that the ion exchanger has a strong, even buffering capacity over a wide pH range.

Operation Principle. It is important that the isoelectric points of the proteins of interest fall roughly in the middle of the pH gradient. For example, let's consider an anion exchanger PBE94 (available from Amersham Pharmacia Biotech for chromatofocusing at pH 4–9). Suppose we want to separate two proteins P_x and P_y, which have pI of 7 and 7.5, respectively. The column will be equilibrated with buffer at pH 9, so that the proteins when loaded would bind to the column. To elute these proteins by pH gradient, the pH of a polybuffer is adjusted to the lower limit of the pH gradient, say, pH 6. As the polybuffer is added, a pH gradient begins to form. At the top of the column, the pH is close to that of the polybuffer. At the bottom of the column, the pH is close to 9. As more and more polybuffer is added, the pH at each point of the column is lowered until the pH of the effluent is 6.

If we look at the two arbitrary proteins, when the pH of the column reaches 7.5, P_y desorbs and travels down the column, whereas P_x will not desorb until the column pH reaches 7 at which point it will begin to travel down the column as fast as the pH gradient. This leads to the eventual separation of the two proteins. Suppose halfway during the process, a second sample of P_x and P_y (say, P_{x2} and P_{y2}) is loaded onto the column. These two proteins travel at the speed of the eluant until the column pH is 7 where

P_{x2} catches up with the first sample of P_x. However, P_{y2} continues at the speed of the eluant until the column pH reaches 7.5 where it catches up with P_y and the two travel down the column at the same pace. This is the *focusing* aspect of *chromatofocusing*. It is claimed (Scopes 1993) that chromatofocusing can be used to resolve proteins differing by 0.5 pI unit.

In summary, successful chromatofocusing requires a linear pH gradient to be formed inside the column. The gradient should not be too steep as poor resolution would be observed nor too flat as sample dilution would result.

Hydrophobic Interaction Chromatography

During the testing of affinity chromatography, when some matrices with nonpolar spacer arms (used to couple the ligand to the matrix) were used as controls, to the surprise of the researchers, these adsorbed as much protein as the affinity ligand bearing–matrices (Shaltiel 1974). As there were no charges on the matrix, it was clear that there was some form of interaction between the protein and the nonpolar spacer arm. This led to what we call hydrophobic interaction chromatography (HIC) a name coined by Shaltiel. Similar observations were reported by Porath et al. (1973).

A process is energetically favorable if the change in free energy (ΔG) is negative. Now ΔG is defined as $\Delta G = \Delta H - T\,\Delta S$, where H is the enthalpy of the process, T is temperature, and S is entropy.

When a protein, P, with nonpolar side chains is dissolved in water, the water molecules become ordered and the process can be represented as

$$P + xH_2O \rightarrow P.xH_2O$$

As we can see, ΔS for this process is negative. This implies that ΔG has increased (become less negative); hence, this process is energetically unfavorable. However, the above process is known to occur; therefore ΔG should be negative, which implies that ΔH is also negative. Hydrophobic interaction, e.g., between two proteins, can be represented as

$$P.xH_2O + P'.yH_2O \rightarrow P - P' + (x + y)H_2O$$

ΔS for this process is positive and large due to the water molecules leaving the more ordered structure around the individual solute molecules for the bulk solution. Hence, ΔG is more negative and the process is energetically favorable. However, the actual sign of ΔG depends on the magnitude of ΔS and ΔH (which is positive in this case). If the process above is conducted in a high-salt solution, the water molecules released would be trapped by the salt ions (hydration), and this drives the process in the forward direction. It is this behavior of proteins that is exploited in HIC, wherein a nonpolar chain is attached to a chromatography matrix and proteins with hydrophobic areas will be adsorbed to the matrix. Organic solvents, polyols, and nonionic detergents all reduce hydrophobic interactions. HIC is also known to be dependent on pH, temperature, and additives (Queiroz et al. 1996).

HIC is often performed under mild conditions (i.e., where irreversible binding is unlikely); thus, biological activity is usually retained. The binding capacity is high and recoveries are very good. As such it is a popular technique (Queiroz et al. 1996). for the large-scale separation of proteins and biopolymers (Staby and Mollerup 1996). It is, however, difficult to improve selectivity by changing experimental conditions. This is made worse if salting out was used in the preliminary stages as both of these processes operate on the sample principle. On the other hand, one does not need to desalt the sample prior to application on HIC.

Operation Principle. The sample must be prepared in a buffer containing a high salt concentration and the column should be equilibrated with the same buffer. The salt used is usually the one employed in salting-out experiments, i.e., ammonium sulfate. When loading is complete, the column should be washed with the equilibration buffer to remove all non-adsorbed proteins. Elution is achieved by passing a buffer of low ionic strength or decreasing polarity. For the latter, solvents like ethylene glycol, ethanol, and so forth, may be mixed with the buffer at desired concentrations.

Affinity Chromatography

Chromatography based on selective interaction of a protein with a ligand molecule is the most powerful separation technique. Affinity chromatography has been described in Chap. 3 and will only be briefly dealt with here. The affinity ligand is covalently coupled to the matrix using a suitable chemical reaction (Porath 1974). On passage of the sample through a column of affinity matrix, only the protein(s) having affinity for the ligand will be retained. After washing of the column, the bound protein is eluted under conditions reducing the protein-ligand interactions. As these interactions are often based on electrostatic, hydrophobic, and hydrogen bonds, agents that weaken these interactions can be used as nonspecific eluants. Alternatively, elution can be effected by the buffer supplemented with the free ligand.

The selectivity of affinity chromatography depends on the specificity of interactions. Some ligands can bind to only one protein whereas others bind to a group of proteins. The ligands used in affinity chromatography are antibodies for antigens, receptor molecules for hormones, substrate analogues and inhibitors for enzymes, lectins for glycoproteins, sugars for lectins, reactive dyes for dehydrogenases and other proteins, metal ions for metal-binding proteins, and so on.

Capacity of the affinity matrix is related to the accessibility of the ligand, strength of ligand-protein interaction, presence of competing molecules, other environmental conditions, and stability of the ligand-matrix linkage.

Covalent Chromatography

In this type of bioselective adsorption, the protein becomes covalently bound to the matrix, i.e., the protein reacts chemically to the matrix. The other components can be washed from the column and the covalent bond is readily cleavable under mild conditions to yield the native target protein. Brocklehurst et al. (1973) were the first to call this type of chromatography "covalent chromatography." Thiols (glutathione, thiopropyl, *N*-acetyl homocysteine derivatives) are commonly used in this type of chromatography.

Most proteins contain the amino acid cysteine in their side chains. From the three-dimensional structure of proteins, we also know that disulfide bonds are common either in the same chain or between neighboring chains. It is this property that is exploited in covalent chromatography. A molecule containing a thiol group is immobilized onto a matrix. These immobilized thiol groups can couple SH groups on proteins. C=O, C=C, and C=N functional groups can similarly be coupled.

The key to this technique lies in the way the thiol is attached to the matrix. For the case of glutathione (GSH), there are two mechanisms in which it can be immobilized onto the matrix.

Mechanism 1

agarose + HOOC H$_2$N ... NH ... CH$_2$SH ... HOOC NH ⟶ agarose —NH ... HOOC ... NH ... CH$_2$SH ... HOOC NH

Mechanism 2

agarose + HSH$_2$C ... NH COOH ... NH ... NH$_2$ COOH ⟶ agarose —S ... NH COOH ... NH ... NH$_2$ COOH

The dominant product is a function of the pH at which the coupling is conducted. Simons and Vander Jagt (1977) observed that the product obtained by mechanism 1 (at higher pH) has less specificity to glutathione S-transferase (GST) (an enzyme used as a fusion partner for the purification of recombinant proteins) than the one obtained at pH 7 (mechanism 2). In fact, to produce product 2, one of the protocols recommends doing the experiment at pH 8.5 (Brocklehurst et al. 1974; Scouten 1981). However, contrary to Simons and Vander Jagt's observation, we observed that product obtained at higher pH showed better specificity and higher capacity than that produced at neutral pH. This difference may be due to the isomers of GST being purified.

Although product 1 can be used to purify some proteins, the proton of SH is not a good leaving group. Therefore, product 1 is reacted with 2, 2′-dipyridyldisulfide (dithiol dipyridine) to introduce a better leaving group (hence activated). Any protein with thiol (SH) functional groups can thus be adsorbed. The process is represented below.

Matrix]-SH → matrix]-S—S (pyridyl)

[Protein]-SH

[Protein]-SH ← Excess glutathione — matrix]-S-S-[protein]

Operation Principle. This section is specific for the GSH matrix activated by reaction with Py-2-S-S-2-Py. As in ion exchange, we recommend that the sample be prepared in the buffer used for equilibrating the column. Tris-HCl buffer, 0.1 M, pH 8 and sodium acetate buffer, 0.1 M, pH 4 both containing 0.3 M NaCl and 1 mM EDTA are recommended buffers (Scouten 1981). Other buffers may also be used as long as they do not interfere with the stability of the matrix and the target protein. During loading, Py-2-SH is eluted from the column, and since it absorbs at 343 nm it can easily be monitored. In fact, the displacement of Py-2-SH can also be used to determine the efficiency of the coupling reaction.

After loading the column is washed with the buffer until absorbance at 280 nm or 343 nm has returned to zero. Elution can be achieved with L-cysteine or GSH in buffer. The column is regenerated by reaction with 30 mM dithiothreitol followed by 1.5 mM 2-Py-S-S-2-Py in buffer.

Gel Filtration Chromatography

In gel filtration chromatography, also called molecular exclusion or molecular sieve chromatography, molecules are separated according to size as they pass through a bed of porous beads. Since proteins differ in size, gel filtration can be used to purify a target protein. When a protein sample is applied to a bed, small proteins penetrate the pores of the beads while the large proteins are excluded (hence the term molecular exclusion). These larger molecules consequently elute earlier than the small molecules. If the resolution of proteins is poor, the only option is to use beads of different pore sizes or increase the retention time by increasing column length or decreasing flow rate. Thus, it is important to keep in mind that the success of gel filtration depends on the pore size of the beads (cutoff points), column packing, and column length. Assuming that the right column, column packing, and column length are in place, gel filtration is easy to run.

The column is equilibrated with an appropriate buffer. A small sample (0.5–5% of bed volume) is usually applied and then washed down the column by the buffer. If an on-line detector is used (which is a standard practice these days), the peaks can be collected and analyzed. Some proteins may exhibit ionic or hydrophobic interactions with the matrix as a result of which they are eluted slowly or may be retained on the column. In such cases, the mobile phase may be modified with respect to ionic strength, pH, polarity, and so forth in order to suppress these interactions.

Gel Filtration as a Separation Tool. A typical example of gel filtration as a separation tool is the purification of IgG from sheep serum. The main constituents of whole serum are IgG (M_w 160 kDa) albumin (M_w 67 kDa). A typical run of sheep whole serum on Sephadex G-100 is shown in Fig. 7. As expected, two peaks are observed. The first corresponds to IgG that is excluded from the pores of the gel. The second corresponds to albumin that could penetrate into the pores of the gel and thus has a longer retention time.

The main limitations of gel filtration are its low capacity and relatively broad distribution of pore size, allowing only small sample volumes to be chromatographed and separation of proteins having a large difference in size. As a result, gel filtration has been found useful as a final polishing stage during downstream processing when the product is present in small volumes. The technique is used for separation of the native target protein from the aggregated forms or other contaminants, for desalting, or for buffer exchange.

Gel Filtration for Molecular Weight Determinations. Since the elution volume V_e depends on the molecular weight and the column para-

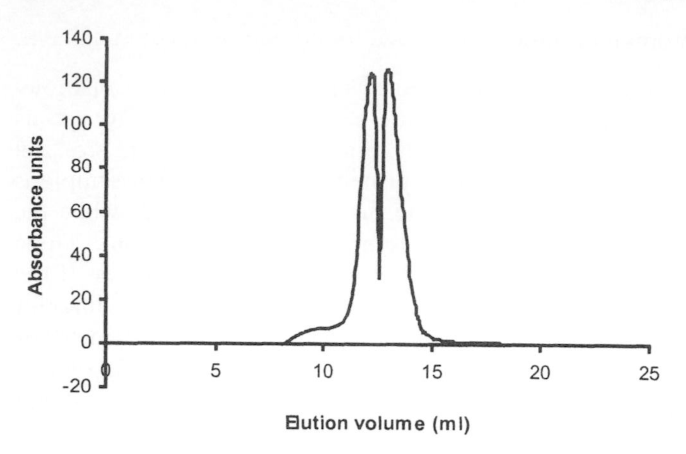

Figure 7 Gel filtration of sheep whole serum on Sephadex G-100. The experiment was conducted on a 30-cm column (9 mm i.d.) and flow rate was 1 ml/min.

meters, it follows that same molecules will elute at the same time on the same column. If the column has been calibrated with proteins of known molecular weight, the molecular weight of unknown protein sample can be estimated. This has been applied with success (Siegel and Monty 1966; Locascio et al. 1969; Ansari and Mage 1977). So far we have ignored the shape of the molecules. It should be obvious that the rate at which a spherical molecule will move down the column would be different from that of a linear molecule. For molecular weight determinations to be accurate, the proteins must all be of the same shape. This is made possible by the addition of urea or guanidine chloride, which transform proteins into a random coil, thus reducing structural differences.

The molecular weight of proteins can also be determined from some relationship that exists between K_{av} and the logarithm of molecular weight (K_{av} is a partition coefficient though not a true one as it is defined differently from the one in Sec. 4.1). Consider a packed column of total volume V_t. The mobile phase will travel in the void volume V_0. Any molecule loaded onto the column will elute after volume V_e. K_{av} is defined as

$$K_{av} = \frac{V_e - V_0}{V_t - V_0}$$

Note that $(V_t - V_0 =)V_p + V_i)$ where V_p is the volume of the particles and V_i is the interstitial volume, i.e., the total volume of the pores. Since V_p is inaccessible to all molecules, K_{av} can not be a true partition coefficient. For a particular matrix it is found that K_{av} has a linear relationship with $\log M_w$. This leads to the construction of selectivity curves that can be used to estimate molecular weights of unknown samples (Giddins 1965; Johansson and Aahsberg 1986).

6. MANUFACTURERS OF CHROMATOGRAPHIC MEDIA AND EQUIPMENT

A variety of chromatographic media are commercially available. We have listed a few and given some examples for each type of matrix (Table 1). It is left to the reader to look for specific matrices from the manufacturers' catalogues. It is also worth mentioning here that most manufacturers pro-

Table 1 Some Commercially Available Matrices for Chromatography of Proteins

Media	Manufacturer/supplier	Examples	Particle size (μm)
Cation exchange matrix	Sigma-Aldrich	CM Sepharose, CM Cellulose, CM Sephadex	45–165
	Amersham Pharmacia Biotech	CM Sephadex C-50,	110–400
		CM Sepharose high performance	34
		SP Sepharose XL	45–165
	Biorad	AG 50W-X8	
		CM Bio-Gel A	
Anion exchange matrix	Sigma-Aldrich	Q Sepharose, DEAE Sepharose	45–165
		QAE Sephadex	40–125
	Amersham Pharmacia Biotech	DEAE Sephacel	40–160
		DEAE Sephadex A-25	45–190
	Biorad	DEAE Bio-Gel A	
Affinity matrix	Sigma-Aldrich	Blue dextran, *S*-Hexylglutathione Cellobiose	
	Amersham Pharmacia Biotech	Heparin Sepharose 6 Fast Flow,	45–165
		IgG Sepharose 6 Fast Flow	45–165
		Protein A Sepharose 4 Fast Flow	45–165
	Biorad	Affi-Gel Protein A	
Hydrophobic matrix	Sigma-Aldrich	Benzylamine, Isopropyl, Phenyl	
	Amersham Pharmacia Biotech	Phenyl Sepharose CL-4B	45–165
		Octyl Sepharose 4 Fast Flow	45–165
	Biorad	Methyl, *t*-Butyl	50

vide activated matrices for attaching specific ligands. We haven't included these activated matrices in our list.

Further to the list provided, to our knowledge Amersham Pharmacia Biotech provides media and equipment for use in expanded bed chromatography. These are sold under the Streamline trade name. Lately other companies, such as Bioprocessing Ltd, Upfront, and Biosepra (Ghose 1999), have come up with matrices for expanded bed absorption chromatography. In the table we have not included any expanded bed matrices or equipment.

The chromatography equipment, including columns, fraction collectors, UV detectors, and pumps, are available from Amersham Pharmacia Biotech, Biorad, and Gilson.

REFERENCES

Adams B.A., Holmes E.L. (1935) Absorptive properties of synthetic resins. J. Soc. Chem. Indust. 54, 1–6T.

Alm R.S., Williams R.J.P., Tiselius A. (1952) Gradient elution. Acta Chem. Scand. 6, 826–836.

Amersham Pharmacia Biotech, Expanded Bed Adsorption: Principles and Methods. Booklet.

Ansari A.A., Mage R.G. (1977) Molecular weight estimation of proteins using Sepharose CL-6B in guanidine chloride. J. Chomatogr. 140, 98–102.

Brocklehust K., Carlsson J., Kierstan M.P.T., Crook E.M. (1973) Covalent chromatography: preparation of fully active papain from dried papaya latex. Biochem. J. 133, 573–584.

Brocklehust K., Carlsson J., Kierstan M.P.T., Crook E.M. (1974) Covalent chromatography by thiol-disulfide interchange. In: Affinity Techniques: Enzyme Purification, Part B (Jakoby W.B., Wilchek M., eds.) Methods in Enzymology, Vol. 34, Academic Press, New York, pp. 531–544.

Chang Y.K., Chase H.A. (1996) Ion exchange purification of G6PDH from unclarified yeast cell homogenates using expanded bed adsorption. Biotechnol. Bioeng. 49, 204–215.

Chase H.A., Draeger N.M. (1992) Affinity purification of proteins using expanded beds. J. Chromatogr. 597, 129–145.

Consden R., Gordon A.H., Martin A.J.P. (1944) Quantitative analysis of proteins. A partition chromatographic method using paper. Biochem. J. 38, 224–232.

Ettre L.S., Zlatkis A. (1975) 75 Years of Chromatography: A Historical Dialogue. Elsevier, Amsterdam.

Giddins J.C. (1965) The Dynamics of Chromatography, Part 1: Principles and Theory. Marcel Dekker, New York.

Ghose S. (1999) Protein adsorption, expanded beds. In: Bioprocess Technology: Fermentation, Biocatalysis and Bioseparation (Flickinger M.C., Drew S.W., eds.), John Wiley, New York, pp. 2124–2133.

Heftmann E. (1973) History of chromatography and electrophoresis. In: Chromatography. A Laboratory Handbook of Chromatographic and Electrophoretic Methods (E. Heftmann, ed.), Van Nostrand-Reinhold, New York, pp. 19–26.

Heftmann E. (1983) Chromatography: Fundamentals and Applications of Chromatographic and Electrophoretic Methods Part A: Fundamentals and Techniques, Elsevier, Amsterdam.

Horvath C.G., Preiss B.A., Lipsky S.R. (1967) Fast liquid chromatography: an investigation of operation parameters and separation of nucleotides on pellicular ion exchangers. Anal. Chem. 39, 1422–1428.

James A.T., Martin A.J.P. (1952) Gas-Liquid partition chromatography: the separation and micro-estimation of volatile fatty acids from formic acid and dedicanoic acid. Biochem. J. 50, 679–690.

Janson J.C. and Rydén L., eds. (1998) Protein Purification: Principles, High Resolution Methods and Applications. John Wiley and Sons, New York.

Johansson B.-L., Aahsberg L. (1986) Column lifetime of Superose at 37°C and basic pH. J. Chromatogr. 351, 136–139.

Kirchner J.G., Mille J.M., Keller G.J. (1951) Separation and identification of some terpenes by a new chromatographic technique. Anal. Chem. 23, 420–425.

Locascio G.A., Tigier H.A., Batle A.M. (1969) Estimation of molecular weight of proteins by agarose gel filtration. J. Chromatogr. 40, 453–457.

Martin A.J.P., Synge R.L.M. (1941) A new form of chromatogram employing two liquid phases. Biochem. J. 35, 1358–1368.

McCreath G.E., Chase H.A., Purvis D.R., Lowe C.R. (1992) Novel affinity separations based on perfluorocarbon emulsions: use of a perfluorocarbon affinity emulsion for the purification of human serum albumin from blood plasma in a fluidized bed. J. Chromatogr. 597, 189–196

Moore S., Stein W.H. (1951) Chromatography of amino acids on sulfonated polystyrene resins. J. Biol. Chem. 192, 663–681.

Porath J. (1974) General methods and coupling procedures. In: Affinity Techniques: Enzyme Purification: Part B (Jakoby W.B., Wilchek M., eds.) Methods in Enzymology, Vol. 34, Academic Press, New York, 13–30.

Porath J., Flodin P. (1959) Gel filtrations: A method for desalting and group separation. Nature 183, 1657–1659.

Porath J., Sundberg L., Fornstedt N., Olsson I. (1973) Salting out in amphiphilic gels as a new approach to hydrophobic adsorption. Nature 245, 465–466.

Porath J., Carlsson J., Olsson I. (1975) Metal chelate affinity chromatography, a new approach to protein purification. Nature 258, 598–599.

Queiroz J.A., Garcia F.A.P., Cabral J.M.S. (1996) Hydrophobic interaction chromatography of *Chromobacterium viscosum* lipase on polyethylene glycol immobilised on Sepharose. J. Chromatogr. A 734, 213–219.

Rieman III, W., Walton H.F. (1970) Ion Exchange in Analytical Chemistry, Pergamon Press, Oxford.

Regnier F.E., Gooding K.M. (1992) Proteins. In: Chromatography, Part B, Journal of Chromatography Library, Vol. 56B (Heftmann E., ed.) Elsevier, Amsterdam, pp. 151–169.

Reichstein T., Shoppee C.W. (1949) Chromatography of steroids and other colourless substances by the method of fractional elution. Faraday Soc. Disc. 7, 305–311.

Scopes R. (1993) Protein Purification: Principles and Practice, Springer-Verlag, New York.

Scouten W.H. (1981) Affinity Chromatography. Bioselective Adsorption on Inert Matrices, John Wiley and Sons, New York.

Shaltiel S. (1974) Hydrophobic chromatography. In: Affinity Techniques: Enzyme Purification: Part B (Jakoby W.B., Wilchek M., eds.) Methods in Enzymology, Vol. 34, Academic Press, New York, pp. 126–140.

Siegel L.M., Monty K.J. (1966) Determination of molecular weights and fractional ratios of proteins in impure systems by use of gel filtration and density gradient centrifugation. Application to crude preparations of sulfite and hydroxyamine reductase. Biochim Biophys. Acta 112, 346–352.

Simons P.C., Vander Jagt D.L. (1977) Purification of glutathione-s-transferases from human liver by glutathione-affinity chromatography. Anal. Biochem. 82, 334–341.

Sluyterman L.A.A., Elgersma O. (1978a) Chromatofocusing: Isoelectric focusing on ion exchange columns 1. General principles. J. Chromatogr. 150, 17–30.

Sluyterman L.A.A., Elgersma O. (1978b) Chromatofocusing: Isoelectric focusing on ion exchange columns II. Experimental verification. J. Chromatogr. 150, 31–34.

Sluyterman L.A.A., Wijdenes J. (1981) Chromatofocusing III: The properties of a DEAE-agarose anion exchanger and its suitability for protein separations. J. Chromatogr. 206, 429–444.

Staby A., Mollerup J. (1996) Solute retention of lysozyme in hydrophobic interaction perfusion chromatography. J. Chromatogr. A 734, 205–212.

Tiselius A. (1937) A new apparatus for the electrophoretic analysis of colloidal mixtures. Faraday Soc. Trans. 33, 524–531.

Tiselius A., Hagdahl L. (1950) A note on "Carrier displacement" chromatography. Acta Chem. Scand. 6, 394–395.

William R., Kirby H.M. (1948) Paper chromatography using capillary accent. Science 107, 481–483.

Zechmeister L., von Cholnoky L. (1937) Die Chromatographische Adsorptionsmethode, Springer-Verlag, Vienna.

3

Selectivity in Affinity Chromatography

Ipsita Roy and Munishwar Nath Gupta
Indian Institute of Technology, New Delhi, India

1. INTRODUCTION

Chromatographic techniques have always occupied a somewhat prominent place among bioseparation protocols for enzymes/proteins (Villamon et al. 1999; Lopuska et al. 1999; Feng et al. 1999). Their continued general popularity as a bioseparation strategy is based on the high resolution one can usually achieve. A widely used book is rightly limited to chromatography and electrophoresis while covering high-resolution methods (Janson and Rydén 1989). In the last decade or so, the needs of the biotechnology industry have created immense pressure to develop techniques that can handle large volumes and deliver high purity. Recent chromatographic techniques have met this challenge quite effectively. Displacement chromatography (Chap. 12; Freitag et al. 1999), perfusion chromatography (a convective chromatography) (Fahrner and Blank 1999; Liao et al. 1999), radial flow chromatography (Wallworth 1996), and expanded bed chromatography (Chap. 10; Hjorth 1997; Chase 1998; Galaev 1998) are capable of handling high volumetric throughput. The requirement of high purity has been met by interfacing chromatography with the affinity approach (Yoshikawa et al. 1999; Guyonnet et al. 1999).

Historically speaking, chromatography has been the first technique to be converted to an affinity-based approach (Jakoby and Wilchek 1974; Scouten 1981; Dean et al. 1984; Dunn 1984; Parikh and Cuatrecasas 1985; Pharmacia 1993), and affinity chromatography is now a well-established technique used by enzymologists (Jakoby and Wilchek 1974; Skorey et al. 1999). Lately, this type of interfacing of the affinity concept has been extended to a large number of separation techniques (Herak and Merrill 1989; Gupta and Mattiasson 1994; Gupta et al. 1996; Zusman 1998) (Fig. 1). Actually, throughout this chapter, most of the discussion on selectivity in affinity chromatography is equally relevant to the techniques mentioned in Fig. 1.

Affinity chromatography is a type of adsorption chromatography in which the molecule to be purified is specifically and reversibly adsorbed by a complementary binding substance ("ligand") immobilized on an insoluble support ("matrix"). Purification can be of the order of 100–1000-fold, and recoveries of the bioactive material are generally very high. Affinity chromatography involves the formation of a reversible complex between the target protein and the ligand. It can, in principle, be applied to a wide variety of macromolecule-ligand systems. For example, specific adsorbents may be used to purify enzymes, antibodies, antigens, nucleic acids, vitamin-binding proteins, transport proteins, repressor proteins, drug or hormone receptors, sulfhydryl-containing proteins, peptides formed by organic synthesis, intact cell populations, specialized polyribosomal complexes, and multienzyme complexes. The target protein from the crude extract interacts with the

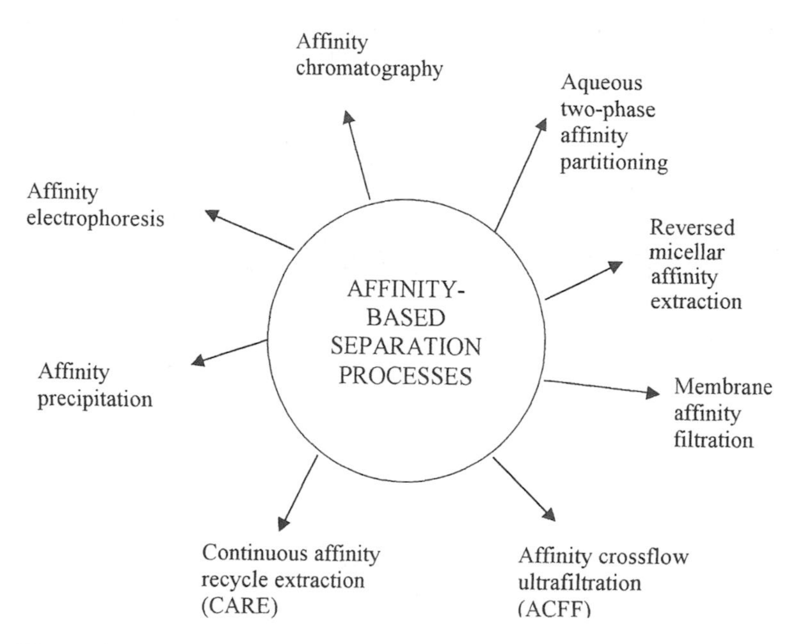

Figure 1 Some affinity-based separation techniques.

immobilized ligand and remains bound to it; the other constituents of the crude extract (referred to as "contaminants") are removed. In some cases, a "spacer arm" is introduced between the ligand and the matrix to improve binding (Table 1). Desorption of the bound target molecule is generally accomplished by a nonspecific or a specific elution method.

Before we go ahead with a discussion on selectivity, it is worthwhile to ponder the nature of affinity interactions. Table 2 provides an illustrative list of the classical (from "precombinatorial" era) affinity ligands, which have been used in affinity chromatography, and other affinity-based approaches. Except in the cases of protein A, textile dyes, and metal ions, the choice of the affinity ligand has its origin in in vivo biochemical phenomena. During such a molecular recognition process, molecules are supposed to show adequate binding because of "biochemical affinity." Conceptually speaking, this so-called biochemical affinity was accepted as a *fait accompli* of nature and simply exploited in an in vitro situation for its capability to pick out one member of an affinity pair using the other. As our understanding of these affinity interactions at a molecular level grew as a result of powerful techniques like X-ray and nuclear magnetic resonance (NMR) (Bolognesi et al. 1999; Geyer et al. 1999), it has been possible to dissect the affinity interactions in terms of individual weak interactions. In many cases, it has been possible to do so in a quantitative manner in terms of thermodynamic quantities associated with each type of non-covalent interactions, e.g., in the case of lectin-sugar interactions (Srinivas et al. 1999). Thus, one can say that the affinity-based recognition between two members of an affinity pair is the result of an optimization of bonds acting in synergy to generate adequate bonding between them. Hence, it should be possible to engineer (protein engineering) and design (combinatorial chemistry) affinity ligands. It is vital to understand this transition in our conceptual understanding of biological affinity as this paved the way for obtaining novel affinity ligands without regard to the structure of the "natural" ligand. Figure 2 traces and underlines this transition.

Table 1 Use of Spacer Arms in Affinity Chromatography

Ligand	Protein	Spacer
Small	Small	Short
Small	Large	Long
Large	Small	None
Large	Large	None

Table 2 Some of the Ligands Used in Affinity Chromatography

Ligand	Specificity
Triazine dyes	Nucleotide-binding proteins, kinases, dehydrogenases
Lectins	Carbohydrates
Benzamidine	Trypsin
Protein A	Fc fragment of antibody
Protein G	Antibodies
Chelated metal ions	Histidine residues
Histones	DNA
NAD(P)	Dehydrogenases
Poly (U)[a]	Poly(A)[a]
Poly(A)[a]	Poly(U)[a]
Lysine	Plasminogen, rRNA, dsDNA
Arginine	Prothrombin, fibronectin
Gelatin	Fibronectin
2′,5′-ADP	$NADP^+$
Heparin	Lipoproteins, RNA, DNA
Boronate	tRNA, plasminogen, *cis*-diols
Calmodulin	Kinases
Polymyxin	Endotoxins

[a]A, adenine; U, uridine.

1.1 Nature of the Affinity Ligand

Ligands are often classified as either mono- or group-specific. Monospecific ligands are more specific and bind to a single enzyme/protein. Vitamins and hormones for their respective receptors, enzyme inhibitors for enzymes, and antibodies for antigens are some illustrative examples of such affinity pairs. Group-specific ligands, on the other hand, bind to specific classes. Coenzymes [like NAD^+ and $NAD(P)^+$] for their dependent dehydrogenases, biomimetic ligands, lectins for their specific glycoproteins, sugars for lectins, etc., belong to this class. A large number of enzymes/proteins bind to heparin, which is a very versatile group-specific affinity ligand (Estable et al. 1999; Poulouin et al. 1999). Immobilized metal affinity chromatography (IMAC) can also be placed in this class (Wisen et al 1999; Wu and Bruley 1999).

Quite a few of the affinity media incorporating both classes of affinity ligands are commercially available. Such affinity media are often available for both low- and medium-pressure chromatography as well as high-performance liquid affinity chromatography.

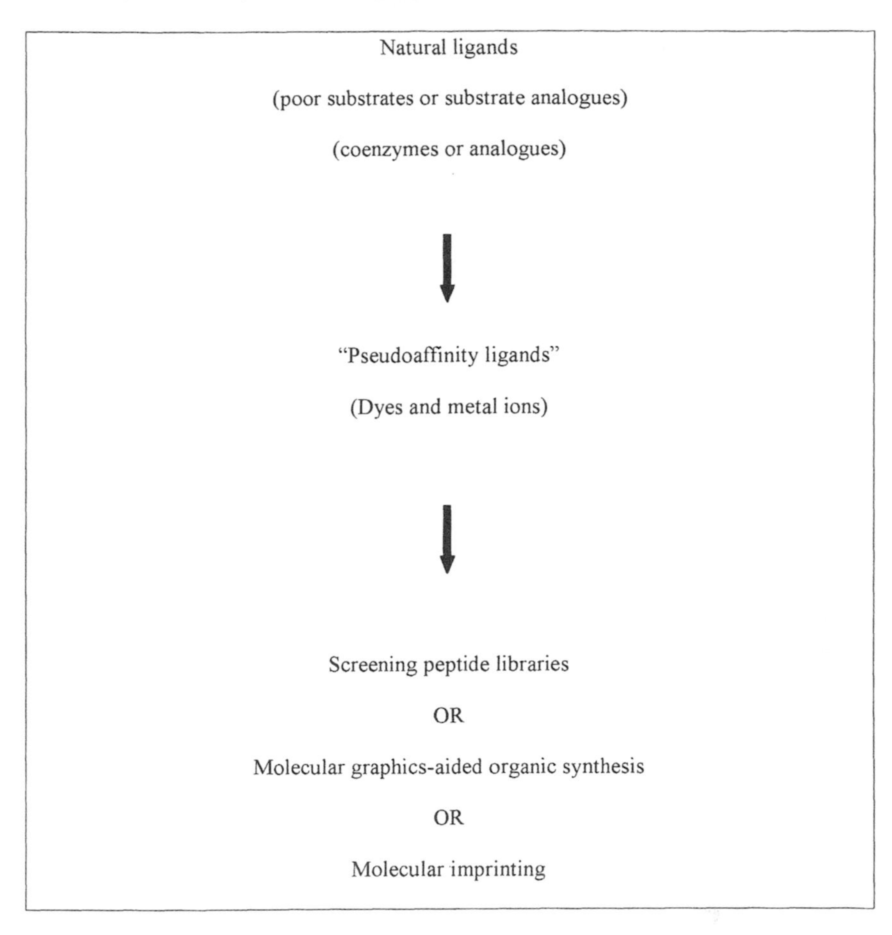

Figure 2 Different stages of development of ligand design.

1.2 Coupling Chemistry

It is generally advisable to covalently couple the affinity ligand to the matrix. This ensures the storage as well as operational stability of the affinity media. Today a very large variety of options are available while choosing the coupling chemistry for linking the affinity ligand to the matrix. Some excellent volumes and many reviews are available on this subject (Wilchek et al. 1984; Mosbach 1987, Mosbach 1988; Carlsson et al. 1989; Gupta and Mattiasson 1994; Liapis and Unger 1994; Burton 1996). Only a brief overview is presented here.

Covalent coupling involves three steps:

Activation of the matrix
Coupling of the affinity ligand
Blocking of residual activated groups by adding a suitable low molecular weight substance

With low molecular weight affinity ligands, it is sometimes necessary to insert a "spacer arm" between the matrix and the affinity ligand to avoid low steric availability of the ligand to the target molecule. This also minimizes the nonspecific interaction between the target enzyme/protein and the matrix. However, spacers, whether hydrophobic like γ-aminobutyric acid, diaminobutane or diaminohexane, or hydrophilic like glycine oligomers, often constitute sites of nonspecific interactions and should be used only if required. Also, glutaraldehyde and bisoxirane, if used for activation, provide inbuilt spacers. Recently, dextrans have been proposed as spacer arms (Penzol et al. 1998). The authors report a 25% increase in the enzyme activity when rennin is immobilized on agarose gel using dextrans as spacer arms instead of 12-carbon-atom (6% increase) or 2-carbon-atom (<3% increase) spacer arms.

The choice of the activating chemistry is of course dictated both by the nature of the matrix and the functionalities present on the affinity ligand (Hermanson et al. 1992; TechNote #205, 1999). The most common situation is a matrix with $-OH$ or $-COOH$ group or, less frequently, an amino group.

Table 3 lists the most commonly used methods for activating such groups. This table is not a comprehensive one but the impression that it conveys, that more methods are available for working with matrices

Table 3 Some Commonly Used Methods for Activating Functional Groups on the Matrix

Functional group (matrix)	Activating reagent	Functional group on the ligand to be coupled
$-OH$	Cyanogen bromide	$-NH_2$
	Bisoxiranes	$-OH$; $-NH_2$; $-SH$
	Epichlorohydrin	$-OH$; $-NH_2$' $-SH$
	Carbonyldiimidazole	$-NH_2$
	Tosyl chloride	$-NH_2$; $-SH$
	Tresyl chloride	$-NH_2$; $-SH$
	Divinylsulfone	$-OH$; $-NH_2$; $-SH$
$-COOH$	Carbodiimides	$-NH_2$
$-NH_2$	Glutaraldehyde	$-NH_2$

containing −OH group, is essentially correct. Methods are also available for activating matrices with other groups like amide, −SH, or −SiOH. An interesting underexploited method is the use of Ugi's reaction (Carlsson et al. 1989). In this, the ligand to be attached contains one or more functional groups out of the amino, carboxyl, carbonyl, or isocyanide groups. The matrix should contain any of the others. The remaining functional groups (from the list of four) have to be added in the form of reagents.

One can also convert the functional group on the matrix to other groups if demanded by the nature of the ligand. For example, $-NH_2$ group on a matrix can be converted to −COOH group by reacting with succinic anhydride. The reverse is easily achievable by attaching a diamine spacer via carbodiimide method. Also, nonfunctionalized matrices can be used after some creative applications of chemistry. For example, polystyrene can be used after nitration, conversion of nitro group to aromatic amine, diazotization, and coupling via carboxyl group on the ligand.

It may be worthwhile to add some pointers regarding different methods that are often overlooked. The buffers should be carefully chosen so as to not interfere with the reactions. Thus, one should avoid buffers like Tris or glycine while using chemistry involving amino group either on the matrix or on the ligand. Similarly, one should avoid phosphate and acetate buffers while using carbodiimides. The efficiency of carbodiimide [two frequently used water-soluble carbodiimides are 1-ethyl-3-(3-dimethylaminopropyl) carbodiimide and 1-ethyl-3-(3-dimethylaminopropyl)carbodiimide hydrochloride] coupling can be increased by adding sulfo-*N*-hydroxysuccinimide. Sometimes it is advisable to incorporate a blocker in the coupling buffer to prevent nonspecific binding of the affinity ligand (and even target enzyme/protein subsequently) to the matrix. This is one aspect that is often overlooked. Bovine serum albumin, polyethylene glycol, and nonionic surfactants like Tween-20 and Triton X-100 are the most frequently used blockers (Hauri and Bucher 1986; Vogt et al. 1987).

Most of the chemistry involved in preparing an affinity media is very similar to what is employed in immobilization of enzymes/proteins via the covalent coupling method. Thus, the experience gained or lessons learned in one area have often been successfully transferred to the other.

1.3 Elution Procedures

Elution of the bound enzyme/protein from an affinity matrix can be carried out by a nonspecific method or in a specific manner. Changes in pH (Gupta et al. 1994; Posewitz and Tempst 1999; Yang et al. 1999); ionic strength (Gupta et al. 1993; Tyagi et al. 1996b); use of chaotropic salts (such as KSCN or KI) (Bradshaw and Sturgeon 1990); use of denaturing agents like urea or guanidine hydrochloride (Greenwood et al. 1994; Diaz-Camino and Villanueva 1999; Govrin and Levine 1999); use of detergents (below critical micellar concentration) like octylglycosides, Triton X-100, or Tween-20 (Le et al., 1994) have all been frequently done successfully. The principle is to decrease the strength of one or more types of bonds that are responsible for effective association. For example, if hydrophobic forces dominate the binding, incorporation of ethylene glycol often helps in eluting the protein (Yarmush et al. 1992; Agarwal and Gupta 1996). Obviously, the use of a parameter (low or high pH) or an eluting agent has to be compatible with the stability of the bound enzyme/protein.

On the other hand, specific elution is generally a mild desorption method that involves the use of a competitive substance (Eccleston and Harwood 1995; Armisen et al. 1999; Zwicker et al. 1999). Use of a competitive inhibitor/coenzyme of an enzyme (Sardar and Gupta 1998) or sugar for eluting lectin (or glycoproteins from a lectin column) (Tyagi et al. 1996a; Anuradha and Bhide 1999) are examples of specific elution.

In both cases, isocratic or gradient elutions are possible and have been used in a wide variety of situations (Miletti et al. 1999; Willoughby et al. 1999).

2. SOME IMPORTANT AFFINITY LIGANDS

A few main classes of ligands are discussed below.

2.1 Biomimetic Ligands

In addition to exploiting biological affinity of the ligand for an enzyme/protein, one can also use biomimetic ligands. Such ligands, while not physiologically relevant, still bind to the enzyme/protein. Textile dyes are the most important example of this class.

Textile Dyes

The principal application of immobilized dye column chromatography has been to purify proteins, a role that has enjoyed spectacular success largely due to the economy, stability, and capacity of immobilized reactive dye columns (Lowe et al. 1992; Stellwagen 1993; Labrou and Clonis 1994). Most applications employ positive immobilized dye chromatography. One case in which negative immobilized dye affinity chromatography has been used is in the removal of albumin (up to 98%) for the isolation of proteins from blood plasma or serum on a column of Cibacron blue-Sepharose

(Travis et al. 1976). Since blue dextran-Sepharose functions as an affinity column for ligand sites constructed by the NAD binding domain (Fig. 3), binding to the column can be used to predict the occurrence of this motif in proteins that have not yet been crystallized. However, mere binding to this column does not confirm the presence of this domain. This column also retains some proteins, known not to contain this domain, e.g., cytochrome

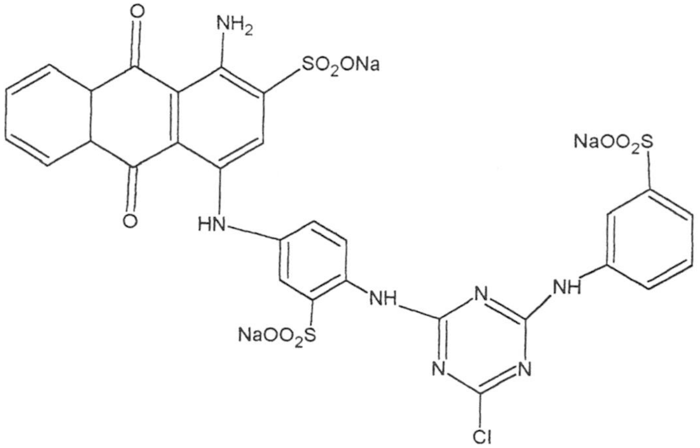

(a)

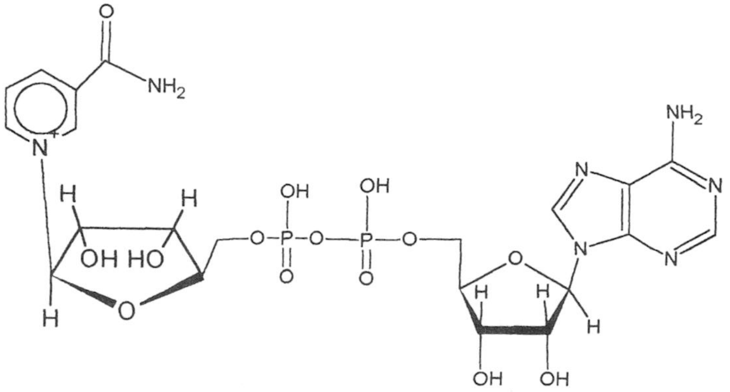

(b)

Figure 3 Structural similarity between the triazine dye Cibacron blue 3GA (a) and nicotinamide adenine dinucleotide (b). The aromatic anthraquinone group of the dye appears to interact with the target protein in a similar fashion to the adenine moiety of NAD^+, and the anionic benzenesulfonates mimic binding of the pyrophosphate group of NAD^+.

c, ribonuclease, and chymotrypsinogen. Elution, however, is carried out by NaCl and this (effectiveness of specific or nonspecific eluents) may be used as a criterion for the presence of NAD binding domain. Recently, Alberghina et al. (1999) isolated G-DNA structures, formed by a 27-mer guanosine-rich oligodeoxyribonucleotide, AACCCGGCGTTCGGGGGG-TACCGGGTT, and studied it by dye-ligand affinity chromatography, using a Reactive Green 19-agarose resin. The results show the selectivity of the approach. The more stable species were those obtained in the presence of K^+, whereas in the presence of Li^+ the oligonucleotide was almost exclusively present as a stem-loop structure recognized by the RG19-agarose affinity resin. The specificity of a Cibacron blue column may be significantly improved by adding bovine serum albumin (BSA) to the affinity column. BSA is thought to selectively mask the sites on the dye, not showing specific affinity for the nucleotide-binding fold of the protein and thus improving the recovery of the target protein using a selective eluent. Elution of bound nitrate reductase increased from almost zero to 45% on addition of 0.1 mM NADH when the crude containing the enzyme was added to blue Sepharose column after pretreatment with BSA (Ramadoss et al. 1983).

Another recent innovative approach to cut down nonspecific interactions in dye-based affinity chromatography has been to use "shielding polymers" (Galaev and Mattiasson 1994a) (Fig. 4). In this case, moderately hydrophobic polymers like polyvinylpyrrolidone (PVP) (Galaev et al. 1994a), polyvinyl alcohol (PVA) (Galaev et al. 1994a), and polyvinyl caprolactam (PVCL) (Galaev et al. 1994b) mask the nonspecific sites on the dye-polymer matrix. Both higher yields and higher purification factors have been reported for several enzymes (Galaev et al. 1994a,b; Galaev and Mattiasson 1994b,c; Garg et al. 1994). An interesting case has been the use of temperature for displacement of the bound enzyme from the column (Galaev et al. 1994b). In this case, PVCL at 40°C fails to effectively "shield" the column; on lowering the temperature to 23°C, PVCL undergoes "globule to coil" transition and while trying to "shield" the column, displaces the bound protein.

Birkenmeier and Kopperschläger (1991) have reported that dyes can be directly used to precipitate proteins. The approach looks very attractive and economical. In our hands, we found that it did not work with the enzymes we tried and entailed loss of almost all enzyme activity in an irreversible fashion (Nihalani 1999).

The use of biomimetic designer ligands (which show structural similarity to the natural cofactors) offers many advantages that increase the specificity and robustness of the system. The first step in designing a biomimetic ligand is to investigate the interactions of the target enzyme and the commercially available textile dye to identify the forces responsible for the interaction. This is followed by the synthesis of analogues of the dye and

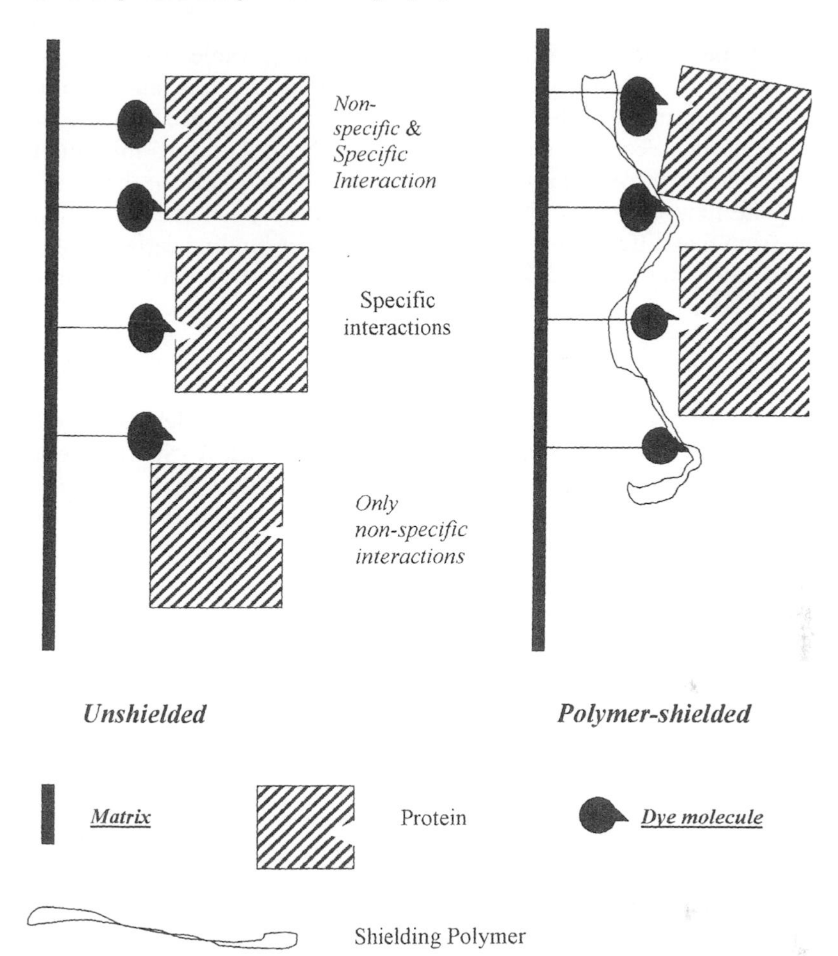

Figure 4 Polymer-shielded dye affinity chromatography. The polymer-pretreated dye column is shielded from nonspecific binding resulting in improved specific binding of the target enzyme/protein from the dye column.

studying altered interaction between the "designer" dye molecule and the target enzyme/protein. Molecular modeling has also been used for the design of a biomimetic chimeric ligand for the purification of L-lactate dehydrogenase (L-LDH) from bovine heart (Labrou et al. 1999). By using the three-dimensional structure of L-LDH as a guide, the authors synthesized an anthraquinone-based monochlorotriazinyl dye comprising two moieties: (1) the ketocarboxyl biomimetic moiety mimicking the natural substrate of the enzyme and (2) the anthraquinone chromophore moiety acting as a pseudo-mimetic of the cofactor NAD^+. The authors, during molecular modeling, tried to mimic the enzyme's polar and hydrophobic regions, and to introduce restriction in flexibility without too much hindrance. The bound enzyme could be selectively eluted with 0.1 mM NAD^+ and 1.5 mM sulfite, resulting in 56% recovery with the enzyme purified 44-fold from the starting extract. The new biomemetric ligand was found to be effective for crude extracts from five different sources, leading to higher binding capacity of the adsorbent and higher specific activity of L-LDH over the non-biomimetic adsorbent.

Protein A Mimic

Computer-aided molecular modeling has also been used to develop a series of nonpeptidyl mimics for *Staphylococcus aureus* protein A (SpA) (Li et al. 1998). One of the members binds IgG competitively with SpA in solution and has been immobilized on agarose beads for the purification of antibodies from human plasma by affinity chromatography. The adsorbent was found to be stable during routine cleaning-in-place procedures.

Surface Template Polymerization

Recently, the technique of surface template polymerization has been used to create artificial biocatalysts that mimic a variety of enzymes (Uezu et al. 1999). An enzyme-mimic polymer was prepared by imprinting a substrate analogue through complex formation between a cobalt ion and oleyl imidazole, the functional host molecule, which showed much higher activity than the control polymer. Complementary specific recognition sites were constructed by the imprint molecule and by the functional host molecule that were specially positioned on the polymer surface.

Metal Ions

In recent years, IMAC has emerged as another powerful tool for bioseparation (Fig. 5). As briefly mentioned earlier, Cu^{2+}, Zn^{2+}, and Ni^{2+} bind to available histidine, cysteine, and tryptophan residues on the surface of the protein. One of the most frequently used IMAC column continues to be one originally described by Porath et al. (1975), namely, iminodiacetic acid linked to agarose and charged with Cu^{2+}. It is normal to include 0.1–1 M NaCl during sorption for suppressing nonspecific ionic interactions. Elution is generally carried out by reduction in pH or inclusion of imidazole in the wash buffer. As the technique and its applications have been adequately reviewed in a number of places (Sulkowski 1985; Kågedal 1989; Arnold 1991; Porath 1992; Lopatin and Varlamov 1995), we will not discuss it in greater detail.

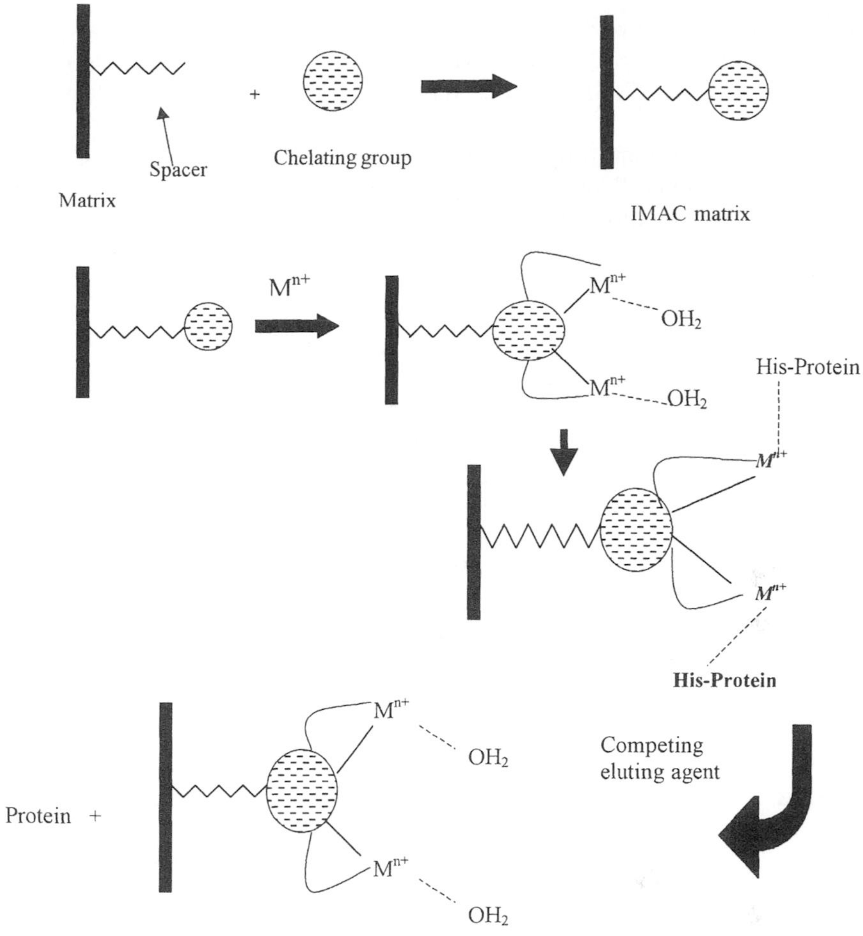
Spacer
Chelating group
Matrix
IMAC matrix
M^{n+}
M^{n+}
OH_2
M^{n+}
OH_2
His-Protein
M^{n+}
M^{n+}
His-Protein
M^{n+}
OH_2
Competing
eluting agent
Protein +
M^{n+}
OH_2

Figure 5 Schematic diagram of immobilized metal ion affinity chromatography. Metal ions are immobilized on the chelating resin, which then binds the target protein reversibly. Elution of the bound protein is carried out with a change in pH, introduction of imidazole in the wash buffer or a metal chelator (e.g., EDTA).

2.2 Lectins

Lectins are carbohydrate-binding proteins or glycoproteins of nonimmune origin that agglutinate cells and/or precipitate glycoconjugates and bear at least two sugar binding sites (Goldstein and Hayes 1978). These proteins are selective in their binding and this has led to their applications in bioseparation, cell-membrane biology, and molecular biology and immunology. These in vitro applications and their importance in vivo have been described at a number of places (Leiner et al. 1986, Basu et al. 1993, van Damme et al. 1997). Numerous data are now available on lectin-sugar interactions.

It is very common to talk of lectin specificity in terms of a sugar that inhibits the blood cell agglutination activity of that particular lectin. Table 4 gives such information for some well-known and/or interesting lectins. A recent handbook is devoted exclusively to lectins of plant origin (van Damme et al. 1997). It lists 13 lectins whose specificities are still unknown! Thus, in terms of selectivity, lectins represent a huge repertoire.

Lectin-sugar affinity pairing is one area wherein considerable insight at the molecular level is now available. Some information, purely illustrative in nature, is given here. Peanut lectin is extensively used as a tool for recognition of T-antigen on the surface of malignant cells and immature thymocytes. This lectin also recognized *N*-acetyllactosamine, which is present at the termini of several cell surface glycoproteins. Based on the X-ray diffraction picture of lectin-sugar complex, it was possible to obtain, by site-directed mutagenesis, two lectin mutants, (L 212N) and (L 212A), which exhibited distinct preferences for T-antigen and *N*-acetyllactosamine, respectively (Sharma et al. 1996). Similarly, affinities of E129A and E129D for C-2 substituted derivatives of galactose vary. "The mutant E129D showed significant binding towards N-acetyl galactosamine, suggest-

Table 4 Characteristics of Some of the Lectins Isolated from Plants

Source	Inhibitory sugar	Blood group specificity
Arachis hypogaea (peanut)	Gal	N. S.
Ricinus communis (Castor bean)		W. H.
Amaranthus caudatus (tassel flower)	GalNAc	ABO
Triticum aestivum (wheat)		N. S.
Allium moly [dwarf onions (flowering)]	Man	N. S.
Canavalia ensiformis (Jack bean)		N. S.
Lathyrus odoratus (sweet pea)	Glu	N. I.
Vicia faba (fava bean)		N. S.
Tetragonolobus purpureas (asparagus pea)	Fuc	O
Ulex europaeus (gorse)		O
Sambucus canadensis (elderberry, Canadian)	Sialic acid	N. S.
Sambucus ebulus (dwarf elder)		A, O

N.S., no specificity observed; W.H., no agglutination observed with human erythrocytes; N.I., no information available.

ing that the residue Glu 129 is crucial in imparting exclusive galactose-specificity upon peanut lectin" (Sharma et al. 1998). An exciting recent finding is that molten globule-like state of the lectin monomer still retains its carbohydrate specificity (Reddy et al. 1999). This contradicts the prevailing view that precise subunit topology is required for biological activities of such lectins.

Another important finding is that the specific binding of concanavalin A (con A, another well-characterized plant lectin) to different sugars is accompanied by differential uptake of water molecules during the binding process. This provides the reason for the observed compensatory behaviour of enthalpies with entropies in lectin-sugar interactions (Swaminathan et al. 1998).

Sharma and Surolia (1997) have carried out an extensive analysis of sequences and structures of several legume lectins. They report that despite the hypervariability of their combining regions, there is a uniform pattern. For example, in the case of mannose/glucose-specific lectins, the size of the binding site loop is invariant.

Thus, the stage is set wherein redesigning and enhancing specificity are possible, at least in the case of this interesting class of proteins.

Somewhat puzzling is the binding of lysozyme (which is devoid of any carbohydrate) to Con A (Vaidya and Gupta 1988). Another interesting example is the purification of human alkaline phosphatase by precipitation with Con A (Latner and Hodson 1980). In this case, affinity of the enzyme for the lectin could be varied by controlling the presence and concentration of metal ions required by the lectin for its activity. Con A is perhaps the oldest lectin still in use. A new method for the isolation of glycoprotein G from viral hemorrhagic septicemia virus (VHSV), a fish rhabdovirus, was developed by using affinity chromatography with immobilized Con A (Perez et al. 1998). The glycoprotein G was isolated from detergent-solubilized concentrated virions and from large-volume virion-free supernatants from VHSV infected cells (soluble form). The purity achieved was higher than 85%.

The difference in the sugar specificity of lectins has been used by Oda et al. (1999) to separate the two components of recombinant thyrotropin hormone receptor by lectin affinity chromatography. In particular, *Galanthus nivalis* lectin, which binds mannose-type sugars, bound the 100-kDa band but not the 120-kDa band, whereas *Datura stramonium* lectin, which binds complex sugars, bound the 120-kDa band but not the 100-kDa band. A minus point is the high cost of lectins, something that efficient downstream processing methods have been able to alleviate, at least in some cases (Tyagi et al. 1996a,b; Senstad and Mattiasson 1989).

2.3 Affinity Tails/Tags

Advances in molecular biology and genetic engineering have made possible the creation of fusion proteins, which show a built-in affinity between the fused protein and the standard adsorbent designed for that portion (see Chapter 4). The technique has finally introduced some order in a field where no two purification protocols are identical. The use of cellulose-binding fusion proteins has achieved the purification of such diverse proteins as biotinylated proteins (Le et al. 1994), alkaline phosphatase (Greenwood et al. 1994), and others. The affinity chromatography is carried out on cellulose or chitin and the cellulose binding domain peptide is subsequently cleaved by a protease. However, current techniques based on enzymatic cleavage are expensive and result in the presence of additional amino acids at either end of the proteins, as well as contaminating proteases in the preparation. Rais-Beghdadi et al. (1998) have evaluated an alternative method to the one-step affinity/protease purification process for large-scale purification. It is based on the cyanogen bromide (CNBr) cleavage at a single methionine placed in between a histidine tag and a *Plasmodium falciparum* antigen. Su et al. (1999) have purified Pen c 1 (an allergen protein) (expressed in *Escherichia coli* as a fusion protein bearing an N-terminal histidine affinity tag) on an affinity column. Full caseinolytic activity could be regenerated in the purified recombinant protein by sulfonation and renaturation, followed by removal of the affinity tag.

2.4 Biotin-Avidin

An analogous approach, which does not utilize recombinant methods, is the exploitation of avidin-biotin interaction, known to be the affinity pair with the highest binding constant. It has given rise to a technology of its own (Wilchek and Bayer 1990). Biotinylated proteins/enzymes binding to avidin/streptavidin columns have been frequently used, with positive results in many cases (Kim et al. 1999a; Pouny et al. 1998). It should be added that fusion proteins with biotin binding domains have also been used for purification purposes (Greenwood et al. 1994). In vivo biotinylation machinery of *Escherichia coli* has also been used to provide an elegant strategy for site-specific biotinylation of recombinant proteins during their production (Cronan 1990).

2.5 Protein A and Protein G

The affinity matrices containing covalently coupled protein A have been widely used for the purificaton of IgG molecules (Surolia et al. 1982).

Protein A is a cell wall component of several strains of *S. aureus* and binds specifically to Fc region of the immunoglobulins. X-ray diffraction has recently shown that it is the B domain of protein A which recognizes the Fc region of IgG (Rongxiu et al. 1998). The strength of the interaction varies with the species and the subclass of immunoglobulin. Another similar ligand for immunoglobulins is protein G, a bacterial cell wall protein from Group A *streptococci* (Åkerström and Björck 1986). It has greater affinity for most mammalian immunoglobulins than protein A and binds also serum albumin (Björck et al. 1987). The IgG-binding protein is eluted from an affinity column with a low pH buffer (0.1 M glycine HCl, pH 2.0). Both of these proteins are now commercially available in recombinant forms. In the case of recombinant protein G, nonspecific binding has been eliminated. For example, recombinant protein G can be used to separate albumin from crude human immunoglobulin preparations. A gene fusion product, protein A/protein G, is also commercially available that combines the binding profiles of protein A and protein G (Eliasson et al. 1988). This protein is purified on an affinity column using 1 M acetate buffer, pH 2.8 as eluant.

These examples vividly illustrate how the specificity of the naturally occurring protein ligands can be curtailed (to make them more specific) or expanded (to make them widely applicable) using recombinant approaches.

2.6 Antigen-Antibody Interactions

Antigen-antibody pair forms one of the oldest recognized bioaffinity pairs (Davies et al. 1988; Gerard and Gerard 1990; van Oss 1997). While early work was carried out using polyclonal sera, monoclonal antibodies now constitute among the most specific affinity ligands. While purification of antibodies using an immobilized antigen has been used frequently, lately many other affinity ligands like protein A/protein G, C1q, mannan-binding protein, and thiophilic gel chromatography have been used more often for this purpose. In this context, a recent interesting observation is that the choice of the affinity ligand while purifying antibodies has great influence on the performance of immunoassays with polyclonal preparations (Choi et al. 1999). The purification of antigens on immobilized antibodies continues to be a powerful method. A major consideration is to immobilize in such a way that antigen binding sites remain free. O'Shannessy and Hofman (1987) had reported site-directed immobilization or oriented immobilization in which coupling was through carbohydrate groups in the Fc region. The immobilization chemistry uses sodium periodate to oxidize glycoproteins, converting vicinal hydroxyl groups to reactive aldehyde groups which then react with hydrazide groups on the matrix to form hydrazone. Pierce Chemical Co (Rockford, IL) offers an immobilized TNB-Thiol™

column, which couples reduced antibody via its hinge region. The linking chemistry is reversible disulfide bond formation. One also has the option of obtaining oriented immobilized antibodies via affinity immobilization on protein A/protein G columns. In addition to improper orientation, steric hindrance and modified conformations have also been proposed as factors responsible for diminished binding capacity of the immobilized antibodies. Spitznagel and Clark (1993) have shown that one can do away with unnecessary protein domains and get better antigen binding by using the fragments. An interesting finding was that "although a fraction of immobilized antibody was inactive at the higher loadings, EPR spectroscopy revealed no significant changes in the conformation of active immobilized antibody."

The nature of interactions involved in antigen-antibody binding has been discussed quite rigorously by van Oss (1997). "The interplay between the macroscopic aspecific repulsion and the microscopic-scale specific attraction between antigens and antibodies is described in light of the long-range non-covalent forces acting on antigens and antibodies (Lifschitz–van der Waals, Lewis acid-base, and electrostatic interactions) and as a function of the distance." The discussion, of course, is relevant to all specific ligand-target molecule pairs in affinity chromatography. Another very relevant reference is a minireview by Davies et al. (1988) who have discussed some interesting issues like: "Is charge neutralization a significant factor in recognition and in stabilization of the antigen-antibody complex? Or is the flexibility observed at the junction between the VH:VL and CH1:CL modules a signal transduction mechanism or is it simply an expression of flexibility between more rigid components?"

The complexity of antigen-antibody interactions also means that dissociation of the complex is not always easy and conditions for elution vary from system to system. One of the best discussions on this can be found in a review by Yarmush et al. (1992). Reducing the pH to 2 or 3 is the most frequently used method. Use of higher pH (between 10 and 11) is avoided because alkaline denaturation sets in. In the case of IgM, alkaline denaturation is less pronounced, and successful elution at higher pH has been reported. Chaotropic salts and 3 M thiocyanate in particular have been frequently effective. High ionic strength, denaturants (urea and guanidine hydrochloride), and organic solvents (ethylene glycol, dimethyl sulfoxide, acetonitrile, dioxane, pyridine, etc.) have also been used. Among the nonchemical approaches, elution with changes in temperature and pressure and electrophoretic elution have been tried with limited success.

A special mention may be made of the effectiveness of the use of divalent cations, e.g., $MgCl_2$ and $CaCl_2$, in some cases. It is believed that the effect is due to the combined influence of ionic strength, mild chaotropic effects, and reduced pH (Yarmush et al. 1992).

Some recent applications are briefly mentioned for illustrative purposes. For more comprehensive data, various older reviews may be referred to (Davies et al. 1988; van Oss 1997). McKercher et al. (1997) have reported the single-column purification of herpes simplex virus type I protease by immunoaffinity employing a monoclonal against the catalytic domain of the protease. High-performance immunoaffinity chromatography has been used to specifically remove β_2-microglobulin from human plasma (Kojima 1997). As such steps can be inserted in an extracorporeal circulation system, this approach is useful for medical applications, such as elimination of any unwanted plasma component. An application of a different kind is the use of a polyclonal antisera-based column for determination of peanut protein in chocolates (Newsome and Abbott 1999). In yet another application, antibodies specific for major epitopes of VPI of foot and mouth disease virus have been purified using recombinant antigens (Bayry et al. 1999). Affinity perfusion chromatography has been used for purification of the product of the oncosuppressor gene *BRCA1* in breast tissues (Hizel et al. 1999). The antibody raised against the protein was the affinity ligand.

A major advance in the area in the coming years is likely to be the use of genetic engineering and combinatorial chemistry to create antibodies that are stable and have binding properties that are more amenable to gentle elution methods. Another trend in the future, of course, is going to be increasing interface of immunoaffinity with emerging efficient downstream processing methods such as affinity precipitation, expanded bed chromatography, and perfusion chromatography.

3. OPTIMAL BINDING CONSTANTS IN AFFINITY CHROMATOGRAPHY

What is the ideal range of dissociation constant (K_d) values for an affinity pair to be exploited for affinity chromatography? Burton (1996), in a very perceptive article writes, "The specificity of protein-ligand interactions ultimately determines the degree of purification that can be achieved. Thus, highly specific antigen-antibody interactions with dissociation constant values in the range of 10^{-8}–10^{-10} M have the potential to deliver higher degrees of purification as compared to immobilized coenzymes, which typically have K_d values in the range of 10^{-6} M. However, very tight binding can be disadvantageous in that the bound protein may be difficult to elute without causing permanent damage. Similarly, very weak binding is disadvantageous as proteins are merely retarded and soon appear in the column eluate. Consequently, there is an optimal K_d range of 10^{-4}–10^{-8} M."

Liapis and Unger (1994) have grouped various ligands according to the dissociation constants. Low molecular weight inhibitors, such as benzamidine (for trypsin), triazine dyes, metal chelates, and so forth, have K_d in the range of 10^{-3} M and sugars (for lectins), hormones, and antibodies have K_d in the range of 10^{-4}–10^{-15} M. The recent work of Bergström et al. (1998), concerning weak monoclonals for affinity chromatography, further illustrates the potential of weak affinity interactions for separation purposes.

There is obviously a trade-off between specificity and binding constants. For example, use of cross-reacting antibodies not highly specific for the target molecule will make gentle elution possible but will not result in very high-fold purification. This concept is illustrated at many other places in this chapter.

However, weaker affinities (10^{-2}–10^{-4} M) have also been exploited in what has been referred to as weak affinity chromatography (Wikström and Ohlson 1992). The weaker affinities enable fast and dynamic separations, and one can operate in high-performance mode. The higher on and off rates observed during this approach have been exploited in a number of systems (Ohlson et al. 1988, 1998).

Of course, the right range of K_d values does not guarantee a successful design. The identical affinity pairs, which are used in affinity chromatography, also form the basis of reversible immobilization via affinity interactions (Mattiasson 1988; Gupta and Mattiasson 1992).

Various methods for the determination of these binding constants (e.g., equilibrium dialysis, UV and visible spectroscopy, fluorescence spectroscopy, etc.) are well established (Yamasaki et al. 1999; Zhou and van Etten 1999; Patel et al. 1999). High-performance affinity capillary electrophoresis (ACE) is another fast-emerging method of choice. Honda et al. (1992) determined association constants of some galactose-specific lectins with lactobionic acid to be around 10^2–10^3 M^{-1}. Gomez et al. (1994) have suggested a method for estimation of binding constants by ACE after comparison of variable electrostatic flux. Dunayevskiy et al. (1998) have introduced ACE-MS as a solution-based approach for screening combinatorial libraries for drug leads. The method allows on-line, one-step selection and structural identification of candidate ligands. In another novel application, Gao et al. (1996) used ACE to calculate the binding affinities between charged ligands and the members of a charge ladder of carbonic anhydrase constructed by random acetylation of the amino groups on its surface. The idea was to evaluate the effect of changes in charges present on the enzyme (but away from the binding interface) on the free energy of binding. This work is of general importance because it applies ACE to quantify the contribution of electrostatics to free energies of molecular recognition.

Some efforts have also been made to bind affinity ligands to capillaries. As an example of what may be possible, one may mention the work of Nashabeh and el Rassi (1991) on the separation of glycopeptides on a Con A-silica column. The difficulty in creating affinity ligand-bound capillaries is in ensuring enough affinity ligand density inside the capillaries for efficient "affinity capture" of the target molecule. The idea of incorporating affinity ligands into gels has not been very successful. However, Nilsson et al. (1994) have reported the use of imprinted polymer particles selective for pentamidine (a drug used for the management of AIDS-related pneumonia) in imprinted polymer capillary electrophoresis (IMPCE).

Yet another technique that is likely to be used more often in the future is biospecific interaction analysis (BIA). Real-time BIA is increasingly being used to assess the rates and binding constants of interaction between affinity pairs. The technique is based on surface plasmon resonance, which utilizes the changes in the reflection of laser light at a metal-liquid interface, caused by a change in the refractive index. BIA has been used for assessing the kinetic parameters for the interaction of monoclonal antibodies with recombinant HIV-1 core protein p24 as a model system (Application Note 301). Antibodies are captured on the sensor chip surface from unfractionated hybridoma culture supernatants, thus eliminating the need for prior purification of antibodies. The affinity constants for p24 antibodies lie in the range of 2.7×10^7 to 6×10^9 M. BIAcore has also been used to study the kinetics of the interaction of endotoxin with polymyxin B and other peptide-based drug analogues in an attempt to determine the structure-activity correlation between them (Thomas and Surolia 1999). A subsequent study has allowed them to define the mechanism of action of polymyxin B in chemical terms and to propose that "the design of molecules as effective antidotes for sepsis should incorporate the ability to sequester endotoxin specifically" (Thomas et al. 1999). BIAcore is now replacing ELISA in supporting preclinical and clinical trials for the detection of antibodies capable of binding to IL-10 and IL-4 (Swanson et al. 1999). Increased throughput and direct detection for binding interactions enables detection of low-affinity antibodies that are not detected by ELISA.

4. MAGNETIC SEPARATION

Affinity ligands attached to magnetic beads constitute yet another option of exploiting biological selectivity. The various applications have been documented at various places (Safarik and Safarikova 1997, 1999). The beads are avaiable from suppliers like Dynal (Norway) or Bangs (USA). The technique is somewhat expensive, although one can make one's own magnetic beads to reduce costs (Tyagi and Gupta 1995; Kubal et al. 1996). The approach has been mostly limited to handling volumes up to few hundred milliliters.

Worlock et al. (1991) have compared the isolation of metabolically labeled human leukocyte antigen (HLA) class I and II molecules from cell lysates by protein A-Sepharose and Dynabeads sheep antimouse IgG. The Dynabeads gave lower nonspecific binding. Karlsson and Platt (1991) reported a similar experience while detecting Tfr protein with the monoclonal antibody.

5. GENERATION OF AFFINITY LIGANDS THROUGH COMBINATORIAL CHEMISTRY

Combinatorial chemistry is an extremely versatile approach for obtaining affinity ligands with adequate association constants. As is often the case with such highly innovative technologies, the concept behind the approach is simple.

Step I: Choose some building blocks
Step II: Create molecular libraries by random or controlled assembly of such blocks
Step III: Screen these assembled molecules for affinity for the chosen target enzyme/protein

In order to appreciate the power of this approach in creating molecular diversity, one merely has to look at some simple calculations. The combinatorial assembly of 20 gene-coded amino acids can generate up to 8000 (20^3) tripeptides, 160,000 tetrapeptides, and 64×10^6 hexapeptides. Once a reasonably sized pool of diverse molecules is available, it is not unlikely that a fortuitous combination of interactions is available (with some members) which effectively simulates ligand-target enzyme–protein interactions.

It is logical that one mostly uses peptides and oligonucleotides in this technique since synthetic methods in both cases have been extensively developed (Birnbaum and Mosbach 1992; Clackson and Wells 1994; Dower 1998). The peptide synthesis in solid phase, for example, has been extensively used with a variety of supports such as polystyrene beads, pins, cellulose disks, and silica chips (Geyson et al. 1984; Lam et al. 1991; Frank 1992; Andres et al. 1999). One critical requirement is the availability of a high-throughput screening method for identifying useful candidate peptides/oligonucleotides. Also, as synthesis invariably leads to a mixture, one is looking at average affinity in this approach. A novel variation that was reported a few years back combines light-directed chemical synthesis with semiconductor-based photolithography (Jacobs and Fodor 1994). Detection of the binding between the ligand peptide and the target molecule is conveniently possible with high selectivity using laser confocal fluorescence microscopy.

Combinatorial chemistry is increasingly being used in drug design (Olivera et al. 1995; Kirkpatrick et al. 1999; Barry et al. 2000). Nevertheless, there are enough reports of successful application of the strategy in designing affinity-based separations. For example, a recent report describes identification of an aptamer (specific oligonucleotide sequences) that selectively binds to human L-selectin (Romig et al. 1999). The use of this aptamer column as the initial purification step resulted in a 1500-fold purification with 83% recovery of the target molecule.

The approach is not limited to obtaining peptides and oligonucleotides. Peptide-like polymers from *N*-acetylated glycines and oligocarbamates have been reported (Simon et al. 1992; Cho et al. 1993). The synthetic polymer chemistry, already extensively developed, is likely to cross-fertilize this approach in the coming years.

It is interesting to record that cone snails, a group of marine predators, appear to use combinatorial strategy to obtain pharmacologically active peptides to incapacitate their prey (Olivera et al. 1995). Thus, the largest genera of marine invertebrates have a 50 million year history of using combinatorial chemistry! The cliché that some of our best science has resulted from copying nature seems a relevant statement.

5.1 Phage Display Libraries

Phase display libraries are the basis of a biological method to generate molecular diversity through a combinatorial approach. The majority of applications have been in obtaining antibodies with defined specificity (de Bruin et al. 1999; van Der Vuurst De Vries and Logtenberg 1999; Kim et al. 2000). While otherwise a very versatile technique, the source of molecular diversity is limited to the gene-coded amino acids. The basic principle of the method is again elegantly simple. Phage particle is engineered to express a peptide in the form of a fusion protein with one of the coat proteins. As each phage encodes a single peptide, pools of 10^8–10^9 different phages/peptides are obtained and can be suitably screened. Amplification of desirable clones after affinity selection and the possibility of multiple rounds of selection make this a rather powerful approach. Again, a large number of applications have been reported (Kim et al. 1999b; Stolz et al. 1999; Winthrop et al. 1999). Recently, a recombinant human monoclonal with antitumor activity was constructed from phage-displayed antibody fragements (Huls et al. 1999).

6. MOLECULAR IMPRINTING

The term *molecular imprinting* has been used by different workers to mean different things! However, while the approaches differ, the common thread has been the affinity of a material for a preselected ligand. The latter is invariably a small molecular weight substance. Let us look at the different approaches.

6.1 Molecularly Imprinted Polymers

Two variations based on ligand recognition via non-covalent and reversible covalent bonds have been described (Fig. 6). In the former case, the print molecule is mixed with a suitable mixture of monomers prior to polymerization (Mosbach and Ramström 1996). Methacrylic acid has been most frequently used as the monomer (Dabulis and Klibanov 1992; Lin et al. 1997; Tomioka et al. 1997), but other monomers, such as itanoic acid, vinylpyridines, and vinylimidazole, have also been employed. After polymerization, the imprint molecule is removed, and the imprinted polymer generally shows reasonable rates for binding and release of the print molecule. Generally, use of less polar solvents during polymerization yields better recognition capability in the imprinted polymer (Mosbach and Ramström 1996). Also, it has been reported that at low temperature polymerization results in higher enantiomeric recognition capabilities.

The second variation, wherein adducts are chemically synthesized and added to the polymerization mixture, results in correct spatial incorporation of functional groups in the recognition site (Wulff 1993; Shea 1994). Initial removal of the print molecule requires harsh conditions and subsequent association-dissociation with the imprinted polymer generally shows rather slow kinetics. However, boronic acid esters have shown reasonable rates (Mosbach and Ramström 1996). Numerous applications of molecularly imprinted polymers have been reported in chromatography as well as in analysis (Andersson 1988; Yu and Mosbach 1998; Haupt and Mosbach 1998).

6.2 Bioimprinting

In this approach, the high rigidity of protein molecules in anhydrous organic solvents has been exploited to imprint them with stereospecific ligands. In one variation, the print molecule is precipitated along with a protein in nearly anhydrous media. The second approach involves lyophilizing the ligand along with the protein (Russell and Klibanov 1988; Braco et al. 1990). The imprint molecule is removed in both cases and the imprinted

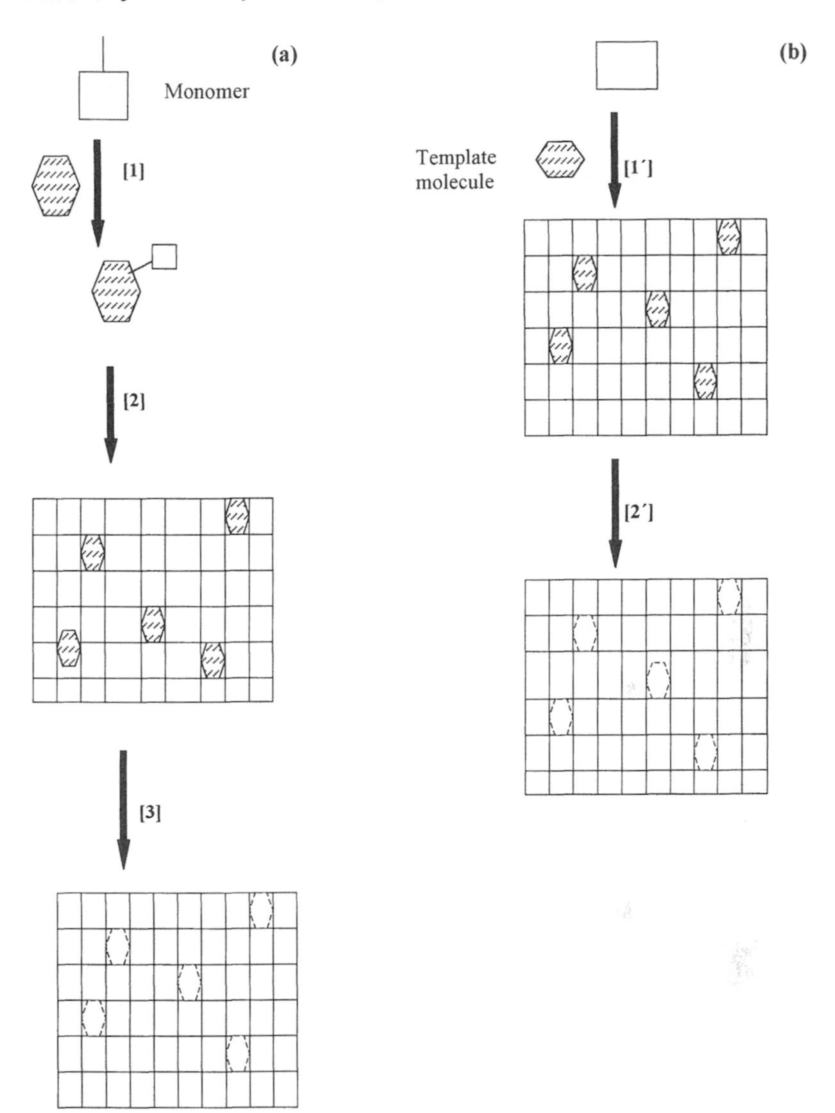

Figure 6 Schematic diagram of molecular imprinting using covalent (a) and non-covalent (b) approaches. In the former case, a covalent bond is formed between the imprint molecule and the monomer [1] followed by polymerization [2]. Extraction of the template from the imprinted polymer [3] is carried out by cleavage of covalent bonds. In the non-covalent approach, the monomer is polymerized in the presence of the print molecule [1′]. Extraction of the template after polymerization [2′] requires mild conditions.

polymer acquires the property of selective binding. Using this method, polymers like dextran have been successfully imprinted. Unlike the former approach, bioimprinting is erased once even a small concentratiaon of water (0.4%) is added to the system. It is for this reason that the approach has not found much practical application.

Soler et al. (1995) have also called their technique "bioimprinting" wherein unfolded immobilized enzymes have refolded in the presence of substrates. This is reported to result in enzyme derivatives with improved catalytic properties. The relevance of this application for the chapter lies in the fact that enzymes, after all, are the best examples of selective affinity. One exciting development in this area is of recent origin. Shi et al. (1999) described a method for imprinting surfaces with protein recognition sites. Template-imprinted nanostructural surfaces were created using radio-frequency glow discharge plasma deposition to form polymeric thin films around proteins coated with disaccharide molecules. The resulting "polysaccharide-like cavity" showed selective recognition for the template proteins such as albumin, IgG, lysozyme, ribonuclease, and streptavidin.

7. CONCLUSION

Before we end, it is pertinent to do some loud thinking about the future directions in this area. Combinatorial chemistry promises affinity ligands and designing of selectivities for any preset purpose. Does that mean that one day in the future all the affinity ligands in use will be peptides? Peptides are costly to produce and are not some of the most rugged molecules. Thus, the considerations of economy, convenience, stability, and scale-up will continue to dictate that traditional ligands remain in vogue.

The nonspecific adsorption by the matrix or part of the affinity ligand continues to be an issue in search of a technological solution. We need more inputs from surface chemists here. Thus, a multidisciplinary approach for the design of an affinity matrix is sorely needed.

One approach, which perhaps is underexploited, is to think of *matrix-less affinity chromatography*. Cellulose beads for cellulase purification (Roy et al., 1999) and alginate beads for amylases (Sardar and Gupta 1998; Roy et al. 2000) possibly represent the tip of an iceberg.

Finally, one should add that selectivity or affinity binding of a molecule to a protein is not relevant to affinity chromatography only. Looking at the phenomenon from a different angle, of equally broad ramifications is the binding of protein to a surface (Norde 1986; Bajpai and Dengre 1999). Both for in vitro and in vivo contexts, these interactions occupy center stage in many phenomena.

ACKNOWLEDGMENTS

The preparation of this manuscript was supported by project funds from Department of Biotechnology, Council for Scientific and Industrial Research (CSIR; Extramural) and CSIR (Technology Mission on Oilseeds, Pulses and Maize) (all Government of India organizations).

ABBREVIATIONS

ACE	affinity capillary electrophoresis
ACE-MS	affinity capillary electrophoresis–mass spectrometry
AIDS	acquired immunodeficiency syndrome
BIA	biospecific interaction analysis
BSA	bovine serum albumin
CH1:CL	module of constant region in heavy chain constant region of light chain
ELISA	enzyme-linked immunosorbent assay
HIV-1 coat protein	human immunodeficiency virus-1 coated protein
HLA class I and II	human leukocyte antigen
IMPCE	imprinted polymer capillary electrophoresis
L-LDH	L-lactate dehydrogenase
PVA	polyvinyl alcohol
PVCL	polyvinyl chloride
PVP	polyvinyl pyrrolidone
Tfr	transferrin receptor
VH1:CL	module of variable region of heavy chain constant region of light chain
VHSV	viral haemorrhagic septicaemia virus
VP1	major epitope of foot and mouth disease virus

REFERENCES

Agarwal R, Gupta MN (1996). Sequential precipitation with reversibly soluble-insoluble polymers as a bioseparation strategy: Purification of beta-glucosidase from *Trichoderma longibrachiatum*. Protein Expr. Purif. 7:294–298.

Åkerström B, Björck L (1986). A physicochemical study of Protein G molecule with unique immunoglobulin G-binding properties. J. Biol. Chem. 261:10240–10247.

Alberghina G, Fisichella S, Renda E (1999). Separation of G structures formed by a 27-mer guanosine-rich oligodeoxyribonucleotide by dye-ligand affinity chromatography. J. Chromatogr. A 840:51–58.

Andersson L (1988). Preparation of amino acid ester-selective cavities formed by non-covalent imprinting by a substrate in highly cross-linked polymers. Bioact. Polym. 9:29–41.

Andres CJ, Denhart DJ, Deshpande MS, Gillman KW (1999). Recent advances in the solid phase synthesis of drug-like heterocyclic small molecules. Comb. Chem. High Throughput Screen 2:191–210.

Anuradha P, Bhide SV (1999). An isolectin complex from *Trichosanthes anguina* seeds. Phytochemistry 52:751–758.

Application Note 301. Pharmacia Biosensor AB. Assessment of antigen-antibody interaction constants by real-time BIA.

Armisen P, Mateo C, Cortes E, Barredo JL, Salto F, Diez B, Rodes L, Garcia JL, Fernandez-Lafuente R, Guisan JM (1999). Selective adsorption of poly-His tagged glutaryl acylase on tailor-made metal chelate supports. J. Chromatogr. A 848:61–70.

Arnold FH (1991). Metal-affinity separations: A new dimension in protein processing. Trends Biotechnol. 9:151–155.

Bajpai AK, Dengre R (1999). Protein adsorption: A fascinating phenomenon. Ind. J. Chem. 38A:101–112.

Barry CE III, Slayden RA, Sampson AE, Lee RE (2000). Use of genomics and combinatorial chemistry in the development of new antimycobacterial drugs. Biochem. Pharmacol. 59:221–231.

Basu J, Kundu M, Chakrabart P (1993). Lectins: Biology, Biochemistry, Clinical Biochemistry, Vol. 9. New Delhi: Wiley Eastern.

Bayry J, Prabhudas K, Bist P, Reddy GR, Suryanarayana VV (1999). Immuno affinity purification of foot and mouth disease virus type specific antibodies using recombinant protein adsorbed to polystyrene wells. J. Virol. Meth. 81:21–30, 1999.

Bergström M, Lundblad A, Pahlsson P, Ohlson S (1998). Use of weak monoclonal antibodies for affinity chromatography. J. Mol. Recognit. 11:110–113.

Birkenmeier G, Kopperschlaeger G (1991). Dye-promoted precipitation of serum proteins. Mechanism and application. J. Biotechnol. 21:93–108, 1991.

Birnbaum S, Mosbach K (1992). Peptide screening. Curr. Opin. Biotechnol. 3:49–54.

Björck L, Kastern W, Lindahl G, Widebäck K (1987). Streptococcal protein G, expressed by streptococci or by *Escherichia coli*, has separate binding sites for human albumin and IgG. Mol. Immunol. 24, 1113–1122.

Bolognesi M, Boffi A, Coletta M, Mozzarelli A, Pesce A, Tarricone C, Ascenzi P (1999). Anticooperative ligand binding properties of recombinant ferric *Vitreoscilla* homodimeric hemoglobin: a thermodynamic, kinetic and X-ray crystallographic study. J. Mol. Biol. 291:637–650.

Braco L, Dabulis K, Klibanov AM (1990). Production of abiotic receptors by molecular imprinting of proteins. Proc. Natl. Acad. Sci. USA 87:274–277.

Bradshaw AP, Sturgeon RJ (1990). The synthesis of soluble ligand complexes for affinity precipitation studies. Biotechnol. Tech. 4:67–71.

Burton SJ (1996). Affinity chromatography. In: Verrall MS, ed. Downstream Processing of Natural Products. New York: John Wiley, pp. 193–207.

Carlsson J, Janson J-C, Sparrman M (1989). Affinity chromatography. In: Janson J-C, Ryden L, eds. Protein Purification: Principles, High Resolution Methods and Applications. New York: VCH Publishers, pp. 275–329.

Chase HA (1998). The use of affinity adsorbents in expanded bed adsorption. J. Mol. Recognit. 11:217–221.

Cho CY, Moran EJ, Cherry S, Stephens J, Fodor SPA, Adams C, Sundaram A, Jacobs JW, Schultz PG (1993). An unnatural biopolymer. Science 261:1303–1305.

Choi J, Kim C, Choi MJ (1999). Influence of the antibody purification method on immunoassay performance: hapten-antibody binding in accordance with the structure of the affinity column ligand. Anal. Biochem. 274:118–124.

Clackson T, Wells JA (1994). In vitro selection from protein and peptide libraries. Trends Biotechnol. 12:173–184.

Cronan JE (1990). Biotinylation of proteins in vivo: a posttranslational modification to label, purify, and study proteins. J Biol. Chem. 265: 10327–10333.

Dabulis K, Klibanov AM (1992). Molecular imprinting of proteins and other macromolecules resulting in new adsorbents. Biotechnol. Bioeng. 39:176–185.

Davies DR, Sherriff S, Padian EA (1988). Antibody-antigen complexes. J. Biol. Chem. 263:10541–10544.

Dean PDG, Johnson WS, Middle FA (1984). Affinity Chromatography—A Practical Approach. Oxford, UK: IRL Press.

de Bruin R, Spelt K, Mol J, Koes R, Quattrocchio F (1999). Selection of high affinity phage antibodies from phage display libraries. Nat. Biotechnol. 17:397–399.

Diaz-Camino C, Villanueva MA (1999). Purification of multiple functional leaf-actin isoforms from *Phaseolus vulgaris L.* Biochem. J. 343:597–602.

Dower WJ (1998). Targeting growth factors and cytokine receptors with recombinant peptide libraries. Curr. Opin. Chem. Biol. 2:328–334, 1998.

Dunayevskiy YM, Lyubarskaya YV, Chu YH, Vouros P, Karger BL (1998). Simultaneous measurement of nineteen binding constants of peptides to vancomycin using affinity capillary electrophoresis-mass spectrometry. J. Med. Chem. 41:1201–1204.

Dunn BM (1984). The study of ligand-protein interactions utilizing affinity chromatography. Appl. Biochem. Biophys. 9;261–284.

Eccleston VS, Harwood JL (1995). Solubilisation, partial purification and properties of acyl-CoA: glycerol-3-phosphate acyltransferase from avocado (*Persea americana*) fruit mesocarp. Biochim. Biophys. Acta 1257:1–10.

Eliasson M, Olsson A, Palmcrantz E, Wiberg K, Inganas M, Guss B, Lindberg M, Uhlen M (1988). Chimeric IgG-binding receptors engineered from staphylococcal protein A and staphylococcal protein G. J Biol Chem 263:4323–4327.

Estable MC, Hirst M, Bell B, O'Shaughnessy MV, Sadowski I (1999). Purification of RBF-2, a transcription factor with specificity for the most conserved cis-element of naturally occurring HIV-1 LTRs. J. Biomed. Sci. 6:320–332.

Fahrner RL, Blank GS (1999). Real-time control of antibody loading during protein A affinity chromatography using an on-line assay. J. Chromatogr. A 849:191–196.

Feng W, Graumann K, Hahn R, Jungbauer A (1999). Affinity chromatography of human estrogen receptor-alpha expressed in *Saccharomyces cerevisiae*. Combination of heparin- and 17beta-estradiol-affinity chromatography. J. Chromatogr. A 852:161–173.

Frank R (1992). Spot synthesis: An easy technique for the positionally addressable, parallel chemical synthesis on a membrane support. Tetrahedron 48:9217–9232.

Freitag R, Vogt S, Modler M (1999). Thermoreactive displacers for anion exchange and hydroxyapatite displacement chromatography. Biotechnol. Prog. 15:573–576.

Galaev IY (1998). New methods of protein purification. Expanded bed chromatography. Biochemistry (Mosc) 63:619–624.

Galaev IY, Mattiasson B (1994a). Shielding affinity chromatography. Bio/Technology 12:1086.

Galaev IY, Mattiasson B (1994b). Poly(N-vinylpyrrolidone) shielding of matrices for dye-affinity chromatography. Improved elution of lactate dehydrogenase from Blue Sepharaose and secondary alcohol dehydrogenase from scarlet Sepharose. J. Chromatogr. 662:27–35.

Galaev IY, Mattiasson B (1994c). Polymer-shielded dye affinity chromatography. In: DL Pyle, ed. Separations for Biotechnology 3. Cambridge, UK: SCI and Royal Society of Chemistry, pp. 179–185.

Galaev IY, Garg N, Mattiasson B (1994a). Interaction of Cibacron blue with polymers: implications for polymer-shielded dye-affinity chromatography of phosphofructokinase from baker's yeast. J. Chromatogr. 684:45–54.

Galaev IY, Warrol C, Mattiasson B (1994b). Temperature-induced displacement of proteins from dye-affinity columns using an immobilized polymeric displacer. J. Chromatogr. 684:37–43.

Gao J, Mammen M, Whitesides GM (1996). Evaluating electrostatic contributions to binding with the use of protein charge ladders. Science 272:535–537.

Garg N, Galaev IY, Mattiasson B (1994). Effect of poly (vinyl alcohol) treatment of the dye matrix on the chromatography of pyruvate kinase. Biotechnol. Tech. 8:645–650.

Gerard NP, Gerard C (1990). Construction and expression of a novel recombinant anaphylatoxin, C5a-N19, as a probe for the human C5a receptor. Biochemistry 29:9274–9281.

Geyer M, Assheuer R, Klebe C, Kuhlmann J, Becker J, Wittinghofer A, Kalbitzer H (1999). Conformational states of the nuclear GTP-binding protein ran and its complexes with the exchange factor RCC1 and the effector protein RanBP1. Biochemistry 38:11250–11260.

Geyson MH, Meloen RH, Barteling SJ (1984). Use of peptide synthesis to probe viral antigens for epitopes to a resolution of a single amino acid. Proc. Natl. Acad. Sci. USA 81:3998–4002.

Goldstein IJ, Hayes CE (1978). The lectins: Carbohydrate-binding proteins of plants and animals. Adv. Carbohydr. Chem. Biochem. 35:127–340.

Gomez FA, Avila LZ, Chu YH, Whitesides GM (1994). Determination of binding constant of ligands to proteins by affinity capillary electrophoresis: compensation for electroosmotic flow. Anal. Biochem. 66:1755–1792.

Govrin E, Levine A (1999). Purification of active cysteine proteases by affinity chromatography with attached E-64 inhibitor. Protein Expr. Purif. 15:247–250.

Greenwood JM, Gilkes NR, Miller RC Jr., Kilburn DG, Warren RAJ (1994). Purification and processing of cellulose-binding domain-alkaline phosphatase fusion proteins. Biotechnol. Bioeng. 44:1295–1305.

Gupta MN, Kaul R, Guoqiang D, Dissing U, Mattiasson B (1996). Affinity precipitation of proteins. J. Mol. Recognit. 9:356–359.

Gupta MN, Guoquiang D, Mattiasson B (1993). Purification of endopolygalacturonase by affinity precipitation using alginate. Biotechnol. Appl. Biochem. 18:321–324.

Gupta MN, Guoquiang D, Mattiasson B (1994). Purification of xylanase from *Trichoderma viride* by precipitation with an anionic polymer Eudragit S-100. Biotechnol. Tech. 8:117–122.

Gupta MN, Mattiasson B (1992). Unique Applications of immobilized proteins in bioanalytical sytems. In: Suelter CH, Kricka L, eds. Bioanalytical Applications of Enzymes, Vol. 36, John Wiley and Sons, New York, pp. 1–34.

Gupta MN, Mattiasson B (1994). Affinity precipitation. In: Street G, ed. Highly Selective Separations in Biotechnology. Chapman and Hall, London, pp. 7–33.

GuyonnetV, Tluscik F, Long PL, Polanowski A, Travis J (1999). Purification and partial characterization of the pancreatic proteolytic enzymes trypsin, chymotrypsin, and elastase from the chicken. J. Chromatogr. A 852:217–225.

Haupt K, Mosbach K (1998). Plastic antibodies: developments and applications. Trends Biotechnol. 16:468–475.

Hauri HP, Bucher K (1986). Immunoblotting with monoclonal antibodies: importance of the blocking solution. Anal Biochem. 159:386–389.

Herak DC, Merrill EW (1989). Affinity cross-flow filtration: Experimental and modeling work using the system of HSA and Cibacron Blue-agarose. Biotechnol. Prog. 5:9–17.

Hermanson GT, Mallia AK, Smith PK (1992). Immobilized Affinity Ligand Techniques. San Diego: Academic Press.

Hizel C, Maurizis J-C, Rio P, Communal Y, Chaassagne J, Favy D, Bignon YJ, Bernard-Gallon DJ (1999). Isolation, purification and quantification of BRCA1 protein from tumour cells by affinity perfusion chromatography. J. Chromatogr. B 721:163–170.

Hjorth R (1997). Expanded bed adsorption in industrial bioprocessing: recent developments. Trends Biotechnol. 15:230–235.

Honda S, Taga A, Suzuki K, Suzuki S, Kakehi K (1992). Determination of the association constant of monovalent mode protein-sugar interaction by capillary zone electrophoresis. J. Chromatogr. 597:377–382.

Huls GA, Heijnen IAFM, Cuomo MA, Koningberger JC, Weigman L, Boel E, van der Vuurst de Vries A, Loyson SAJ, Helfrich W, van Berge Henegouwen GP, van Meijer M, de Kruif J, Logtenberg T (1999). A recombinant, fully human monoclonal antibody with antitumor activity constructed from phage-displayed antibody fragments. Nature Biotechnol. 17:276–281.

Jacobs JW, Fodor SPA (1994). Combinatorial chemistry—application of light-directed chemical synthesis. Trends Biotechnol. 12:19–26.

Jakoby W, Wilchek M (1974). Meth. Enzymol. 34:265–267.
Janson JC, Rydén L (1989). Protein Purification: Principles, High Resolution Methods, and Applications. New York: VCH Publishers.
Kågedal L (1989). Immobilized metal ion affinity chromatography. In: Janson JC, Rydén L, eds. Protein Purification: Principles, High Resolution Methods, and Applications. VCH Publishers, New York, pp. 227–251.
Karlsson GB, Platt FM (1991). Analysis and isolation of human transferrin receptor using the OKT-9 monoclonal antibody covalently crosslinked to magnetic beads. Anal. Biochem. 199:219–222.
Kim HO, Durance TD, Li-Chan EC (1999a). Reusability of avidin-biotinylated immunoglobulin Y columns in immunoaffinity chromatography. Anal. Biochem. 268:383–397.
Kim JH, Lee JH, Kim J, Kim JK (1999b). Identification of phage-displayed peptides which bind to the human HnRNPA1 protein-specific monoclonal antibody. Mol. Cells 9:452–454.
Kim S, Titlow CC, Margolies MN (2000). An approach for preventing recombination-deletion of the 40–50 anti-digoxin antibody V(H) gene from the phage display vector pComb3. Gene 241:19–25.
Kirkpatrick DL, Watson S, Ulhaq S (1999). Structure-based drug design: combinatorial chemistry and molecular modeling. Comb. Chem. High Throughput Screen 2:211–221.
Kojima K (1997). Selective removal of plasma components by high performance immunoaffinity chromatography. Ther. Apher. 1:169–173, 1997.
Kubal BS, Godbole SS, D'Souza SF (1986). Preparation and characterization of magnetic hen egg white beads containing coimmobilized glucose oxidase, magnetite and MnO_2. Ind. J. Biochem. Biophys. 23:240–241.
Labrou NE, Clonis YD (1994). The affinity technology in downstream processing. J. Biotechnol. 36:95–119.
Labrou NE, Eliopoulos E, Clonis YD (1999). Molecular modeling for the design of a biomimetic chimeric ligand. Application to the purification of bovine heart L-lactate dehydrogenase. Biotechnol. Bioeng. 63:322–332.
Lam KS, Salmon SE, Hersh EM, Hruby VJ, Kazimeirski WM, Knapp RJ (1991). A new type of synthetic peptide library for identifying ligand-binding activity. Nature 354:82–84.
Latner AL, Hodson AW (1980). Differential precipitation with Concanavalin A as a method for the purification of glycoproteins: human alkaline phosphatase. Anal. Biochem. 101:483–487.
Le KD, Gilkes NR, Kilburn DG, Miller RC Jr., Saddler JN, Warren RAJ (1994). A streptavidin-cellulose-binding domain fusion protein that binds biotinylated proteins to cellulose. Enzyme Microb. Technol. 16:496–500.
Leiner IE, Sharon N, Goldstein IJ (1986). The Lectins: Properties, Functions and Applications in Biology and Medicine. Academic Press, San Diego.
Li R, Dowd V, Stewart DJ, Burton SJ, Lowe CR (1998). Design, synthesis and application of a Protein A mimetic. Nature Biotechnol. 16:190–195.

Liao CH, Revear L, Hotchkiss A, Savary B (1999). Genetic and biochemical characterization of an exopolygalacturonase and a pectate lyase from *Yersinia enterocolitica*. Can J. Microbiol. 45:396-403.

Liapis AI, Unger KK (1994). The chemistry and engineering of affinity chromatography. In: Street G, ed. Highly Selective Separations in Biotechnology. Chapman and Hall, London, pp. 121–162.

Lin JM, Nakagama T, Uchiyama K, Hobo T (1997). Temperature effect on chiral recognition of some amino acids with molecularly imprinted polymer filled capillary electrochromatography. Biomed. Chromatogr. 11:298–302.

Lopatin SA, Varlamov VP (1995). New trends in immobilized metal ion affinity chromatography of proteins. Appl Biochem. Microbial. 3:212–217.

Lopuska A, Polanowska J, Wilusz T, Polanowski A (1999). Purification of two low-molecular-mass serine proteinase inhibitors from chicken liver. J. Chromatogr. A 852:207–216.

Lowe CR, Burton SJ, Burton NP, Alderton WK, Pitts JM, Thomas JA (1992). Designer dyes: "Biomimetic" ligands for the purification of pharmaceutical proteins by affinity chromatography. Trends Biotechnol. 10:442–448.

Mattiasson B (1988). Affinity immobilization. Meth. Enzymol. 137:647–656.

McKercher G, Bonneau PR, Lagace L, Thibeault D, Massariol MJ, Krogsrud R, Lawetz C, McDonald PC, Cordingley MG (1997). Improved purification protocol of the HSV-I protease catalytic domain, using immunoaffinity. Biochem Cell Biol. 75:795–801.

Miletti LC, Marino C, Marino K, de Lederkremer RM, Colli W, Alves MJ (1999). Immobilized 4-aminophenyl 1-thio-beta-D-galactofuranoside as a matrix for affinity purification of an exo-beta-D-galactofuranosidase. Carbohydr. Res. 320:176–182.

Mosbach K. Meth. Enzymol. 135:1987.

Mosbach K. Meth. Enzymol. 137:1988.

Mosbach K, Ramström O (1996). The emerging technique of molecular imprinting and its future impact on biotechnology. Bio/Technology 14:163–170.

Nashabeh W, el Rassi Z (1991). Capillary zone electrophoresis of alpha 1-acid glycoprotein fragments from trypsin and endoglycosidase digestions. J. Chromatogr. 536:31–42.

Newsome WH, Abbott M (1999). An immunoaffinity column for the determination of peanut protein in chocolate. J AOAC Int 82:666–668.

Nihalani I (1999). Protein-dye interaction and its application in protein bioseparation. M. Tech. thesis. Indian Institute of Technology, Delhi.

Nilsson K, Lindell J, Norrlöv O, Sellergren B (1994). Imprinted polymers as antibody mimetics and new affinity gels for selective separations in capillary electrophoresis. J. Chromatogr. 680:57–61.

Norde W (1986). Adsorption of proteins from solution at the solid-liquid interface. Adv. Colloid Interface Sci. 4:267–340.

Oda Y, Sanders J, Roberts S, Maruyama M, Kiddie A, Furmaniak J, Smith BR (1999). Analysis of carbohydrate residues on recombinant human thyrotropin receptor. J. Clin. Endocrinol. Metab. 84:2119–2125.

Ohlson S, Bergström M, Leickt L, Zopf D (1998). Weak affinity chromatography of small saccharides with immobilized wheat germ agglutin and its application to monitoring of carbohydrate transferase activity. Bioseparation 7:101–105.

Ohlson S, Lundblad A, Zopf D (1988). Novel approach to affinity chromatography using "weak" monoclonal antibodies. Anal. Biochem. 169:204–208.

Olivera BM, Hillyard DR, Marsh M, Yoshikami D (1995). Combinatorial peptide libraries in drug design: lesons from venomous cone snails. Trends Biotechnol. 13:422–426.

O'Shannessy DJ and Hofman WI (1987). Site-directed immobilization of glycoproteins on hydrazide-containing supports. Biotechnol. Appl. Biochem. 9:488–496.

Parikh I, Cuatrecasas P (1985). Affinity chromatography. Chem. Eng. News 17–32.

Patel AB, Srivastava S, Phadke RS (1999). Interaction of 7-hydroxy-8-(phenylazo)1,3-naphthalenedisulfonate with bovine plasma albumin. Spectroscopic studies. J. Biol. Chem. 74:21755–21762.

Penzol G, Armisén P, Fernández-Lafuente R, Rodés L, Guisán JM (1998). Use of dextrans as long and hydrophilic spacer arms to improve the performance of immobilized proteins acting on macromolecules. Biotechnol Bioeng 60:518–523.

Perez L, Estapa A, Coll JM (1998). Purification of the glycoprotein G from viral haemorrhagic septicaemia virus, a fish rhabdovirus, by lectin affinity chromatography. J. Virol. Meth. 76:1–8.

Pharmacia (1993). Affinity Chromatography: Principles and Methods.

Porath J (1992). Immobilized metal ion affinity chromatography. Protein Expr. Purif. 3:263–281.

Porath J, Carlsson J, Olsson I, Belfrage G (1975). Metal chelate affinity chromatography, a new approach to protein fractionation. Nature 258:598–599.

Posewitz MC, Tempst P (1999). Immobilized gallium(III) affinity chromatography of phosphopeptides. Anal. Chem. 71:2883–2892.

Poulouin L, Gallet O, Rouahi M, Imhoff JM (1999). Plasma fibronectin: three steps to purification and stability. Protein Expr. Purif. 17:146–152.

Pouny Y, Weitzman C, Kaback HRI (1998). In vitro biotinylation provides quantitative recovery of highly purified active lactose permease in a single step. Biochemistry 37:15713–15719.

Rais-Beghdadi C, Roggero MA, Fasel N, Reymond CD (1998). Purification of recombinant proteins by chemical removal of the affinity tag. Appl. Biochem. Biotechnol. 74:95–103.

Ramadoss CS, Steczko J, Uhlig JW, Axelrod B (1983). Effect of albumin on binding and recovery of enzymes in affinity chromatography on Cibacron blue. Anal. Biochem. 130:481–484.

Reddy GB, Srinivas VR, Ahmad N, Surolia A (1999). Molten globule-like state of peanut lectin monomer retains its carbohydrate specificity. J. Biol. Chem. 274:4500–4503.

Romig TS, Bell C, Drolet DW (1999). Aptamer affinity chromatography. Aptamer combinatorial chemistry applied to protein purification. J. Chromatogr. B Biomed. Sci. Appl. 731:275–284.

Rongxiu L, Dowd V, Stewart DJ, Burton SJ, Lowe CR (1998). Design, synthesis and application of a protein A mimetic. Nature Biotechnol. 16:190–195.

Roy I, Pai A, Lali A, Gupta MN (1999). Comparison of batch, packed bed and expanded bed purification of *A. niger* cellulase using cellulose beads. Bioseparation 8:317–326.

Roy I, Sardar M, Gupta MN (2000). Exploiting unusual affinity of usual polysaccharides for separation of enzymes on fluidized beds. Enzyme Microb. Technol. 27:53–65.

Russell AJ, Klibanov AM (1988). Inhibitor-induced enzyme activation in organic solvents. J. Biol. Chem. 263:11624–11626.

Safarik I, Safarikova M (1997). Overview of magnetic separations used in biochemical and biotechnological applications. In: Hafeli U, Schut W, Teller J, Dorowski M, eds. Scientific and Clinical Applications of Magnetic Carriers. Plenum Press, New York, pp 323–340.

Safarik I, Safarikova M (1999). The use of magnetic techniques for the isolation of cells. J. Chromatogr. B 722:33–53.

Sardar M, Gupta MN (1998). Alginate beads as an affinity material for alpha amylases. Bioseparation 7:159–165.

Scouten WH (1981). Affinity chromatography: Bioselective adsorption on inert matrices. New York: John Wiley and Sons.

Senstad C, Mattiasson B (1989). Purification of wheat germ agglutinin using flocculation with chitosan and a subsequent centrifugation or floatation step. Biotechnol. Bioeng. 34:387–393.

Sharma V, Srinivas VR, Adhikari P, Vijayan M, Surolia A (1998). Molecular basis of recognition by Gal/GalNAc specific legume lectins: Influence of Glu 129 on the specificity of peanut agglutinin (PNA) towards C-2 substituents of galactose. Glycobiology 8:1007–1012.

Sharma V, Surolia A (1997). Analyses of carbohydrate recognition by legume lectins: Size of the combining site loops and their primary specificity. J. Mol. Biol. 267:433–445.

Sharma V, Vijayan M, Surolia A (1996). Imparting exquisite specificity to peanut agglutinin for the tumour-assisted Thomsen-Friedenreich antigen by redesign of its combining site. J. Biol. Chem. 271:21209–21213.

Shea KJ (1994). Molecular imprinting of synthetic network polymers: the de novo synthesis of macromolecular binding and catalytic sites. Trends Polym. Sci. 19:9–14.

Shi H, Tsai WB, Garrison MD, Ferrari S, Ratner BD (1999). Template-imprinted nanostructured surfaces for protein recognition. Nature 398:593–597.

Simon RJ, Kama RS, Zuckermann RN, Huebner VO, Jewell DA, Banville SC, Ng S, Wang I, Rosenberg S, Marlowe CK, Spellmeyer D, Tan R, Frankel AD, Sann DV, Cohen FF, Bartlett PA (1992). Peptoids: A modular approach to drug discovery. Proc. Natl. Acad. Sci. USA 89:9367–9371.

Skorey KL, Johnson NA, Huyer G, Gresser MJ (1999). A two-component affinity chromatography purification of *Helix pomatia* arylsulfatase by tyrosine vanadate. Protein Exp. Purif. 15:178–187.

Soler G, Blanco RM, Fernandez-Lafuente R, Rosell CM, Guisan JM (1995). Design of novel biocatalysts by "bioimprinting" during unfolding-refolding of fully dispersed covalently immobilized enzymes. Ann. NY Acad. Sci. 750:349–356.

Spitznagel TM, Clark DS (1993). Surface-density and orientation effects on immobilized antibodies and antibody fragments. Bio/Technology 11:825–829.

Srinivas VR, Reddy GB, Surolia A (1999). A predominantly hydrophobic recognition of H-antigenic sugars by winged bean acidic lectin: A thermodynamic study. FEBS Lett. 450;181–185.

Stellwagen E (1993). Affinity chromatography with immobilized dyes. In: Ngo TT, ed. Molecular Interactions in Bioseparation. Plenum Press, New York, pp. 247–275.

Stolz J, Ludwig A, Stadler R, Biesgen C, Hagemann K, Sauer N (1999). Structural analysis of a plant sucrose carrier using monoclonal antibodies and bacteriophage lambda surface display. FEBS Lett. 53:375–379.

Su NY, Yu CJ, Shen HD, Pan FM, Chow LP (1999). Pen c 1, a novel enzymic allergen protein from *Penicillium citrinum*. Purification, characterization, cloning and expression. Eur. J. Biochem. 261:115–123.

Sulkowski E (1985). Purification of proteins by IMAC. Trends Biotechnol. 3:1–7.

Surolia A, Khan MI, Pain D (1982). Protein A: Nature's universal antibody. Trends Biochem. Sci. 7:74–76.

Swaminathan CP, Surolia N, Surolia A (1998). Role of water in the specific binding of mannose and mannooligosaccharides to concanavalin A. J. Am. Chem. Soc. 120:5153–5159, 1998.

Swanson SJ, Jacobs SJ, Mytych D, Shah C, Indelicato SR, Bordens RW (1999). Applications for the new electrochemiluminescent (ECL) and biosensor technologies. Dev. Biol. Stand. 97:135–147.

Technote #205 (1999). Covalent coupling. Bangs Laboratories.

Thomas CJ, Surolia A (1999). Kinetics of the interaction of endotoxin with Polymyxin B and its analogs: a surface plasmon resonance analysis. FEBS Lett. 445:420–424.

Thomas CJ, Surolia N, Surolia A (1999). Surface plasmon resonance studies resolve the enigmatic endotoxin neutralizing activity of Polymyxin B. J. Biol. Chem. 274:29624–29627.

Tomioka Y, Kudo Y, Hayashi T, Nakamura H, Niizeki M, Hishinuma T, Mizugaki M (1997). Phenobarbital molecularly-imprinted polymer selectively binds phenobarbital. Biol. Pharm. Bull. 20:397–400.

Travis J, Bowen J, Tewksbury D, Johnson D, Pannell R (1976). Isolation of albumin from whole human plasma and fractionation of albumin-depleted plasma. Biochem. J. 157:301–306.

Tyagi R, Gupta MN (1995). Purification and immobilization of *Aspergillus niger* pectinase on magnetic latex beads. Biocat. Biotrans. 12:293–298.

Tyagi R, Agarwal R, Gupta MN (1996a). Purification of peanut lectin using guar gum as an affinity ligand. J. Biotechnol. 46:79–83.

Tyagi R, Kumar A, Sardar M, Kumar S, Gupta MN (1996b). Chitosan as an affinity macroligand for precipitation of N-acetyl glucosamine binding proteins/enzymes. Isol. Purif. 2:217–226.

Uezu K, Yoshida M, Goto M, Furusaki S (1999). Molecular recognition using surface template polymerization. Chemtech 29:12–18.

Vaidya S, Gupta MN (1988). Concanavalin A interacts with lysozyme—A model for understanding lectin-glycoprotein interaction. Biochem. Int. 17:647–653.

van Damme EJM, Peumans W J, Pusztai A, Bardocz S (1997). Handbook of Plant Lectins: Properties and Biomedical Applications. New York: John Wiley and Sons.

van Der Vuurst De Vries AR, Logtenberg T (1999). A phage antibody identifying an 80-kDa membrane glycoprotein exclusively expressed on a subpopulation of activated B cells and hairy cell leukemia B cells. Eur. J. Immunol. 29:3898–3907.

van Oss CJ (1997). Hydrophobic and hydrophilic interactions in antigen-antibody binding. Int. J. Biochromatogr. 3:1–8.

Villamon E, Gozalbo D, Martinez JP, Gil ML (1999). Purification of a biologically active recombinant glyceraldehyde 3-phosphate dehydrogenase from *Candida albicans*. FEMS Microbiol. Lett. 179:61–65.

Vogt RF, Phillips DL, Henderson LO, Whitfield W, Spierto FW (1987). Quantitative differences among various proteins as blocking agents for ELISA microtiter plates. J. Immunol. Methods 101:43–50.

Wallworth DM (1996). The use of radial flow chromatography for the purification of biomolecules. In: M Verrall, ed. Downstream Processing of Natural Products. John Wiley and Sons, Chichester, pp. 209–221.

Wikstrom M, Ohlson S (1992). Computer simulation in weak affinity chromatography. J Chromatogr. 597:83–92.

Wilchek M, Miron T, Kohn J (1984). Affinity chromatography. Meth. Enzymol. 104:3–55.

Wilchek M, Bayer E (1990). Introduction to avidin-biotin technology. Meth. Enzymol. 184.

Willoughby NA, Kirschner T, Smith MP, Hjorth R, Titchener-Hooker NJ (1999). Immobilised metal ion affinity chromatography purification of alcohol dehydrogenase from baker's yeast using an expanded bed adsorption system. J. Chromatogr. A 840:195–204.

Winthrop MD, DeNardo SJ, DeNardo GL (1999). Development of a hyperimmune anti-MUC-1 single chain antibody fragments phage display library for targeting breast cancer. Clin. Cancer Res. 5:3088s–3094s.

Wisen S, Jiang F, Bergman B, Mannervik B (1999). Expression and purification of the transcription factor NtcA from the cyanobacterium *Anabaena* PCC 7120. Protein Expr. Purif. 15:351–357.

Worlock AJ, Sidgwick A, Horsburgh T, Bell PRF (1991). The use of paramagnetic beads for the detection of major histocompatibility complex class I and class II antigens. Biotechniques 10:310–315.

Wu H and Bruley DE (1999). Homologous human blood protein separation using immobilized metal affinity chromatography: Protein C separation from prothrombin with application to the separation of factor IX and prothrombin. Biotechnol. Prog. 15:928–931.

Wulff G (1993). The role of binding-site interactions in the molecular imprinting of polymers. Trends Biotechnol. 11:85–87.

Yamasaki K, Maruyama T, Yoshimoto K, Tsutsumi Y, Narazaki R, Fukuhara A, Kragh-Hansen U, Otagiri M (1999). Interactive binding to the two principal ligand binding sites of human serum albumin: effect of the neutral-to-base transition. Biochim. Biophys. Acta 1432:313–323.

Yang J, Moyana M, Xiang J (1999). Enzyme-linked immunosorbent assay-based selection and optimization of elution buffer for TAG72-affinity chromatography. J. Chromatogr. B Biomed. Sci. App. 731:299–308.

Yarmush ML, Antonsen KP, Sundaram S, Yarmush DM (1992). Immunoadsorption: Strategies for antigen elution and production of reusable adsorbents. Biotechnol. Prog. 8:168–178.

Yoshikawa K, Umetsu K, Shinzawa H, Yuasa I, Maruyama K, Ohkura Y, Yamashita K, Suzuki T (1999). Determination of carbohydrate-deficient transferrin separated by lectin affinity chromatography for detecting chronic alcohol abuse. FEBS Lett. 458:112–116.

Yu C, Mosbach K (1998). Insights into the origins of binding and the recognition properties of molecularly imprinted polymers prepared using an amide as the hydrogen-bonding functional group. J. Mol. Recognit. 11:69–74.

Zhou M, van Etten R (1999). Structural basis of the tight binding of pyridoxal 5′-phosphate to a low molecular weight protein tyrosine phosphatase. Biochemistry 38:2636–2646.

Zusman I (1998). Gel fiberglass membranes for affinity chromatography columns and their application to cancer detection. J. Chromatogr. B 715:297–306.

Zwicker N, Adelhelm K, Thiericke R, Grabley S, Hanel F (1999). Strep-tag II for one-step affinity purification of active bHLHzip domain of human c-Myc. Biotechniques 27:368–375.

4

Genetic Approaches to Facilitate Protein Purification

Stefan Ståhl, Sophia Hober, Joakim Nilsson, Mathias Uhlén, and Per-Åke Nygren
Royal Institute of Technology, Stockholm, Sweden

This chapter will describe how genetic approaches can be used to simplify recovery of recombinant proteins. Gene fusion technology resulting in the expression of fusion proteins, having the combined properties of the parental gene products, has found widespread use in biotechnology. Applications of fusion proteins include facilitated purification of the target protein; means to decrease proteolysis of the target protein; display of proteins on surfaces of cells, phages, or viruses; construction of reporter molecules for the monitoring of gene expression and protein localization; and strategies to increase the circulation half-life of protein therapeutics. However, as pioneered by Uhlén and coworkers (1983), the most common application of gene fusions has been for the purpose of affinity purification of recombinant proteins. Recently, gene technology in the format of combinatorial protein engineering has allowed in vitro selection of novel proteins that selectively bind a desired target molecule. This new emerging technology enables the de novo creation of purpose-designed ligands suitable for affinity chromatography applications. A general overview of commonly used affinity tags will be followed by some examples of specific applications for affinity tags and a description of some novel trends in how genetic approaches can be employed to facilitate purification.

1. INTRODUCTION

There is a great interest in developing methods for fast and convenient purification of proteins. For example, to facilitate functional and structural studies of proteins derived from the rapidly growing number of genes coming out of genome programs such as the Human Genome Project (Caskey and Rossiter 1992), efficient and robust production and purification strategies are necessary (Larsson et al. 1996). For industrial production of recombinant proteins, simple and fast purification methods introduced as an early unit operation can drastically improve on the overall economy of the process.

A powerful purification technique made possible by the introduction of genetic engineering is to purify the target protein by the use of a genetically fused affinity fusion partner. Such fusion proteins can often be purified to near-homogeneity from crude biological mixtures by a single, and fusion partner–specific, affinity chromatography step. The strong, specific, and well-characterized interaction between the Fc part of immunoglobulin G (IgG) and *Staphyloccus aureus* protein A (SPA) was the first biointeraction exploited for the creation of an affinity fusion system allowing affinity chromatography purification of expressed recombinant gene products (Uhlén et al. 1983).

To date, a large number of different gene fusion systems have been described involving fusion partners that range in size from one amino acid to whole proteins capable of selective interaction with a ligand immobilized onto a chromatography matrix (For reviews, see Ford et al. 1991; Uhlén et al. 1992; Kaul and Mattiasson 1992; Flaschel and Friehs 1983; Lavallie and McCoy 1995; Makrides 1996; Nilsson et al. 1997b). In such systems, different types of interactions have been utilized such as enzymes-substrates, bacterial receptors–serum proteins, polyhistidines-metal ions, and antibodies-antigens. The conditions for purification differs from system to system, and the environment tolerable by the target protein is an important factor for deciding which affinity fusion partner to choose. Also other factors, including costs for the affinity matrix, buffers, and the possibility of removing the fusion partner by site-specific cleavage, are important to consider.

More recently, other strategies based on DNA technology have also emerged for the purpose of facilitating recovery of recombinant proteins. Targeting of an expressed gene product to the extracellular space by means of secretion systems represents a convenient strategy to simplify the initial recovery of a recombinant protein (Hansson et al. 1994). The efficiency of using a secretion strategy, which can be employed with all different host cells, depends on the inherent properties of the target protein (Ståhl et al.

1997a). For the production of gene products inefficiently secreted but still remaining soluble when expressed intracellularly, heat release can in certain cases be a very efficient way of obtaining selective product release to facilitate the initial product recovery (Jonasson et al., 2000).

A technology that most probably will have a significant impact on the design of future recovery processes of "biotech" products is the use of combinatorial protein engineering to construct protein libraries from which target-specific ligands suitable for affinity chromatography can be selected using in vitro techniques (Nygren and Uhlén 1997; Nord et al. 1997, 2000).

This chapter will present some commonly used affinity fusion partners, followed by some relevant specific examples of applications for affinity fusion strategies. Furthermore, we will describe how phage display technology can be used to select novel protein domains, "affibodies," with completely new binding specificities and discuss how such affibodies might become useful ligands in bioseparation applications. Potential future trends in the use of affinity fusions and in vitro–selected ligands will be discussed.

2. AFFINITY FUSION PARTNERS

Some of the most commonly used affinity fusion systems are listed in Table 1, together with their respective elution condition(s) and agents suitable for specific detection of expressed fusion proteins. Using gene fusion strategies for the production of native proteins, efficient means for site-specific cleavage of the fusion protein and subsequent removal of the affinity fusion partner are needed. The introduction of a recognition sequence for a chemical agent or a protease between the fusion partner and the target protein allows for site-specific cleavage of the fusion protein to remove the affinity fusion partner (for reviews, see Carter 1990; Nygren et al. 1994; Nilsson et al. 1997b). If employing enzymatic strategies for cleavage, strategies for the removal of the protease itself must be developed. Special considerations must be taken if the target protein itself also has to be further processed to give the desired final product (Nilsson et al. 1996a; Jonasson et al. 1996, 1999). Careful upstream design of the fusion protein construct using genetic strategies can greatly facilitate the subsequent purification of the target protein and also allow for integrated systems involving coprocessing of the protein and efficient removal of the affinity fusion partner as well as the protease used for cleavage (Nilsson et al. 1996a; Gräslund et al. 1997). However, for many applications cleavage might not be necessary and the target protein is instead suitable for use as fused to the affinity tag: (1) as immunogens for the generation and purification of antibodies (Ståhl et al.

Table 1 Commonly Used Affinity Fusion Systems

Fusion partner	Size	Ligand	Elution conditions	Supplier
Protein A	31 kDa	hIgG	Low pH	Amersham Pharmacia Biotech
Z	7 kDa	hIgG	Low pH	Amersham Pharmacia Biotech
ABP	5–25 kDa	HSA	Low pH	—
Poly His	≈ 1 kDa	Me^{2+} chelator	Imidazole/ low pH	Novagen, Qiagen
GST	25 kDa	Glutathione	Reduced glutathione	Amersham Pharmacia Biotech
MBP	41 kDa	Amylose	Maltose	New England Biolabs
FLAG	1 kDa	MAbs M1/M2	EDTA/ Low pH	Sigma-Aldrich
PinPoint[a]	13 kDa	Streptavidin/avidin	Biotin[c]	Promega
Bio[b]	13 aa	Streptavidin/avidin	Biotin[c]	—

[a] A 13-kDa subunit of the transcarboxylase complex from *Propionibacterium shermanii* which is biotinylated in vivo by the *E. coli* cells.
[b] A 13 amino acid peptide which is selected from a combinatorial library and found to be in vivo biotinylated.
[c] If monomeric avidin is used as ligand, free biotin can be used for mild elution.
aa, amino acids; ABP, albumin-binding protein; GST, glutathione *S*-transferase; hIgG, human immunoglobulin G; HSA, human serum albumin; MAb, monoclonal antibody; MBP, maltose-binding protein.

1989; Kaslow and Shiloach 1994; Larsson et al. 1996); (2) when the biological activity of the target protein is unaffected by the affinity fusion partner (Gandecha et al. 1992; Kunz et al. 1992; Walker et al. 1994; Nilsson et al. 1994); (3) for directed immobilization of the target protein (Ljungquist et al. 1989a,b; Kashlev et al. 1993; Stempfer et al. 1996).

2.1 Affinity Tags Based on Protein A or Protein G

Staphylococcal protein A (SPA) is an immunoglobulin-binding receptor present on the surface of the gram-positive bacterium *Staphylococcus aureus*. The strong and specific interaction between SPA and the constant part (Fc) of certain immunoglobulins (IgG) has made it useful for the purification and detection of immunoglobulins in a variety of different applications (Langone 1982). SPA is capable of binding to the constant part of IgG from a large number of different species, including humans (Langone 1982; Eliasson et al. 1988). In addition to this interaction, SPA also binds to a limited number of Fab fragments containing certain (variable heavy chain) VH sequences (Vidal and Conde 1985; Potter et al. 1996). The cloning and sequencing of the SPA gene revealed a highly repetitive organization of the protein that could be divided into a signal sequence followed by five homologous domains (E, D, A, B, and C) and a cell surface anchoring sequence denoted XM (Fig. 1) (Löfdahl et al. 1983; Uhlén et al. 1984). It was later shown that all five domains of SPA were separately capable of binding IgG (Moks et al. 1986), both Fc and Fab (Jansson et al. 1998). The isolation of the gene made it possible to take advantage of different extensions of the receptor for use as affinity fusion partners, allowing one-step IgG affinity purification of target proteins (Table 1) (Uhlén et al. 1983; Nilsson and Abrahmsén 1990; Ståhl et al. 1997a and b).

Several additional properties of SPA (or fragments thereof) have made it particularly suitable as fusion partner for production of recombinant proteins: (1) SPA is highly stable against proteolysis in various hosts (Ståhl et al. 1997a and b). (2) The N and C termini of the three-helix bundle structure of an individual IgG-binding SPA domain are solvent exposed (Deisenhofer 1981; Torigoe et al. 1990; Gouda et al. 1992; Jendeberg et al. 1996), which favors independent folding of a fused target protein and the SPA fusion partner. (3) SPA does not contain any cysteine residues that otherwise could interfere with the disulfide formation within a fused target protein (Uhlén et al. 1984). (4) SPA is highly soluble and renatures efficiently after having been subjected to denaturants such as urea and guanidinium-HCl, which facilitates the refolding of SPA fusion proteins from inclusion bodies or the "reshuffling" of disulfide bridged misfolded or multimeric forms of target proteins (Samuelsson et al. 1991, 1994; Hober et al. 1994; Jonasson et al. 1996; Nilsson et al. 1996a). In addition, SPA fusion proteins can be produced to high levels within the *E. coli* cell and remain soluble (Nilsson et al. 1994, 1996a; Larsson et al. 1996). (5) It has been demonstrated to be possible to introduce different protease recognition sequences

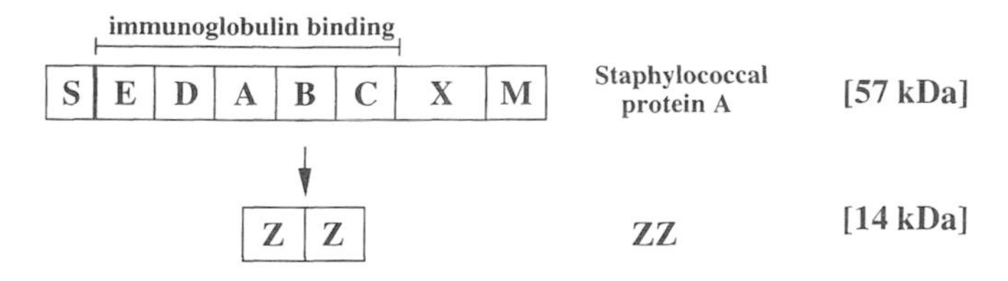

Figure 1 Schematic representation of staphylococcal protein A (SPA) and the single-domain SPA analogue Z, in its commonly used divalent form. S, signal peptide (processed during secretion); E, D, A, B, and C, immunoglobulin binding domains; XM, cell wall anchoring region.

accessible for site-specific cleavage of SPA fusion proteins to release the target protein (Forsberg et al. 1992). (6) The SPA fusion partner is secretion competent, and SPA fusions can be efficiently secreted to the *E. coli* periplasm and in some cases also to the culture medium, using the SPA promoter and signal sequence (Abrahmsén et al 1985, 1986; Moks et al. 1987a,b; Ståhl et al. 1990; Hansson et al. 1994).

To allow chemical cleavage of SPA fusion proteins with both CNBr and hydroxylamine, an engineered domain was designed based on domain B lacking methionine. In this new domain, designated Z (Table 1), the glycine residue in a hydroxylamine-sensitive Asn-Gly sequence was replaced by alanine, making it resistant to both of these chemical agents (Fig. 1) (Nilsson et al. 1987). In most of the recently constructed SPA affinity fusion vectors, different multiplicities of this Z domain have replaced the original SPA fusion partner. Analysis of different repeats of Z binding to IgG showed that there was no advantage in using more than two Z domains (Ljungquist et al., 1989b). A number of different expression vectors for *E. coli* production of Z or ZZ fusions have been developed where the fusion protein is either exported (Löwenadler et al. 1987; Hammarberg et al. 1989; Ståhl et al. 1989; Sjölander et al. 1991) or kept intracellularly (Köhler et al. 1991; Murby et al. 1994; Nilsson et al. 1994, 1996a; Jonasson et al. 1996; Larsson et al. 1996). In addition to *E. coli*, several other hosts have also been used for successful expression of SPA and ZZ fusion proteins, such as gram-positive bacteria, yeast, plant cells, CHO cells, and insect cells (Nygren et al. 1994; Ståhl et al. 1997a and b).

Streptococcal protein G (SPG) is a bifunctional receptor present on the surface of certain strains of streptococci that is capable of binding to both IgG and serum albumin (Björck and Kronvall 1984; Björck et al. 1987; Nygren et al. 1988). The structure is highly repetitive, with several structurally and functionally different domains (Fahnestock et al. 1986; Guss et al. 1986). The regions responsible for the affinities to serum albumin and IgG, respectively, are structurally separated, which allows for the subcloning of fragments displaying serum albumin binding only (Fig. 2) (Nygren et al. 1988). The serum albumin binding region has been proposed to contain three binding motifs (about 5 kDa each) of which one has been structurally determined showing a three-helix bundle domain, surprisingly similar in structure to an IgG binding domain of SPA (Falkenberg et al. 1992; Kraulis et al. 1996). The region binds to serum albumins from different species with various affinities and shows strong binding to human serum albumin (HSA) (Nygren et al. 1990). Expression in *E. coli* has demonstrated that the region shows a high proteolytic stability, is highly soluble, and can be efficiently secreted (Nygren et al. 1988; Ståhl et al. 1989; Nord et al. 1995; Larsson et al. 1996).

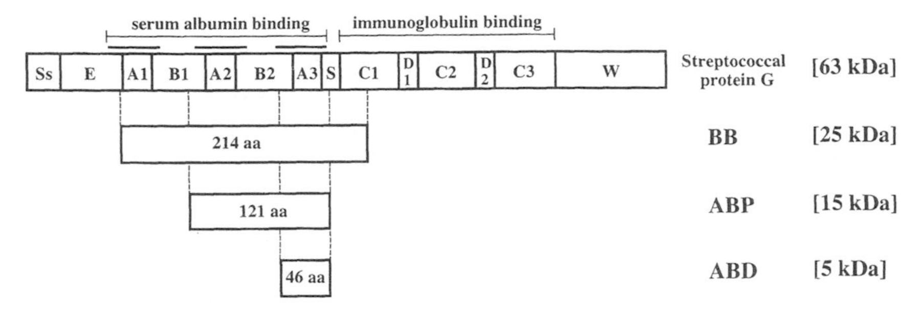

Figure 2 Schematic representation of streptococcal protein G (SPG) and the extensions of various serum albumin binding subfragments described in the text. S, signal peptide (processed during secretion); W, cell wall anchoring region. The regions responsible for serum albumin and immunoglobulin binding, respectively, are marked. BB (25 kDa) was the first isolated serum albumin–binding affinity tag. More recently, the new tags ABP (15 kDa) and ABD (5 kDa), containing two or one serum albumin–binding motifs, respectively, were designed on the basis of postulated borders for the albumin-binding protein domains. The three postulated minimal albumin-binding motifs are indicated as bars above the SPG.

Different parts of the region, denoted BB, ABP (albumin-binding protein), or ABD (albumin-binding domain) (Fig. 2), have been utilized as fusion partners allowing one-step HSA affinity chromatography purification of fusion proteins expressed either intracellularly or as secreted proteins (Table 1) (Nygren et al. 1988; Ståhl et al. 1989; Lorca et al. 1992; Nilsson et al. 1994; Murby et al. 1995; Larsson et al. 1996; Jonasson et al. 2000). Other hosts, such as COS and CHO cells (Makrides et al. 1996) and gram-positive bacteria (Hansson et al. 1992; Samuelson et al. 1995), have in addition to *E. coli* successfully been used to produce BB or ABP fusion proteins.

The SPG-based function partners have inherent properties that could make such fusion proteins of particular interest for delivery of protein therapeutics, as well as for administration of subunit vaccines. The strong and specific binding to serum albumin has been proposed to be responsible for the increased in vivo stability seen for otherwise rapidly cleared proteins when injected into the bloodstream of a laboratory animal (Nygren et al. 1991; Makrides et al. 1996). Serum albumin, which is the most abundant protein in serum, has a long half-life itself (19 days in humans) (Carter and Ho 1994) and could therefore be an ideal "stabilizer" of sensitive proteins via a serum albumin binding fusion partner. Nygren and coworkers (1991) showed that the half-life in mice of CD4 (the target receptor for human immunodeficiency virus) could be significantly increased by fusion to BB.

Recently, the same strategy was employed to increase the serum half-life of human soluble complement receptor type 1 (sCR1) in rats, by fusion to different serum albumin–binding fragments of SPG (Makrides et al. 1996).

The serum albumin binding region of SPG seems to have immunopotentiating properties, when used as a carrier protein genetically fused to a protein immunogen used for immunization (Sjölander et al. 1993). Recently, it was demonstrated that a BB fusion protein (BB-M3) containing a malaria peptide (M3) induced significant antibody responses in mice strains that were nonresponders to the malaria peptide alone, thus suggesting that BB has the ability to provide T-cell help for antibody production (Sjölander et al. 1997). Furthermore, a fusion protein BB-G2N, comprising a 101-amino-acid sequence from human respiratory synctial virus (RSV) was shown to induce protective immunity in mice to RSV challenge (Power et al. 1997; Lisbon et al. 1999). It was shown that by inclusion of the BB part, a more potent G2N-specific B-cell memory response was evoked (Libon et al. 1999). This indicates that the SPG-derived BB can function both as an affinity tag to facilitate purification and as a carrier protein with immunopotentiating properties. To date, it is not fully elucidated whether this capacity is due solely to strong T-cell epitopes or is also related to the serum albumin binding capacity resulting in a prolonged exposure of the immunogen to the immune system.

Protocols for Protein A-IgG Affinity Chromatography

The affinity chromatography purification of SPA fusions has been shown to be extremely specific, resulting in a highly purified product after a single-step procedure. Even for highly expressed proteins, yields of up to 100% have been demonstrated in the chromatography step (Moks et al. 1987b; Hansson et al. 1994), and the fast kinetics of the binding allows sample loading onto the IgG-Sepharose columns with relatively high flow rates. It has also been demonstrated that the columns can potentially be regenerated for repeated use up to 100 times (Uhlén and Moks 1990). The human polyclonal IgG used as ligand can be replaced by recombinant Fc fragments (Jendeberg et al. 1997) avoiding the use of a human serum protein in the purification protocol. Alternative matrices to Sepharose can also be used. For example, IgG-coated paramagnetic beads (Dynal AS, Oslo, Norway) could be suitable for certain applications.

Typical protocol for the affinity purification of SPA fusion proteins using IgG-Sepharose is as follows:

1a. If a secretion strategy is employed, samples containing the SPA fusion products are collected from the periplasm after an osmotic shock procedure (Nossal and Heppel 1966), or directly from the culture medium after sedimentation of the cells by centrifugation (10,000*g*).

1b. For intracellularly produced proteins, the samples are collected from the supernatant of centrifuged (20,000*g*) cell disintegrates. If the product is precipitated in inclusion bodies, procedures to solubilize the precipitated proteins would be necessary (see below).

2. The pH-adjusted (pH 7) samples are loaded directly onto IgG-Sepharose (Amersham Pharmacia Biotech, Uppsala, Sweden) columns, previously equilibrated with a washing buffer (50 mM Tirs-HCl, pH 7.4, 0.15 M NaCl, 0.05% Tween-20). To avoid clogging of the columns, samples should preferably be filtered (0.45 μm) prior to loading.
3. The columns are subsequently washed with five column volumes of 5 mM NH_4Ac (pH 6.0) to lower the buffer capacity and remove salt before elution with 0.5 M HAc, pH 3.3.
4. Fractions are collected, and the protein content is suitably estimated by $A_{280\ \mathrm{nm}}$ measurements.

Proteins of low solubility and proteins containing hydrophobic transmembrane regions are extremely difficult to secrete and often precipitate intracellularly in *E. coli* into so-called inclusion or refractile bodies. For such proteins, an intracellular production strategy has to be used. Intracellular expression has become an increasingly attractive alternative due to recent advances for in vitro renaturation of recombinant proteins from intracellular precipitates (Rudolph 1995). Production by the inclusion body strategy has the main advantages that the recombinant product normally is protected from proteolysis and that it can be produced in large quantities. Recently, it was demonstrated that affinity purification indeed can be useful also for the recovery of proteins with a strong tendency to precipitate during renaturation from inclusion bodies following standard protocols. Murby and coworkers (1994) developed an alternative recovery scheme for the production of ZZ fusions to various fragments of the fusion glycoprotein (F) from the human RSV. Earlier attempts to produce these labile and precipitation-prone polypeptides in *E. coli* had failed, but several different ZZ-F fusions could by this novel strategy be produced and recovered as full-length products with substantial yields (20–50 mg/L). Since it was demonstrated that the IgG-Sepharose was resistant to 0.5 M guanidine hydrochloride, efficient recovery from inclusion bodies of the ZZ-F fusions could be achieved by affinity chromatography on IgG-Sepharose in the presence of the chaotropic agent throughout the purification process (Murby et al. 1994). In contrast, exclusion of guanidine hydrochloride during the procedure resulted in precipitation of the fusion proteins on the affinity column. The described strategy has so far been successfully evaluated for a number of proteins of low solubility and should be of interest for efficient recovery of other heterologous proteins that form inclusion bodies when expressed in a bacterial host.

Renaturation and IgG affinity recovery of protein A fusion proteins from inclusion bodies takes place as follows:

1. Insoluble material after sonication is pelleted by centrifugation and precipitated intracellular proteins are recovered by an initial solubilization in 7 M guanidine hydrochloride (GuaHCl, Sigma) and 25 mM Tris-HCl, pH 8.0 followed by incubation at 37°C for 2 h. When cysteines are present in the fusion proteins, 10 mM dithiothreitol (DTT) or 100 mM β-mercaptoethanol is included.
2. The solubilization mixture is centrifuged and the supernatant is slowly pipetted into 100 ml of renaturation buffer containing 1 M GuaHCl, 25 mM Tris-HCl, pH 8.0, 150 mM NaCl, and 0.05% Triton X-100.
3. The mixture is incubated at 4°C under slow stirring for 15 h.
4. The renaturation mixture is centrifuged, filtered (0.45 μm), and applied to an affinity column containing 5 ml IgG-Sepharose (Amersham Pharmacia Biotech, Uppsala, Sweden) at 4°C at a low flow rate (0.5 ml/min). The IgG-Sepharose had previously been treated with alternating pulses (five column volumes) of TSTG buffer (50 mM Tris-HCl, pH 8, 200 mM NaCl, 0.05% Tween-20, 0.5 mM EDTA, and 0.5 M GuaHCl) and 0.3 M HAc, pH 3.3 containing 0.5 M GuaHCl.
5. After sample loading, the column is washed with 100 mL TSTG and 25 ml 5 mM NH_4Ac, pH 5.5 containing 0.5 M GuaHCl.
6. Bound proteins are eluted by 20 ml 0.3 M HAc, pH 3.3 with 0.5 M GuaHCl.
7. The chaotropic agent is removed by dialysis against 2 L of 0.3 M HAc, pH 3.3, twice.

Mild Elution of Protein A and Protein G Fusions

Protein A (Z) and ABP fusion proteins are most conveniently eluted from affinity columns using low pH (e.g., HAc of pH $\approx$ 3). The eluant is usually removed either by vaporization or by the use of a desalting column. However, a low pH can, in some instances, be destructive for target protein function and must in such cases be avoided. In addition, the use of low-pH eluants following an initial capture of a fusion protein at physiological pH could result in passage of the isolectric point (pI) of the immobilized protein, which in some cases might lead to aggregation. Alternative elution conditions reported for ABP fusion proteins include high pH (pH 12; Makrides et al. 1996), lithium *diiodosalicate* (0.25 M; Lorca et al. 1992), and heat (>70°C; Nilsson et al. 1997a).

A general concept based on competition for mild elution of Z fusion proteins, which also should be applicable to ABD fusions, was developed recently (Nilsson et al. 1994). To obtain an efficient elution, a target protein fused to a single Z domain is competitively eluted using a divalent ZZ competitor protein having a more than 10-fold higher apparent affinity for hIgG. In addition, to be able to specifically remove the competitor after elution the ZZ domains were fused to a second affinity tag using the divalent serum albumin binding regions BB. Thus, the competitor present in the eluate can be captured through a simple HSA affinity chromatography step. The strategy thus allows also for the potential reuse of the ZZ-BB competitor after elution of the fusion protein from the column. Furthermore, the ZZ-BB competitor should also be useful for effective competitive elution of monovalent albumin binding domain fusion proteins.

A conceptually different approach to allow milder elution of SPA fusion proteins was recently evaluated by Gülich and coworkers (1999). In their approach, protein engineering was employed to generate two novel SPA domain analogues. The second loop in the three-helix bundle SPA domain was subjected to site-directed changes by insertion or substitution mutagenesis. Both resulting SPA domain variants were shown to be significantly destabilized, and efficient elution from an hIgG matrix could be obtained at as high a pH as 4.5. These new SPA domains would thus be suitable as fusion partners for pH-sensitive target proteins, or alternatively they could be used to obtain a "protein A"-Sepharose that would allow recovery of IgG antibodies under milder conditions.

2.2 Polyhistidine Tags

A different concept for affinity purification of proteins was presented by Porath and coworkers (1975). The method is based on the interaction between the side chains of certain amino acids, particularly histidines, on a protein surface and immobilized transition metal ions and is today known as immobilized metal ion affinity chromatography (IMAC). The metal ions are immobilized by the use of a chelating agent, capable of presenting the metals for binding to the protein. Numerous gene fusion systems employing histidine-rich tags for purification of recombinant proteins on immobilized metal ions have since then been described (Table 1) (Flaschel and Friehs 1993). Tags, either N- or C-terminal, consisting of consecutive histidine residues binding selectively to immobilized Ni^{2+} ions were described by

Hochuli and coworkers (1988). Adsorption of the poly-His-tagged proteins to a metal-chelate adsorbent was performed at neutral or slightly alkaline pH at which the imidazole residues of the histidines are not protonated (Hochuli, 1990). The expressed fusion proteins were recovered with a purity of more than 90% in a single step using an Ni^{2+}-nitrolotriacetic acid (NTA) adsorbent and elution with low pH or by competition using imidazole (Hochuli et al. 1988; Hochuli 1990).

Other research groups have described the use of alternative chelators such as iminodiacetic acid (IDA) (Smith et al. 1988; Ljungquist et al. 1989b; Van Dyke et al. 1992; Witzgall et al. 1994; Nilsson et al. 1996b) or TALON [Clontech], as well as other His-containing tag sequences such as variants containing Trp (Smith et al. 1988; Kasher et al. 1993) or multiple copies of the peptide Ala-His-Gly-His-Arg-Pro (Ljungquist et al. 1989b). For IMAC purification of various His-Trp containing fusion proteins the use of Co^{2+} and Zn^{2+} rather than Ni^{2+} as immobilized metal ions has been reported to lead to less contamination of *E. coli* proteins (Kasher et al. 1993). General expression vectors for both intracellular (Hoffmann and Roeder 1991; Van Dyke et al. 1992; Kroll et al. 1993; Chen and Hai 1994; Nilsson et al. 1996b; Braciak et al. 1996) and secreted production [Skerra 1994; Hayashai et al. 1995) of His_6 fusion proteins in *E. coli* have been developed. Furthermore, His_6 sequences have also been employed in the production and purification of dual-affinity fusions in *E. coli* in combination with a modified S-peptide of ribonuclease A (Kim and Raines 1993, 1994), the albumin-binding protein ABP (Gräslund et al., 1997), or the GST fusion partner (Panagiotidis and Silverstein 1995). The use of poly-His peptides as fusion partners for purification has also been demonstrated for recombinant fusion proteins produced in *Saccharomyces cerevisiae* (Bush et al. 1991; Kaslow and Shiloach 1994), mammalian cells (Janknecht et al. 1991; Witzgall et al. 1994), or baculovirus-infected *Spodoptera frugiperda* (Kuusinen et al. 1995).

An important advantage with the His_6 affinity tag is the possibility of purifying proteins under denaturing conditions. Thus, proteins that have aggregated into inclusion bodies can be dissolved in a suitable agent, such as urea or GuaHCl, and then be purified by IMAC. Refolding of the target protein can then be performed without interference from other proteins (Hochuli 1990; Sinha et al. 1994). Furthermore, small affinity tags such as the His_6 peptide can easily be genetically fused to a target gene by polymerase chain reaction (PCR) techniques (Nygren et al. 1994; Hindges and Hübscher 1995). Furthermore, the strong interaction between poly-His tags and immobilized Ni^{2+} ions has been demonstrated to result in a directed immobilization of the fused target protein, allowing protein-protein or protein-DNA interaction studies (Hoffman and Roeder 1991; Kashlev et al. 1993; Lu et al. 1993).

2.3 In vivo Biotinylated Affinity Tags

The strong binding ($K_d \sim 10^{-15}$ M) between biotin and avidin or streptavidin (Green 1975; Wilchek and Bayer 1990) is frequently used in biochemistry and molecular biology for immobilization and detection purposes. For use in such applications biotinylation of proteins is usually achieved through covalent coupling of biotin to the protein by the use of biotin-ester reagents that preferentially modify lysine residues (Savage et al. 1992). However, such coupling is difficult to direct to a specific position and can occur at several residues, some of which might be important for the protein structure and/or activity. An elegant strategy for site-specific biotinylation of recombinant proteins during their production was demonstrated by Cronan in 1990, utilizing the in vivo biotinylation machinery of *E. coli*. Analysis of various extensions of the naturally biotinylated 1.3S subunit of *Propionibacterium shermanii* transcarboxylase (Samols et al. 1988) showed that a 75-residue fragment was sufficient to serve as substrate for the biotinylation. Thus, fusion proteins containing this domain could be purified by affinity chromatography employing immobilized monovalent avidin (Kohanski and Lane 1990) and free biotin for mild elution (Cronan 1990). The stability of the avidin ligand allows the capture of proteins also under denaturing conditions, such as in the presence of 1% SDS or 8 M urea (Cronan 1990).

Few proteins are naturally modified by biotin incorporation. In *E. coli* only one such protein, the biotin carboxyl carrier protein (BCCP) of acetyl-CoA carboxylase, has been found (Fall 1979). The biotin is covalently attached to a specific lysine residue (Samols et al. 1988), a reaction catalyzed by biotin ligase (BirA) (Barker and Campbell 1981; Cronan 1990). Recently, the C-terminal 101 residues of the BCCP protein was utilized as an affinity handle for production and purification of a Fab antibody fragment in *E. coli* (Weiss et al. 1994). The fusion protein was secreted into the culture medium and purified on streptavidin-agarose beads, resulting in a simultaneous directed immobilization of the antibody fragment allowing one-step immunoaffinity purification of a recombinant human tumor necrosis factor-α (TNF-α) fusion protein. If desired, biotinylated proteins bound to streptavidin can be eluted by boiling in an SDS-urea solution (Swack et al. 1978). In addition to the BCCP and the *P. shermanii* protein, similar proteins from *Klebsiella pneumoniae* and *Arabidopsis thaliana* have been used to produce biotin-modified fusion proteins in *E. coli* (Cronan 1990; Choi et al. 1995). The 1.3S subunit of *P. shermanii* transcarboxylase has also been described to function as a biotinylated fusion partner in recombinant Semliki Forest virus–infected BHK and CHO cells (Lundstrom et al. 1995).

Recently, from a combinatorial library of short peptides, sequences capable of mimicking the normal BirA substrate were found using an elegant selection strategy based on fusion of the peptides to a plasmid binding lac repressor protein (Schatz 1993). From the series of sequences found to be in vivo biotinylated at a central invariant lysine residue, a consensus 13-amino-acid stretch was defined. Thus, this sequence could replace the larger in vivo biotinylated fusion partners used earlier by other groups. Recently, this tag was used in combination with a His_6 peptide and ABP into a novel affinity fusion partner (See below; Nilsson et al. 1996b). An alternative small affinity tag that binds to streptavidin was recently found in a peptide library screened for peptides that mimic biotin (Schmidt and Skerra 1993). A 9-amino-acid streptavidin-binding peptide termed strep-tag was selected from the library and used as an affinity fusion partner for purification of an *E. coli*-expressed antibody Fv fragment by streptavidin affinity chromatography. The fusion protein was competitively eluted with the biotin analogue iminobiotin, which allowed for column regeneration by washing with the equilibration buffer. More recently, such mild elution conditions (diaminobiotin) were used in the streptavidin affinity chromatography purification of various membrane proteins captured via strep-tag-fused Fv fragments (Kleymann et al. 1995).

A common complication associated with avidin/streptavidin affinity chromatography of in vivo biotinylated proteins is that any free biotin should preferentially be removed before loading the cell lysate onto the column because otherwise irreversible blocking of the ligand can be expected. In addition, if not removed in a first purification step, biotinylated host proteins may be copurified with the biotinylated target protein. Recently, this problem was circumvented using a combined affinity tag that enabled removal of biotin and BCCP by employing an HSA affinity chromatography step prior to immobilization of the affinity fusion protein on a streptavidin surface (Nilsson *et al.* 1996b).

2.4 Glutathione *S*-transferase

The glutathione *S*-transferases (GST) are a family of enzymes that modify toxic substances, such as nitro and halogenated compounds, by adding sulfur from glutathione, leading to their excretion in the urine (Mehler 1993). Many mammalian GSTs can be purified by affinity chromatography using the immobilized cofactor glutathione, followed by competitive elution with an excess of reduced glutathione (Simons and Vander Jagt 1981). Based on this specific interaction, a gene fusion system for *E. coli* expression was developed by Smith and Johnson (1988), using GST from the parasitic helminth *Schistosoma japonicum* (Table 1). They demonstrated that after fusion to the C terminus of GST, a number of eukaryotic proteins could efficiently be purified from a crude *E. coli* lysate by glutathione affinity chromatography. In addition, cleavage sites for the proteases thrombin and blood coagulation factor Xa had been introduced to allow removal of the GST fusion partner (Smith and Johnson 1988). It has been demonstrated that GST fusion proteins can be renatured from inclusion bodies after solubilization using 6 M Gua-HCl and subsequently purified by glutathione affinity chromatography (Zhao and Siu 1995). Recently, a series of vectors was described allowing either N- or C-terminal fusion of a target protein to GST for production in *E. coli* (Sharrocks 1994). Furthermore, two vectors have been developed that allow, in addition to N- or C-terminal fusion to GST, site-specific in vitro ^{32}P labeling of the fusion protein after glutathione affinity purification, employing a protein kinase recognizing an introduced substrate sequence (Jensen et al. 1995). The GST fusion partner has also been used as an N-terminal constituent in a dual-affinity fusion approach, in combination with a His_6 tag at the C terminus of a tripartite fusion protein (Panagiotidis and Silverstein 1995). Furthermore, expression vectors have been constructed for production of GST fusion proteins in insect cells (Davies et al. 1993; Peng et al. 1993), yeast (Ward et al 1994), and COS-7 cells (Chatton et al. 1995). A possible complication associated with the GST fusion system is the use of reduced glutathione (a reducing agent) for elution, which can affect target proteins containing disulfides (Sassenfeld, 1990).

2.5 FLAG Peptide

An affinity gene fusion system that has become popular in recent years is the so-called FLAG system, based on the fusion of an 8-amino-acid peptide to a target protein for immunoaffinity chromatography using immobilized monoclonal antibodies. The FLAG peptide sequence (DYKDDDDK) is hydrophilic and contains an internal enterokinase cleavage recognition sequence (DDDDK) (Hopp et al. 1988). Monoclonal antibodies to the FLAG sequence were obtained after immunization of mice with a FLAG fusion protein. One monoclonal antibody (M1) was found to bind to FLAG fusion proteins in a calcium-dependent manner, which allowed gentle elution by the addition of a chelating agent (Hopp et al. 1988; Prickett et al. 1989). A drawback with the M1 antibody is that it can only bind the FLAG peptide when it is located at the extreme N terminus of the fusion protein. This limits the use of the M1 antibody to the purification of fusions exposing an N-terminal FLAG peptide, such as after removal of a signal peptide from a secreted FLAG fusion. Recently, it was found that the M1 antibody binds with almost the same affinity to a shortened version of the FLAG peptide, consisting only of the first four amino acids (DYKD) (Knappik and Plückthun 1994). Alternatively, a different monoclonal antibody (M2) is available, capable of binding to the FLAG peptide also if preceded by a methionine at the N terminus or fused to the C terminus of the target protein (Brizzard et al. 1994). However, the M2 antibody interacts with the FLAG peptide in non-calcium-dependent manner, and a fusion protein can therefore not be eluted with a chelating agent. Instead low pH or competition with an excess of synthetically produced FLAG peptide is used for elution of FLAG fusion proteins bound to an M2 affinity matrix (Table 1) (Brizzard et al. 1994).

Different hosts have been used for production of FLAG fusion proteins such as *E. coli*, *S. cerevisiase* (Hopp et al. 1988), recombinant baculovirus-infected *Spodoptera frugiperda* (Kuusinen et al. 1995), and COS-7 cells (Kunz et al., 1992). A major drawback with the FLAG system is that the M1 and M2 affinity matrices are expensive.

2.6 Maltose-Binding Protein

E. coli maltose-binding protein (MBP), the product of the malE gene, is exported to the periplasmic space where it binds specifically with high affinity to maltose or maltodextrins for subsequent transport of these sugars across the cytoplasmic membrane (Duplay et al. 1984). MBP can be purified in a single step by affinity chromatography on resins containing cross-linked amylose followed by competitive elution with maltose (Kellerman and Ferenci 1982). The low costs for amylose resins together with the fact that MBP does not contain any cysteine residues that can interfere with disulfide bond formation in the target protein led to the development of an expression system for MBP fusion proteins in *E. coli* allowing affinity purification under mild conditions (Table 1) (Guan et al. 1988). Vectors for both intracellular (Guan et al. 1988) and secreted production (Guan et al. 1988; Bedouelle and Duplay 1988) have been developed. The vectors designed by Guan and coworkers were further developed by the introduction of a cleavage recognition sequence for the factor Xa protease at the C terminus of MBP to allow release of a target protein fused to MBP (Maina et al. 1988). Later, expression vectors were constructed allowing translational fusions to MBP in all reading frames (Aitken et al. 1994). The possibility of renaturing an MBP fusion protein from inclusion bodies followed by amylose affinity purification has also been demonstrated (Chen and Gouax 1996).

3. INTRACELLULAR PRODUCTION VIA INCLUSION BODIES

As discussed earlier, production by the inclusion body strategy has the main advantages that the recombinant product normally is protected from proteolysis and that it can be produced in large quantities. Levels of up to 50% of total cell protein content have been reported (Rudolph 1995). The mechanism for inclusion body formation is not fully elucidated but fusions to certain proteins, such as TrpE and, to a lesser extent, β-galactosidase, may result in inclusion body formation (Ståhl et al. 1997a). In addition, expression from strong promoters such as the T7 promoter, leading to very high expression levels, seems to increase the tendency for intracellular aggregation of the produced recombinant protein (Ståhl et al. 1997a). Gene fusion technology has been demonstrated to greatly facilitate the recovery also of proteins with low solubility (Murby et al. 1994).

The recovery of biologically active or native proteins by in vitro refolding from insoluble inclusion bodies is often hampered by the aggregation of the product during the procedure, leading to low overall yields. To make the procedure more efficient, several improved protocols have been developed, including the addition of different "folding enhancers" (Nygren et al. 1994). Alternatively, the protein itself can be engineered to facilitate refolding. The presence of hydrophilic peptide extensions during the refolding can dramatically improve the folding yield, probably by conferring a higher overall solubility of the protein (Rudolph 1995). Samuelsson and coworkers (1991) showed that the reshuffling of misfolded disulfides in recombinant insulin-like growth factor-I (IGF-I) was greatly facilitated by fusion to the highly soluble ZZ fusion partner, derived from staphylococcal protein A. Compared with unfused IGF-I, the fusion ZZ-IGF-I could be successfully refolded at a 100-fold higher concentration (1–2 mg/ml), without the formation of precipitates (Samuelsson et al. 1994).

4. COMBINED AFFINITY TAGS

A common problem in heterologous gene expression is proteolytic degradation. Using an affinity fusion strategy, eventual degradation products of the target protein are purified together with the full-length fusion protein. Gene fusions can in some cases be used to stabilize labile proteins, but single fusions have shown limited stabilizing effects (Murby et al. 1991). However, employing a dual-affinity approach, as first described for the production of the peptide hormone IGF-II (Hammarberg et al. 1989), two different affinity fusion partners (ZZ and BB) were fused at each end of the target protein. Thus, after two successive affinity purification steps only proteins containing both tags, and therefore also the central target protein, will by definition be obtained (Hammarberg et al. 1989). Several other combinations of affinity tags have thereafter been used in various dual-affinity fusion concepts (Jansson et al. 1990; Kim and Raines 1994; Panagiotidis and Silverstein 1995; Gräslund et al. 1997). The dual-affinity concept demonstrates the advantage for certain applications to use a combination of different affinity tags. In addition, it has been observed that a dual-affinity approach can have a stabilizing effect on several proteolytically sensitive proteins (Murby et al. 1991).

Although a multitude of systems have been described, no single-affinity fusion strategy is ideal for all expression or purification situations. For example, if secretion of the product into the periplasm or culture medium is desired, a fusion to affinity tails derived from normally intracellular proteins (e.g., the β-galactosidase system) is not applicable, and if the purification has to be performed under denaturating conditions, protein ligands like monoclonal antibodies (mAbs), e.g., anti-FLAG M1 and M2 mAbs, are unsuitable. Therefore, to obtain general expression vectors for affinity gene fusion strategies, applicable in several situations, a combination of affinity fusion domains could be introduced into a single fusion partner. Such composite fusion partners consisting of several independent affinity domains could potentially be used in different detection, purification, and immobilization situations, employing the affinity function that is most suitable for the situation. However, the included affinity domains should be carefully chosen not to functionally interfere with each other, and each moiety should be able to withstand the purification conditions dictated by the affinity domain requiring the most harsh affinity chromatography conditions. For instance, if denaturing conditions are used, as in IMAC purification of proteins containing a hexahistidyl sequence from inclusion bodies solubilized by urea or GuaHCl, the other included affinity domains should not be irreversibly denatured during this step. This could otherwise cause aggregation of the eluted proteins. Another advantage of using combined affinity tags is that different methods for the detection and immobilization of the fusion protein can also be envisioned.

This strategy was utilized in designing an affinity tag combination consisting of an in vivo biotinylated peptide (Bio), a hexahistidyl peptide (His_6), and the ABP that could allow for flexible use in different detection, purification, and immobilization situations (Nilsson et al. 1996b). Fusions to the Bio-His_6-ABP affinity tag could be used for easy detection using commercial streptavidin conjugates, for purification under both native and denaturing conditions, and for immobilization purposes with an option to choose between three strong affinity interactions. In addition, the ABP part of the affinity tag is soluble and effectively refolded after denaturation, which can be an advantage during refolding of the target protein fused to the Bio-His_6-ABP affinity tag.

5. AFFINITY-TAGGED PROTEASES

For general use in site-specific cleavage of fusion proteins comprising relatively large target proteins, proteases such as trypsin recognizing basic arginine and lysine residues are not ideal, since the occurrence of these amino acids generally is proportional to protein size. Alternative enzymes for broader use in biotechnological applications should preferably be highly specific proteases, easy to produce by recombinant means in large scale. Interesting candidate enzymes can be found in human picoRNA viruses, whose maturation relies on the site-specific cleavage of a large polyprotein precursor to yield the viral components. Some of these proteases have been described to be functionally produced at high levels in bacterial expression systems (Miyashita et al. 1992; Walker et al. 1994), which also would facilitate the production of variants constructed by protein engineering. Recently, a new general strategy was described where the 3C protease of rhinovirus was fused to an affinity tag and subsequently used for specific cleavage of a target protein (Walker et al. 1994). This facilitated removal of the affinity-tagged protease and the cleaved affinity fusion partner by simple affinity chromatography steps. In addition, it was possible to add any desired amount of protease to ensure efficient cleavage, since the protease could be easily removed after cleavage (Walker et al. 1994).

Another affinity-tagged protease, consisting of coxsackievirus 3C ($3C^{pro}$) fused to the serum albumin binding ABP at the N terminus and His_6 at the C terminus, was produced in *E. coli* and used for the production of a truncated *Taq* DNA polymerase (ΔTaq) (Gräslund et al. 1997). The heat-stable polymerase was produced as an ABP–ΔTaq fusion protein having a $3C^{pro}$ cleavage site introduced between the two protein moieties. After affinity purification of ABP–ΔTaq the fusion was efficiently cleaved using the affinity-tagged protease ABP-$3C^{pro}$, which allowed for the recovery of nonfused, fully active ΔTaq after passage of the cleavage mixture over an HSA column (Gräslund et al. 1997).

Furthermore, an affinity fusion protease can also facilitate the on-column cleavage of a fusion protein immobilized onto an affinity column in cases where the affinity ligand is insensitive to the protease (Walker et al. 1994; Gräslund et al. 1997). Recently, a novel system for simultaneous affinity purification and on-column cleavage of affinity fusion proteins was described (Chong et al. 1996; Impact I System, New England Biolabs]. In this system the target protein is fused to a fusion partner consisting of an intein domain from *S. cerevisiae* and a chitin binding domain. After affinity immobilization on chitin, the target protein is released through the proteolytic activity of the intein domain.

6. IN VIVO ASSOCIATION TO ENABLE AFFINITY CAPTURE

A completely new concept to improve the fraction of correctly folded recombinant IGF-I was recently presented (Samuelsson et al. 1996). It was demonstrated that coexpression of a specific binding protein, IGF-binding protein-1 (IGFBP-1), significantly increased the relative yield of IGF-I having native disulfide bridges when expressed in a secreted form in *E. coli.* A glutathione redox buffer was added to the growth medium to enhance formation and breakage of disulfide bonds in the periplasm of the bacteria. In the presented example, both IGF-I and IGFBP-1 were produced as affinity fusions, which facilitated the purification of in vivo assembled heterodimers by alternative purification methods (Samuelsson et al. 1996).

For general use of the concept, an attractive strategy would be to express the target protein as a nonfused gene product and the specific binding protein in an affinity-tagged configuration. Correctly folded target protein would thus be affinity captured as a heterodimer via the tagged binding protein. This would employ the benefits of high specificity for affinity chromatography without introducing requirements of proteolytic processing to recover the native target protein. The binding proteins to be used in these kinds of bioprocesses could be naturally existing ligands, as in the described example (Samuelsson et al. 1996), or novel proteins specific for correctly folded molecules optionally selected from protein libraries produced employing combinatorial protein chemistry.

A somewhat similar strategy was employed by Hentz and Daunert (1996) in their study of the interaction between SPA and DnaK. A calmodulin-SPA fusion protein was expressed in *E. coli* and allowed to interact in vivo with DnaK. Subsequently, the complex of the fusion protein and DnaK was captured on a phenothiazine column, which specifically bound the calmodulin moiety. An interesting aspect of this approach is that the column could be efficiently regenerated before reuse (Hentz and Daunert 1996).

7. IN VIVO SELECTED PROTEIN LIGANDS FOR SPECIFIC CAPTURE

Affinity capture is an attractive procedure for isolation of biomolecules from complex mixtures because it can offer a single-step increase in purity and product concentration (Jones et al. 1995). However, for affinity chromatography to be generally available for bioprocess design, methodology to obtain affinity ligands meeting industrial demands on stability and selectivity has to be improved. Ideally, an affinity ligand should bind its target reversibly with high enough affinity and selectivity, be easy to produce and immobilize onto a suitable chromatographic support, and be able to withstand repeated column sanitation protocols, preferably using alkaline solutions.

Hybridoma technology used for the generation of monoclonal antibodies (mAbs) has the potential to supply selective ligands to almost any given target (Jack 1994). However, subsequent production of mAb ligands in mammalian cells is often associated with a concern of viral contamination of the product, leading to extensive process validation procedures. In addition, mAbs are considered to be relatively fragile to the cleaning-in-place (CIP) conditions often required to achieve efficient column sanitation (Girot et al. 1990). Although extraordinary achievements have been reported concerning both the selection of novel antibody specificities using, say, phage display technology (Griffiths and Duncan 1998; Hoogenboom et al. 1998) and high-level production of antibody fragments in nonmammalian expression systems (Verma et al. 1998), methodology for generating robust and efficiently produced ligands of more simple overall structures for use in bioseparation applications could prove to be advantageous.

An example of a protein fulfilling the criteria of a suitable ligand is SPA, widely used for affinity recovery of antibodies, e.g., from cell cultures (Bottomley et al. 1995). SPA is remarkably stable to alkaline solutions, allowing repeated washings with 0.5 M sodium hydroxide without significant loss of binding capacity (Girot et al. 1990). The Z domain from SPA thus has several features, including proteolytic stability, high solubility, small size, compact and robust structure, which make it interesting as a scaffold for protein library constructions. The residues responsible for the IgG Fc binding activity have been identified from the crystal structure of the complex between its parent B domain and human IgG Fc, showing that these residues are situated on the outer surfaces of two of the helices making up the domain structure and do not contribute to the packing of the core (Deisenhofer 1981). The Fc binding surface covers an area of approximately 600 $Å^2$, similar in size to the surfaces involved in many antigen-antibody interactions. Taken together, these data suggest that if random mutagenesis of this binding surface was performed, novel domains could be obtained with the possibility of finding variants capable of binding molecules other than Fc of IgG. In addition, these domains would theoretically share the overall stability and α-helical structure of the wild-type domain.

Recently, an attempt to obtain such ligands with completely new affinities was initiated, employing phage display technology (Nord et al. 1995). A genetic library was created in which 13 surface residues (involved in the IgG Fc binding) of the Z domain were randomly and simultaneously substituted with any of the 20 possible amino acids. This Z library has, after genetic fusion to the gene for coat protein 3 of filamentous phage M13, resulted in a phage library adapted for selection of novel specificities by biopanning (Nord et al. 1997). The library has been subjected to such affinity selection against different molecules for investigation as a source of novel binders. Novel Z variants, so-called affibodies, have successfully been selected to diverse targets, such as *Taq* DNA polymerase, human insulin, and a human apolipoprotein variant (Nord et al. 1997).

In a recent study, such affibodies were investigated as ligands (Fig. 3) in affinity chromatography purification of their target proteins, being bacterial *Taq* DNA polymerase and human apolipoprotein A-1, respectively, from *E. coli* cell lysates (Nord et al. 2000). Using affinity columns prepared by coupling of dimeric versions of the affibody ligands to chromatographic media, both target proteins were efficiently recovered from applied cell lysates with high selectivities, as analyzed by SDS-PAGE and N-terminal sequencing. No loss of column capacity or selectivity was observed for repeated cycles of sample loading, washing, and low pH elution. Interestingly, column sanitation protocols involving exposure of

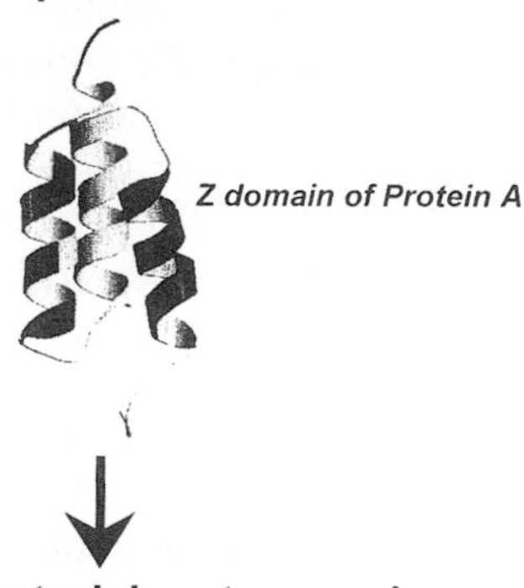

Figure 3 Outline of the rationale to obtain robust affinity ligands. From a library of proteins constructed by combinatorial mutagenesis of the surface of a robust protein domain used as scaffold, e.g., the Z domain of staphylococcal protein A, variants capable of binding to a desired target are selected. Such ligands could subsequently be used as ligands in affinity chromatography applicants.

the affibody ligands to 0.5 M sodium hydroxide could successfully be repeatedly used without observable loss of purification performance (Nord et al. 2000).

To make an affinity ligand suitable for affinity purification applications, the affinity for the target protein should be high enough to enable efficient capture from crude process samples. Although affibodies with affinities in the micromolar range ($K_D = 10^{-6}$) can be readily selected from medium-sized naive libraries of the Z domain (Nord et al. 1997; Hansson et al. 1999), in some applications ligands of higher affinities could prove important. Gunneriusson and coworkers (1999) recently demonstrated that affibodies with affinities in the nanomolar range can indeed be obtained by affinity maturation strategies (Fig. 4) in combination with affibody dimerization. These results suggest that combinatorial approaches (Nygren and Uhlén 1997) using robust protein domains can be a general tool in the process of designing purification strategies for biomolecules.

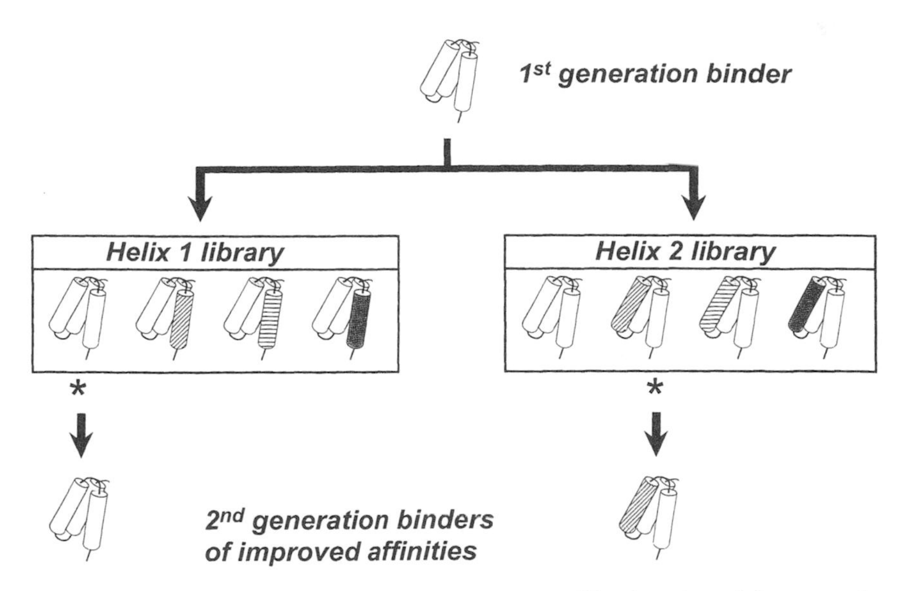

Figure 4 The helix shuffling strategy. A primary affibody, selected from a naive library of the three-helix bundle Z domain constructed through combinatorial mutagenesis of 13 surface-located positions in helices 1 and 2, is subjected to affinity maturation using a helix shuffling strategy. From a secondary library, constructed through selective rerandomization of one of the helices involved in the binding interaction, second-generation affibodies can be selected.

8. CONCLUDING REMARKS

Affinity fusion technology has during the past decade become increasingly important in many fields of research including biochemistry, applied microbiology, immunotechnology, vaccinology, and biotechnology. Affinity fusion tags have been used as research tools for detection, purification, and immobilization of recombinant proteins expressed in bacteria, yeast, insect cells, plants, and mammalian cells. However, on an industrial scale the use of affinity fusions to facilitate bioprocessing of recombinant proteins has not yet been widely used. The main reason for this is most likely the fact that there has been a demand to produce heterologous proteins in a native form in the "first generation" of recombinant products. Affinity fusion strategies thus introduce an additional problem into the downstream processing, since a site-specific cleavage is needed. This chapter points out new strategies based on affinity fusions, such as the use of affinity-tagged proteases and affinity-assisted in vivo folding, which could be of importance in circumventing such problems. The use of an affinity purification step as an early unit operation in a bioprocess is likely to decrease the overall number of unit operations and thus the overall cost. As pointed out by the affibody examples (Sec. 6), one alternative to the use of affinity fusions to the target protein would be to create specially designed affinity matrices for recovery of correctly folded forms of the target protein. Such strategies could potentially be implemented in biopharmaceutical industry in the near future. We believe that the applications described in this chapter make it evident that genetic approaches for the purpose of facilitating purification will find broad use in biotechnology.

REFERENCES

Abrahmsén L., Moks T., Nilsson B., Hellman U., and Uhlén M. (1985) Analysis of the signals for secretion in the staphylococcal protein A gene. EMBO J. 4, 3901–3906.

Abrahmsén L., Moks T., Nilsson B. and Uhlén M. (1986) Secretion of heterologous gene products to the culture medium of *Escherichia coli*. Nucl. Acids. Res. 14, 7487–7500.

Aitken R., Gilchrist J., and Sinclair M.C. (1994) Vectors to facilitate the creation of translational fusions to the maltose-binding protein of *Escherichia coli*. Gene 144, 69–73.

Barker D.F., and Campbell A.M. (1981) The *birA* gene of *Escherichia coli* encodes a biotin holoenzyme synthetase. J. Mol. Biol. 146, 451–467.

Bedouelle H., and Duplay P. (1988) Production in *Escherichia coli* and one-step purification of bifunctional hybrid proteins which bind maltose: Export of the Klenow polymerase into the periplasmic space. Eur. J. Biochem. 171, 541–549.

Björck L., and Kronvall G. (1984) Purification and some properties of streptococcal protein G, a novel IgG-binding reagent. J. Immunol. 133, 969–974.

Björck L., Kastern W., Lindahl G., and Widebäck K. (1987) Streptococcal protein G, expressed by streptococci or by *Escherichia coli*, has separate binding sites for human albumin and IgG. Mol. Immunol. 24, 1113–1122.

Bottomley, S.P., Sutton, B.J., and Gore, M.G. (1995) Elution of human IgG from affinity columns containing immobilised variants of protein A. J. Immunol. Meth. 182, 185–192.

Braciak T.A., Northemann W., Chong D.K., Schroeder J.-A., and Gauldie J. (1996) Vector-derived expression of recombinant rat interleukin-6. Prot. Expr. Purif. 7, 269–274.

Brizzard B.L., Chubet R.G., and Vizard D.L. (1994) Immunoaffinity purification of FLAG epitope-tagged bacterial alkaline phosphatase using a novel monoclonal antibody and peptide elution. BioTechniques 16, 730–734.

Bush G.L., Tassin A.-M., Fridén H., and Meyer D.I. (1991) Secretion in yeast: Purification and in vitro translocation of chemical amounts of prepro-α-factor. J. Biol. Chem. 21, 13811–13814.

Carter P. (1990) Site-specific proteolysis of fusion proteins. In: Protein Purification: From Molecular Mechanism to Large-Scale Processes (M.R. Ladish, R.C. Willson, C.-D.C. Painton, and S.E. Builder, eds.), Vol. 427, American Chemical Society Symposium Series. pp. 181–193.

Carter D.C., and Ho J.X. (1994) Structure of serum albumin. Adv. Prot. Chem. 45, 153–202.

Caskey C.T., and Rossiter B.J. (1992) The human genome project. Purpose and potential. J. Pharm. Pharmacol. 44, 198–204.

Chatton B., Bahr A., Acker J., and Kedinger C. (1995) Eukaryotic GST fusion vector for the study of protein-protein associations in vivo: application to interaction of ATFa with Jun and Fos. BioTechniques 18, 142–145.

Chen B.P.C., and Hai T. (1994) Expression vectors for affinity purification and radiolabelling of proteins using *Escherichia coli* as host. Gene 139, 73–75.

Chen, G.-Q., and Gouaux, J.E. (1996) Overexpression of bacterio-opsin in *Escherichia coli* as a water-soluble fusion to maltose binding protein: Efficient regeneration of the fusion protein and selective cleavage with trypsin. Prot. Sci. 5, 456–467.

Choi J.K., Yu F., Wurtele E.S., and Nikolau B.J. (1995) Molecular cloning and characterization of the cDNA coding for the biotin-containing subunit of the chloroplastic acetyl-coenzyme A carboxylase. Plant Physiol. 109, 619–625.

Chong S., Shao Y., Paulus H., Benner J., Perler F.B., and Xu M.-Q. (1996) Protein splicing involving the *Saccharomyces cerevisae* VMA intein: The steps in the splicing pathway, side reactions leading to protein cleavage, and establishment of an in vitro splicing system. J. Biol. Chem. 271, 22159–22168.

Cronan J.E. (1990) Biotinylation of proteins in vivo: A post-translational modification to label, purify, and study proteins. J. Biol. Chem. 265, 10327–10333.

Davies A.H., Jowett J.B., and Jones I.M. (1993) Recombinant baculovirus vectors expressing glutathione-S-transferase fusion proteins. Biotechnology (N Y) 11, 933–936.

Deisenhofer J. (1981) Crystallographic refinement and atomic models of a human Fc fragment and its complex with fragment B of protein A from *Staphylococcus aureus* at 2.9- and 2.8-Å resolution. Biochemistry 20, 2361–2370.

Duplay P., Bedouelle H., Fowler A., Zabin I., Saurin W., and Hofnung M. (1984) Sequences of the malE gene and of its product, the maltose-binding protein of *Escherichia coli* K12. J. Biol. Chem. 259, 10606–10613.

Eliasson M., Olsson A., Palmcrantz E., Wiberg K., Inganäs M., Guss B., Lindberg M., and Uhlén M. (1988) Chimeric IgG-binding receptors engineered from staphylococcal protein A and streptococcal protein G. J. Biol. Chem. 9, 4323–4327.

Fahnestock S.R., Alexander P., Nagle J., and Filpula D. (1986) Gene for an immunoglobulin-binding protein from a group G *Streptococcus.* J. Bacteriol. 167, 870–880.

Falkenberg C., Björck L., and Åkerström B. (1992) Localization of the binding site for streptococcal protein G on human serum albumin. Identification of a 5.5-kilodalton protein G binding albumin fragment. Biochemistry 31, 1451–1457.

Fall R.R. (1979) Analysis of microbial biotin proteins. Meth. Enzymol. 62, 390–398.

Flaschel E., and Friehs K. (1993) Improvement of downstream processing of recombinant proteins by means of genetic engineering methods. Biotech. Adv. 11, 31–78.

Ford C.F., Suominen I., and Glatz C.E. (1991) Fusion tails for the recovery and purification of recombinant proteins. Prot. Expr. Purif. 2, 95–107.

Forsberg G., Baastrup B., Rondahl H., Holmgren E., Pohl G., Hartmanis M., and Lake M. (1992) An evaluation of different enzymatic cleavage methods for recombinant fusion proteins, applied on Des(1-3)insulin-like growth factor I. J. Prot. Chem. 11, 201–211.

Gandecha A.R., Owen M.R.L., Cockburn B., and Whitelam G.C. (1992) Production and secretion of a bifunctional staphylococcal protein A antiphytochrome single-chain Fv fusion protein in *Escherichia coli.* Gene 122, 361–365.

Girot, P., Moroux, Y., Duteil, X.P., Nguyen, C., and Boschetti, E. (1990) Composite affinity adsorbents and their cleaning in place. J. Chromatogr. 510, 213–223.

Gouda H., Torigoe H., Saito A., Sato M., Arata Y., and Shimada I. (1992) Three-dimensional solution structure of the B domain of staphylococcal protein A: Comparison of the solution and crystal structure. Biochemistry 31, 9665–9672.

Gräslund T., Nilsson, J., Lindberg M., Uhlén M., and Nygren P.-Å. (1997) Production of a thermostable DNA polymerase by site specific cleavage of a heat-eluted affinity fusion protein. Prot. Expr. Purif. 9, 125–132.

Green N.M. (1975) Avidin. Adv. Protein Chem. 29, 85–133.

Griffiths, A.D., and Duncan, A.R. (1998) Strategies for selection of antibodies by phage display. Curr. Opin. Biotechnol. 9, 102–108.

Guan C.-D., Li P., Riggs P.D., and Inouye H. (1988) Vectors that facilitate the expression and purification of foreign peptides in *Escherichia coli* by fusion to maltose-binding protein. Gene 67, 21–30.

Gülich, S., Uhlén, M., and Hober, S. (2000) Protein engineering of an IgG-binding domain allows milder elution conditions during affinity chromatography. J. Biotechnol. 76, 233–244.

Gunneriusson, E., Nord, K., Uhlén, M., and Nygren, P.-Å. (1999) Affinity maturation of a *Taq* DNA polymerase specific affibody by helix shuffling. Proten Eng. 12, 873–878.

Guss B., Eliasson M., Olsson A., Uhlén M., Frej A.-C., Jörnvall H., Flock J.-I., and Lindberg M. (1986) Structure of the IgG-binding region of streptococcal protein G. EMBO J. 5, 1567–1575.

Hammarberg B., Nygren P.-Å., Holmgren E., Elmblad A., Tally M., Hellman U., Moks T., and Uhlén M. (1989) Dual affinity fusion approach and its use to express recombinant human insulin-like growth factor II. Proc. Natl. Acad. Sci. USA 86, 4367–4371.

Hansson M., Ståhl S., Nguyen T.N., Bächi T., Robert A., Binz H., Sjölander A., and Uhlén M. (1992) Expression of recombinant proteins on the surface of the coagulase-negative bacterium *Staphylococcus xylosus*. J. Bacteriol. 174, 4239–4245.

Hansson M., Ståhl S., Hjorth R., Uhlén M., and Moks T. (1994) Single-step recovery of a secreted recombinant protein by expanded bed adsorption. Bio/Technology 12, 285–288.

Hansson, M., Ringdahl, J., Robert, A., Power, U., Goetsch, L., Nguyen, T. N., Uhlén, M., Ståhl, S. and Nygren, P.-Å. (1999) An in vitro selected binding protein (affibody) shows conformation-dependent recognition of the respiratory syncytial virus (RSV) G protein. Immunotechnology 4, 237–252.

Hayashi N., Kipriyanov S., Fuchs P., Welschof M., Dörsam H., and Little M. (1995) A single expression system for the display, purification and conjugation of single-chain antibodies. Gene 160, 129–130.

Hentz, N.G., and Daunert, S. (1996) Bifunctional fusion proteins of calmodulin and protein A as affinity ligands in protein purification and in the study of protein-protein interactions. Anal. Chem. 68, 3939–3944.

Hindges R., and Hübscher U. (1995) Production of active mouse DNA polymerase ∂ in bacteria. Gene 158, 241–246.

Hober S., Hansson A., Uhlén M., and Nilsson B. (1994) Folding of insulin-like growth factor I is thermodynamically controlled by insulin-like growth factor binding protein. Biochemistry 33, 6758–6761.

Hochuli E., Bannwarth W., Döbeli H., Gentz R., and Stüber D. (1988) Genetic approach to facilitate purification of recombinant proteins with a novel metal chelate adsorbent. Bio/Technology 6, 1321–1325.

Hochuli E. (1990) Purification of recombinant proteins with metal chelate adsorbent. Genet. Eng. 12, 87–98.

Hoffmann A., and Roeder R.G. (1991) Purification of his-tagged proteins in non-denaturating conditions suggests a convenient method for protein interaction studies. Nucl. Acids Res. 19, 6337–6338.

Hoogenboom, H.R., de Bruine, A.P., Hufton, S.E., Hoet, R.M., Arends, J.W., and Roovers. R.C. (1998) Antibody phage display technology and its applications. Immunotechnology 4, 1–20.

Hopp T.H., Prickett K.S., Price V.L., Libby R.T., March C.J., Ceretti D.P., Urdal D.L., and Conlon P.J. (1988) A short polypeptide marker sequence useful for recombinant protein identification and purification. Bio/Technology 6, 1204–1210.

Jack, G.W. (1994) Immunoaffinity chromatography. Mol. Biotechnol. 1, 59–86.

Janknecht R., Martynoff G.D., Lou J., Hipskind R.A., Nordheim A., and Stunnenberg H.G. (1991) Rapid and efficient purification of native histidine-tagged proteins expressed by recombinant vaccinia virus. Proc. Natl. Acad. Sci. USA 88, 8972–8976.

Jansson B., Palmcrantz C., Uhlén M., and Nilsson B. (1990) A dual affinity gene fusion system to express small recombinant proteins in a soluble form: expression and characterization of protein A deletion mutants. Prot. Eng. 2, 555–561.

Jansson, B., Uhlén, M., and Nygren, P.-Å. (1998) All individual domains of staphylococcal protein A show Fab binding. FEMS Immunol. Med. Microbiol. 20, 69–78.

Jendeberg L., Tashiro M., Tejero R., Lyons B.A., Uhlén M., Montelione G.T., and Nilsson B. (1996) The mechanism of binding staphylococcal protein A to immunoglobin G does not involve helix unwinding. Biochemistry 35, 22–31.

Jendeberg, L., Nilsson, P., Larsson, A., Denker, P., Uhlén, M., Nilsson, B., and Nygren, P.-Å. (1997) Engineering of Fc1 and Fc3 of human immunoglobulin G to analyse subclass specificity for staphylococcal protein A. J. Immunol. Meth. 201, 25–34.

Jensen T.H., Jensen A., and Kjems J. (1995) Tools for the production and purification of full-length, N- or C-terminal 32p-labeled protein, applied to HIV-1 gag and rev. Gene 162, 235–237.

Jonasson P., Nilsson J., Samuelsson E., Moks T., Ståhl S., and Uhlén M. (1996) Single-step trypsin cleavage of a fusion protein to obtain human insulin and its C-peptide. Eur. J. Biochem. 236, 656–661.

Jonasson, P., Nygren, P.-Å, Jörnvall, H., Johansson, B.-L., Wahren, J., Uhlén, M. and Ståhl, S. (2000) An integrated bioprocess for production of the human proinsulin C-peptide via heat release of an intracellular heptameric fusion protein. J. Biotechnol. 76, 215–226.

Jones, C., Patel, A., Griffin, S., Martin, J., Young, P., O'Donnell, K., Silverman, C., Porter, T., and Chaiken, I. 1995) Current trends in molecular recognition and bioseparation. J. Chromatogr. A 707, 3–22.

Kasher M.S., Wakulchik M., Cook J.A., and Smith M.C. (1993) One-step purification of recombinant human papillomavirus type 16 E7 oncoprotein and its binding to the retinoblastoma gene product. BioTechniques 14, 630–641.

Kashlev M., Martin E., Polyakov A., Severinov K., Nikiforov V., and Goldfarb A. (1993) Histidine-tagged RNA polymerase: dissection of the transcription cycle using immobilized enzyme. Gene 130, 9–14.

Kaslow D.C., and Shiloach J. (1994) Production, purification and immunogenecity of a malaria transmission-blocking vaccine candidate: TBV25H expressed in yeast and purified using nickel-NTA agarose. Bio/Technology 12, 494–499.

Kaul, R., and Mattiasson, B. (1992) Secondary purification. Bioseparation 2, 1–26.

Kellerman O.K., and Ferenci T. (1982) Maltose binding protein from *Escherichia coli*. Meth. Enzymol. 90, 459–463.

Kim J.-S., and Raines R.T. (1993) Ribonuclease S-peptide as a carrier in fusion proteins. Prot. Sci. 2, 348–356.

Kim J.S., and Raines R.T. (1994) Peptide tags for a dual affinity fusion system. Anal. Biochem. 219, 165–166.

Kleymann G., Ostermeier C., Ludwig B., Skerra A., and Michel H. (1995) Engineered Fv fragments as a tool for the one-step purification of integral multisubunit membrane protein complexes. Bio/Technology 13, 155–160.

Knappik A., and Plückthun A. (1994) An improved affinity tag based on the FLAG peptide for the detection and purification of recombinant antibody fragments. BioTechniques 17, 754–761.

Kohanski R.A., and Lane M.D. (1990) Monovalent avidin affinity columns. Meth. Enzymol. 184, 195–200.

Köhler K., Ljungquist C., Kondo A., Veide A., and Nilsson B. (1991) Engineering proteins to enhance their partition coefficients in aqueous two-phase systems. Bio/Technology 9, 642–646.

Kraulis P.J., Jonasson P., Nygren P.-Å., Uhlén M., Jendeberg L., Nilsson B., and Kördel J. (1996) The serum albumin-binding domain of streptococcal protein G is a three-helical bundle: A heteronuclear NMR study. FEBS Lett. 378, 190–194.

Kroll D.J., Abdel-Malek Abdel-Hafiz H., Marcell T., Simpson S., Chen C.Y., Gutierrez-Hartmann A., Lustbader J.W., and Hoeffler J.P. (1993) A multi-functional prokaryotic protein expression system: Overproduction, affinity purification, and selective detection. DNA Cell Biol. 12, 441–453.

Kunz D., Gerard N.P., and Gerard C. (1992) The human leukocyte platelet-activating factor receptor: cDNA cloning, cell surface expression and construction of a novel epitope bearing analog. J. Biol. Chem. 267, 22676–22683.

Kuusinen A., Arvola M., Oker-Blom C., and Keinanen K. (1995) Purification of recombinant GluR-D glutamate receptor produced in Sf21 insect cells. Eur. J. Biochem. 233, 720–726.

Langone J.J. (1992) Protein A of *Staphylococcus aureus* and related immunoglobulin receptors produced by streptococci and pneumonococci. In: Advances in Immunology (F.J. Dixon and H.G. Kunhel, eds.), Vol. 32, Academic Press, New York, pp. 157–252.

Larsson M. Brundell E., Nordfors L., Höög C., Uhlén M., and Ståhl S. (1996) A general bacterial expression system for functional analysis of cDNA-encoded proteins. Prot. Expr. Purif. 7, 447–457.

Lavallie E.R., and McCoy J.M. (1995) Gene fusion expression systems in *Escherichia coli*. Curr. Opin. Biotechnol. 6, 501–505.

Libon, C., Corvaïa, N., Haeuw, J.-F., Nguyen, T.N., Ståhl, S., Bonnefoy, J.-F. and Andréoni, C. (1999) The serum albumin-binding region of streptococcal protein G (BB) potentiates the immunogenicity of the G130–230 RSV-A protein. Vaccine 17, 406–414.

Ljungquist C., Jansson B., Moks T., and Uhlén M. (1989a) Thiol-directed immobilization of recombinant IgG-binding receptors. Eur. J. Biochem. 186, 557–561.

Ljungquist C., Breitholtz A., Brink-Nilsson H., Moks T., Uhlén M., and Nilsson B. (1989b) Immobilization and affinity purification of recombinant proteins using histidine peptide fusions. Eur. J. Biochem. 186, 563–569.

Löfdahl S., Guss B., Uhlén M., Philipson L., and Lindberg M. (1983) Gene for staphylococcal protein A. Proc. Natl. Acad. Sci. USA 80, 697–701.

Lorca T., Labbé J.-C., Devault A., Fesquet D., Strausfeld U., Nilsson J., Nygren P.-Å., Uhlén M., Cavadore J.-C., and Dorée M. (1992) Cyclin A-cdc2 kinase does not trigger but delays cyclin degradation in interphase extracts of amphibian eggs. J. Cell. Sci. 102, 55–62.

Löwenadler B., Jansson B., Paleus S., Holmgren E., Nilsson B., Moks T., Palm G., Josephson S., and Uhlén M. (1987) A gene fusion system for generating antibodies against short peptides. Gene 58, 87–97.

Lu T., Van Dyke M., and Sawadogo M. (1993) Protein-protein interaction studies using immobilized oligohistidine fusion proteins. Anal. Biochem. 213, 318–322.

Lundstrom K., Vargas A., and Allet B. (1995) Functional activity of a biotinylated human neurokinin 1 receptor fusion expressed in the Semliki Forest virus system. Biochem. Biophys. Res. Commun. 208, 260–266.

Maina C.V., Riggs P.D., Grandea III A.G., Slatko B.E., Moran L.S., Tagliamonte J.A., McReynolds L.A., and diGuan C.-D. (1988) An *Escherichia coli* vector to express and purify proteins by fusion to and separation from maltose-binding protein. Gene 74, 365–373.

Makrides, S. (1996) Strategies for achieving high-level expression of genes in *Escherichia coli*. Microbiol. Rev. 60, 512–538.

Makrides S., Nygren P.-Å., Andrews B., Ford P., Evans K.S., Hayman E.G., Adari H., Levin J., Uhlén M. and Toth C.A. (1996) Extended in vivo half-life of human soluble complement receptor type I fused to serum albumin-binding receptor J. Pharm. Exp. Therap. 277, 534–539.

Mehler A.H. (1993) Glutathione S-transferases function in detoxification reactions. In: Textbook of Biochemistry, 3rd ed. (T.M. Devlin, ed.), Wiley-Liss, New York, pp. 523–524.

Miyashita K., Kusumi M., Utsumi R., Komano T., and Satoh N. (1992) Expression and purification of recombinant 3C proteinase of coxsackievirus B3. Biosci. Biotechnol. Biochem. 56, 746–750.

Moks T., Abrahmsén L., Nilsson B., Hellman U., Sjöquist J., and Uhlén M. (1986) Staphylococcal protein A consists of five IgG-binding regions. Eur. J. Biochem. 156, 637–643.

Moks T., Abrahmsén L., Holmgren E., Bilich M., Olsson A., Pohl G., Sterky C., Hultberg H., Josephsson S., Holmgren A., Jörnvall H., Uhlén M., and Nilsson

B. (1987a) Expression of human insulin-like growth factor I in bacteria: Use of optimized gene fusion vectors to facilitate protein purification. Biochemistry 26, 5239–5244.

Moks, T., Abrahmsén, L., Österlöf, B., Josephson, S., Östling, M., Enfors, S.-O., Persson, I., Nilsson, B., and Uhlén,M. (1987b) Large-scale affinity purification of human insulin-like growth factor I from culture medium of *Escherichia coli.* Bio/Technology 5, 379–382.

Murby M., Cedergren L., Nilsson J., Nygren P.-Å., Hammarberg B., Nilsson B., Enfors S.-O., and Uhlén M. (1991) Stabilization of recombinant proteins from proteolytic degradation in *Escherichia coli* using a dual affinity fusion strategy. Biotechnol. Appl. Biochem. 14, 336–346.

Murby M., Nguyen T.N., Binz H., Uhlén M., and Ståhl S. (1994) Production and recovery of recombinant proteins of low solubility. In: Separations for Biotechnology, 3rd ed. (D.L. Pyle, ed.), Bookcraft Ltd, Bath, UK, pp. 336–344.

Murby M., Samuelsson E., Nguyen T.N., Mignard L., Power U., Binz H., Uhlén M., and Ståhl S. (1995) Hydrophobicity engineering to increase solubility and stability of a recombinant protein from respiratory syncytial virus. Eur. J. Biochem. 230, 38–44.

Nilsson B., Moks T., Jansson B., Abrahmsén L., Elmblad A., Holmgren E., Henrichson C., Jones T.A., and Uhlén M. (1987) A synthetic IgG-binding domain based on staphylococcal protein A. Prot. Eng. 1, 107–113.

Nilsson B., and Abrahmsén L. (1990) Fusions to staphylococcal protein A. Meth. Enzymol. 185, 144–161.

Nilsson J., Nilsson P., Williams Y., Pettersson L., Uhlén M., and Nygren P.-Å. (1994) Competitive elution of protein A fusion proteins allows specific recovery under mild conditions. Eur. J. Biochem. 224, 103–108.

Nilsson J., Jonasson P., Samuelsson E., Ståhl S., Uhlén M. and Nygren P.-Å. (1996a) Integrated production of human insulin and its C-peptide. J. Biotechnol. 48, 241–250.

Nilsson J., Larsson M., Ståhl S., Nygren P.-Å., and Uhlén M. (1996b) Multiple affinity domains for the detection, purification and immobilization of recombinant proteins. J. Mol. Recogn. 9, 585–594.

Nilsson J., Bosnes M., Larsen F., Nygren P.-Å., Uhlén M., and Lundeberg J. (1997a) Heat-mediated activation of affinity immobilized *Taq* DNA polymerase. BioTechniques 22, 744–751.

Nilsson, J., Ståhl, S., Lundeberg, J., Uhlén, M., and Nygren, P.-Å. (1997b) Affinity fusion strategies for detection, purification and immobilization of recombinant proteins. Prot. Expr. Purif. 11, 1–16.

Nord K., Nilsson J., Nilsson B., Uhlén M., and Nygren P.-Å. (1995) A combinatorial library of an α-helical bacterial receptor domain. Prot. Eng. 8, 601–608.

Nord, K., Gunneriusson, E., Ringdahl, J., Ståhl, S., Uhlén, M. and Nygren, P.-Å (1997) Binding proteins selected from combinatorial libraries of an α-helical bacterial receptor domain. Nat. Biotechnol. 15, 772–777.

Nord, K., Gunneriusson, E., Uhlén, M., and Nygren, P.-Å. (2000) Ligands selected from combinatorial libraries of protein A for use in affinity capture of apolipoprotein A-1M and Tag DNA polymerase. J. Biotechnol. 80, 45–54.

Nossal, N.G., and Heppel, L.A. (1966) The release of enzymes by osmotic shock from *Escherichia coli* in exponential phase. J. Biol. Chem. 241, 3055–3062.

Nygren, P.-Å., Eliasson, M., Palmcrantz, E., Abrahmsén, L., and Uhlén, M. (1988) Analysis and use of the serum albumin binding domains of streptococcal protein G. J. Mol. Recognit. 1, 69–74.

Nygren P.-Å., Ljungquist C., Trømborg H., Nustad K., and Uhlén M. (1990) Species-dependent binding of serum albumins to the streptococcal receptor protein G. Eur. J. Biochem. 193, 143–148.

Nygren P.-Å., Flodby P., Andersson R., Wigsell H., and Uhlén M. (1991) In vivo stabilization of a human recombinant CD4 derivative by fusion to a serum albumin-binding receptor. In: Vaccines 91, Modern Approaches to Vaccine Development (R.M. Chanock, ed.), Cold Spring Harbor Laboratory Press, New York, pp. 363–368.

Nygren P.-Å., Ståhl S., and Uhlén M. (1994) Engineering proteins to facilitate bioprocessing. Trends Biotechnol. 12, 184–188.

Nygren, P.-Å., and Uhlén M. (1997) Scaffolds for engineering novel binding sites in proteins. Curr. Opin. Struct. Biol. 7, 463–469.

Panagiotidis C.A., and Silverstein S.J. (1995) pALEX, a dual-tag prokaryotic expression vector for the purification of full-length proteins. Gene 164, 45–47.

Peng S., Sommerfelt M., Logan J., Huang Z., Jilling T., Kirk K., Hunter E., and Sorscher E. (1993) One-step affinity isolation of recombinant protein using the baculovirus/insect cell expression system. Prot. Expr. Purif. 4, 95–100.

Porath J., Carlsson J., Olsson I., and Belfrage G. (1975) Metal chelate affinity chromatographay, a new approach to protein fractionation. Nature 258, 598–599.

Potter K.N., Li Y. and Capra D. (1996) Staphylococcal protein A simultaneously interacts with framework region 1, complementarity-determining region 2, and framework region 3 on human V_H3-encoded Igs. J. Immunol. 157, 2982–2988.

Power, U.F., Plotnicky-Gilquin, H., Huss, T., Robert, A., Trudel, M., Ståhl, S., Uhlén, M., Nguyen, T.N. and Binz, H. (1997) Induction of protective immunity in rodents by vaccination with a prokaryotically expressed recombinant fusion protein containing a respiratory syncytial virus G protein fragment. Virology 230, 15–166.

Prickett K.S., Amberg D.C., and Hopp T.P. (1989) A calcium-dependent antibody for identification and purification of recombinant proteins. BioTechniques 7, 580–589.

Rudolph, R. (1995) Successful protein folding on an industrial scale. In: Principles and Practice of Protein Folding (J.L. Cleland and C.S. Craik, eds.), John Wiley and Sons, New York, pp. 283–298.

Samols D., Thornton C.G., Murtif V.L., Kumar G.K., Haase F.C., and Wood H.G. (1988) Evolutionary conservation among biotin enzymes. J. Biol. Chem. 263, 6461–6464.

Samuelson P., Hansson M., Ahlborg N., Andreoni C., Götz F., Bächi T., Nguyen T.N., Binz H., Uhlén M., and Ståhl S. (1995) Cell surface display of recombinant proteins on *Staphylococcus carnosus*. J Bacteriol. 177, 1470–1476.

Samuelsson E., Wadensten H., Hartmanis M., Moks T., and Uhlén M. (1991) Facilitated in vitro refolding of human recombinant insulin-like growth factor I using a solubilizing fusion partner. Bio/Technology 9, 363–366.

Samuelsson E., Moks T., Nilsson B. and Uhlén M. (1994) Enhanced in vitro refolding of insulin-like growth factor I using a solubilizing fusion partner. Biochemistry 33, 4207–4211.

Samuelsson, E., Jonasson, P., Viklund, F., Nilsson, B., and Uhlén, M. (1996) Affinity-assisted in vivo folding of a secreted human peptide hormone in *Escherichia coli*. Nat. Biotechnol. 14, 751–755.

Sassenfeld H.M. (1990) Engineering proteins for purification. Trends Biotechnol 8, 88–93.

Savage M.D., Mattson G., Desai S., Nielander G.W., Morgensen S., and Conklin E.J. (1992) Biotinylation reagents. In: Avidin-Biotin Chemistry: A Handbook, Pierce Chemical Company, Rockford, IL, pp. 25–88.

Schatz P.J. (1993) Use of peptide libraries to map the substrate specificity of a peptide-modifying enzyme: A 13 residue consensus peptide specifies biotinylation in *Escherichia coli*. Bio/Technology 11, 1138–1143.

Schmidt T.G.M., and Skerra A. (1993) The random peptide library-assisted engineering of a C-terminal affinity peptide, useful for the detection and purification of a functional Ig Fv fragment. Prot. Eng. 6, 109–122.

Sharrocks A. (1994) A T7 expression vector for producing N- and C-terminal fusion proteins with glutathione S-transferase. Gene 138, 105–108.

Simons P.C., and Vander Jagt D.L. (1981) Purification of gluthatione S-transferases by gluthatione-affinity chromatography. Meth. Enzymol. 77, 235–237.

Sinha D., Bakhshi M., and Vora R. (1994) Ligand binding assays with recombinant proteins refolded on an affinity matrix. BioTechniques 17, 509–514.

Sjölander A., Lövgren K., Ståhl, S., Åslund L., Hansson M., Nygren P.-Å., Wåhlin B., Berzins K., Uhlén M., Morein B., and Perlmann P. (1991) High antibody responses in rabbits immunized with inflenza virus ISCOMs containing a repeated sequence of the *Plasmodium falciparum* antigen Pf155/RESA. Vaccine 9, 443–450.

Sjölander, A., Ståhl, S. and Perlmann, P. (1993) Bacterial expression systems based on protein A and protein G designed for the production of immunogens, applications to *Plasmodium falciparum malaria* antigens. ImmunoMethods 2, 79–92.

Sjölander, A., Nygren, P.-Å., Ståhl, S., Berzins, K., Uhlén, M., Perlmann, P., and Andersson, R. (1997) The serum albumin binding region of streptococcal protein G: a bacterial fusion partner with carrier-related properties. J. Immunol. Meth. 201, 115–123.

Skerra A. (1994) A general vector, pASK84, for cloning, bacterial production, and single-step purification of antibody Fab fragments. Gene 141, 79–84.

Smith D.B., and Johnson K.S. (1988) Single-site purification of polypeptides expressed in *Escherichia coli* as fusions with glutathione S-transferase. Gene 67, 31–40.

Smith, M.C., Furman, T.C., Ingolia, T.D. and Pidgeon, C. (1988) Chelating peptide-immobilized metal ion affinity chromatography. J. Biol. Chem. 263, 7211–7215.

Ståhl S., Sjölander A., Nygren P.-Å., Berzins K., Perlmann P., and Uhlen M. (1989) A dual expression system for the generation, analysis and purification of antibodies to a repeated sequence of the *Plasmodium falciparum* antigen Pf155/RESA. J. Immunol. Meth. 124, 43–52.

Ståhl S., Sjölander A., Hansson L., Nygren P.-Å., and Uhlén M. (1990) A general strategy for polymerization, assembly and expression of epitope-carrying peptides applied to the *Plasmodium falciparum* antigen Pf155/RESA. Gene 89, 187–193.

Ståhl S., Nygren P.-Å., and Uhlén M. (1997a) Strategies for gene fusion. In: Recombinant Gene Expression Protocols. Methods in Molecular Biology Series, Vol. 62 (R.S. Tuan, ed.), Humana Press, Totowa, NJ, pp. 37–54.

Ståhl S., Nygren P.-Å., and Uhlén M. (1997b) Detection and isolation of recombinant proteins based on binding affinity of reporter: protein A. In: Recombinant Proteins Protocols: Detection and Isolation. Methods in Molecular Biology Series, Vol. 63 (R. S. Tuan, ed.), Humana Press, Totowa, NJ, pp. 103–118.

Stempfer G., Höll-Neugebauer B., and Rudolph R. (1996) Improved refolding of an immobilized fusion protein. Nat. Biotechnol. 14, 329–334.

Swack J.A., Zander G.L., and Utter M.F. (1978) Use of avidin-Sepharose to isolate and identify biotin polypeptides from crude extracts. Anal. Biochem. 87, 114–126.

Torigoe H., Shimada I., Saito A., Sato M., and Arata Y. (1990) Sequential ^{1}H NMR assignments and secondary structure of the B domain of staphylococcal protein A: Structural changes between the free B domain in solution and the Fc-bound B domain in crystal. Biochemistry 29, 8787–8793.

Uhlén M., Nilsson B., Guss B., Lindberg M., Gatenbeck S., and Philipson L. (1983) Gene fusion vectors based on staphylococcal protein A. Gene 23, 369–378.

Uhlén M., Guss B., Nilsson B., Götz F., and Lindberg M. (1984) Expression of the gene encoding protein A in *Staphylococcus aureus* and coagulase-negative staphylococci. J. Bacteriol. 159, 713–719.

Uhlén, M. and Moks, T. (1990) Gene fusions for purpose of expression: An introduction. Meth. Enzymol. 185, 129–143.

Uhlén M., Forsberg G., Moks T., Hartmanis M., and Nilsson B. (1992) Fusion proteins in biotechnology. Curr. Opin. Biotechnol. 3, 363–369.

Van Dyke M.W., Sirito M., and Sawadogo M. (1992) Single-step purification of bacterially expressed polypeptides containing an oligo-histidine domain. Gene 111, 99–104.

Verma, R., Boleti, E., and George, A.J. (1998) Antibody engineering: comparison of bacterial, yeast, insect and mammalian expression systems. J. Immunol. Meth. 216, 165–181.

Vidal M.A., and Conde F.P. (1985) Alternative mechanism of protein A-immunoglobulin interaction: The V_H-associated reactivity of a monoclonal human IgM. J. Immunol. 135, 1232–1238.

Walker P.A., Leong L.E.-C., Ng P.W.P., Han Tan S., Waller S., Murphy D., and Porter A.G. (1994) Efficient and rapid affinity purification of proteins using recombinant fusion proteases. Bio/Technology 12, 601–605.

Ward A.C., Castelli L.A., Macreadie I.G., and Azad A.A. (1994) Vectors for Cu(2+)-inducible production of glutathione S-transferase- fusion proteins for single-step purification from yeast. Yeast 10, 441–449.

Weiss E., Chatellier J., and Orfanoudakis G. (1994) In vivo biotinylated recombinant antibodies: Construction, characterization, and application of a bifunctional Fab-BCCP fusion protein produced in *Escherichia coli*. Prot. Expr. Purif. 5, 509–517.

Wilchek M., and Bayer E.A. (1990) Introduction to avidin-biotin technology. Meth. Enzymol. 184, 5–13.

Witzgall R., O'Leary E., and Bonventre J.V. (1994) A mammalian expression vector for the expression of GAL4 fusion proteins with an epitope tag and histidine tail. Anal. Biochem. 223, 291–298.

Zhao X., and Siu C.H. (1995) Colocalization of the homophilic binding site and the neuritogenic activity of the cell adhesion molecule L1 to its second Ig-like domain. J. Biol. Chem. 270, 29413–29421.

[illegible]

5
Solid–Liquid Separation

Ricardo A. Medronho
Federal University of Rio de Janeiro, Rio de Janeiro, Brazil

1. INTRODUCTION

Separation and purification of biological products present special characteristics and difficulties when compared with separation and purification of other chemical products. The main difference is related to the fact that biological products are, in general, large and unstable molecules diluted in a complex solid-liquid mixture. The first step to be performed in a separation and purification protocol is usually a solid-liquid separation. This separation step can also be used at the end of a separation and purification protocol if the product has been concentrated, e.g., by precipitation or crystallization. In this chapter, some common unit operations used for solid-liquid separation in biotechnological processes are discussed, taking into account the peculiarities of biological particles. These unit operations are centrifugal separation and cake filtration.

Centrifugal separations are processes in which a centrifugal field is applied to promote the separation of particles, drops, or bubbles from a liquid, or particles and drops from a gas. This chapter will deal with the separation of biological particles from liquid media. The use of centrifuges to promote cell separations is widespread both in laboratory and in industry. A detailed description of the theory and types of centrifuges is seen in Sec. 2. This includes an item on separation efficiency that provides the theoretical basis for the understanding of centrifuge performance. In spite of the emergence of new separation methods, such as ultrasound-enhanced sedimentation (Hawkes et al. 1997) and dielectrophoresis separation (Pethig and Markx 1997), it is expected that centrifuges will continue to play an important role in the separation of bioparticles in the near future.

Hydrocyclones also use centrifugal field to promote separation but have a great advantage over centrifuges. They have no moving parts. As such, they are extremely simple equipments. Recent works (Lübberstedt 2000a,b) have shown that they are able to separate mammalian cells. This opens the possibility of using them in perfusion systems or as a pretreatment in harvesting systems. A brief description of hydrocyclones is given in Sec. 2 of this chapter.

Conventional cake filtration is one of the oldest ways of carrying out solid-liquid separation. This is why it has been used since the beginning of the biotechnology industry. The cell high compressibility and the presence of submicrometer particles suspended in the liquid make the filtration of culture media a difficult task. The use of filter aids increases the cake permeability and porosity, facilitating the filtration of broths that would be otherwise difficult to filter. The theory of filtration, the use of filter aids and a description of the main types of filters are seen in Sec. 3 of this chapter.

The aggregation of bioparticles is an important step in many solid-liquid separation processes. Section 4 deals with flocculation, which is a common way of increasing the separability degree of suspensions with difficult settling properties or poor filterability.

Apart from two exceptions, all equations given in this chapter are dimensionally consistent. Therefore, they may be used with any coherent system of units. The two exceptions are Eqs. (46) and (52), which must be used only with the International System of Units (SI).

2. CENTRIFUGAL SEPARATION

Separation in a centrifugal field occurs only if there is a density difference between the particle and the liquid. Unfortunately, this difference is usually small when dealing with biological particles. Apart from that, their sizes are relatively small. These two facts explain why these particles attain very low terminal settling velocities when settling individually in the gravitational field, as shown in Table 1. Therefore, it is necessary to use high centrifugal accelerations and, usually, also high residence times in order to obtain high separation efficiencies of biological particles. Table 2 gives both the size and density ranges of some biological particles (Cann 1999; Datar 1984; Kildeso and Nielsen 1997; Linz et al. 1990; Lin et al. 1991; Ling et al. 1997; Pons and Vivier 1998; Bendixen and Rickwood 1994; Yuan et al. 1996b; Taylor et al. 1986; Werning and Voss, 1993), and Fig. 1 shows the typical size range of

Table 1 Typical Settling Velocities of Some Biological Particles Falling in Water at 20°C Under the Gravitational Field

Biological particle	Typical Settling Velocity (mm h^{-1})
Influenza virus	0.004
Cell debris of *Escherichia coli*	0.02
Interferon-γ as inclusion body	0.4
Escherichia coli	0.2
Saccharomyces cerevisiae	7
HeLa cell	30

some biological particles and the operational range of common separation processes.

Equation (1), known as Stokes' law, gives the settling velocity v of a spherical particle settling in the laminar region (Stokes region) under the influence of a gravitational or centrifugal field.

$$v = \frac{(\rho_s - \rho) b d^2}{18\mu} \tag{1}$$

where $b = g$ for gravitational field, $b = \omega^2 r$ for centrifugal fields, ρ_s and ρ are the particle and liquid densities, d is the particle diameter, μ is the liquid

Table 2 Ranges of Typical Sizes and Densities of Some Biological Particles

Biological particle	Size (μm)	Density (g cm^{-3})
Cells	0.5–400	1.005–1.14
Bacteria	0.5–10	1.05–1.11
Yeast	3–25	1.05–1.13
Mammalian cells	8–40	1.06–1.14
Insect cells	13–30	—
Plant cells	20–400	1.05–1.09
Nuclei from mammalian cells	2.5–5.0	—
Inclusion bodies	0.4–3.0	1.03–1.18
Cell debris	0.1–5	1.01–1.20
Virus	0.02–0.2	1.09–1.39

Data from Cann 1997; Datar 1984; Kildeso and Nielsen 1997; Linz et al. 1990; Lin et al. 1991; Ling et al. 1997; Pons and Vivier 1998; Bendixen and Rickwood 1994; Yuan et al. 1996b; Taylor et al. 1986; Werning and Voss 1993).

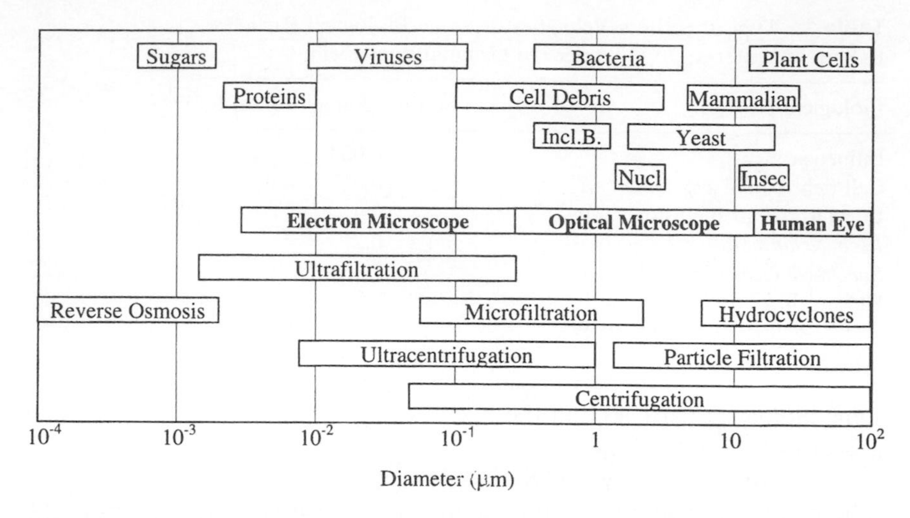

Figure 1 Typical sizes of some biological particles and the operational range of common separation processes. (Mammalian, mammalian cells; Inc.B., inclusion bodies; Nucl, mammalian cells nuclei; Insec, insect cells.)

viscosity, g is the gravity acceleration, ω is the angular velocity, and r the radial position of the particle.

Equation (1) is valid for a particle settling without the interference of other particles, i.e., for diluted systems. The particle settling velocity decreases as particle concentration increases. This phenomenon is known as *hindered settling*. Several equations can be found in the literature to account for this phenomenon. Nevertheless, a convenient method to calculate the hindered-settling velocity is by using Eq. (1) with the apparent density and viscosity of the suspension replacing the liquid density and viscosity, respectively (Heiskanen 1993).

The ratio between the settling velocity of a particle under a centrifugal field v_c and this velocity under the gravitational field v_g is known as g-factor ζ and is given by Eq. (2).

$$\zeta = \frac{v_c}{v_g} = \frac{\omega^2 r}{g} \tag{2}$$

The g factor is also known as g-number, centrifugation factor, acceleration factor, and, improperly, g force and relative centrifugal force. As Eq. (2) shows, ζ is also the ratio between the centrifugal and the gravitational accelerations. The angular velocity in Eq. (2) must be used in radians per second (units: s^{-1}). As a full revolution means 2π radians, the following equation may be used to convert revolutions per minute (rpm) to radians per second:

$$\omega_{rad/s} = \frac{\pi \omega_{rpm}}{30} \tag{3}$$

For example, if a tubular centrifuge of 10.5 cm diameter is rotating at 15,000 rpm, from Eq. (3), its angular velocity is 1571 radians per second and, from Eq. (2), the g factor at the internal lateral wall of the bowl is 13,205. Therefore, the centrifugal acceleration at the wall is 13,205 times greater than the gravitational acceleration, i.e., 13,205 × g.

2.1 Separation Efficiency

When promoting a separation of particles from liquid, the degree of separation is an important parameter to be evaluated. This can be done through the use of some concepts as total efficiency, grade efficiency, and cut size. These concepts are applicable to any equipment whose performance remains constant if the operational conditions do not change. They are valid, therefore, for equipment such as sedimenting centrifuges, hydrocyclones, cyclones, gravitational classifiers, elutriators, and so forth.

Figure 2 shows a schematic diagram of a typical separator. Q, Q_o, and Q_u are volumetric flow rates, W_S, W_{So}, and W_{Su} are mass flow rates of solids, and y, y_o, and y_u are cumulative size distributions (undersize) of the feed, overflow and underflow, respectively.

The total efficiency E_T of a separator is defined by Eq. (4) and gives the mass fraction of solids recovered in the underflow. Authors who work with centrifuges usually refer to the fraction of unsedimented solids, which is equal to $1 - E_T$.

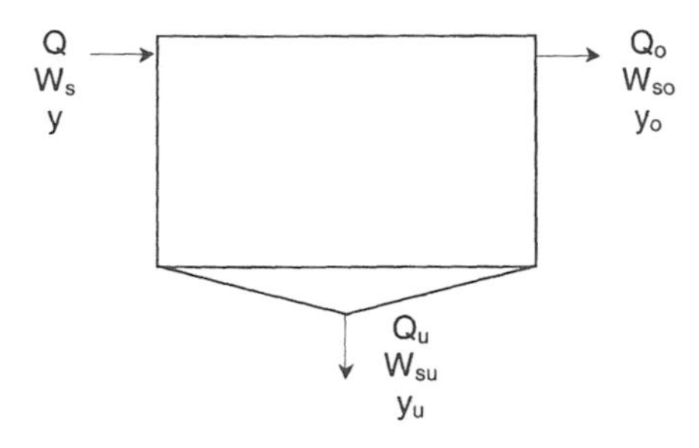

Figure 2 Schematic diagram of a typical separator.

$$E_{\mathrm{T}} = \frac{W_{\mathrm{Su}}}{W_{\mathrm{S}}} \tag{4}$$

Assuming that there is no agglomeration or comminution inside the separator, a mass balance for the particles smaller than a given diameter d produces Eq. (5), which gives a relationship between the total efficiency and the particle size distributions.

$$E_{\mathrm{T}} = \frac{y_{\mathrm{o}} - y}{y_{\mathrm{o}} - y_{\mathrm{u}}} \tag{5}$$

The fraction of fluid that is discharged in the underflow is called flow ratio R_{f}, as given by Eq. (6).

$$R_{\mathrm{f}} = \frac{Q_{\mathrm{Lu}}}{Q_{\mathrm{L}}} \tag{6}$$

where Q_{Lu} and Q_{L} are the volumetric flow rate of fluid in the underflow and in the feed, respectively.

Separators can operate with or without a fluid flow rate in the underflow. For instance, tubular centrifuges operate normally with $R_{\mathrm{f}} = 0$ and scroll-type centrifuges with $R_{\mathrm{f}} > 0$. A flow ratio equal to R_{f} means that this fraction of feed fluid is leaving the separator through the underflow. Since the fluid carries solid particles with it, some particles will be discharged into the underflow due not to the centrifugal action of the separator but to entrainment. In spite of some controversy regarding hydrocyclones (Frachon and Cilliers 1999; Roldán-Villsana et al. 1993; Del Villar and Finch 1992), this bypass is normally assumed to be equal to the flow ratio (Heiskanen 1993). Therefore, R_{f} is the minimal efficiency at which a separator will operate even if no centrifugal action takes place.

The reduced total efficiency E'_{T}, also called centrifugal efficiency, is used for separators with $R_{\mathrm{f}} > 0$, and gives the separation efficiency taking into account only those particles that will be separated or not due to the centrifugal field intensity. Hence, E'_{T} does not consider the particles that are separated due to the flow ratio. The reduced total efficiency is defined by Eq. (7).

$$E'_{\mathrm{T}} = \frac{E_{\mathrm{T}} - R_{\mathrm{f}}}{1 - R_{\mathrm{f}}} \tag{7}$$

The total efficiency and the reduced total efficiency are useful when, for instance, analyzing the influence of operational conditions on a given solid-liquid separation. However, these figures only have meaning when accompanied by many other data, such as particle size distribution of the feed, densities of the solids and liquid, feed concentration of solids, feed flow rate, etc. If, however, the efficiency is calculated for every particle size present in the feed, the resulting curve, known as grade efficiency curve, is usually independent of the solids size distribution and density and is constant for a particular set of operating conditions (Svarovsky 1990). The grade efficiency, G, gives the mass fraction of solids of a given diameter recovered in the underflow. Its definition is, thus, similar to the total efficiency but is applied to a given particle size, as shown in Eq. (8). For a better comprehension, the grade efficiency is the total efficiency that a separator would give if fed only with particles of a given size.

$$G = \frac{W_{Su}|_d}{W_S|_d} \tag{8}$$

As the particle size frequency x gives the fraction of particles of size d:

$$G = \frac{W_{Su} x_u}{W_S x} \tag{9}$$

By definition, the frequency x is equal to the derivative of the cumulative size distribution y in relation to the diameter d. Therefore:

$$G = E_T \frac{dy_u}{dy} \tag{10}$$

Based on the mass balance given by Eq. (5), it is possible to obtain expressions similar to Eq. (10) based on y_o and y or on y_o and y_u.

$$G = 1 - (1 - E_T)\frac{dy_o}{dy} \tag{11}$$

$$\frac{1}{G} = 1 + \left(\frac{1}{E_T} - 1\right)\frac{dy_o}{dy_u} \tag{12}$$

Equations (10), (11), and (12) show that it is possible to obtain the grade efficiency curve based in the total efficiency and in two of the three size distributions.

Figure 3a shows a typical grade efficiency curve for a tubular centrifuge working without flow ratio. This curve is also known as partition or classification curve. The particle size, which corresponds to a grade efficiency of 50%, is known as cut size, d_{50}. A typical grade efficiency curve for a scroll-type centrifuge working with a 10% flow ratio is shown in Fig. 3b. As can be seen, the curve starts at the flow ratio value because the very fine particles present in the feed follow the flow and are therefore split in the same ratio as the fluid.

Similar to the total efficiency, the grade efficiency can also be reduced, producing the reduced grade efficiency G' as given by Eq. (13). Like the reduced total efficiency, G' does not consider the particles that are only

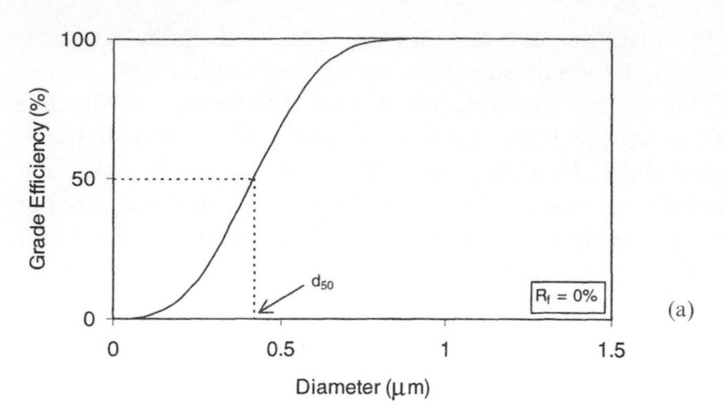

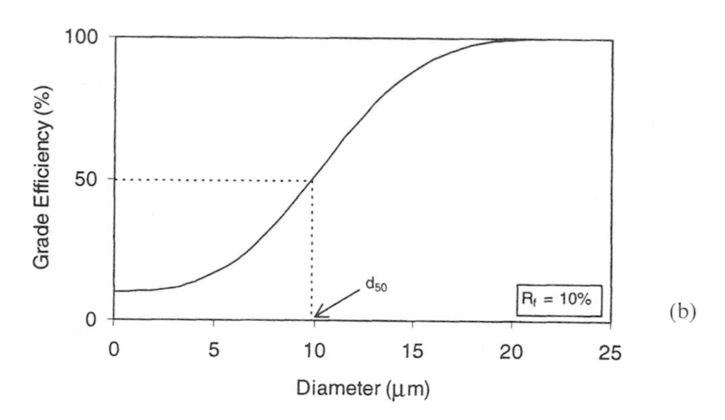

Figure 3 A typical grade efficiency curve for (a) a tubular centrifuge and (b) a scroll-type centrifuge, showing the cut size (d_{50}).

separated due to the flow ratio. A typical reduced grade efficiency curve is shown in Fig. 4. This curve was plotted based on the data from Fig. 3b. In analogy with the grade efficiency curve, the particle size that corresponds to $G' = 50\%$ is known as reduced cut size d'_{50}.

$$G' = \frac{G - R_f}{1 - R_f} \tag{13}$$

The grade efficiency curve (when $R_f = 0$) and the reduced grade efficiency curve (when $R_f > 0$) are usually presented as a function of d/d_{50} or d/d'_{50}, respectively (see an example in Fig. 5). Grade efficiency curves and reduced grade efficiency curves usually have an S shape for equipment that uses either screening (cake filters, microfilters, etc.) or particle dynamics in

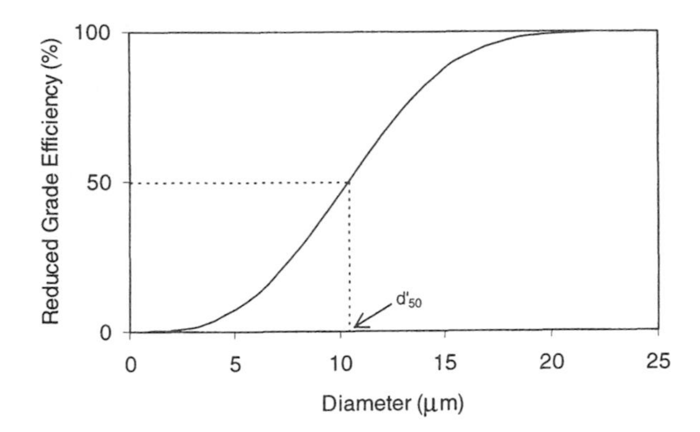

Figure 4 A typical reduced-grade efficiency curve and the reduced cut size (d'_{50}). This plot was based on data from Fig. 3b.

which body forces acting on the particles, such as gravity or centrifugal forces, are opposed by drag forces (elutriators, centrifuges, hydrocyclones, etc.) (Svarovsky 1990).

Equation (14), which can be obtained from Eq. (10), shows that it is possible to estimate the total efficiency when the grade efficiency curve and the feed size distribution are known.

$$E_T = \int_0^1 G\,dy \tag{14}$$

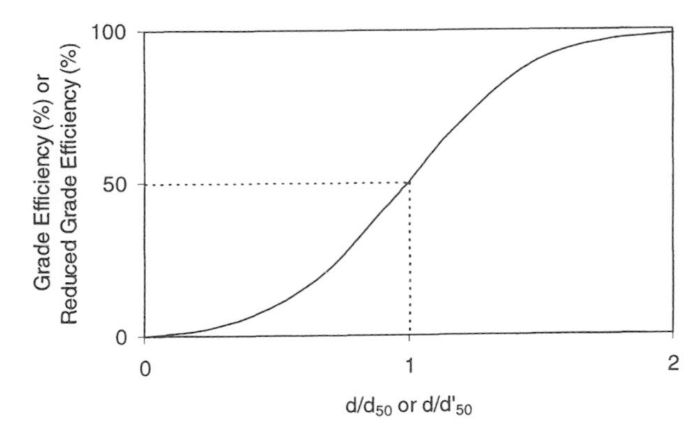

Figure 5 A typical grade efficiency curve as a function of the ratio between particle size and cut size (d/d_{50}). This S-shaped curve is usually found also for the reduced grade efficiency as a function of the ratio between particle size and reduced cut size (d/d'_{50}).

Based on Eqs. (7) and (14), a similar equation can be obtained for the reduced total efficiency:

$$E'_T = \int_0^1 G' dy \tag{15}$$

2.2 Centrifuges

Centrifugation is an important tool in the biotechnology industry. It is used for separation of whole cells, cell debris, protein precipitates, blood plasma fractionation, etc. In the laboratory, centrifuges are used in similar applications and also in some others, such as density gradient centrifugation (Patel et al. 1998) and centrifugal elutriation (Wilton and Strain 1998). An alternative application is the use of centrifuges in liquid-liquid extraction, where they may have several advantages over more classical methods (Ersson et al. 1998).

There are basically two families of centrifuges: sedimenting centrifuges and filtering centrifuges. The former uses centrifugal force to move the particles radially either outward or inward through a liquid, according to whether they are heavier or lighter than the suspending liquid. The principle of separation of these centrifuges is, therefore, sedimentation in a centrifugal field and, as such, they will be seen in this item. The latter, also known as centrifugal filter, uses the centrifugal field to promote the necessary pressure to force the mother liquor through both the filter media and the cake. The principles of filtration will be seen in item 3.

Types of Centrifuges

Some characteristics of the main industrial sedimenting centrifuges are shown in Table 3. For comparison, the data for ultracentrifuges are also included. Three different types of bowl can be seen in Fig. 6: the tubular bowl, the multichamber bowl, and the disk bowl.

Tubular Centrifuge. Figure 7 shows an industrial tubular centrifuge used for pharmaceutical and biotechnology applications. Tubular centrifuges use tubular bowls (see Fig. 6) with length-to-diameter ratios varying from 5 to 7 and diameters usually varying from 75 to 150 mm. They are the most efficient of all industrial sedimenting centrifuges due to their high angular velocities and thin settling zone. They can be used for separation of whole cells, cell debris, protein precipitates, and also for plasma fractionation, and polishing of solutions containing fines. In their laboratory version, speeds up to 50,000 rpm can be achieved leading to *g*

Table 3 Characteristics of the Main Industrial Centrifuges[a]

Type of centrifuge	*g* factor	Velocity (rpm)	Flow rate (m^3 h^{-1})	Conc. (% vol)	Solids removal[b]	Dewatering capacity[c]
Tubular	13000–20000	13000–18000	0.5–4	0–1	B	P
Multi-chamber	6000–11500	5000–10000	2–15	0–5	B	P
Disk	5000–13000	3000–15000	0.3–300	0–25	B, I, C	TS, TkS, P[d]
Scroll	1500–5000	1500–6000	0.5–120	3–60	C	TkS-P
Ultra-centrifuges	4×10^4–10^6	2×10^4–10^5	—	—	B	—

[a] For comparison, data for ultracentrifuges are also included.
[b] B = batch; I, intermittent; C, continuous.
[c] P, paste; TS, thin slurry; TkS, thick slurry.
[d] TS, nozzle type; TkS, solids-ejecting type; P, solid bowl type.

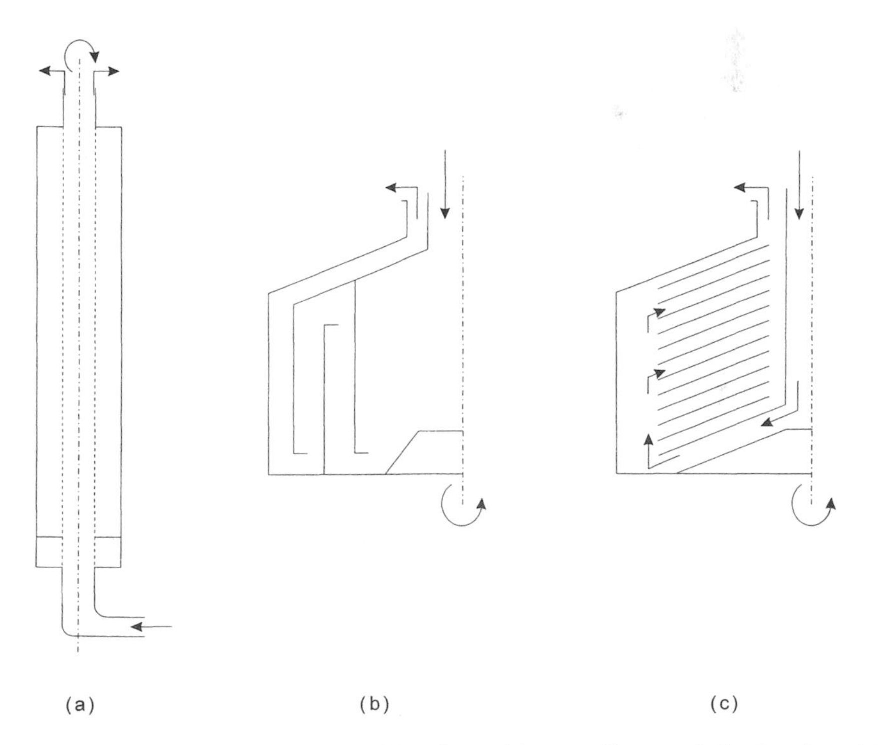

Figure 6 Three different types of bowls used in centrifuges: tubular bowl, multi-chamber bowl, and disk bowl.

Figure 7 A Sharples supercentrifuge used for pharmaceutical and biotechnology applications. (Courtesy of Alfa Laval Separation AB, Sweden.)

factors as high as 62,000. Their main disadvantage is the small sludge holding space (2–10 L for the industrial version and maximum of 200 ml for the lab version), which makes their utilization viable only for dilute suspensions (concentrations lower than 1% by volume). Another disadvantage is that the separated solids must be manually removed.

Multichamber Centrifuge. As shown in Fig. 8, multichamber centrifuges have a certain number of concentric vertical compartments con-

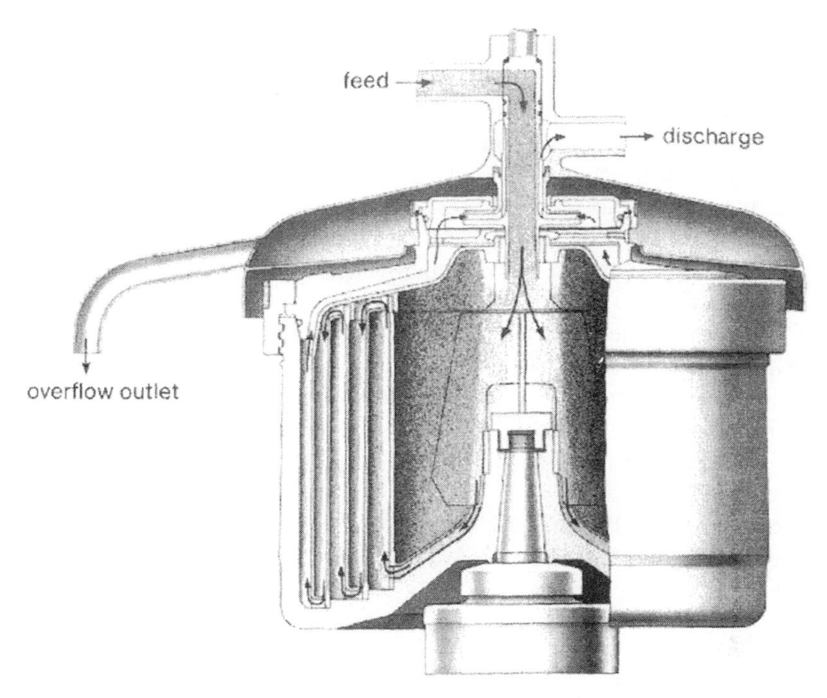

Figure 8 Multichamber centrifuge, vertical cut. (Courtesy of Westfalia Separator AG, Germany.)

nected in such a way that the suspension, which enters in the center, is forced to flow in series through the annuli between the cylinders. Thus, clarification occurs in the thin, axially flowing cylindrical layer of liquid. The thickness of the layers in the chambers decreases the further they are from the center, which implies a decreasing settling distance from the central chamber outwards. Since the centrifugal acceleration is directly proportional to the radius, the centrifugal force rises from the central chamber outward. Therefore, the coarser particles will be separated in the inner chambers and the finer particles in the outer (Hemfort 1984). These centrifuges use the same principle as the tubular ones but present the advantage of having a larger solids holding capacity of up to 65 L. They have usually two to six chambers, and their diameters range from 125 to 530 mm, with a length-to-diameter ratio of about 1. Their disadvantage is the relatively time-consuming process of solids removal. They are used for polishing liquids in the beverage, chemical, and pharmaceutical industries, and some machines can be refrigerated for special applications, such as human blood fractionating.

Disk-Stack Centrifuges. In a disk-stack centrifuge, the settling capacity is increased by means of a stack of conical disks, as shown in Fig. 6. A disk can be seen in Fig. 9. Both the bowl and the disk stack rotate at the same speed. The liquid is fed to the equipment through the center, flows underneath the disk stack, gaining angular velocity, and is split into many thin layers as it passes through the spaces between the disks, toward the overflow annulus at the top center. The centrifugal action makes the particles to move radially outward until they reach the upper conical surface of the individual separating space between two disks. Once settled, the particles slide down in a cohesive layer toward the disk periphery and then into the bowl sludge space. Typical diameters of these centrifuges are between 140 and 1000 mm, and two adjacent disks are usually 0.3–2 mm apart (Hemfort 1984). Disk-stack centrifuges are versatile devices and find application in the dairy, starch, food, and pharmaceutical industries.

There are basically three types of disk-stack centrifuges and they are related to the method of solids removal. These are the solid bowl type, the

Figure 9 A disk with slanted caulks from a disk-stack centrifuge. (Courtesy of Alfa Laval Separation AB, Sweden.)

solids-ejecting type, and the nozzle type. In the solid bowl type, like the one shown in Fig. 6, the solids settle on the wall of the bowl and must be removed manually. Since the solids-holding capacity of these centrifuges is usually 5–20 L, their application is limited to suspensions with low concentrations (less than 1% by volume).

Figure 10 shows a solids-ejecting centrifuge. The intermittent solids ejection is achieved with an automated, periodic partial discharge of the sediment (Axelsson 1999). They can handle suspensions with concentrations up to 25% in volume. Solids-ejecting centrifuges are normally preferred for cell debris removal or for classifying centrifugation to separate inclusion bodies from cell debris. As inclusion bodies usually have higher settling velocities, they settle before the debris. These machines have a wide application, such as in breweries to separate yeast from the fermented broth, in the

Figure 10 Solids-ejecting centrifuge. (Courtesy of Alfa Laval Separation AB, Sweden.)

wine industry, in antibiotic and vaccine production, and in the recovery of bioproducts obtained from manipulated cells.

In the nozzle type, a concentrated suspension is continuously discharged through nozzles, as shown in Fig. 11. They are, therefore, fully continuous machines. The nozzle diameters range from 0.5 to 3 mm, the number of nozzles ranges from 4 in small bowls to 20 in large ones, and the underflow-to-throughput ratio varies from 5% to 50%. As the feed liquid splits in overflow (clarified) and underflow (concentrated), the concepts of reduced total efficiency and reduced grade efficiency may be used here. Nozzle-type machines can process feed suspensions with high concentrations of up to 30% by volume. Nozzle centrifuges are used, for instance, in alcohol distilleries that employ the batchwise Melle-Boinot process, in baker's yeast production, in rDNA yeast processes, and in citric acid production.

Bacteria are normally harvested using disk-stack centrifuges. The dry solid contents by weight achieved by these centrifuges are usually in the 10–15% range for solids-ejecting types and 5–10% for nozzle types. If the bacteria can be flocculated, the fermented broth can be preconcentrated in a

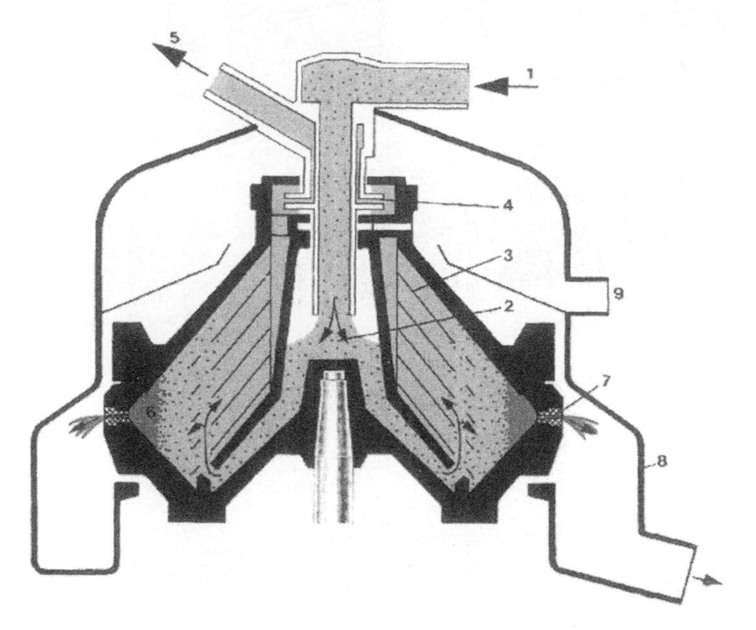

Figure 11 Nozzle type centrifuge. 1 feed, 2 inlet chamber, 3 disks, 4 centripetal pump, 5 clarified liquid discharge, 6 compaction zone, 7 nozzles, 8 concentrate catcher, 9 hood overflow. (Courtesy of Westfalia Separator AG, Germany.)

nozzle centrifuge followed by a further concentration in a scroll-type centrifuge. In this case, dry solids of 25–35% can be obtained.

Scroll-Type Centrifuges. Figure 12 shows a scroll-type machine, also known as a decanter centrifuge. It is equipped with a screw conveyor that continuously discharges the deposited solids. The bowl and the screw conveyor rotate in the same direction, but the latter rotates with a velocity slightly lower or higher than the former. This speed difference is usually 5–100 rpm (Svarovsky 1990). The bowl diameters range from 150 to 1200 mm, and the length/diameter ratios are usually 4–5. The scroll-type machine was developed to deal with large quantities of solids; therefore, they can handle suspensions with concentrations in the 3–60% range. Their conventional mechanical design limits the *g* factor to a maximum of 5000. Thus, their application in biotechnology is limited to the dewatering of biological sludges. Alfa Laval, a well-known centrifuges producer, has claimed that due to design innovations they were able to achieve *g* factors as high as 10,000. These innovations include suspended main bearings, a gear box dynamically decoupled from the bowl, and a swimming conveyor.

New Types of Centrifuges. Shear stress is always present in centrifugal separations in a relatively high degree. This poses a problem when dealing for instance with mammalian cells, which are shear sensitive. The Centritech centrifuge was developed aiming at shear stress minimization. It uses a sterilized and flexible plastic bag that is placed in a rotor and a "principle of the inverted comma" that requires no seals between rotating and no rotating parts. The operational *g* factor is only 100, which is enough to separate mammalian cells in these machines. The feed enters at one top end of the plastic bag and the overflow exits at the other top end. The cell-concentrated suspension is withdrawn intermittently through the

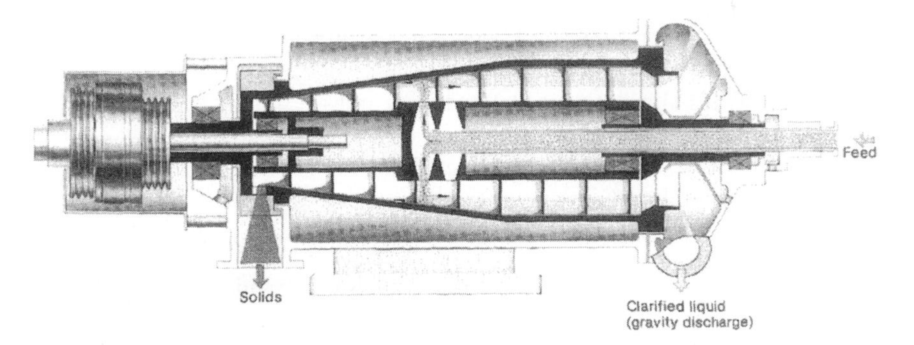

Figure 12 Decanter centrifuge. (Courtesy of Westfalia Separator AG, Germany.)

underflow pipe situated at the bottom end of the plastic bag. It has been reported that a high viable cell concentration of around 1.4×10^7 cells ml^{-1} could be achieved in a 50-L bioreactor performing perfusion culture of CHO cells (Apelman and Björling 1991). In another work, the growth rates and monoclonal antibody production were comparable to those obtained with a filtration-perfusion system when operating the centrifuge in an intermittent fashion (Johnson et al. 1996).

Another new design is the inverted chamber bowl used in the Powerfuge. These centrifuges operate with an intermittent solids discharge and are available in four different sizes, with bowl diameters varying from 150 to 600 mm. They can process from 0.01 to 6 $m^3\ h^{-1}$ of feed suspension and have a solids holding capacity from 0.9 to 60 L. The suspension is fed in the top of the machine and the clarified liquid is discharged through a centrate port. The separated solids are periodically removed during a fully automated scraping cycle. These centrifuges can operate with g factors of up to 20,000, except the largest size that can achieve $15{,}000 \times g$. The main applications of the Powerfuge are cell separation, blood plasma fractionation, and recovery of vaccines.

Grade Efficiency and Cut Size for Tubular Centrifuges

In order to derive an expression for the grade efficiency of tubular centrifuges, the following assumptions are made:

- The particles are homogeneously distributed at the entrance, i.e., at the bottom ($z = 0$) of the annulus formed by the R_1 and R_2 radii (see Fig. 13).
- The particle is a sphere with a smooth surface and the interaction between particles is neglected, i.e., there is no hindered settling.
- The particles settle within the Stokes region, and the end effects can be neglected.
- The particle is separated when it reaches the lateral wall of the centrifuge.
- There is no slip between the particles and the liquid flow in the axial direction.
- The liquid rotates at the same speed as the bowl.
- The liquid velocity profile in the liquid shell is uniform, i.e., the liquid flows as plug flow.

Figure 13 shows a schematic diagram of a tubular centrifuge. If a particle entering at $r = R_1$ and $z = 0$ reaches the wall at $z = L$, it will be separated with 100% efficiency. This particle is the smallest size to be separated with 100% grade efficiency and is known as critical size d_{100}. This

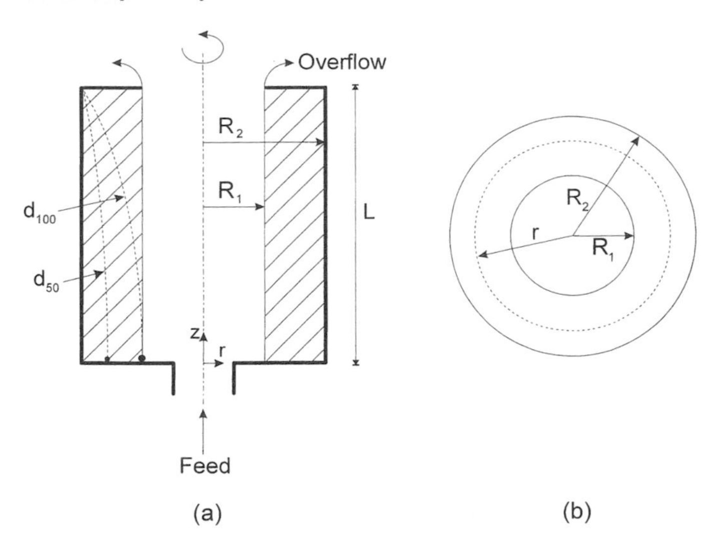

Figure 13 Schematic diagram of a tubular centrifuge showing the cut size d_{50} and the critical size d_{100}.

means that if this particle enters in any other position it will be more easily separated or that larger particles entering in $r = R_1$ and $z = 0$ will be separated before reaching $z = L$. Another interesting situation is the case of a particle that enters at the radius $r = R_{50}$ that divides in two equal parts the area of the annulus formed by R_1 and R_2. If this particle is collected in $z = L$, it will have 50% grade efficiency, since the particles of the same size that enter in $R_{50} < r \leq R_2$ will be collected and those entering in $R_1 \leq r < R_{50}$ will escape. This particle is known as cut size d_{50}. The trajectories of these two particles have been drawn in Fig. 13a. For the same reason, the particle that is collected with 20% efficiency is the one that will be collected in $z = L$ when entering the centrifuge in a position r that divides the area of the annulus in 20% between r and R_2 and 80% between R_1 and r. Therefore, the grade efficiency G of a particle of size d that is collected in $z = L$ when entering in a radius r is the fraction of the total annulus area lying between r and R_2, as given by Eq. (16).

$$G = \frac{R_2^2 - r^2}{R_2^2 - R_1^2} \tag{16}$$

The settling velocity of this particle is given by Eq. (1). As this velocity is a function of radial position, it is better written in a differential form. Therefore, replacing v by dr/dt and b by $\omega^2 r$ in Eq. (1):

$$\frac{dr}{dt} = \frac{(\rho_s - \rho)\omega^2 r d^2}{18\mu} \tag{1b}$$

The residence time t_r of this particle in the bowl is given by:

$$t_r = \frac{V_s}{Q} \tag{17}$$

where Q is the feed flow rate and V_s is the effective volume of the centrifuge, i.e., the volume of the liquid shell, as given by Eq. (18).

$$V_s = \pi(R_2^2 - R_1^2)L \tag{18}$$

Thus, this particle will have to travel the distance between r and R_2 in the available residence time. So, from Eq. (1b):

$$\int_r^{R_2} \frac{dr}{r} = \frac{(\rho_s - \rho)\omega^2 d^2}{18\mu} \int_0^{t_r} dt \tag{19}$$

Integrating Eq. (19), and replacing r and t_r as given by eqs. (16) and (17), respectively:

$$G = \frac{R_2^2}{R_2^2 - R_1^2}\left\{1 - \exp\left[-\frac{(\rho_s - \rho)\omega^2 V_s d^2}{9\mu Q}\right]\right\} \qquad \text{for } d \le d_{100} \tag{20a}$$

$$G = 1 \qquad \text{for } d > d_{100} \tag{20b}$$

The critical size d_{100} can be obtained from Eq. (20a), when $G = 1$:

$$d_{100} = \left[\frac{18\mu Q}{(\rho_s - \rho)\omega^2 V_s} \ln\left(\frac{R_2}{R_1}\right)\right]^{1/2} \tag{21}$$

And the cut size d_{50} can also be obtained from Eq. (20a), when $G = 0.5$:

$$d_{50} = \left[\frac{9\mu Q}{(\rho_s - \rho)\omega^2 V_s} \ln\left(\frac{2R_2^2}{R_2^2 + R_1^2}\right)\right]^{1/2} \tag{22}$$

From Equations (20) and (22) it is possible to obtain an expression for the curve $G(d/d_{50})$:

$$G = \frac{1}{1 - (R_1/R_2)^2}\left\{1 - \left[\frac{2}{1 + (R_1/R_2)^2}\right]^{-(d/d_{50})^2}\right\} \qquad \text{for } d \le d_{100} \tag{23a}$$

$$G = 1 \qquad \text{for d} > d_{100} \tag{23b}$$

If the size distribution of the biological particle in a given suspension and the desired total efficiency are known, the needed cut size can be obtained from Eqs. (14) and (23). Thus, the flow rate that will give this total efficiency can be obtained from Eq. (22). Unfortunately, this theoretical flow rate is always too high in comparison with the experimental value that truly gives the desired total efficiency. There are several reasons for such a deviation, and many of them are related to the initial assumptions made at the beginning of this item. Possibly, the main reason is the hypothesis of plug flow. According to Hemfort (1984), most of the liquid filling the shell seen in Fig. 13 is stationary with respect to the bowl. Only a thin cylindrical layer flows upward from the bottom and is discharged over the bowl overflow weir. This means that Eq. (17) overestimates the residence time and, consequently, the grade efficiency. A correction factor for the theoretical flow rate will be seen in the next item.

The Sigma Factor and the Scale-up of Centrifuges

The terms in Eq. (22) can be rearranged to give:

$$Q = 2\left[\frac{(\rho_s - \rho) g d_{50}^2}{18\mu}\right] \frac{\omega^2 V_s}{g \ln\left(\dfrac{2R_2^2}{R_2^2 + R_1^2}\right)} \tag{24}$$

The term inside the square brackets is the cut size settling velocity under the gravitational field, v_{g50} [see Eq. (1)], and its neighbor term is constant for a given centrifuge operating at a constant speed. Equation (24) can then be rewritten as:

$$Q = 2 v_{g50} \Sigma \tag{25}$$

where Σ is known as sigma factor, machine parameter, or theoretical capacity factor. As Eq. (24) shows, Σ is a function of the centrifuge geometry and its speed. Thus, the sigma factor for tubular centrifuges is given by:

$$\Sigma_{\text{tubular}} = \frac{\pi\left(R_2^2 - R_1^2\right) L \omega^2}{g \ln\left(\dfrac{2R_2^2}{R_2^2 + R_1^2}\right)} \tag{26}$$

The sigma factor has the dimension of an area and represents the area of a settling tank capable of giving the same total efficiency under the gravitational field. That explains why it is also known as equivalent clarification area. According to Svarovsky (1990), this interpretation of Σ is false since, due to Brownian diffusion and convection currents among others, the settling tank would hardly perform as well as a centrifuge.

Through a similar procedure, the sigma factor can also be derived for multichamber, disk-stack, and scroll-type centrifuges.

$$\Sigma_{\text{multichamber}} = \frac{\pi\omega^2 L}{3g} \sum_{i=0}^{i=n} \frac{R_{2i+2}^3 - R_{2i+1}^3}{R_{2i+2}^3 - R_{2i+1}^3} \tag{27}$$

where L is the height of the chambers, $n+1$ is the number of chambers, and indexes of R (radius) with even and odd numbers are related to inner and outer radii of the chamber, respectively.

$$\Sigma_{\text{disc}} = \frac{2\pi\omega^2 n\left(R_2^3 - R_1^3\right)}{3g\ tg\theta} \tag{28}$$

where θ is half cone angle of the disks, n is the number of disks, and R_1 and R_2 are the minimal and maximal radii of the disk, respectively.

$$\Sigma_{\text{scroll}} = \frac{\pi\omega^2}{4g}\left[2L_1\left(3R_2^2 + R_1^2\right) + L_2\left(R_2^2 + 3R_2R_1 + 4R_1^2\right)\right] \tag{29}$$

where L_1 and L_2 are the length of the cylindrical and the conical parts of the bowl, respectively, R_1 is the inner radius of liquid and R_2 is the inner radius of the cylindrical part.

An analysis of Eq. (25) shows that, when working with the same centrifuge at the same speed (constant Σ):

$$\frac{Q_2}{Q_1} = \frac{\left(v_{g_{50}}\right)_2}{\left(v_{g_{50}}\right)_1} = \left[\frac{(d_{50})_2}{(d_{50})_1}\right]^2 \qquad \text{for constant } \Sigma \tag{30}$$

Equation (30) shows that, when working with the same suspension at the same temperature, a reduction in flow rate leads to a decrease in cut size. This generates a higher grade efficiency for each diameter present in the feed size distribution [see, for instance, Eq. (23)]. According to Eq. (14), higher grade efficiencies produce higher total efficiencies. In other words, changes in feed flow rate have a great influence in performance. For instance, if $Q_2/Q_1 = 1/4$, $(d_{50})_2/(d_{50})_1 = 1/2$. Such a reduction in cut size will produce most probably a great increase in the collected solids (or a great decrease in the fraction of unsedimented solids). How large the increase is will be a function of the feed size distribution and the grade efficiency curve. As cells usually have narrow size distributions, small reductions in cut size can produce relatively high increments in efficiency. For instance, Higgins et al. (1978) obtained an increase in separation efficiency of *Escherichia coli* from 58% to 99.5% when decreasing the flow rate to 25% of its initial value. In resume, reductions in flow rate lead to higher efficiencies. This effect can also be understood through the following way: according to Eq. (17), a lower flow rate increases the residence time of the particles in the bowl, i.e., the particles have more time to settle.

Another way of increasing efficiency is to increase the particle settling velocity. According to equation (1), it is possible through increases in density difference, angular velocity and particle diameter (through, for instance, flocculation) or by viscosity reduction (achievable, for instance, through higher temperatures)

As mentioned at the end of last section, the theoretical flow rate calculated with Eq. (25) is always overestimated. Axelsson (1999) suggested the introduction of an efficiency factor to correct the theoretical flow rate given by Eq. (25). In his equation, instead of using v_{g50}, he uses v_{g100}, i.e., the settling velocity of the critical size d_{100}. Axelsson's equations are:

$$Q_{\text{theor}} = v_{g_{100}} \Sigma \tag{31}$$

and

$$Q_{\text{actual}} = \eta v_{g_{100}} \Sigma \tag{32}$$

The efficiency factors η suggested by Axelsson (1999) are within the ranges 0.90–0.98, 0.45–0.73, and 0.54–0.67 for tubular, disk-stack, and scroll-type centrifuges, respectively. As $(d_{50}/d_{100})^2 = 0.5$ for disk centrifuges (Svarovsky 1990), the theoretical flow rates calculated with Eqs. (25) and (31) are identical, for the same centrifuge and speed (constant Σ). That is not the case, for instance, for tubular centrifuges where, according to Eqs. (21) and (22), the ratio between the flow rates given by Eqs. (25) and (31) is a function of R_1 and R_2 and, for the same centrifuge, is usually in the 0.6–0.9 range.

Equation (25) or (31) can be used for scale-up purposes. For instance, data from a pilot centrifuge can be used to predict the flow rate of an industrial centrifuge in order to get the same performance. Equation (25) when applied to both centrifuges becomes:

Pilot centrifuge: $$Q_1 = 2\left(v_{g_{50}}\right)_1 \Sigma_1 \tag{33}$$

Industrial centrifuge: $$Q_2 = 2\left(v_{g_{50}}\right)_2 \Sigma_2 \tag{34}$$

To get the same performance, the cut size d_{50} in both centrifuges should be the same. Therefore, $(v_{g50})_1 = (v_{g50})_2$, hence:

$$\frac{Q_1}{\Sigma_1} = \frac{Q_2}{\Sigma_2} \tag{35}$$

The flow rate in the pilot machine should be adjusted to produce the desired total efficiency; then it is possible to calculate the flow rate for the industrial centrifuge, if both sigma factors are known. The same Eq. (35) can be obtained if Eq. (31) is used with the condition of both centrifuges producing the same critical size d_{100}.

Scale-up of centrifuges of the same type using Eq. (35) is fairly reliable (Svarovsky 1990), and the results are better when geometrically similar machines are compared (Axelsson 1999). These limitations are mainly due to the assumption that to produce the same performance both centrifuges should give the same cut size. This is a necessary condition, but not sufficient. As Eq. (14) shows, a way to guarantee the same total efficiency is by having the same size distribution y and the same grade efficiency G. The first condition is attended when working with the same suspension. To satisfy the second condition, not only must the cut size be the same, but the centrifuges must be geometrically similar [see Eq. (23a)]. In resume, Eq. (35) can be used to compare the performances of geometrically similar centrifuges. If this is not the case, but the centrifuges are of the same type, it can be used with caution. However, it should not be used to predict performance of machines of different types.

Since Svarovsky (1990) stated that the only way to fully describe the performance of a sedimenting centrifuge is by the grade efficiency curve, only a few groups have been using this concept (Wong et al. 1997; Maybury et al. 1998). It would also be good to see the manufacturers of centrifuges adopting this concept.

Test Tube Centrifugation

At the early stage of a bioprocess development, the amount available of biological material is usually not enough to permit tests in a pilot centrifuge. At this stage, test tube centrifugation can be carried out as a preliminary test to determine the degree of separability of a given biological particle suspension. This spin test is usually done at different settling times and speeds, and the result is one or more sets of values that give an acceptable supernatant. It also gives an indication of solids compressibility and rheology, and dry solids content after centrifugation (Axelsson 1999).

Figure 14 shows a schematic view of a swing-out test tube (bottle) centrifuge. As the settling velocity is a function of radial position, from Eq. (1b):

$$\frac{dr}{dt} = \frac{(\rho_s - \rho)\omega^2 r d^2}{18\mu} = \frac{\omega^2 r}{g} v_g \tag{1b}$$

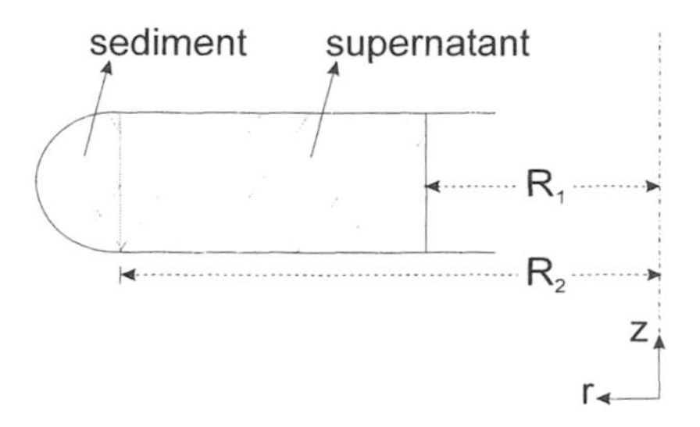

Figure 14 Schematic view of a test tube of a swing-out test tube (bottle) centrifuge.

Supposing R_2 is the radius at the sediment surface and the particles are well distributed in the test tube, a particle with cut size d_{50} that is located in the test tube surface ($r = R_1$) at the beginning of the centrifugation ($t = 0$) will have to travel half the distance between R_1 and R_2 in the centrifugation time (t_c). Thus:

$$\int_{R_1}^{(R_1+R_2)/2} \frac{dr}{r} = \frac{\omega^2}{g} v_{g_{50}} \int_0^{t_c} dt \tag{36}$$

Then:

$$v_{g_{50}} = \frac{g}{\omega^2 t_c} \ln\left(\frac{R_1 + R_2}{2R_1}\right) \tag{37}$$

From Eq. (25):

$$\frac{Q}{\Sigma} = \frac{2g}{\omega^2 t_c} \ln\left(\frac{R_1 + R_2}{2R_1}\right) \tag{38}$$

An equation based on the critical size d_{100} can also be formulated. This particle will have to travel from R_1 until R_2 in the centrifugation time; therefore, from Eq. (36):

$$v_{g_{100}} = \frac{g}{\omega^2 t_c} \ln\left(\frac{R_2}{R_1}\right) \tag{39}$$

From equation (31):

$$\frac{Q}{\Sigma} = \frac{g}{\omega^2 t_c} \ln\left(\frac{R_2}{R_1}\right) \tag{40}$$

Q can be understood here as the ratio V/t_c, where V is the volume of material in the centrifuge tube.

For the same centrifuge and speed, i.e., for the same Σ, Eq. (38) usually gives flow rates from 5% to 15% higher than Eq. (40).

A similar procedure applied to test tube centrifuge with angle head (see Fig. 15) yields equations (41) and (42) based on d_{50} and (43) and (44) based on d_{100}.

$$v_{g_{50}} = \frac{g}{\omega^2 t_c} \ln\left(1 + \frac{D}{2R_1 \cos\gamma}\right) \tag{41}$$

$$\frac{Q}{\Sigma} = \frac{2g}{\omega^2 t_c} \ln\left(1 + \frac{D}{2R_1 \cos\gamma}\right) \tag{42}$$

$$v_{g_{100}} = \frac{g}{\omega^2 t_c} \ln\left(1 + \frac{D}{R_1 \cos\gamma}\right) \tag{43}$$

$$\frac{Q}{\Sigma} = \frac{g}{\omega^2 t_c} \ln\left(1 + \frac{D}{R_1 \cos\gamma}\right) \tag{44}$$

where D is the test tube diameter and γ is angle of the bottle relative to the vertical axis.

For the same centrifuge and speed, i.e., for the same Σ, Eq. (42) usually gives flow rates from 5% to 15% higher than Eq. (44).

Maybury et al. (1998) adapted the sigma theory aiming at predicting the performance of a disk-stack centrifuge based on tests with a bench-top centrifuge. Taking into account the acceleration and deceleration stages present in the test tube centrifuge, they found a "corrected" equation similar to Eq. (38). Based on their equation, they could predict well the performance of the industrial centrifuge when treating suspensions of polyvinyl acetate.

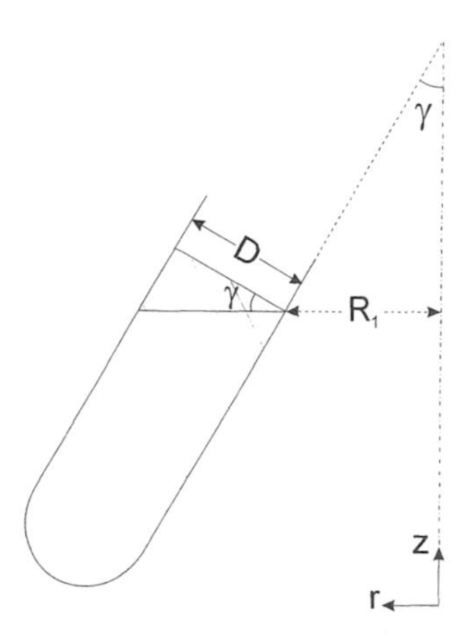

Figure 15 Schematic view of a test tube centrifuge with angle head.

For yeast cell debris, the predictions were good in the high-capacity zone of the industrial centrifuge. However, there was some evidence of overprediction for small capacities. For protein precipitates, the test tube centrifuge always overpredicted the efficiency given by the disk-stack centrifuge. This difference was attributed to the most likely shear-related damage of sensitive precipitate particles in the high shear stress regions of the disk-stack centrifuge.

2.3 Hydrocyclones

A hydrocyclone is a very simple equipment, as shown in Fig. 16. It consists of a conical section joined to a cylindrical portion, which is fitted with a tangential inlet and closed by an end plate with an axially mounted overflow pipe, also called vortex finder. The end of the cone terminates in a circular

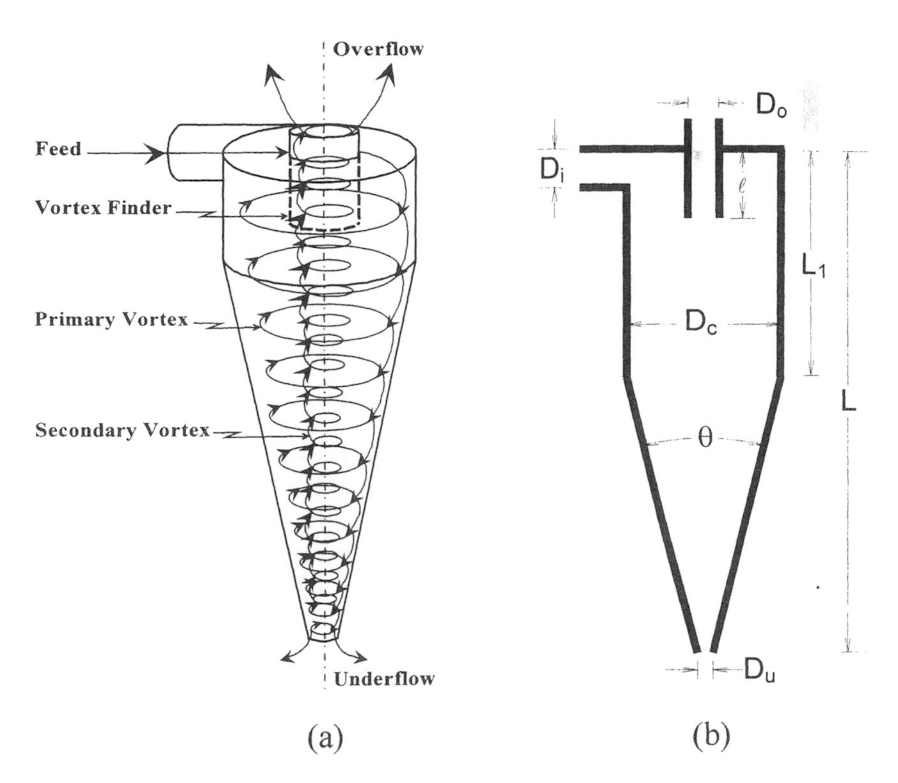

Figure 16 Perspective view of a hydrocyclone showing the internal flow (a), and a schematic view showing the main dimensions (b).

apex opening, called underflow orifice. Despite its simplicity, hydrocyclones are very efficient when promoting solid-liquid separations.

Unlike centrifuges, which use the same separation principle (sedimentation in a centrifugal field), hydrocyclones have no moving parts and the vortex motion is performed by the fluid itself. As Fig. 16a shows, the feed is introduced tangentially through the inlet duct into the upper part of the cylindrical section, acquiring a strong swirling downward movement. In hydrocyclones designed for solid-liquid separations, the underflow orifice is smaller than the overflow diameter; therefore, only a fraction of the feed liquid escapes through the underflow, carrying the coarser (or denser) particles. Most of the flow reverses its vertical direction and goes up in an even stronger vortex motion and out through the overflow pipe, carrying the smaller (or lighter) particles.

Based on correlations published in the literature and reviewed in books (Bradley 1965; Svarovsky 1984; Heiskanen 1993), it is possible to establish the effect of geometrical (see Fig. 16b) and operational variables on capacity and cut size of hydrocyclones, as shown in Table 4 (Matta and Medronho 2000). For instance, an increase in underflow diameter promotes a small increase in capacity and a relatively high reduction in cut size. As a reduction in cut size leads to an increase in total efficiency, it is possible to change the hydrocyclone efficiency by simply changing the underflow diameter, without great changes in flow rate.

Hydrocyclones were originally designed to promote solid-liquid separations, but nowadays they are also used for solid-solid, liquid-liquid, and gas-liquid separations. Potential new applications, such as cell separation, are being developed, and some works can be found in the literature.

Table 4 Effect of Increases in the Values of Geometrical and Operational Variables on Capacity and Cut Size of Conventional Hydrocyclones[a]

	D_c	D_i	D_o	D_u	L	ℓ	θ	ΔP	C_v
Capacity	+ + +	+ +	+ +	+	+	—	—	+ +	—[c]
Cut size[b]	+ +	+ +	+ +	− −	−	+	+	− −	+ +

[a] +, increase; −, reduction. D_c, hydrocyclone diameter; D_i, feed inlet diameter; D_o, overflow diameter; D_u, underflow diameter; L, hydrocyclone length; ℓ, vortex finder length; θ, angle of the hydrocyclone cone; ΔP, pressure drop; C_v, feed volumetric concentration.

[b] Reductions in cut size mean higher efficiencies and vice versa.

[c] Flow rate increases slightly with feed concentration from 0% to 3–5% and then decreases continuously.

These works refer to separation of mammalian cells (Lübberstedt et al. 2000a,b), separation of yeast either from fermented broths or from water suspensions (Rickwood et al. 1992; Yuan et al. 1996a,b; Harrison et al. 1994; Cilliers and Harrison 1996, 1997), separation of yeast from filter aids (Rickwood et al. 1996; Matta and Medronho 2000), and separation of biological sludge in wastewater treatments (Thorwest and Bohnet 1992; Ortega-Rivas and Medina-Caballero 1996; Bednarski 1996; Marschall 1997; Müller 2000).

The use of hydrocyclones for separating mammalian cells from the culture medium opens the possibility of using them to perform perfusion in bioreactors. In contrast to most solid particles, which are insensitive to shear, mammalian cells can be damaged by the relatively high levels of shear stress existing in hydrocyclones. Figure 17 shows the influence of pressure drop on the viability of HeLa cells in the underflow and overflow of three different hydrocyclones (Lübberstedt et al. 2000a). It can be seen that in the underflow of the three hydrocyclones the viability did not decrease with pressure drops up to 4 bar, and in the overflow, it started to decrease for pressure drops higher than 3.5 bar. Therefore, in spite of the relatively high values of shear stresses generated inside hydrocyclones, the cells appear to resist up to a certain limit of pressure drop. This is probably due to the extremely low average residence time of the cells inside the equipment, which lies in the range between 0.03 and 0.1 s. Since hydrocyclones are low-cost equipment and, for this application, their maintenance costs would be virtually nonexistent, their use in perfusion would decrease both capital and operational costs (Lübberstedt et al. 2000b). They could be also used, in some applications, as a pretreatment for a harvesting process.

The performance of hydrocyclones can be estimated with the following equations (Coelho and Medronho 2001):

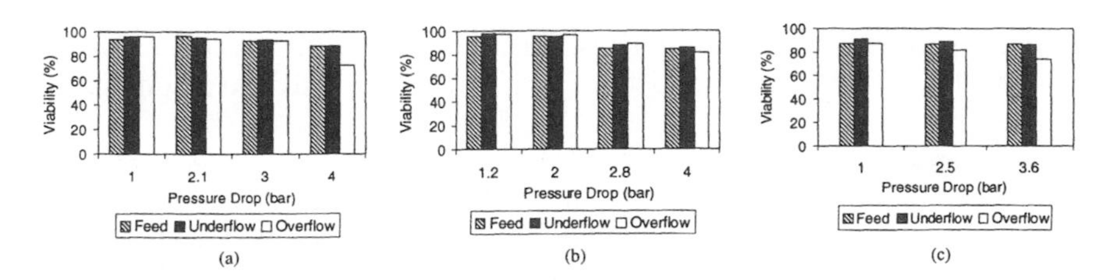

Figure 17 Influence of pressure drop on cell viability in the underflow and overflow of the following hydrocyclones: Mozley (a), Dorr-Oliver (b), and Bradley (c) with diameters of 10 mm, 10 mm, and 7 mm, respectively (Lübberstedt et al. 2000a).

$$\text{Stk}_{50}\text{Eu} = 0.12\left(\frac{D_c}{D_o}\right)^{0.95}\left(\frac{D_c}{L-\ell}\right)^{1.33}[\ell n(1/R_f)]^{0.79}\exp(12.0\,C_v) \tag{45}$$

$$\text{Eu} = 43.5D_c^{0.57}\left(\frac{D_c}{D_i}\right)^{2.61}\left(\frac{D_c}{D_o^2+D_u^2}\right)^{0.42}\left(\frac{D_c}{L-\ell}\right)^{0.98}\text{Re}^{0.12}\exp(-0.51C_v) \tag{46}$$

$$R_f = 1.18\left(\frac{D_c}{D_o}\right)^{5.97}\left(\frac{D_u}{D_c}\right)^{3.10}\text{Eu}^{-0.54} \tag{47}$$

$$G' = 1 - \exp\left[-0.693\left(\frac{d}{d'_{50}}\right)^3\right] \tag{48}$$

where the symbol definitions can be found in Table 4. Stk_{50}Eu is the product between the Stokes number Stk_{50} and the Euler number Eu, Re is the Reynolds number, and R_f is the flow ratio given by Eq. (6). Stk_{50}Eu, Eu, and Re are given by Eqs. (49) to (51), respectively.

$$\text{Stk}_{50}\text{Eu} = \frac{\pi(\rho_s-\rho)\Delta P\,D_c\left(d'_{50}\right)^2}{36\mu\rho Q} \tag{49}$$

$$\text{Eu} = \frac{\pi^2\,\Delta P D_c^4}{8\rho Q^2} \tag{50}$$

$$\text{Re} = \frac{4\rho Q}{\pi\mu D_c} \tag{51}$$

Equations (45), (47), and (48) are totally based on dimensionless groups and dimensionless variables, and Eqs. (49) to (51) are the definitions of well-known dimensionless groups. Therefore, these equations can be used with any coherent system of units as, for instance, the International System of Units (SI). This is not the case for Eq. (46), where only the SI units may be employed.

Equations (45) to (48) can be used for performance prediction of hydrocyclones. For instance, the reduced cut size and the flow rate can be calculated based on Eqs. (45) and (49); and (46), (50), and (51), respectively, leading to Eqs. (52) and (53).

$$d'_{50} = \frac{1.17D_c^{0.64}}{D_o^{0.48}(L-\ell)^{0.67}}\left[\frac{\mu\rho Q}{(\rho_s-\rho)\Delta P}\right]^{0.50}\left[\ell n\left(\frac{1}{R_f}\right)\right]^{0.40}\exp(6.0C_v) \tag{52}$$

$$Q = 0.18_D c^{-0.22} D_i^{1.23} (D_o^2 + D_u^2)^{0.20} (L - \ell)^{0.46} \mu^{0.057} \rho^{-0.53} \Delta P^{0.47} \exp(0.24 C_v) \quad (53)$$

Equation (52) can be used with any coherent system of units as, for instance, the SI units. However, Eq. (53) must be used only in SI units.

If the hydrocyclone dimensions and the suspension data (liquid and particle densities, liquid viscosity, concentration of solids, and particle size distribution) are known, the hydrocyclone performance can be predicted according to the following steps:

1. Select the pressure drop to be used.
2. Calculate the flow rate using Eq. (53).
3. Calculate Eu and R_f using Eqs. (50) and (47), respectively.
4. Calculate d'_{50} using Eq. (52).
5. Calculate the predicted reduced total efficiency E'_T using the particle size distribution of the feed suspension and Eqs. (15) and (48).
6. Calculate the total efficiency E_T using Eq. (7).

Manufacturers of hydrocyclones produce only a limited range of cyclone diameters and, in order to be able to cover a wide range of cut sizes and flow rates, each cyclone of a given size can be operated with different openings sizes (inlet, overflow and underflow) through the use of interchangeable parts. This approach requires accurate knowledge of how geometrical variables affect the equipment performance. An alternative approach is to use a custom-made hydrocyclone based on a geometrically similar family. Although a hydrocyclone is a very simple apparatus to build, this approach is not widely used. A simple procedure to design two well-known families of geometrically similar hydrocyclones (Castilho and Medronho 2000) has been recently proposed.

3. FILTRATION

Filtration is the separation of most of the solids present in a solid-fluid mixture by forcing the fluid through a porous barrier known as filter medium or septum, which can be a cloth, membrane, paper, screen, or a bed of solids. The conventional filtration of biological particles suspended in a liquid will be treated in this chapter. The use of membranes for protein isolation will be seen in Chap. 7. The conventional filtration can be divided into two main types—deep-bed filtration and cake filtration—as shown in Fig. 18. In the former, the particles are smaller than the porous diameters of the filter medium. Therefore, they penetrate into the pores of the bed and are

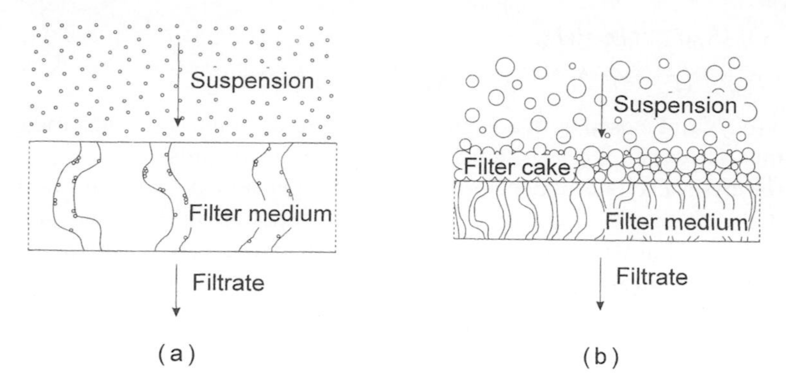

Figure 18 Schematic view of a deep-bed filtration (a) and of a cake filtration (b).

collected due to a series of mechanisms, like gravity, diffusion, and inertia (Svarovsky 1990). In cake filtration, the particles larger than the porous diameters are collected over the surface of the filter medium forming a cake, which has pores even smaller than those of the medium. Therefore, the cake acts as a porous medium for the subsequent incoming suspension, capturing even smaller particles. Cake filtration has a wider application in biotechnology than deep-bed filtration. Hence, only the former will be seen here. More detailed information on filtration can be found in more specialized books (Meltzer and Jornitz 1998; Dickenson 1997; Rushton et al. 1996; Svarovsky 1990).

Cake filtration of fermented broths is not an easy task. Biological particles are normally highly compressible. The broth usually also contains colloidal and submicrometer particles, which usually play an important role in determining the filtration properties. Therefore, the main problems of filtering fermented broths are the blinding of the filter medium due the presence of submicrometer particles and the blinding of the cake due to the microorganism's high compressibility. To overcome these problems, some powders with special properties are added to the system, improving the filtration characteristics of the original suspension. These powders are called filter aids and will be described in detail in Sec. 3.5.

If necessary, after filtration, the cake may be washed to remove liquid containing dissolved materials in order either to clean the cake from soluble contaminants or to recover product retained in the cake pores.

3.1 Theory of Filtration

In spite of filtration being a straightforward procedure, the theory underlying it is very complex. The great majority of biological solids produce compressible cakes that deform themselves continuously under pressure. Therefore, cake porosity ε and permeability k are a function of time and position, i.e., in a given point, ε and k decrease with time, having lower values near the filter medium than away from it. Apart from that, the cake is a porous medium that grows as filtration proceeds. Hence, filtration is an unsteady problem with moving boundaries. Regarding the filter medium, its resistance can increase with time due to some penetration of fine solids into the medium. This resistance may also change with changes in pressure drop, due to the compression of the fibers that compose the medium. However, the medium resistance can always be considered approximately constant, since these small variations in resistance are negligible in comparison with the total resistance offered by the cake and medium together. Use of the conservation equations (continuity and motion equations) to solve such a problem is extremely difficult not only due to the resulting nonlinear equations but also due to the unreliability of the constitutive equations. A way to overcome these difficulties is through the simplified theory of filtration, which incorporates the following assumptions:

Plane deformation of the compressible cake.
One-directional flow of incompressible Newtonian fluid.
"Darcyan" flow, i.e., low fluid velocities.
Fluid velocities inside the cake are much larger than solid velocities; consequently, they are independent of position.
In the motion equation, the acceleration, the viscous force, and the gravitational force are negligible in comparison with the pressure force and the force the fluid exerts on the solids.
Cake porosity and permeability in a given point are a function of the pressure on the solids P_s, which is defined as the difference between the pressures in the cake face and in that point ($P_s = P_\ell - P$).

Figure 19 gives a detailed schematic view of a cake filtration. In this figure, ΔP_c and ΔP_m are the pressure drops across the cake and the filter medium, respectively, ℓ_c and ℓ_m are the cake and medium thickness, A is the filtration area, and dx is a thin slice of the cake.

The equation of motion for the fluid is given by:

$$\rho\left\{\varepsilon\frac{\partial}{\partial t}\left(\frac{\vec{q}}{\varepsilon}\right)+\left[\nabla\left(\frac{\vec{q}}{\varepsilon}\right)\right]\vec{q}\right\} = -\nabla P + \nabla\cdot\vec{\tau} + \frac{\mu}{k}\vec{q} + \rho\vec{q} \qquad (54)$$

where ρ and μ are the fluid density and viscosity, respectively, ε and k are the cake porosity and permeability, respectively, t is time, q is the average fluid velocity ($q = Q/A$), Q is the volumetric flow rate, τ is the stress tensor, and g is the gravity acceleration.

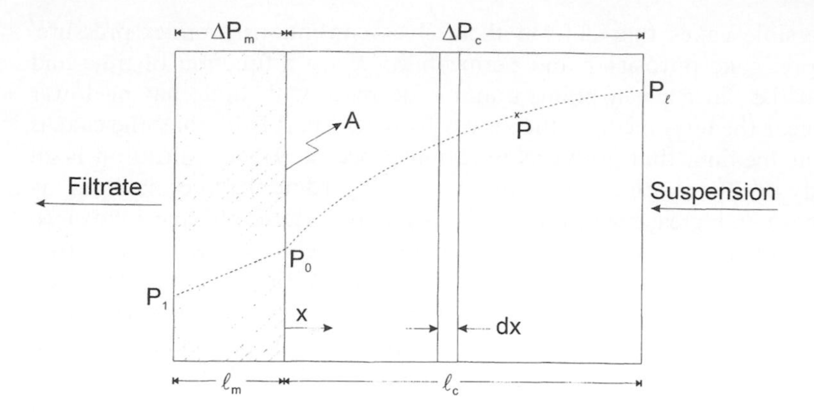

Figure 19 Detailed schematic view of a cake filtration. (Symbol definitions can be found in the text.)

Equation (55) is obtained if the assumptions listed above are applied to equation (54). This equation is known as Darcy's law.

$$\frac{dP}{dx} = \frac{\mu q}{k} \tag{55}$$

The mass of solids dm in a cake slice of thickness dx is given by equation (56), where ρ_s is the density of the solids.

$$dm = \rho_s(1 - \varepsilon)\mathrm{A}dx \tag{56}$$

Replacing dx in equation (55) by equation (56), assuming that P_ℓ is constant, replacing dP by $-d(P_\ell - P)$, and integrating the resulting equation from $P = P_0$ to $P = P_\ell$:

$$\Delta P_c = \frac{\langle\alpha\rangle \mu q m}{A} \tag{57}$$

where m is the total mass of solids in the cake and $< \alpha >$ is the average specific cake resistance given by Eq. (58).

$$\langle\alpha\rangle = \frac{\Delta P_c}{\displaystyle\int_0^{\Delta P_c} \frac{dP_s}{\alpha}} \tag{58}$$

and $\alpha = [k\rho_s(1 - \varepsilon)]^{-1}$ is the specific cake resistance.

Assuming that the volume of liquid entrapped in the cake is negligible in comparison with the volume of filtrate:

$$m \cong \rho c V \tag{59}$$

where V is the volume of filtrate and c is the solid concentration in the feed expressed as mass of solids by mass of liquid.

Substitution of Eq. (59) into Eq. (57) gives:

$$\Delta P_c = \frac{\langle\alpha\rangle \mu q \rho c V}{A} \tag{60}$$

Equation (61) can be obtained if Eq. (55) is applied to the filter medium and integrated from $P = P_0$ to $P = P_1$ for $x = 0$ to $x = -\ell_m$.

$$\Delta P_m = R_m \mu q \tag{61}$$

where $R_m = \ell_m / k_m$ and k_m are the resistance and permeability of the filter medium, respectively.

The total pressure drop ΔP can be obtained adding the cake pressure drop with the pressure drop in the filter medium, both expressed by Eqs. (60) and (61), respectively.

$$\Delta P = \mu q \left(\frac{\langle\alpha\rangle \rho c V}{A} + R_m \right) \tag{62}$$

but:

$$q = \frac{Q}{A} = \frac{1}{A}\frac{dV}{dt} \tag{63}$$

then:

$$\frac{dt}{dV} = \frac{\mu}{A \Delta P} \left(\frac{\langle\alpha\rangle \rho c V}{A} + R_m \right) \tag{64}$$

Equation (64) is the general filtration equation for compressible cakes. The same result is obtained if the cake is incompressible. In this case, as the specific cake resistance is constant, $< \alpha >= \alpha$.

For compressible cakes, the specific cake resistance increases with pressure drop according to the following empirical equation:

$$\langle\alpha\rangle = \alpha_o (\Delta P)^n \tag{65}$$

where α_o is the average specific cake resistance at unit applied pressure drop and n is a compressibility index, which is zero for incompressible cakes and varies between 0.3 and 1.0 for most cakes of biological particles. Table 5 gives the values of α_o and n for some microorganisms. Nakanishi et al. (1987) found that the $\langle\alpha\rangle$ value is not only a function of the microorganism size, but also more strongly of its shape. They explain this dependency as being related to the difference in porosity obtained as a function of shape.

Table 5 Values of the Average Specific Cake Resistance at Unit Applied Pressure Drop α_o and Compressibility Index n for Some Microorganisms (Nakanishi et al. 1987)

Microorganisms	Morphology	α_o at 1 bar (10^{12} m kg^{-1})	n
Baker's yeast	Elliptical	0.4	0.45
Micrococcus glutamicus	Elliptical	3.8	0.31
Bacillus circulans	Rod shaped	30.0	1.00
Rhodopseudomonas spheroides	Rod shaped	54.0	0.88
Escherichia coli	Rod shaped	670.0	0.79

3.2 Constant Pressure Filtration

Filtration is normally carried out at constant pressure, as is the case in vacuum filtration, or at constant flow rate (to be seen in the next item). If the filtration is conducted at constant pressure drop, Eq. (64) can be integrated to give:

$$\frac{t}{V} = \frac{\mu}{A\Delta P}\left(\frac{\langle\alpha\rangle\rho c V}{2A} + R_m\right) \tag{66}$$

Equation (66) is a straight line, if the experimental points of t/V are plotted against V, with the following gradient and intercept:

$$\text{Gradient} = \frac{\mu\langle\alpha\rangle\rho c}{2A^2\Delta P} \tag{67}$$

$$\text{Intercept} = \frac{\mu R_m}{A\Delta P} \tag{68}$$

The average specific cake resistance $\langle\alpha\rangle$ and the filter medium resistance R_m can, therefore, be easily calculated. Filtration at the desired pressure drop should be carried out and the volumes of filtrate at different times measured. With the experimental values of t and V, the values of t/V should be calculated; then a linear regression should be done using t/V as the dependent variable and V as the independent variable. The values of $\langle\alpha\rangle$ and R_m could be then calculated using Eqs. (67) and (68). Figure 20 shows a plot of t/V against V for the filtration of a pectinase extract from fermented solids (solid-state fermentation of wheat bran by *Aspergillus niger*). The filtration was conducted (Ferreira et al. 1995) in a vacuum filter leaf with an area of 78.5 cm^2, under a pressure drop of 0.6 bar, and using perlite as filter aid (precoat-bodyfeed method, as described in Sec. 3.5). The solids concentra-

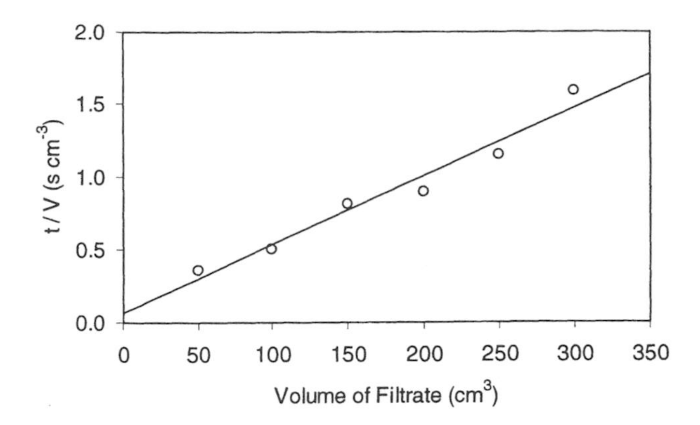

Figure 20 Filtration of pectinase extract from fermented solids (solid-state fermentation of wheat bran by *Aspergillus niger*) using diatomaceous earth as filter aid (precoat-bodyfeed method) in a vacuum filter leaf, under a constant pressure drop of 0.6 bar (Ferreira et al., 1995).

tion was 0.084 g of solids per gram of extract. The calculated resistances were $\langle\alpha\rangle = 4.2 \times 10^{10}$ cm g^{-1} and $R_m = 3.1 \times 10^8$ cm^{-1}.

It should be pointed out that at the very beginning of a constant pressure filtration, since there is no cake formation yet, the resistance to the flow is minimal. Therefore, at this point ($t = 0$), the filtrate flow rate is maximal. As soon as the flow starts, cake begins to be formed on the filter medium, continuously increasing the flow resistance. The net effect is, thus, a continuously decreasing flow rate.

3.3 Constant Rate Filtration

If a positive displacement pump is used to promote filtration, the flow rate through the filter will be constant. In this case, dt/dV in Eq. (64) is constant, so that

$$\frac{dt}{dV} = \frac{t}{V} = \frac{1}{Q} \tag{69}$$

Under this condition, Eq. (64) can be rearranged as follows:

$$\Delta P = \frac{\mu Q}{A}\left(\frac{\langle\alpha\rangle \rho c Q t}{A} + R_m\right) \tag{70}$$

Equation (70) is a straight line, if the experimental points of pressure drop ΔP are plotted against time, with the following gradient and intercept:

$$\text{Gradient} = \frac{\mu\langle\alpha\rangle\rho c Q^2}{A^2} \tag{71}$$

$$\text{Intercept} = \frac{\mu Q R_m}{A} \tag{72}$$

The resistances $\langle\alpha\rangle$ and R_m can then be calculated through a filtration experiment, in a way similar to that in the constant-pressure filtration case.

In a constant rate filtration, the pump delivers a constant flow rate against an increasing cake resistance due to the continuous growth of the cake thickness. In order to give a constant flow rate, the pressure delivered by the pump must also rise with time. Equation (70) shows that this growth in ΔP is linear with time. Since this pressure can rise considerably, most filters that use constant-rate filtration have a pressure relief valve in their pumping system. Figure 21 shows a plot of ΔP against t for yeast filtration at constant flow rate using perlite as filter aid (precoat-bodyfeed method). The filter used was a plate-and-frame press. The calculated resistances were $\langle\alpha\rangle = 6.6 \times 10^{10}$ cm g^{-1} and $R_m = 8.6 \times 10^9$ cm^{-1}.

3.4 Filtration Using a Centrifugal Pump

Most pressure filtration systems use a centrifugal pump for delivering the slurry to the filter. In this situation, the filtration follows neither a constant-pressure filtration nor a constant-rate filtration. This occurs because the flow rate delivered by the pump is a function of the resistance offered by the system, as shown schematically in Fig. 22. At the beginning of the filtration,

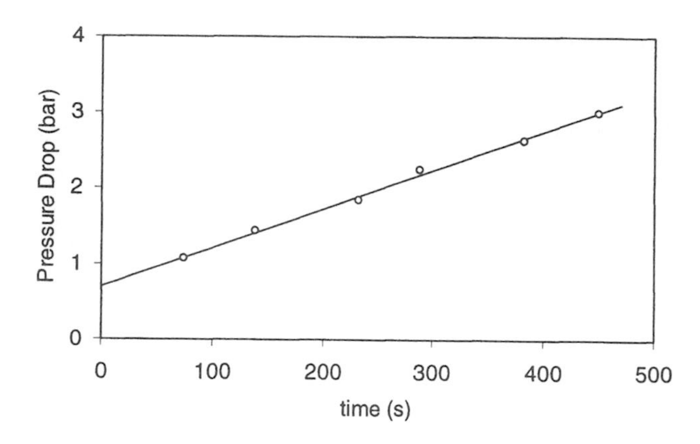

Figure 21 Filtration of yeast using perlite as filter aid (precoat-bodyfeed method) in a plate-and-frame press at constant flow rate.

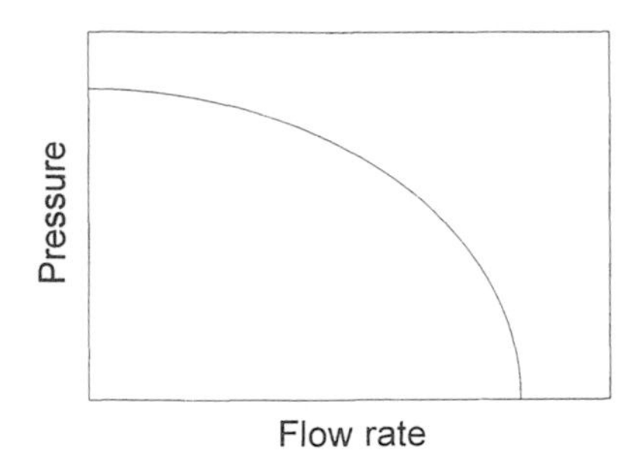

Figure 22 Typical centrifugal pump characteristic.

there is no cake, so the system resistance is at its minimum and the flow rate is at its maximum. As cake starts to build up, the resistance grows and, thus, the flow rate starts to decrease continuously as the pressure drop increases, following the pump characteristic. In other words, the time to filter a given volume of filtrate is a function of the flow rate, which is a function of the pressure drop.

Assuming that the cake volume is negligible compared to the volume of filtrate V, it is possible to write the instantaneous flow rate as:

$$Q = \frac{dV}{dt} \tag{73}$$

Therefore:

$$t = \int_0^V \frac{dV}{Q} \tag{74}$$

The relationship between V and Q given by equation (64) can be rewritten as:

$$\frac{\Delta P}{Q} = \frac{\mu \langle \alpha \rangle \rho c}{A^2} V + \frac{\mu R_m}{A} \tag{75}$$

where the relationship between ΔP and Q is given by the pump characteristics. Thus, through the use of equations (74) and (75) and the pump characteristics, it is possible to calculate the time needed for filtering a desired volume V of filtrate. This is not a straightforward procedure, since the integral expressed by equation (74) must be, in general, numerically solved. However, if a good centrifugal pump is being used (one that possesses a relatively flat pump characteristic curve in the low Q region) and if the pressure drop rapidly attains a constant value, the assumption of pressure constant filtration can be used, simplifying the problem. Figure 23

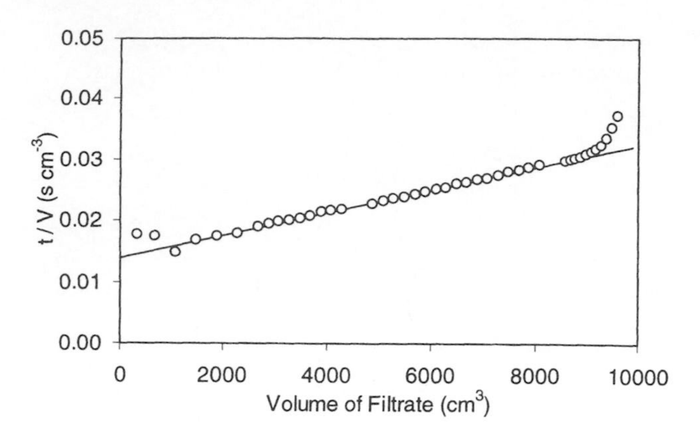

Figure 23 Filtration of yeast using a centrifugal pump and diatomaceous earth as filter aid (precoat-bodyfeed method) in a plate-and-frame press.

shows a plot of t/V against V for yeast filtration using diatomaceous earth as filter aid (precoat-bodyfeed method). A plate-and-frame filter press with a filtration area of 894 cm^2 was employed (pilot filter with square plates and frames of 15.24 cm), at a pressure drop of 3.0 bar. The solids concentration was 0.15 g L^{-1}. As can be seen in this figure, the initial and the final points do not lie in a straight line and must be discarded from the linear regression. The initial points do not fit to a straight line because at the beginning the pressure drop is not constant, so Eq. (66) does not describe this initial stage. At the end, the cake fills all the frames and, consequently, it becomes more and more difficult to filter a given ΔV. In other words, the time to filter a given ΔV increases exponentially. The calculated resistances, discarding the initial and final points, were $\langle\alpha\rangle = 6.6 \times 10^{10}$ cm g^{-1} and $R_m = 8.6 \times 10^9$ cm^{-1}.

The procedure described above is only valid for filtrations where the pressure drop reaches a constant level in a reasonably low time, which should be not superior to 10% of the filtration time. If that is not the case, Eq. (64) may be integrated from a point t_1, V_1 at the beginning of the truly constant period:

$$\int_{t_1}^{t} dt = \frac{\mu\langle\alpha\rangle\rho c}{A^2\Delta P}\int_{V_1}^{V} V dV + \frac{\mu R_m}{A\Delta P}\int_{V_1}^{V} dV \tag{76}$$

Then:

$$\frac{t - t_1}{V - V_1} = \frac{\mu\langle\alpha\rangle\rho c}{2A^2\Delta P}(V + V_1) + \frac{\mu R_m}{A\Delta P} \tag{77}$$

Equation (77) is a straight line when $(t - t_1)/(V - V_1)$ is plotted against $V + V_1$. The resistances $\langle\alpha\rangle$ and R_m may, then, be evaluated through a linear regression.

3.5 Filter Aids

Most fermented broths tend to blind the filter medium due to the presence of submicrometer particles and/or to blind the cake due to the high compressibility of microorganisms. The best way to solve these problems is to use filter aids, which are particles that can protect the filter medium and/or increase the permeability and porosity and decrease the compressibility of the cake. They are used extensively in the chemical industry to aid the filtration of solids that tend to form compressible cakes. In biotechnology, filter aids find application in the antibiotic, beer, and wine industries; in serum purification; in *in vitro* preparations; in enzyme isolation; in plasmid DNA recovery; and so on.

Filter aids are used in two ways: precoat and/or body feed. The precoat method is used to protect the filter medium against fouling due to colloidal particles, since such particles would penetrate into the filter medium, increasing its resistance and eventually blinding it. This can be avoided by initially filtering a diluted suspension (e.g., 0.5% by weight) containing only filter aid, which will deposit on the filter medium. Recirculating the filter aid slurry is recommended since the coarser particles will be first retained on the septum followed by the small ones. Average precoating velocities should be less than 7×10^{-4} m s^{-1} and the pressure drop should be at least 7 kPa. For higher viscosity liquids, much lower rates should be used. This precoat will act as the filtering medium rather than the filter cloth itself, and this filter action involves both sieve retention and adsorptive removal. An additional advantage of precoating the septum is that it facilitates the removal of the cake at the end of the filtration period. When used in pressure filtrations, precoat employs usually only a thin layer of about 1.5–3 cm thickness, equivalent to 0.5–1.0 kg/m^2 of filter area. However, when filtering slimy or gelatinous fermentation broth, a rotary-drum vacuum filter, using a thick layer of filter aid as precoat, is recommended. In this case, a thin layer of filter aid should be removed on every rotation together with the deposited solids. The usual range of precoat thickness is from 5 to 15 cm, and the blade advance varies from 25 to 250 μm per drum revolution. A new precoat is usually needed when the thickness of the old one reaches 0.5–1.5 cm. The recommended concentration for the precoat formation in vacuum filtration is from 2% to 5% by weight. A 5- to 10-cm precoat thickness may be achieved in 1 h, except for very fine filter aids where the time for forming a 5-cm precoat can be as high as 2 h.

The second way of using filter aids is as body feed. In this case, the filter aid is added to the fermented broth at the end of the fermentation. Actually, the body-feed method is usually employed together with the precoat method. The use of filter aid as body feed increases the porosity and permeability of the cake and decreases its compressibility, consequently reducing the cake resistance. The amount of filter aid to be used as body feed should be at least equal to the weight of biomass present. Nevertheless, up to 10 times more filter aid would be necessary if the bioparticles form highly compressible cakes. Filter aids are normally used in a one-time basis, although it is possible to recover them through resuspension of the cake followed by separation of the filter aid from the biological particles using a hydrocyclone (Matta and Medronho 2000).

Within a variety of filter aids, practically only three types have found wide application in the fermentation industry: (1) diatomaceous earth, also known as kieselguhr; (2) perlites; and (3) to a much smaller extent, cellulose. Diatomaceous earths are the most commonly used type of filter aid. They are porous and friable sedimentary rock composed by skeletal remains of single-celled aquatic plants called diatoms. These microscopic algae can extract silica from water to form their skeletal structure (see Fig. 24). They have approximately 90% silica and offer an enormous surface for the adsorption of colloidal particles (Fig. 25). Perlite is a glassy volcanic rock chemically composed primarily of aluminum silicates. It also has in its composition from 2% to 6% combined water. When quickly heated to above 870°C, the crude rock pops as the combined water vaporizes and creates countless tiny glass bubbles. This light-weight material is then milled, sized, and packed. Cellulose is not as frequently used in the biotechnology industry as diatomaceous earth or even perlites. Their greater application is, perhaps, in mixtures with diatomaceous earth or perlite to improve precoat

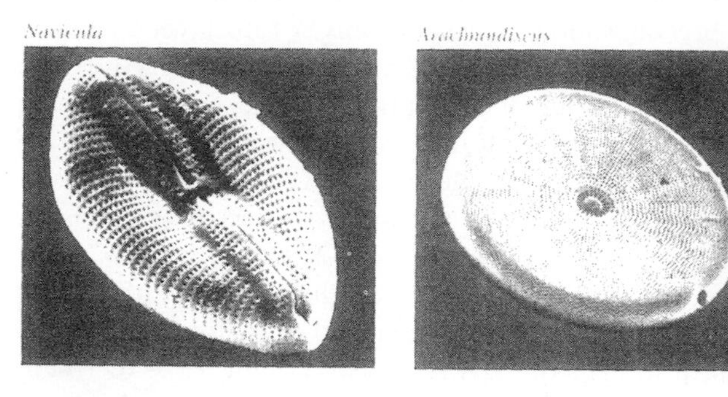

Figure 24 Skeletal structure of two different diatoms. (Courtesy of Celite Corporation.)

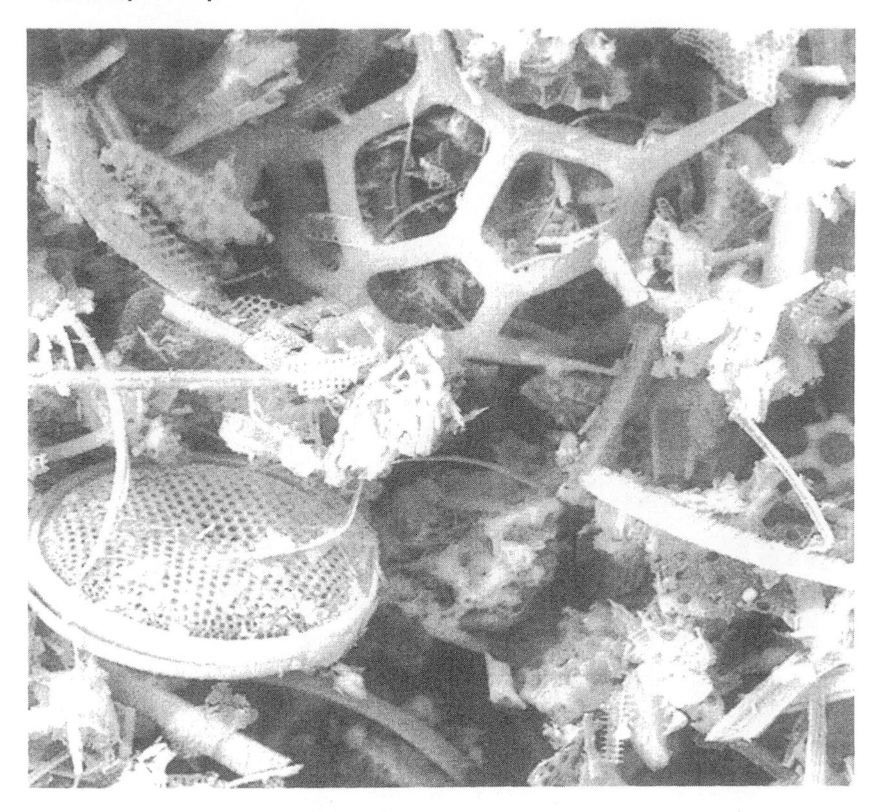

Figure 25 Scanning electron micrographs (1000×) of a commercial grade (Standard Super-Cell) of diatomaceous earth. (Courtesy of Celite Corporation.)

stability. Cellulose also finds application where the presence of trace amounts of silica is undesirable, in the filtration of hot caustic solutions, or when incineration of the cake is desirable. Table 6 gives a comparison between these three filter aid types, which are commercialized by different producers with a variety of brand names such as Celite 500, Filter Cell, Standard Super-Cell, Hyflo Super-Cell, Hyflo HV, Dicalite 215, Superaid, Speedflow, Speedplus, Speedex, Clarcel (all diatomaceous earths), Harbolite, Europerl, Dicalite 428, Grefco 436, Sil-Kleer, Clarcel-Flo (all perlites), and Fibra-Cell, Dicacel, Solka-Floc, and Clar-O-Cel (all celluloses). Figure 26 shows a scanning electron micrograph of 10-μm latex beads in a cake of Hyflo Super-Cell.

The use of filter aids usually improves the filtration of fermentation broths. However, a disadvantage may be the possible adsorption of product

Table 6 A Comparison Between the Main Properties of the Three Main Types of Filter Aids Used in the Biotechnology Industry

Filter aid	Wet density (g cm^{-3})	Permeability (10^{-8} cm^2)	Median pore size (μm)	Compressibility
Diatomite	0.27–0.44	0.02–25	0.5–22	Low
Perlite	0.08–0.32	0.4–8.0	6–18	Medium
Cellulose	0.10–0.38	0.4–8.3	—	High

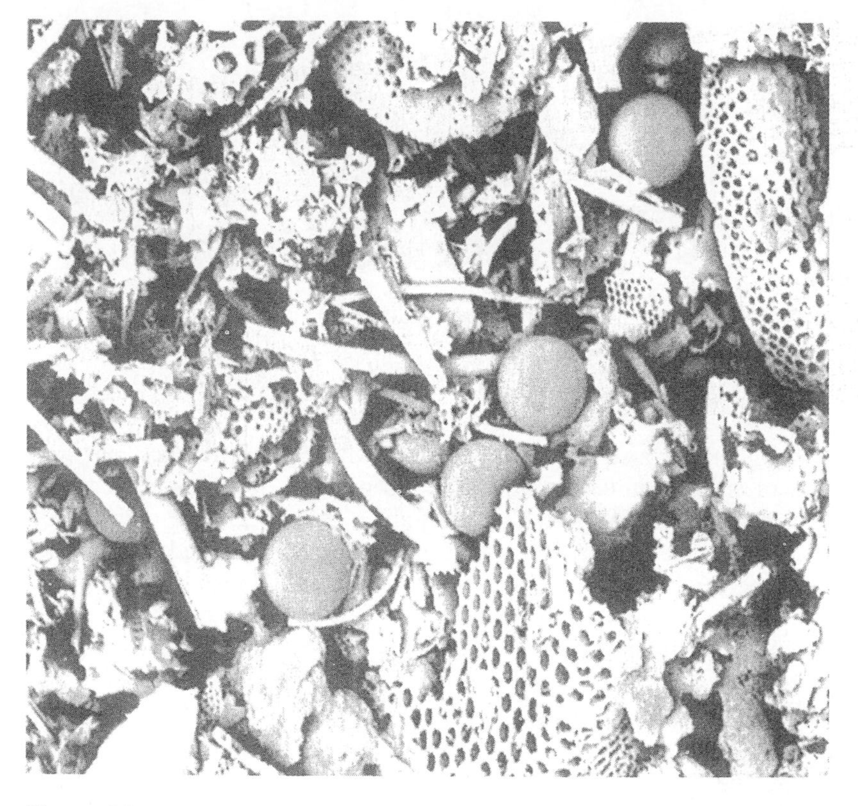

Figure 26 Scanning electron micrograph (1000×) of 10-μm latex beads in a cake of Hyflo Super-Cell. (Courtesy of Celite Corporation.)

to the filter aid. For instance, plasmid genes (Theodossiou et al. 1999), as well as aminoglycoside antibiotics and proteins (Wagman et al. 1975) can bind to diatomite and cellulose.

3.6 Types of Filters

There is an enormous variety of filter types. For instance, Nutsche filters, filter presses, liquid bag filters, external-cake tubular filters, and pressure-leaf filters are batch cake filters. Rotary drum filters (pressure or vacuum), disk filters, and horizontal vacuum filters are continuous filters. Each of these filter types has a great number of variations. For example, there are more than 100 design variations of filter presses. Filters have filtration areas varying from 1000 m^2 in some large units used in the chemical industry, down to 0.1 mm^2 in some microfilters used in conjunction with microreactors (Ehrfeld et al. 2000). Obviously, it is not possible to cover all of these filters in this chapter. Therefore, only the two most important filters will be summarized here, i.e., the filter press and the rotary vacuum filter. The readers can find more detailed descriptions of filters in other sources (Perry and Green 1997; Smith 1998; Svarovsky 1990).

Filter Presses

Filter presses have been used since the early days of the biotechnology industry, but nowadays they are increasingly being replaced by membrane filters. In spite of that, they still find applications ranging from fine pharmaceuticals and beverages to large volumes of waste products (Smith 1998). They are very simple and low-capital-cost equipments. They are available in a wide range of materials, such as metal, plastics, and wood. They are flexible, since their capacity can be easily adjusted by adding or removing plates and frames. When properly operated, a drier cake is obtained than that obtained with most other filters. However, they also have disadvantages. The main disadvantage is the high labor requirement, although a reasonable degree of automation can be obtained in some types of filter presses. In addition, they often present problems of leaking that render them unsuitable for the processing of hazardous liquids.

There are two main types of filter presses: recessed-plate and plate-and-frame. The former is composed only of plates, and the latter of plates and frames. In the recessed-plate press, both faces of each plate have a recess that acts as a chamber where the cake is formed between adjacent plates. These filters can be automated but inspection must be done after the end of each filtration cycle to ensure cake release. Some designs have a rubber diaphragm between plates that, at the end of filtration, compresses the cake to squeeze out more liquid. Compression pressures of up to 16 bar can be achieved, and the bioparticle cake compressibility helps in deliquoring of the cake (Mackay 1996).

The plate-and-frame press (Fig. 27) is an alternate assembly of hollow frames, where the cake is formed, and plates, which are covered on both sides with a filter medium (usually a cloth). When compared with a recessed-plate press, it has the advantage of forming more uniform cakes. But the recessed-plate press design has fewer parts and, therefore, about half as many joints, resulting in tighter closure. Therefore, leaking is a larger problem in plate-and-frame than in recessed-plate presses. In plate-and-frame presses, the feed is pumped into the frames, and the filtrate is driven from the filter through special orifices in the plates. Frames and plates are usually rectangular, but they can also be, for instance, circular. There are basically two different procedures for washing the cake: simple washing and thorough washing. In the former, the wash liquid is pumped at the end of filtration through the cake using the same entrance in the frames as was used for the suspension. This kind of washing is only recommended for very uniform and highly permeable cakes (Perry 1997), which is usually not the case for cakes formed by bioparticles. The second type of washing uses special plates, sometimes called washing plates or three-button plates. The filter is

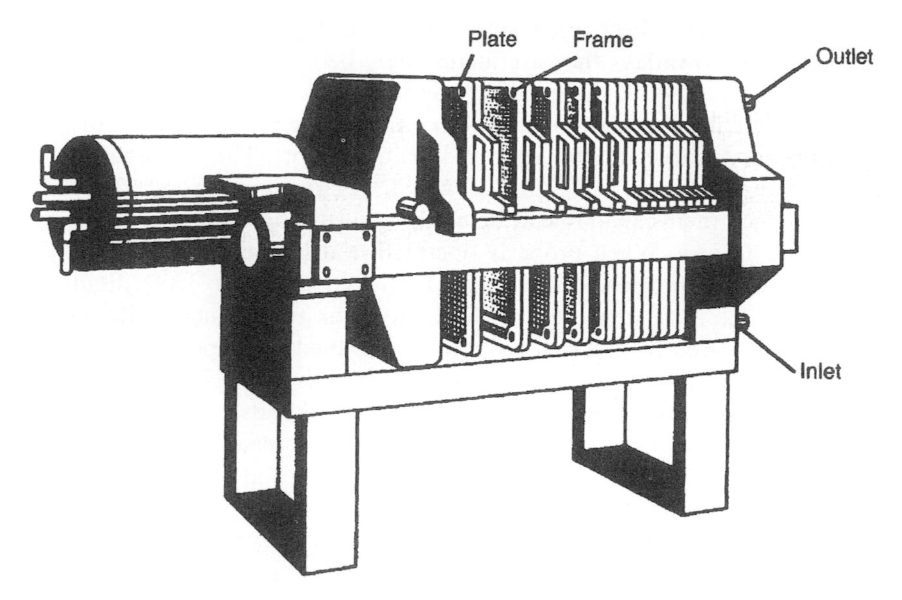

Figure 27 A perspective view of a plate-and-frame press. (Courtesy of Celite Corporation.)

assembled using the normal plates (one button), the frames (two buttons), and the washing plates (three buttons) in the following order: 1-2-3-2-1-2-3-2-1. In other words, the washing plate is always placed between two frames containing the cake. The wash liquid is then injected through the washing plates, passes through the entire cake, and exits through the one-button plates.

If the washing is done at the same pressure drop as at the end of filtration, the washing flow rate will be constant and equal to the flow rate at the end of filtration. This is true because during the washing the cake resistance does not increase because only liquid is passing through it. Therefore, Eq. (56) can be used to calculate the washing flow rate. As $dt/d\mathrm{V} = 1/Q$, where Q is the flow rate for simple washing:

$$Q_\ell = \frac{V_\ell}{t_\ell} = \frac{A\,\Delta P}{\mu\left(\dfrac{\langle\alpha\rangle \rho c V}{A} + R_\mathrm{m}\right)} \tag{78}$$

If the filter has washing plates, the flow rate for thorough washing can be proved to be one-fourth of the simple washing flow rate. Therefore, the time to wash the cake with a volume V_ℓ of washing liquid will be four times greater in the thorough washing.

The scale-up of plate-and-frame press is possible if the following information is obtained with a pilot plate-and-frame press, operating under the same conditions and with the same suspension as the industrial filter (index p is used here for the pilot filter variables and i for the industrial filter variables):

The average specific cake resistance $\langle\alpha\rangle$, the filter medium resistance R_m, and the average cake porosity $\langle\varepsilon\rangle$

The filtration time t_p needed to occupy with cake all volume within the frames of thickness e_p, and the volume of filtrate V_p obtained after t_p

The required ratio between the volume of washing liquid and cake volume ($\beta' = V_\ell/V_\mathrm{t}$)

A complete filtration cycle involves three different times: filtration time t, washing time t_ℓ (if required), and time t_d for cake removal, cleaning, and filter reassembling. Therefore, the filtrate production p will be given by:

$$p = \frac{V}{t + t_\ell + t_\mathrm{d}} \tag{79}$$

Applying Eqs. (64) and (78) to both pilot and industrial filter, it is possible to deduce the scale-up Eqs. (80) to (82), for filtration and washing conducted at the same constant pressure.

$$A_{\mathrm{i}} = A_p \frac{V_{\mathrm{i}}}{V_{\mathrm{p}}} \frac{e_{\mathrm{p}}}{e_{\mathrm{i}}} \tag{80}$$

$$t_{\mathrm{i}} = t_{\mathrm{p}} \left(\frac{e_{\mathrm{i}}}{e_{\mathrm{p}}} \right)^2 \tag{81}$$

$$t_{\ell \mathrm{i}} = k \frac{\beta'}{\beta} t_{\mathrm{i}} \tag{82}$$

where $k = 2$ for simple washing and $k = 8$ for thorough washing, e is the frame thickness, and β is the ratio between the filtrate volume V and the cake volume V_{t}. β can be obtained through Eq. (83).

$$\beta = \frac{V}{V_{\mathrm{t}}} = \frac{(1 - \langle \varepsilon \rangle)\rho_{\mathrm{s}}}{\rho c} \tag{83}$$

where

$$V_{\mathrm{t}} = \left(\frac{A}{2} \right) e \tag{84}$$

In Eq. (84), the filtration area is divided by 2 because the filtration area of each frame is twice the frame face area.

With the above equations it is possible to select a plate-and-frame press for an industrial operation, based on the results obtained in the laboratory using a pilot filter. The scaled-up filtration area should be increased by 25% as a factor of uncertainty (Perry 1997). It should be pointed out that the ratios given by β and β' hold for both filters and that the volume of filtrate V_{p} for the pilot filter is the one at which the cake fills all frames. This volume can be estimated through Eq. (84) or through a plot of t/V against V. For instance, for the filtration shown in Fig. 23, the volume of filtrate and the time of filtration at the point where the frames are full are 9000 cm^3 and 280 s, respectively.

The Rotary Vacuum Filter

The rotary vacuum filters are widely used for continuous large-scale filtrations. They are not truly continuous filters but can be operated for long batch times ranging from 1 to 10 days. They are especially useful for highly compressible particles, such as those found in fermented broths, which blind the cake after forming just a very thin layer. In such a case, a thick precoat of filter aid, as described in Sec. 3.5, is normally employed. Rotary vacuum filters are used for the filtration of waste sludges and have been used for yeast separations for more than a century. They find also wide application in the pharmaceutical industry, although here they are being replaced by other separation methods, such as membrane filtration.

Despite the existence of many design variations, the rotary vacuum filter consists basically of a horizontal-axis rotating drum, as shown schematically in Fig. 28. The cylindrical area of the drum is a grid structure covered by a filter medium with filtration areas varying from 0.4 to 190 m^2. The drum rotates at low speeds, usually between 0.1 and 10 rpm, and is partially submerged in the suspension. A typical submergence of 30–35% of the drum diameter is normally applied. Vacuum is applied inside the drum, so liquid is sucked through the filter medium as the drum rotates. The solids are thus deposited over the filter cloth, forming a cake. At each full rotation of the drum, a cycle of filtration is completed, which includes filtration, cake washing (when needed), cake drying, and cake removal. Vacuum from 15 to 60 cm of mercury is usually employed, with the lower range (15–30 cm Hg) being used for coarser particles and porous cakes, such as those formed by filter aids, and the higher range being used for thicker cakes and when a reduced moisture is important. There are two main types of drum: single compartment and multicompartment. The single compartment is usually a hollow cylinder whereas the multicompartment has a more complicated mechanical design, with internal pipes and valves. Different methods of cake discharge have been developed over the years: knife discharge, precision knife discharge (used in precoat filtration), roller discharge (for sticky cakes), string discharge (for fibrous cakes), belt discharge (for sticky cakes), and so forth. The proper cake-discharge method is mainly a function of the cake characteristics.

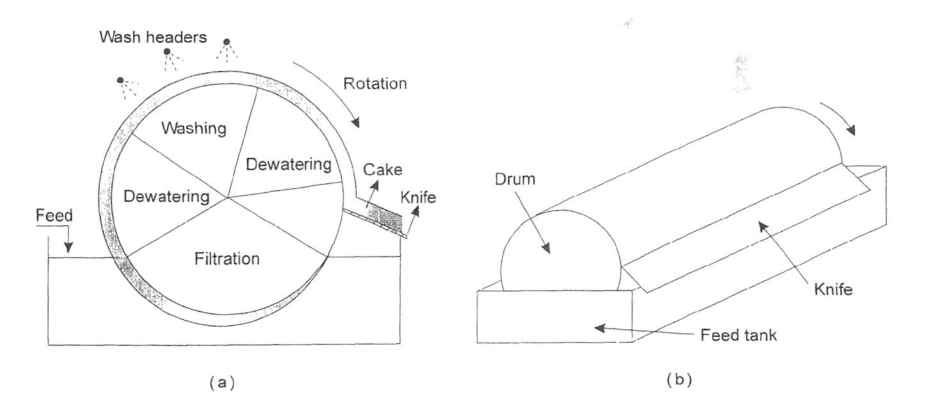

Figure 28 Rotary vacuum filter: schematic front view (a) and simplified perspective view (b).

Selection or performance predictions of rotary vacuum filters can be carried out using the so-called leaf tests. With these tests, it is possible to determine the flux of filtrate, expressed as filtrate flow rate per filtration area, and therefore the area of the filter. Detailed explanation about leaf tests as well as other laboratory tests used for vacuum or pressure filters can be found elsewhere (Rushton et al. 1996). It should be emphasized that the Buchner funnel is not an adequate substitute for a leaf test because of the significant resistance to flow produced by the drainage surface. It may be used for comparative flocculation evaluations, although the capillary suction time test is more appropriate for such application (Rushton et al. 1996).

4. FLOCCULATION

Small particles, such as cell debris or whole cells, may be aggregated to form flocs large enough to be removed by an appropriate process, such as gravity settling, centrifugation, or filtration. The scientific basis of aggregation is complex. It is based on the interfacial electrochemistry of small particles and its relation with colloid stability. Detailed information on flocculation can be found in more specialized books (Dobias 1993; Ives 1978).

An aggregation process may be distinguished as coagulation or flocculation, and the definition of both may vary with the area of application. For workers in the water treatment area, coagulation is the step of mixing an appropriate chemical to destabilize a colloid, and flocculation is the posterior stage of soft mixing that will promote aggregation through particle-particle collisions. For many colloid researchers, coagulation is the process whereby aggregation is caused by the addition of ions and flocculation by the addition of polymers. Since neither of these definitions describes properly the aggregation process, only the term "flocculation" will be adopted here.

Many flocculants found in the market are not approved for food and pharmaceutical use, and most of those approved are expensive. However, some alternative and cheap biocompatible flocculants, such as borax, may be used (Tsoka et al. 2000).

4.1 Colloidal Interactions

The principles that apply to colloidal solutions are also valid for larger particles, but in this case other forces, such as gravitational forces, may become more important. In any case, aggregation may be applied to particles from colloidal size up to the visible size range. The main types of interactions occurring when two particles approach each other are van der

Waals attraction, electrical interaction, solvation forces, steric interaction, and polymer bridging (Gregory 1986). Specific forces may also act in special cases, such as magnetic forces (Ives 1978). The universal van der Waals attractive forces are inversely proportional to the distance between the colloidal particles, are directly proportional to the size of the particle, and are almost independent of ionic strength for most particles (not always valid for bioparticles). Electrical interactions appear as a consequence of positive or negative charges that particles normally present at their surface when dispersed in water. These charges occur due either to preferential adsorption of certain ions from solution or to ionization of groups at the particle surface. As the system can have no net charge, the solution near a charged surface will develop a countercharge. The particle charge and the countercharge together form an electrical double layer. For biological surfaces, the most important groups to suffer ionization are carboxylic acid and amine groups, characteristic of amino acids and proteins. There is usually a characteristic pH value at which these charges at the particle surface are neutralized. Many bioparticles have this point of zero charge in the acid region of pH; consequently, their surfaces hold a negative charge at neutral pH (Gregory 1986). The van der Waals and electrostatic forces are "long range" forces, i.e., they can act at several nanometers of distance. Solvation (hydration) forces are short range forces that play a role in particle-particle interaction, usually adding an extra repulsion. They appear due to a different structure that water assumes close to solid surfaces. Steric interaction can occur if the particles are covered by an adsorbed layer, particularly of polymeric material. In a collision, compression or interpenetration of the adsorbed layers may occur. Compression always generates repulsion and interpenetration can produce either repulsion (more frequent) or attraction, depending on the existing interaction between the polymer and the solvent. In biological systems, polymers produced by microorganisms play an important role in controlling stability (Gregory 1986). Polymer bridging occurs when a polymer promotes an opposite effect that occurs in a steric stabilization. Thus, instead of stabilizing the particles, the polymer promotes flocculation. This difference in behavior can be understood if one looks at the particle surface. Polymer bridging occurs only at very diluted polymer concentrations. Therefore, the same polymer molecule, usually with a very high molecular weight, can have more than one particle attached to it, i.e., one or more polymer molecules bridge a floc. In the steric stabilization, which uses much higher polymer concentrations, the polymer covers all or most of the particle surface, thus, there is no surface left to bind to a polymer molecule already bound to another particle. Therefore, if polymer bridging is desired, tests must be carried out to determine the point of optimal concentration. As shown in Fig. 29, flocculation usually increases with polymer concentra-

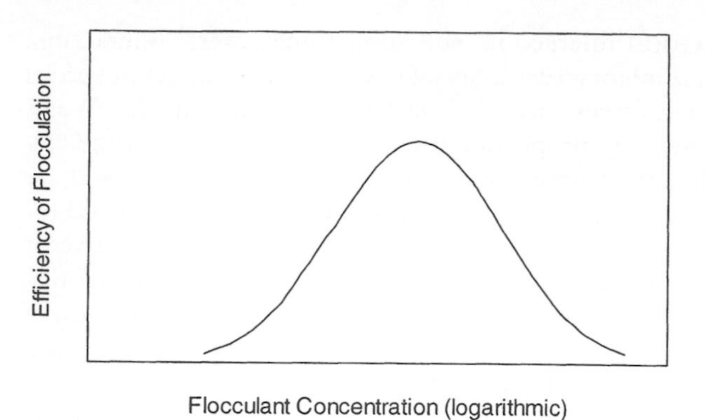

Figure 29 Efficiency of flocculation as a function of flocculant concentration.

tion, passes through a maximum, and then decreases due to the continuous adsorption of more polymer molecules by a given particle. The optimal surface coverage for bridging flocculation has been estimated to be around 50% (Cumming et al. 1994). It is important to point out that bridging is not the only way of action of polymeric flocculants. When using polyelectrolytes (charged polymers), charge neutralization may also be responsible for flocculation (Gregory 1986).

4.2 Aggregation Mechanisms

The presence of several forces, some attractive and some repulsive, that show different behavior according to the medium conditions (pH, temperature, salt concentration, etc.) makes aggregation a very complex theoretical matter. Nevertheless, there are several ways of promoting it, such as addition of chemicals to reduce the stability of the original suspension, neutralization of the electrical forces, addition of polymers to link particles (bridging), and so on (Ives 1978). In any situation, aggregation will only occur if the particles collide with each other. Collisions can occur due to Brownian motion, velocity gradients within the liquid, or differential settling of the particles. Brownian motion of the particles happens due to the thermal energy of the liquid. Velocity gradients can be obtained through stirring the liquid, and differential settling occurs when the particles have different sedimentation velocities. The mechanism of flocculation due to Brownian motion is called *perikinetic flocculation*, and the one due to velocity gradients is known as *orthokinetic flocculation*. Differential settling is usually treated as a special case of orthokinetic flocculation, since the particle motion relative to the liquid also creates velocity gradients. It has been shown (Ives 1978) that, under most significant cases of stirring, the perikinetic rates dominate for particle sizes under 1 μm and the orthokinetic rates dominate for larger particles.

4.3 Flocculation Agents

The surface charge of most bioparticles, such as whole cells and cells debris, depends on the pH of the medium and is usually negative at neutral pH. Therefore, a continuous reduction in pH will lead to a continuous reduction in these negative charges, until zero charge is reached. At this point, if the particles collide, the van der Waals forces may be strong enough to flocculate the particles. Actually, this flocculation can occur not only at the point of zero charge but also in a narrow range around it. It is possible to estimate the point of zero charge by measuring the ζ potential as a function of pH using microelectrophoresis and, for the majority of bioparticles, it is located in the 4.5–6.5 pH range (Mackay 1996). Many fermentation broths have buffering properties. If this is the case, the amount of acid required to reach the proper pH range needed for flocculation may be high (Mackay 1996).

The addition of an electrolyte to a culture medium containing charge-stabilized bioparticles may lead to flocculation. This happens because electrolytes increase the ionic strength of the medium, which decreases the electrical double-layer thickness and, consequently, reduces particle-particle repulsion. When this electrostatic repulsion is reduced, van der Waals forces may predominate and, thus, the particles can flocculate. This effect is more prominent for dissolved salts containing high-valence ions (usually the cation) of opposite charge to the bioparticle. Aluminum and iron salts, usually sulfates or chlorides, are the most common inorganic flocculant agents. In some cases, the addition of flocculation aids, which are insoluble particles that act as nucleating sites, improves the quality and size of the flocs. Common flocculation aids are activated silica, activated carbon, clay, metal oxides, and paper pulp (Daniels 1993).

Polymeric flocculants are organic polymers with a linear or branched chain. They can both decrease the electrical double-layer thickness and hence the strength of repulsion and/or adsorb on different particles bridging them. They can be natural or synthetic and, regarding electrical charge, anionic, cationic, or nonionic. Natural flocculants, such as alginates, starches, gelatines, celluloses (e.g., carboxymethylcellulose), and chitins (e.g., chitosan), have been used since ancient times and still find applications, for instance, in the beer and wine industries. However, the great majority of polyelectrolytes are synthetic, and of these polyacrylamide is the most widely used. It is obtained through polymerization of acrylamide monomer, and polymers with high molecular weight of up to around 2×10^7 can be obtained. In spite of being theoretically nonionic, polyacrylamide usually has a small anionic character due to unintentional hydrolysis of some amide groups to form carboxyl groups (Gregory 1986). Cationic polyelectrolytes can be obtained through copolymerization of acrylamine with a positively charged monomer (e.g., amine, imine, or quaternary ammonium group). Anionic polyelectrolytes are obtained by deliberate hydrolysis of polyacrylamide or through copolymerization of acrylamine with acrylic acid. Other examples of polymeric flocculants are polyethylene oxide (nonionic), polyvinyl alcohol (nonionic), polyethylenimine (cationic), polydiallyldimethylammonium chloride (cationic), and sodium polystyrene sulfonate (anionic).

5. MAIN MANUFACTURERS OF EQUIPMENT

5.1 Centrifuges

Alfa Laval: http://www.alfalaval.com/
Baker Hughes: http://www.bakerhughes.com
Beckman Coulter: http://www.beckman.com/
CARR Separations: http://www.carrsep.com
Eppendorf: http://www.eppendorf.com/
Flottweg: http://www.flottweg.com/
Hettich-Zentrifugen: http://www.hettich-zentrifugen.de/
Kendro: http://www.kendro.com
Rousselet Robatel: http://robatel.com
Sigma-Aldrich: http://www.sigma-aldrich.com
The Western States Machine: http://www.westernstates.com
Westfalia Separators: http://www.westfalia-separator.com

5.2 Hydrocyclones

AKW: http://www.akwauv.com
Dorr-Oliver: http://www.dorr-oliver.com
Krebs Engineers: http://www.krebsengineers.com
Richard Mozley: http://www.mozley.co.uk

5.3 Filters

Alsop Engineering: http://www.alsopengineering.com
Baker Hughes: http://www.bakerhughes.com

BHS-Sonthofen: http://www.bhs-sonthofen.de
D. R. Sperry: http://www.drsperry.com
Dorr-Oliver: http://www.dorr-oliver.com
Giovanola Freres: http://www.giovanola.ch/
Larox: http://www.larox.com
Netzsch Filtrationstechnik: http://www.netzsch.com/

5.4 Filter Aids

Ceca: http://www.ceca.fr
Dicalite: http://www.dicalite-europe.com
Grefco Minerals: http://www.grefco.com
Solka-Floc: http://www.solkafloc.com
World Minerals: http://www.worldminerals.com/

NOMENCLATURE

A filtration area (m^2)
b intensity of the field of forces ($m\ s^{-2}$)
c concentration of solids as mass of solids by mass of liquid
C_v feed volumetric concentration
C_{vu} underflow volumetric concentration
d particle diameter (m)
D test tube diameter (m)
D_c hydrocyclone diameter (m)
D_i feed inlet diameter (m)
D_o overflow diameter (m)
D_u underflow diameter (m)
d_{50} cut size (m)
d'_{50} reduced cut size (m)
e thickness of the frame (m)
E_T total efficiency
E'_T reduced total efficiency
Eu Euler number
g gravity acceleration ($m\ s^{-2}$)
G grade efficiency
G' reduced grade efficiency
k cake permeability (m^2)
k_m filter medium permeability (m^2)
ℓ vortex finder length (m)
L length (m)

ℓ_c cake thickness (m)
ℓ_m filter medium thickness (m)
m mass of solids in the cake (kg)
p filtrate production ($m^3\ s^{-1}$)
P pressure (Pa)
q average fluid velocity ($m\ s^{-1}$)
Q volumetric flow rate of the feed suspension ($m^3\ s^{-1}$)
Q_ℓ volumetric flow rate of cake washing ($m^3\ s^{-1}$)
Q_L volumetric flow rate of water in the feed ($m^3\ s^{-1}$)
Q_{Lu} volumetric flow rate of water in the underflow ($m^3\ s^{-1}$)
Q_u volumetric flow rate of the underflow ($m^3\ s^{-1}$)
r radius (m)
R radius (m)
Re Reynolds number
R_f flow ratio
R_m filter medium resistance (m^{-1})
Stk_{50} Stokes number
t time (s)
t_c centrifugation time (s)
t_d time for cake removal, cleaning, and filter reassembling (s)
t_f filtration time (s)
t_ℓ washing time (s)
t_r residence time (s)
v particle settling velocity ($m\ s^{-1}$)
V volume of filtrate (m^3)
v_g particle settling velocity under the gravitational field ($m\ s^{-1}$)
V_ℓ volume of washing liquid (m^3)
V_s volume of the liquid shell in the centrifuge (m^3)
V_t cake volume (m^3)
x particle size frequency in the feed
x_u particle size frequency in the underflow
W_S mass flow rate of the feed ($kg\ s^{-1}$)
W_{Su} mass flow rate of the underflow ($kg\ s^{-1}$)
y cumulative particle size distribution (undersize) of the feed suspension
y_o cumulative particle size distribution (undersize) of the overflow
y_u cumulative particle size distribution (undersize) of the underflow
α specific cake resistance ($m\ kg^{-1}$)
$\langle\alpha\rangle$ average specific cake resistance ($m\ kg^{-1}$)
α_o average specific cake resistance at unit-applied pressure drop ($m\ kg^{-1}$)
β ratio between the filtrate volume and the cake volume
β' ratio between the volume of washing liquid and the cake volume

γ angle of the bottle
ΔP pressure drop (Pa)
ε cake porosity
ζ g factor
η efficiency factor
θ angle of the hydrocyclone cone
μ viscosity of liquid (Pa s)
ρ density of liquid (kg m^{-3})
ρ_s density of solids (kg m^{-3})
Σ sigma factor (m^2)
τ stress tensor (Pa)
ω angular velocity (s^{-1})

REFERENCES

Apelman, S. and Björling, T. (1991) New centrifugal separator. Biotech Forum Europe, 8, 356–358.

Axelsson, H. (1999) Cell separation, centrifugation. In: Encyclopedia of Bioprocess Technology: Fermentation, Biocatalysis, and Bioseparation, Flickinger, M.C. and Drew, S.W. (eds.), John Wiley and Sons, New York, 513–531.

Bednarsky, S. (1996) The thickening of biological solids in hydrocyclones. In: Hydrocyclones ’96, Claxton, D., Svarovsky, L. and Thew, M. (eds.), Mechanical Engineering Publications, London & Bury Saint Edmunds, 151–159.

Bendixen, B. and Rickwood, D. (1994) Effects of hydrocyclones on the integrity of animal and microbial cells, Bioseparation, 4, 21–27.

Bradley, D. (1965) The Hydrocyclone, Pergamon Press, Oxford.

Cann, A.J. (ed.) (1999) Virus Culture—A Practical Approach, Oxford University Press, Oxford.

Castilho, L.R. and Medronho, R.A. (2000) A simple procedure for design and performance prediction of Bradley and Rietema hydrocyclones, Minerals Engineering, 13, 183–191.

Cilliers, J.J. and Harrison, S.T.L. (1996) The effect of viscosity on the recovery and concentration of micro-organisms using mini-hydrocyclones. In: Hydrocyclones ’96, Claxton, D., Svarovsky, L. and Thew, M. (eds.), Mechanical Engineering Publications, London & Bury Saint Edmunds, 123–133.

Cilliers, J.J. and Harrison, S.T.L. (1997) The application of mini-hydrocyclones in the concentration of yeast suspensions, Chemical Engineering Journal, 65, 21–26.

Coelho, M.A.Z. and Medronho, R.A. (2001) A model for performance prediction of hydrocyclones, Chemical Engineering Journal, 84, 7–14.

Cumming, R.H., Robinson, P.M., and Martin, G.F. (1994) Towards a mathematical model for flocculation of *Escherichia coli* with cationic polymers. In: Pyle,

D.L. (ed.), Separations for Biotechnology 3, The Royal Society of Chemistry, Cambridge 127–133.

Daniels, S.L. (1993) Flocculation. In: Unit Operations Handbook, Vol. 2, McKetta, J.J. (ed.), Marcel Dekker, New York.

Datar, R. (1984) Centrifugal and membrane filtration methods in biochemical separation, Filtration and Separation, 21, 402–406.

Del Villar, R. and Finch, J.A. (1992) Modelling the cyclone performance with a size dependent entrainment factor, Minerals Engineering, 5, 661–669.

Dickenson, T.C. (1997) Filters and Filtration Handbook, 4th ed., Elsevier, Oxford.

Dobiáš, B. (ed.) (1993) Coagulation and Flocculation, Marcel Dekker, New York.

Ehrfeld, W., Hessel, V. and Löwe, H. (2000) Microreactors—New Technology for Modern Chemistry, Wiley-VCH, Weinheim.

Ersson, B., Rydén, L. and Janson, J.-C (1998) Introduction to protein purification. In: Janson, J.-C and Rydén, L. (eds.), Protein Purification, 3–40, Wiley-Liss, New York.

Ferreira, J.S.G., Castilho, L.R., and Paiva, S.P. (1995) Produção de pectinases para a indústria de bebidas, Final Report, School of Chemistry, Federal University of Rio de Janeiro, Rio de Janeiro.

Frachon, M. and Cilliers, J.J. (1999) A general model for hydrocyclones partition curves, Chemical Engineering Journal, 73, 53–59.

Gregory, J. (1986) Flocculation. In: Progress in Filtration and Separation 4, Wakeman, R.J. (ed.), Elsevier, Amsterdam.

Harrison, S.T.L., Davies, G.M., Scholtz, N.J., and Cilliers, J.J. (1994) The recovery and dewatering of microbial suspensions using hydrocyclones. In: Separations for Biotechnology 3, Pyle, D.L. (ed.), SCI, The Royal Society of Chemistry, Cambridge, 214–220.

Hawkes, J.J., Limaye, M.S. and Coakley, W.T. (1997) Filtration of bacteria and yeast by ultrasound-enhanced sedimentation, Journal of Applied Microbiology, 82, 39–47.

Heiskanen, K. (1993) Particle Classification, Chapman & Hall, London.

Hemfort, H. (1984) Separators: Centrifuges for clarification, separation and extraction processes. Technical Scientific Documentation No. 1, Westfalia Separator AG, Oelde.

Higgins, J.J., Lewis, D.J., Daly, W.H., Mosqueira, F.G., Dunnill, P., and Lilly, M.D. (1978) Investigation of the unit operations involved in the continuous flow isolation of β-galactosidase from *Escherichia coli*. Biotechnology and Bioengineering, 20, 159–182.

Ives, K.J. (ed.) (1978) The Scientific Basis of Flocculation, Sijthoff & Noordhoff, Alohen aan den Rijn.

Johnson, M., Lantier, S., Massie, B., Lefebvre, G., and Kamen, A.A. (1996) Use of the Centritech lab centrifuge for perfusion culture of hybridoma cells in protein-free medium. Biotechnology Progress, 12, 855–864.

Kildeso, J. and Nielsen, B.H. (1997) Exposure assessment of airborne microorganisms by fluorescence microscopy and image processing, Ann Occup Hyg, 41, 201–216.

Lin, A.A., Nguyen, T., and Miller, W.M. (1991) A rapid method for counting cell nuclei using a particle sizer/counter, Biotechnology Techniques, 5, 153–156.

Ling, Y., Wong, H.H., Thomas, C.J., Williams, D.R.G., and Middelberg, A.P.J. (1997) Pilot-scale of PHB from recombinant *E. coli* by homogenization and centrifugation, Bioseparations, 7, 9–15.

Linz, F., Degelau, A., Friehs, K., Eghtessagi, F., and Scheper, Th. (1990) The use of cytometry for controlling biotech processes, DECHEMA Biotechnology Conferences, 4, Part B, 941–944.

Lübberstedt, M., Medronho, R.A., Anspach, F.B., and Deckwer, W.-D. (2000a) Separation of mammalian cells using hydrocyclones, Proceedings of the World Congress on Biotechnology, Biotechnology 2000, Vol. 1, 460–462, Berlin.

Lübberstedt, M., Medronho, R.A., Anspach, F.B., and Deckwer, W.-D. (2000b) Abtrennung tierischer Zellen mit Hydrozyklonen, Chemie Ingenieur Technik, 72, 1089–1090.

Mackay, D. (1996) Broth conditioning and clarification. In: Downstream Processing of Natural Products, Verrall, M.S. (ed.), John Wiley and Sons, Chichester, 11–40.

Marschall, A. (1997) Biomasseabtrennung mit Hydrozyklonen, Thesis, Technical University of Braunschweig, Brunswick.

Matta, V.M. and Medronho, R.A. (2000) A new method for yeast recovery in batch ethanol fermentations: filter aid filtration followed by separation of yeast from filter aid using hydrocyclones, Bioseparation, 9, 43–53.

Maybury, J.P., Mannweiler, K., Hooker, N.J.T., Hoare, M., and Dunnill, P. (1998) The performance of a scaled down industrial disc stack centrifuge with a reduced feed material requirement, Bioprocess Engineering, 18, 191–199.

Meltzer, T.H. and Jornitz, M.W. (eds.) (1998) Filtration in the Biopharmaceutical Industry, Marcel Dekker, New York.

Müller, M. (2000) Selektive Trennung von Anaerobschlamm mit Hydrozyklonen, Thesis, Technical University of Braunschweig, Brunswick.

Nakanishi, K., Tadokoro, T., and Matsuno, R. (1987) On the specific resistance of cake of microorganisms, Chemical Engineering Communications, 62, 187–201.

Ortega-Rivas, E. and Medina-Caballero, H.P. (1996) Wastewater sludge treatment by hydrocyclones, Powder Handling and Processing, 8, 355–359.

Patel, D., Ford, T.C., and Rickwood, D. (1998) Fractionation of cells by sedimentation methods. In: Cell Separation—A Practical Approach, Fischer, D., Francis, G.E., and Rickwood, D. (eds.), Oxford University Press, Oxford, 43–896.

Perry, R.H. and Green, D.W. (eds.) (1997) Chemical Engineers' Handbook, 7th ed., McGraw-Hill, New York.

Pethig, R. and Markx, G.H. (1997) Applications of dielectrophoresis in biotechnology, Trends in Biotechnology, 15, 426–432.

Pons, M.N. and Vivier, H. (1998) Beyond Filamentous Species ..., Advances in Biochemical Engineering Biotechnology, 60, 61–93.

Rickwood, D., Freeman, G.J., and McKechnie, M. (1996) An assessment of hydrocyclones for recycling kieselguhr used for filters in the brewing industry. In: Hydrocyclones '96, Claxton, D., Svarovsky, L. and Thew, M. (eds.), Mechanical Engineering Publications, London & Bury Saint Edmunds, 161–172.

Rickwood, D., Onions, J., Bendixen, B., and Smyth, I. (1992) Prospects for the use of hydrocyclones for biological separations. In: Hydrocyclones: Analysis and Applications, Svarovsky, L. and Thew, M.T. (eds.), Kluwer Academic, Dordrecht, 109–119.

Roldán-Villasana, E.J., Williams, R.A. and Dyakowski, T. (1993) The origin of the fish-hook effect in hydrocyclone separator, Powder Technology, 77, 243–250.

Rushton, A., Ward, A.S. and Holdich, R.G. (1996) Solid-Liquid Filtration and Separation Technology. VHC, Weinheim.

Smith, G.R.S. (1998) Filter aid filtration. In: Filtration in the Biopharmaceutical Industry, Meltzer, T.H. and Jornitz, M.W. (eds.), Marcel Dekker, New York.

Svarovsky, L. (1984) Hydrocyclones, Holt, Rinehart and Winston, London.

Svarovsky, L. (ed.) (1990) Solid-Liquid Separation, 3rd ed., Butterworths, London.

Taylor, G., Hoare, M., Gray, D.R., and Marston, F.A.O. (1986) Size and density of protein inclusion body, Bio/Technology, 4, 553–557.

Theodossiou, I., Thomas, O.R.T., and Dunnill, T.P. (1999) Methods of enhancing the recovery of plasmid genes from neutralized cell lysate, Bioprocess Engineering, 20, 147–156.

Thorwest, I. and Bohnet, M. (1992) Aufkonzentrierung von Biomasse mit Hydrozyklonen, Chemie Ingenieur Technik, 64, 1123–1125.

Tsoka, S., Ciniawskyj, O.C., Thomas, O.R.T., Titchener-Hooker, N.J., and Hoare, M. (2000) Selective flocculation and precipitation for the improvement of virus-like particle recovery from yeast homogenate, Biotechnology Progress, 16, 661–667.

Wagman, H., Bailey, V., and Weistein, M. (1975) Binding of antibiotics to filtration materials, Antimicrobial Agents and Chemotherapy, 7, 316–319.

Werning, H. and Voss, H. (1993) Verfahrenstechnische Ansätze lassen hoffen, Chemische Industrie, 10, 31–35.

Wilton, J.C. and Strain, A.J. (1998) Centrifugal separation. In: Cell Separation—A Practical Approach, Fischer, D., Francis, G.E., and Rickwood, D. (eds.), Oxford University Press, Oxford, 91–129.

Wong, H.H., O'Neill, B.K., and Middelberg, A.P.J. (1997) Cumulative sedimentation analysis of *Escherichia coli* debris size, Biotechnology and Bioengineering, 55, 556–564.

Yuan, H., Rickwood, D., Smyth, I.C., and Thew, M.T. (1996a) An investigation into the possible use of hydrocyclones for the removal of yeast from beer, Bioseparation, 6, 159–163.

Yuan, H., Thew, M.T. and Rickwood, D. (1996b) Separation of yeast with hydrocyclones. In: Hydrocyclones '96, Claxton, D., Svarovsky, L. and Thew, M. (eds.), Mechanical Engineering Publications, London & Bury Saint Edmunds, 135–149.

6
Membranes for Protein Isolation and Purification

Roland Ulber and Kerstin Plate
University of Hannover, Hannover, Germany

Oskar-Werner Reif and Dieter Melzner
Sartorius AG, Göttingen, Germany

1. INTRODUCTION

The use of membranes during the purification and separation of proteins is becoming increasingly important both in academic sciences and in industrial process engineering. Thus, membranes play an important role in different downstream processes, starting with cell separation from the bioprocess broth up to final polishing of the product. A pressure driving force and a semipermeable membrane are used to effect a separation of components in a solution or colloidal dispersion. The separation is based mainly on molecular size, but also to a lesser extent on shape and charge. As a pressure-driven process the membrane filtration allows processing of proteins at ambient temperatures and without exposing them to thermal stress or alterations of its chemical environment. Membrane processes are easily scaled up, and the opportunity of using the same materials and configurations in different sizes from laboratory to process scale reduces the validation effort enormously. However, filtration processes are limited in selectivity. Fractionation of proteins can only be achieved at large differences in molecular weight of the proteins, and it is important to keep in mind that a certain difference in molecular weight between two proteins does not mean the same degree of difference in molecular size. Proteins that differ in molecular weight by 10 times may differ in size by only 3 times when in the globular or folded form.

The function of membranes has now been enhanced to more than its role as a selective barrier for filtration of molecules. Selective adsorption of molecules to the membranes for their separation based on different chemical behavior is being increasingly applied as an integrated downstream processing operation.

2. MEMBRANE FILTRATION

2.1 Classification of Membrane Types

During the last 30 years, a number of membrane processes have been developed for molecular separation. These filtration techniques can be divided in four major groups: reverse osmosis (RO; hyperfiltration), nanofiltration (NF), ultrafiltration (UF), and microfiltration (MF). The dimensions of the components involved in these separations are given in Fig. 1 (Datar and Rosén 1993; Lewis, 1996).

RO membranes or NF membranes are used when low molecular weight substances (salts, sugars, amino acids) have to be separated from a solvent. RO membranes have a higher retention for monovalent ions like Na^+ and Cl^- than NF membranes, but a lower flux. Applications are mainly concentration of solutions (dewatering).

UF membranes have a pore size between 1 nm and 0.05 μm. They are typically used for the retention of macromolecules and colloids from a solution. They are commonly classified not by pore size but by nominal molecular weight cutoff (MWCO). The cutoff is the molecular weight of that macromolecule, for which the retention of the membrane is 90%. This MWCO number is not an absolute value for membrane retention because it is also influenced by the nature of the protein, by interaction of the protein with the membrane or other components in the process solution, and by the operating conditions. Therefore, for a proper selection of a membrane for a specific application the optimal membrane has to be selected by tests or with the help of application specialists.

MF membranes have a pore size between 0.02 and 10 μm. They are used to remove colloids and particles from solutions. The most important applications are cold sterilization of solutions (products or feed solutions for a bioreactor using 0.2 μm pore size membranes), particle (or cell) removal, cell harvesting, and preparation of ultrapure water. Hydrophobic MF membranes are used for sterile filtration of gases, e.g., aeration of bioreactors or venting of storage tanks.

To remove low molecular weight solutes (e.g., salts) from a protein solution, *diafiltration* is used. Here, the protein solution is concentrated by

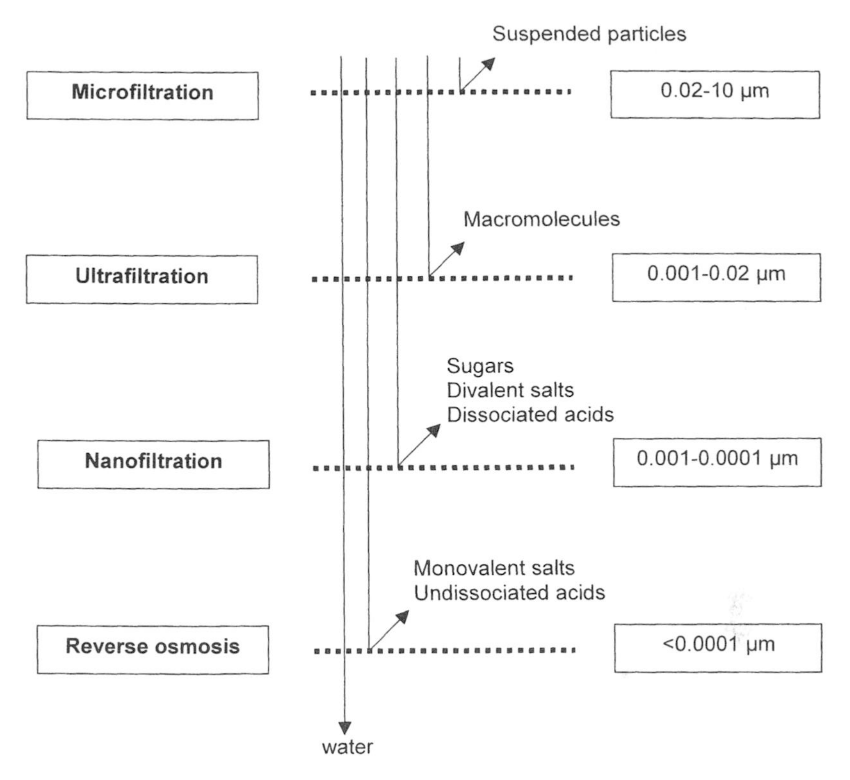

Figure 1 Pressure-driven membrane processes and their separation characteristics (Datar and Rosén 1993; Lewis 1996).

UF, pure solvent is added, and the protein is concentrated again. The dilution step has to be repeated until the desired purity of the protein is reached.

2.2 Fundamentals of Filtration Techniques

During a membrane separation process the low-viscosity feed suspension is usually applied on one side of a membrane. The stream that passes through the membrane under the influence of the pressure force is termed the *permeate* (filtrate). After removal of the required amount of permeate, the remaining material is termed *retentate* (concentrate). The extent of the concentration is characterized by the *concentration factor* (f), which is the ratio of the feed volume to the final volume [Eq. (1)].

$$f = \frac{V_F}{V_c} \tag{1}$$

where V_F is feed volume and V_c is final retentate volume.

Thus, a process with a concentration factor of 10 would have a volume reduction of 90% (Lewis, 1996).

Another important factor is the *retention factor* (R) of the target product, which is defined as the distribution of components between the permeate and retentate.

$$R = \frac{c_F - c_p}{c_F} \tag{2}$$

where c_F is concentration in feed solution and c_p is concentration in permeate.

Retention factors vary between 0 and 1, where R of 1 means that the component is retained completely in the feed and R of 0 means that the component is freely permeating. According to this, an ideal reverse-osmosis membrane would give a retention factor of 1 for all components, whereas an ideal ultrafitration membrane, used to remove a low molecular weight component, would give a retention factor of 0.

2.3 Processes and Process Design

Figure 2 shows the modes in which membrane filtration is operated. In *dead-end filtration* mode the total process fluid stream flows through the membrane. The retained solids accumulate on the membrane and build up a filter

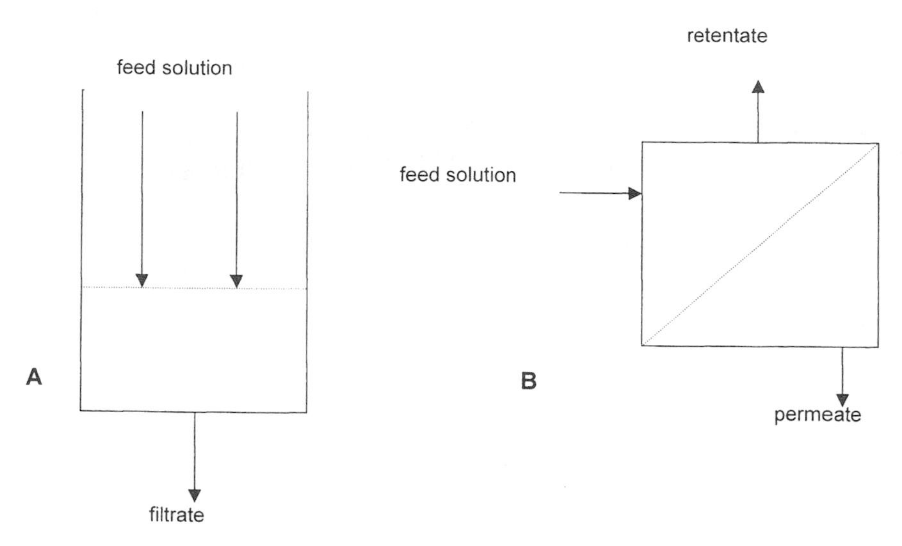

Figure 2 Schematic presentation of (A) dead-end and (B) cross-flow filtration.

cake. The membrane has to be changed when the membrane pores are clogged by solids. Dead-end mode is only applicable when liquids with very low particle content are processed and no concentration polarization takes place. Typically it is used in microfiltration or prefiltration steps.

Dead-end systems are mainly used in laboratory scale separations. The membranes are available as ready-to-use devices with different membrane areas for one-time application (see also Table 1). For process applications, pleated-membrane cartridges are used, and to increase the economics of the process the process stream is prefiltered before entering the final sterile filter.

When the feed flow is directed parallel to the membrane surface, the term *cross-flow filtration* is used to describe such applications. The tangential flow of liquid removes retained molecules or particles from the membrane surface and a stable flux for a longer time period results. This mode of filtration is applied for RO, UF, and MF with high solids content. From time to time the membrane has to be cleaned because there is a slight flux decline with time caused by deposits on the membrane. To achieve the tangential flow a cross-flow unit has a recirculation loop. Cross-flow filtration can be operated in batch or continuous mode.

2.4 Membranes and Formats

Membrane filtration has played an important role in concentration and purification of proteins in laboratory and process scale. A broad range of membranes from different materials and products in various configurations are available on the market. The structure of the membranes differs from type to type. However, some general characteristics of micro- and ultrafiltration membranes will be discussed shortly within this chapter.

There are several methods for manufacturing membranes. Some of the methods are applicable to a variety of polymers and others are material specific. Each of these methods result in different ultrastructures, porosity, and pore size distribution. The source of the polymers used in membrane production may be naturally available materials such as cellulose or synthetic polymers such as polycarbonate, polyethylene, or polysulfone. During production the polymer is combined with a suitable solvent and a swelling agent or nonsolvent to form a casting dope, which is then sent to the casting machine. The casting machine is a critical factor in the quality of the membrane. Casting equipment is designed specifically for each kind of membrane and is usually fabricated by the membrane manufacturers themselves (Cheryan 1998).

The first commercially available membranes were made of cellulose acetate and are termed first-generation membranes. These were followed in the mid-1970s by other polymeric membranes (second-generation mem-

Table 1 Manufacturers of Membranes for Protein Separation

Filter material	Typical applications	Macherey & Nagel	Millipore	Osmonics	Pall-Gelman	Sartorius	Schleicher & Schüll	Whatman
Cellulose nitrate	Cleaning and sterilization	NC (0.2, 0.45 μm); porablot NCP (0.45 μm); porablot NCL (0.45 μm)	Immobilon-NC pure	—	BioTrace NT	Typ 113 (0.1, 0.2, 0.3, 0.45, 0.65, 0.8, 1.2, 3.0, 5.0, 8.0 μm); Polyamid (0,2, 0.45 μm)	AE/NC (0.025, 0.05, 0.1, 0.2, 0.45, 0.6, 5.0, 8.0, 12.0 μm)	Cellulose nitrate (0.1, 0.2, 0.45, 0.65, 0.8, 1.0, 1.2, 3.0, 5.0 μm)
Cellulose acetate	Protein and enzyme separation	CA (0.2, 0.45, 0.8, 1.2 μm)	—	—	Sepraphore III: Super Sepraphore; Optiphor-10	Typ 111 (0.2, 0.45, 0.65, 0.8 μm)	ST/OE (0.15, 0.2, 0.45, 0.8, 1.2 μm)	Cellulose acetate (0.2, 0.45 μm)
Mixed cellulose ester	Same as Cellulose acetate	CM (0.2, 0.45 μm); CM with spacer (0.45, 0.65, 0.80, 1.20 μm); MV (0.20, 0.45, 0.65, 0.80 μm)	MF-Millipore (0.025, 0.05, 0.1, 0.22, 0.3, 0.45, 0.65, 0.8, 1.2, 3.0, 5.0, 8.0 μm)	MCE without (0.1, 0.22, 0.45, 0.65, 0.8, 1.2, 5.0, 8.0 μm) and with spacer (0.22, 0.45, 0.7 μm)	GN-4/6 Metricel (0.45, 0.8 μm)	—	ME (0.2, 0.3, 0.45, 0.6, 0.8, 1.2, 3.0 μm)	Mixed ester (0.45 μm)
Polyether-sulfone/ Polysulfone (PES/PS)	As cellulose, and protein and nucleic acid separation; sterilization	—	Millipore Express (0.22 μm)	PES (0.1, 0.2, 0.45, 0.6 μm)	Nova (1–100 kDa); Omega (1–1000 kDa, 0.16, 0.3, 0.5 μm); Supor (0.1, 0.2, 0.45, 0.8 μm); HAT Tuffryn	Polysulfone-ultrafilter	—	—

Source: http://biotech.fr.belwue.de

branes), with polyamides and polysulfones. Until now more than 150 organic polymers have been investigated for membrane applications. Inorganic membranes based on sintered and ceramic materials are also now available. The physical structure of these membranes is complex and most of them are used for microfiltration (Grandison and Lewis 1996). The membranes themselves are thin and in most cases require support against high pressure. The support material itself should be porous. The membrane and its support are normally referred to as a module. Two major configurations that have withstood the test of time are the tubular and the flat-plate configurations.

Tubular membranes come in range of diameters. In general, tubes offer no dead spaces, do not block easily and are easy to clean. However, as the tube diameter increases, the membranes occupy a larger space, have a higher hold-up volume, and incur higher pumping costs. The two major types are the hollow fiber (Fig. 6b), with a fiber diameter of 0.001–1.2 mm, and a wider tube with diameters up to 25 mm (Grandison and Lewis 1996).

The flat-plate module can take the form of a plate-and-frame-type geometry or a spirally wound geometry (Fig. 3). The plate-and-frame system employs membranes stacked together, with appropriate spacers and collec-

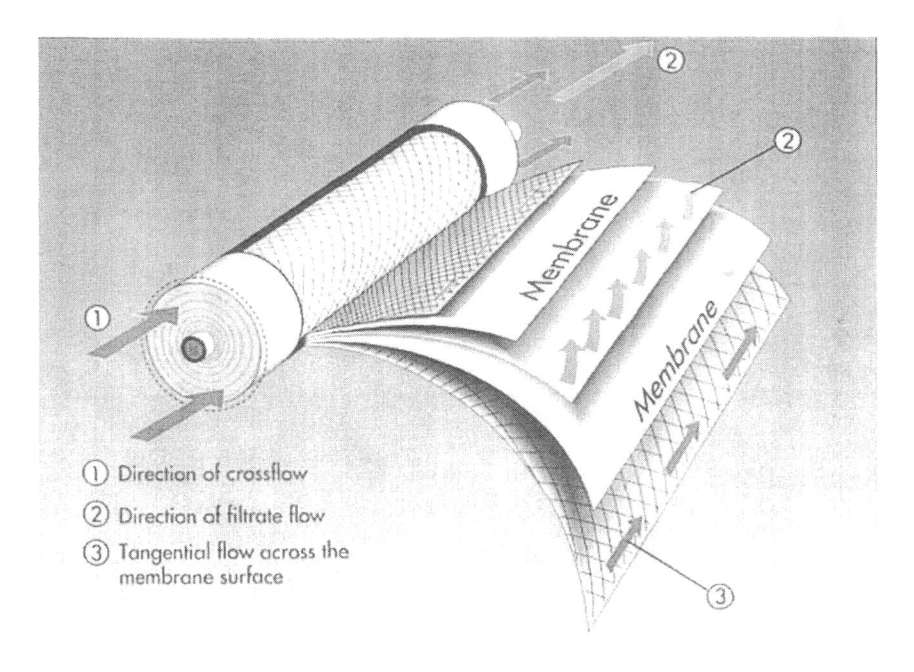

Figure 3 Spiral wound membrane module. (Schematics courtesy of Sartorius AG, Göttingen, Germany).

tion plates for permeate removal. Flow may be streamline or turbulent, and the feed may be directed over the plates in a parallel or series configuration.

The spiral wound system is now widely used and costs for membranes are quite low. In this case a sandwich is made from several sheet membranes, which enclose a permeate spacer mesh. This is attached at one end to a permeate removal tube and the other three sides of the sandwich are sealed. The typical dimensions of one spiral membrane unit would be about 12 cm in diameter and about 1 m in length (Grandison and Lewis 1996).

The main terms used to describe membranes are *microporous* and *asymmetrical*. Microporous membranes have a uniform porous structure throughout, although the pore size may not be uniform across the thickness of the membrane. They are usually characterized by a normal pore size, and no particle larger than this will pass through the membrane.

In microfiltration the membranes are isotropic and the internal structures are cohesive systems of open-celled foams, where the walls of the vacuoles form a three-dimensional network. There are no regular cylinders or capillaries inside, and the upper and lower surface show pores of variable size and shape (Fig. 4).

The exclusion limit defines the "pore size" and is determined by a Bacteria Challenge Test, e.g. with *Brevundimonas diminuta* for a 0.2-μm

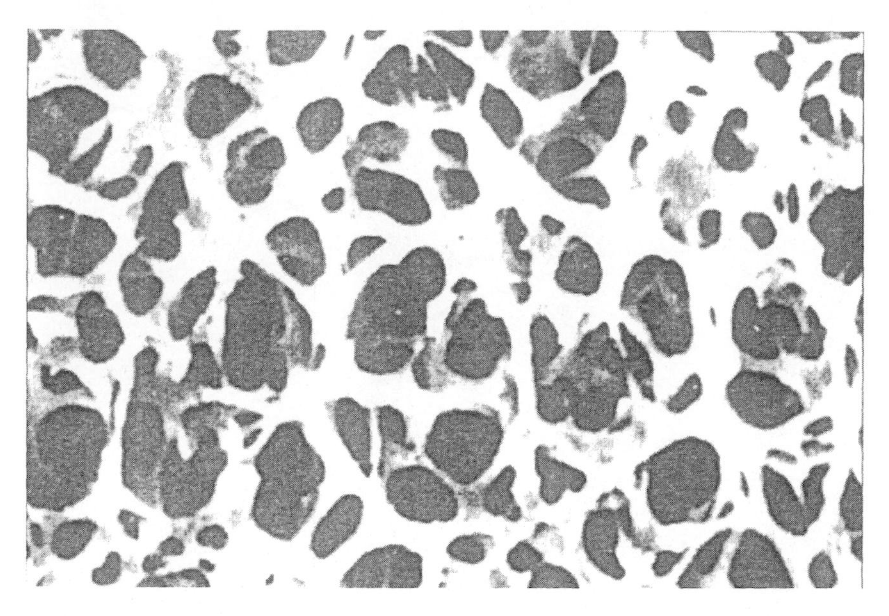

Figure 4 Microfiltration membrane, internal structure. Cross-section through a microporous membrane filter (SEM) (Sartorius).

membrane or with *Serratia marcescens* for a 0.45-μm membrane. The pore size distribution can be analyzed by Mercury Intrusion Test or with a Coulter porometer (Fig. 5). Ultrafiltration membranes are anisotropic, consisting of a thick substrate layer and a thin skin layer; the skin surface is smooth. The macroporous substrate layer serves as drain and mechanical support, and the microporous skin layer serves as the separating membrane (Fig. 6a and 6b). The pores in the skin are formed by cylindrical capillaries across the skin and exclusion limit ("cutoff") is determined by challenging the membrane with marker molecules. A UF membrane retaining the marker (nearly) completely has an exclusion limit according to the size of the marker molecule, and therefore the exclusion limit is termed nominal molecular weight limit (NMWL) and is expressed in daltons. It is equivalent to the pore size.

For a very detailed description monographs about membranes are available (Mulder 1996; Cheryan 1998; Walter 1998; Grandison and Lewis 1996). An overview of common commercially available membranes is given in Table 1, and a schematic view of membrane applications in protein downstream processing is given in Fig. 7.

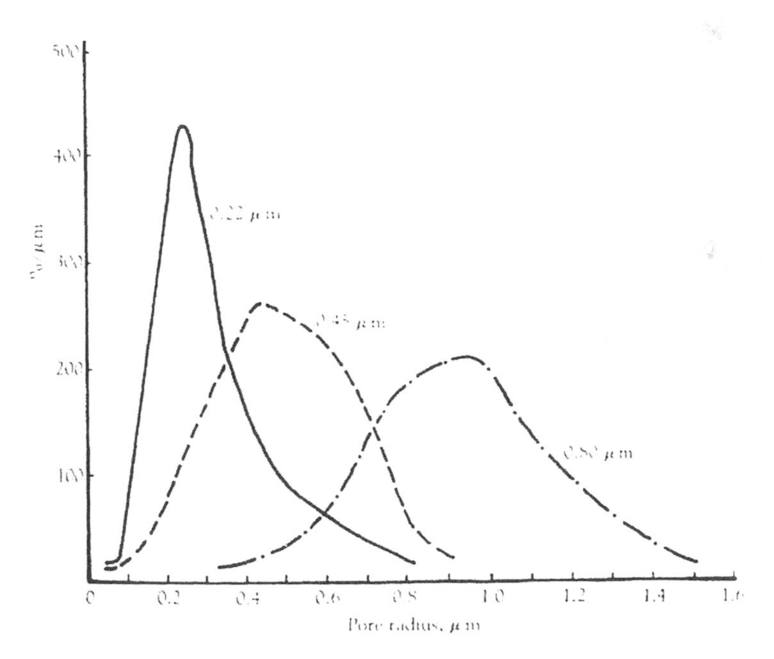

Figure 5 Pore size distribution of microporous membranes (obtained by mercury intrusion test) (Sartorius).

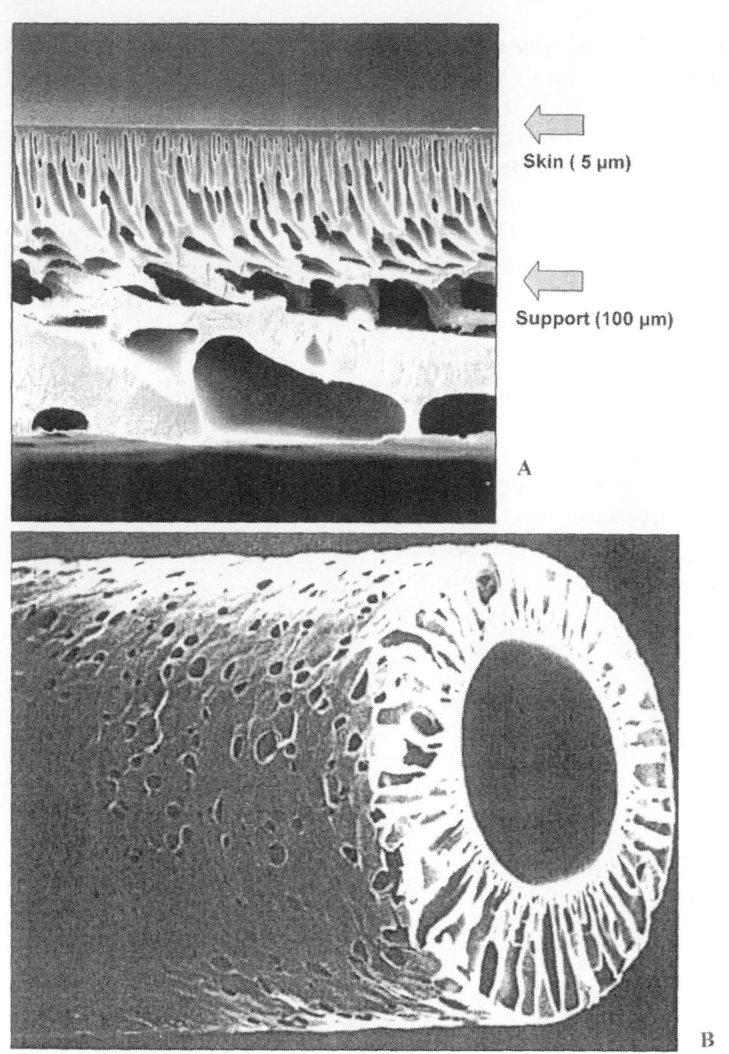

Figure 6 Cross-section through ultrafiltration membrane (SEM). (A) Flat sheet membrane and (B) hollow fiber membrane (Sartorius).

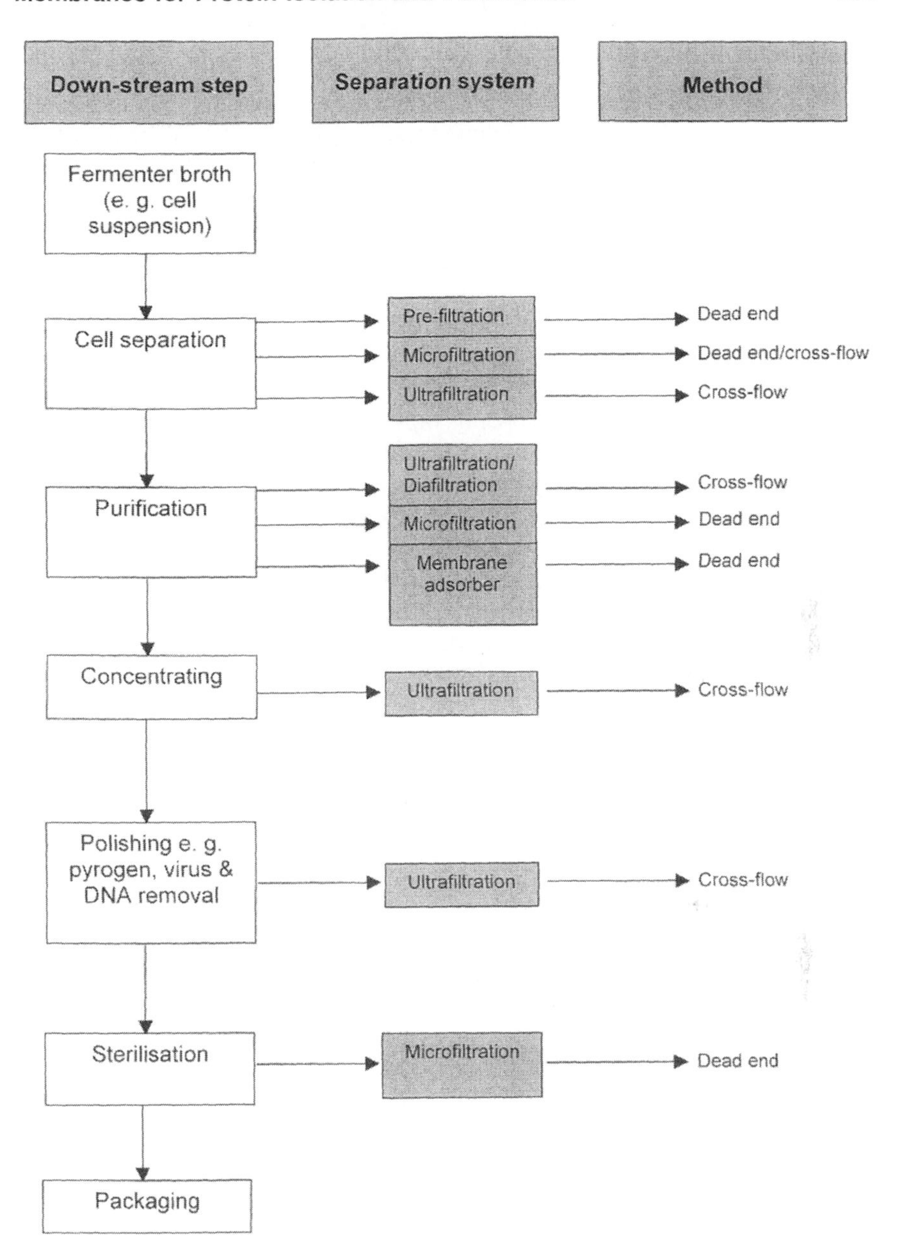

Figure 7 Flow chart for the use of membranes during protein isolation and purification.

2.5 Membrane Operation

As the cross-flow filtration technique plays a more important role in protein isolation and purification, this technique will be described in more detail. As was mentioned, the big advantage of cross-flow filtration is the formation of a nearly constant permeate flux over time (Fig. 8).

During dead-end filtration a continuous formation of a filter cake occurs. With the thickness of this filter cake the pressure drop of the membrane increases whereas the permeate flux decreases. Better results can be obtained by using the cross-flow filtration technique. During cross-flow MF and UF invariably a "secondary membrane," often called the "dynamic membrane," is formed on top of the actual separating membrane ("primary membrane"). This formation depends on the amount of colloidal suspended particles (such as lipids) and other macromolecules (such as proteins) interacting with the primary membrane.

The intrinsic properties of the primary membrane typically control the process only during the early stage of cross-flow filtration. As filtration proceeds, formation of the secondary membrane continuously alters the flux and retention characteristics of the filtration process (Datar and Rosén 1993). Detailed information about different types of these secondary membranes can be found in Tanny (1978). However, as shown in Fig. 9, the permeate flux can also drastically decrease in cross-flow filtration during the formation of a secondary membrane while the transmembrane pressure (TMP) increases. In this case the process parameters must be changed. One possible problem can be a slow retentate flux, which allows formation of a thick secondary membrane. Another possibility is a strong interaction of the sample with the membrane material. This unspecific protein adsorption is dependent on several factors:

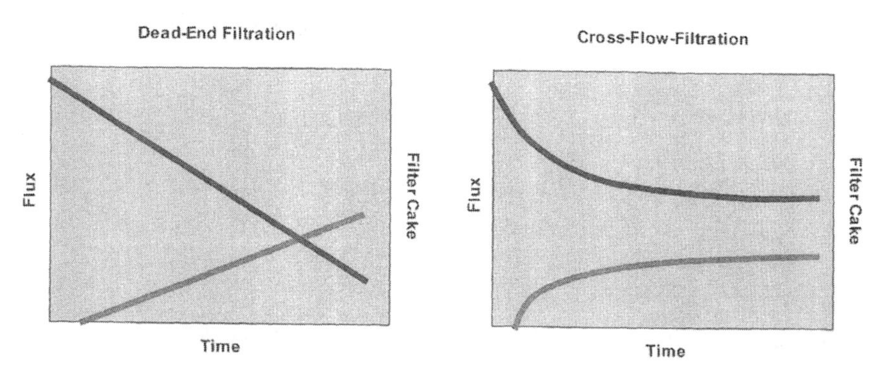

Figure 8 Dependence of permeate flux on the formation of a filter cake during dead-end and cross-flow filtration (Sartorius).

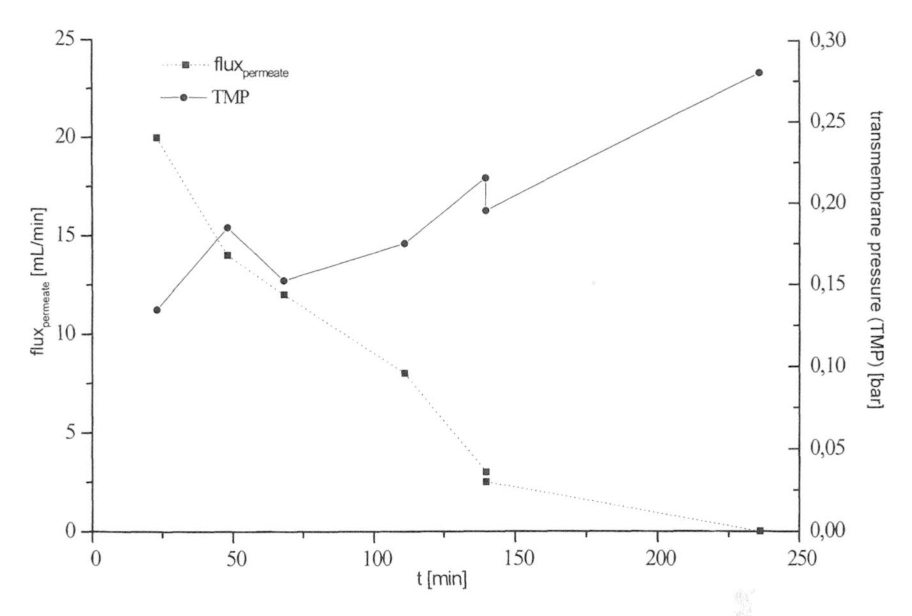

Figure 9 Effect of secondary membrane on the cross-flow filtration of sweet whey using a cellulose acetate membrane (0.2 μm; Sartorius); decrease of flux and increase of transmembrane pressure.

1. Proteins adhere to all wetted surfaces.
2. The amount of adsorbed protein depends on:
   - surface tension of the surface
   - molecular interaction between product and surface
   - protein concentration
   - pH and ionic strength
   - area of wetted surface (membranes!)

The most important problems are that

1. The binding is unspecific, and
2. Binding is irreversible

The adsorption is caused by hydrophobic interaction such as van der Waals forces and weak interactions. Thus, different membrane types must be screened to minimize the unspecific binding of the sample if a new filtration procedure should be developed. An easy change of the flow conditions (e.g., flow rates, feed temperature) should be possible to minimize unspecific binding. Thus, cross-flow filtration devices, at least in research, should be very flexible in their setup. Some typical unspecific binding capacities of different membrane materials are given in Table 2.

A typical cross-flow filtration setup is shown in Fig. 10. The system consists of a feed tank, which can, in the case of cell harvesting, be a bioreactor. The content of the feed tank is passed through the filtration unit using a circulation pump. This pump should allow different flow rates to overcome the problems described above. By regulation of the flow rate, shear forces that prevent the deposition of solute layers can be controlled.

As cross-flow filtration is a pressure driven procedure, control of the transmembrane pressure is the most important part. For this, several pressure sensors are used—one in front of the filtration unit, one for the retentate, and one on the permeate side of the membrane. The transmembrane pressure can also be regulated using a valve on the retentate side of the filtration unit, as shown in Fig. 10. Using this setup the filtration performance can either be regulated by changing the feed flow (increasing or decreasing the pump rate) or by changing the transmembrane pressure using the valve. However, the performance of a cross-flow filtration is limited by several factors as described above.

The following factors influence the molecular separation by membranes:

size of components
type of membrane
type of module
running conditions (parameters)
configuration of cross-flow filtration system

The type of membrane module is selected based on the characteristics of solution. The concentration of solids is one important factor—it is better to use open channels for high solids concentration. The viscosity of the solution also influences the performance of a filtration procedure. The best results for highly viscous solutions are obtained by spacer-filled channels. The channel is the space in a membrane module where the feed

Table 2 Unspecific Binding Capacities of Some Membrane Materials

Membrane material	Adsorbed IgG ($\mu g/cm^2$)
Polyethersulfone	4.7
Cellulose triacetate	2.7
Hydrosart	2.6

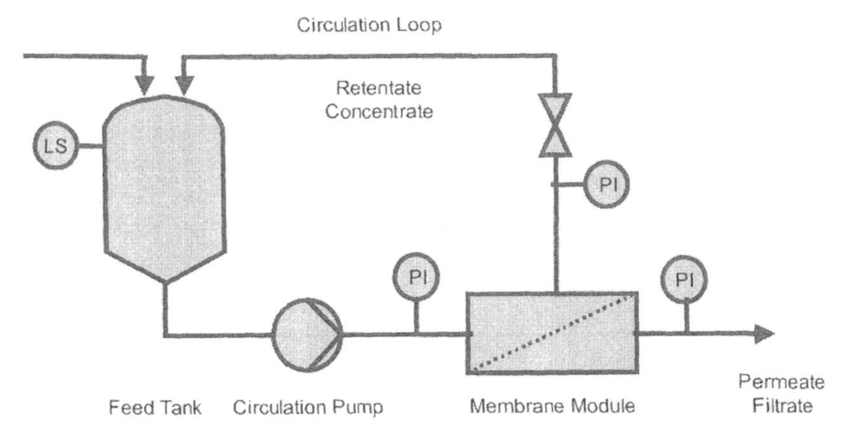

Figure 10 Schematic setup of a cross-flow filtration.

flows (e.g., the inside of hollow fibers or tubes or the space between parallel plates).

When using membrane operations for the separation of proteins, some other aspects are also important:

Avoid high temperature to minimize risk of unfolding proteins.
Avoid foaming because large areas of gas-liquid interfaces are the main source of inactivation.
Avoid high shear forces that might be a problem for mammalian cells or for very large (aggregated) molecules. However, most proteins are not shear sensitive.
Use low-shear pumps; avoid too high cross-flow velocity.

2.6 Regeneration of Membranes

A major limiting factor in membrane technology is "fouling" of the membrane. Fouling manifests itself as a decline in flux with time of operation. In its strictest sense, the flux decline should occur when all other operating parameters are kept constant (e.g., pressure, flow rate, temperature). Fouling problems can occur for several reasons:

Changes in membrane properties (physical or chemical deterioration of the membrane)
Change in feed properties
Surface topography (grade of roughness)

Charge of the membrane (most membranes are negatively charged due to their production procedure)
Pore size
Proteins (!); they are the major foulants in membrane processing, considering the multiplicity of functional groups, the charge density, the varying degrees of hydrophobicity, and the complex structure that allows a protein to interact with other feed components as well as the membrane itself.
Salts
pH
Temperature
Lipids, fats, and oils

As one can imagine from the above list, cleaning of the membranes is one of the most crucial steps in membrane operations since nearly all parameters can cause fouling of the membrane. Cleaning is the removal of foreign material from the surface and body of the membrane and associated equipment. There are some important aspects to be considered during the cleaning of membranes (Cheryan 1998):

Membrane materials and chemistry: One has to take care that the cleaning procedure does not influence the chemical structure of the membrane (e.g., cellulose acetate is very sensitive to highly acidic or alkaline conditions; polyamides are very sensitive to chlorine).
Fluid mechanics: The cleaning solutions should be pumped through the system under turbulent flow conditions.
Temperature: Chemical reaction rates double with a 10°C increase in temperature. Thus, the temperature of the cleaning solution should be as high as possible.

Several companies sell chemical cleaning compounds specifically for membranes (e.g., Ecolab-Klenzage, Diversery, Pfizer, Sigma). Many membrane manufacturers also supply their own cleaning reagents, either as powders or as liquids. In general, the following recommendations can be given for cleaning (Cheryan 1998):

Fats, oils and proteins:
0.5 M NaOH for 30-60 min at 25–55°C or
0.1% SDS, 0.1% Triton X-100 for 30 min to overnight at 25–55° C
Proteins:
Enzyme detergents for 30 min to overnight at 30–40°C
Fats, oils and grease:
20–50% ethanol for 30–60 min at 25–50°C

DNA, mineral salts:
0.1–0.5 M acid (acetic, citric, nitric) for 30–60 min at 25–35°C

Table 3 shows the results of different cleaning methods after a cell-harvesting step with a microfiltration module (Sartorius Hydrosart Cassette). The cleaning efficiency is significantly increased by washing with isopropanol, which removes the antifoam agent and the proteins more efficiently. It is also possible to recover 72% of the water flux after cleaning the UF membrane with 1 M NaOH for 1 h at 50°C. In general, it is very important to flush new filtration units with a large amount of purified water (approx. 10 L) to flush all traces of storage solution from the cassette. In addition, it is necessary to build up the pressure drop over the membrane in a very slow manner to avoid membrane blocking.

2.7 Practical Applications

In the following section some practical applications will be shown for the use of membrane systems in protein recovery, isolation, and concentration.

Cell Harvesting and Concentration of α-amylase with Cross-Flow Filtration Techniques

The use of microorganisms for the production of enzymes for industrial applications is very common today. Application of cross-flow techniques is very useful for achieving fast cell harvesting and product concentration.

α-Amylase is produced as an extracellular enzyme during the cultivation of *Bacillus subtilis*. The cultivation is performed at 20°C (pH range 7–8) up to a dry mass of 8% in batch mode (60 L). The downstream processing needs to be optimized in terms of process time while maintaining the activity of the enzyme. MF and UF are used to recover the enzyme from the culture broth (Fig. 11). The cell harvesting from the broth is performed using a

Table 3 Cleaning Efficiency of Different Chemicals for a Hydrosart 0.2 μm Membrane After a Cell Harvesting Procedure (Sartorius AG, Göttingen, Germany)

Cleaning	Water flux (L/hm^2bar) at 20°C	Recovery (%)
New membrane	1.380	100
1 M NaOH, 1 h, 50°C	1.000	72
Isopropanol 30%, 1 h, 45°C	1.350	97

0.2-μm cellulose acetate membrane (Hydrosart; Sartorius). During this step it is very important to start under most gentle conditions. Therefore, the transmembrane pressure (TMP) should not exceed 0.5 bar while the filtrate valve is closed. After 5 min the filtrate valve can be slowly opened to increase the pressure from 0.5 to 2.0 bar. The filtration is done under a constant TMP. The filtration performance is shown in Fig. 12a. A significant flow rate drop can be seen during the first hour. The performance plateau is reached at 17 L/hm^2. The increase of TMP during the filtration step does not improve the performance significantly in relation to the increased energy input.

The sharp decrease of flux is due to the presence of antifoam agent in the solution and pore clogging by the small enzyme. The filtration was stopped after recovering 60 L filtrate, which took approximately 5 h. After 210 min the enzyme was completely transferred into the permeate of 42 L. No enzyme activity was found in the retentate.

During the second filtration step the α-amylase, in the permeate obtained above, is concentrated using a UF membrane with a cutoff of 10 kDa (Hydrosart UF 10 kDa; Sartorius) as the enzyme has a molecular weight of approximately 25 kDa. The performance is shown in Fig. 12b. It is possible to obtain an almost constant flow rate during the filtration. The concentration of 40 L is complete after approximately 45 minutes with a concentration factor of 20 (i.e., reduction in volume of the enzyme solution from 42 to 2 L). The flux decrease at the end of the trial is produced by the need to reduce the TMP to avoid membrane blocking. During the concentration, temperature of the retentate increases from 21°C to 28°C. This increase can influence the stability of the product and must be taken into account. However, in the case of α-amylase this increase is not considered to be a problem for the product quality as the enzyme is stable up to 37°C. The

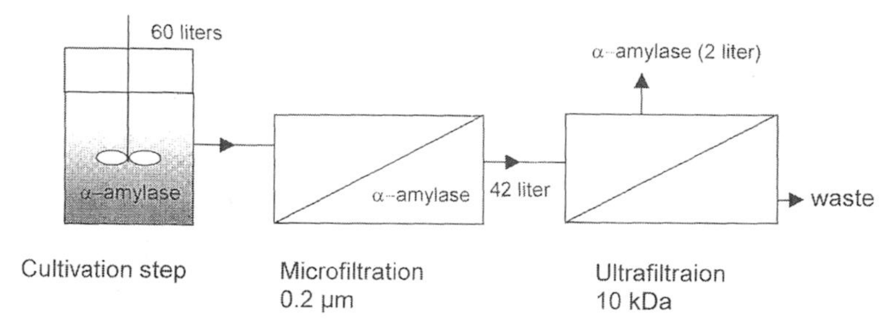

Figure 11 Process scheme for the production of α-amylase from *Bacillus subtilis* using microfiltration and ultrafiltration steps.

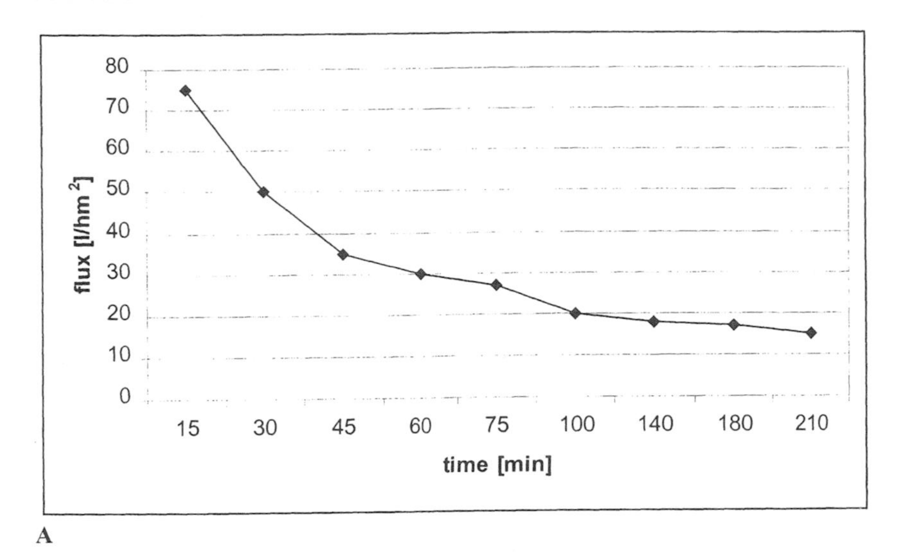

A

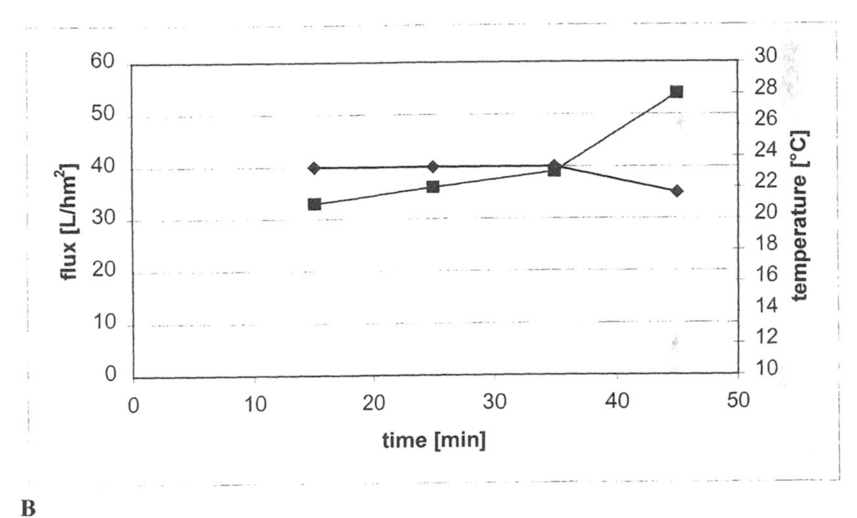

B

Figure 12 Clarification and concentration of amylase from culture broth of *Bacillus subtilis*. Filtration performance during (A) cell harvesting by microfiltration with Hydrosart 0.2-μm open channel (Sartorius), and (B) ultrafiltration with Hydrosart 10-kDa (Sartorius). ■, temperature (°C); ◆, flux (L/h m^2).

product recovery after this concentration step is 70%. No enzyme activity has been detected in the permeate.

The concentration step can easily be performed during one day, including preparation of the setup and cleaning of the system and modules.

Microfiltration and Subsequent Concentration by Ultrafiltration of an Antigen

In this example, a 15.5-kDa antigen is produced by bacterial fermentation (*E. coli* mutant) in 8-L scale. As in the above example, this antigen is removed from the biomass and concentrated using MF and UF steps.

In the first step the biomass was removed using a 0.45-μm Hydrosart slice cassette (Sartorius). With the permeate valve closed and the retentate valve open, the pump was turned on and the inlet pressure adjusted using the pump speed to $P_{in} = 2.0$ bar. This was maintained for 3–5 min to allow the formation of a loose filter cake at the membrane surface. The permeate valve was then opened and the operating pressures were set as follows: $P_{in} =$ 2.0 bar, $P_{ret} = 0.0$ bar, and $P_{perm} = 0.0$ bar. The permeate flux rate was then determined at 5-min intervals. The flux during this MF step is shown in Fig. 13a.

It is also important to evaluate the influence temperature of the fermentation broth on the filtration efficiency. As shown in Table 4, higher temperature can lead to a better concentration factor and higher fluxes due to lower viscosity of the fermentation broth. However, the limiting factor is the temperature stability of the product.

The second filtration step (UF with single 10-kDa nominal molecular weight cutoff polysulfone slice cassette, Sartorius) was performed as follows. The product was kept at 20°C for the duration of the process. The filtration was run under the procedure described above. After 1 h P_{in} was adjusted to 2.5 bar, P_{ret} adjusted to 0.5 bar, and P_{perm} remained unchanged at 0.0 bar. The performance of this filtration procedure is shown in Fig. 13b. Within 90 min filtration time, a 6044-ml antigen solution was concentrated to 425 ml.

Plasma Protein Fractionation with Advanced Membrane Systems: Optimization of the Ultrafiltration Step in Cohn's Process

Human albumin is usually obtained from plasma according to Cohn's process (developed in 1947), which is based on differential precipitation of the plasma proteins with ethanol. The pH and temperature are also varied to cause individual proteins to precipitate so that they can be separated. Until today, Cohn's process (Fig. 14) was utilized with only minor modifications in large centralized fractionation facilities connected to blood collection centers all over the world. The present annual world production of

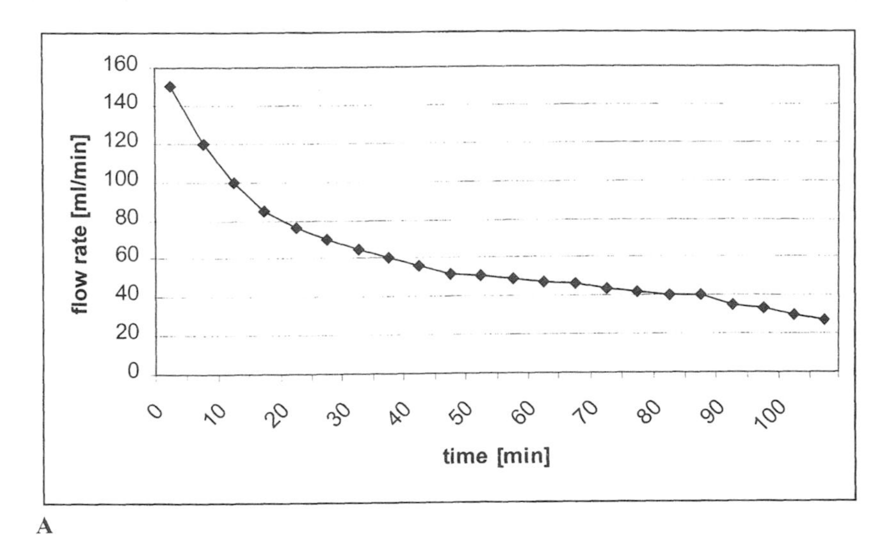

A

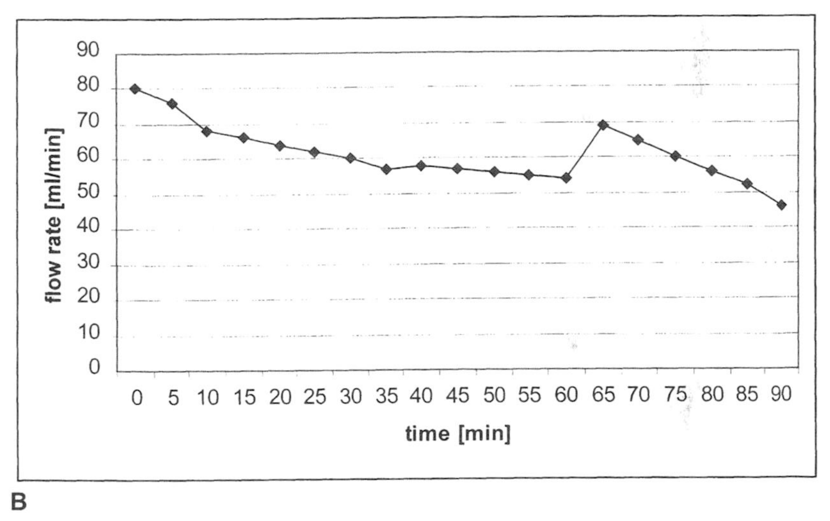

B

Figure 13 Clarification and concentration of an antigen from *E. coli* culture broth. Flux during (A) microfiltration (0.45-m Hydrosart slice cassette at 32°C) (Sartorius); and (B) ultrafiltration (10-μkDa nominal molecular weight cutoff polysulfone slice cassette) (Sartorius).

Table 4 Influence of Temperature on Filtration Efficiency

Temp. (°C)	Total volume (ml)	Retentate volume (ml)	Permeate volume (ml)	Concentration factor	Total time (min)
32	6970	650	6320	9.72	105
37	7003	475	6528	13.74	90

human serum albumin (HSA) is estimated at 300 tons derived from about 17 million liters of plasma.

In the last step, the albumin concentration in supernatant IV is approximately 13 g/L and the alcohol content of this solution is 40 vol %. To be able to handle this solution in UF steps, this solution has been further processed in four stages:

Dilution to an alcohol concentration of 20 vol % to avoid damaging of polysulfone UF cassettes

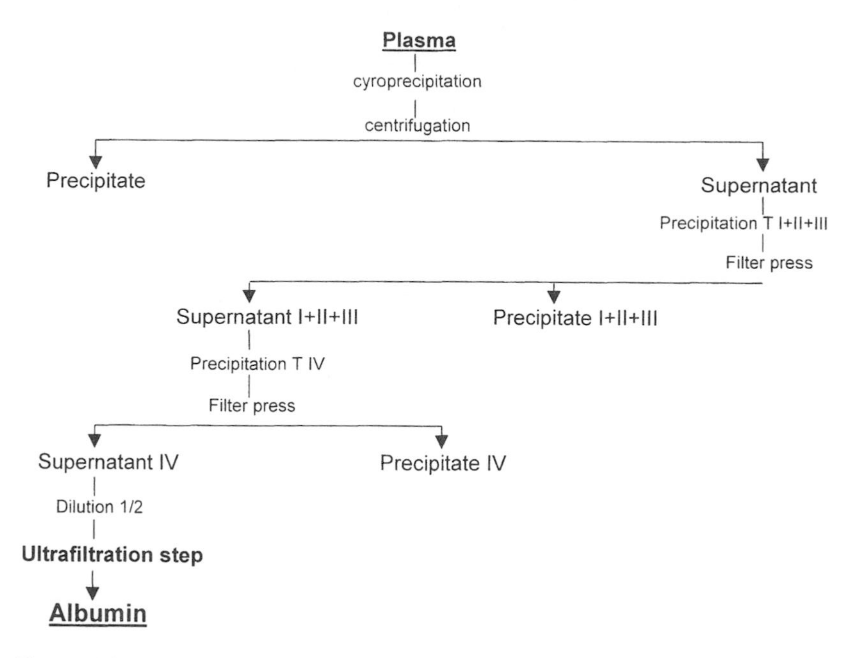

Figure 14. Scheme of Cohn's process for human serum albumin production from plasma.

Further concentration of the albumin concentrate to approximately 80 g/L. This step is the longest of the entire process (Jaffrin and Charrier 1994)
Diafiltration with water to remove alcohol
Concentration to an albumin content of approximately 200 g/L

To avoid this very complex and time-consuming UF procedure, UF was performed directly on supernatant IV having the alcohol concentration of 40 vol %. This was possible using Sartorius Hydrosart ultrafiltration cassettes (SM 305). This reduces the entire albumin processing time.

Excellent cleaning of the membrane system was achieved with 2 M NaCl. This is an additional factor demonstrating that the cassette has low adsorption of albumin. The systems used had an original flux of 48 L/hm^2 for water. This flux was attained each time the cassette had been flushed with the NaCl solution.

As shown in Fig. 15, all solutions tested could be filtered without any difficulty. In all cases, it was determined that the flux shows a strong linear dependence on the transmembrane pressure and is not drastically influenced by the content of ethanol in the sample. Thus, Hydrosart cassettes show a very low tendency to become fouled when alcohol-albumin solutions are filtered through them. This can also be manifested by a rapid stabilization (10–15 min) of the flux to a constant value after the operating conditions have been changed.

3. MEMBRANE ADSORPTION TECHNIQUES

Chromatographic processes based on adsorption with ion exchange and (immuno-) affinity media or by size exclusion have been developed since the early 1980s (Gebauer et al. 1997). The use of membrane adsorption techniques offers in relation to conventional column chromatographic procedure some advantages that lead to a better process performance, such as

Lower manufacturing costs
Non-diffusion-controlled exchange kinetics, making higher fluxes possible
Easier handling in various module forms
Easier upscaling

Raising the separation efficiency by maximizing mass transfer is the basic idea when using modified microporous membranes as the stationary matrix in liquid chromatography. Membranes can be converted into efficient adsorbers by attaching functional groups to the inner surface of synthetic

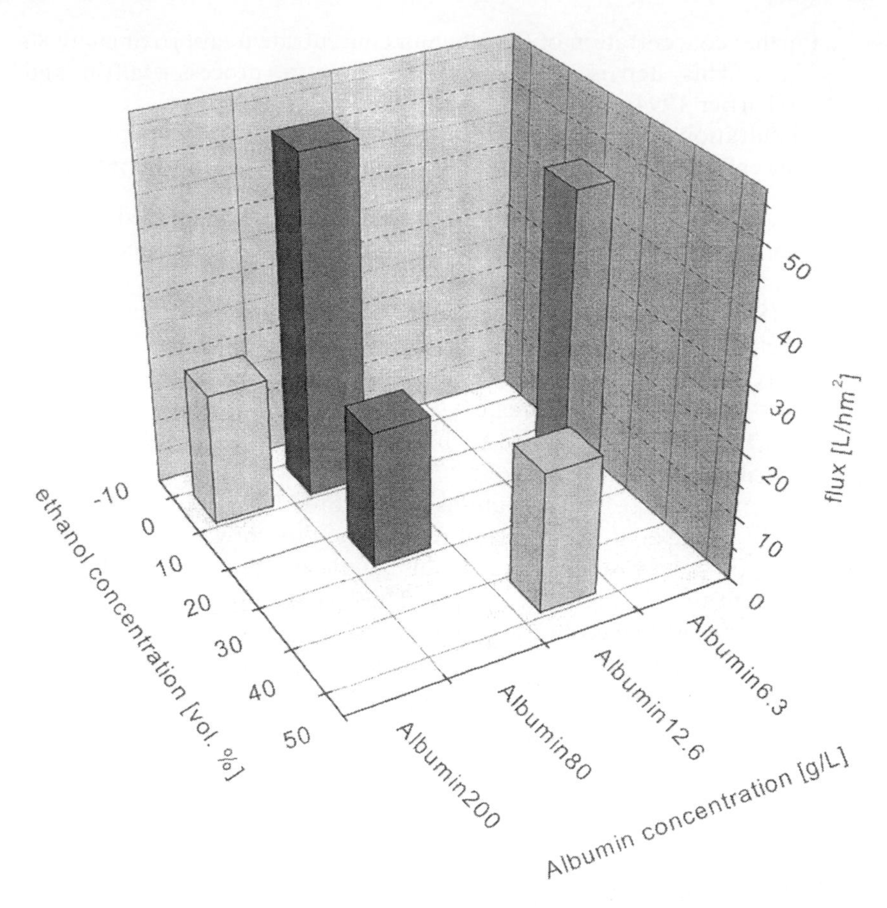

Figure 15 Recovery of human serum albumin by ultrafiltration in Supernatant IV (see Fig. 14) (different model solutions) using Satorius Hydrosart ultrafiltration cassettes (SM 305, Sartorius). The figure shows the obtained flux by combination of different albumin concentrations with different ethanol concentrations; the flux is not drastically influenced by the content of ethanol within the sample.

microporous membranes. Affinity adsorption, ion exchange, or immobilized metal affinity chromatography (IMAC) can be obtained by these membranes. Commercially available are membrane ion exchangers of strongly acidic (sulfonic acid), strongly basic (quaternary ammonium), weakly acidic (carboxylic acid), and weakly basic (diethylamine) types. A chelating membrane based on the iminodiacetate (IDA) group is applicable for IMAC. The membranes are available in products for laboratory and process scale. For process applications the modules and systems can be adopted to the special needs of the specific separation process to achieve optimal conditions.

For production and large-scale application the Sartobind Factor-Two Family of membrane adsorber modules has been developed. The modules consist of a Sartobind membrane reeled up like a paper roll to form a cylindrical module sealed at both ends with polyoxymethylene (POM) caps. For scaling up, the modules have areas between 0.12 and 8 m^2.

The different module heights can be purchased with 15, 30, or 60 layers of membrane. In combination with the different heights a variety of 15 large-scale modules are available. Since the direction of flow is from the inside to the outside of the membrane adsorber cylinder, a solid core of the appropriate size is inserted into the module to keep hold-up volume as small as possible. The solid POM cores are also available in lengths of 3, 6, 12, 25, and 50 cm and the thickness varies with the number of membrane layers used. The module is inserted in a specific housing, which consists of a top and base plate, the housing tube, and the solid core. For operating the system the unit is filled first with starting buffer. The feed solution enters the unit at the top (Fig. 16). The central cylindrical core distributes the fluid to the inside of the module. The flow is directed from the inner channel radial through the module to the outer channel (see arrows in Fig. 16). The permeate leaves the housing at the bottom plate.

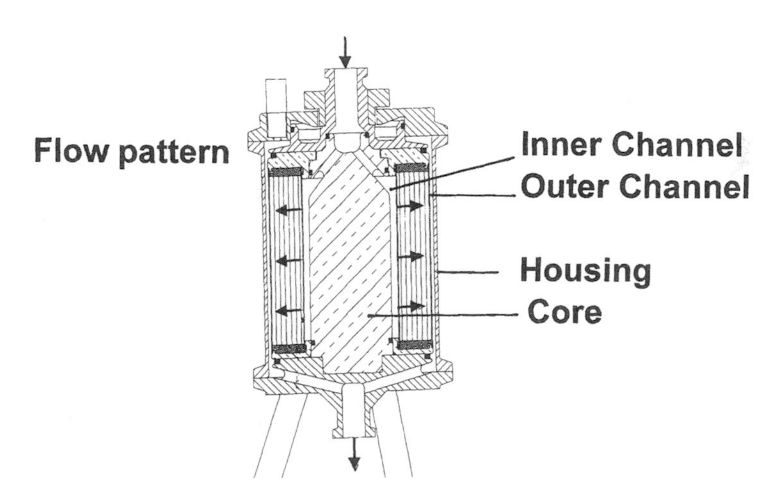

Figure 16 Scheme of Sartobind Factor-Two Family module (Sartorius); arrows indicate the flow direction of the feed.

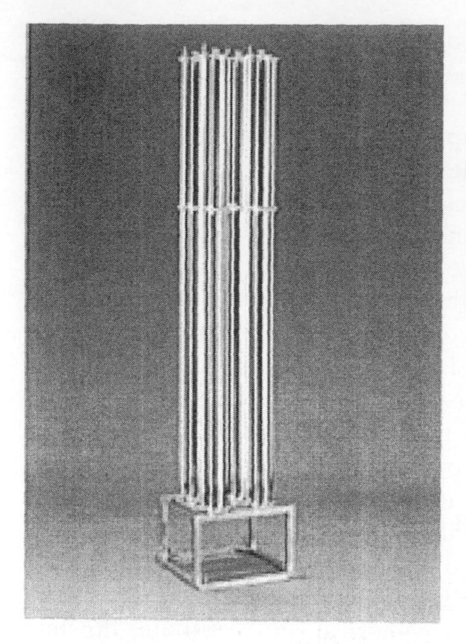

Figure 17 Different types of Sartobind Factor-Two Family modules for large-scale operation (Sartorius).

For large-scale protein isolation, adsorber modules of different sizes can be combined to achieve a desired yield and productivity. The modules offer unprecedented high flow rates and short cycle times in the range of minutes. For process plants, parallel running modular units can be combined with modules in series (Fig. 17).

Typical applications of the membrane adsorber technology are concentration of proteins and monoclonal antibodies, removal of contaminants (e.g., DNA, endotoxins), and reduction of virus content.

The membrane adsorber technology has several major advantages compared with classical separation methods. Due to the membrane structure the binding of proteins is not limited by diffusional processes; therefore, the loading and elution can be performed at very high fluxes, resulting in very short cycle times. Compressibility of the membrane under normal operation conditions can be neglected, channeling can not occur, and the pressure distribution inside the modules is designed to have plug flow through the module altogether leading to sharp breakthrough curves. Scale-up is very easy, materials and systems allow CIP (cleaning in place), and the validation of the process is made easier due to usage of standard products and validation service of suppliers.

3.1 Examples

Advanced Membrane Adsorbents in Plasma Protein Fractionation

Gebauer et al. (1997) explained in detail the application of high-capacity membrane adsorbents as a stationary phase for preparative chromatographic purification of HSA [termed *membrane adsorption chromatography* (MAC); see also Brandt et al. 1988; Hou and Mandaro 1986]. Currently other therapeutically important proteins are isolated from plasma such as IgG (10 tons), coagulation factor VIII, and protease inhibitors (Burnouf 1995). Although Cohn-related methods are employed for the production of these products as well, considerations of purity and safety require chromatographic steps in the final processing.

For the isolation of HSA from serum, Gebauer et al. (1997) designed a MAC process analogous to an established and well-documented two-step ion exchange procedure based on Sepharose (Berglöf and Erikson 1989). A schematic outline of the setup is shown in Fig. 18.

Pretreated serum was applied in a dead-end mode to the anion exchange module having a dynamic capacity of 18 mg protein per milliliter adsorbent. Albumin is bound to the membrane at pH 5.4, whereas the IgG fraction passes through this separation module. After washing the Sartobind

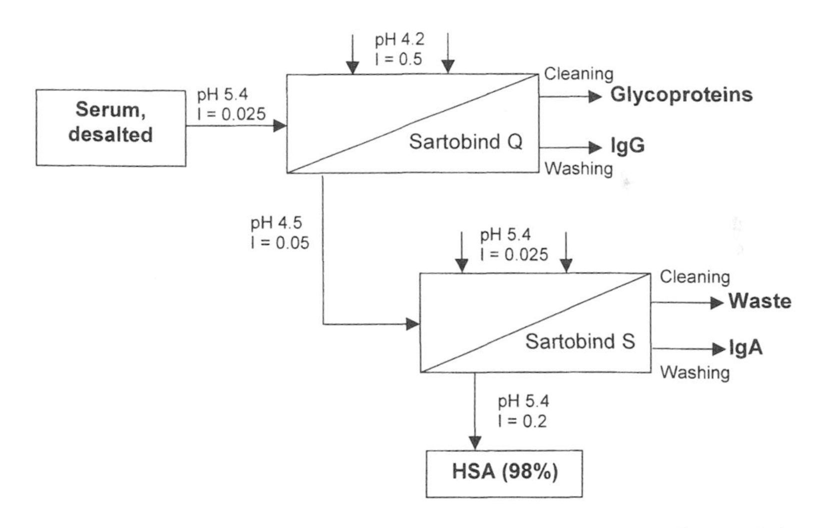

Figure 18 Tandem ion exchange plasma fractionation according to Gebauer et al. (1997); I: ionic strength (sodium acetate/acetic acid buffer); membrane system: Sartobind Q (strongly basic anion exchanger) and Sartobind S (strongly acid cation exchanger) with a total membrane area of 75 cm^2.

Table 5 Results of HSA Isolation from Human Serum by Tandem Ion Exchange Chromatography

	Fast-flow Sepharose (DEAE/CM)	Sartobind MA550 (Q/S)
Flow rate (ml/min)	2.150	820
Anion exchange step:		
HSA productivity (g/L h)	26	79
Total cycle time (min)	70	5.6
Pressure drop (bar)	3	0.2
Yield (%)	99	94
Dyn. capacity HSA (g/L)	31	7.5
Cation exchange step:		
HSA productivity (g/L h)	26	40
Total cycle time (min)	70	4.5
Pressure drop (bar)	3	0.2
Yield/purity (%)	95/97	99/82
Dyn. capacity HSA (g/L)	30	3.0

Data from Gebauer et al., 1997.

Q system the albumin was eluted by decreasing the pH down to 4.5 while slightly increasing the ionic strength (fraction A). Better binding glycoproteins can be removed at a high salt concentration.

The second adsorbent unit, a cation exchanger (Sartobind S), has a capacity of 14 mg/ml of adsorbent. The albumin of fraction A is bound to the membrane while other proteins (IgA) are removed. After elution of HSA the membrane is regenerated at high pH and increasing salt concentration.

Table 5 shows the results of HSA isolation by MAC process in comparison to conventional particle-based chromatography on fast-flow Sepharose. It is obvious that the membrane adsorbents result in much higher productivity. Even though the loading capacity of the membrane system is much lower than that of Sepharose, 10-fold shorter cycle times makes the Sartobind Q/S system is up to nine times more efficient with regard to the productivity.

Isolation of Protein from Whey

About 10 million tonnes of whey, a by-product in dairy industry, was produced in Germany in 1996. Further processing involves production of whey powder, lactose, and whey protein concentrates (WPCs). Due to the enormous volume, whey is concentrated about five times with evaporators before transportation to the whey-refining industry. Besides the major whey proteins (α-lactalbumin, β-lactoglobulin, BSA, and bovine IgG), minor whey proteins, such as lactoperoxidase, lacto(trans)ferrin, and other minor compounds, are present. The glycoprotein lactoferrin (LF) belongs to the transferrin protein family with threefold stronger binding properties to iron than transferrin. Although the protein was already isolated more than 35 years ago, its function is still not totally obvious. Beside the well-known biological functions as an antimicrobial and antiviral agent, the protein shows immunomodulatory functions in the host defense system.

To date, minor protein components like bovine lactoferrin (bLF) are isolated by standard column chromatographic procedures. Now Sartobind adsorber (Sartorius, Göttingen, Germany) is used for the direct removal of bLF from sweet whey. Due to its high isoelectric point of 8.5–9.0, bLF can be bound to strong cationic membrane adsorber (Sartobind S). However, using such membrane absorber technique one requires a pretreatment of whey to remove the insoluble particles and lipids that otherwise will block the membrane. Thus, one must set-up a continuous cross-flow filtration step. The permeate coming from this filtration step can be directly pumped through the membrane adsorber. Figure 19 depicts the setup of the whole downstream process.

For the cross-flow filtration of cheese whey various membranes of different geometries and pore sizes were used, as shown in Table 6. The experiments were carried out in the recycled batch modus, in which the retentate and the permeate were brought back into the batch reactor to keep the concentration of bovine lactoferrin constant over the whole filtration time. The filtration temperature was 50°C. To assess the cleaning effect of the filtration, optical transmission measurements of the retentate and permeate samples were made at 600 nm against water as blank. Permeates showing transmission values of at least 30 did not cause blocking effects on the ion exchange membranes.

The tubular modules provided the highest permeate fluxes and constantly high permeation rates of bLF. Based on their geometry, tubular modules allow extremely high retentate fluxes, the resulting shear rates of about 9000 L/s prevent fouling on the membrane's surface that limits flux and permeation rates. The pore size of 1 μm did not lead to permeates of adequate quality (transmission values of less than 30), so for the following whey filtration a Microdyn 0.2-μm tubular module was used.

For the isolation of bLF strongly acidic ion exchange membranes from Sartorius were used. The spirally wounded modules are available in various heights and layers, resulting in membrane areas ranging from 0.1 to 8.0 m^2 per module. The results presented were obtained with two 1m^2

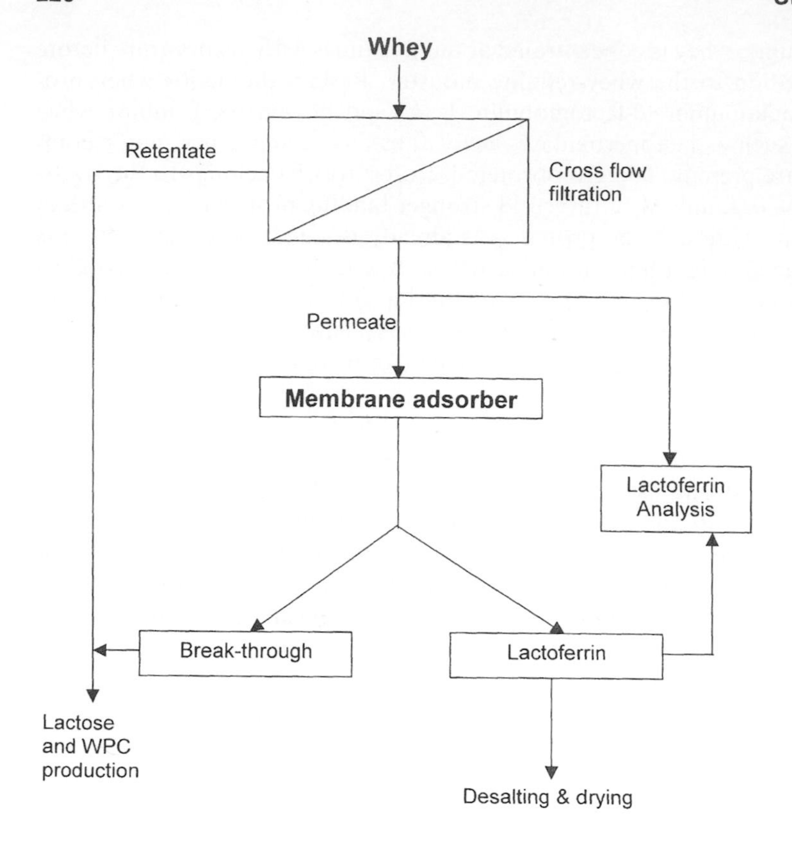

Figure 19 Schematic view of the downstream process for recovery of bovine lactoferrin (bLF) from sweet whey using a membrane adsorber system.

Table 6 Results of the Cross-Flow-Filtration of Cheese Whey with Various Modules of Different Geometries and Pore Sizes

Module	Geometry/ pore size (μm)	Permeate flux (L/m^{2}h)	Transmission 600 nm	Permeation of lactoferrin [%]
Microdyn	Tubular / 0.2	450	80	90
	Capillary / 0.4	300	80	55
	Tubular / 1.0	1150	23	85
DSS	Flat sheet / 0.1	45	70	5
	Flat sheet / 0.2	50	73	5
	Flat sheet / 0.45	85	20	25
APV	Ceramic / 0.8	60	80	20
Koch-Abcor	Spiral / 1.2	45	54	35

modules (type S-10k-15-25) in series. The flow rates varied between 1.0 and 3.0 L/min; the process was monitored by a UV detector (Model 662 UV Analyzer, Wedgewood Technol., San Carlos, CA) at 280 nm. Prior to use, the membranes were pre-equilibrated with a 20 mM sodium chloride solution at pH 7.0; afterward bLF and lactoperoxidase as byproducts were eluted at increasing sodium chloride concentrations.

More than 90% of the bound lactoperoxidase was eluted from the modules at a salt concentration of 0.2 M NaCl, together with about 15% of the bound bLF. At salt concentrations of more than 0.5 M NaCl, no other proteins could be detected in the eluates. To get two protein fractions as pure as possible the salt concentration of the used eluants was optimized. The best result was achieved with a three-step salt gradient, which led to a lactoperoxidase fraction of about 85% purity and an bLF fraction of about 95% purity (Fig. 20).

In previous experiments the dynamic binding capacity of the cation exchange membranes for bLF was found to be 0.2 mg/cm^2, so that it was possible to isolate 4 g of the target protein per process cycle. Several cycles were carried out successively without a cleaning procedure, as the loading step of each new process cycle led to a self-regeneration of the ion exchange groups on the surface of the modules.

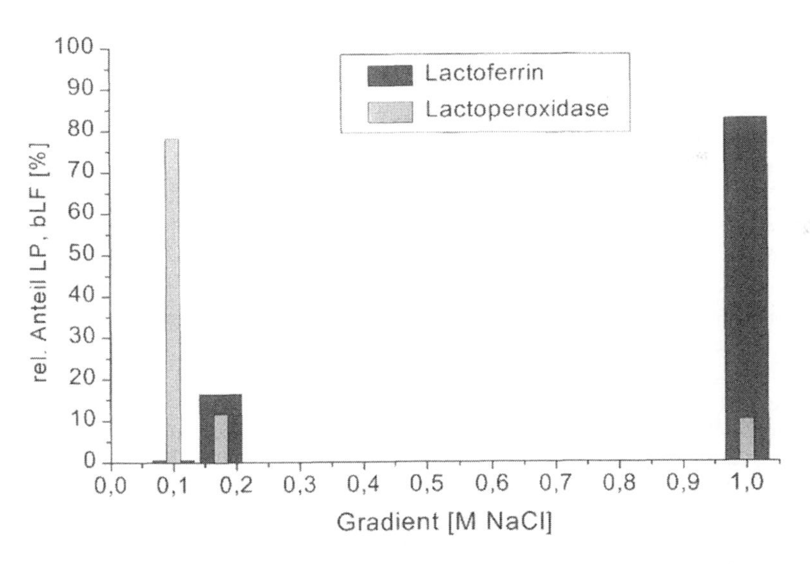

Figure 20 Optimized three-step salt gradient (0.1 M, 0.175 M, 1 M NaCl) for the elution of lactoperoxidase and lactoferrin from Sartobind S (Sartorius).

The desalting and concentration of the protein-containing eluate was carried out by cross-flow ultrafiltration. The collected bLF-containing permeates of five process cycles, about 12 L, were desalted within 2 h. The loss of protein that was bound to the membrane's surface was 10%. By rinsing the module with distilled water after finishing the ultrafiltration the loss could be reduced to 6%.

Using a two-step downstream process consisting of a cross-flow filtration and a membrane adsorber, it is possible to isolate bLF from sweet whey in a very suitable manner. It is very important to produce within the cross-flow filtration a permeate with a high permeation of bLF (about 90%), whereas the transmission at 600 nm should be higher than 30. Using a Microdyn tubular system this can also be performed at high permeate fluxes (450 L/h m^2). Using a strong cationic membrane adsorber the bLF could be isolated from the permeate produced during the cross-flow filtration within 15 min (loading 8 min, elution 4 min, washing 2 min). Thus, this system offers great advantages in comparison with chromatography columns such as non-diffusion-controlled kinetics resulting in shorter cycle times.

REFERENCES

Berglöf, J. H., Erikson, S. E. (1989) Plasma fractionation by chromatography of albumin and IgG. In: Biotechnology of Plasma Proteins (Stoltz, J. F., and Rivat, C., eds.). INSERM, Paris, pp. 207–216.

Brandt, S., Goffe, R. A., Kessler, S. B., O'Conner, J. L., Zale, S. E. (1988) Membrane-based affinity technology for commercial scale purifications. Bio/Technology 6, 779–782.

Burnouf, T. (1995) Chromatography in plasma fractionation: Benefits and future trends. J. Chromatogr. 664, 3–15.

Cheryan, M. (1998) Ultrafiltration Handbook. Technomic, Basel.

Datar, R. V., Rosén, C.-G. (1993) Cell and cell debris removal: Centrifugation and cross flow filtration. In: Biotechnology, 2nd ed. (Rehm, H.-J., Reed, G., eds.). VCH, Weinheim, pp. 472–502.

Gebauer, K. H., Thömmes, J., Kula M.-R. (1997) Plasma protein fractionation with advanced membrane adsorbents. Biotechnol. Bioeng. 53, 181–189.

Grandison, A. S., Lewis, M. J. (eds.) (1996) Separation Processes in the Food and Biotechnology Industries. Woodhead Publishing Ltd, Cambridge, UK.

Hou, K. C., Mandaro, R. M. (1986) Bioseparation by ion exchange cartridge chromatography. Biotechniques 4, 358–367.

Jaffrin, M. Y., Charrier, J. Ph. (1994) Optimisation of ultrafiltration and diafiltration processes for albumin production. J. Membr. Sci. 97, 71–81.

Lewis, M. J. (1996) Pressure-activated membrane processes. In: Separation Processes in the Food and Biotechnology Industries (Grandison, A. S., and Lewis, M. J., eds.) Woodhead Publishing, Cambridge, UK, pp. 65–96.

Mulder, M. (1996) Basic Principles of Membrane Technology. Kluwer Academic, Dordrecht.

Tanny, G. (1978) Dynamic membranes in ultrafiltration and reverse osmosis. Sep Purif Meth 7(2), 183–200.

Walter, J. K. (1998) Strategies and considerations for advanced economy in downstream processing of biopharmaceutical proteins. In: Bioseparation and Bioprocessing, Vol. 2 (Subramanian, G., ed.). Wiley-VCH, Weinheim.

[illegible]

7

Precipitation of Proteins: Nonspecific and Specific

Ashok Kumar, Igor Yu. Galaev, and Bo Mattiasson
Lund University, Lund, Sweden

1. INTRODUCTION

Precipitation of proteins from biological fluid, e.g., the precipitation of casein from milk by dilute acid, is known since old times. For some time, precipitation was the only practical way of separating different types of proteins by causing part of a mixture to precipitate through altering some property of the solvent. Precipitation still remains an important operation for the laboratory and industrial scale recovery and purification of proteins—often used in the early stages of downstream processing for both product stream concentration and fractionation. Though chromatography is the workhorse of protein recovery and purification operations, precipitation remains an indispensable unit operation. The efficiency of chromatographic separations is largely dependent on initial steps involving precipitation procedures (Englard and Seifer 1990). Up to 80% of published protein purification protocols include at least one precipitation step, ranging from primary isolation [separating cellular debris from protein product in recombinant cell cultures (McGregor 1983; Nakayama et al. 1987)] to finishing operations [producing solids prior to drying and formulation (Hoare et al. 1983; Paul and Rosas 1990; Niederauer and Glatz 1992)], to yield a partially purified product of reduced volume (Englard and Seifer 1990; Chen et al. 1992). Selective precipitation of target protein from a crude mixture is more attractive as the separate protein-enriched phase forms from a homogeneous solution during the process, and the mechanical separation of the protein-enriched phase (pellet) from the protein-depleted phase (supernatant) is easily achieved by well-established and simple techniques such as filtration or centrifugation. The main function of precipitation is to concentrate the target protein. The protein precipitated from a large volume of the crude extract could be dissolved afterward in a small buffer volume. The technique also allows rapid isolation of protein products from denaturing conditions and proteolytic enzymes (Chen et al. 1992). It is easily adapted to large scale, uses simple equipment, has a large number of inexpensive alternatives, and can be done without denaturation of biological products (Glatz 1990).

The most important aspect of protein precipitation is its specificity, i.e., how a particular protein is precipitated selectively from the other proteins in the crude mixture. While most widely used precipitation modes are nonspecific in nature, much of the emphasis has been given lately to impart specificity in precipitation. Traditionally precipitation of target protein is achieved by the addition of large amounts of salts such as ammonium sulfate, polymers such as polyethylene glycol (PEG), or organic solvents miscible with water such as acetone or ethanol (Scopes 1994). Precipitation of the target protein occurs because of changing bulk parameters of the medium, the driving force being integral physicochemical and surface properties of the protein macromolecule. The macromolecular nature of protein molecules combined with a general principle of their folding—hydrophilic amino acids at the surface, hydrophobic inside the core—make proteins rather similar in their surface properties. One does not expect high selectivity to be achieved by traditional precipitation techniques, as the selectivity of precipitation is limited to the differences in integral surface properties of protein molecules. Nevertheless, the techniques are widely used as the first step of protein purification, combining capture of the target protein with some purification. The introduction of higher specificity to precipitation techniques is certainly of great importance. The present trend in downstream processing of proteins is the introduction of highly selective affinity steps at the early stage of purification protocol when using robust, biologically and chemically stable ligands (Kaul and Mattiasson 1992). The present chapter describes various aspects of specific and nonspecific precipitation for protein isolation and purification. An attempt to introduce the highly selective affinity technique at the very beginning of protein purification is discussed.

2. Precipitation of Proteins: Nonspecific

Original classifications of proteins mainly depended on their solubility behavior in aqueous solution. Best known examples are serum albumin and immunoglobulins. Globulins are proteins of generally low solubility in aqueous media that tend to have a substantial hydrophobic amino acid content at their surface. Conversely, there are many true albumins, with high aqueous solubility and low hydrophobicity. The distribution of charged and hydrophobic residues at the surface of the protein molecule is the feature that determines solubility in various solvents. In a protein molecule, a substantial number of hydrophobic groups reside on the surface, in contact with the solvent, and are as important in determining the behavior of the molecules as are charged and other polar groups. The solubility behavior of the protein can be changed drastically as the solvent properties of water are manipulated, causing the protein to precipitate out from the medium. Protein precipitants include inorganic cations and anions, NH_4^+, K^+, Na^+, SO_4^{2-}, PO_4^{3-}, acetate$^-$, Cl^-, Br^-, NO_3^-, ClO_4^-, I^-, SCN^- for salting out (England and Seifer 1990; Shih et al. 1992; Scopes 1994; Baldwin 1996); bases or acids, H_2SO_4, HCl, NaOH for isoelectric precipitation (Chan et al. 1986; Fisher et al. 1986); organic solvents such as ethanol, acetone, methanol, *n*-propanol (Englard and Seifer 1990; Scopes 1994); nonionic polymers such as PEG (Ingham 1990) and polyelectrolytes (polyacrylic acid, carboxymethylcellulose, polyethyleneimine) (Clark and Glatz 1987; Chen and Berg 1993; Li et al. 1996). Heat- and pH-induced denaturing perturbations are also used to precipitate contaminant proteins (Scopes 1994).

Most of these protein precipitation methods involve the following steps. First, the protein environment is altered by the addition of a precipitating agent, causing the solution to become unstable. Second, a solid phase appears as small spherical "primary" particles of solid protein, which grow by diffusional transport of protein molecules to the solid surface. Third, primary particles aggregate as a result of convective transport and lead to floc formation. Finally, the aggregate (floc) size is limited by hydrodynamic disruption of the aggregates, generating smooth and uniform precipitate particles. Subsequent separation of solids depends on maximizing the aggregate's size and density. This is schematically presented in Fig. 1. Nelson and Glatz (1985), when studying soy protein precipitates, have shown that the precipitant, precipitation conditions, and reactor configuration significantly affect the primary particle size. However, in most of the precipitating modes using salt, solvent, isoelectric or metal ion, the mechanism of precipitate aggregation is the same (Chan et al. 1986). In all of the precipitation methods, protein fractionation occurs at the primary particle level.

2.1 Salt Precipitation

The most common type of precipitation for proteins is salt-induced precipitation. At low concentration of the salt, solubility of the proteins usually

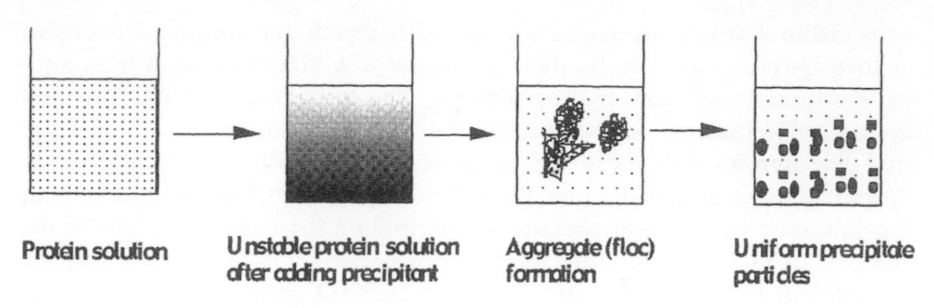

Figure 1 Schematic presentation of different steps involved in nonspecific precipitation.

increases slightly (*salting in*). But at high concentrations of salt, the solubility of the proteins decreases sharply (*salting out*) and the proteins precipitate out (Fig. 2). The addition of salt in high concentration diminishes electrostatic repulsion between similarly charged groups at the protein surface and disturbs the structure of water molecules around the protein globule. Salt ions compete with protein globules for water and, eventually, at a sufficiently high concentration, strip the latter of aqueous shell. Aqueous salt solution becomes a poor solvent for proteins, which precipitate out. The number and distribution of charges, nonionic polar groups, and hydropho-

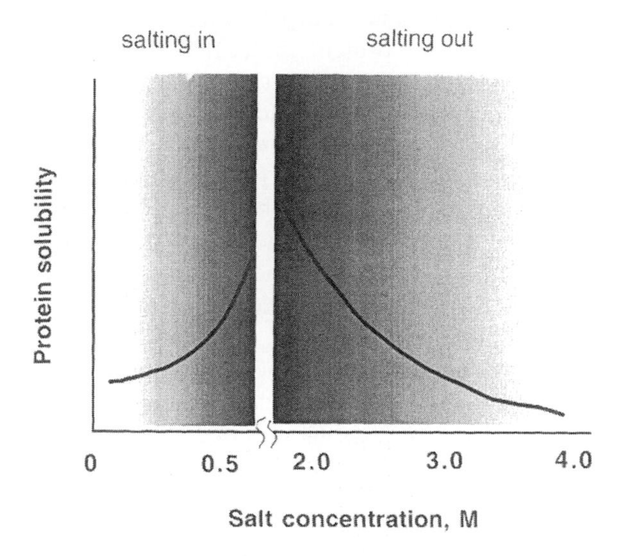

Figure 2 Dependence of protein solubility on salt concentration: *Salting in* and *salting out* of proteins. (Adapted from Scopes 1994.)

bic residues on the surface of the protein determines the concentration of the salt needed to cause precipitation of the protein. The size and shape of the protein also contribute to the precipitating property of the protein. This makes the basis for the particular group of proteins with closely related properties to precipitate from the mixture of proteins and is not specific to a particular protein. For practical considerations, one must consider physical characteristics of the salt, such as solubility, besides its effectiveness in causing precipitation. The medium conditions, such as pH and temperature, are also important (Shih et al. 1992). The most effective salts are those with multiple-charged anions such as sulfate, phosphate, and citrate; the cation is less important. Salting-out ability of anions follows the Hofmeister series. In order of decreasing effectiveness, this series is citrate > sulfate > phosphate > chloride > thiocynate. The tendency for a salt to cause denaturation of a protein is inversely related to its position in the Hofmeister series (Bell et al. 1983). For cations, monovalent ions should be used, with $NH_4^+ > K^+ > Na^+$ in precipitation effectiveness. Because of higher solubility in the wide temperature range of 0–30°C and lower density of the saturated solution (in comparison with the other salts), ammonium sulfate is preferred over other useful salts. The solubility of the protein as a function of the ionic strength follows the empirical formula:

$$\log S = A - m(\text{salt concentration})$$

where A is a constant dependent on temperature and pH, and m is a constant dependent on the salt employed but independent of temperature and pH. The more theoretical and practical aspects of the salt precipitation of proteins are described in the reviews by Englard and Seifter (1990) and Scopes (1994).

Various models have been proposed that deal with protein solubility with respect to salts. The first model was proposed in 1925 by Cohen. Many years later, Melander and Horvath (1977) suggested protein solubility as a function of salt concentration by relating the solubility behavior to hydrophobic effects. Recently, a molecular-thermodynamic model was proposed for salt-induced protein precipitation (Chiew et al. 1995; Kuehner et al. 1996). The protein molecules are considered to interact in a manner described by a set of spherically symmetrical two-body potentials of mean force. These include screened Coulombic repulsion, dispersion (van der Waals) attraction, osmotic attraction, and an attractive square-well potential intended to model specific protein-protein chemical interactions (including the hydrophobic effect and protein self-associations). More recently, Agena et al. (1999) described protein solubility model as a function of salt concentration and temperature for a four-component system consisting of a protein, pseudosolvent (water and buffer), cation, and anion (salt). Two different systems, lysozyme with sodium chloride and concanavalin A with ammonium sulfate, were investigated. Comparison of the modeled and experimental protein solubility data results in an average root mean square deviation of 5.8%, demonstrating that the model closely follows the experimental behavior.

A new versatility for the ammonium sulfate precipitation that offered operational ease and simplified tedious manual operations was recently introduced (Ito 1999). A novel chromatographic system internally generates a concentration gradient of ammonium sulfate along a long channel to fractionate proteins according to their solubility in the salt solution. The principle of the method is based on centrifugal precipitation chromatography and is depicted in Fig. 3. The separation column consists of a pair of disks with mutually mirror-imaged spiral channels that are separated by a semipermeable membrane. The disk assembly is mounted on a seamless continuous flow centrifuge. A concentrated ammonium sulfate solution is introduced into the upper channel while water solution is passed through the lower channel in the opposite direction in the rotating column. A mixture of proteins injected into the water channel moves along a salt gradient of increasing concentration that has been established in the water solution. Each protein species can precipitate at a different salt concentration along the gradient. The group of proteins with lowest solubility in the salt solution is the first to be precipitated and does so in an early portion of the lower

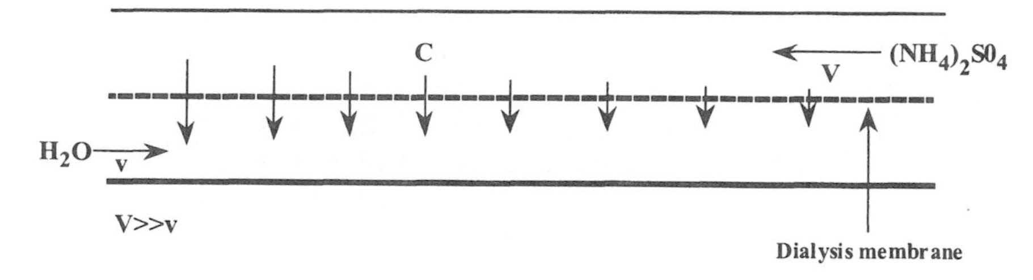

Figure 3 Principle of centrifugal precipitation chromatography applied to fractionation of proteins with ammonium sulfate. Two channels partitioned with a dialysis membrane. Ammonium sulfate solution and water flow countercurrent through these channels to produce an exponential concentration gradient of salt in the water channel. Salt transfer takes place from the upper channel to the lower channel through the dialysis membrane at every portion of the column. As the process continues, the two liquids soon establish a steady-state equilibrium where a salt concentration gradient is formed in each channel along the entire length. Since the flow rate of salt solution through the upper channel (V) is much greater than that of water through the lower channel (v), the ammonium sulfate concentration (C) in the upper channel is not significantly altered. (Reproduced from Ito 1999 with permission.)

channel, whereas others continue to advance through the lower channel until they reach their own critical points. Thus, proteins are precipitated along the distal wall of the lower channel according to their solubility. After all proteins in the sample are precipitated in this manner, the concentration of the ammonium sulfate fed into the upper channel is now gradually decreased. This leads to the decrease in the gradient of salt concentration in the lower channel at every point. This dissolves the once precipitated proteins, but they will again be precipitated at a slightly more advanced location of the channel. Thus, the proteins are subjected to a repetitive process of dissolution and precipitation until they are eluted out from the channel unit. The eluate is continuously monitored and fractions are collected to isolate the desired proteins. This salt fractionation method also promises some specificity in protein isolation when a specific affinity ligand is introduced in the protein sample solution. For example, a recombinant enzyme, ketosteroid isomerase, from crude *E. coli* lysate was selectively precipitated using β-estradiol-17-methyl-PEG 5000 in the protein lysate. The ligand effects the precipitation behavior of the target protein; thus, the protein-ligand complex showed lower solubility than the protein alone, thereby precipitating selectively earlier in the channel unit (Fig. 4).

2.2 Precipitation with Organic Solvents

Precipitation by addition of an organic solvent has been especially important on the industrial scale, in particular in plasma protein fractionation (Curling 1980). Its use in laboratory scale purifications has been less extensive, though there are some advantages compared with salting-out precipitation. Addition of solvent such as ethanol or acetone to an aqueous extract containing proteins has a variety of effects which together lead to protein precipitation. The principal effect is the reduction in water activity. There is a medium decrease in the dielectric constant with the addition of an organic solvent leading to the decrease in the solvating power of water for a charged, hydrophilic protein molecule, and thus protein solubility decreases and precipitation occurs. This is described by the following empirical equation:

$$\text{Log } S = A/\varepsilon^2 + \log S_0$$

where S is solubility in the presence of a solvent and S_0 is the original solubility. A is a constant depending on the temperature and protein employed, and ε is the dielectric constant depending on the type of solvent used.

The ordered water structure around hydrophobic areas on the protein surface can be displaced by organic solvent molecules, leading to a relatively higher solubility of these areas. So some extremely hydrophobic proteins,

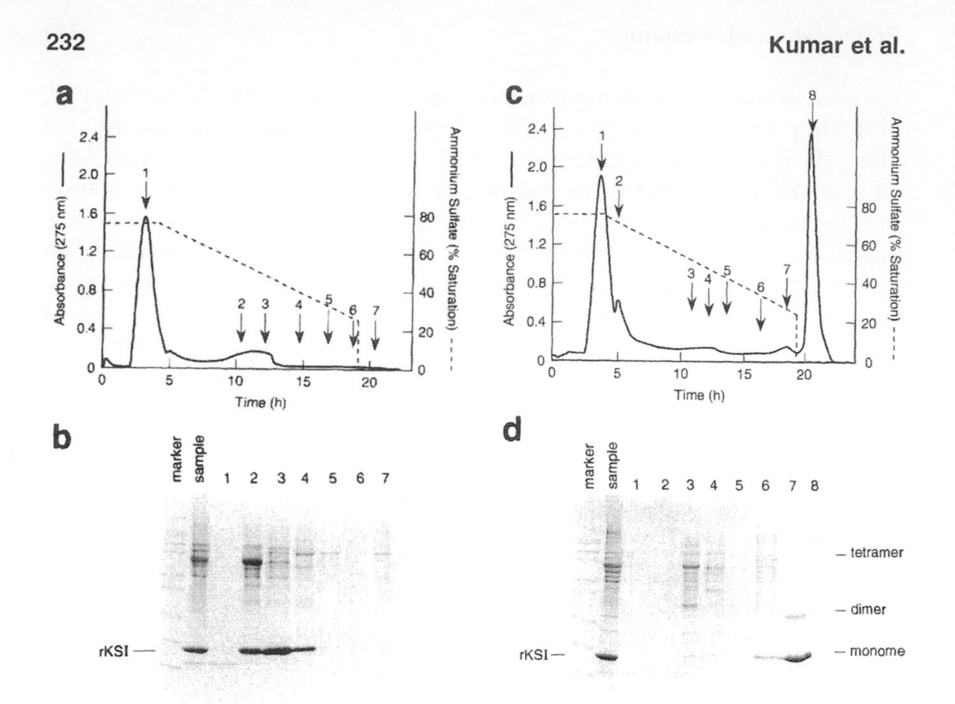

Figure 4 Purification of recombinant ketosteroid isomerase from crude *E. coli* lysate using centrifugal precipitation chromatography in the absence (a and b) and in the presence of affinity ligand, β-estradiol-17-methyl-PEG-5000 (c and d). (Reproduced from Ito 1999 with permission.)

normally located in membranes, may be soluble in nearly 100% organic solvent. As a rule, the net effect on cytoplasmic and other water-soluble proteins is a decrease in solubility to the point of aggregation and precipitation. The principal causes of aggregation are likely to be electrostatic and dipolar van der Waals forces. The size of the protein molecule is also an important factor for aggregation: the larger the molecule, the lower the percentage of organic solvent required to precipitate it (Fig. 5).

The two most widely used solvents are ethanol and acetone, whereas others, such as methanol, *n*-propanol, *i*-propanol, and dioxane, are also frequently used. Other more exotic alcohols, ethers, and ketones are used under specific conditions. The solvents used must be completely water miscible, unreacting with proteins, and have a good precipitating effect. The precipitation temperature should be kept very low, otherwise there are high chances of protein denaturation in presence of solvents. One obvious advantage with solvent precipitation is that it can be carried out at subzero tem-

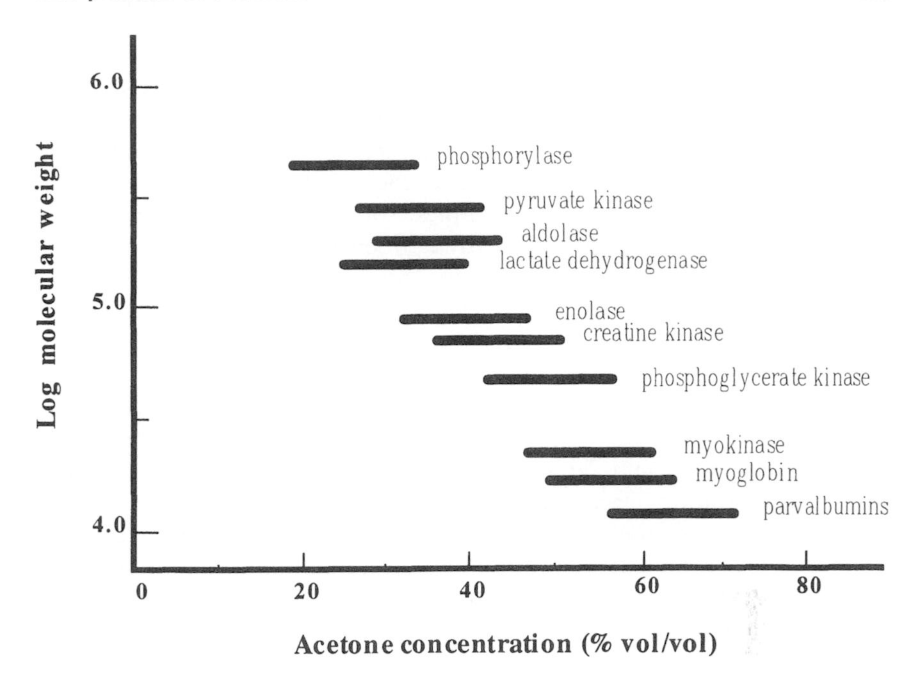

Figure 5 Approximate precipitation ranges in acetone at 0°C, pH 6.5, $I = 0.1$, of some proteins found in muscle tissue extracts. (Adapted from Scopes 1994.)

peratures, since all the miscible solvents form mixtures with water that freeze well below 0°C. Most of the proteins are precipitated with acetone or ethanol in the concentration range of 20–50% vol/vol. At 50% solvent only proteins of molecular weight less than 15,000 are likely to remain in solution.

Recently, a very attractive precipitation process of protein powders has been developed using supercritical CO_2 as an antisolvent (supercritical antisolvent, SAS) to induce rapid protein precipitation from an organic solvent (Yeo et al. 1993; Winters et al. 1996). SAS precipitation is a continuous process in which the organic solution is typically sprayed through an orifice, and is simultaneously expanded and extracted upon contact with continuum of concurrently flowing supercritical CO_2. Uniform protein precipitate particles (1–5 μm size range), suggest these microparticles to be well suited for controlled release and pulmonary aerosol drug delivery applications. While CO_2-induced precipitation is a promising method for the micronization of protein particles (Winters et al. 1996), its application for the fractional separation of proteins has just been investigated (Winters et al.

1999). CO_2 was used to induce the precipitation of alkaline phosphatase, insulin, lysozyme, ribonuclease, and trypsin from dimethylsulfoxide (DMSO). These proteins have high solubilities in pure DMSO. Because of the high miscibility of DMSO and high pressure CO_2, isothermal addition of vapor-phase CO_2 was used to expand batches of protein solutions in liquid DMSO. The volumetric expansion of these solutions reduced the concentration of DMSO, lowering its solvent strength towards dissolved proteins. The proteins precipitated at specific pressures and were collected with an isobaric filtration step. Except alkaline phosphatase the activities of the other proteins were almost quantitatively recovered after redissolution of the precipitates.

Vapor phase precipitants offer several advantages over conventional precipitation techniques, as they are easily separated and recovered from protein precipitates and liquid solvents by pressure reduction in a flash vessel. Furthermore, purer products with less solvent inclusion are expected to precipitate from the gas-expanded liquid solvents, which have higher diffusion coefficients and lower viscosities than liquids at ambient pressures. The density, solvent power, and solubility of vapor phase precipitants are continuously adjustable with only moderate changes in pressure, allowing for fractional separation and accurate control over precipitate characteristics (e.g., particle size, particle size distribution, and residual solvent). With these potential advantages, however, the main drawback in some cases can be the loss of the biological activity because of the marked increase in the β-sheet content and substantial loss of α helicity. Because of this, the process may not show general applicability. Nevertheless, the method will become increasingly attractive for purifying proteins that show activity recovery following SAS processing.

2.3 Isoelectric (pI) Precipitation

A third method is precipitation by changing the pH of the protein solution. This effect is due to the different ionic groups of a protein molecule. At the *isoelectric point* (pI) where the net charge on a protein is zero, the electrostatic repulsions between molecules are at a minimum and result in aggregation due to predominating hydrophobic interactions. The pI is different for different proteins, and logically speaking this could have imparted selectivity in isoelectric precipitation. However, several of the proteins have very close pI_s, and in a protein mixture different proteins with similar properties coprecipitate and aggregate. Thus, most isoelectric precipitates are aggregates of many different proteins and may include particulate fragments and protein-nucleic acid complexes. If the initial composition of the solution is changed, a desired protein may not exhibit the same apparent solubility behavior if its partners in precipitation are absent. Since mostly isoelectric precipitations are carried out at the pH away from the physiological, it is necessary to make sure that the protein is stable at the desired pH. Often isoelectric precipitation is more useful when combined with other modes of precipitation where other solutes are added. Organic solvents or PEG together with the pH adjustment can precipitate the desired protein, as shown for the purification of yeast phosphofructokinase (Welch and Scopes 1981). Specific enzymes were precipitated from a mixture of proteins by the simultaneous addition of a soluble aluminate and acid. The pH of the mixture was at least one pH unit from the isoelectric point of the particular enzyme (Nielsen 1994).

Recently, more thrust in these precipitation modes has been given for application in the recovery of recombinant proteins. A model system where canola protein extract was spiked with recombinant T4 lysozyme was developed for the separation of the recombinant protein (Zaman et al. 1999). The minimal solubility of canola proteins occurs at pH 5.0 with precipitation of about 70% of the protein. The minimal solubility of T4 lysozyme in the extract was observed at pH 7.0 with precipitation of around 40% of the protein. This produced a native canola concentrate and a substantially enriched "recombinant" (i.e., T4 lysozyme) fraction. The same system was also studied with polyelectrolyte precipitations, but with less success.

The choice of the acid or base for the pH adjustment in pI precipitation depends on the individual protein systems. However, while milder acids or bases are mainly used to bring change in pH, in some systems abrupt changes are necessarily achieved by using strong acids or bases. Another important aspect can be the way of addition of these acids or bases. The isoelectric precipitation of soy proteins by acid addition (two extremes in the speed of acid addition or, alternatively, in mixing during acid addition) does not affect the fraction composition, microstructure, or overall protein yield (Fisher et al. 1986). However, other properties are affected. Rapid acid addition creating high supersaturation gives primary particles that are larger and a particle size distribution showing stronger aggregation. Rapid precipitation also reduces the opportunity for protein molecules to arrange, achieving minimal chemical potential and resulting in a higher surface potential on the primary particles. These potentials affect the aggregate strength so that the rapid precipitation produces larger aggregates with higher fraction of the precipitating agent entrapped.

2.4 Heat/pH-Induced Precipitation

Heat- and pH-induced denaturation of proteins is also used as subtractive adjunct precipitation method. The proteins exhibit a wide range of stabilities to perturbations in heat and pH; thus, some proteins, such as adenylate kinase and trypsin and certain proteins of thermophilic organisms are relatively heat stable in comparison with the majority of the other enzymes. Similarly some proteins are stable and even biologically active outside the pH range of pH 5–10, where many of the proteins undergo denaturative changes in confirmation. The precipitation strategy therefore involve the option to use heat and pH to denature and precipitate the unwanted proteins while the desired protein remains unaffected (Englard and Seifer 1990). This approach is now commonly used for single-step purification of thermostable enzymes expressed in a mesophilic host.

2.5 Precipitation with Polymers

Nonionic Water-Soluble Polymers

In addition to the above commonly used precipitation modes, proteins can also be precipitated by the addition of non-ionic, water-soluble polymers. There are several such polymers, which are effective in causing precipitation, however, the high viscosity of most of the solutions made their use as protein precipitants rather difficult. The one exception is PEG, which is available in a variety of degrees of polymerization, and solutions up to 20% (w/v) are not too viscous. PEG fractional precipitation of plasma proteins was first introduced by Polson et al. (1964). Since then this precipitation mode has become widely useful in purification of proteins. The advantages of PEG as a fractional precipitating agent stem primarily from its well-known benign chemical properties. Unlike several other precipitating agents, PEG has little tendency to denature or otherwise interact with proteins even when present at high concentrations and elevated temperatures. One of the main advantages of using PEG precipitation over salt or ethanol precipitation has been the shorter time required for the precipitated proteins to equilibrate and achieve a physical state suitable for large-scale centrifugation. PEG with a nominal average molecular weight of 4000–6000 is commonly used, although higher molecular weights are equally effective. Very high molecular weight PEG makes more viscous solutions difficult to handle, while PEGs with lower than 4000 molecular weight require higher concentration to achieve protein precipitation.

The extensive study on the mechanism of precipitation of proteins by PEGs was carried out by Atha and Ingham (1981). The apparent solubilities of various proteins (14,000–670,000) were measured in the presence of PEGs of different molecular weights. It was well understood that the solubilities of proteins decrease exponentially with increasing concentration of polymer according to the equation:

$$\log S = \log S_o - \beta C$$

where S is the solubility in the presence of PEG at concentration C (%, w/v) and S_o is the apparent intrinsic solubility obtained by extrapolation to zero PEG. Plots of log S vs. PEG concentration exhibit striking linearity over a wide range of protein concentration, the slope for a given protein being relatively insensitive to pH and ionic strength, but markedly dependent on the size of the PEG up to about 6000. Thus, the increment in PEG concentration required to achieve a given reduction in solubility is unique for a given protein-polymer pair, being insensitive to solution conditions but primarily dependent on the size of the protein and the polymer. Manipulation of the solution conditions is expected to improve the separation of a given pair of proteins to the extent that their intrinsic solubilities diverge. The behavior of the proteins during PEG precipitation is somewhat similar to their behavior in precipitation by organic solvents, and indeed the PEG molecule can be regarded as polymerized organic solvent, although the percentage required to cause a given amount of precipitation is lower.

Since PEG precipitation has been very selective for fractionating plasma/serum proteins, it has found wide applications in clinical diagnostic testing. The desired protein is precipitated from the blood by adding various concentrations of PEG. Prolactin is measured in sera after PEG precipitation as a screening method for macroprolactinemia (Fahie-Wilson 1999). Selective precipitation and estimation of protein S with PEG was shown to be comparable to more tedious monoclonal antibody–based ELISA method for prothrombotic investigations in patients (Murdock et al. 1997).

In some cases PEG precipitation method has been more advantageous for purification and fractionation of proteins than even more specific affinity methods or other precipitation methods, e.g., for fractionating thyroid stimulating immunoglobulin (Jap et al. 1995).

A simple and effective method to purify apolipoprotein H with higher yield was PEG precipitation followed by heparin affinity chromatography (Cai et al. 1996). Precipitation by the polymer retained more integrated structure of the protein, which resulted in higher bioactivity of the protein. Earlier purification methods involving acid extraction of the protein induced disordered structure of the protein molecule and hence loss of bioactivity. Precipitation and fractionation of more closely related proteins (e.g., collagens type I, II, and III) was achieved by PEG at neutral pH (Ramshaw et al. 1984). The method can be used to obtain collagens and procollagens from tissue culture media both at analytical and preparative scale. Another potential example has been the fractional precipitation of closely related seven acid phosphatases from *Aspergillus ficuum* culture filtrate

(Hamada 1994). The proteins were sequentially fractionated with 4, 9, 15 19, 24, 30, and 36% PEG with 93% activity recoveries. Further purification of the individual PEG precipitate fractions was possible with high recovery using high-performance liquid chromatography. Similar application of PEG fractional precipitation was also seen when milk proteins α-lactalbumin and β-lactoglobulin were isolated from bovine whey (Ortin et al. 1992). In crude protein extracts with high lipid contents, chloroform precipitation prior to PEG precipitation has been quite useful. For example, immunoglobulin Y is recovered with higher yields and activity from egg yolks by chloroform-PEG precipitation than the PEG alone (Polson 1990). The efficiency of PEG precipitation for the purification of lipase was compared with the other precipitation modes in a recent study (Table 1) (Boominathan et al. 1995). Besides precipitation of protein, PEG has also found wide application in nucleic acid precipitation and purifications (Ferreira et al. 1999; Kresk and Wellington 1999).

It is generally thought that PEG is not as easy to remove from protein fraction as either salt or organic solvent. However, since PEG precipitation is usually used in the early stages of purification scheme, the polymer is removed in the subsequent chromatographic steps. In general, PEG has no tendency to adsorb to the chromatographic matrices; however, in some specific columns it may affect the performance of the matrix. In such situations the polymer can be first removed from protein fraction by ultrafiltration or salt-induced phase separations (Ingham 1990).

Table 1 Precipitation of Lipase by Different Reagents from the Culture Filtrate of *Humicola lanuginosa* (Boominathan et al. 1995)

Precipitating agent	Total activity in ppt (U)	Total protein in ppt (mg)	Specific activity (U/mg protein)	Recovery of enzyme (%)	Fold purification
Ammonium sulfate	4850	170	28.50	97	4.07
Acetone	4600	119	38.65	92	5.25
Ethanol	3950	98	40.30	79	5.75
Isopropanol	4200	106	39.62	84	5.66
Acetic acid	3600	90	40.00	72	5.71
PEG 4000	4200	106	39.62	84	5.66
PEG 6000	3800	106	35.84	76	5.12
PEG 20000	3500	108	32.40	70	4.60

Initial total activity of the enzyme = 5000 U and protein = 710 mg. U = micromoles fatty acid released per minute.

Synthetic and Natural Polyelectrolytes

The charge-based electrostatic nonspecific binding of polyanions and polycations to proteins below and above their isoelectric points, respectively, is well established (Xia and Dubin 1994). These interactions may result in soluble complexes (Sacco et al. 1988), complex coacervation (Burgess and Carless 1984; Dubin et al. 1987), or formation of amorphous precipitates (Sternberg and Hershberger 1974; Kokufuta et al. 1981; Nguyen 1986). All these states of protein-polyelectrolyte complexes may be achieved by the selection of the polyelectrolyte, choice of the ionic strength and pH, and control of the concentration of the macromolecular components.

The practical advantages of these phase changes may govern the use of polyelectrolytes for protein separation (Morawetz and Hughes 1952; Strege et al. 1990; Bozzano et al. 1991) and immobilization or stabilization of enzymes in polyelectrolyte complexes (Burgess and Jendrisak 1975; Margolin et al. 1984). For protein separation and purification, aggregate formation of the complex is desirable. Protein precipitation by polyelectrolytes may lead to closely packed aggregates that are conveniently separated by settling or can generate open-textured aggregates more feasible for separation by filtration. However, in both cases the aggregation should be essentially reversible. The precipitated proteins are recovered from the insoluble protein-polyelectrolyte complex aggregates by redissolution achieved by pH or ionic strength adjustment. Use of polyelectrolytes as precipitating agents offers several advantages even though their costs may be high. Generally, very low concentrations of the polyelectrolytes are required that can also be recycled (Hughes and Lowe 1988; Patrickios et al. 1994), and the fractionation potential appears promising.

Several techniques have been employed to study the protein-polyelectrolyte complexes and the application of these complexes to protein separation. These include turbidimetric titrations, viscometry, light scattering, spectroscopy, electron spin resonance, size exclusion chromatography, and other techniques (Xia and Dubin 1994). These techniques have helped to elucidate some details of the mechanism of complex formation and factors influencing protein-polyelectrolyte complexation. Most of the research in the field of polyelectrolyte precipitation is focused on the screening of polymers with varying charge density and chain length or with choice of conditions (pH, ionic strength, concentration of polyelectrolyte, composition of the mixtures, etc.) (for reviews, see Clark and Glatz 1990; Kokufta 1994; Shieh and Glatz 1994; Xia and Dubin 1994). The nature of the polyelectrolyte plays a major role in the precipitation efficiency. Both the charges on the polymer and the charge density are important. Steric factors or the flexibility of the polyelectrolyte may also influence the effectiveness of the precipitation.

Solution pH is an important determinant in the precipitation efficiency, and the optimum pH will vary with both the protein and the polyelectrolyte. Similar effect can be seen with the change in the ionic strength in the medium. With increasing ionic strength, the reduction in the protein-polyelectrolyte interactions is generally observed. This also gives evidence that complex results from electrostatic interactions. Furthermore, the protein precipitation increases with polyelectrolyte dosage to an optimum, then decreases with further addition of the polymer. The dosage requirement depends on the nature of the polyelectrolyte, the degree of ionization, and the protein nature. The schematic presentation of the principle of polyelectrolyte complex and protein-polyelectrolye complex formation is presented in Fig. 6.

In general, polyelectrolyte precipitations have inherent selectivity with regard to differently charged species. The change of medium conditions also exerts such selectivity. Table 2 summarizes some examples of polyelectrolyte precipitations of different proteins. If the protein mixture contains closely related proteins with respect to charge, specifically precipitating the desired protein is somewhat difficult. However, by changing the medium conditions like pH or ionic strength, relative charge differences can be increased among the proteins, thus imparting more specificity in precipitation. Lysozyme and hemoglobin were both precipitated with poly(acrylic acid) (PAA) (Sternberg and Hershberger 1974), the former precipitating quantitatively in the pH range of 4.5–6.5, and the latter at 4.25–5.25. The broader pH range of lysozyme precipitation is due to its more basic character in comparison with hemoglobin.

One recent example has shown how the impurities in the protein extract can adversely effect the polyelectrolyte precipitation of the specific protein. Recombinant protein T4 lysozyme (exhibits nine positive net charges at pH close to 7) was quantitatively precipitated by oppositely charged polyelectrolytes, PAA or polyphosphate glass H (Zaman et al. 1999). When the same protein was spiked into the canola protein extract and subjected to precipitation by adding polymers, the precipitation was adversely affected and only 40% of the recombinant protein was precipitated using nine times higher polymer dosage than with pure enzyme. The interference from canola components, which may affect the stoichiometry of polyelectrolyte to protein charge, offers the likely explanation for poorer selectivity observed here. However, in another case, lysozyme, a protein of similar size with charge of 5.8 per molecule at pH 7.5 was readily precipitated with PAA from egg white protein extract (Fisher and Glatz 1988). The stoichiometry of polyelectrolyte to protein charge was 1 in this case, as compared with 1.7 in case of T4 lysozyme.

Often polyelectrolyte precipitation is used for clearing-up bacterial cell lysates, where the impurities are precipitated from the extract leaving the

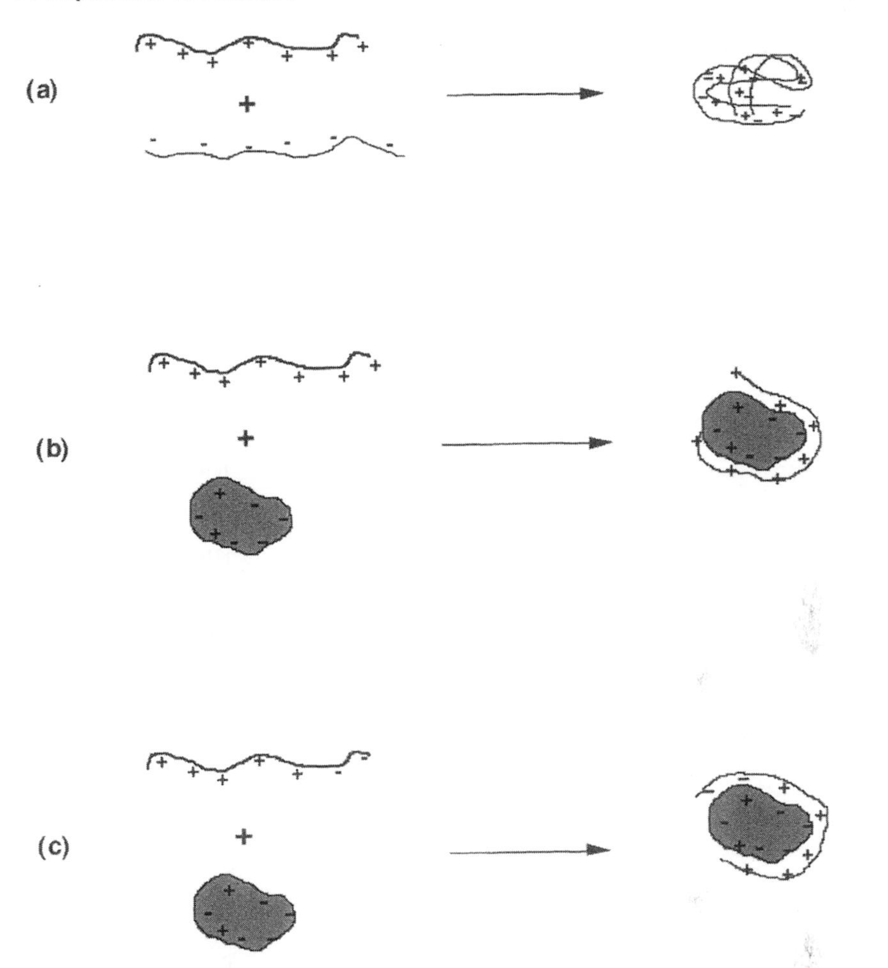

Figure 6 Schematic presentation of polyelectrolyte complexation: (a) polycation and polyanion; (b) polycation and protein; and (c) polyampholyte and protein.

target proteins in solution. Poly(ethyleneimine) (PEI), a positively charged polymer under neutral and acidic conditions, is very efficient for precipitating nucleic acids and acidic proteins (bacterial proteins are predominantly acidic). The method was successfully used as the first-step isolation of cysteine proteinase inhibitor stefin B (Jerala et al. 1994). Addition of 0.2% PEI to bacterial lysate at pH 8.0 produced bulky precipitate composed mainly of nucleic acids along with other 90% contaminating proteins (with pI $\leq$ 6.0), leaving about 80% pure recombinant inhibitor stefin B (pI 7.7) in

Table 2 Examples of Protein Separation by Polyelectrolytes by Nonspecific Precipitation

Proteins	Polyelectrolytes	Conditions	Ref.
BSA	PMAA, PVAm	pH 3.95–8.70	Morawetz and Hughes 1952; Bozzano et al. 1991
Catalase and BSA	PAA	pH 5.2	Xia and Dubin 1994
Lysozyme from egg white	PAA	pH 5.45	Sternberg and Hershberger 1974; Chen et al. 1992
Lysozyme/ovalbumin	CMC	pH 4.2, 5.8, 7.5	Clark and Glatz 1990
RNA polymerase	PEI	pH 7.0 I = 0.2 NaCl	Jendrisak 1987
Whey proteins	CMC	pH 2.5–4.0	Hill and Zadow 1978
Amyloglucosidase/lactase/α-amylase/β-amylase/calf rennet/lipoxygenase	PAA	pH 3.2–4.8	Sternberg 1976
β-Galactosidase	PEI		Zhao et al. 1990
Penicillin acylase	30% Sulfomethylated (PAm)	pH 7.8	Bryjak and Noworyta 1994
T4 lysozyme	PAA/Glass H	pH 7.4/7.1	Zaman et al. 1999
Cysteine proteinase inhibitor Stefin B	PEI	pH 8.0	Jerala et al. 1994
Trypsin inhibitor	pectin/CMC-sodium salt/DS	pH 3.2/3.5/7.0	Kiknadze and Antonov 1998
Soybean trypsin inhibitor/ribonuclease A/lysozyme	Triblock methacrylic polyampholytes	pH 4.0–9.0	Patrickios et al. 1994

PAm, poly(acrylamide); PAA, poly(acrylic acid); PEI, poly(ethyleneimine); PMAA, poly(methacrylic acid); CMC, carboxymethylcellulose; DS, dextran sulfate; PVAm, poly(vinyl acrylamide).

the solution. One more chromatographic step was sufficient to get a final homogeneous preparation. PEI is also very efficient and mostly used in the isolation of enzymes involved in nucleic acid metabolism. The dependence of protein precipitation on the polymer charge and nature was shown in the case of trypsin inhibitor precipitation from albumin fraction of alfalfa (*Medicago sativa* L.) leaf juice using different polyelectrolytes (Kiknadze and Antonov 1998). Weak polyelectrolytes, pectin, and CMC-sodium salt caused significantly low precipitation of the inhibitor protein, whereas the strong polyelectrolyte, dextran sulfate (DS), precipitated about 50% of the inhibitor. The protein-polyelectrolyte complex was insoluble in water but could be completely solubilized in high-ionic-strength buffered saline (1 M NaCl in 50 mM Tris-HCl, pH 7.8), suggesting the electrostatic nature of the complex.

The removal of polyelectrolyte from the protein complex is rather convenient. For example, PAA can be conveniently removed from the protein complex by first changing the pH or in some cases using high ionic strength to dissolve the complex. To separate the solubilized polyacrylate from the protein, the mixture is treated with $CaCl_2$ solution and pH changed away from the precipitation range of protein-polyelectrolyte. The white insoluble calcium-polyelectrolyte salt formed can be removed by filtration or centrifugation.

In some situations, the limitations of polyelectrolyte precipitation can be the relatively narrow pH range at which optimal precipitation will take place. In some cases, the pH where precipitation has to be made can be damaging to the functional property of the particular protein.

More versatility in polyelectrolyte precipitation is achieved by using synthetic block *polyampholytes* that bear a net positive or negative charge like the protein (a biological ampholyte) (Patrickios et al. 1994; Kudaibergenov and Bekturov 1989). These polyelectrolytes, consisting of di- or triblock copolymers of different charged groups, show the advantage of self-aggregation around their isoelectric point. The self-aggregation will provide the ability for polymer removal and recycling at the end of the protein separation process. Polyampholytes interact strongly both with themselves and with proteins, and precipitate within a pH range determined by the polymer and protein net charges. The kinetics of precipitation both of the pure polyampholytes and the protein-polymer complex are quite fast. Also very low concentrations of polymers (0.01%) are typically being used in these precipitations, making the process quite economical. The resulting protein-polyampholyte precipitate can be removed from the system and redissolved at a different pH. Finally, protein and polyampholytes can be separated from each other by precipitating the polyampholyte at its isoelectric point. Synthetic polyampholytes constructed as triblock copolymers of methacrylic acid, dimethylaminoethyl methacrylate, and methyl methacrylate (the ratio of the three govern the net charge on the polymer) were used for the separation of one acidic protein (soybean trypsin inhibitor) and two basic proteins (ribonuclease A and lysozyme) (Patrickios et al. 1994).

The higher the discriminating power between the target protein and protein impurities during protein-polyelectrolyte complex formation, the more potential it has for bioseparation. Increase in selectivity of precipitation by polyelectrolytes has been shown by the use of tailor-made polyelectrolytes having affinity ligands specific for the target protein coupled to the polymer (Izumrudov et al. 1999). This is discussed in the section, "Affinity Precipitation Using Polyelectrolytes."

3. PRECIPITATION OF PROTEINS: SPECIFIC

The lattice theory of specific precipitation was first proposed by Marrack in the early twentieth century and later became the basis for studies of the mechanism of such precipitations (Boyd 1973). It is assumed that divalent antibody interacts with multivalent antigen, the bimolecular compound thus formed interacts with another molecule of antigen or antibody (or with an antibody-antigen compound already formed), and this process continues until the resulting aggregates are so large that they perforce separate out of the solution as a precipitate. This forms the basis of *immunoprecipitation*. It is realized and well accepted that such interactions involve high affinity; hence, there is a direct relationship between specificity and affinity. The factors that lead to high-affinity binding are a good fit between the surfaces of the two molecules and charge complementarity, and the same factors give high specificity for a target molecule. This concept was recently discussed in detail by Eaton et al. (1995) in his review, wherein it was assumed that selection for high-affinity binding automatically leads to high specific binding. Affinity interactions can be tailored to separate proteins from a complex mixture with very high efficiency and selectivity. However, one needs to keep in mind that affinity interactions offer selectivity and specificity, but not necessary always in a form that can lead to precipitation.

When the concept of affinity interactions is combined with precipitation methods, much more is achieved in downstream processing. This led to the development of *affinity precipitation* technique, which has shown tremendous potential and is now considered as a powerful technique for protein purification (Mattiasson and Kaul 1993; Gupta and Mattiasson 1994; Galaev and Mattiasson 1997). As a rule of thumb, there are five basic steps in affinity precipitation: (1) carrying out affinity interactions in free solution, (2) precipitation of the affinity reagent-target protein complex from the solution, (3) recovery of the precipitate, (4) dissociation and recovery of the target molecule from the complex, and, finally, (5) recovery of the affinity reagent. The basic idea of affinity precipitation is the use of a macroligand, i.e., a ligand with two or more affinity sites. The macroligands could be synthesized either by covalent linking of two low molecular weight ligands (directly or through a short spacer) or by covalent binding of several ligands to a water-soluble polymer, which provide two main approaches (homo-bifunctional and hetero-bifunctional modes) of affinity precipitation.

3.1 Affinity Precipitation: Homo-Bifunctional Mode

If the protein has a few ligand binding sites, complexation with a macro-ligand results in a formation of poorly soluble large aggregates (Fig. 7), which could be separated by filtration or centrifugation from the supernatant containing soluble impurities, immunoprecipitation of antigens with divalent antibodies being a good illustration of this principle. Optimal concentrations of the two molecules are needed to form the aggregated complexes and if this ratio is deviated from, the complex starts to dissolve. The homo-bifunctional mode of affinity precipitation was introduced by Larsson and Mosbach (1979; also Larsson et al. 1984) who performed quantitative precipitation of purified lactate dehydrogenase (tetrameric enzyme) and glutamate dehydrogenase (hexamer) from solution using bis-NAD [N^2,$N^{2'}$-adipohydrazidobis(N^6carbonylmethyl)-NAD]. The precipitate was easily soluble in the presence of NADH, which forms stronger bonds with lactate dehydrogenase and replaces the enzyme from complex with bis-NAD. The effectiveness of the precipitation strongly depends on the length of the spacer linking the two molecules of ligand and on the ratio enzyme/bis derivative; excess of the latter decreases the effectiveness of precipitation down to completely soluble complex.

Lactate dehydrogenase of more than 95% purity was isolated in a similar way from homogenate of bovine heart with 90% yield (Larsson et al. 1984). The same enzyme was also precipitated with bis derivatives of the triazine dye Cibacron blue F3GA (II) and recovered from rabbit muscle extract with 100-fold purification (Hyet and Vijaylakshimi 1986; Riahi and Vijaylakshimi 1989). Precipitation of the enzyme with polymeric derivatives of this dye in polyvinyl alcohol was less effective (Fisher et al. 1989; Morris and Fisher 1990). This is because of the interaction between the dye molecules bound to one polymer molecule, which loosen the binding of the enzyme with the polymeric derivative by about one order of magnitude compared with that with the bis derivative (Morris and Fisher 1990).

Avidin (a tetramer) from egg white was successfully precipitated with iminobiotin conjugate obtained by covalent linking of *N*-hydroxysuccinimi-nobiotin with copolymer of acrylamide and *N*-(3-aminopropyl)methacrylamide. However, the main problem encountered was to isolate the small amount of precipitate from the large volumes of solution (Morris et al. 1993). A more successful approach was to use phospholipid dimyristoyl-L-

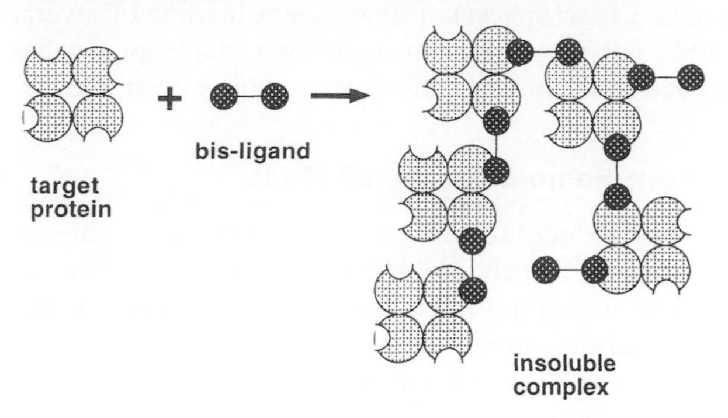

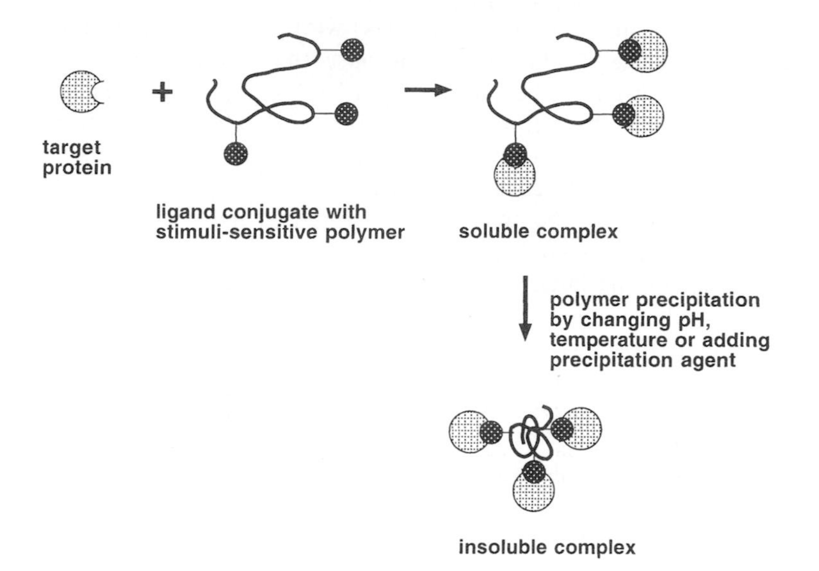

Figure 7 Homo- and hetero-bifunctional mode of affinity precipitation.

α-phosphatidylethanolamine modified with biotin (Powers et al. 1992). The precipitation resulted from specific binding of avidin with biotin ligand and the subsequent aggregation of the avidin/biotin-phospholipid complexes due to hydrophobic interactions between hydrophobic tails of the phospholipid. Avidin was isolated both from model mixtures with myoglobin, bovine serum albumin, and lysozyme, and from egg white with 80–90% yields.

By definition, homo-bifunctional mode of affinity precipitation may be used only for proteins with two or more ligand binding centers; this significantly limits its application. An exception, however, is metal affinity precipitation, which can exploit different metal-binding amino acid residues on the same polypeptide chain (Arnold 1991). It is also necessary to thoroughly control the conditions of precipitation, especially the ratio of purified protein to ligand. It should be noted that the suggested mechanism of aggregation due to the binding of protein molecules by polyligand in insoluble aggregates in many cases is speculative and not experimentally proven. Certain proteins are specifically precipitated in the presence of monoligands representing the components of the polyligands usually used in the homo-bifunctional mode of affinity precipitation. For example, the precipitation with nonmodified triazine dyes Cibacron blue F3GA and reactive blue 2 was used to purify L-lactate dehydrogenase (Pearson et al. 1989) and serum proteins (Birkenmeier 1989), and the selective precipitation with transition metals was used to purify lectin, concanavalin A (Agarwal and Gupta 1994), bovine serum albumin, or bovine γ-globulins (Iyer and Przybycien 1995).

3.2 Affinity Precipitation: Hetero-Bifunctional Mode

Hetero-bifunctional mode of affinity precipitation is a more general approach. The precipitation is induced by a moiety not directly involved in the affinity interactions and is represented by a reversibly soluble-insoluble polymer. The affinity ligands, for interaction with the target protein, are covalently coupled to the polymer. Thus, one part of the *macroligand* has the affinity for the target protein/enzyme while the other controls the solubility of the complex. Macroligands composed of soluble/insoluble polymers dominate affinity precipitation as compared with latexes or liposomes, and we are mainly referring to the former systems when discussing heterofunctional affinity precipitation.The ligand-polymer conjugate first forms a complex with the target protein and phase separation of the complex is triggered by small changes in the environment resulting in transition of polymer backbone into an insoluble state. The target protein is then either eluted directly from the insoluble macroligand-protein complex or the precipitate is dissolved, the protein dissociated from the macroligand, and the ligand-polymer conjugate reprecipitated without the protein, which remains in the supernatant in a purified form.

Smart Polymers

The polymers with reversed solubility are often combined under the names "smart polymers," "intelligent polymers," or "stimuli-responsive" polymers. Smart or intelligent materials have the capability to sense changes in their environment and respond to the changes in a preprogrammed and pronounced way (Gisser et al. 1994). Referring to water-soluble polymers and hydrogels, this definition can be formulated as follows: Smart polymers undergo fast and reversible changes in microstructure triggered by small changes of medium property (pH, temperature, ionic strength, presence of specific chemicals, light, electric or magnetic field). This microscopic changes of polymer microstructure manifest themselves at the macroscopic level as a precipitate formation in a solution or as manifold decrease/increase of the hydrogel size and hence of water content. Macroscopic changes in the system with smart polymer/hydrogel are reversible, and elimination of the trigger changes in the environment returns the system to its initial state.

In general, all smart polymer/hydrogel systems can be divided into three groups (for details, see Galaev et al. 1996; Galaev and Mattiasson 2000). The first group consists of the polymers for which poor solvent conditions are created by decreasing net charge of polymer. The net charge can be decreased by changing pH to neutralize the charges on the macromolecule and hence to reduce the repulsion between polymer segments. For instance, copolymers of methyl methacrylate and methacrylic acid precipitate from aqueous solutions on acidification to pH around 5, whereas copolymers of methyl methacrylate with dimethylaminethyl methacrylate are soluble at acidic pH but precipitate in slightly alkaline media. Figure 8 presents a pH-dependent precipitation curve of a random copolymer of methacrylic acid and methacrylate (commercialized as Eudragit S-100 by Röhm Pharma GMBH, Weiterstadt, Germany) as well as *p*-aminophenyl-α-D-glucopyranoside-modified Eudragit S-100 (Linné-Larsson and Mattiasson 1994).

The charges on the macromolecule can also be neutralized by addition of an efficient counterion, e.g., low molecular weight counter ion or a polymer molecule with the opposite charges. The latter systems are combined under the name *polycomplexes*. The complex formed by poly(methacrylic acid) (polyanion) and poly(*N*-ethyl-4-vinylpyridinium bromide) (polycation) undergoes reversible precipitation from aqueous solution at any desired pH value in the range 4.5–6.5 depending on the ionic strength and polycation/polyanion ratio in the complex (Dainiak et al. 1999). Polyelectrolyte complexes formed by poly(ethyleneimine) and polyacrylic acid undergo soluble-insoluble transition in even broader pH range, from pH 3 to pH 11 (Dissing and Mattiasson 1996).

The second group consists of thermosensitive smart polymers. On raising the temperature of aqueous solutions of smart polymers to a point higher than the critical temperature [lower critical solution temperature (LCST) or "cloud point"], separation into two phases takes place. A poly-

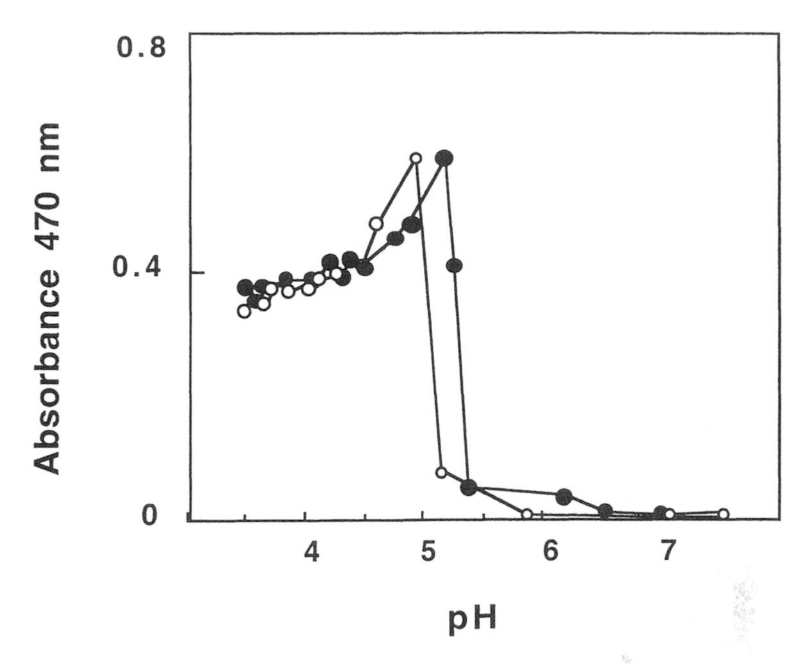

Figure 8 Precipitation curves of Eudragit S-100 (○) and *p*-aminophenyl-α-D-glucopyranoside-modified Eudragit S-100 (●) measured as turbidity at 470 nm. Some decrease in turbidity at lower pH values is caused by flocculation and sedimentation of polymer precipitate. (Reproduced from Linné-Larsson and Mattiasson 1994 with permission.)

mer-enriched phase and an aqueous phase containing practically no polymer are formed, which can be easily separated. This phase separation is completely reversible, and the smart polymer dissolves in water on cooling. Three thermosensitive smart polymers are most widely studied and used, poly(*N*-isopropylacrylamide) [poly(NIPAM)] with LCST 32–34°C, poly(vinyl methyl ether) with LCST 34°C, and poly(*N*-vinylcaprolactam) [poly(VCL)] with LCST 32–40°C. The modern polymer chemistry provides polymers with different transition temperatures from 4–5°C for poly(N-vinylpiperidine) to 100°C for poly(ethylene glycol) (Galaev and Mattiasson 1993a,b).

Contrary to pH-sensitive smart polymers, which contain carboxy or amino groups that could be used for the covalent coupling of ligands, the thermosensitive polymers do not have inherent reactive groups. Thus, a copolymer containing reactive groups should be synthesized. *N*-Hydroxyacrylsuccinimide (Liu et al. 1995) or glycidyl methacrylate (Mori et al. 1994) were used as active comonomers in copolymerization, with NIPAM allowing further coupling of amino group–containing ligands to the synthesized copolymers. An alternative strategy is to modify the ligand with acryloyl group and then copolymerize the modified ligand with NIPAM (Maeda et al. 1993; Umeno et al. 1998).

The third group of smart polymers combines systems with reversible non-covalent cross-linking of separate polymer molecules into insoluble polymer network. The most familiar systems of this group are Ca-alginate (Charles et al. 1974; Linné et al. 1992) and boric acid-polyols (Wu and Wisecarver 1992; Kokufuta and Matsukawa 1995; Kitano et al. 1993) or boric acid polysaccharides (Bradshaw and Sturgeon 1990). These types of polymers found limited application as carriers in affinity precipitation, but they are more promising for the development of "smart" drug delivery systems capable of releasing drugs in response to the signal, e.g., release of insulin when glucose concentration is increasing (Lee and Park 1996).

Properties of the Polymer Precipitate

When designing a polymer for affinity precipitation one should consider also the properties of polymer precipitate (or polymer-enriched phase) formed. When phase separation of the polymer takes place, three distinct types of the precipitate can be recognized (Fig. 9).

The first type of polymer-enriched phase has high water content as in pH-induced precipitation of chitosan (Senstad and Mattiasson 1989a,b; Tyagi et. al. 1996) or Ca-induced precipitation of alginate (Linné et al. 1992). The loose structure of such gel makes it difficult to separate the gel from the supernatant, and the entrapped solution containing impurities decreases the purification factor. The attractive feature, however, is their relatively low hydrophobicity and the fact that large protein molecules and even cells can be processed without a risk of denaturing.

The second type of polymer phase with high polymer content occurs during thermoprecipitation of polymers like poly(*N*-vinylcaprolactam) (Galaev and Mattiasson 1993a,b) or poly(ethylene glycol-*co*-propylene glycol) random and block copolymers (Lu et al. 1996). The polymer phase is easily separated from supernatant with practically no entrapment of impurities from solution. High polymer content renders the polymer phase very hydrophobic and makes it an unfavorable environment for proteins. In fact, unbound proteins are repelled from the polymer phase. Such nature of polymer though disadvantageous for affinity precipitation could be quite useful in another mode of protein purification, namely, partitioning in aqueous two-phase polymer systems (Alred et al. 1994). Affinity ligands with strong binding to target protein are required to achieve specific co-precipitation of the

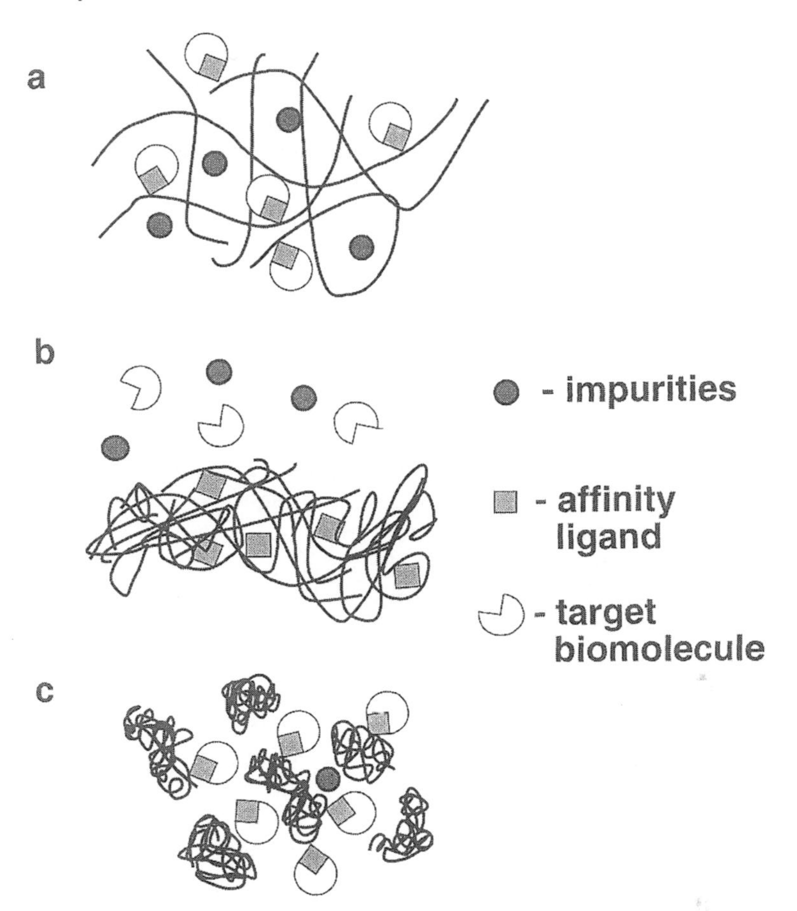

Figure 9 Types of the polymer enriched phase formed after the phase separation of the polymer during affinity precipitation. a, loose gel with high water content; b, compact hydrophobic phase with low water content; and c, suspension of compact polymer.

protein with the polymer (Galaev and Mattiasson 1992, 1993a). Some of the thermosensitive polymers are able to form aqueous two-phase systems with dextrans. When a specific ligand is coupled to the polymer, the target protein partitions preferentially into the aqueous phase formed by this polymer. After mechanical separation from the dextran phase, the target protein can be recovered by thermoprecipitation of the polymer. The latter forms a compact polymer-enriched phase, whereas the purified target protein remains in the supernatant (Franco et al. 1997; Harris et al. 1991).

The third case looks to be the most preferable for affinity precipitation. Compact particles of aggregated polymer are easily separated from the supernatant accompanied with only minimal entrapment of the supernatant and impurities. Moreover, the ligands with bound target protein are exposed to the solution and hence are in a comfortable environment. The examples are the compact precipitation formation by pH-sensitive polymer, Eudragit, or the temperature-sensitive polymer poly-*N*-isopropylacrylamide.

Ligands Used in Hetero-Bifunctional Affinity Precipitation

Affinity precipitation, in contrast to affinity chromatography, is a one-plate process. The protein molecule once dissociated from the ligand on precipitating the polymer has practically no chance to interact with the ligand again as the polymer is removed from solution after precipitation. Hence, in order to be successful, the affinity precipitation requires stronger protein-ligand interactions than affinity chromatography, where the protein molecule once dissociated from the ligand has several chances to be bound again as it moves along the column matrix with its high concentration of the ligands available. The rough estimation is that the binding constant for a single ligand-protein interaction in affinity precipitation should be at least 10^{-5} M (Galaev and Mattiasson 1993a). Ligand coupling to a polymer usually results in 100- to 1000-fold decrease in affinity. Thus, free ligand must bind to the protein of interest with constants of about 10^{-7} to 10^{-8} M. Such strong affinity is mostly restricted to systems with protein ligands, e.g., antibodies. On the other hand, flexibility of the polymer chain allows interaction of a few ligands with multi-subunit proteins and this multipoint attachment promotes binding strength, thereby also making relatively weak interactions useful. This fact allows exploitation of relatively weak binding pairs as sugar-lectin for the efficient affinity precipitation of tetrameric lectin, concanavalin A using pH-sensitive conjugate of Eudragit S-100 with *p*-aminophenyl-*N*-acetyl-D-galactosamine (Linné-Larsson et al. 1996) or *p*-aminophenyl-α-D-glucopyranoside (Linné-Larsson and Mattiasson 1994).

The affinity precipitation is designed to be applied at the first stages of purification protocol dealing with crude unprocessed extracts. Hence, the ligands used should be robust to withstand both harmful components present in the crude extracts and precipitating/eluting agents. Triazine dyes, which are robust affinity ligands for many nucleotide-dependent enzymes, were successfully used in conjugates with Eudragit S to purify dehydrogenases from various sources by affinity precipitation (Guoqiang et al. 1993, 1994a,b, 1995b; Shu et al. 1994). Sugar ligands constitute another attractive alternative for use in bioseparation of lectins (Linné-Larsson and Mattiasson 1994, 1996; Linné-Larsson et al. 1996; Hoshino et al. 1998), and ligands with immobilized metal are efficient for affinity precipitation of proteins having histidine residues at the surface (Galaev et al. 1997, 1999; Kumar et al. 1998a,b, 1999; Mattiasson et. al. 1998), *p*-aminobenzamidine ligands were used for affinity precipitation of proteolytic enzymes (Nguyen and Luong 1989; Pécs et al. 1991).

More complex structures and even proteins have been used as ligands in affinity precipitation. Poly(NIPAM) conjugate with $(dT)_8$ was used for a model separation of a complementary oligonucleotide $(dA)_8$ from a mixture of $(dA)_8$ and $(dA)_3(dT)(dA)_4$. Affinity precipitation resulted in recovery of 84% of $(dA)_8$ while 92% of $(dA)_3(dT)(dA)_4$ remained in solution (Umeno et al. 1998). Restriction endonuclease Hind III was isolated using poly(NIPAM) conjugate with λ-phage DNA (Maeda et al. 1993), and C-reactive protein was isolated using poly(NIPAM) conjugate with *p*-aminophenylphosphorylcholine (Mori et al. 1994).

Protein ligands offer good selectivity and high binding strength needed for affinity precipitation, provided they are stable enough under conditions of polymer precipitation and elution. Such proteins include soybean trypsin inhibitor (Galaev and Mattiasson 1992; Kumar and Gupta 1994; Chen and Jang 1995), immunoglobulin G (Kamihira et al. 1992; Mattiasson and Kaul 1993, 1994), and protein antigen (for binding monoclonal antibodies) (Dainiak et al. 1999).

The selectivity of affinity precipitation as a protein purification technique depends on the nature of the ligands used as well as the polymer. Precipitation of noncharged thermosensitive polymers shows very low nonspecific coprecipitation of proteins as compared with pH-sensitive polymers where ionic interactions can contribute to the unwanted protein-polymer interactions. In the worst case, a nearly quantitative nonspecific coprecipitation of proteins took place with the native polymer (Kumar et al. 1994). However, in the subsequent study it was shown that using the polymer-ligand conjugate, the nonspecific protein binding decreases significantly (Kumar and Gupta 1996). Also, whatever little protein was bound nonspecifically could not be eluted under the specific dissociation conditions for the target protein and ligand, thus preventing the contamination of the purified target protein. When polyelectrolyte complexes precipitate, the charges of one polyanion are compensated by the charges of the oppositely charged polyanion. Strong cooperative interactions of oppositely charged polyions resulted in efficient displacement of nonspecifically bound proteins; the amount of nonspecifically bound protein does not exceed a few percentage points of that present in solution (Dainiak et al. 1999). On the other hand, the affinity of the proteins to nonmodified polymer could be harnessed for purification of the proteins. Endopolygalacturonase was purified by coprecipitation with alginate (Gupta et al. 1993), xylanase from *Trichoderma viride* by coprecipitation with Eudragit S 100 (Gupta et al. 1994), β-glucosidase from *Trichoderma longibrachiatum* by sequential precipitation of chitosan (to remove coprecipitated cellulase activity) and coprecipitation with Eudragit S-100 (Agarwal and Gupta 1996), lysozyme and lectins specific for *N*-acetylglucosamine (e.g., lectins from rice, tomato, and potato) by coprecipitation with chitosan (Tyagi et al. 1996).

3.3 Applications of Affinity Precipitation

Affinity precipitation of proteins using smart polymers emerged in the early 1980s. Since then it has evolved to a technique capable of simple, fast, and efficient purification of a variety of proteins. A number of examples are now available in which affinity precipitation is used as a selective and specific precipitation technique for protein purification. Table 3 summarizes published protocols of protein isolation using affinity precipitation technique. The authors consider it relevant to discuss here some of their recent works carried out in this area as model examples.

Affinity Precipitation Using Polyelectrolytes

The main problem encountered in polyelectrolyte precipitation of proteins is the selectivity of polymer-protein interaction (discussed earlier under "Nonspecific Precipitations"). Introduction of affinity interactions in the system by coupling an affinity ligand to one of the components make the polyelectrolyte precipitations very specific for the target protein. This has been proven when lactate dehydrogenase from beef heart extract was specifically purified using Cibacron blue 3GA coupled to positively charged polymer poly(ethyleneimine) (Dissing and Mattiasson 1996). The polyelectrolyte complex was formed by adding negatively charged polymer, poly(acrylic acid). The precipitated complex with the target protein was removed from the solution containing impurities. LDH was obtained with a yield of 85% and a purification factor of 11-fold.

In one of the more recent examples, when antigen, inactivated glyceraldehyde-3-phosphate dehydrogenase, from rabbit was covalently coupled to a polycation, the resulting complex was used for purification of antibodies from 6G7 clone specific for the protein. The crude extract was incubated with the polymer complex and the precipitation of the latter was carried at 0.01 M NaCl and pH 4.5, 5.3, 6.0, and 6.5 using complexes with polycation/polyanion ratio of 0.45, 0.3, 0.2, and 0.15, respectively. Purified antibodies were eluted at pH 4.0 where polyelectrolyte complexes of all compositions used were insoluble. Quantitative recoveries were

Table 3 Published Protocols of Protein Isolation Using Affinity Precipitation

Target protein	Polymer	Ligand	Method used to precipitate polymer	Protein recovery (%)	Purification fold	Ref.
Wheat germ agglutinin	Chitosan	—[a]	ΔpH	55–70	10	Senstad and Mattiasson 1989b
Lysozyme, lectins specific for *N*-acetylglucosamine	Chitosan	—[a]	ΔpH	—[b]	—[b]	Tyagi et al. 1996
Trypsin	Chitosan	Soybean trypsin inhibitor	ΔpH	93	5.5	Senstad and Mattiasson 1989a
Trypsin	Copolymer NIPAM and *N*-acrylhydroxy-succinimide	*p*-amino-benzamidine	Δ*T*	74–82	—[b]	Nguyen and Luong 1989
Trypsin	Alginate	Soybean trypsin inhibitor	addition of Ca^{2+}	30–36	—[b]	Linne et al. 1992
Trypsin	Poly(*N*-vinyl caprolactam)	Soybean trypsin inhibitor	Δ*T*	54	28	Galaev and Mattiasson 1992
Protein A	Hydroxypropyl-methyl-cellulose succinate	IgG	ΔpH	50–70 53–90	4.9–5.4 32	Taniguchi et al. 1990 Taniguchi et al. 1989
Human IgG	Galactomannan	Protein A	Addition of tetraborate	—[b]	—[b]	Bradshaw and Sturgeon 1990
Alkaline protease	Poly-NIPAM	*p*-aminobenz-amidine	Δ*T*	—[b]	—[b]	Pécs et al. 1991
β-Glucosidase	Chitosan	—[a]	ΔpH	—[b]	—[b]	Homma et al. 1993

Table 3 Continued

Target protein	Polymer	Ligand	Method used to precipitate polymer	Protein recovery (%)	Purification fold	Ref.
Restriction endonuclease Hind III	Poly-NIPAM	Phage λ DNA	ΔT	—[b]	—[b]	Maeda et al. 1993
Xylanase from *Trichoderma viride*	Eudragit S-100	—[a]	ΔpH	70	4.6	Gupta and Mattiasson 1994
D-Lactate dehydrogenase from *Lactobacillus bulgaricus*	Eudragit S-100	—[a]	ΔpH	85	2.4	Guoqiang et al. 1993
Endopoly-galacturonase from *Aspergillus niger*	Eudragit S-100	—[a]	ΔpH	86	8.8	Gupta et al. 1993
Trypsin	Eudragit S-100	Soybean trypsin inhibitor	ΔpH	74	1.8	Kumar and Gupta 1994
Concanavalin A	Eudragit S-100	*p*-aminophenyl-α-D-glucopyranoside	ΔpH	83–91	—[b]	Linne-Larsson and Mattiasson 1994
D-Lactate dehydrogenase from *Leuconostoc mesenteroides* ssp. *cremoris*	Eudragit S-100	Cibacron blue	ΔpH	56	4.3	Shu et al. 1994
Lactate dehydrogenase, and pyruvate kinase from porcine muscle	Eudragit S-100	Cibacron blue	ΔpH	63 59	7.8 4.4	Guoqiang et al. 1994b
Alcohol dehydrogenase from bakers yeast	Eudragit S-100	Cibacron blue in the presence of Zn^{2+}	ΔpH	40–42	8.2–9.5	Guoqiang et al. 1995b

β-Glucosidase from *Trichoderma longibrachiatum*	I. chitosan II. Eudragit S-100	—[a]	ΔpH	82–86	10–14	Agarwal and Gupta 1996
α-Amylase inhibitor from wheat	co-polymer of NIPAM and 1-vinylimidazole	Immobilized Cu ions	Addition of salt	89	4	Kumar et al. 1998a
α-Amylase inhibitors I-1 and I-2 from seeds of ragi (Indian finger millet)	co-polymer of NIPAM and 1-vinyl imidazole	Immobilized Cu ions	Addition of salt	84 (I-1) 89 (I-2)	13 (I-1) 4(I-2)	Kumar et al. 1998b
Monoclonal antibodies	Polyelectrolyte complex formed by poly(methacrylic acid) and poly(N-ethyl-4-vinylpyridinium bromide)	Antigen (inactivated glyceraldehyde-3-phosphate dehydrogenase)	ΔpH	96–98	1.4–1.8	Dainiak et al. 1999
Lactate dehydrogenase, from porcine muscle	Carboxymethyl-cellulose	Cibacron blue	Addition of Ca^{2+} and PEG	87	21	Lali et al. 1999
α-Glucosidase from *Saccharomyces cerevisiae*	Poly(*N*-acryloyl-piperidine)	Maltose	ΔT	54–68	170–200	Hoshino et al. 1998

[a] Target protein has the affinity for the polymer itself.
[b] Not presented.

achieved under optimal conditions (Table 4). Relatively low values of purification factors were attributable to the relative purity of the preparation of antibodies used in this work. Precipitated polyelectrolyte complexes could be dissolved at pH 7.3 and used repeatedly (Dainiak et al. 1999).

The successful affinity precipitation of antibodies in the above example indicates that the ligand is exposed to the solution. This fact was used to develop a new method of the production of monovalent Fab fragments of antibodies containing only one binding site. Traditionally, Fab fragments are produced by proteolytic digestion of antibodies in solution followed by separation of Fab fragments. In the case of monoclonal antibodies against inactivated subunits of glyceraldehyde-3-phosphate dehydrogenase, digestion with papain resulted in significant damage of binding sites of the Fab fragment. Proteolysis of monoclonal antibodies in the presence of the antigen-polycation conjugate followed by (1) precipitation induced by addition of polyanion, poly(methacrylic) acid and pH-shift from 7.3 to 6.5 and (2) elution at pH 3.0 resulted in 90% immunologically competent Fab fragments. Moreover, papain concentration required for proteolysis was 10 times lower than that for free antibodies in solution. The antibodies bound to the antigen-polyelectrolyte complex were digested by papain to a lesser extent, suggesting that binding to the antigen-polycation conjugate not only protected binding sites of monoclonal antibodies from proteolytic damage but also facilitated the proteolysis probably by exposing antibody molecules in a way convenient for proteolytic attack by the protease (Table 5) (Dainiak et al. 2000).

Metal Chelate Affinity Precipitation

The metal ions selectively precipitate proteins from solution by coordinating the lone-pair electrons of heteroatoms on the side chains of the amino acids, predominantly surface-exposed histidines and, to a lesser extent, cysteines and tryptophans (Gurd and Wilcox 1956). Bis complexes of transition

Table 4 Antibody Purification by Affinity Precipitation Using Glyceraldehyde-3-phosphate Dehydrogenase Bound to Polyelectrolyte Complex

Polycation/polyanion ratio	0.45	0.35	0.2	0.15
pH of incubation	7.3	6.5	5.5	4.6
pH of precipitation	6.5	6.0	5.3	4.5
Recovery, %	98	96	60	60
Purification factor (SDS-PAGE data)	1.8	1.4	1.4	1.4
Purification factor (ELISA data)	1.5	1.5	1.0	—

Table 5 Yield of Fab Fragments After Digestion of Antibodies at Different Conditions

	Free antibodies (%)		Antibodies on PEVP (%)		Antibodies on N-PEC (%)	
Factor	Low papain	High papain	Low papain	High papain	Low papain	High papain
Antibodies,[a] %	90	0	10	0	50	0
Fab fragments (active),[a] %	0	0[b]	90	70[b]	50	65[b]

[a] Initial concentration of immunologically competent antibodies was taken as 100%.

[b] It was observed that the part or all (in the case of free antibodies) Fab fragments lost affinity to antigen after digestion at high concentration of papain.

Low papain, 0.04 mg/mg antibodies; high papain, 1 mg/mg antibodies.

metals (III) were used for affinity precipitation of proteins in a homofunctional mode. Human hemoglobulin (26 histidine residues) and whale myoglobin were quantitatively precipitated with Cu-bis complex in model experiments (Van Dam et al. 1989). In another study, Lilius et al. (1991) reported the purification of genetically engineered galactose dehydrogenase with polyhistidine tail by metal affinity precipitation, using a bis-zinc complex with ethylene glycol-bis-(β-aminoethyl ether)-N,N,N',N'-tetraacetic acid, EGTA–Zn_2. The requirement of multibinding functionality of the target protein and slow precipitation rate pose some limitations to this approach of metal precipitation.

However, the heterofunctional format of metal affinity precipitation is expected to deliver better performance. Extensive work is carried out in our laboratory in this direction using the thermosensitive polymers poly-NIPAM and poly-VCL. For metal coordination, imidazole was used as metal ligand and copolymerized with the polymer backbone (Galaev et al. 1997). In a copolymer of NIPAM and vinylimidazole, poly(VI/NIPAM), the incorporation of relatively hydrophilic imidazole moieties hindered the hydrophobic interactions of the native poly-NIPAM and resulted in increase in the precipitation temperature (Fig. 10). The effect was more evident at lower pH values where imidazole moieties were protonated and hence rendered the polymer more hydrophilic. Poly(VI/NIPAM) (15.6 mol % VI), for example, did not precipitate at all on heating up to 70°C at pH 4 and 6, whereas precipitation of homopolymer, poly(NIPAM), was independent of pH (Fig. 10) (Galaev et al. 1999).

Poly(VI/NIPAM) copolymers are capable of binding metal ions via imidazole moieties (Galaev et al. 1999). When loaded with Cu(II), all the Cu-loaded copolymers did not precipitate on heating up to 70°C. The increase in ionic strength dramatically facilitated precipitation of Cu(II)-poly(VI/NIPAM) (Fig. 11), sodium sulfate being the most efficient salt additive followed by ammonium sulfate and NaCl. At NaCl concentration of 0.6 M at room temperature, all the polymer was instantaneously precipitated and flocculated in a clump, the remaining solution being transparent. The efficient precipitation of Cu(II)-poly(VI/NIPAM) by high salt concentrations at mild temperatures is very convenient for metal affinity precipitation. High salt concentration does not interfere with protein-metal ion-chelate interaction (Porath and Olin 1983). On the other hand, it reduces the possibility of nonspecific binding of foreign proteins to the polymer both in solution and after precipitation.

Cu(II) ion can form a complex with up to four imidazoles in solution. The flexible copolymer poly(VI/NIPAM) can adopt in solution a conformation where two to three imidazole molecules are close enough to form a complex with the same metal ion (Kumar et al. 1999) (Fig. 12). With about

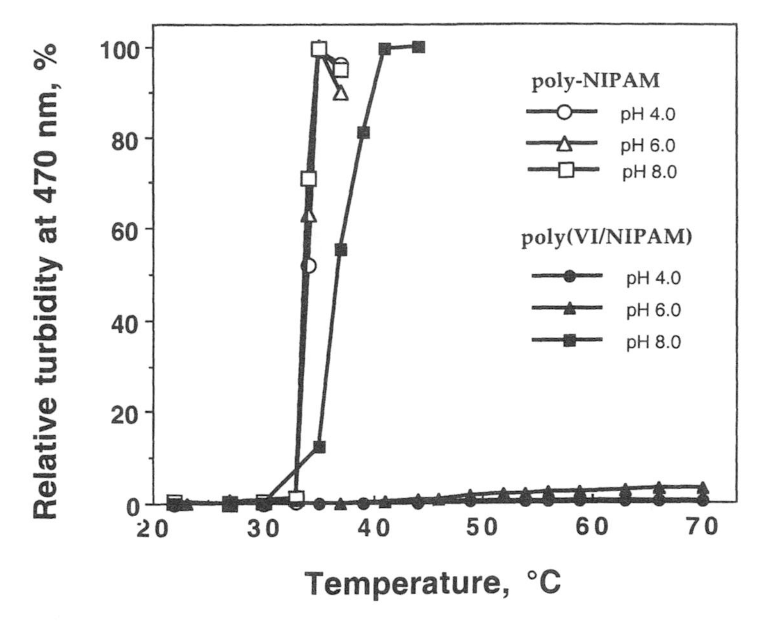

Figure 10 Thermoprecipitation of poly-NIPAM and poly(VI/NIPAM) (15.6 mol % VI) from aqueous solution at various pH values monitored as turbidity at 470 nm. Polymer concentration 1.0 mg/ml. (Reproduced from Galaev et al. 1999 with permission.)

two to three imidazole ligands bound to the metal ion, e.g., Cu(II), one could expect binding strength of log $K = 6.0–9.0$ (Gold and Greger 1960; Liu and Gregor 1965), providing significant strength of interaction. It is clear that not all available coordination sites of the metal ion are occupied by imidazole ligands of the polymer. The unoccupied coordination sites of the metal ion could be used for complex formation with the protein molecule via histidine residues on its surface. This thermosensitive, metal-chelated affinity reagent was used successfully for the purification of Kunitz soybean trypsin inhibitor from soybean meal (Galaev et al. 1997), α-amylase inhibitor from wheat (Kumar 1998a), and α-amylase/trypsin inhibitor from ragi seeds (Kumar et al. 1998b). These protein inhibitors were purified with more than 80% yields and with significant purification factors.

Other Applications

We have studied the rheological properties of the polymer phase formed after pH-induced precipitation of Eudragit S-100. The neutralization of the

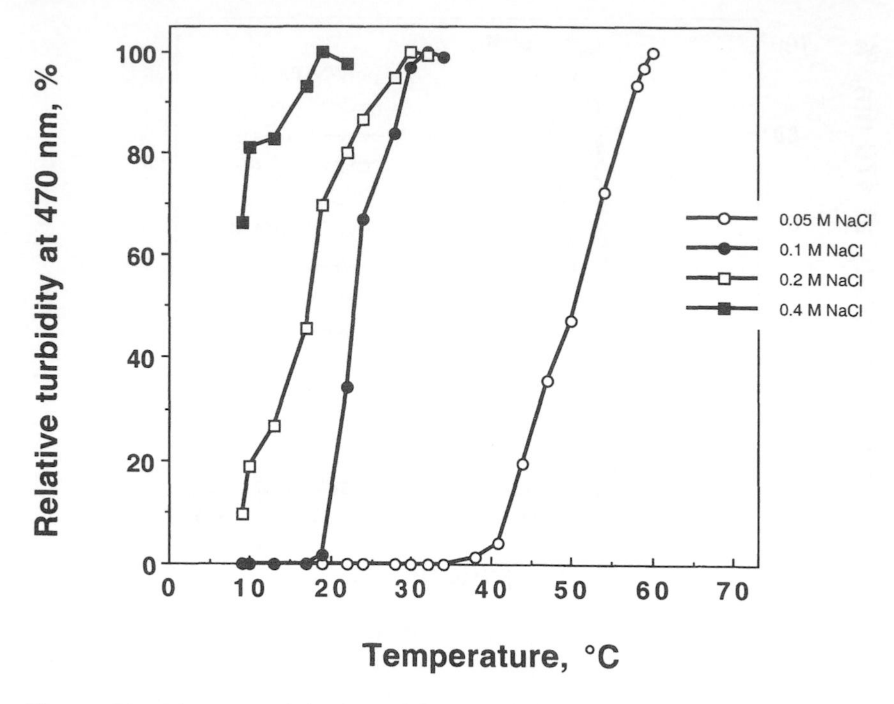

Figure 11 Thermoprecipitation of Cu(II)-poly(VI/NIPAM) (15.6 mol % VI) at various concentrations of NaCl at pH 6.0 monitored as turbidity at 470 nm. Polymer concentration 1.0 mg/ml. (Reproduced from Galaev et al. 1999 with permission.)

negatively charged carboxy groups results in the increase in hydrophobicity of the polymer molecules. The latter aggregate due to the hydrophobic interactions and form a three-dimensional network with high viscosity and strong shear-thinning behavior. Upon further decrease in pH and increase in hydrophobicity of the polymer molecules, the latter rearrange from three-dimensional network into a suspension of particles of aggregated polymer chains. This rearrangement is accompanied by significant drop in viscosity and much less pronounced shear-thinning behavior, indicating formation of the polymer precipitate of the third type (Linné-Larsson and Mattiasson 1996). Not surprisingly, quite a few successful affinity precipitation procedures were reported when using this polymer (Kamihira et al. 1992; Gupta et al. 1994; Guoqiang et al. 1993, 1994a,b; 1995a,b; Mattiasson and Kaul 1994; Shu et al. 1994) or similar polymer Eudragit L-100 (Chern et al. 1996). An alternative pH-sensitive polymer used successfully for affinity precipitation is hydroxypropylmethylcellulose acetate succinate (Taniguchi

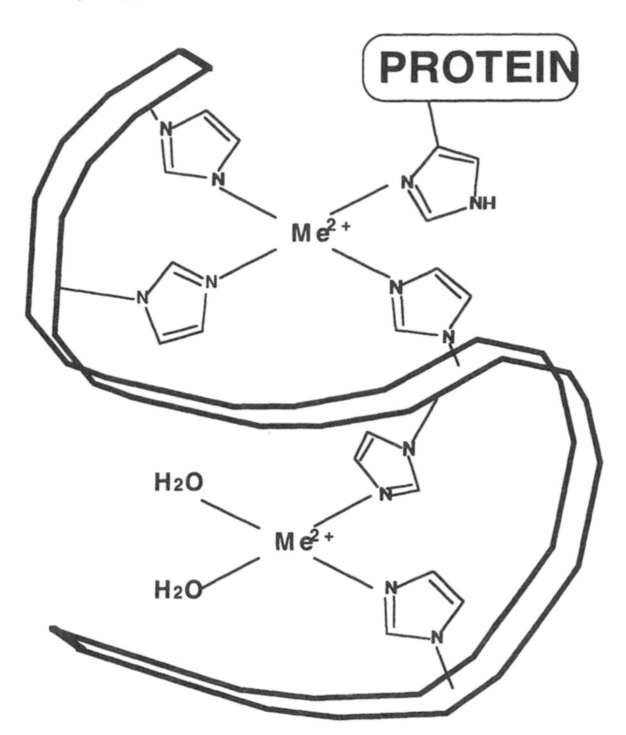

Figure 12 Imidazole-metal complex formation of flexible poly (VI/NIPAM) copolymer with surface His-containing protein. (Reproduced from Kumar et al. 1999 with permission.)

et al. 1989, 1990), which is also used for enteric coating like Eudragit polymers.

Affinity precipitation is readily combined with other protein isolation techniques, e.g., partitioning in aqueous two-phase polymer systems (Kamihira et al. 1992; Guoqiang et. al. 1994a; Mattiasson and Kaul 1994; Chen and Jang 1995). Partitioning of protein complexed with ligand-polymer conjugate is usually directed to the upper hydrophobic phase of aqueous two-phase polymer systems formed by PEG and dextran/hydroxypropyl starch, whereas most of the proteins present in crude extracts or cell homogenates partition into lower hydrophilic phase. Then the precipitation of the protein-polymer complex is promoted by changing pH. Trypsin was purified using conjugate of soybean trypsin inhibitor with hydroxypropylcellulose succinate acetate (Chen and Jang 1995), lactate dehydrogenase, and protein A using a conjugate of Eudragit S-100 with the triazine dye

Cibacron blue (Guoqiang et al. 1994a) and immunoglobulin G (Kamihira et al. 1992; Mattiasson and Kaul 1994), respectively. Combination of partitioning with affinity precipitation improves yield and purification factor and allows easier isolation of protein from particulate feed streams.

An interesting example of the use of poly(*N*-acryloylpiperidine) terminally modified with maltose for affinity precipitation was presented recently (Hoshino et al. 1998). Use of the polymer with extremely low critical temperature (soluble below 4°C and completely insoluble above 8°C) made it possible to use the technique for purification of thermolabile α-glucosidase from cell-free extract of *Saccharomyces cerevisiae* achieving 206-fold purification with 68% recovery.

Yet another carrier, polymerized liposome was developed for salt-induced affinity precipitation (Sun et al. 1995). Polymerized liposomes, prepared using a synthesized phospholipid with a diacetylene moiety in the hydrophobic chain and an amino group in the hydrophilic head, showed a reversible precipitation on salt addition and removal, respectively. Polymerized liposomes with immobilized soybean trypsin inhibitor were used to resolve a model mixture of trypsin and chymotrypsin.

Thermosensitive latexes composed of thermosensitive polymers or with a layer of thermosensitive polymer at the surface present another example of polymeric carriers. These are insoluble but reversibly suspended in response to increasing/decreasing temperature. Until recently, thermosensitive latexes were used mainly as carriers for reversibly soluble biocatalyst (Kondo et al. 1994; Shiroya et al. 1995a,b; Kondo and Fukuda 1997; Okubo and Ahmad 1998). Their potential for affinity precipitation remains to be evaluated.

pH-sensitive latex based on the copolymer of poly(methacrylate) and poly(methyl-methacrylic acid) was compared with a soluble polymer of the similar composition, Eudragit L-100, as carrier for precipitation of positively charged protein, lysozyme. Latex particles precipitated much faster but their capacity was about an order of magnitude lower than that of the soluble polymer.

4. CONCLUSION

Precipitation is a well-established technique for separating biomolecules from reaction fluids. Experience with large-scale processes have established that there are no problems in scaling up. There are, however, two issues that need improvement among the classical precipitation protocols: selectivity and environmental load. The former deals with the challenge on how to precipitate one protein among many others such that selective enrichment is achieved. The different concepts of affinity precipitation represent promising steps in a direction of highly selective precipitation processes. So far, precipitation has mainly been carried out in bulk solution, but the introduction of smart polymers may also open for other configurations, e.g., precipitation on the surface of adsorbant particles.

Precipitation processes have for a long period been dominated by the use of ammonium sulfate. It is by far the most used and a very well-functioning process. However, one needs to also bring in the concept of environmental load into the process evaluation, and then a precipitation mode involving reagents that can be recycled would be more attractive. Efforts to recycle salts from precipitation processes are ongoing at the research level. Alternatively, polymers used for precipitation may be efficiently recycled.

LIST OF ABBREVIATIONS

CMC	carboxymethylcellulose
DMSO	dimethyl sulfoxide
DS	dextran sulfate
pI	isoelectric point
PEG	poly(ethylene glycol)
PEI	poly(ethylene imine)
PAm	poly(acrylamide)
PAA	poly(acrylic acid)
PMAA	poly(methacrylic acid)
PVAm	poly(vinyl acrylamide)
Poly(NIPAM)	poly(*N*-isopropylacrylamide)
Poly(VCL)	poly(*N*-vinyl caprolactam)
Poly(VI/NIPAM)	vinylimidazole-*co*-*N*-isopropyl acrylamide
Poly(VI/VCL)	vinylimidazole-*co*-vinyl caprolactam
LCST	lower critical solution temperature
LDH	lactate dehydrogenase
SAS	supercritical antisolvent

REFERENCES

Agarwal, R. and Gupta, M. N. (1994) Copper affinity precipitation as an initial step in protein purification. Biotechnol. Techniq. 8, 655–658.

Agarwal, R. and Gupta, M. N. (1996) Sequential precipitation with reversibly soluble insoluble polymers as a bioseparation strategy: Purification of β-glucosidase from *Trichoderma longibrachiatum*. Protein Expr. Purif. 7, 294–298.

Agena, S. M., Pusey, M. L. and Bogle, I. D. L. (1999) Protein solubility modeling. Biotechnol. Bioeng. 64, 144–150.

Alred, P. A., Kozlowski, A., Harris, J. M. and Tjerneld, F. (1994) Application of temperature-induced phase partitioning at ambient temperature for enzyme purification. J. Chromatogr. A 659, 289–298.

Arnold, F. H. (1991) Metal-affinity separations: a new dimension in protein processing. Bio/Technology 9, 151–156.

Atha, D. H. and Ingham, K. C. (1981) Mechanism of precipitation of proteins by polyethylene glycols. Analysis in terms of excluded volume. J. Biol. Chem. 256, 12108–12117.

Baldwin, R. L. (1996) How Hofmeister ion interactions affect protein stability. Biophys. J. 71, 2056–2063.

Bell, D. J., Hoare, M. and Dunnill, D. (1983) The formation of protein precipitates and their centrifugal recovery. Adv. Biochem. Eng. Biotechnol. 26, 1–72.

Birkenmeier, G. (1989) Dye serum protein interaction-analysis and application. In: Vijaylakshimi, M. A. and Bertrand, O. (eds.), Protein-Dye Interactions. Developments and Applications. Elsevier Applied Science, London, pp. 253–254.

Bradshaw, A. P. and Sturgeon, R. J. (1990) The synthesis of soluble polymer-ligand complexes for affinity precipitation studies. Biotechnol. Techn. 4, 67–71.

Bryjak, J. and Noworyta, A. (1994) Storage stabilization and purification of enzyme by water-soluble synthetic polymers. Enzyme Microb. Technol., 16, 616–621.

Boominathan, R., Mishra, P. and Chand, S. (1995) Isolation of lipase from the culture filtrate of *Humicola lanuginosa*. Bioseparation, 5, 235–239.

Boyd, W. C. (1973) The mechanism of specific precipitation: another look. Experientia 12, 1565–1566.

Bozzano, A. G., Andrea, G. and Glatz, C. E. (1991) Separation of proteins from polyelectrolytes by ultrafiltration. J. Membr. Sci., 55, 181–198.

Burgess, D. J. and Carless, J. E. (1984) Microelectrophoretic studies of gelatin and acacia for the prediction of complex coacervation. J. Colloid Interf. Sci., 98, 1–8.

Burgess, R. R. and Jendrisak, J. J. (1975) A procedure for the rapid, large-scale purification of *Escherichia coli* DNA-dependent RNA polymerase involving polymin P precipitation and DNA-cellulose chromatography. Biochemistry 14, 4634–4638.

Cai, G., Guo, Y. and Shi, J. (1996) Purification of apolipoprotein H by polyethylene glycol precipitation. Prot. Expr. Purif. 8, 341–346.

Chan, M. Y. Y., Hoare, M. and Dunnill, P. (1986) The kinetics of protein precipitation by different reagents. Biotechnol. Bioeng. 28, 387–393.

Charles, M., Coughlin, R. W., and Hasselberger, F. X. (1974) Soluble-insoluble enzyme catalysts. Biotechnol. Bioeng. 16, 1553–1556.

Chen, W. and Berg, J. C. (1993) The effect of polyelectrolyte dosage on floc formation in protein precipitation by polyelectrolytes. Chem. Eng. Sci. 48, 1775–1784.

Chen, J.-P. and Jang, F.-L. (1995) Purification of trypsin by affinity precipitation combining with aqueous two-phase extraction. Biotechnol. Techn. 9, 461–466.

Chen, W., Walker, S. and Berg, J. C (1992) The mechanism of floc formation in protein precipitation by polyelectrolytes. Chem. Eng. Sci. 47, 1039–1045.

Chern, C. S., Lee, C. K. and Chen, C. Y. (1996) Biotin-modified submicron latex particles for affinity precipitation of avidin. Colloids Surf B Biointerfaces 7, 55–64.

Chiew, Y. C., Kuehner, D., Blanch, H. W. and Prausnitz, J. M. (1995) Molecular thermodynamics for salt-induced protein precipitation. AIChE J. 41, 2150–2159.

Clark, K. M. and Glatz, C. E. (1987) Polymer dosage considerations in polyelectrolyte precipitation of proteins. Biotechnol. Prog. 3, 241–247.

Clark, K. M. and Glatz, C. E. (1990) Protein fractionation by precipitation with carboxymethyl cellulose. In: Downstream Processing and Bioseparation, Hammel, J-F., Hunter, J. B. and Sikdar, S. K. (eds.). Vol. 419, ACS Symposium Series. Washington DC, pp. 170–187.

Curling, J. M. (ed.) (1980) Methods of Plasma Protein Fractionation. Academic Press, New York.

Dainiak, M. B., Izumrudov, V. A., Muronetz, V. I., Galaev, I. Yu., and Mattiasson, B. (1999) Affinity precipitation of monoclonal antibodies by nonstoichiometric polyelectrolyte complexes. Bioseparation 7, 231–240.

Dainiak, M. B., Muronetz, V. I., Izumrudov, V. A., Galaev, I. Yu., and Mattiasson, B. (2000) Production of Fab fragments of monoclonal antibodies using polyelectrolyte complexes. Anal. Biochem. 277, 58–66.

Dissing, U. and Mattiasson, B. (1996) Polyelectrolyte complexes as vehicles for affinity precipitation of proteins. J. Biotechnol. 52, 1–10.

Dubin, P. L., Ross, T. D., Sharma, I. and Yegerlehner, B. (1987) Coacervation of polyelectrolyte-protein complexes, In: Ordered Media in Chemical Separations. Hinze, W. L. and Armstrong, D. W. (eds.), American Chemical Society. Washington, DC, pp. 162–169.

Eaton, B. E., Gold, L. and Zichi, D. A. (1995) Let's get specific: the relationship between specificity and affinity. Chem. Biol. 10, 633–638.

Englard, S. and Seifter, S. (1990) Precipitation techniques. Meth. Enzymol. 182, 285–300.

Eudragit (1993) Röhm Pharma GMBH Information Materials.

Fahie-Wilson, M. N. (1999) Polyethylene glycol precipitation as a screening method for macroprolactenemia. Clin. Chem. 45, 436–437.

Ferreira, G. N., Cabral, J. M. and Prazeres, D. M. (1999) Development of process flow sheets for the purification of supercoiled plasmids for gene therapy applications. Biotechnol. Prog. 15, 725–731.

Fisher, R. R., Machiels, B., Kyracou, K. C. and Morris, J. E. (1989) Affinity ultrafiltration and affinity precipitation using a water-soluble complex of poly(vinyl alcohol) and cibacron blue F3GA. In: Vijaylakshimi, M. A. and Bertrand, O. (eds.), Protein-Dye Interactions: Developments and Applications. Elsevier Applied Science, London, pp. 190–196.

Fisher, R. R. and Glatz, C. E. (1988) Polyelectrolyte precipitation of proteins: the effects of reactor conditions. Biotechnol. Bioeng. 32, 777–785.

Fisher, R. R., Glatz, C. E. and Murphy, P. A. (1986) Effects of mixing during acid addition on fractionally precipitated protein. Biotechnol. Bioeng. 28, 1056–1063.

Franco, T. T., Galaev, I. Yu., Hatti-Kaul, R., Holmberg, N., Bülow, L. and Mattiasson, B. (1997) Aqueous two-phase system formed by thermoreactive vinyl imidazole/vinyl caprolactam copolymer and dextran for partitioning of a protein with a polyhistidine tail. Biotechnol. Techn. 11, 231–235.

Galaev, I. Yu., Gupta, M. N. and Mattiasson, B. (1996) Use smart polymers for bioseparations. Chemtech, December 19–25.

Galaev, I. Yu., Kumar, A., Agarwal, R., Gupta, M. N. and Mattiasson, B. (1997) Imidazole – a new ligand for metal affinity precipitation. Precipitation of Kunitz soybean trypsin inhibitor using Cu(II)-loaded copolymers of 1-vinylimidazole with N-vinylcaprolactam and N-isopropylacrylamide. Appl. Biochem. Biotechnol. 68, 121–133.

Galaev, I. Yu., Kumar, A. and Mattiasson, B. (1999) Metal-copolymer complexes of N-isopropylacrylamide for affinity precipitation of proteins. J. Macromol. Sci. 36, 1093–1105.

Galaev, I. Yu. and Mattiasson, B. (1992) Affinity thermoprecipitation of trypsin using soybean trypsin inhibitor conjugated with a thermo-reactive polymer, poly(N-vinyl caprolactam). Biotechnol. Techn. 6, 353–358.

Galaev, I. Yu. and Mattiasson, B. (1993a) Affinity thermoprecipitation: Contribution of the efficiency of ligand-protein interaction and access of the ligand. Biotechnol. Bioeng. 41, 1101–1106.

Galaev, I. Yu. and Mattiasson, B. (1993b) Thermoreactive water-soluble polymers, nonionic surfactants, and hydrogels as reagents in biotechnology. Enzyme Microb. Technol. 15, 354–366.

Galaev, I. Yu. and Mattiasson, B. (1997) New methods for affinity purification of proteins. Affinity precipitation: a review. Biochemistry (Moscow) 62, 571–577.

Galaev, I. Yu. and Mattiasson, B. (2002) Affinity precipitation of proteins using smart polymers. In: Smart Polymers for Bioseparation and Bioprocessing. Galaev, I. Yu. and Mattiasson, B. (eds.), Taylor & Francis, London, pp. 55–77.

Gisser, K. R. C., Geselbracht, M. J., Capellari, A., Hunsberger, L., Ellis, A. B., Perepezko, J. and Lisensky, G. C. (1994) Nickel-titanium memory metal. A "smart" material exibiting a solid-state phase change and superelasticity. J. Chem. Educ. 71, 334–340.

Glatz, C. E. (1990) Precipitation. In: Downstream Processing in Biotechnology. Asenjo, J. (ed.), Marcel Dekker, New York.

Gold, D. H. and Gregor, H. P. (1960) Metal-polyelectrolyte complexes. VIII. The poly-N-vinylimidazole-copper(II) complex. J. Phys. Chem. 64, 1464–1467.

Guoqiang, D., Batra, R., Kaul, R., Gupta, M. N. and Mattiasson, B. (1995a) Alternative modes of precipitation of Eudragit S 100: a potential ligand carrier for affinity precipitation of proteins. Bioseparation 5, 339–350.

Guoqiang, D., Benhura, M. A. N., Kaul, R. and Mattiasson, B. (1995b) Affinity precipitation of yeast alcohol dehydrogenase through metal ion promoted binding with Eudragit bound Cibacron blue 3 GA. Biotechnol. Prog. 11, 187–193.

Guoqiang, D., Kaul, R. and Mattiasson, B. (1993) Purification of *Lactobacillus bulgaricus* D-lactate dehydrogenase by precipitation with an anionic polymer. Bioseparation 3, 333–341.

Guoqiang, D., Kaul, R. and Mattiasson, B. (1994a) Integration of aqueous two-phase extraction and affinity precipitation for the purification of lactate dehydrogenase. J. Chromatogr. A 668, 145–152.

Guoqiang, D., Lali, A., Kaul, R. and Mattiasson, B. (1994b) Affinity precipitation of lactate dehydrogenase and pyruvate kinase from porcine muscle using Eudragit bound Cibacron blue. J. Biotechnol. 37, 23–31.

Gupta, M. N., Guoqiang, D. and Mattiasson, B. (1993) Purification of endo-polygalacturonase by affinity precipitation using alginate. Biotechnol. Appl. Biochem. 18, 321–327.

Gupta, M. N., Guoqiang, D., Kaul, R. and Mattiasson, B. (1994) Purification of xylanase from *Trichoderma viride* by precipitation with anionic polymer Eudragit S 100. Biotechnol. Techn. 8, 117–122.

Gupta, M. N. and Mattiasson, B. (1994) Affinity precipitation. In: Street, G. (eds.). Highly Selective Separations in Biotechnology. Blackie Academic & Professional, London, pp. 7–33.

Gurd, F. R. N. and Wilcox, P. E. (1956) Complex formation between metallic cations and proteins, peptides, and amino acids. Adv. Prot. Chem. 11, 311–427.

Hamada, J. S. (1994) Use of polyethylene glycol and high-performance liquid chromatography for preparative separation of *Aspergillus ficuum* acid phosphatases. J. Chromatogr. A 658, 371–380.

Harris, P. A., Karlström, G. and Tjerneld, F. (1991) Enzyme purification using temperature-induced phase formation. Bioseparation 2, 237–246.

Hill, R. D. and Zadow J. G. (1978) The precipitation of whey proteins with water soluble polymers. N. Z. J. Diary Sci. Technol., 113, 61–64.

Hoare, M., Dunnill, P. and Bell, D. J. (1983) Reactor design for protein precipitation and its effect on centrifugal separation. Ann. NY. Acad. Sci. 413, 254–269.

Homma, T., Fujii, M., Mori, J., Kawakami, T., Kuroda, K. and Taniguchi, M. (1993) Production of cellobiose by enzymatic hydrolysis: removal of β-glucosidase from cellulase by affinity precipitation using chitosan. Biotechnol. Bioeng. 41, 405–410.

Hoshino, K., Taniguchi, M., Kitao, T., Morohashi, S. and Sasakura, T. (1998) Preparation of a new thermo-responsive adsorbent with maltose as a ligand and its application to affinity precipitation. Biotechnol. Bioeng. 60, 568–579.

Hughes, P. and Lowe, C. R. (1988) Purification of proteins by aqueous two-phase partition in novel acrylic co-polymer systems. Enzyme Microb. Technol. 10, 115–122.

Hyet, M. and Vijaylakshmi, M. A. (1986) Affinity precipitation of proteins using bis-dyes. J. Chromatgr. 376, 157–161.

Ingham, K. C. (1990) Precipitation of proteins with polyethylene glycol. Meth. Enzymol. 182, 301–306.

Ito, Y. (1999) Centrifugal precipitation chromatography applied to fractionation of proteins with ammonium sulfate. J. Liq. Chrom. Rel. Technol. 22, 2825–2836.

Iyer, H. V. and Przybycien, T. M. (1995) Metal affinity protein precipitation: effects of mixing, protein concentration, and modifiers on protein fractionation. Biotechnol. Bioeng. 48, 324–332.

Izumrudov, V. A., Galaev, I. Y. and Mattiasson, B. (1999) Polycomplexes—potential for bioseparation. Bioseparation 7, 207–220.

Jap, T. S., Jenq, S. F., Wong, M. C. and Chiang, H. (1995) Polyethylene glycol method is superior to ammonium sulfate and protein-A Sepharose-4B method in fractionating thyroid stimulating immunoglobulin in Grave's disease. Chung Hua I Hsueh Tsa Chih (Taipei), 55, 19–24.

Jendrisak, J. (1987) The use of polyethyleneimine in protein purification. In: Protein Purification: Micro to Macro. Burgess, R. (ed.), Alan R, Liss, New York, pp.75–97.

Jerala, R., Kroon-Zitka, L. and Turk, V. (1994) Improved expression and evaluation of polyethyleneimine precipitation in isolation of recombinant cysteine proteinase inhibitor Stefin B. Prot. Expr. Purif. 5, 65–69.

Kamihira, M., Kaul, R. and Mattiasson, B. (1992) Purification of recombinant protein A by aqueous two-phase extraction integrated with affinity precipitation. Biotechnol. Bioeng. 40, 1381–1387.

Kaul, R. and Mattiasson, B. (1992) Secondary purification. Bioseparation 3, 1–26.

Kiknadze, E. V. and Antonov, Yu. A. (1998) Use of polyelectrolytes for isolation of trypsin inhibitor from industrial waste of Alfalfa leaf protein fractionation. Appl. Biochem. Microbiol. 34, 508–512.

Kitano, S., Hisamitsu, I., Koyama, Y., Kataoka, K., Okano, T., Yokoyama, M. and Sakurai, Y. (1993) Preparation of glucose-responsive polymer complex system having phenylboronic acid moiety and its application to insulin-releasing device. In: Takagi, T., Takahashi, K., Aizawa, M. and Miyata, S. (eds.), Proceedings of the First International Conference on Intelligent Materials, Technomic Publishing, Lancaster, pp. 383–388.

Kokufuta, E. (1994) Complexion of proteins with polyelectrolytes in a salt-free system. In: Dubin, P. L. (ed.), Macromolecular Complexes in Chemistry and Biology. Springer-Verlag, Berlin, pp. 301–325.

Kokufuta, E. and Matsukawa, S. (1995) Enzymatically induced reversible gel-sol transition of a synthetic polymer system. Macromolecules 28, 3474–3475.

Kokufuta, E., Shimizu, H. and Nakamura, I. (1981) Salt linkage formation of poly(diallyldimethylammonium chloride) with acidic groups in the polyion complex between human carboxyhemoglobin and potassium poly(vinyl alcohol) sulfate. Macromolecules 14, 1178–1180.

Kondo, A. and Fukuda, H. (1997) Preparation of thermo-sensitive magnetic hydrogel microspheres and application to enzyme immobilization. J. Ferment. Bioeng. 84, 337–341.

Kondo, A., Imura, K., Nakama, K. and Higashitani, K. (1994) Preparation of immobilized papain using thermosensitive latex perticles. J. Ferment. Bioeng. 78, 241–245.

Kresk, M. and Wellington, E. M. (1999) Comparison of different methods for the isolation and purification of total community DNA from soil. J. Microb. Meth. 39, 1–16.

Kudaibergenov, S. Y. and Bekturov, Y. A. (1989) Influence of the coil-globule conformational transition in polyampholytes affecting the sorption and desorption of polyelectrolytes in human serum albumin. Polym. Sci. USSR 31, 2870–2874.

Kuehner, D. E., Blanch, H. W. and Prausnitz, J. M. (1996) Salt-induced protein precipitation: phase equilibria from an equation of state. Fluid Phase Equilibria 116, 140–147.

Kumar, A., Agarwal, R., Batra, R. and Gupta, M. N. (1994) Effect of polymer concentration on recovery of the target proteins in precipitation methods. Biotechnol. Techn. 8, 651–654.

Kumar, A., Galaev, I. Yu. and Mattiasson, B. (1998a) Affinity precipitation of alpha-amylase inhibitor from wheat meal by metal chelate affinity binding using Cu(II)-loaded copolymers of 1-vinylimidazole with N-isopropylacrylamide. Biotechnol. Bioeng. 59, 695–704.

Kumar, A., Galaev, I. Yu. and Mattiasson, B. (1998b) Isolation of α-amylase inhibitors I-1 and I-2 from seeds of ragi (Indian finger millet, *Elusine coracana*) by metal chelate affinity precipitation. Bioseparation 7, 129–136.

Kumar, A., Galaev, I. Yu. and Mattiasson, B. (1999) Metal chelate affinity precipitation: a new approach to protein purification. Bioseparation 7, 185–194.

Kumar, A. and Gupta, M. N. (1996) An assessment of nonspecific adsorption to Eudragit S-100 during affinity precipitation. Mol. Biotechnol. 6, 1–6.

Kumar, A. and Gupta, M. N. (1994) Affinity precipitation of trypsin with soybean trypsin inhibitor linked Eudragit S-100. J. Biotechnol. 37, 185–189.

Lali, A., Balan, S., John, R. and D'Souza, F. (1999) Carboxymethyl cellulose as a new heterofunctional ligand carrier for affinity precipitation of proteins. Bioseparation 7, 195–205.

Larsson, P.-O. and Mosbach, K. (1979) Affinity precipitation of enzymes. FEBS Lett. 98, 333–338.

Larsson, P.-O., Flygare, S. and Mosbach, K. (1984) Affinity precipitation of dehydrogenases. Adv. Enzymol., 104, 364–369.

Lee, S. J. and Park, K. (1996) Glucose-sensitive phase-reversible hydrogels. In: Hydrogels and Biodegradable Polymers for Bioapplications. Ottenbrite, R.M., Huang, S.J. and Park, K. (eds.), American Chemical Society, Washington, DC, pp. 2–10.

Li, Y., Mattison, K. W., Dubin, P. L., Havel, H. A. and Edwards, S. L. (1996) Light scattering studies of the binding of bovine serum albumin to a cationic polyelectrolyte. Biopolymers 38, 527–533.

Lilius, G., Persson, M., Bülow, L. and Mosbach, K. (1991) Metal affinity precipitation of proteins carrying genetically attached polyhistidine affinity tails. Eur. J. Biochem. 198, 499–504.

Linné, E., Garg, N., Kaul, R. and Mattiasson, B. (1992) Evaluation of alginate as a carrier in affinity precipitation. Biotechnol. Appl. Biochem. 16, 48–56.

Linné-Larsson, E., Galaev, I. Y., Lindahl, L. and Mattiasson, B. (1996) Affinity precipitation of concanavalin A with *p*-amino-α-D-glucapyranoside modified Eudragit S-100. I. Initial complex formation and build-up of the precipitate. Bioseparation 6, 273–282.

Linné-Larsson, E. and Mattiasson, B. (1994) Isolation of concanavalin A by affinity precipitation. Biotechnol. Techn. 8, 51–56.

Linné-Larsson, E., and Mattiasson, B. (1996) Evaluation of affinity precipitation and a traditional affinity chromatography for purification of soybean lectin from extracts of soya flour. J. Biotechnol. 49, 189–199.

Liu, F., Liu, F. H., Zhuo, R. X., Peng, Y., Deng, Y. Z. and Zeng, Y. (1995) Development of a polymer-enzyme immunoassay method and its application. Biotechnol. Appl. Biochem. 21, 257–264.

Liu, K.-J. and Gregor, H. P. (1965) Metal-polyelectrolyte complexes. X. Poly-N-vinylimidazole complexes with zinc (II) and with copper(II) and nitrilotriacetic acid. J. Phys. Chem. 69, 1252–1259.

Lu, M., Albertsson, P.-Å., Johansson, G. and Tjerneld, F. (1996) Ucon-benzoyl dextran aqueous two-phase systems: protein purification with phase component recycling. J. Chromatogr. B 680, 65–70.

Maeda, M., Nishimura, C., Inenaga, A. and Takagi, M. (1993) Modification of DNA with poly(N-isopropylacrylamide) for thermally induced affinity precipitation. Reactive Funct Polym 21, 27–35.

Margolin, A. L., Sherstyuk, S. F., Izumrudov, V. A., Zezin, A. B., Kabanov, V. A. (1984) Enzymes in polyelectrolyte complexes. The effect of phase transition on thermal stability. Eur. J. Biochem. 146, 625–632.

Mattiasson, B. and Kaul, R. (1993) Affinity precipitation. In: Molecular Interactions in Bioseparations. Ngo, T., (ed.), Plenum Press, New York, pp. 469–477.

Mattiasson, B. and Kaul, R. (1994) "One-pot" protein purification by process integration. Bio/Technology 12, 1087–1089.

Mattiasson, B., Kumar, A. and Galaev, I. Yu. (1998) Affinity precipitation of proteins: design criteria for an efficient polymer. J. Mol. Recogn. 11, 211–216.

McGregor, W. C. (1983) Large-scale isolation and purification of proteins from recombinant *E. coli.* Ann. N. Y. Acad. Sci. 413, 231–237.

Melander, W. and Horvath, C. (1977) Salt effects on hydrophobic interactions in precipitation and chromatography of proteins: an interpretation of the lyotropic series. Arch. Biochem. Biophys. 183, 200–215.

Morawetz, H. and Hughes, W. L. Jr. (1952) The interaction of proteins with synthetic polyelectrolytes. I. Complexing of bovine serum albumin. J. Phys. Chem. 56, 64–69.

Mori, S., Nakata, Y. and Endo, H. (1994) Purification of rabbit C-reactive protein by affinity precipitation with thermosensitive polymer. Prot. Expr. Purif. 5, 151–156.

Morris, J. E. and Fisher, R. R. (1990) Complications encountered using Cibacron Blue F3G-A as a ligand for affinity precipitation of lactate dehydrogenase. Biotechnol. Bioeng. 36, 737–743.

Morris, J. E., Hoffman, A. S. and Fisher, R. R. (1993) Affinity precipitation of proteins by polyligands. Biotechnol. Bioeng. 41, 991–997.

Murdock, P. J., Brooks, S., Mellars, G., Cheung, G., Jacob, D., Owens, D. L., Parmar, M. and Riddell, A. (1997) A simple monoclonal antibody based ELISA for free protein S. Comparison with PEG precipitation. Clin. Lab. Haematol. 19, 111–114.

Nakayama, A., Kawamura, K., Shimada, H., Akaoka, A., Mita, I., Honjo, M. and Furutani, Y. (1987) Extracellular production of human growth hormone by a head portion of the prepropeptide derived from *Bacillus subtilis*. J. Biotechnol. 5, 171–179.

Nelson, C. D. and Glatz, C. E. (1985) Primary particle formation in protein precipitation. Biotechnol. Bioeng. 27, 1434–1444.

Nguyen, T. Q. (1986) Interactions of human haemoglobin with high-molecular weight dextran sulfate and diethylaminoethyl dextran. Makromol. Chem. 187, 2567–2578.

Nguyen, A. L. and Luong, J. H. T. (1989) Syntheses and application of water-soluble reactive polymers for purification and immobilization of biomolecules. Biotechnol. Bioeng. 34, 1186–1190.

Niederauer, M. Q. and Glatz, C. E. (1992) Selective precipitation. In: T. Provender (ed.), Advances in Biochemical Engineering/Biotechnology-Bioseparation, Vol. 47, Springer-Verlag, New York, pp. 159–188.

Nielsen Niels Viktor. (1994) Method for purification of an aqueous enzyme solution. Patent abstract, PCT Pub. No. WO94/01537, Novo Nordisk A/S.

Okubo, M., and Ahmad, H. (1998) Enzymatic activity of trypsin adsorbed on temperature-sensitive composite polymer particles. J. Polym. Sci A Polym. Chem. 36, 883–888.

Ortin, A., Cebrian, J. A. and Johansson, G. (1992) Large scale extraction of alpha-lactalbumin and beta-lactoglobulin from bovine whey by precipitation with polyethylene glycol and partitioning in aqueous two-phase systems. Prep. Biochem. 22, 53–66.

Patrickios, C. S., Herler, W. R. and Hatton, T. A. (1994) Protein complexion with acrylic polyampholytes. Biotechnol. Bioeng. 44, 1031–1039.

Paul, E. L. and Rosas, C. B. (1990) Challenges for chemical engineers in the pharmaceutical industry. Chem. Eng. Prog. 22, 17–25.

Pearson, J. C., Clonis, Y. D. and Lowe, C. R. (1989) Preparative affinity precipitation of L-lactate dehydrogenase. J. Biotechnol., 11, 267–274.

Pécs, M., Eggert, M. and Schügerl, K. (1991) Affinity precipitation of extracellular microbial enzymes. J. Biotechnol. 21, 137–142.

Polson, A. (1990) Isolation of IgY from the yolks of eggs by chloroform-polyethylene glycol procedure. Immunol. Invest. 19, 253–258.

Polson, A., Potgieter, G. M., Largier, J. F., Mears, G. E. F. and Joubert, F. J. (1964) The fractionation of protein mixtures by linear polymers of high molecular weight. Biochim. Biophys. Acta 82, 463–475.

Porath, J. and Olin, B. (1983) Immobilized metal ion affinity adsorption and immobilized metal ion affinity chromatography of biomaterials. Serum protein affinities for gel-immobilized iron and nickel ions. Biochemistry 22, 1621–1630.

Powers, D. D., Willard, B. L., Carbonell, R. G. and Kilpatrick, P. K. (1992) Affinity precipitation of proteins by surfactant-solubilized ligand-modified phospholipids. Biotechnol. Prog. 8, 436–453.

Ramshaw, J. A., Bateman, J. F. and Cole, W. G. (1984) Precipitation of collagens by polyethylene glycols. Anal. Biochem. 141, 361–365.

Riahi, B. and Vijaylakshmi, M. A. (1989) Affinity precipitation of (NAD dependent dehydrogenase) lactate dehydrogenase using cibacron blue dimers. In: Protein-Dye Interactions: Developments and Applications, Vijaylakshimi, M. A. and Bertrand, O. (eds.), Elsevier Applied Science, London, pp. 197–204.

Sacco, D., Bonneaux, F. and Dellacherie, E. (1988) Interaction of haemoglobin with dextran sulphates and the oxygen-binding properties of the covalent conjugates. Int. J. Biol. Macrmol. 10, 305–310.

Scopes, R. P. (1994) Protein Purification: Principles and Practice. Springer-Verlag, New York.

Senstad, C. and Mattiasson, B. (1989a) Affinity precipitation using chitosan as a ligand carrier. Biotechnol. Bioeng. 33, 216–220.

Senstad, C. and Mattiasson, B. (1989b) Purification of wheat germ agglutinin using affinity flocculation with chitosan and a subsequent centrifugation or flotation step. Biotechnol. Bioeng. 34, 387–393.

Shieh, J. Y. and Glatz, C. E. (1994) Precipitation of proteins with polyelectrolytes: role of polymer molecular weight. In: Macromolecular Complexes in Chemistry and Biology. Dubin, P. L. (ed.), Springer-Verlag, Berlin, pp. 272–284.

Shih, Y. C., Prausnitz, J. M. and Blanch, H. W. (1992) Some characteristics of protein precipitation by salts. Biotechnol. Bioeng. 40, 1155–1164.

Shiroya, T., Tamura, N., Yasui, M., Fujimoto, K. and Kawaguchi, H. (1995a) Enzyme immobilization on thermosensitive hydrogel microspheres. Colloids Surf. B. Biointerfaces 4, 267–274.

Shiroya, T., Yasui, M., Fujimoto, K. and Kawaguchi, H. (1995b) Control of enzymatic activity using thermosensitive polymers. Colloids Surf. B. Biointerfaces 4, 275–285.

Shu, H.-C., Guoqiang, D., Kaul, R. and Mattiasson, B. (1994) Purification of the D-lactate dehydrogenase from *Leuconostoc mesenteroides* ssp. *cremoris* using a sequential precipitation procedure. J. Biotechnol. 34, 1–11.

Sternberg, M. (1976) Purification of industrial enzymes with polyacrylic acids. Process Biochem. 11, 11–12.

Sternberg, M. and Hershberger, C. (1974) Separation of proteins with polyacrylic acids. Biochim. Biophys. Acta 342, 195–206.

Strege, M. A., Dubin, P. L., West, J. S. and Flinta, C. D. (1990) Protein separation via polyelectrolyte complexation, In: Protein Purification. Ladisch, M. R., Willson, R. C., Painton, C. C. and Builder, S. E. (eds.), Vol. 427, ACS Symposium Series, Washington, DC, pp. 66–79.

Sun, Y., Yu, K., Jin, X. H. and Zhou, X. Z. (1995) Polymerized liposome as a ligand carrier for affinity precipitation of proteins. Biotechnol. Bioeng. 47, 20–25.

Taniguchi, M., Kobayashi, M., Natsui, K. and Fujii, M. (1989) Purification of staphylococcal protein A by affinity precipitation using a reversibly soluble-insoluble polymer with human IgG as a ligand. J. Ferment. Bioeng. 68, 32–36.

Taniguchi, M., Tanahashi, S. and Fujii, M. (1990) Purification of staphylococcal protein A by affinity precipitation: Dissociation of protein A from the adsorbent with chemical reagents. J. Ferment. Bioeng. 69, 362–264.

Tyagi, R., Kumar, A., Sardar, M., Kumar, S. and Gupta, M. N. (1996) Chitosan as an affinity macroligand for precipitation of N-acetyl glucosamine binding proteins/enzymes. Isol. Purif. 2, 217–226.

Umeno, D., Mori, T. and Maeda, M. (1998) Single stranded DNA-poly(N-isopropylacrylamide) conjugate for affinity precipitation separation of oligonucleotides. Chem. Commun. 1433–1434.

Van Dam, M. E., Wuenchell, G. E. and Arnold, F. H. (1989) Metal affinity precipitation of proteins. Biotechnol. Appl. Biochem. 11, 492–502.

Welch, P. and Scopes, R. K. (1981) Rapid purification and crystallization of yeast phosphofructokinase. Anal. Biochem. 110, 154–157.

Winters, M. A., Frankel, D. Z., Debenedetti, P. G., Carey, J., Devaney, M. and Przybycien, T. M. (1999) Protein purification with vapor-phase carbon dioxide. Biotechnol. Bioeng. 62, 247–258.

Winters, M. A., Knuston, B. L., Debenedetti, P. G., Sparks, H. G., Przybycien, T. M., Stevenson, C. L. and Prestrelski, S. J. (1996) Precipitation of proteins in supercritical carbon dioxide. J. Pharm. Sci. 85, 586–594.

Wu, K.-Y. A. and Wisecarver, K. D. (1992) Cell immobilization using PVA crosslinked with boric acid. Biotechnol. Bioeng. 39, 447–449.

Xia, J. and Dubin, P. L. (1994) Protein-polyelectrolyte complexes. In: Macromolecular Complexes in Chemistry and Biology. Dubin, P., Bock, J., Davis, R., Schulz, D. N. and Thies, C. (eds.), Springer-Verlag, Berlin, pp. 247–272.

Yeo, S. D., Lim, G. B., Debenedetti, P. G. and Bernstein, H. (1993) Formation of microparticulate protein powders using a supercritical fluid antisolvent. Biotechnol. Bioeng. 41, 341–346.

Zaman, F., Kusnadi, A. R. and Glatz, C. E. (1999) Strategies for recombinant protein recovery from canola by precipitation. Biotechnol. Prog. 15, 488–492.

Zhao, J. Y., Ford, C. F., Glatz, C. E., Rougvie, M. A. and Gendel, S. M. (1990) Polyelectrolyte precipitation of beta-galactosidase fusions containing poly-aspartic acid tails. J. Biotechnol. 14, 273–283.

8
Protein Crystallization for Large-Scale Bioseparation

E. K. Lee
Hanyang University, Ansan, Korea

Woo-Sik Kim
Kyunghee University, Suwon, Korea

1. OBJECTIVES OF PROTEIN CRYSTALLIZATION

In general, protein crystallization is performed for one of two distinct reasons: (1) X-ray crystallography for structural biology studies and (2) purification as a means of bioseparation. In the first application, the protein crystals are used to ascertain the three-dimensional structure of the molecule, which is then used to study the various properties of a protein such as optical, spectroscopic, electrical, thermal, and mechanical. The information on the biological functions and on the exact bioactive sites can be exploited for improved drug identification and design.

The second application exploits the fact that crystallization has been one of the commonly employed separation techniques not only for low molecular weight substances but also for macromolecular biological substances. Compared with other protein purification/separation techniques, it has an inherent advantage of not denaturing the protein of interest and often providing some stabilization effect. As such, crystallization can be used as a large-scale, high-yield purification process of industrially important proteins. Properly executed, crystallization is a powerful and economical protein purification method since high-purity proteins can be obtained from a single-step operation. However, research articles dedicated to large-scale protein purification by crystallization are relatively scarce.

The purity and the size of a single protein crystal from a batch are of utmost importance for the crystallization intended for X-ray crystallography. Usually, only a minute amount is needed and the crystal yield does not matter so much. In the large-scale protein purification process, however, recovery yield as well as purification factor is the key issue. In the well-designed and executed process, higher than 95% yields were reported for various enzymes, including β-amylase (Visuri and Nummi 1972). The purifying power of crystallization is usually better than any other large-scale methods exploiting the protein solubility characteristics such as extraction, precipitation, etc.

The basics involved in the crystallization of proteins are essentially the same as those for small-molecule crystallizations such as inorganic salts and simple organics. In each case the solution is allowed to become supersaturated with a solute, and at some point solute crystals start to appear and grow in size. Proteins differ from small molecules only in the degree of supersaturation required to induce and allow a sufficiently high rate of crystallization. Other than that, the mechanisms involved in crystal nucleation, growth, and cessation are identical.

2. STEPS OF PROTEIN CRYSTALLIZATION

In general, crystallization is defined as one way to reach a more stable, lower energy state from a metastable supersaturated state by reducing a solute concentration (Weber 1991). Common to all systems, crystallization goes through the three stages of nucleation, growth, and cessation.

Nucleation is the process in which thermodynamically stable aggregate with a regularly repeating lattice is produced from a solution. By crystallization, free energy of a protein is lowered by about 3–6 kcal/mol relative to the solution state (Drenth and Haas 1992). For a successful nucleus, aggregate must first exceed a certain size attribute (the critical size) defined by the competition of the ratio of the surface area to its volume (Feher and Kam 1985; Boistelle and Astier 1988). Once the critical size is exceeded, the aggregate becomes a supercritical nucleus capable of further growth. If not, spontaneous dissolution would occur much faster than crystallization. Also, the formation of noncrystalline aggregate proceeds more quickly because it does not involve the competition between surface area and volume. The size of a nucleus is also dependent upon the degree of supersaturation. The more supersaturated, the larger and more stable aggregates can be produced due to increased probability of collision among the diffusing molecules. At lower solute concentrations the formation of a large number of small, single crystals is favored. Therefore, for successful nucleation we need to identify and establish the "optimal" degree of supersaturation.

Crystal growth proceeds in continuation of nucleation. The growth rate is determined by a combination of the crystal's surface characteristics and the solute's diffusion rate. Boistelle and Astier (1988) used periodic bond chain theory to propose three different and characteristic types of growing crystal faces (flat, stepped, and kinked). It was demonstrated that the crystal growth rate is strongly influenced by the type of the faces; the smoother the face, the slower the rate would be. Another factor influencing the growth rate is the rate of convective diffusion of protein molecules reaching the crystal surface. As in nucleation, increased solubility results in an increased rate of growth. Ultraviolet microscopy was used to demonstrate that the protein concentration in the adjacent layer, or halos, surrounding the growing crystals was lower than that of the bulk solution (Feher and Kam 1985). The rate of protein diffusion across these halos could be a rate-limiting factor.

Cessation of crystal growth occurs for many reasons. The most obvious is the decrease in solute concentration in the liquid phase. Crystal growth can be extended by the addition of more solutes. Other major reasons for the cessation include lattice strain effects and surface poisoning. The former, demonstrated by Feher and Kam (1985) using hen egg white lysozyme, suggests that the long-range propagation of strain effectively prevents addition of molecules to the surface once a certain critical volume is reached. The latter occurs when impurity molecules are incorporated into the crystal face. Since the growth of crystal lattices typically selects for perfect or target molecules over damaged or impurity molecules, surface poisoning is more likely to occur in the later period of growth when the crystal size increases. Using laser scattering tomography, Sato et al. (1992) were able to visualize lattice defects of the large crystals of orthogonal hen egg white lysozyme. They showed the poisoning occurs not only at the surface but also within the bulk of the crystal itself (see Section 5.1 for more details).

3. SOLUTION FOR PROTEIN CRYSTALLIZATION

3.1 Solubility and Supersaturation

In a given solvent the crystallization purification of a biological macromolecule (i.e., a solute) is driven by the supersaturation, which means the concentration of a solute in solution is above its solubility. Therefore, to design a crystalization process we need to first understand the solubility of a solute in a solvent. Since a solute is dissolved in a solvent by various intermolecular interactions between solute-solvent, solute-solute, and solvent-

solvent, the solubility of a solute depends primarily on the physicochemical properties of the solvent and the solute. The interactions between the molecules are thermodynamically defined as cohesive energy, which is composed of dispersion, polar, and hydrogen bonding energies. To dissolve a solute in a solvent, intermolecular interaction energy between solvent-solute should be similar to or greater than the intramolecular interaction energies between solvent-solvent and solute-solute. Even though estimation of the molecular interactions between a solute and a solvent is possible from the thermodynamic data, the selection of the best solvent for a crystallization of biological macromolecule is a completely different problem because a tiny difference in solvent properties may cause the significant changes of crystallization behavior.

In general, the solvent is classified as polar or nonpolar based on their dielectric constants. In nonpolar solvents having a low dielectric constant, the molecular interaction of a solvent relies mostly on the van der Waals force, therefore, the nonpolar solute having a similar molecular interaction with the solvent can be dissolved in it. The polar solvents are subdivided into the polar protic one, of which the molecular interaction results from the hydrogen bonding, and the dipolar aprotic one, which causes the strong molecular interaction by the dipole-dipole intermolecular bonding. To be dissolved in the polar protic solvents, a solute molecule should break the hydrogen bonding of solvent molecules to form a stable molecular interaction between solute-solvent molecules. For this reason, the protic solute, which can generate the hydrogen bonding between solute-solvent molecules, is usually preferred to have a reasonable solubility in the protic solvent. In the dipolar aprotic solvents, a dipolar and aprotic solute is well dissolved because it interacts easily with solvent molecules to form a stable dipole-dipole bond. Based on the molecular interactions the solvents can be classi-

Table 1 Classification of Solvents Based on the Intermolecular Interaction

Polar solvents		Nonpolar solvents
Polar protic	Dipolar aprotic	
Examples	*Examples*	*Examples*
Water	Nitrobenzene	Benzene
Methanol	Acetonitrile	*n*-Hexane
Formic acid	Furfural	Ethyl ether
Ethanol	Acetone	
Acetic acid		

fied as shown in Table 1. Among these solvents, water is frequently used as an exclusive solvent for the crystallization of biological macromolecular solutes because it is capable of dissolving many kinds of, and reasonable amounts of, the solutes (e.g., organic and inorganic solutes), and inexpensive and innocuous as well.

In thermodynamics, the solubility of a solute in a solvent is defined as the equilibrium concentration of the solute dissolved in a solution and depends on the solution conditions, such as temperature, pH, composition, and the types and amounts of impurities present. To crystallize the solute out of solution, the supersaturation, which means that the solute concentration is higher than its solubility limit, is required. Accordingly, the crystallization process can be classified by the method employed to create the supersaturation. For example, if the solubility is sensitive to temperature, a temperature drop can be used to generate the supersaturation. This process is called *cooling crystallization*. In this case, the influence of temperature on solubility is commonly expressed as a polynomial equation:

$$X_S = A + BT + CT^2 + \cdots \tag{1}$$

or

$$X_S = A' + B'T^{-1} + C' \log T + \cdots \tag{2}$$

where X_S is the mole fraction of solute at equilibrium, T is the temperature, and A, B, C, A', B', C', etc. are constants. It is rarely needed to resort to higher order polynomial for the empirical relation. For selecting the cooling crystallization to separate biological macromolecules such as proteins, the changes in other chemical and physical properties of the solute with respect to the temperature change must be considered because protein denaturation may be caused by wide thermal history.

To induce the crystallization of biological macromolecules in aqueous solution, a salting-out process (salting-out crystallization) is frequently used. In this type of crystallization, the solubility change of the biological macromolecule is caused by addition of a small amount of a salt(s). To predict the solubility change of proteins relative to salt concentration, the semiempirical Cohn equation is used (Esdal 1947):

$$\log \frac{C_s'}{C_s} = \beta + K_s I \tag{3}$$

where C_s and C_s' are the protein solubilities in pure water and a salt solution, respectively, β is the solubility ratio of a protein at isoionic point, and K_s is a constant. The ionic strength of the solution, I, depends on the salt concentration as:

$$I = \frac{1}{2}\sum_{i} z_i^2 m_i \tag{4}$$

where z_i is the charge number of an ion and m_i is the molar salt concentration. In Eq. (3), if the constant K_s is positive, the protein solubility increases with increasing salt concentration. It is called "salting in." However, if K_s is negative, the protein solubility decreases with increasing salt concentration, which is called "salting out." This correlation of the protein solubility relative to the salt concentration is commonly employed to characterize the protein crystallization from aqueous solution using inorganic electrolytic salt.

The above linear relationship between the solubility and the salt concentration in aqueous solution is usually valid in the limited concentration range of the salt. Deviation of the protein solubility from the correlation in a wide range of salt concentration is mainly due to the electrostatic and hydrophobic interactions of a solute (Melander and Horvath 1977). By adding the salt into the solution, the salting-out effect based on the hydrophobic interaction and the salting-in effect based on the electrostatic repulsion occur at the same time. Therefore, a realistic expression for the protein solubility in a salt solution can be introduced as:

$$\log\frac{C_s'}{C_s} = K_{in}\sqrt{I} - K_{out}I \tag{5}$$

where K_{in} and K_{out} are the salting-in and salting-out constants, respectively. If the salting-out effect is greater than the salting-in effect, the protein is crystallized out of the solution as the salt concentration increases. The dependence of the protein solubility on the salt concentration can be predicted as shown in Fig. 1.

Actually, the hydrophobic interaction of a protein with a salt determines the dehydration behavior of a protein. It means that the salting-out effect of a salt may vary with the salt activity coefficient as well as the salt concentration. As shown in Fig. 2, the solubility of lysozyme is reduced with increasing the activity coefficient of the salt although the concentrations of each salt are the same (Ataka 1986).

3.2 Metastable Solution

When a solute concentration in a solution is above its solubility, the solution is supersaturated and unstable. If any crystals of the solute are not initially present in the solution, the crystallization has to be started by primary nucleation. For the primary nucleation to occur, a high level of

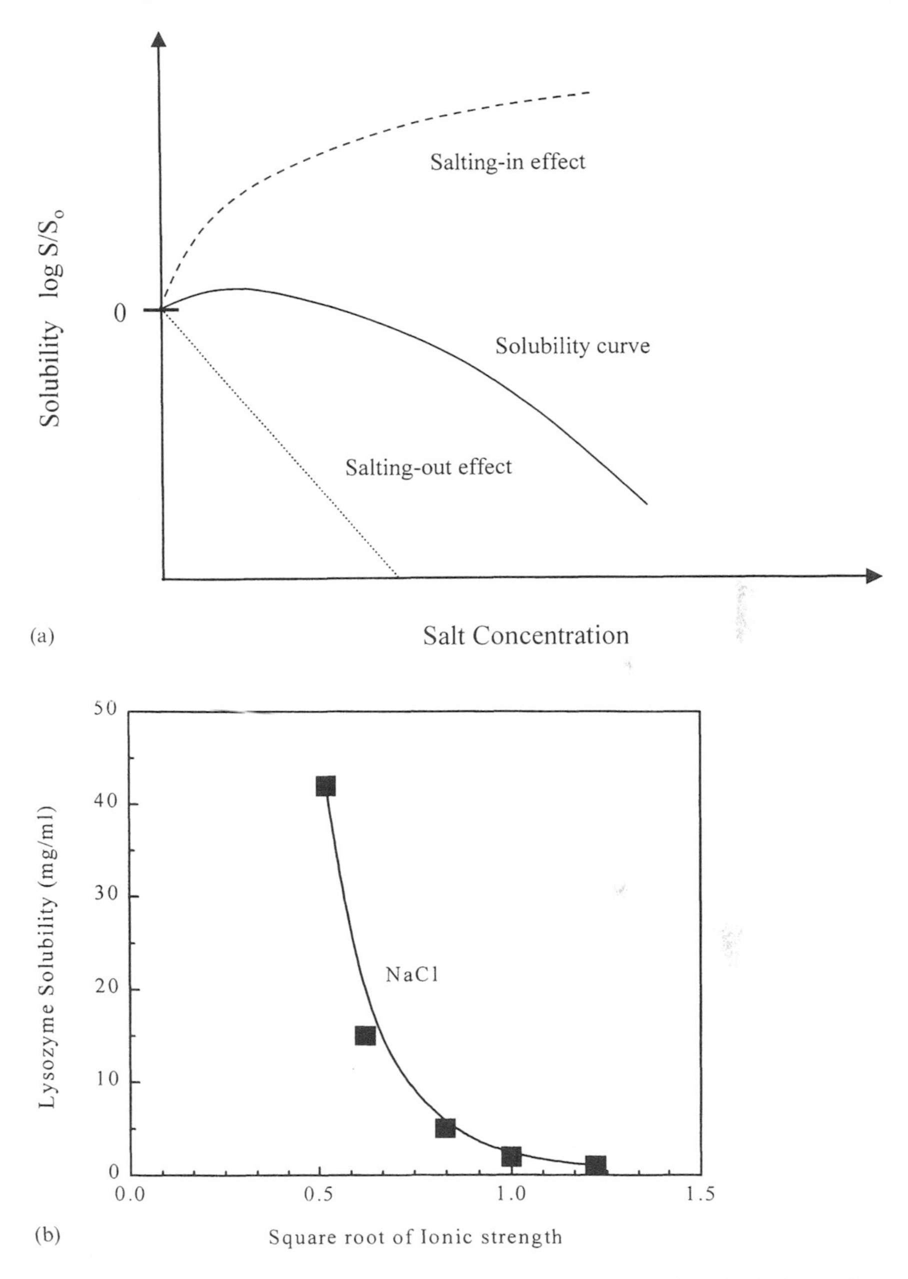

Figure 1 Dependence of protein solubility on salt concentration. (a) Effects of salt concentration on protein solubility. (b) Solubility change of lysozyme protein relative to the ionic strength of salt (Ries-Kautt and Ducruix 1992).

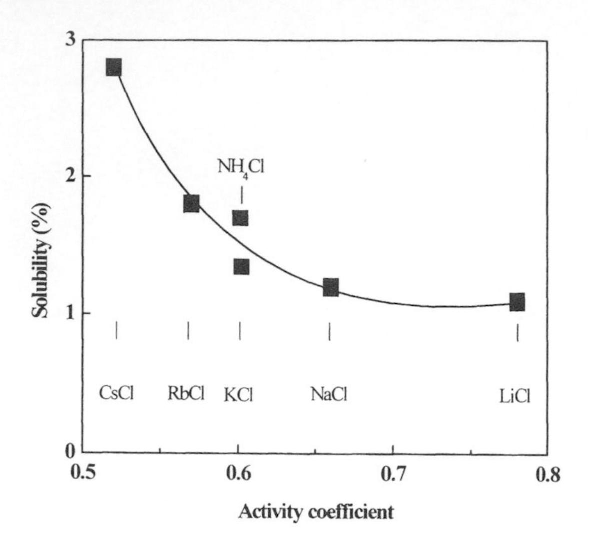

Figure 2 Effect of ionic activities of salts on the solubility of lysozyme protein (Ataka 1986).

supersaturation is required because of the high activation energy. If the supersaturation level is not high enough to overcome the activation energy barrier against the nucleation, no crystal can be formed in the solution. Then, the solution will remain clear without crystalization and appear stable even if it is supersaturated. This solution is defined as "metastable," and the maximal supersaturation level not allowing the primary nucleation to occur is called "metastable zone." It should be noted that the crystal growth might occur in the metastable zone if the solute crystals are added intentionally to the solution. As shown in Fig. 3, as the solution temperature decreases in cooling crystallization, the solute solubility also decreases. Then the solution becomes supersaturated and remains stable in the metastable zone. In that zone, the solute crystals may appear suddenly. According to Mersmann (1995) and Mullin (1993), the metastable zone width can be estimated as:

$$\Delta C_{\mathrm{max}} = \left(\frac{dC_{\mathrm{s}}}{dT}\right)\Delta T_{\mathrm{max}} \tag{6}$$

where ΔC_{max} is the maximum metastable supersaturation defined as $(C_{\mathrm{meta}} - C_{\mathrm{s}})$, and ΔT_{max} is the maximal allowable undercooling temperature.

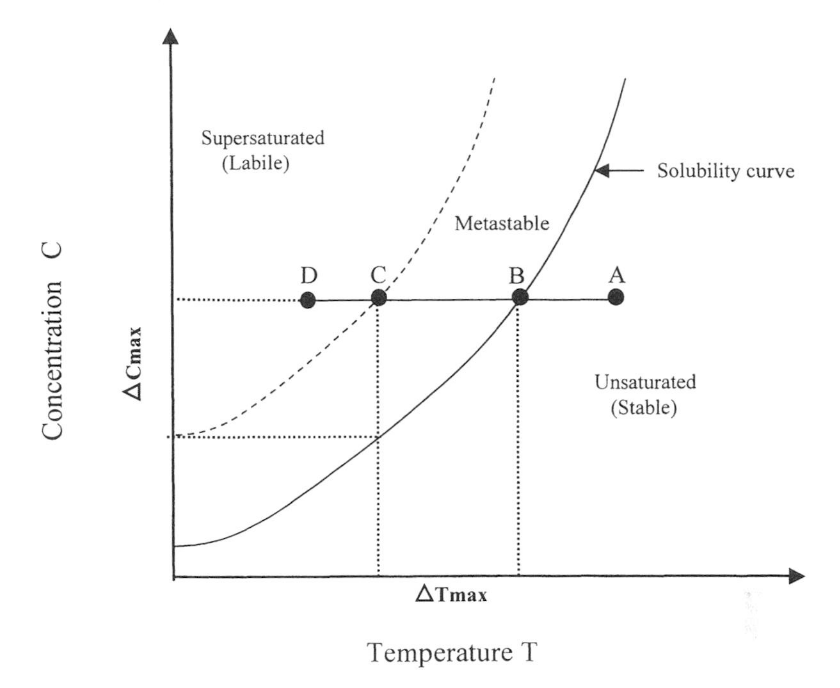

Figure 3 Change of solution condition relative to temperature.

3.3 Crystal Size and Solubility

In general, the solubility is determined by the physicochemical properties of the solute and the solvent. However, it is often observed that even under equilibrium the small-size crystals disappear and the large-size crystals grow simultaneously and continuously. This dynamic phenomenon is called Thomson-Ostwald ripening, and its effect appears more pronounced in the crystallization of highly soluble proteins. To thermodynamically describe the ripening effect, a Thomson-Ostwald equation was developed as (Mullin 1993; Mersmann 1995):

$$\frac{C_s(\mathrm{L})}{C_s(\infty)} = \exp\left(\frac{2M\sigma}{RT\rho L}\right) \tag{7}$$

where $C_s(\mathrm{L})$ is the size-dependent solubility of the crystal, $C_s(\infty)$ is the normal solubility of the solute, R is the universal gas constant, ρ is the density of the crystal, M is the molecular weight of the solute, σ is the interfacial tension of the crystal in contact with the solution, and L is the crystal size. According to Eq. (7), the equilibrium concentration of the crystals increases with decrease in the crystal size. When the crystals in a wide range of their size distribution are in the normal equilibrium solution, the solution will be subsaturated with respect to small crystals and supersaturated with respect to large crystals. Therefore, it will result in the dissolution of small crystals and the growth of large crystals.

From Eq. (7), it is unrealistic that the crystal solubility reaches infinity as the crystal size approaches zero. To overcome this, Knapp (1922) modified the solubility effect on the crystal size as:

$$\frac{C_s(\mathrm{L})}{C_s(\infty)} = \exp\left(\frac{A}{L} - \frac{B}{L^4}\right) \tag{8}$$

where A and B are constants depending on the physical properties of the solute. Using Eq. (8), the solubility effect on the crystal size is depicted in Fig. 4.

4. KINETICS OF PROTEIN CRYSTALLIZATION

4.1 Nucleation

Under labile conditions, a solute can be transformed from liquid to solid (crystal) phase by internal or external stimuli such as supersaturation, seeds,

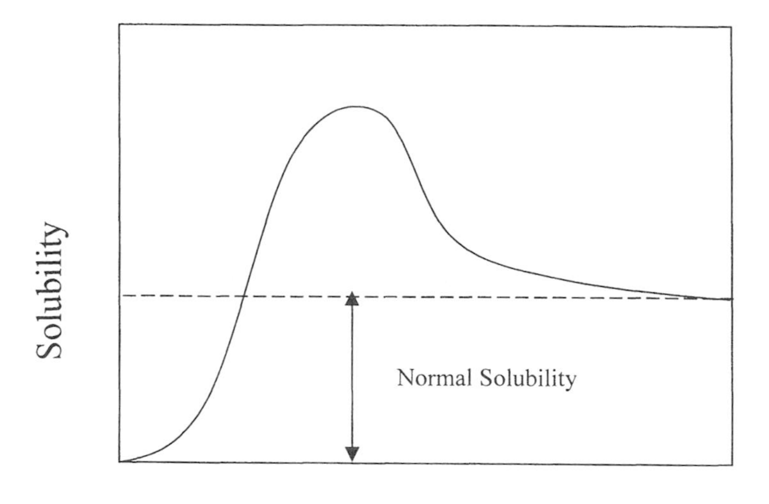

Figure 4 Description of crystal solubility relative to crystal size (Knapp 1922).

foreign surface, agitation, and impurities. Before the solid crystal appears in solution, the so-called nucleation process first generates the crystal nuclei, which are the tiniest crystals. This process can occur by many different ways depending on the solution conditions. Based on the mechanisms creating the nuclei, the nucleation processes can be classified into primary and secondary nucleation, as displayed in Fig. 5. In the primary nucleation, the nucleation occurs in the absence of its own crystalline matter. If the nuclei are generated through embryo, which is the molecular cluster of a solute formed in the homogeneous solution phase, it is called "homogeneous" nucleation. However, if the nucleus is formed with the aid of foreign surfaces, such as reactor wall, baffle, impeller surface, etc., it is called "heterogeneous" nucleation. In the heterogeneous nucleation, the foreign surface plays a role like a catalyst reducing the energy barrier to transform the embryos to nuclei, and the nuclei can easily be produced at a lower supersaturation level than in the homogeneous nucleation. It should be noted that homogeneous nucleation is more sensitive to the supersaturation level than the heterogeneous nucleation process. Frequently, most crystalization separation of biological macromolecules starts with the heterogeneous nucleation process because of the difficulties and limitations involved in generating the supersaturation state.

In the presence of its own crystalline matter, it is often observed that nuclei are continuously produced even at a very low supersaturation level. This is called "secondary nucleation." For example, in the agitated suspension the fluid shear on the crystal surface can produce small fragments of crystals playing like a nucleus; this is called "shear nucleation." In addition, the attrition, contact, breakage, abrasion, and needle fraction

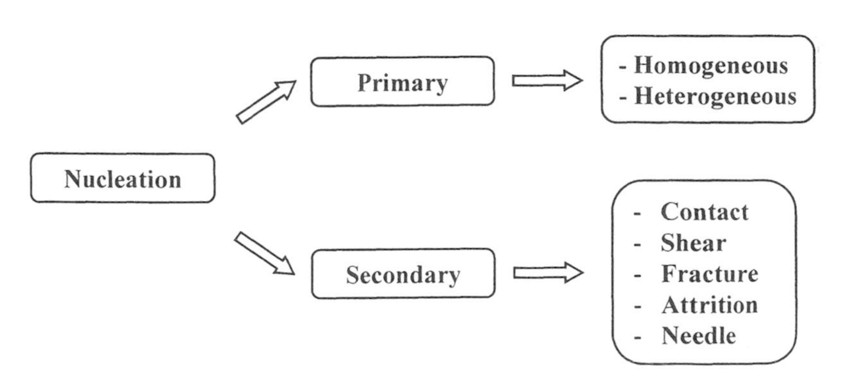

Figure 5 Classification of crystal nucleation processes based on crystal nucleation mechanisms.

are the plausible reasons for the secondary nucleation. The nucleation mechanism relative to the supersaturation level is classified in Fig. 6.

In the supersaturated solution the solute molecules collide to form molecular embryos, which are thermodynamically unstable clusters. These embryos can either be developed to stable crystal nuclei by continuous molecular addition or be dismissed into liquid phase. Once the embryos reach the minimal size of stable nuclei, they can grow naturally in the supersaturated solution. This is the way the solute molecules transform from the liquid phase to the crystal phase. From the thermodynamic point of view, the Gibbs free energy change related to the crystallization at constant temperature and pressure is expressed as:

$$\Delta G = V\Delta G_{\mathrm{V}} + A\Delta G_{\mathrm{A}} \tag{9}$$

where V is the volume of embryo, A is the surface area of embryos, and ΔG_{V} and ΔG_{A} are the Gibbs free energy changes due to changes of embryo volume and embryo surface area, respectively. The surface free energy of embryo is originated from the interfacial energy difference between the crystal and the liquid. The volume free energy of embryo is related to the solutes' molecular affinity, which depends on the solute concentration in the solution:

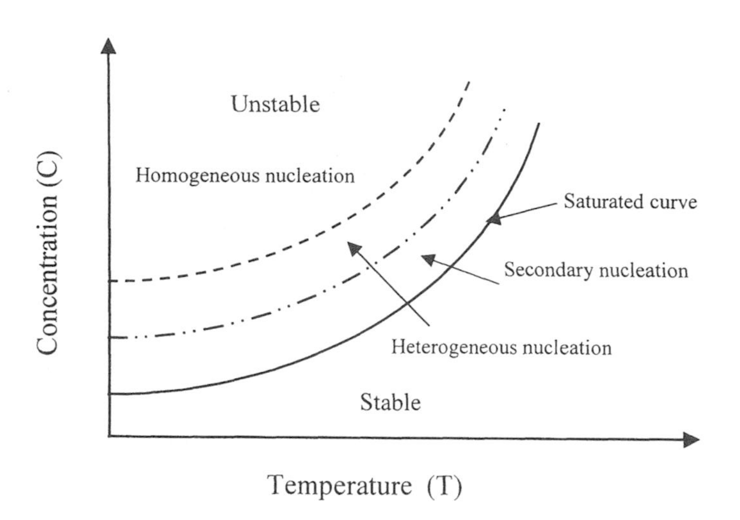

Figure 6 Classification of crystal nucleation mechanisms relative to the supersaturation level of the solution.

$$\Delta G_A = \sigma \quad (10)$$

$$\Delta G_V = -(kT/V_M)\ln(a/a_0) \quad (11)$$

where σ is the surface tension, k is Boltzmann's constant, T is temperature, V_M is the volume of a solute molecule, a and a_0 are the activities of the dissolved and the pure solute in equilibrium saturation, respectively. For simplicity, the activity coefficient of a solute is frequently assumed to be unity and then,

$$\Delta G = -V(kT/V_M)\ln(C/C_s) + A\sigma \quad (12)$$

where C and C_s are the solute concentration in a solution and at equilibrium, respectively. If the embryo is spherical, the Gibbs free energy change is described as function of the embryo size as:

$$\Delta G = -4/3\pi r^3(kT/V_M)\ln(C/C_s) + 4\pi r^2\sigma \quad (13)$$

where r is the radius of the embryo. Since the surface tension, embryo radius, a single molecular volume of the solute, temperature, and the Boltzmann constant all assume positive values physically, Eq. (13) suggests the Gibbs free energy is always increasing relative to the embryo radius, if the solute concentration is below the equilibrium concentration ($C/C_s \leq 1$). This means that in undersaturated solution crystallization cannot occur spontaneously, and external work should be provided to make the embryo grow. However, if the solute concentration is above the equilibrium concentration ($C/C_s > 1$), the volume change of the embryo causes the reduction in free energy. It may result in maximization of the free energy change relatively to the embryo size, as shown in Fig. 7. The maximal Gibbs free energy change, ΔG^*, relative to the embryo radius can be obtained by derivation of Eq. (13) as:

$$\Delta G^* = 16\pi\sigma^3/3[(kT/V_M)\ln(C/C_s)]^2 \quad (14)$$

at

$$r_c = 2\sigma/(kT/V_M)\ln(C/C_s) \quad (15)$$

Here the embryo radius at the maximal free energy change is defined as the critical radius (r_c). The maximal Gibbs free energy at the critical embryo radius is called "critical activation energy," which is an energy barrier to be overcome to create a stable nucleus. As shown in Fig. 7, below the critical radius the Gibbs free energy increases with an increase in the embryo radius. It also indicates that the dissolution of the embryo is reversible. However, above the critical radius, the embryo growth is irreversible because the Gibbs free energy decreases relative to the embryo size. The embryo of the critical size is called "nucleus," which is the minimum-sized, stable

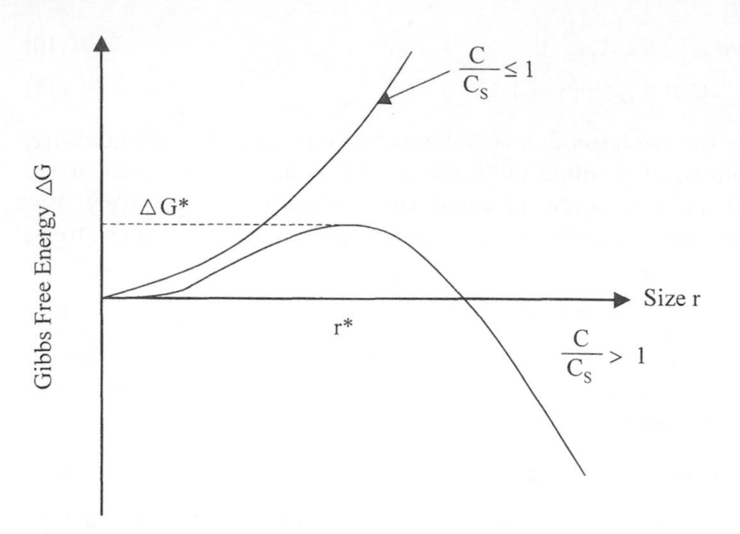

Figure 7 Gibbs free energy change for formation of crystal in solution.

crystal existing in the supersaturated solution. Equations (14) and (15) indicate that the critical free energy and the critical radius are reduced as the supersaturation level increases.

When the embryos are growing by the molecular addition in the supersaturated solution, the nucleation rate, B^0, is expressed as:

$$\begin{aligned} B^0 &= (2D/d^5)\exp(-\Delta G^*/kT) \\ &= (2D/d^5)\exp\left(-16\pi\sigma^3/3\left\{[(kT/V_M)\ln(C/C_S)]^2 kT\right\}\right) \end{aligned} \tag{16}$$

where D is the diffusion coefficient and d the diameter of the solute molecule. As shown in Eq. (16), the nucleation rate increases exponentially with an increase in the supersaturation level. This type of nucleation rate expression was first suggested by Volmer and Weber (1926) and later analytically developed by Nielsen (1964). This expression is very similar to that of the Arrhenius equation for the chemical reaction rate constant. Here the critical Gibbs free energy corresponds to the energy barrier for the chemical reaction, which is called "activation energy." Although the physical parameters in Volmer's expression, such as diffusion coefficient, solute molecule size, volume of single solute molecule, temperature, etc., are measurable and available in the literature, the surface tension of nucleus is very difficult to measure or obtain from the literature because the nucleus is too tiny to allow measurement of the interfacial contact angle and its size varies with the supersaturation level. Nielsen (1964) showed that the surface tension of a nucleus, which was estimated by comparing the predicted nucleation rate with the experimental data, has the same order of magnitude as that of the macroscopic crystal. Despite the fact that the above nucleation rate expression may include intrinsic errors coming from many assumptions and approximations used to describe the nucleation rate in measurable parameters, it may give important information to estimate the nucleation rate within a quite wide range of the supersaturation level [up to $\ln(C/C_s) = 10$)].

When the supersaturation level is low, the heterogeneous nucleation becomes dominant. In the heterogeneous nucleation, the foreign surface such as reactor wall, impeller, baffle surface, and so forth are favored to generate the nuclei because of its low surface energy. In general, the critical activation energy for the heterogeneous nucleation, $\Delta G^{*'}$, is less than that for the homogeneous nucleation and is simply described as:

$$\Delta G^{*'} = \phi \Delta G^{*} \tag{17}$$

where ϕ is the proportional factor that depends on the interfacial tension and usually less than unity. The interfacial tension between a foreign surface and the solution is frequently determined by the contact angle, and ϕ can be derived as

$$\phi = (2 + \cos\theta)(1 - \cos\theta)^2/4 \tag{18}$$

where θ is the contact angle between the solid and the liquid. This equation indicates that when θ is 180° $\Delta G^{*'}$ is equal to ΔG^{*}, which means the critical activation energy is not reduced by the foreign surface. When it is less than 180°, however, the foreign surface has a role in reducing the critical activation energy and promoting the nucleation. A simple power-law expression on the nucleation rate is adopted for the application in an industrial crystallization separation.

$$B^0 = k_n(C - C_s)^n \tag{19}$$

where k_n and n are the nucleation rate constant and the power index, respectively. Even though the above expression is not analytical, the nucleation rate is fairly well approximated in various conditions and can be applied to the homogeneous as well as the heterogeneous nucleation. Of course, the nucleation rate constant and the power index may vary with nucleation mechanisms, i.e., homogeneous or heterogeneous.

The secondary nucleation occurs at much lower supersaturation levels than the primary nucleation and is driven by many different mechanisms such as contact, shear, attrition, fracture, and so forth. Therefore, it will be difficult to develop the secondary nucleation rate based on the kinetic theory such as Volmer's expression for the primary nucleation. Fortunately, the simple power-law expression involving the major influencing factors is normally used to estimate the secondary nucleation rate as follows:

$$B^0 = k_n'(C - C_s)^i(\mathrm{MF})^j \tag{20}$$

where k_n' is the secondary nucleation rate constant, MF is the major influencing factor on the secondary nucleation except the supersaturation, and i and j are the power indices. For example, when the secondary nucleation is predominantly controlled by the agitation speed, the MF in Eq. (20) may be agitation speed, and when the attrition and fracture are the dominant mechanisms for the secondary nucleation, the solid density of the suspension can be a major influencing factor. In some cases the multiple influencing factors can be considered and described by multipower indices as like $(\mathrm{RPM})^{j1}(M_T)^{j2}(\mathrm{AREA})^{j3}$, where M_T means the solid density of the suspension, RPM means the agitation speed, and AREA means the surface area of the crystals.

4.2 Crystal Growth

In order for the crystal to grow in a supersaturated solution, the solute molecule dissolved in the solution has to be transformed into the solid crystal. This transformation is a very complex process consisting of the multiple consecutive steps such as:

1. Diffusion of a solute to crystal surface
2. Adsorption of the solute on to crystal surface
3. Dehydration of the adsorbed solute
4. Surface diffusion of the growth unit to energetically favorable lattice site
5. Incorporation of the growth unit into the lattice

First, the solute molecule in the solution is transported from the bulk by the concentration difference to the crystal surface to be adsorbed. The adsorbed solute releases the water molecules associated with it and then it diffuses on the surface to find the lowest energy site, which is called "kink" or "growth site." If the solute molecule is dissociated into ions in the solution, the crystal growth is much more complicated because the dissociated ions are separately diffused and rejoined on the crystal surface. Among these steps, the diffusion in solution and the integration on the crystal are the rate-limiting steps for crystal growth.

Diffusion-Controlled Crystal Growth Rate

When the crystal growth steps occurring on the crystal surface are fast enough, the mass transfer of the solute by diffusive and convective transport mechanism driven by the chemical potential difference becomes rate limiting. In this case, it may be assumed that the crystal growth rate is proportional to the supersaturation level as follows:

$$dm/dt = Ak_{\mathrm{m}}(C - C_{\mathrm{s}}) \tag{21}$$

where $1/A \cdot (dm/dt)$ is defined as a mass growth rate of the crystal (G_{m}), m and A are the mass and surface area of a single crystal, respectively, and k_{m} is the mass transfer coefficient. Assuming the constant morphology for all the crystals, the mass and the surface area of the single crystal are defined simply as:

$$\begin{aligned} m &= k_{\mathrm{v}}\rho L^3 \\ A &= k_{\mathrm{A}}L^2 \end{aligned} \tag{22}$$

where k_{v} and k_{A} are the geometrical shape factors of the crystal for its volume and surface area, respectively, ρ is the density of the crystal, and L is the characteristic length of the crystal. When the crystal shape is spherical and the characteristic length is a diameter of the sphere, the geometrical shape factors of the volume and the surface area are $\pi/6$ and π, respectively. From Eqs. (21) and (22), the linear growth rate of the crystal ($G_{\mathrm{L}} = dL/dt$) is derived as:

$$G_{\mathrm{L}} \equiv \frac{dL}{dt} = \frac{k_{\mathrm{A}}}{k_{\mathrm{v}}}\frac{k_{\mathrm{m}}}{3\rho}(C - C_{\mathrm{s}}) \tag{23}$$

It is also linearly proportional to the supersaturation level.

The mass transfer coefficient is strongly dependent on the hydrodynamic conditions around the crystal. When the fluid velocity around the crystal is high, the boundary layer becomes thinner and mass transfer is promoted. It results in the enhancement of the crystal growth rate. Therefore, the agitation speed of the suspension is considered as one of the important influencing parameters to determine the crystal growth rate. The mass transfer coefficient relative to the fluid velocity in a stirred-tank reactor is provided by the semiempirical correlation expressed as a function of impeller Reynolds number.

Often, the hydrodynamic condition on a crystal surface determines the crystal shape. Around the cubic-shape crystal the hydrodynamic condition on its surface is uneven. In the agitated suspension, as shown in Fig. 8, the mass transfer rate on the edge and the corner of the cubic crystal is higher than on the planes because the fluid velocity on the edge and the corner is relatively

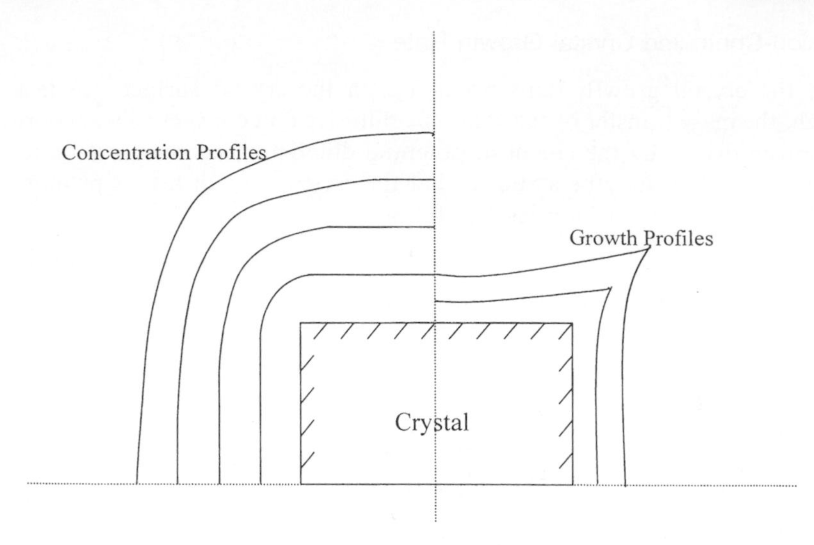

Figure 8 The concentration profiles around crystal (left-hand side) and the contour of crystal growth rate around crystal (right-hand side) growing with a diffusion-controlled growth process (Nielsen 1964).

higher than that on the plane. Then, the edge and the corner are favored for growth that can lead to the dendritic or star-like shape (Nielsen 1964).

Integration-Controlled Crystal Growth Rate

If the mass transfer of the solute from the bulk solution to the crystal surface is fast enough, the crystal growth is predominantly determined by the steps occurring on the crystal surface, such as surface adsorption, surface diffusion, dehydration, lattice integration, etc. Sometimes, all of the steps occurring on the crystal surface are collectively called "surface integration." Since the growth steps on the crystal surface are complicatedly dependent on the supersaturation, it is very difficult to fully describe the surface integration phenomena with mathematics. Fortunately, however, the rate of the overall surface integration is controlled by one step and then it may be simply expressed by a power-law equation.

$$dm/dt = Ak_r(C - C_s)^r \tag{24}$$

where k_r is the surface integration coefficient and r is the order of the surface integration. Although the power-law expression is too simple to analytically predict all of the steps, it is often used to estimate the crystal growth in industrial crystallization.

Using Eq. (22), the linear growth rate of the crystal can be written as:

$$G_L \equiv \frac{dL}{dt} = \frac{k_A}{k_v}\frac{k_r}{3\rho}(C - C_s)^r \tag{25}$$

In general, it is believed that the surface integration rate is dependent on the chemical conditions of the solution such as supersaturation, pH, temperature, impurity, etc., but is independent of the physical conditions such as mixing, hydrodynamics, etc. The temperature dependence of the surface integration rate is described on the basis of the Arrhenius law:

$$k_r = k_{r0} \exp\left(-\frac{\Delta E_r}{RT}\right) \tag{26}$$

where k_{r0} is the surface integration coefficient at the reference temperature and ΔE_r is the activation energy. Equation (26) suggests that the surface integration coefficient is promoted as temperature is increased. However, if the solubility of the solute increases sensitively to the temperature, the influence of temperature on the crystal growth rate should be marginal.

Diffusion- and Integration-Controlled Crystal Growth Rate

For industrial crystallization, the crystal growth process is frequently described as a simple mechanism of diffusion and integration control. Karpinski (1985) developed this two-step growth model consisting of the mass transfer and the surface integration steps. In the model, the solute is transported from the bulk to the crystal interface by a mass transfer process (diffusion step) and then the solute is arranged into the crystal growth site by the molecular integration process (surface integration step). Each step can be expressed as

$$dm/dt = Ak_m(C - C_i) \tag{27}$$

$$dm/dt = Ak_r(C_i - C_S)^r \tag{28}$$

where C_i is the supersaturation at the crystal interface. Using Eq. (22), the above equation can be rearranged as:

$$C - C_s = \frac{k_v}{k_A}\frac{3\rho}{k_m}G_L + \left(\frac{k_v}{k_A}\frac{3\rho}{k_r}G_L\right)^{1/r} \tag{29}$$

When the surface integration coefficient is infinitely large, i.e., $k_r \to \infty$, it becomes the diffusion-controlled growth. If there is no limitation in the mass transfer process, i.e., $k_m \to \infty$, it is equivalent to the rate equation for the integration-controlled growth. The rate equations for the crystal growth are summarized in Table 2.

Table 2 Model Equations for the Crystal Growth Mechanism

Model	Equation
Diffusion limitation	$G_m = k_m(C - C_s)$
Surface integration limitation	$G_m = k_r(C - C_s)^r$
Two-step crystal growth	$G_m = k_r(\Delta C - G_m/k_m)^r$ where $\Delta C = C = C_s$
in case of $r = 1$	$G_m = \dfrac{k_m k_r}{k_m + k_r} \Delta C$
in case of $r = 2$	$G_m = k_m \Delta C + \dfrac{k_m}{2k_r} - \sqrt{\dfrac{k_m^4}{4k_r^2} + \dfrac{k_m \Delta C}{k_r}}$

4.3 Aggregation, Agglomeration, and Breakage

The aggregation, agglomeration, and breakage of crystals are the most important process in determining the crystal size and their distribution, which in turn are the important parameters in the downstream processing of solid-liquid separation for crystal recovery. Frequently, the aggregation, agglomeration, and breakage critically influence the crystal purity because of the impurity incursion into the crystals. The conceptual diagram of the aggregation, agglomeration, and breakage of crystals is demonstrated in Fig. 9.

According to Macy and Cournil (1991), the aggregation is carried out by the crystal collision and physical adhesion, since the binding force between the individual crystals in aggregate is rather weak. However, the aggregate becomes a concrete body by the molecular crystal growth process. This is called "agglomeration." According to their study, the agitation and the dispersion medium are important factors in the aggregation and agglomeration of crystals because the agitation and the dispersion medium generally determine the collision of crystals and the physical interactive forces between crystals, respectively. If the agglomerate is strong enough to withstand the turbulent shear force induced by agitation, its size will increase beyond the expected hydrodynamic limit. The crystal size distribution is thus strongly influenced by the aggregation and agglomeration of the crystals. Due to the adhesion of small crystals to large ones the bimodal distribution occurs frequently.

The importance of the supersaturation to the crystal agglomeration has to be considered also. The agglomeration is promoted by increasing the supersaturation because the molecular crystal growth is enhanced.

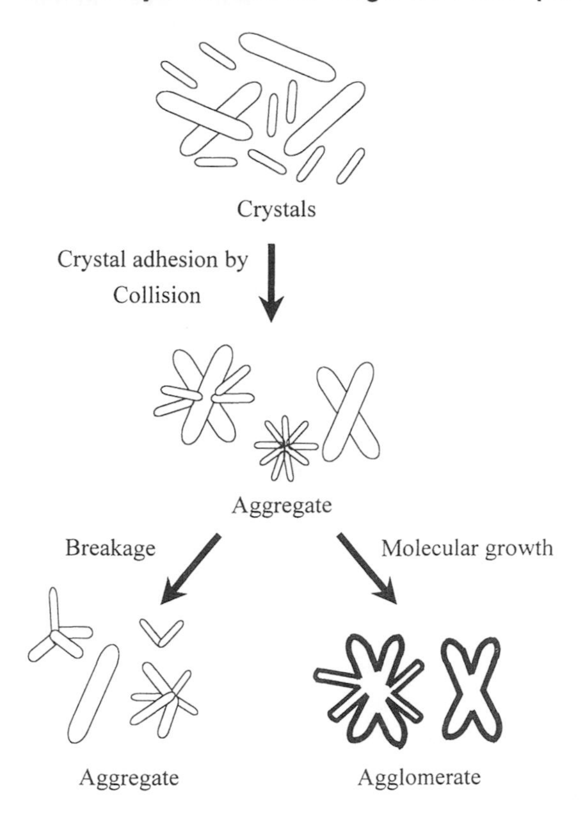

Figure 9 Schematic diagram of the crystal aggregation, agglomeration, and breakage.

Therefore, higher supersaturation tends to produce larger agglomerates. For example, the agglomerate size increases with an increase in the feed concentration in a continuous stirred tank reactor (Wojcik and Jones 1997).

Even though agitation of the suspension promotes the crystal collision frequency resulting in the promotion of the crystal aggregation, its opposite effect, i.e., aggregate breakage by turbulent fluid motion, is sometimes observed. Although the breakage and the aggregation occurs simultaneously and reversibly, too strong agitation rather inhibits the agglomeration more significantly because the turbulent fluid motion can disrupt the aggregate.

For the quantitative description of the turbulent agitation effect on the crystal aggregation and breakage, the turbulent dissipation energy per unit mass of suspension, ϵ, is generally used (McCabe and Smith 1976).

$$\epsilon = \frac{N_p n^3 D_a^5 \rho_s}{g_c m_s} \tag{30}$$

where n is the agitation speed (rps), D_a is the impeller diameter, ρ_s is the density of the suspension, m_s is the mass of the suspension, and N_p is the Power number, which can be correlated with the Reynolds number. A stirred-tank crystallizer is widely used and the Power number depends uniquely on the tank geometry and impeller type. The standard Rushton tank equipped with four baffle and a six-paddle impeller is most widely used for batch and continuous crystallization. The turbulent dissipation energy of the Rushton tank relative to the agitation speed is summarized in Table 3.

From the turbulent dissipation energy the important characteristics of the turbulent fluid motion such as Kolmogorov microscale, λ, and turbulent shear stress, τ, are calculated as follows (Ayazi Shamlou and Titchener-Hooker 1993):

$$\lambda = \left(\frac{\nu^3}{\epsilon}\right)^{1/4} \tag{31}$$

$$\tau = \mu\left(\frac{\epsilon}{\nu}\right)^{1/2} \tag{32}$$

where μ and ν are dynamic viscosity and kinematic viscosity defined as μ/ρ, respectively. If the aggregation occurs by crystal collision and adhesion and the breakage is mainly due to the turbulent shear force, the aggregation and the breakage rates in the viscous sublayer can be related with the turbulent dissipation energy:

Table 3 Power Number and Turbulent Dissipation Energy Corresponding to Agitation Speed in Standard Rushton Tank

rpm	N_{Re}	N_p	ϵ (m^2/s^3)
100	1818	5	1.3×10^{-4}
200	3636	6	1.2×10^{-3}
300	5454	6	4.2×10^{-2}
500	9090	6	1.94×10^{-1}
700	12726	6	5.33×10^{-1}
900	16362	6	1.131
1200	21816	6	2.68
1500	27270	6	5.24

$$r_{agg} \propto \epsilon^{1/2} \tag{33}$$

$$r_{break} \propto \epsilon \tag{34}$$

Then, the net rate of the agglomerate size change relative to the turbulent dissipation energy can be expressed as the sum of the crystal aggregation and breakage rates as follows:

$$\frac{dL}{dt} = k_1\epsilon^{1/2} - k_2\epsilon \tag{35}$$

where k_1 and k_2 are the rate constants of aggregation and breakage, respectively. Here k_1 and k_2 depend on the suspension properties such as crystal size, crystal population, and crystal collision efficiency. Assuming that the suspension properties are independent of the turbulent dissipation energy, the agglomerate size produced for a given time period can be approximated as:

$$\begin{aligned} L &= \left(\int_0^t k_1 dt\right)\epsilon^{1/2} - \left(\int_0^t k_2 dt\right)\epsilon \\ &= K_1\epsilon^{1/2} - K_2\epsilon \end{aligned} \tag{36}$$

According to Eq. (36), the agglomerate size profile relative to the turbulent dissipation energy increases first and then decreases after the maximum, as shown in Fig. 10. Both the aggregation and the breakage are enhanced by the increase in the turbulent dissipation energy. In the range of low dissipation energy, the crystal aggregation rate is higher than that of the breakage rate, and the agglomerate size increases with the dissipation energy. However, if the breakage rate surpasses the aggregation rate after the optimal dissipation energy level is reached, the agglomerate size is reduced with an increase in the dissipation energy.

From Eq. (36), the maximal agglomerate size and the optimal dissipation energy can be estimated as

$$L_{max} = \frac{1}{4}\frac{K_1^2}{K_2} \tag{37}$$

at

$$\epsilon_{max} = \frac{1}{4}\left(\frac{K_1}{K_2}\right)^2 \tag{38}$$

Equations (37) and (38) indicate that the maximal agglomerate size and the optimal turbulent dissipation energy are determined by the intrinsic properties of the solute and the solvent as well as the operational conditions, such as batch time and mean residence time in a crystallizer.

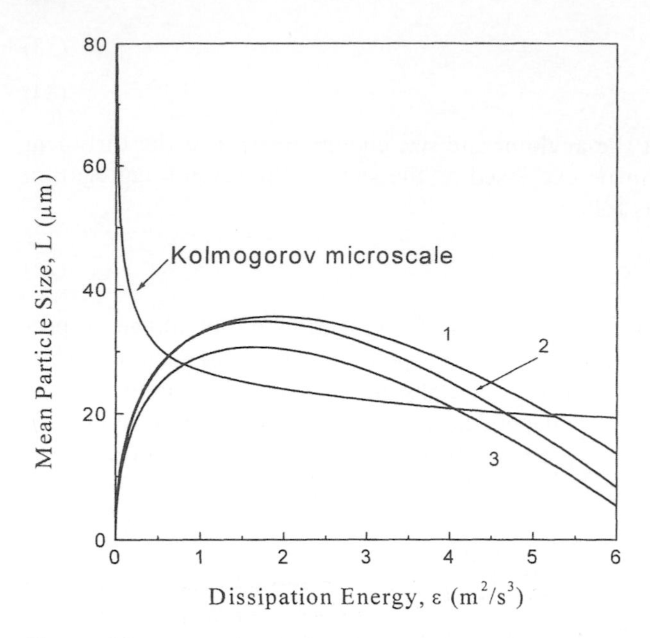

Figure 10 Description of effect of turbulent dissipation energy on the crystal size determined predominantly by aggregation, agglomeration, and breakage processes. Each curve is determined by its values for K_1 and K_2 of Eq. (36).

In actual processes of crystallization separation of biological macromolecules, the crystal breakage occurs much more severely as a result of the collision between the crystals and the impeller blade because the biomolecular crystals are not as rigid as inorganic crystals. Since the collision directly affects nuclei attrition and crystal breakage, the shape and the material of the impeller must be carefully chosen and the impeller speed has to be considered as an important operational parameter. Mersmann (1995) reported that crystal breakage is determined primarily by the collision intensity between the crystal and the impeller. The major factors influencing the collision intensity are summarized in Table 4.

5. FACTORS INFLUENCING PROTEIN CRYSTALLIZATION PERFORMANCE

Protein crystallization is a highly variable process. Many factors are known to influence the crystallization performance, i.e., the yield and the crystal purity. They can be classified into three categories: biological, chemical, and

Table 4 Major Influencing Factors Determining Collision Intensity for Crystal Breakage

Influencing factor	Examples
Crystallizer geometry	Volume of crystallizer
	Diameter of impeller
	Number of blades
	Breadth and thickness of blade, etc.
Operational parameter	Impeller speed
	Pumping capacity of impeller
Physical properties of crystal	Crystal density
	Liquid density
	Elastic modulus of crystal
	Shear modulus of crystal
	Dynamic viscosity of liquid

physical. The biological factors include, among others, the presence of impurities such as macromolecules, ligands, cofactors, and inhibitors; post-translational modifications; and aggregation behavior. Ionic strength, pH, precipitant type and concentration, and redox potential are the chemical factors of importance. The physical factors include temperature, pressure, viscosity, diffusivity, and so forth. In addition to these factors, the high instability of protein molecules in solution (unfolding, hydration requirements, temperature/pH sensitivity, etc.) makes it difficult and fruitless to predict the protein crystallization performance by de novo calculation.

Protein crystallization can be simply understood as a purely physicochemical process by which the solute is crystallized out of the solution. However, in actual crystallization, the feed solution includes not only the target protein to be precipitated out but also many other noncrystallizing substances. It means that the crystallization is usually carried out in a multicomponent solution and thus is frequently influenced strongly by foreign substances. This may be the somewhat unique characteristic of the crystallization for bioseparation as compared with that for crystallography. The foreign substances are defined as "impurities" in case of unintentional presence or as "additives" in the case of their being added for the purpose of crystallization modification. Even though it is hard to generalize and quantify the influence of the impurities and the additives on crystallization, the inorganic impurities and additives seem to exhibit more dramatic influence than the organic ones. Sometimes a tiny amount of an inorganic impurity may cause a completely different crystallization. Therefore, the ability of the impurities and additives to change the crystal morphology and the kinetics of nucleation, growth, and agglomeration, and their influence on the crystallization process has been a subject of intensive investigations (Nyvilt and Ulrich 1995; Rauls et al. 2000). However, until now the influence of the impurities on the nucleation had not been clearly elucidated. Empirically, it is observed that the presence of the impurities and/or additives induces primary nucleation at a lower supersaturation level and also reduces the nucleation rate.

5.1 Effects of Protein Solution Purity and Impurities

Probably the most important factor for obtaining high-quality protein crystals is the purity of the protein solution to be crystallized. Although they tend to be excluded during crystal growth, even the presence of very low concentration of impurities will lead to contamination of the crystal lattice and, ultimately, poor-quality crystals. To assure the homogeneity for laboratory scale crystallization, Lin et al. (1992) performed protein liquid chromatography (FPLC) as a pretreatment step. They reported several cases where the FPLC pretreatment improved the crystal quality and the process reproducibility. In large-scale crystallization, ultrafiltration is often used in combination with diafiltration as the pretreatment step to achieve the homogeneity (Park et al. 1997).

The presence of impurities, whether or not they are structurally similar to the target protein, can substantially reduce the crystal growth rates. Few generalities exist in elucidating the exact mechanism by which the impurities affect the growth rate. One model involves adsorption of impurities onto the crystal surface; once located on the surface, the impurity molecules form a barrier for the target protein molecule from the bulk solution to adsorb to the surface. This model was used to relate the crystal growth rate to an impurity concentration in solution through an adsorption isotherm (Rousseau et al. 1976; Rousseau and Woo 1980). Other models call for the impurity molecules to either occupy active growth sites or be integrated into the crystal structure, which leads to reduction in growth rate and production of impure crystals.

The modification of crystal growth by the impurities and additives adsorbed on the crystal surface has been extensively studied (Myerson et al. 1993; Cabrera et al. 1958). In most cases, the impurities and additives hinder the surface integration step because they may preoccupy the active growth site to delay the lattice integration. The inhibition of crystal growth is proportional to the coverage of the crystal surface by the impurities and additives. Based on this mechanism, the influence of the impurities and additives on the crystal growth rate is suggested as (Punin and Franke 1998):

$$\frac{G_{L0} - G_L(C_{add})}{G_{L0} - G_L(G_{add}^{sat})} = \Theta \tag{39}$$

were G_{L0} and G_L are the crystal growth rate in the pure and impure solutions, respectively, and C_{add} and C_{add}^{sat} are the impurity concentrations in the bulk and the saturated state, respectively. Here, the surface coverage by the adsorbed impurity, Θ, can be described by the Langmuir isotherm as:

$$\Theta = \frac{KC_{add}}{1 + KC_{add}} \tag{40}$$

where K is the adsorption constant.

In an effort to quantify the inhibiting effect of the impurities, Kubota (1996) introduced the effectiveness factor of impurity, α:

$$\frac{G_{L0} - G_L}{G_{L0}} = \alpha \frac{KC_{add}}{1 + KC_{add}} \tag{41}$$

The higher α value indicates the more effective inhibition of crystal growth. It decreases with an increase in the supersaturation. When the effectiveness factor is greater than unity, the crystal growth process can be completely inhibited by the impurity even though the crystal surface is not saturated by the impurity. If the effectiveness factor is unity, the crystal growth rate becomes zero, suggesting that the crystal is saturated with the adsorbed impurity. In most cases, the impurity effectiveness factor is less than unity, and the growth rate approaches to a finite asymptote as the surface becomes saturated by the impurity.

When the surface integration step, which is usually the rate limiting of the crystal growth process, is hindered by the impurities and additives adsorbed on the crystal surface, the crystal growth rate relative to the impurity concentration in the solution can be estimated as:

$$k_r = k_{r0}(1 - \Theta) \tag{42}$$

and

$$\frac{k_{r0}}{k_{r0} - k_r} = 1 + \frac{1}{KC_{add}} \tag{43}$$

where k_{r0} is the surface integration coefficient in pure solution. The above equations describing the influence of the impurity on the crystal growth are highly consistent with the experimental results performed with inorganic impurities and additives.

Another important effect of impurities is that they may change the crystal habit. Habit change is considered to result from the unequal growth rates at different faces because impurities and additives change the surface energy of each face in different ways. Davey (1979) reviewed the role of impurities in the general context of habit modification. In particular, a surfactant was observed as an effective habit modifier (Michaels and Colville 1967).

The impurity adsorption on a crystal surface is due to an interaction between crystal and impurity molecules. Thus, if a certain face of a crystal has relatively high interaction energy with an impurity, the impurity will be preferentially adsorbed on the crystal face, which in turn strongly inhibits the crystal growth resulting in the crystal habit modification. The factors influencing the preferential adsorption of impurity may include the steric arrangement, charge, and dipole moment of impurity molecules as well as the electric field on the crystal surface. As an example, Michaels and Tausch (1961) observed that a detergent significantly changes the morphology of adipic acid crystals, as displayed in Fig. 11. When adipic acid was crystallized in an anionic detergent–added solution, the crystal morphology was shifted from a hexagonal prism, which was a typical shape crystallized in the

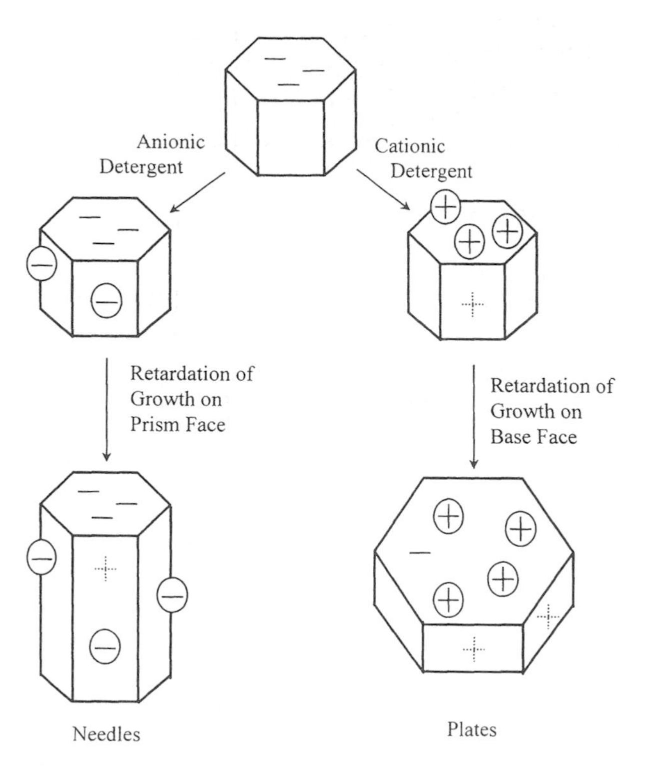

Figure 11 Modification of crystal shape by the ionic additive in adipic acid crystallization (Michaels and Tausch 1967).

additive-free solution, to a needle shape because of the preferential adsorption of the detergent molecule on the prism face to retard the growth. Meanwhile, a plate shape was produced when a cationic detergent was added.

5.2 Effects of Chemical Precipitants (Inorganic Salts and PEG)

Chemical precipitants are widely used to achieve supersaturation of proteins. These precipitants in general influence the bulk solvent (e.g., water) properties rather than those of solutes. Inorganic ionic salts have been the most common and effective precipitants used for protein crystallization, although in crystallography they tend to decrease the signal-to-noise ratio due to increased mean electron density. Because of the strong electric field around these ions, a large number of water molecules are loosely bound around the ions in a sphere of hydration. This, in turn, reduces the amount of water that is free to keep the protein molecules in solution. Thus, the supersaturation level can be effectively increased. The efficacy of a particular salt as a crystallization inducer is proportional to the square of the valences of the associated ions and generally follows the Hofmeister series. Sehnke et al. (1988) listed the effective salt precipitants, their maximal concentrations, and effective concentration ranges. It is interesting to note that as the molecular weight of the molecules to be crystallized increases the effective concentration ranges decreases.

Although some chaotropic salts, such as KSCN (potassium thiocyanate), are used at low concentrations (~ 10 mM) to increase the rate of crystallization, these salts are not frequently used because they disrupt the secondary and tertiary structures of proteins. Addition of nonchaotropic salts increases the surface tension of water molecules surrounding the protein molecules, thus dehydrating the protein's surface and creating the excluded volume effects. Solutions of these charged salts increase the solvent dielectricity that in turn promotes protein-protein interactions. These salts are also known to stabilize the proteins in solution (Scopes 1994). Because of these, the inorganic salts are probably the most favored precipitants used in large-scale protein crystallization process.

Of the high molecular weight linear polymers, polyethylene glycol (PEG) was the most effective in both precipitating ability and cost-effectiveness. Like the salts, PEG competes with proteins for water and exerts the excluded volume effects. However, unlike salts, PEG decreases the effective dielectricity of the solution, which increases the effective distance among protein molecules and discourages the interaction of heavy-atom compounds with the protein crystals. The PEGs with molecular weights less than 1000 are generally used at concentrations above 40% (v/v), whereas higher molecular weight PEGs are used in the 5–10% range. In a large-scale crystallization of *Geotrichum candidum* lipase, it was observed that the size of the crystals was dependent on the molecular weight of the PEG use (Hedrich et al. 1991).

When the crystal growth process is controlled by the mass transfer step, the influence of the polymer additives on the crystal growth may be different in their mechanism. According to Kim and Tarbell (1993), the polymer additive adsorbed on the crystal surface forms a layer to resist the solute transport from the bulk to the surface. Thus, the crystal growth rate is reduced with an increase in the polymer additive concentration in the bulk solution because its density in the polymer additive layer on the crystal surface increases. They expressed the mass transfer coefficient relative to the density of the polymer additive layer as follows:

$$k_{\mathrm{m}} = k_{\mathrm{m0}}\left(\frac{1-\varphi}{1+\varphi}\right)^2 \tag{44}$$

where k_{m0} indicates the mass transfer coefficient in additive-free solution and φ is the volume fraction of the polymer additive in the adsorbed layer. Using the Langmuir adsorption, φ relative to the polymer additive concentration in bulk was developed by Heller (1966):

$$\varphi = \varphi_{\infty}\frac{KC_{\mathrm{add}}}{1+KC_{\mathrm{add}}} \tag{45}$$

where φ_{∞} is the volume fraction of the additive in the adsorbed layer at saturation.

In some cases, upon the addition of a chemical precipitant the protein may form amorphous precipitation or one- or two-dimensional poor-quality crystals. In that case, a combination of precipitants may be used, with one type increasing the dielectricity and the other decreasing it. Both types may effectively maximize the size exclusion and the water competition effects while preventing the intermolecular charge repulsion by increasing the solvent dielectricity, leading to improved yield and quality of crystals. Using this, crystallization in aqueous two-phase system (e.g., immiscible biphasic system created by mixing high concentrations of low molecular weight salts and PEGs) was tried (Kuciel et al. 1992; Ray and Bracker 1986; Garavito et al. 1986). In another case, it was reported the crystal habit varied depending on the precipitant used; needle-shaped crystals of bacterial luciferase from *Vibrio harveyi* were obtained in ammonium sulfite whereas rhombic crystals were obtained in ammonium sulfate (Lang et al. 1992).

Sometimes, in a laboratory scale, dialysis against deionized water can be used as a method of crystallization. It exploits the Donnan effect, which states that a cloud of charged ions should surround a protein molecule to make it stay soluble. If this cloud layer is removed by dialysis, the protein will attempt to surround itself with whatever charged species are present, which in this case will be other protein molecules.

5.3 Effects of Solution pH and Temperature

Other important factors in protein crystallization are solution pH and temperature. Because pH changes the ionic attraction forces among amino acids, it can directly influence the crystallization performance. Although there are some exceptions, it is far more typical for protein crystallization to occur over a fairly narrow range of pH (< 1 pH unit). Crystal morphology is also directly related to pH; it is gradually improved as the proper pH range is approached and falls off on either side. In some cases, pI of a protein can be the optimal crystallization pH. Most proteins are crystallized at either 4°C or ambient temperature (around 22°C). Lower temperature tends to act as a preservative for sensitive proteins but it increases the protein solubility, requiring more salt precipitant. Also, at the lower temperatures the crystals tend to grow more slowly. Due to these reasons and the energy cost involved in the low-temperature operation, the ambient temperature crystallization is more widely practiced in large-scale operations if the protein stability is not affected.

5.4 Effects of Protein Concentration

Initial protein concentration that determines the supersaturation level can play a key role in the yield and rate of crystallization. In general, higher concentration or solubility is likely to yield a higher growth rate (Bourne 1980). If the protein concentration is too low, it may never be crystallized regardless of the amount of precipitant added. As the concentration increases, the crystal yield increases before it plateaus off. When the initial protein concentration is too high, amorphous precipitation is likely to occur instead of crystalline precipitation. Therefore, a certain optimal concentration range should be identified for any successful crystallization. It is interesting to note that protein crystallization may still occur when the initial concentration is substantially below the solubility (Shih et al. 1992). For α-chymotrypsin and bovine serum albumin (BSA), it was reported that the solubility was strongly affected by their initial concentration prior to salting out. Higher solubility was obtained from higher concentration and vice versa. A similar result was observed for alkaline protease (Park et al. 1997).

6. DESIGN AND OPERATION OF PROTEIN CRYSTALLIZATION PROCESSES

6.1 Crystal Population and Material Balances

To design the crystallization process, the crystal population and the material balances have to be understood. Although crystallization is among the oldest and the most widely used industrial chemical process for separation and purification, the mathematical description for crystallization was not firmly established until Randolph and Larson did so (1971). According to them, the population balance for the crystals can be based on the following relationship: accumulation rate = input rate − output rate + net generation rate. At a fixed subregion of crystal phase space, the mathematical expression for the population balance is obtained as:

$$\frac{\partial n}{\partial t} + V \cdot (v_e n) + \nabla \cdot (v_i n) + D - B = 0 \tag{46}$$

when n is the population density of crystals, D and B are the death and birth rates of crystals, which are originated from the crystal breakage and aggregation/agglomeration, respectively. If the Lagrangian space (x, y, z) is taken for the external coordinate, the velocity of crystal, v_e is equivalent to moving velocity of crystals in suspension, v_x v_y, and v_z. Meanwhile, the crystal length is usually counted as the internal coordinate for the crystal space, and then the internal coordinate velocity, v_i, is defined as dL/dt, which is equivalent to the linear crystal growth rate, G_L. Equation (46) describes the change of particle population in internal as well as external spaces and is called "microscopic population balance."

When the suspension in the system is well mixed, it can be assumed that all terms except the external coordinate velocity in Eq. (46) are independent of volume integral over system. However, the volume integral of external coordinate velocity is derived as:

$$\int_V \nabla \cdot v_e n dV = \sum_k Q_k n_k + n \frac{dV}{dt} \tag{47}$$

where V is the volume of the system, Q_k, which means the flow stream, is taken as positive for the flow out of system and negative for the flow into system, and n_k is the population density in the flow stream. Then the population balance in the system can be developed as:

$$\frac{\partial n}{\partial t} + \nabla \cdot G_L n + n \frac{d \ln V}{dt} + \sum_k \frac{Q_k n_k}{V} + D - B = 0 \tag{48}$$

In Eq. (48), the crystal population is averaged in the external coordinate space and distributed in the internal coordinate space. It is called "macroscopic population balance," which is useful to describe the crystal size distribution in a well-mixed tank crystallizer.

To solve the population balance, which is a first-order partial differential equation, Hulbert and Katz (1964) suggested the analytical method of moments. According to this method, the population balance is first transformed into a set of ordinary differential equations and then the population density function is analytically reconstructed from the moments using Laguerre polynomials. Therefore, when the moment is defined as

$$m_j \equiv \int_0^\infty L^j n \, dL \tag{49}$$

where j is the power index of moment, the population balances of Eq. (46) and (48) can be rearranged by the moment transformation, respectively, as

$$\frac{\partial m_j}{\partial t} + \nabla \cdot v_e m_j - 0^j B^0 - jG_L m_{j-1} + \bar{D} - \bar{B} = 0 \tag{50}$$

$$\frac{dm_j}{dt} + m_j \frac{d \ln V}{dt} - 0^j B^0 - jG_L m_{j-1} + \sum_k \frac{Q_k m_{j,k}}{V} + \bar{D} - \bar{B} = 0 \tag{51}$$

where B^0 is the birth rate of crystal nuclei. It should be mentioned that in the moment transformation of Eq. (46) and (48) the McCabe's ΔL law, which means the size independent growth, is adopted. To solve the above equations the transformation functions of $\bar{D}$ and $\bar{B}$ have to be defined. However, there is no general way to formulate the death and birth rates due to their complicated behaviors resulting from aggregation, breakage, and agglomeration of crystals. For a binary collision system, fortunately, Hulbert and Katz (1964) developed the function for the crystal aggregation/agglomeration.

Since the analytical expressions for crystal nucleation and growth rates, B^0 and G_L, are suggested as functions of supersaturation of solute (see Sec. 4), the material balance should be included in the design of crystallization process. The method to generate the supersaturation depends uniquely on the crystallization system characteristics, such as crystallizer type used (e.g., batch, semibatch, plug flow, or stirred-tank crystallizer) and type of driving force applied (e.g., cooling, evaporation, reaction, vacuum, salting-out and drowning-out crystallizations). Therefore, the material balance has to be developed on the basis of the crystallization system. In general, the material balance for the solute in the crystallizer can be described as

$$\frac{dV(C + M_T)}{dt} = Q_{in}(C_{in} + M_{T,in}) - Q_{out}(C_{out} + M_{T,out}) + VR \tag{52}$$

$$M_T \equiv k_v \tilde{\rho}_s \int_0^\infty nL^3 dL = k_v \tilde{\rho}_s m_3 \tag{53}$$

where m_T is the magma density meaning the crystal molar mass per unit volume of suspension, k_v is the volumetric shape factor, $\tilde{\rho}_s$ is the molar density of crystal, Q_{in} and Q_{out} are the flow-in and flow-out streams for the crystallizer, respectively, and R indicates the rate of generation or disappearance of solute by reaction. It should be noted that the volume fraction of crystals in the suspension is assumed to be negligible.

6.2 Batch Crystallizer

The batch crystallization system, the most frequently used for the biological macromolecular product in practice, is conceptually depicted in Fig. 12. If the suspension in a batch crystallizer is well mixed and the averaged values for the external coordinates are assumed, the macroscopic population balance can be adopted to describe the crystallization. With an additional assumption of no agglomeration and breakage, the population balance and the moment equation can be simplified, respectively, as

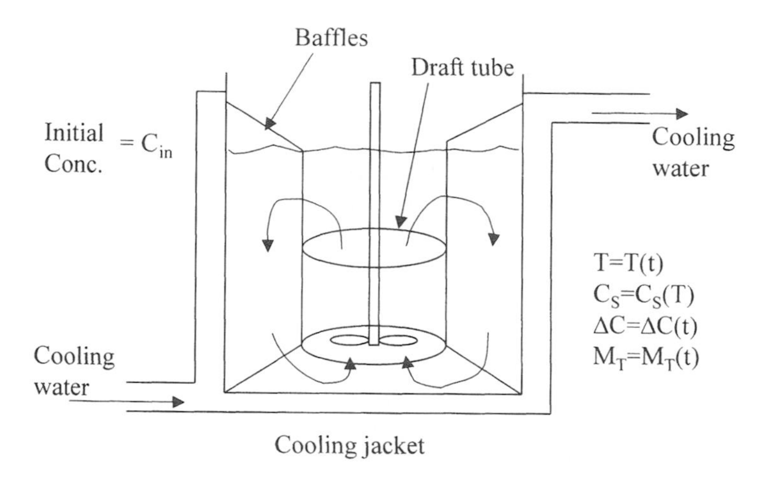

Figure 12 Schematic diagram for the cooling crystallization in a batch crystallizer.

$$\frac{\partial n}{\partial t} + G_{\mathrm{L}} \frac{dn}{dL} = 0 \tag{54}$$

$$\frac{dm_j}{dt} - 0^j B^0 - jG_{\mathrm{L}} m_{j-1} = 0 \tag{55}$$

In a cooling crystallization, the supersaturation, by which the crystal nucleation and growth rates are determined, can be predicted by material balance as

$$\frac{dM_{\mathrm{T}}}{dt} = -\frac{d\,\Delta C}{dt} + \frac{dC_{\mathrm{s}}}{dt} \tag{56}$$

where C_{s} is the solubility of the solute and ΔC is equal to $C - C_{\mathrm{s}}$. In the cooling crystalization, both the supersaturation (ΔC) and solubility (C_{s}) vary simultaneously with time because the solution temperature decreases continuously. Fortunately, the solubility change relative to the temperature is frequently available in literature or easily measurable in a laboratory. Then the material balance can be rearranged as

$$k_{\mathrm{v}} \tilde{\rho}_{\mathrm{s}} \frac{dm_3}{dt} = -\frac{d\,\Delta C}{dt} + \frac{dC_{\mathrm{s}}}{dT}\frac{dT}{dt} \tag{57}$$

By programming the cooling rate (dT/dt) in the crystallization operation, we can solve the material balance by combining the set of moment equations in Eq. (55). After obtaining the supersaturation profile with time, the crystal population balance of Eq. (54) can be numerically solved by the finite difference method, as suggested by Kim and Tarbell (1991).

6.3 Continuous MSMPR Crystallizer

A continuous mixed suspension mixed product removal (MSMPR) crystallizer is also widely used in industrial crystallizations for large-scale separation and in laboratory experiments for crystallization kinetics analysis. Agitation of suspension with baffles is usually enough to attain the MSMPR condition inside a crystallizer. In an industrial crystallizer a draft-tube baffle is commonly used. Then we can suppose that the suspension is homogeneous in terms of the external space, and it allows us to adopt the macroscopic population balance and the moment equation to describe the crystallization behavior. If the crystal agglomeration and breakage are negligible and no crystal is included in the inlet feed flows, the population balance and the moment equation at steady state can be expressed as:

$$G_{\mathrm{L}}\frac{dn}{dt}+\frac{n}{\tau}=0 \tag{58}$$

$$\frac{m_j}{\tau}=0^j B^0+jG_{\mathrm{L}}m_{j-1} \tag{59}$$

where τ is the mean residence time in the crystallizer (V/Q).

For a steady-state cooling crystallizer displayed in Fig. 13, the material balance is simplified as

$$C_{\mathrm{in}}=\Delta C+C_{\mathrm{s}}+k_{\mathrm{v}}\tilde{\rho}_{\mathrm{s}}m_3 \tag{60}$$

Based on the McCabe's ΔL law, the population balance of Eq. (58) can be analytically solved as

$$n=n_0\exp\left(-\frac{L}{G_{\mathrm{L}}\tau}\right) \tag{61}$$

where n_0 is the population density of crystal nuclei defined as

$$\left.\frac{dN}{dL}\right|_{L\to 0}$$

where N is the crystal population. Since the population density distribution in Eq. (61) is linear on a semilog plot, as shown in Fig. 14, the values of the

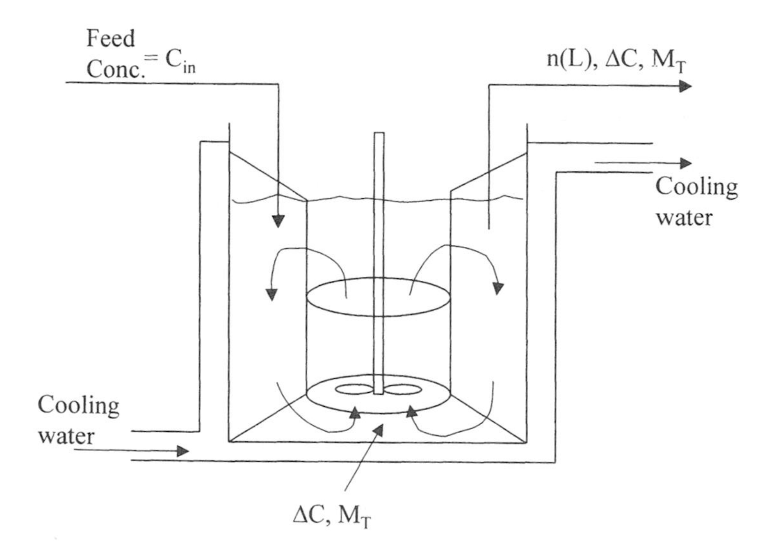

Figure 13 Schematic diagram for cooling crystallization in an MSMPR crystallizer.

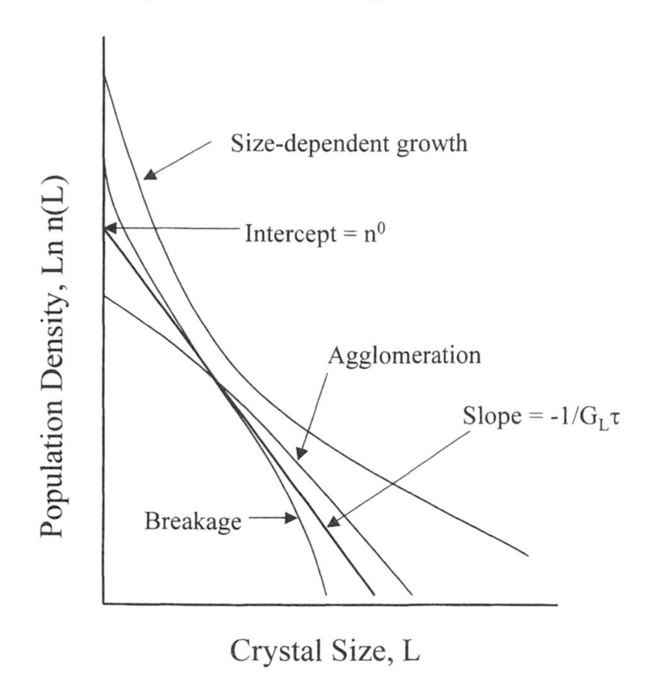

Figure 14 Crystal population density distributions in ideal and nonideal MSMPR crystallizers.

nucleus density, n_0, and linear growth rate, G_L, can be obtained from the intercept and the slope of the linear plot. Meanwhile, the definition of the crystal nucleation rate is related with the nucleus density and linear growth rate of crystal as

$$B^0 \equiv \lim_{L \to 0} \frac{dN}{dt} = \lim_{L \to 0} \left[\frac{dN}{dL} \frac{dL}{dt} \right] = n^0 G_L \tag{62}$$

Therefore, from the crystal size distribution in a steady-state MSMPR crystallizer, the information on the kinetics of crystal growth and nucleation rates can be obtained. This is why the continuous MSMPR crystallizer is frequently used in lab studies for crystallization.

Using the population density distribution [Eq. (61)], the mean crystal size based on crystal numbers and magma density in the MSMPR crystallizer can be calculated as

$$\bar{L} \equiv \frac{\int_0^\infty nLdL}{\int_0^\infty ndL} = \frac{m_1}{m_0} = G_L\tau \tag{63}$$

$$M_T = k_v\tilde{\rho}_s m_3 = 6k_v\tilde{\rho}_s B^0 G_L^3\tau^4 \tag{64}$$

By substituting Eq. (64) with Eq. (60), the supersaturation in the MSMPR crystallizer is also obtained as

$$\Delta C = \Delta C_{in} - 6k_v\tilde{\rho}_s B^0 G_L^3\tau^4 \tag{65}$$

where ΔC_{in} is equal to $C_{in} - C_s$.

It should be noted that the above analysis is based on the ideal mixing of suspension inside a crystallizer to guarantee a spatial homogeneity for external coordinates. However, in cases of high magma density in a large-scale crystallizer, the high agitation speed is required to suspend the crystals and that may result in the crystal breakage and agglomeration by collisions. Then, the crystal size distribution on the semilog plot is deviated from the linearity as displayed in Fig. 14. More deviations in the crystallization behavior between the ideal and the industrial MSMPR crystallizer can occur due to dead volume of mixing, fine dissolution, classified product removal size, size-dependent growth, and growth rate dispersion.

6.4 Operational Modes

Crystallization can be performed in single or multistage or in batch operations. It is known that for small molecules crystallization the staged operation is more effective to produce uniform and/or larger crystals (Moyer and Rousseau 1987). For natural products purification a series of fractional crystallization, in which additional crystals are recovered from mother liquor, could be performed for maximal yield (Mullin 1972). In protein crystallization, however, addition of proper precipitant is preferred over evaporation and/or vacuum cooling to generate the supersaturation; thus, suspension-type batch crystallization is predominantly employed. The batch operation tends to produce a narrower and more uniform crystal size distribution than the continuous well-mixed operation and is often more economical for low-capacity crystallization (less than 50 kg/h). The batch crystallizer is usually a stirred tank equipped with either jacketed or forced-type heat exchanger for cooling, an agitator with Rushton impellers for uniform distribution and collision of crystals, and a screen or "chunk breaker" installed at the bottom to avoid plugging of the drain line.

7. STEPS OF PROCESS DEVELOPMENT FOR LARGE-SCALE PROTEIN CRYSTALLIZATION

Large-scale crystallization of proteins is not as straightforward as it might seem. There are a number of process conditions to be analyzed, and often it is not easy to precisely maintain their conditions especially during batch crystallization. The process development work generally goes through the following stages: (1) initial screening of the key factors and optimization of their conditions; (2) reproducibility analysis and fine tuning of the conditions; and, finally, (3) process scale-up and implementation.

The critical factors to be identified in the initial screening stage would include but not be limited to precipitant type and concentration, pH, initial protein concentration (or supersaturation level), and temperature. The screening usually uses an iterative process based on two-dimensional matrices in which two different factors are tested as the two independent axes. It starts with large grids over wide ranges. The results are graded based on the yield and purity of the crystals. Once the best range of the conditions is identified, the test is repeated with narrower ranges until the initial optimal set of the two conditions with the adequate ranges is established. This condition then can be used to test for any additional factors and their conditions. Due to the repetitive nature, robotic automation could be applied (Cox and Weber 1988). This method is straightforward and works well if the protein can be crystallized early, but it consumes relatively large amounts of samples. Also, once a new and unexpected factor is discovered, one might need to repeat the whole process of the screening because the new factor may alter and/or interfere with the previously established conditions.

In this stage of identifying the "operating window," 5–50 mg/ml protein solution can be used at as low as 500 μl volume. Of course, these values may vary depending on the solubility and availability of the target protein. When the quantity of the protein sample is the limiting factor, several screening methods have been proposed. Dialysis button was used by which protein could be easily reused by dialysis against a buffer to remove precipitant. Carter et al. (1988) used this in an incomplete factorial approach. Jancarik and Kim (1991) described the use of a sparse set of conditions, which was later developed commercially. The idea was to provide a broad sampling by random combination of conditions initially and then improve on them later.

The protein purity used for the crystallization for structural studies should be 99% plus, e.g., single band on sodium dodecyl sulfate and isoelectric focusing gel electrophoresis using silver staining (Bauer and Stubbs, 1999). For large-scale bioseparation of industrial enzymes such as alkaline protease it could be as low as 50–60%, although higher purity solution tends to result in the crystals with more uniform shape and size distribution. Crystal habit is observed usually by using optical, binocular microscope with 50–100× magnification. Crystals larger than 0.2 mm in each direction are regarded as good crystals for the subsequent recovery step of solid-liquid separation (Bauer and Stubbs 1999).

After a large window of operating conditions is initially identified, more experiments are performed to narrow down the conditions to the optimal point. Alternative precipitants and additives, different levels of supersaturation and method of equilibration, and/or addition of nucleation seeds can be tested. For some labile proteins, addition of complexers, such as substrates and/or inhibitors, can be considered. Process reproducibility or consistency in terms of crystal yield and purity is closely examined in this stage. For an industrial application, process robustness is evaluated in this stage. Usually, 5–20 L of protein solution is used in a batch-type crystallizer to monitor the key parameters such as temperature distribution and control, change in the level of supersaturation, crystal formation kinetics, crystal size and habits, and batch-by-batch variations. If the process is too delicate and sensitive, the operating conditions are changed so as to render increased ruggedness and simplicity at the expense of yield and purity.

Despite a few theories for crystallization kinetics and design methods that are available for crystallizer configurations, no clear-cut guidance has been provided for scale-up (Moyer and Rousseau 1987). This is because of the complexities involved in rationally describing and interrelating the kinetics equations with the process configuration and the mechanical features of the crystallizer. Many factors contribute to the nonideal behaviors of large-scale crystallization process, and unfortunately not all can be easily explainable. Therefore, the scale-up is done usually by extending the bench or pilot data straightforwardly, focusing on how to control the supersaturation levels, maintain the crystallization time, match the vessel configuration, and duplicate the hydraulic regimes, among others. For large-scale batch crystallization, maintaining the uniform distribution of temperature and precipitant concentration inside the vessel can be a major problem, which can be solved by providing adequate degree of mixing.

8. CONCLUDING REMARKS

Crystallization is a very powerful method for large-scale bioseparation of proteins. Well designed and executed, it can provide high yield and purity with very attractive process economics. However, protein crystallization itself is a highly variable process. Numerous factors of biological, chemical, and physical natures influence the crystallization behavior, often interactively. Furthermore, high instability and diversity of the protein in the solution phase make it very difficult to theoretically analyze and predict the correct outcome of the protein crystallization process. This is why, despite the abundant information available on small-molecule crystallization, there are no set rules or guidelines directly applicable to large-scale protein crystallization. Thus, the design and scale-up have been heavily dependent on the extension of experimental data rather than theoretical simulations. However, as the experimental data are being accumulated very rapidly in this field, we need to constantly thrive to combine the two areas, i.e., experimental observations on protein crystallization and theoretical analyses on small-inorganics crystallization, together to formulate and derive any theories and guidelines universally applicable to successful protein crystallization on a large scale. We hope that this chapter will be of help in that effort.

REFERENCES

Ataka M. (1986) Growth of large single crystals of lysozyme. Biopolymers 25, 337–349.

Ayazi Shamlou P., Titchner-Hooker N. (1993) Turbulent aggregation and breakup of particles in liquid in stirred vessel. In: Processing of Solid-Liquid Suspension (Ayazi Shamlou, ed.), Butterworth-Heinemann, Oxford, UK, pp. 1–25.

Bauer M.M.T. and Stubbs M.T. (1999) Crystallization of proteinases. In: Proteolytic Enzymes: Tools and Targets (Sterchi E.E. and Stoecker W., eds.), Springer-Verlag Berlin Heidelberg, Germany, pp 124–147.

Bourne J.R. (1980) The influence of solvent on crystal growth kinetics. AIChE Symp. Ser. No. 193, 76, 59–64.

Boistelle R., Astier J.P. (1988) Crystallization mechanisms in solution. J. Cryst. Growth 90, 14–30.

Cabrera N., Vermileya D. (1958) Growth and Perfection of Crystals (Doremus R.H. and Turnbull D., eds.), Wiley, New York, pp. 393–425.

Carter C.W. Jr., Baldwin E.T., Frick L. (1988) Statistical design of experiments for protein crystal growth and the use of a precrystallization assay. J. Cryst. Growth, 90, 60–73.

Cox M.J., Weber P.C. (1988) An investigation of protein crystallization parameters using successive automated grid searches (SAGS). J. Cryst. Growth, 90, 318–324.

Davey R.J. (1979) The control of crystal habit. In: Industrial crystallization '78 (de Jong E.I. and Jančič S.J., eds.), North-Holland, Amsterdam, pp. 169–188.

Drenth J., Haas C. (1992) Protein crystals and their stability. J. Cryst. Growth 122, 107–109.

Esdal J.T. (1947) The plasma proteins and their fractionation. Adv. Protein Chem. 3, 383–389.

Feher G., Kam Z. (1985) Nucleation and growth of protein crystals: general principles and assays. Meth. Enzymol. 114, 77–111.

Garavito R.M., Markovic-Housley Z., Jenkins J.A. (1986) The growth and characterization of membrane protein crystals. J. Cryst. Growth 76, 701–709.

Hedrich H.C., Spener F., Menge U., Hecht H.J., Schmid R.D. (1991) Large-scale purification, enzymic characterization, and crystallization of the lipase from *Geotrichum candidum*. Enzyme Microb. Technol. 13, 840–847.

Heller W. (1966) Effect of macromolecular components in dispersion systems. Pure Appl. Chem. 12, 249–250.

Hulbert H.M., Katz S. (1964) Some problems in particle technology: A statistical mechanical formulation. Chem. Eng. Sci. 19, 555–574.

Jancarik J., Kim S-H.J. (1991) Sparse matrix sampling: a screening method for crystallization of proteins. Appl. Cryst. 24, 409–411.

Karpinski P.H. (1985) Importance of the two-step crystal growth model. Chem. Eng. Sci., 40, 641–649.

Kim W.S., Tarbell J.M. (1991) Numerical technique for solving population balance in precipitation processes. Chem. Eng. Comm., 101, 115–129.

Kim W.S., Tarbell J.M. (1993) Effect of PVA and gelatin additives on barium sulfate precipitation in an MSMPR reactor. Chem. Eng. Comm., 120, 119–137.

Knapp L.F. (1922) The solubility of small particles and the stability of colloids. Trans. Faraday Soc. 17, 457–465.

Kubota N., Yokota M., Mullin J.W. (1996) Kinetic models for the crystal growth from aqueous solution in the presence of impurities; Steady and unsteady state impurity actions. In: Proceedings of the 13th Symposium on Industrial Crystallization, Toulouse, France, pp. 111–116.

Kuciel R., Jakob L., Lebioda L., Ostrowski W.S. (1992) Crystallization of human prostatic acid phosphatase using biphasic systems. J. Cryst. Growth 122, 199–203.

Lang D., Erdmann H., Schmid R.D. (1992) Bacterial luciferase of *Vibrio harveyi* MAV: purification, characterization and crystallization. Enzyme Microb. Technol., 14, 479–483.

Lin S-X., Sailofsky B., Lapointe J., Zhou M. (1992) Preparative fast purification procedure of various proteins for crystallization. J. Cryst. Growth 122, 242–245.

Macy J.C., Cournil M. (1991) Using a turbidimetric method to study the kinetics of agglomeration of potassium sulfate in a liquid medium. Chem. Eng. Sci., 46, 693–701.

McCabe W.L., Smith J.C. (1976) Unit Operations of Chemical Engineering. 5th ed., McGraw-Hill, Koshaido Printing Co., Tokyo, Japan.

Melander W., Horvath C. (1977) Salt effects on hydrophobic interaction in precipitation and chromatography of proteins: an interpretation of the lyotrophic series. Arch. Biochem. Biophys. 183, 200–209.

Mersmann A. (1995) Crystallization Technology Handbook, Marcel Dekker, New York.

Michaels, A., Tausch Jr., F.W. (1961) Modification of growth rate and habit of adipic acid crystals with surfactants. J. Phys. Chem. 65, 1730–1737.

Michaels A.S., Colville A.R. (1967) The effect of surface active agents on crystal growth rate and crystal habit. J. Phys. Chem. 64, 13–19.

Moyers C.G., Jr., Rousseau R.W. (1987) Crystallization operations. In: Handbook of Separation Process Technology (Rousseau R.W., ed.), Wiley, New York, pp. 1–58.

Mullin J.W. (1972) Crystallisation techniques. In: Crystallisation, 2nd ed., Butterworths, London, pp. 233–257.

Mullin J.W. (1993) Crystallization, 3rd ed., Butterworth-Heinemann, Oxford, UK.

Myerson A.S., Weisinger Y., Grinde R. (1993) Crystal shape, the role of solvents and impurities. In: Industrial Crystallization '93 Symposium Proceedings (Rojkowski Z., ed.), pp. 3–135, Warsaw, Poland.

Nielsen A.E. (1964) Kinetics of Precipitation, Pergamon Press, New York, USA.

Nyvilt J., Ulrich J. (1995) Admixtures in Crystallization, VCH, Weinheim, Germany.

Park D.H., Lee H.J., Lee E.K. (1997) Crystallization of alkaline protease as a means of purification process. Korean J. Chem. Eng. 14, 64–68.

Punin Y.O., Franke V.D. (1998) Effect of carbamide adsorption on the growth kinetics of the ammonium chloride crystals. Crystal Res. Tech. 33, 166–172.

Randolph A.D., Larson M.A. (1971, 1988) Theory of Particulate process: Analysis and Techniques of Continuous Crystallization, 1st and 2nd eds., Academic Press, London, UK.

Rauls M., Bartosh K., Kind M., Kuch St., Lacmann R., Mersmann A. (2000) The influence of impurities of crystallization kinetics—a case study on ammonium sulfate. J. Crystal Growth 213, 116–128.

Ray W.J. Jr., Bracker C.E. (1986) Polyethylene glycol: catalytic effect on the crystallization of phosphoglucomutase at high salt concentration. J. Cryst. Growth, 76, 562–576.

Ries-Kautt M., Ducruix A. (1992) Phase Diagram, Crystallization of Nucleic Acids and Proteins: Practical Approach (Ducruix A., Giege R., eds.), Oxford Press, Oxford, UK, pp. 195–218.

Rousseau R.W., Tai C.Y., McCabe W.L. (1976) The influence of quinoline yellow on potassium alum growth rates. J. Cryst. Growth, 32, 73–82.

Rousseau R.W., Woo R. (1980) Effects of operating variables on potassium alum crystal size distribution. AIChE Symp. Ser. No. 193, 76–27–35.

Sato K., Fukuba Y., Mitsuda T., Hirai K., Moriya K. (1992) Observation of lattice defects in orthorhombic hen egg white lysozyme crystals with laser scattering tomography. J. Cryst. Growth 122, 87–94.

Scopes R.K. (1994) Protein Purification: Principles and Practice, Springer-Verlag, New York.

Sehnke P.C., Harrington M., Hosur M.V., Li Y., Usha R., Tucker R.C., Bomu W., Stauffacher C.V., Johnson J.E. (1988) Crystallization of viruses and virus proteins, J. Cryst. Growth 90, 222–230.

Shih Y.-C., Prausnitz J.M., Blanch H.W. (1992) Some characteristics of protein precipitation by salts. Biotech. Bioeng 40, 1155–1164.

Visuri K., Nummi M. (1972) Purification and characterization of crystalline β-amylase from Barley. Eur. J. Biochem. 28, 555–565.

Volmer M., Webber A. (1926) Keimbildung in ubersttigten Gebilden. Z. Physik. Chem. 119, 71–73.

Weber P.C. (1991) Physical principles of protein crystallization. Adv. Prot. Chem. 41, 1–36.

Wojcik J.A., Jones A.G. (1998) Particle disruption of calcium carbonate crystal agglomerates in turbulently agitated suspensions. Chem. Eng. Sci. 53, 1097–1101.

9
Extraction for Rapid Protein Isolation

M. T. Cunha, M. R. Aires-Barros, and J. M. S. Cabral
Instituto Superior Técnico, Lisbon, Portugal

1. INTRODUCTION

The production of foreign proteins using a selected host with the necessary posttranslational modifications is one of the key successes in modern biotechnology. This methodology allows the industrial production of proteins that were otherwise produced in small quantities. However, the separation and purification of these proteins from the fermentation media constitutes a major bottleneck for the widespread commercialization of recombinant proteins. The major production costs (50–90%) for a typical biological product resides in the purification strategy. There is a need for efficient, effective, and economic large-scale bioseparation techniques to achieve high purity and high recovery while maintaining the biological activity of the molecule.

The recombinant protein may be accumulated in the cytoplasm or the periplasm, or secreted to the extracellular medium. The intracellular protein can also form aggregates with highly dense structures, called inclusion bodies. The choice of the purification scheme depends on the location of the target protein and on the desired purity of the product, which is also determined by the further utilization of the protein. As a first step, after cell lysis, if the protein is not secreted, the cellular material is separated from the liquid medium, using a centrifugation or a filtration step, followed by chromatographic steps.

It has been demonstrated that it is possible to use liquid-liquid extraction technology to integrate the first two steps into one so as to simultaneously obtain separation and concentration of target protein. Two different systems have been proposed and are presently under study. The first one uses two immiscible aqueous phases of simple electrolytes and water-soluble polymers (like polyethylene glycol, PEG) or of incompatible water-soluble polymers (e.g., dextran/PEG) (Kula et al. 1982; Huddleston et al. 1991; Sebastião et al. 1996; Cunha et al. 1997). The second one, developed more recently, uses a water-in-oil microemulsion in equilibrium with an aqueous phase to achieve the desired separation/concentration of the product (Göklen and Hatton 1985, 1987; Rahaman et al. 1988; Pires and Cabral 1993; Carneiro-da-Cunha et al. 1996; Pires et al. 1996; Krieger et al. 1997). In both cases, it has been claimed that the ease of scale-up and the high partition coefficients obtained allow its potential application in large-scale downstream processing of proteins produced by fermentation.

2. EXTRACTION IN AQUEOUS TWO-PHASE POLYMER SYSTEMS

The use of aqueous two-phase systems (ATPSs) for the purification of proteins is a well-documented procedure. Detailed discussion of basic and applied aspects can be found in several monographs and reviews (Kula et al. 1982; Kula 1985; Albertsson 1986; Diamond and Hsu 1992; Zaslavski 1995). This technique is very powerful for the primary downstream processing steps. The main advantages have been summarized by Albertsson (1986) and are given below:

1. Both phases of the system are of aqueous nature.
2. Rapid mass transfer and mixing until equilibrium requires little energy input.
3. Technique facilitates the processing of solid-containing streams.
4. Polymers stabilize the proteins.
5. Separation can be made selective.
6. Scale-up from small laboratory experiments is easy and reliable.
7. Continuous operation is possible.
8. Technique is cost effective.

The simplest procedure of this technique is the one-step extraction. The phase system is prepared and the mixture to be separated is added. After mixing, phase separation is accomplished either by settling under gravity or by centrifugation. The phases are separated and analyzed or used to recover the separated components of the initial mixture. The target protein should be concentrated in one of the phases and the contaminants in the other.

The theoretical yield in the top phase, Y_t, can be given in relation to the volume ratio of the phases, VR (volume top/volume bottom), and the partitioning coefficient of the target protein ($K = C_{top}/C_{bottom}$):

$$Y_t = \frac{100}{1 + (1/\text{VR})(1/K)} (\%)$$

The theoretical concentration factor in the top phase, CF_t, of a protein is defined as the ratio between the target protein concentration in the top phase and the target protein concentration in the input mixture. This can be given as a function of the theoretical yield, volume ratio, and weight percentage of media added to the separation system:

$$\text{CF}_t = \frac{Y_t}{100} \frac{\%\,\text{media}}{100} (1 + 1/\text{VR})$$

The theoretical purification factor of a target protein purified in the top phase is defined as follows:

$$\text{PF}_t = \frac{\left(\frac{m_{\text{target protein}}}{m_{\text{total protein}}}\right)_{\text{top}}}{\left(\frac{m_{\text{target protein}}}{m_{\text{total protein}}}\right)_{\text{feed}}} = \frac{Y_t}{100} \times \left(1 + \frac{1}{K_{\text{total protein}} \times \text{VR}}\right)$$

In the former equation, m stands for mass and it is assumed that the target protein represents a minor part of the total protein, and the partitioning of all proteins in the sample is similar.

In Figs. 1–3, theoretical yield, concentration, and purification factors are depicted against the volume ratio for several K values of the target and

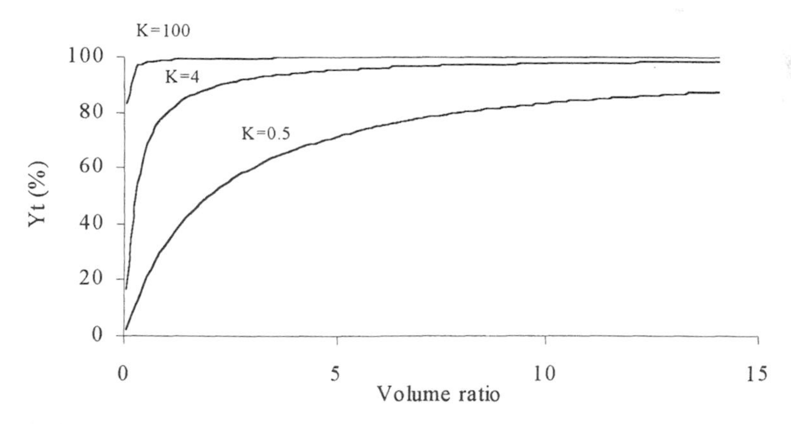

Figure 1 Theoretical product yield in the top phase for different partitioning coefficients of the target protein.

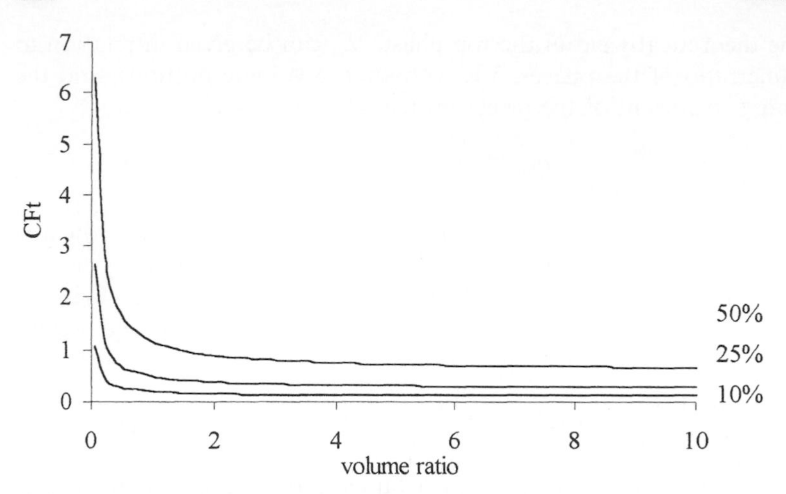

Figure 2 Theoretical product concentration factor, in the top phase, for different mixture loads (weight percentage) considering a partitioning coefficient for the target protein of 20.

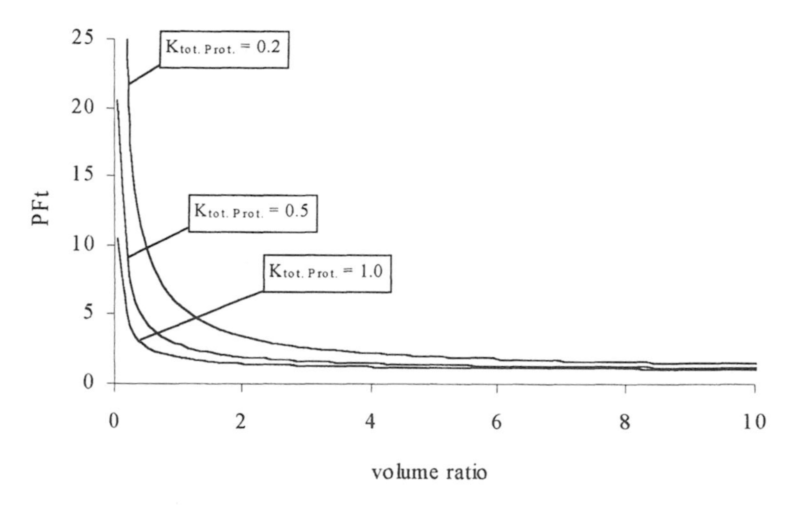

Figure 3 Theoretical product purification factor, in the top phase, for different total protein partitioning coefficients, and considering a partitioning coefficient for the target protein of 20.

total protein, and different mixture loads, respectively. For high volume ratios, higher yields can be achieved. However, this is accomplished at a cost of increased dilution of the input mixture to be separated, lower purification factors, and less usage of the chemicals per unit weight of mixture, which is economically unfavorable.

2.1 Phase Separation and System Properties

The phase components in ATPS may be either two different hydrophilic polymers, such as PEG and dextran; or one polymer and one low molecular weight solute, such as potassium phosphate. Above certain critical concentrations of these components, phase separation occurs. Separation is dependent on molecular weight of the polymers, additives, and temperature. Each of the phases is enriched in one of the components. The composition of each phase can be determined for the total system composition from the phase diagram (Fig. 4).

In Fig. 4, the three systems A, B, and C differ in the initial compositions and in the volume ratios. However, they all have the same top phase equilibrium composition (X_t, Y_t) and the same bottom phase equilibrium composition (X_b, Y_b). This is because they are lying on the same tie line, whose end points determine the equilibrium phase compositions and lie in a convex curve, termed binodal, which represents the separation between the two immiscible phases.

It should be noted that commercial polymers are usually polydisperse and their molecular weight distributions may vary from lot to lot, even when

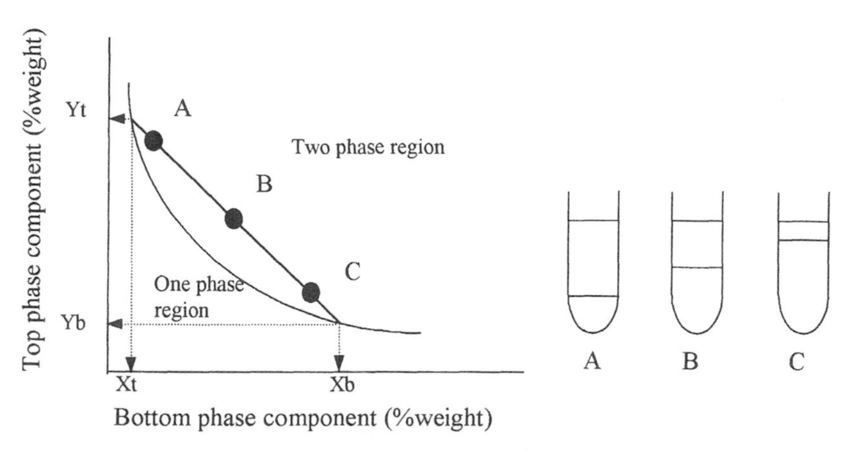

Figure 4 Two-phase diagram for Y-X water system.

obtained from the same manufacturer. The phase diagrams for systems formed by different lots differ accordingly.

The length of the tie line (TLL) and the slope of the tie line (STL) can be related to the equilibrium phase composition as follows:

$$\mathrm{TTL} = \sqrt{(X_\mathrm{b} - X_\mathrm{t})^2 + (Y_\mathrm{t} - Y_\mathrm{b})^2}$$

$$\mathrm{STL} = \frac{(Y_\mathrm{t} - Y_\mathrm{b})}{(X_\mathrm{t} - X_\mathrm{b})} = \frac{\Delta Y}{\Delta X}$$

Table 1 shows some common examples of ATPS. For more extensive lists of ATPS, the reader is referred to the references: Albertsson (1986), Diamond and Hsu (1992), and Zaslavsky (1995).

Table 1 Polymer Systems Capable of Phase Separation in Water Solutions

Polymer	Polymer	Ref.
Polyethylene glycol	Polyvinyl alcohol	Albertsson (1986)
	Dextran (Dex)	Albertsson (1986)
	Hydroxypropyl starch (HPS)	Tjerneld et al. (1986)
	Ficoll	Albertsson (1986)
Polypropylene glycol	Methoxypolyethylene glycol	Albertsson (1986)
	Polyethylene glycol	Albertsson (1986)
	Dextran	Albertsson (1986)
Ethylhydroxyethylcellulose	Dextran	Albertsson (1986)
	Hydroxypropyl starch	Tjerneld (1989)
Ethylene oxide–propylene oxide	Dextran	Harris et al. (1991)
	Hydroxypropyl starch	Harris et al. (1991)
	Low molecular weight solute	
Polypropylene glycol	Potassium phosphate	Albertsson (1986)
	Glucose	Albertsson (1986)
Polyethylene glycol	Inorganic salts, e.g., K^+, Na^+, Li^+, $(NH_4)^+$, PO_4^{3-}, SO_4^{2}, etc.	Albertsson (1986); Ananthaphadmanabhan and Goddard (1987)

Two-Polymer Systems

The most important factor for phase separation is the chemical nature of both polymers. In two-polymer ATPSs, the phase separation is due to small repulsive interactions between the two types of monomers in the solution. The total interaction between the two polymers is large because each one is composed of several monomers.

Molecular Weight. The higher the molecular weights (MWs) of the polymers, the lower is the polymer concentration required for phase separation (Albertsson 1986; Diamond and Hsu 1989a; Abbot et al. 1990), i.e., the binodal is depressed. Forciniti et al. (1991a) evaluated the effect of polymers MW and temperature on the phase composition of a PEG/dextran ATPS. The TLL, the $\Delta_{Dextran}$, and Δ_{PEG} were found to increase with the MW, and this effect was higher, the higher the difference in molecular size between the two polymers, with a consequent increase of the diagrams' asymmetry.

Temperature. The concentration of phase polymers required for phase separation usually increases with increasing temperature. The experiments of Sjöberg and Karlström (1989) suggest that at temperatures below 90°C, a change in the temperature has only a minor effect in the phase diagram of PEG/dextran aqueous two-phase system.

For the PEG/dextran ATPS, the effect of the polymer's MW is further increased with increasing temperatures. The STL increased with the temperature due to the fact that $\Delta_{Dextran}$ decreased with increasing temperature whereas Δ_{PEG} remained nearly constant (Forciniti 1991a).

Inorganic Salts. The hydrophobic (water structure breaking) salts (e.g., $KClO_4$, KI, KSCN) generally elevate the binodal of an aqueous two-polymer two-phase system as does the temperature increase, while the hydrophilic (water structure making) salts (e.g., K_2SO_4, KF) depress the binodal of the system (Zaslavsky 1995).

The PEG/dextran system is much less susceptible to the salt effects, when compared with PVP/dextran or Ficoll/dextran ATPSs. These effects, on phase separation of PEG/dextran, seem to be similar to the ones observed on the cloud point temperatures (see Sec. 2.3) in the dextran-free aqueous solutions of PEG (Zaslavsky 1995).

The addition of a given salt affects the polymer composition of the two phases depending on the type and total amount of salt. The salt composition of the phases is also influenced by the total polymer concentration of the system (Zaslavsky et al. 1988). Bamberger et al. (1984) found that PEG rejects phosphate, sulfate, and, to a lesser extent, chloride, while the effect of dextran on the distribution of either salt is much smaller. The magnitude of PEG's effect on the salt distribution behavior was found to be proportional to the polymer concentration.

Zaslavsky et al. (1988) have established the following empirical relationship between the partition coefficient of the salt (P_{salt}) and the polymer concentration difference of PEG (PVP or Ficoll) in both phases:

$$\ln P_{salt} = B_{salt} \Delta C_{PEG}$$

where B_{salt} is a constant depending on the type of the phase polymers and the type and total concentration of salt additive.

Hydrophobic salts were found to favor the PEG (PVP or Ficoll)–rich phase, whereas hydrophilic salts favored the dextran-rich phase. The STL was related to the total salt concentration in the ATPS (Zaslavsky et al. 1988).

In ATPSs, anions and cations distribute unequally across the interface. To keep the electroneutrality between the phases, a potential difference results (Müller 1985; Albertsson 1986; Diamond and Hsu 1992). Water structure–making ions (Li^+, Na^+, NH_4^+, Ca^{2+}, Mg^{2+}, F^-, SO_4^{2-}, CO_3^{2-}, PO_4^{3-}, CH_3COO^-) favor the more hydrophilic phase, whereas water structure–breaking ions (K^+, Rb^+, Cs^+, Cl^-, Br^-, I^-, SCN^-, NO_3^-, ClO_4^-) favor the more hydrophobic phase.

In summary, salt additives partition between the two phases and lead to a redistribution of the polymers between the phases, i.e., a change in the phase polymer composition. Therefore, when evaluating the partitioning of proteins in two-polymer ATPSs with salt additives, the interrelationship between the polymer and the ionic composition of the coexisting phases should be taken into account.

One-Polymer Systems

These systems can either be polymer/water two-phase systems or polymer/low molecular weight solute/water two-phase systems. The polymer/water ATPS is formed due to temperature changes and is described further in Sec. 2.3.

Phase separation of nonionic polymer/inorganic salt is known to occur at room temperatures, and most are composed with hydrophilic salts. The higher the valency of the anion, the lower is the salt concentration required for phase formation. The higher the polymer MW, the lower the salt concentration required for phase separation. The STL usually increases with increasing polymer molecular weight (Zaslavsky 1995). The higher the temperature, the lower is the concentration of polymer and salt required for phase separation, i.e., the binodal is depressed.

Physical Properties of ATPS

The physical properties of the ATPS (density, viscosity, and interfacial tension) determine the phase separation and influence the solute partitioning.

Low-density differences between the phases or highly viscous phases give rise to long separation times. Systems with short TLL have low viscosity but also low density difference; therefore, they will take a long time to separate. At the other extreme, i.e., large TLL, the density difference is high but so are the phases' viscosity, again leading to long separation times. An intermediate choice will allow a minimal separation time.

Changing the volume ratios, i.e., the dispersed and continuous phases, affects the separation times. When the more viscous phase is dispersed in the less viscous continuous phase (volume of the more viscous phase is the smallest), the separation time is shorter and vice versa.

The interfacial tensions in aqueous polymer systems are very difficult to measure because they are in the range of 0.5–500 mN m^{-1}, i.e., two to three orders of magnitude lower than those typical for water-organic solvent systems (Albertsson 1985). Low interfacial tensions give rise to small drop diameters, which enable high mass transfer rates. However, very fine sizes of dispersion lead to long separation times.

2.2 Factors Affecting Protein Partitioning

To date, mainly intracellular enzymes have been extracted and purified with ATPSs. The method might also be useful for the purification of extracellular proteins. However, care should be taken when choosing the system composition to avoid extreme dilution of the target protein (Fig. 2). This is particularly important when the expression levels in the microorganism are very low.

High level expression of recombinant protein expressed in prokaryotic hosts often results in accumulation of the protein both in soluble and insoluble form. Clarification by centrifugation is usually the first recovery step, with a common loss of protein either accumulated in the cells of the discarded pellet or soluble protein in discarded supernatant. A method developed by Genentech (Palo Alto, CA) integrates in situ solubilization of inclusion bodies with aqueous two-phase extraction (Hart et al. 1994). This strategy facilitated recovery of about 90% of the solubilized insulin-like growth factor-I (IGF-I) in the light phase of the ATPS. Furthermore, it was reproducible at scales from 10 to 1000 L.

Cells and cell debris usually partition to the bottom phase, both in PEG/salt and PEG/dextran systems. Cell material contributes to the phase formation, with a consequent decrease of the volume ratio with increasing cell concentration. In PEG/salt systems, nucleic acids and polysaccharides partition strongly into the salt-rich phase. In contrast, most colored byproducts of fermentation broths are hydrophobic and partition into the top phase (Hustedt et al. 1985).

The partitioning coefficient of a biomolecule in an ATPS was found to be a function of many variables (Albertsson 1986).

$$\ln K = \ln K^0 + \ln K_{elec} + \ln K_{hfob} + \ln K_{size} + \ln K_{biosp} + \ln K_{conf}$$

where the subscripts of the several contributions denote the following: elec, electrochemical; hfob, hydrophobic; biosp, biospecific; and conf, conformational. These contributions are both from the protein structural properties (size; net charge; hydrophobicity; other surface properties; primary, secondary, tertiary, and quaternary structures), and from the surrounding environment conditions (salts and concentration, pH, type of polymers, polymer molecular weights and concentrations, temperature). K^0 include other factors.

Polymers Molecular Weight and Concentration

Polymer MW affects the phase compositions, as previously mentioned. The trend usually observed for protein partitioning is that an increase in MW of the polymer, which is enriched in one of the phases, leads to the partitioning of the biomolecule into the other phase. The magnitude of this effect is larger for larger proteins (Albertsson 1986; Forciniti et al. 1991b; Diamond and Hsu 1992).

Protein partition coefficients become more one sided when the TLL is increased, and the effect of the TLL is larger for the larger proteins than for the smaller ones (Albertsson 1986; Forciniti et al. 1991b). The effect of polymer MW on the solute partition may not be separated from that of the polymer concentration. In addition to Δ_{PEG}, the STL is needed to describe the effect of the polymer MW on the solute partition (Zaslavsky 1995).

Temperature

The effect of temperature has not been given enough attention, and there are very few studies to allow any generalization. The temperature influences the phase composition, the electrostatic and hydrophobic interactions, proteins' conformational state, and it can also induce protein denaturation, self-association, or dissociation. The separation of some of these effects is very complex.

According to Forciniti et al. (1991c), the influence is highly dependent on the protein and total polymer concentration, but it does not depend on the polymer MW. Partition of hydrophobic and large proteins is more sensitive to temperature changes.

Diamond and Hsu (1990, 1992) have developed a model for the effect of temperature on partitioning in PEG/dextran systems, based on a modified form of the Flory-Huggins theory of polymer solution thermodynamics.

Salt and pH

Changing the pH is a common practice in the partitioning studies of proteins due to the net charge of the protein and its interaction with the surrounding environment. The balance of these interactions may reverse the predictions exclusively based on the protein net charge. Furthermore, considering that the pH can induce structural and conformational changes of the protein, the best way to study the pH effect is to do it experimentally. In addition, the different pH values are accomplished with different buffers; therefore, the salt composition of the ATPS is also altered, and the partitioning of proteins may change accordingly.

Partition coefficients of cutinase (pI 7.8) in PEG/phosphate systems at different pH values illustrated typical pH effects and its usefulness to steer protein partitioning as needed (Table 2). Adding extra salts to the ATPS in order to direct the partitioning of proteins is also very common.

Table 2 Partition Coefficients of *Fusarium solani pisi* Cutinase in Aqueous Two-Phase Polymer Systems

	Buffer[a]	Partition coefficient K[c]	
pH	EOPO system	PEG/phosphate[b]	EO50PO50/Hydroxypropyl starch[a]
4	Na-citrate		0.13
5	Na-citrate		0.77
6	Na-citrate	0.44	0.92
7	Na-citrate		1.0
8		0.67	
9	Na-Borax	2.8	0.80

[a]System composition: 6% wt EO50PO50-3900–12% wt Hps-200 with 10 mM buffer.
[b]System composition: 20% wt PEG 3350–11% phosphate (buffer potassium phosphate).
[c]Protein concentration in the systems: 0.125 g/L, room temperature.
Data from Sebastião et al. 1993a and Cunha et al. 1999.

Salt additives affect the composition and the properties of the phases of a given system, as well as the properties of the protein (size, conformational state).

Regarding protein-salt interactions, low salt concentrations enable the establishment of favorable electrostatic interactions between the salt ions and the charged residues of the protein. At higher salt concentrations, most salt ions are excluded from the protein's domain due to unfavorable interactions between the salt ions and the hydrophobic residues of the protein, producing salting-out behavior (Curtis et al. 1998).

Taking advantage of the protein-ion interactions and of the uneven distribution of the different ions between the two phases, which established a small electric potential across the interface (see section "Two-Polymer Systems"), the proteins are steered as needed. Albertsson has developed the following equations for the protein partition (K_p) with respect to the interfacial potential (Ψ), which is created by the salt whose ions have the charges Z^+ and Z^-:

$$\ln K_p = K_p^0 + (Z_p F/RT)\Psi$$

$$\Psi = \frac{RT}{Z^+ + Z^-} F \ln \frac{K_-}{K_+}$$

where R is the gas constant, F is the Faraday constant, T is the absolute temperature, and K_- and K_+ are hypothetical partition coefficients of the ions in the absence of a potential. K_p^0 is the value of the protein partition coefficient when the interfacial potential is zero or when the protein net charge Z_p is zero. There are several experimental verifications of this prediction; however, it is also important to note that deviations from this prediction are often found in the literature as other factors besides the protein net charge may change with pH (Abbott et al. 1990).

In Fig. 5 these effects are illustrated for cutinase partitioning in EO50PO50/hydroxypropyl starch ATPS, with the salts sodium perchlorate and trimethylammonium sulfate (Cunha et al. 1999).

How to Find a Suitable System for Protein Extraction

As emphasized previously, protein partitioning is governed by many parameters. Therefore, understanding the relative importance of the factors affecting their partitioning behavior is critical for an effective separation. A good strategy for the system optimization should be carefully designed. The experiments may be carried out in 10-mL graduated centrifuge tubes. Following addition of the polymers (stock solutions are usually used), buffers, and protein mixture, the systems are thoroughly mixed, followed by

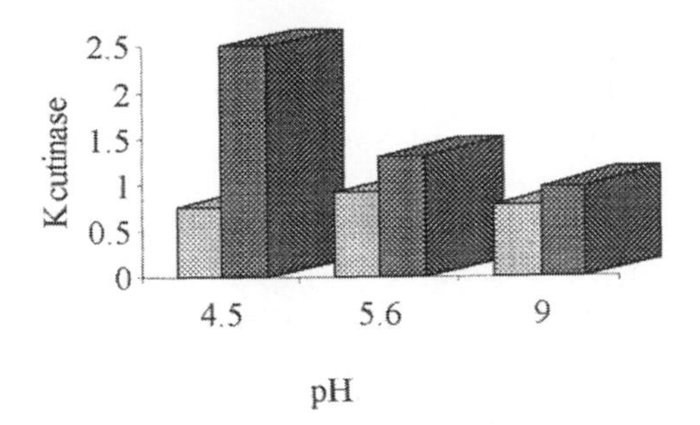

Figure 5 Partition coefficients of *Fusarium solani pisi* cutinase in EO50PO50/Hps two-phase system at different pH values. System composition: 6% wt EO50PO50-3900–12% wt Hps-200. Buffer: Na-citrate 10 mM (pH 4.5 and 5.6) and Na-Borax 10 mM (pH 9.0); extra salts: 100 mM $NaClO_4$ (pH 4.5 and 5.6) and 100 mM $(Et_3NH)_2SO_4$ (pH 9.0). (Data from Cunha et al. 1999.)

low-speed centrifugation. Influence of the phase components on the quantification assays should be checked. Protein determination can be performed using Bradford's procedure (Bradford, 1976) after suitable dilution of the phases.

Multifactorial analysis strategy indicates major trends and possible interactions among the factors studied (Box et al. 1978; Barker 1985). Furthermore, it is economically beneficial, as it requires relatively few runs per factor studied and saves time. Factors that can be studied include the TLL, the system pH, the additive concentration, the protein mixture load, the volume ratio, and the top and bottom phase polymer concentrations (Hart et al. 1995; Zijlstra et al. 1998). Parameters that can be evaluated are partition coefficient of the target protein, partition coefficient of contaminants, yield, purification factor, and concentration factor.

Affinity Extraction

Aqueous two-phase affinity partitioning has been mostly applied to PEG/dextran systems containing PEG derivatives (Andrews et al. 1990). Only a few affinity two-phase studies have been performed in PEG/salt systems due to the fact that biospecific interactions are usually obstructed by high salt concentrations.

Several ligands have been developed and applied to the purification of proteins. The high purity required for molecular biology diagnostic and therapeutic proteins has spurred efforts to design and discover new powerful ligands. Affinity chromatography (or chromatography based) has dominated the affinity separation technologies. However, there are some evident advantages of aqueous two-phase partitioning over liquid-solid affinity chromatography. These include improved availability of ligand for binding the protein, stabilization by phase polymers, and absence of nonspecific sorption as no solid matrix is present (Zaslavsky 1995).

Two major categories of ligands have been distinguished (Labrou and Clonis 1994): high affinity (polyclonal and monoclonal antibodies) and general affinity ligands. The latter exhibit a wide spectrum of interactions with protein molecules; therefore, their selectivity is reduced and they can be used in different purification cases. Some of these include textile and biomimetic triazine dyes, chelated metal cations, carbohydrate, coenzymes, fatty acids, amino acid sequences, and protein ligands. Some examples of affinity partitioning of proteins in aqueous two-phase systems are given in Table 3.

Triazine dyes are the most common choice as ligands, since they are commercially available as low-cost chemicals and are easily coupled to PEG. They are also safe, chemically and biologically stable, and display moderate to high specificity toward many proteins. These characteristics also make them the most promising candidates for large-scale purification. Cordes and Kula (1986) have successfully proven the efficiency of Procion red HE3B in the purification of formate dehydrogenase from cell homogenate of *Candida boidinii*.

Hydrophobic affinity (fatty acid esters) is mostly used on an analytical scale and for cell separations. The small amounts of modified PEG-fatty acid esters usually required have no significant effect on the physical properties of the phases. The different intensities of the hydrophobic interactions between the proteins and the ligand have led to successful separation of proteins.

Iminodiacetic acid (IDA) has been bound to PEG and loaded with divalent ions (Cu^{2+}, Zn^{2+}, and Ni^{2+}) for the extraction of proteins with surface histidine residues (Walter et al. 1991; Suh and Arnold 1990; Plunkett and Arnold 1990). Charged groups, like trimethylamine, are also very effective. By changing the system pH, the partitioning coefficients can be shifted by several orders of magnitude (Johansson et al. 1973). The immobilization of affinity ligands to magnetic particles that selectively partition in a PEG/phosphate system has been applied for the purification of protein A from crude extract of recombinant *E. coli* (Suzuki et al. 1995).

Monoclonal antibodies (Elling et al. 1991) have been used as ligands to carry out the separation of the complex antigen-antibody and antigen in the so-called immunoaffinity partitioning procedure. Affinity partitioning is usually performed by addition of the polymer-bound ligand (e.g., PEG-X), to the mixture containing the target protein and the other phase com-

Table 3 Proteins Purification in ATPS Composed of Affinity Polymers

Protein	Ligand	Ligand carrier	ATPS	Ref.
Dehydrogenases	Coenzyme (NADH); triazine dyes	PEG	PEG/crude dextran	Kula et al. (1982)
β-galactosidase, BSA, lysozyme, β-lactoglobulin	Valeryl, benzoyl	Dextran	PEG/dextran	Lu et al. (1994)
Dehydrogenase	Triazine dyes	UCON	UCON/dextran	Alred et al. (1992)
Hemoglobins	Cu(II)IDA	PEG	PEG/sodium sulfate	Plunkett et al. (1990)
Cytocrome *c*, myoglobins	Cu(II)IDA	PEG	PEG/dextran	Suh and Arnold (1990)
Peroxidase	Peroxidase Mab	PEG	PEG/dextran	Elling et al. (1991)
Interferon-α_1	Phosphate ester	PEG	PEG/phosphate	Guan et al. (1996)
BSA, lysozyme, myoglobin, β-lactoglobulin, ovalbumin	Fatty acid (palmitate)	PEG	PEG/dextran	Shanbhag and Axelsson (1975)
Penicillin acylase	Trimethylamine	PEG	PEG/phosphate	Guan et al. (1992)
IgG	Dye	PEG	PEG/dextran	Zilstra et al. (1998)
Protein A	Human IgG	Eudragit-Mag	PEG/phosphate	Suzuki et al. (1995)

ponents (e.g., PEG and dextran). After mixing and phase separation, the target protein is enriched in the phase containing the polymer-bound ligand (top phase). A washing step may follow, by equilibrating the phase containing the ligand and product (top phase) with pure phase (bottom). To isolate the target protein from the ligand another partition step is performed, with a fresh (bottom) phase, under dissociating conditions. Usually pH and salt variations are used or, if an ethylene oxide–propylene oxide copolymer is used, temperature is increased above the cloud point of the polymer solution. The protein partitions spontaneously to the phase (bottom) opposite to that containing the ligand.

Protein Design

Partitioning in a two-phase system depends on the surface properties of the protein. Recently, the contribution of the surface-exposed amino acid residues towards partition coefficient of a protein was demonstrated using variants of the enzyme cutinase differing in one or several amino acid residues (Berggren et al. 2000b). In this study, the effect on partitioning could be described only taking solvent accessibility (determined by computer analysis) and the type of amino acid into account.

Fusion of a peptide/protein tag to recombinant proteins has been used as a tool to direct partitioning to the desired phase in a two-phase system. β-Galactosidase was found to be an adequate affinity tag in the purification of proteins in PEG/salt ATPS (Köhler et al. 1991a,b). Its strong affinity for the PEG-rich phase was attributed to the high content of tryptophan residues that were exposed on the surface of the enzyme. Tryptophan-rich peptides have been used as fusion tags in a number of studies to improve the partitioning of recombinant proteins to a PEG or ethylene oxide–propylene oxide (EOPO) phase (Köhler et al. 1991b; Berggren et al. 1999, 2000a). However, fusion proteins with Trp-rich peptide tags give rise to reduced protein yields and have shown increased sensitivity towards proteolysis (Köhler et al. 1991b; Berggren et al. 1999). The partitioning data of the free peptides obtained in a recent study by Berggren et al. (2000b) showed that Tyr and Phe are good alternatives to Trp. While Phe resembled Trp in exhibiting poor solubility in aqueous solutions, no solubility problems were encountered for Tyr-containing peptides. The authors suggest that Tyr-rich fusion tags could be effective in partitioning a recombinant protein to an EOPO-rich phase.

Luther and Glatz (1994, 1995) have studied the effect of charged peptides (e.g., polyarginine and polyaspartic tails) as fusion tags in the partitioning of proteins in systems of PEG/dextran with small amounts of DEAE-dextran (diethylaminoethyl-dextran).

2.3 Temperature-Sensitive Polymers

Some polymers exhibit a lower critical solution temperature (LCST) in water. This means that the solution separates into two phases above this temperature: a polymer-rich and a polymer-poor phase. LCST is often called the cloud point temperature (CPT), and its value is dependent on the polymer concentration and molecular weight.

PEG is a thermoseparating polymer. The CPT is around 180°C for the lower molecular weights and decreases with increasing molecular weight, reaching a value of approximately 95°C for molecular weights of 200,000 or more (Bailey and Koleske 1976). Other examples of thermoseparating polymers are the random copolymers of EOPO, often referred to as EOPO polymers, which have much lower CPT values. The CPT of this type of polymer solution is a linear function of the mass fraction of the PO in the copolymer (Fig. 6). A PO content of 0% represents the PEG polymer.

The cloud point diagrams of thermoseparating polymers in aqueous solution have a typical shape, as represented in Fig. 7. The addition of another component to the polymer/water system may change the clouding behavior. Hydrophilic additives, including salts (Florin et al. 1983; Ananthapadmanabhan and Goddard 1986; Louai et al. 1991; Cunha et al. 1998), glycine (Johansson et al. 1997), and sugars (Sjöberg et al. 1989), lower the CPT and strongly partition to the water phase after thermoseparation. Cunha et al. (1997) determined the polymer and salt content in the phases of a 10% EO50PO50 water solution with and without potassium phosphate after thermoseparation at 50°C and 60°C, respectively. The polymer content in the polymer-rich phase (bottom) reached values between 40% and 65% (w/w), and in the water-rich phase, values between 1% and 3% (w/w), depending on the system temperature. Hydrophobic additives, like phenol

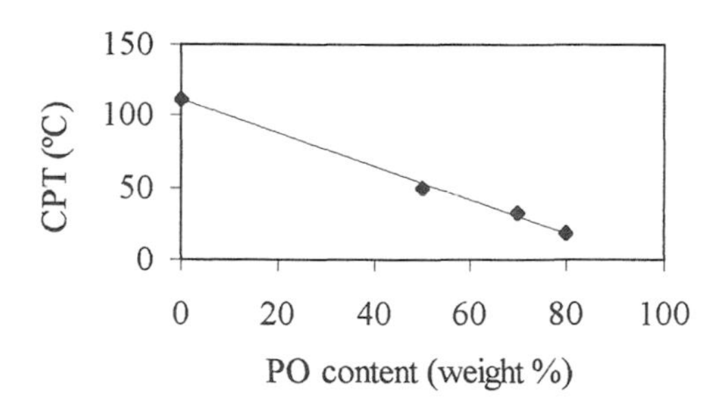

Figure 6 Cloud point temperature of 10% EOPO polymer/90% water system with the PO content. (Data from Alred et al. 1994; Johansson et al. 1993, 1996; Cunha et al. 1998.)

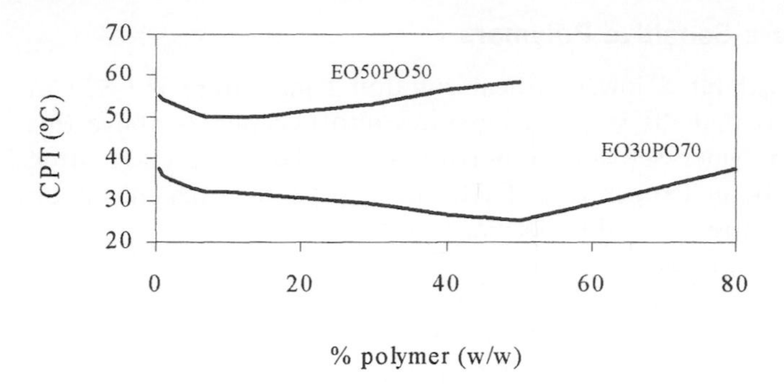

Figure 7 Cloud point diagram for the systems EO50PO50/water and EO30PO70/water. (Data from Johansson et al. 1996; Cunha et al. 1998.)

and *n*-butanol, lower the CPT and partition to the polymer-rich phase (Louai et al. 1991; Johansson et al. 1997). Amphiphilic additives like ionic surfactants (e.g., SDS) increase the CPT, and nonionic surfactants (e.g., Tween-20) change the CPT almost not at all (Cunha et al. 1998).

There seems to be evidence that the phase separation in polymer/salt/water systems and clouding of polymer solutions upon heating are one and the same phenomenon (Zaslavsky 1995).

EOPO polymers have CPT values that facilitate their use in the extraction of proteins. Two-polymer systems such as EOPO/dextran or EOPO/hydroxypropyl starch (HPS), have been used for the purification of proteins (Table 4). In the first extraction step, the target protein partitions to the top

Table 4 Proteins Purification in ATPS Composed of Thermoseparating Polymers

Protein	ATPS	Ref.
3-Phosphoglycerate kinase hexokinase	EO50PO50/dex. or HPS	Harris et al. (1991)
3-Phosphoglycerate kinase hexokinase, glucose-6-phosphate dehydrogenase	EO20PO80/dex.	Alred et al. (1994)
BSA, lysozyme	EO50PO50 or EO30PO70/dex.	Johansson et al. (1996)
Cutinase	EO50PO50/HPS	Cunha et al. (1999)
Peptides	EO30PO70/dex.	Berggren et al. (1999)
Amino acids	EO50PO50	Johansson et al. (1995)

phase, which is enriched in the EOPO polymer. In the second extraction step, the former top phase is heated to a temperature somewhat higher than the CPT of the EOPO solution, and a new phase system is formed after 30 min. The target protein quantitatively partitions to the water-rich phase, and the polymer-rich phase can be recycled and used for subsequent extraction.

Hydrophobic proteins strongly partition to the EOPO-rich phase. Persson et al. (1998) successfully purified a mutant of apolipoprotein A-I out of *E. coli.* in an EO50PO50/Hps system with 2.5 M of urea, purification factor of about 3 and a yield of 80% after thermoseparation were obtained.

When dealing with less hydrophobic proteins, salts can be used to direct the target protein to the desired phase. Lysozyme was purified from hen egg white with a purification factor of 8 and a yield of 60% after thermoseparation at 45°C. The system used was EO30PO70/Hps at pH 8.6, with 100 mM sodium perchlorate with 50% (w/w) of hen egg white (Cunha and Tjerneld 1995).

2.4 Polymers and Salts Removal

Removal of PEG from the target protein can be accomplished by shifting the desired protein into the salt-rich phase, which contains low amounts of PEG. The residual PEG can be separated in subsequent purification steps when removing the salt by diafiltration or ultrafiltration. Alternatively, ion exchange chromatography can be used, adsorbing the protein and washing off PEG. Cooling down the salt phase to 6°C may lead to salt precipitation. The addition of alcohols (e.g., ethanol) to the bottom phase of a PEG/salt system allowed recycling of 95% of the salt (Greve and Kula 1991).

Recycling the PEG can reduce the process costs, and this can be performed by directly reusing the PEG-rich phase after shifting the desired protein into the salt-rich phase (Hustedt 1986). Extraction of PEG with organic solvent followed by evaporation may also be considered.

The thermoseparating polymers offer the possibility of both polymer recycling and easy separation of polymer from biomolecules.

3. EXTRACTION IN DETERGENT-BASED AQUEOUS TWO-PHASE SYSTEMS

Bordier (1981) reported, for the first time, the use of Triton X-114 for separation of hydrophilic and integral membrane proteins. Subsequently, several publications have followed, mostly using Triton X-114 in the characterization of membrane proteins. Also, increasing attention was paid to the phase separation phenomena of surfactants and its potential for the purification of proteins (Table 5).

A surfactant monomer is an amphiphilic molecule, composed of a hydrophilic head group and a hydrophobic tail group. The surfactants are distinguished between ionic and nonionic owing to the nature of their head groups. The surfactant molecules have a strong tendency to self-aggregate into micelles in an aqueous environment due to the strong attractive hydrophobic interaction between the hydrocarbon chains (Picullel and Lindman 1992), which sequester themselves inside the aggregate while the polar groups orient to the aqueous phase. The concentration at which the micelles start to form is defined as the critical miceller concentration (CMC). At low detergent concentrations only monomers occur, whereas at high concentrations both monomers and micelles exist in equilibrium (Helenius and Simons 1975). Surfactant aggregates, in contrast to polymer molecules, are not fixed assemblies and the aggregation number of a micelle, which is the number of monomers aggregated, may vary with the experimental conditions. The micelles can then have variable sizes and shapes. In addition, monomers may form mixed aggregates; thus, surfactant-based affinity ligands may be incorporated to selectively purify target biological molecules.

The majority of the examples found in the literature for the purification of proteins in detergent-based aqueous two-phase systems are with polyoxyethylene detergents (Table 5 and 6). These surfactants are considered to be nondenaturing to proteins (Helenius and Simons 1975). The structure of these surfactants is normally abbreviated to C_nE_m, where C_n

Table 5 Proteins Purification in Detergent-Based Aqueous Two-Phase Systems

Protein	Surfactant	Ref.
Mycoplasma hyopneumoniae surface proteins	Triton X-114	Wise and Kim (1987)
Escherichia coli penicillin-binding protein 4	Triton X-114	Mottl and Keck (1991)
Interleukin-2	Triton X-114	Bergmann et al. (1991)
Lipases	$C_{14}E_6$	Terstappen et al. (1992)
Cholesterol oxidase	Triton X-114	Ramelmeier et al. (1991)
Cholesterol oxidase	$C_{12\text{-}18}E_5$	Minuth et al. (1996)
Phenol oxidases	Triton X-114	Sánchez-Ferrer et al. (1994)
Bacteriorhodopsin, cytochrome *C*	C_9 or C_{10}-APSO4	Saitoh and Hinze (1991)
Cytocrome *c*, ovalbumin, catalase	$C_{10}E_4$, C_8-lecithin	Liu et al. (1996)

Table 6 Thermoseparating Polyoxyethylene Surfactants and Respective Cloud Point Temperatures

Surfactant	Trade name	CPT (°C)	Ref.
C_5E_2		36	Chakhovskoy (1956)
C_8E_3		8	Chakhovskoy (1956)
C_8E_4		35.5	Chakhovskoy (1956)
$C_{10}E_4$		18.8	Liu et al. (1996)
$C_{10}E_5$		44	Lang and Morgan (1980)
$C_{12}E_4$		6.9	Fujimatsu et al. (1988)
$C_{12}E_5$		24.8	Fujimatsu et al. (1988)
$C_{12}E_6$		50.5	Fujimatsu et al. (1988)
$C_{14}E_6$		35	Terstappen et al. (1992)
$C_8\Phi E_{7\text{-}8}$	Triton X-114	20	Sepulveda et al. (1968)
$C_9\Phi E_{7.5}$	PONPE-7.5	18	Kenjo (1966)
$C_9\Phi E_{12.5}$		2	Kenjo (1966)

represents the length of the alkyl group and E_m the number of oxyethylene units. The general formula is $C_nH_{2n-1}(OCH_2CH_2)_mOH$. If there is a phenyl ring between the alkyl group and the polyoxyethylene chain, it can be abbreviated to $C_n\Phi E_m$. An extensive list of these types of surfactants can be found in Galaev and Mattiasson (1993). A solution of such nonionic surfactants, when heated, separates into two phases—one surfactant-rich phase and the other phase containing a very low concentration of surfactant approximately equal to the CMC. CPT varies slightly with the surfactant concentration and with the inclusion of additives (Evans 1994). Hydrophilic salts like sodium citrate, sodium phosphate, and sodium sulfate normally lower the CPT, whereas hydrophobic salts as sodium nitrate and sodium iodide raise the CPT (Ramelmeir et al. 1991). Aliphatic alcohols, fatty acids, and phenols lower the CPT substantially (Sánchez-Ferrer et al. 1994). The addition of more hydrophilic surfactants (deoxycholate, Triton X-100) can raise the CPT and vice versa.

Advantages of these systems, among others, include (1) ability to concentrate solutes, (2) safety and cost benefits (low amounts of nonvolatile and nonflammable surfactants are required), and (3) easy disposal of surfactant (burnt in the presence of waste acetone or ethanol) (Saitoh and Hinze 1991). However, commercial nonionic surfactants are usually less pure and exhibit high background absorbance in the ultraviolet (UV) region.

The proteins with sufficient surface hydrophobicity partition to the surfactant-rich phase, whereas the others partition to the other phase. From the surfactant-rich phase, the protein can be recovered either by 2-

butanol extraction of the surfactant (Minuth et al. 1996) or by ion exchange chromatography, although the sample has to be diluted due to its high viscosity (Ramelmeier et al. 1991).

More recently, systems composed of ionic surfactants have been applied to the purification of proteins (Table 5). Zwitterionic dialkylphosphatidylcholine surfactants, based on phosphatidylcholines, often called lecithins, as well as synthetic dialkylmethylammonium surfactants have been used. The solutions of these surfactants become cloudy when the temperature is lowered, and a two-phase system is formed. Thus, these systems form two phases under (rather than over) the cloud point temperature ($CPT_{C9\text{-}APSO4} = 65°C$; Saitoh and Hinze 1991). Some of the advantages that have been claimed are minimal background absorbance at UV detection wavelengths and the possibility of extracting thermally labile proteins. However, these surfactants exhibit higher CMC values than nonionic surfactants; therefore, higher amounts of the material are required to form an extraction system, and higher losses in the surfactant-depleted phase are obtained. There is the possibility of recovering this surfactant by adding a high concentration of salt (Saitoh and Hinze 1991). On the other hand, high CMC values permit dialysis across a membrane where rapid removal or displacement of detergent is desired.

Aqueous mixtures of cationic and anionic surfactants at concentrations much higher than CMCs can separate into two immiscible aqueous phases and have been applied to the partitioning of proteins (Zhao and Xiao 1996).

Detergent based aqueous two-phase systems have very-low-density differences, low interfacial differences, and complex rheological behavior of the surfactant-rich phase, which makes processing very difficult. Minuth and coworkers (1997) have successfully applied these systems for the extraction of a cell membrane–bound enzyme from a fermentation broth, in pilot scale, using both gravity settling and centrifugal separation for phase separation.

4. EXTRACTION IN REVERSED MICELLAR SYSTEMS

Reversed micelles are aggregates of surfactants in apolar media. The surfactant molecules are amphiphiles that can be arranged in such a way that the hydrophobic tails are in contact with the apolar bulk solution and the polar head groups are turned toward the interior of the aggregate, forming a polar inner core. The amount of water solubilized in the reversed micelles is commonly referred to as w_0, the molar ratio of water to surfactant ($w_0 = [H_2O]/[\text{surfactant}]$). Usually, the reversed micellar system is a ternary system of water, organic solvent, and surfactant, with extra components added when advantageous or necessary. The cationic surfactants require a cosurfactant for stabilization of the reversed micelles, which is usually an aliphatic alcohol that partitions between the micelle interphase and the continuous phase. Depending on the relative concentrations of various components of the system, phase structure can change dramatically, including normal and reversed micelles, liquid crystals, and lamellar phases.

4.1 Factors Affecting Protein Transfer

The distribution of proteins between a micellar organic phase and an aqueous solution is largely determined by mechanisms based on electrostatic and hydrophobic interactions that depend on the conditions in the aqueous bulk phase, namely, pH, ionic strength, type of salt, and temperature. The parameters related to the organic phase, such as concentration and type of surfactant, presence of cosurfactant, and type of solvent, also influence the partition of a protein. In addition, changes in temperature can affect the solubilization of biomolecules. The phase transfer depends on the specific characteristics of the proteins, namely, isoelectric point, size and shape, hydrophobicity, and charge distribution.

Surfactant Type and Concentration

The most often used surfactant in protein extraction by reversed micelles is AOT (sodium di-2-ethylhexylsulfosuccinate), primarily due to the fact that a cosurfactant is not required for protein solubilization. AOT aggregates spontaneously in hydrocarbon solvents, forming water pools with radii greater than 170 Å (Maitra 1984). Despite the large amount of work on AOT reversed micellar systems, it is difficult to separate the proteins from the surfactant subsequent to its extraction, and phase separation takes a long time.

Several analogues of AOT have been synthesized, but their poor solubility in aliphatic solvents is a limitation for practical applications (Leydet et al. 1994). Hu and Gulari (1996) used sodium bis(2-ethylhexyl)phosphate (NaDEHP), an anionic surfactant that has the same hydrocarbon tail as AOT but a different polar head, for the extraction and back extraction of cytochrome *c* and α-chymotrypsin. Under appropriate conditions, more than 90% of the α-chymotrypsin and nearly 100% of the cytochrome *c* can be transferred into the NaDEHP reverse micellar phase. Overall recoveries of 98% for cytochrome *c* and 67% for α-chymotrypsin were achieved (Hu and Gulari 1996). The surfactant NaDEHP can be easily broken by converting the sodium salt NaDEHP to a non-surface-active divalent salt,

$M(DEHP)_2$. Moreover, the phase separation of NaDEHP is much faster than that of AOT, and the surfactant can be readily recycled.

Goto et al. (1997) developed a new anionic surfactant, dioleyl phosphoric acid (DOLPA), for protein extraction. The structure of the surfactants, in particular the hydrophobic moiety of the surfactant, strongly affects the degree of protein extraction. The protein extraction appears to be governed by the nature of interfacial complex formed between the protein and the surfactant, with the solubilization favored when the interfacial complex has a high hydrophobicity. Also, surfactants that can form a close-packed complex with the protein are excellent protein-solubilizing agents (Goto et al. 1997).

The surfactant nature can also have a strong effect on retention of protein activity due to direct interactions with the hosted molecules (Melo et al. 1996, 1998). The effect of surfactant charge and the presence of alcohols as cosurfactants on the structure and activity of a recombinant cutinase were investigated (Melo et al. 1998). With the anionic surfactant AOT a fast deactivation of cutinase occurs due to a reversible denaturation process. The deactivation and denaturation of cutinase is slower in small cationic (CTAB/hexanol) reversed micelles and dependent on micelle size. The higher stability of cutinase in CTAB/hexanol reversed micelles is at least partly due to the presence of the cosurfactant, hexanol (Carvalho et al. 1997, 1999; Melo et al. 1998).

Increasing the surfactant concentration favors protein solubilization in the organic phase (Aires-Barros and Cabral 1991; Hentsch et al. 1992; Carneiro-da-Cunha et al. 1994; Krieger et al. 1997), resulting in broader pH-solubilization peaks (Göklen 1986). On the other hand, higher surfactant concentrations make difficult the backward transfer of proteins into a second aqueous phase (Castro and Cabral 1988; Hentsch et al. 1992; Carneiro-da-Cunha et al. 1994a). Therefore, the optimal surfactant concentration in a double-phase transfer corresponds to the minimum limit for achieving maximal transfer to the organic phase.

The selectivity of the reversed micellar phase for a given protein can be changed by choosing an appropriate surfactant concentration in the organic phase. For example, smaller proteins are more easily extracted into the organic phase than proteins with higher molecular weight at low surfactant concentration (about 50 mM). These results also point to a limitation of the size of the protein that can be extracted from the aqueous phase. Although this problem can be solved by increasing the surfactant concentration when dealing with large proteins ($>$ 100,000 Da), this has a spurious effect on the phase separation by decreasing it (Göklen 1986; Ishikawa et al. 1990).

A recent study (Imai et al. 1997) relates the protein hydrophilicity with the AOT concentration and the water level required for complete extraction.

The minimal AOT concentration required for a 100% extraction was obtained for ribonuclease A, lysozyme, and cytochrome *c* (Ishikawa et al. 1992). The hydrophilic surroundings, defined as the ratio of water to protein in the micelle organic phase at a certain pH and salt concentration, were found to be linearly related to Fisher's polarity ratio, *p* (total volume of polar residues/total volume of nonpolar residues). In order to solubilize more hydrophilic proteins (those with a larger *p* value), more water and AOT molecules are necessary in the reversed micelle organic phase. Also, different location of protein in the AOT reversed micelles leads to different water uptake in the organic phase (similar or larger than in the protein-free system). It can be considered that the balance between the hydrophilic surroundings and protein hydrophobicity, which is intermediate by the amphiphilic molecules, is important for protein solubilization.

Optimization of the surfactant concentration is important not only in terms of achieving high extraction yields, but also in the surfactant partitioning to the aqueous phase, which may limit the applicability of reversed micelle protein extraction. Contamination with surfactants is a crucial issue for high-purity proteins or proteins in pharmaceutical use. The commonly used surfactant concentrations for extraction with reversed micelles are in the range of 50–100 mM for AOT. However, some reports (Ishikawa et al. 1992; Adachi and Harada 1993; Paradkar and Dordick 1994) have shown that the optimal surfactant concentration for the extraction of proteins into reversed micelle phases may be much lower. Hentsch et al. (1996) demonstrated that quite low surfactant concentrations, in the order of 2–20 mM for AOT, are required for extracting a maximum of protein at a concentration of 15 μM, when using equal volumes of both phases. The same authors also showed that AOT does not strongly adsorb on proteins because simple dialysis was sufficient to drastically reduce the AOT content. Therefore, an efficient AOT removal should be possible by classical processes used in protein purification, such as diafiltration and gel permeation (Hentsch et al. 1996), and the toxicity of AOT does not represent a problem with respect to pharmaceutical proteins.

Another important aspect related to surfactant partitioning to the aqueous phase is the loss of AOT during forward extraction (2–4% under optimal conditions) (Hentsch et al. 1996). The loss of AOT into the aqueous phase depends mainly on the initial AOT concentration in the nonpolar phase and the salt concentration in the aqueous phase. Lowering the salt concentration favored the partitioning of AOT in the aqueous phase (Hentsch et al. 1996), therefore, most of the surfactant losses occurred during forward extraction. Extensive losses of surfactants can be expected particularly in continuous extraction processes (Dekker et al. 1986). Precipitation of proteins by surfactants (Adachi and Harada 1993) and solubilization of surfactants by proteins (Paradkar and Dordick 1994), both in the aqueous phase, may also represent other problems at low surfactant/protein ratio. Low surfactant concentrations are beneficial for protein stripping from the reversed micelle phase as well as for accelerating the process owing to improved phase separation characteristics.

Cosurfactant Type

Most of the available surfactants have a limited solubility in aliphatic organic solvents, and addition of a cosurfactant is necessary to help surfactants to be dissolved in organic solvents and to form suitable reversed micelles to host target proteins. The use of cosurfactants increases the complexity of phase diagrams, making it difficult to develop a physical picture of the reversed micellar system and their hosted solutes (Goto et al. 1997). However, cosurfactants have been used to increase the selectivity of protein recovery (Woll and Hatton 1989; Coughlin and Baclaski 1990; Kelley et al. 1993b; Chen and Jen 1994; Paradkar and Dordick 1994; Chang et al. 1997).

Cosurfactants usually interpose between the surfactant chains, increasing the interface flexibility and the interdroplet interaction (Hou et al. 1988). On the other hand, the micelle's shape and diameter may be altered either to bigger or to smaller values depending on the alcohol chain length and polarity (Hayes and Gulari 1990; Forland et al. 1998). The introduction of a cosurfactant can improve the solubility capacity of the organic phase when using cationic surfactants (Dekker et al. 1989) probably due to an increase in the micelle size.

A cationic surfactant, Aliquat 336, in isooctane with the addition of six different kinds of straight-chain *n*-alcohols as the cosurfactants, was used to extract α-amylase using a full forward and backward extraction cycle (Chang et al. 1997). The most suitable cosurfactant for the full extraction cycle was *n*-butanol, with an α-amylase activity recovery of 80%. With the other alcohols the recovery of enzyme activity was relatively low. Circular dichroism (CD) spectra analysis showed that the conformation of α-amylase in the stripping solution could be restored almost completely to its native state after an extraction cycle. Probably due to high polarity of *n*-butanol, the interaction between the enzyme and the reversed micelles may be weaker (Chang and Chen 1995), therefore not affecting the structure and activity of α-amylase. For low-polarity cosurfactants, this interaction is stronger leading to enzyme denaturation.

Interactions between the proteins solubilized inside the reversed micelles and the cosurfactant seems to be very important in achieving a maximal activity retention of the extracted proteins (Sebastião et al. 1993b; Carneiro-da-Cunha et al. 1994a, 1996; Melo et al. 1998; Carvalho et al. 1999).

pH

The pH of the aqueous solution determines the net charge of proteins. Electrostatic interactions between the protein and the surfactant head groups can favor the transfer of protein into the organic phase. This trend has been observed using anionic (Göklen 1986; Luisi and Magid 1986; Göklen and Hatton 1987; Hatton 1989; Dekker et al. 1989; Camarinha-Vicente et al. 1990; Aires-Barros and Cabral 1991; Cabral and Aires-Barros 1993; Chang et al. 1994; Carneiro-da-Cunha et al. 1994a; Hu and Gulari 1996; Pires et al. 1996; Krieger et al. 1997) and cationic surfactants (Krei et al. 1995), with solubilization favored when the protein and the surfactant head groups have opposite charges. This effect is not always observed. Luisi et al. (1979) obtained identical pH profiles for α-chymotrypsin (pI 8.3) and pepsin (pI < 1.1) using TOMAC, a cationic surfactant in cyclohexane. Similar results were reported by Göklen (1986) with another cationic surfactant, didodecyldimethylammonium bromide (DDAB), and a mixture of proteins with isoelectric points ranging from 5.5 to 7.8. In both studies using quaternary ammonium salts, the transfer of proteins into the reversed micellar phase is more effective in the alkaline pH range (less protonated residue on the protein), which could be explained by ion pairing between the anionic surfaces of the protein and the positively charged surfactant. Luisi et al. (1979) found sharp decreases in the transfer at pH values higher than 13 for α-chymotrypsin and pepsin, attributed to the competition of the proteins with OH^- ions. Protein denaturation and changes in the ionization state of the surfactant at extreme pH values make the interpretation of the phase transfer pattern extra difficult. Studies on the extraction of α-amylase by reversed micelles of a cationic surfactant, BDBAC (benzyldodecylbishydroxyethylammonium chloride) reported that the solubilization mechanism of this enzyme might be due to a micelle as well as ion-pair binding depending on the ratio of micelle and protein radius (Krei et al. 1995).

Ionic Strength

Increasing ionic strength of the aqueous phase reduces electrostatic interactions as a result of the Debye screening effect, which is stronger for larger ions. Thus, at a higher ionic strength, the interactions between hydrophilic biomolecules and the surfactant polar groups are reduced, and smaller micelles are formed. As a consequence of this effect, the solubilization capacity of the organic phase for water and biomolecules decreases. Complete solubilization of three proteins—cytochrome *c*, lysozyme, and ribonuclease A—using an AOT/isooctane system, was obtained at low ionic strength (0.1 M KCl), and no solubilization occurred at high ionic strength (1.0 M KCl)

(Göklen and Hatton 1987). Inversion of the partition coefficients is achieved in a very narrow range of pH, which depends on the protein species and does not correlate with the isoelectric point of the protein. Ribonuclease A is the protein with the lower isoelectric point, but a sharp change of solubilization behavior occurs at a KCl concentration (0.5 M) lower than that in the case of lysozyme (0.7 M) (Göklen and Hatton 1987). The role of electrostatic interactions in phase transfer is supported by narrowing of the pH peaks when ionic strength is increased, as observed for the extraction of cytochrome *c* (Göklen 1986). These results with AOT are consistent with the phase transfer behavior of α-amylase (Dekker et al. 1986) in a system including trioctylmethylammonium chloride (TOMAC), a cationic surfactant.

Studies with a lipase from *Chromobacterium viscosum* (Aires-Barros and Cabral 1991) using AOT as surfactant also referred to a strong effect of the ionic strength in lipase solubilization. A sharp decrease in solubilization was observed at 250 mM KCl ($w_0 < 10.8$), and no protein solubilization occurred with a KCl concentration greater than 350 mM ($w_0 < 9.6$) (Aires-Barros and Cabral 1991).

The ionic strength of the aqueous phase also affected the extraction and back extraction of a recombinant cutinase using AOT/isooctane reversed micelles by changing the radii of the water pool. The extraction is favored at low ionic strength (100 mM KCl) and the back extraction at higher ionic strengths (500 mM KCl) (Carneiro-da-Cunha et al. 1994a).

Extraction studies of cytocrome *c* and α-chymotrypsin with an anionic surfactant, sodium bis(2-ethylhexyl)phosphate (NaDEHP), also show that the extraction process is a strong function of the NaCl concentration in the aqueous phase (Hu and Gulari 1996).

Additional effects of the ionic strength in the solubilization of proteins in reversed micellar biphasic systems are the competition with ionic species for transfer into the reversed micelles and changes in the electrostatic state of the micelles and/or proteins. It has been noticed that when ionic strength of the aqueous phase is below a certain limit, reversed micelles and phase separation do not occur, a stable microemulsion being formed. Therefore, the transfer of proteins between phases requires a certain minimum ionic strength of the aqueous solution. A minimal concentration of KCl around 0.1 M was found to transfer cytochrome *c* into the AOT/isooctane system (Göklen and Hatton 1985). Often the buffer in the aqueous phase can supply enough electrolyte.

Type of Electrolyte

The salt composition of the aqueous phase can also influence the efficiency of protein transfer between phases, apart from the expected different contribution to the ionic strength of the solution. Specific interactions of the protein and/or the surfactant with the ions in solution can account for a change in the solubilization pattern of proteins under given conditions in the presence of different ionic species (Leser et al. 1986; Göklen 1986; Castro and Cabral 1988; Marcozzi et al. 1991). The buffering system itself can influence the solubilization behavior of proteins (Göklen 1986; Meier et al. 1984).

Protein Charge Distribution

The yield of extraction depends on the total charge of the protein surface, since it strongly depends on the pH of the aqueous phase. The importance of charge distribution on the protein surface was addressed by Wolbert et al. (1989), who found that proteins with higher charge asymmetry are more easily extracted into an organic phase containing TOMAC surfactant and obtained a correlation between the yield of extraction and the degree of charge asymmetry (Barlow and Thornton 1986) of the protein. This was not a general observation because the same was not valid when using AOT reversed micelles.

The mechanism of extraction of cytochrome b_5 into a reversed micellar phase of AOT in octane was studied using protein engineering of surface residues (Pires et al. 1994). The modified proteins with substitutions of glutamic acid by lysine residue at positions 44 (E44K), 56 (E56K), and 92 (E92K) provide a net charge increase of +1 to +2, depending on the pH value. The percent area of positive potential caused by basic amino acid residues was calculated at the surface of the protein, based on a protein titration program (TITRA) used to compute the average charge distribution of the protein at a given pH. A higher positive surface charge led to higher extraction yields, although the cytochrome was also extracted at an unfavorable pH due to hydrophobic interactions.

Solvent Type

The type of organic solvent affects the size of the reversed micelles and consequently the water solubilization capacity (Göklen 1986; Mat and Stuckey 1993). A different micellar structure resulting from the solvent effect could be exploited to achieve significant changes in the solubilization of α-chymotrypsin with different solvents (Luisi et al. 1979). When comparing the results obtained with several proteins, in a system including cyclohexane as organic solvent, the transfer efficiencies were also very different. Several solvents, including isooctane, octane, heptane, kerosene, hexane, and cyclohexane, were used to form AOT reversed micelles to extract trypsin (Chang and Chen 1995). Recovery of trypsin activity after a full cycle of forward and backward extraction is nearly the same for isooctane, octane, heptane, and kerosene (about 15%), compared with hexane (about 10%) and cyclohexane (about 5%), which are less efficient solvents. Aliphatic solvents are preferable for use in reversed micelle extraction because they are less likely to denature proteins as compared with aromatic, alcohol, and chloric solvents (Goto et al. 1997). However, due to the high hydrophilicity of anionic head groups, such as sulfonic or phosphoric acid, a surfactant with a relatively long hydrophobic tail is required for dissolving in such hydrophobic solvents.

Temperature

Changes in temperature have drastic effects on physicochemical properties of the reversed micellar system. The temperature facilitates the transfer of proteins to a reversed micelle phase by increasing the hydrophobic interactions of the hydrophobic surface of the protein with the apolar tails of the surfactant (Pires and Cabral 1993; Carneiro-da-Cunha et al. 1994a). The extraction yield of a recombinant cytochrome b_5 was highly influenced by the temperature, and increased solubilization of the protein was observed at higher temperatures (40°C), a pH = pI of the protein, and low ionic strength (Pires and Cabral 1993). The hydrophobic interactions are strongly dependent on the temperature, becoming less important at lower temperatures, which leads to a decrease in the amount of recombinant cutinase extracted (100% to 35%) as the temperature is lowered from 40°C to 4°C (Carneiro-da-Cunha et al. 1994a). The only drawback of temperature increase is the partial denaturation of protein (Carneiro-da-Cunha et al. 1994a). Also, higher temperatures facilitate the transfer of species to the reversed micelle phase leading to an increase of the interdroplet exchange rate and reducing the free energy barrier to droplet fusion (Fletcher et al. 1987).

Based on the effect of temperature on the phase diagrams of oil/water/surfactant, Huang and Chang (1995) used a temperature cycle for forward and backward extractions of α-chymotrypsin so as to enhance the efficiency of the global process. The temperature for forward extraction was 10°C, which led to an extraction yield of 98%, and that for backward extraction was 40°C at a low KCl concentration of 0.75 M, giving an enzyme activity recovery of 91.3%. Usually, high salt concentrations are added during the backward extraction, which requires additional purification after extraction.

Biospecific Ligands

Significant enhancements in the selectivity of the protein extraction process with reversed micellar systems can be achieved by the introduction of affinity ligands in the organic phase (Woll et al. 1989; Coughlin and Baclaski

1990; Paradkar and Dordick 1991; Kelley et al. 1993a; Chen and Jen 1994; Poppenborg and Flaschel 1994; Adachi et al. 1996).

Increased partitioning in the transfer of concanavalin A into reversed micelles was obtained when the biosurfactant octyl-β-D-glycopyranoside (2–20% of total surfactant concentration) was included in the organic phase, while glucose inhibited the solubilization of concanavalin A due to competition with the ligand for two binding sites in the protein (Woll et al. 1989). The transfer of the model protein, ribonuclease A, was not changed by the presence of the biological detergent. High values for protein/ligand binding constants in the organic and the aqueous phases, and number of binding sites in the protein favor the protein solubilization in reversed micelles, whereas a low protein/ligand constant in the aqueous solution is desired.

Increased selectivity for the extraction of avidin into reversed micelles was also achieved by using a biotinylated affinity ligand, incorporated into the micelles as a cosurfactant (Coughlin and Baclaski 1990).

Kelley and coworkers (1993a) used affinity cosurfactants, consisting of hydrophilic ligands derivatized with hydrophobic tails, to purify concanavalin A, a membrane-associated protein and chymotrypsin. The addition of a small concentration of the affinity cosurfactants to the micellar phase extends the operating range of pH and salt concentration over which proteins can be extracted. The length of the affinity cosurfactant tail also influences protein uptake (Kelley et al. 1993b).

More recently, the use of nonionic surfactants to form reversed micelles has been suggested (Adachi et al. 1996, 1997; Shioi et al. 1996). Introduction of an affinity ligand to these systems could result in the solubilization of the desired protein only by biospecific interaction, as the reversed micelles composed of only nonionic surfactants had no ability to extract proteins (Poppenborg and Flaschel 1994; Adachi et al. 1996; Shioi et al. 1996). The nonionic surfactant, tetraoxyethylenedecylether (C10E4), was used to extract concanavalin A using alkylglucosides as affinity ligands, and a higher efficiency was obtained compared with the surfactant AOT (Adachi et al. 1996). The same nonionic surfactant (C10E4) as the reverse micellar phase containing trypsin inhibitor as an affinity ligand was applied for separation of trypsin from a mixture of several contaminant proteins (Adachi et al. 1997). No loss of activity of the recovered trypsin was observed after the global extraction process.

Recently, a new reversed micellar system composed of lecithin modified with the affinity ligand Cibacron blue F-3GA was used for lysozyme recovery (Sun et al. 1998). This is a promising system for purification of diagnostic and therapeutic proteins produced by recombinant DNA technology. The biocompatibility of the reversed micellar system and the weak electrostatic interactions between the reversed micelles and the protein lead to a high activity recovery without alteration of the secondary structure. Furthermore, due to the weak electrostatic interactions, the selectivity of the target protein can be significantly improved by introduction of the biospecific ligands (Sun et al. 1998).

4.2 Interfacial Transport of Proteins Into and Out of Reversed Micelles

The majority of studies on protein extraction using reverse micellar systems have focused on the thermodynamic behavior of the microemulsion phases. Equilibrium studies are very important for the optimization of reversed micellar extraction and the understanding of the parameters that affect protein solubilization in these systems. However, studies of the kinetics of protein transfer are also vital for the design of liquid-liquid extraction systems. It is important to have information about the rates of forward and backward transfer of the protein into and out of the reverse micelles across the oil-water interface. Knowledge of these rates is required to determine whether interfacial transport is rate limiting relative to bulk transport processes. The nature of transfer of proteins in reversed micelles is at present unknown, being dependent on different forces (chemical, electrostatic, hydrophobic, and flux dynamics interactions) acting between the solute (protein) and the interface.

Exhaustive studies of interfacial transport in reversed micellar systems focused on the importance of the kinetics associated with forward and backward extraction by measuring the interfacial mass transfer coefficients and transport rates for two model proteins, α-chymotrypsin and cytochrome *c* (Dungan et al. 1991; Bausch et al. 1992). The measurements were made using a stirred-cell technique that allows the importance of the interfacial resistance to protein transfer in these systems to be established. The kinetics of protein transfer into the micellar phase was up to three orders of magnitude (overall mass transfer coefficients of 10^{-3} cm/s) faster than rates measured for small solute transfer (Albery et al. 1987). This indicates that the protein facilitates its own solubilization process by a significant influence on the interfacial formation process due to the large size of proteins A strong decrease in forward transfer rates with an increase in the pH and ionic strength of the aqueous phase was also observed (Dungan et al. 1991).

Back-transfer rates were found to be three orders of magnitude (2×10^{-5} cm/s for α-chymotrypsin and 6×10^{-6} cm/s for cytochrome *c*) slower then those for forward transfer, as well as being much lower than water bulk transfer rates. The protein desolubilization rates are dependent on aqueous pH value, suggesting that coalescence of the protein-filled micelle with the bulk interface dominates the desolubilization kinetics (Dungan et al. 1991).

5. SCALE-UP AND LARGE-SCALE APPLICATIONS

Large-scale protein recovery in two liquid phase systems requires two operations: (1) mixing of the phase components and (2) phase separation.

In batch processes, equilibration is usually done in agitated vessels and in mixer-settler devices. Phase separation is performed either by settling under gravitational force for fast-settling systems like polymer-salt and micelle organic-water, or with common centrifugal separators. Alternatively, column-type extractors can be used to improve extraction efficiency. The selection of a particular column depends on the needs of the operation, the properties of the biomolecules, and the type of the two-phase system involved. Columns can be operated more efficiently in a continuous countercurrent mode or by keeping one phase stationary. When operating an extraction column with ATPS the problem is not the generation of sufficient exchange surface between the phases, but the avoidance of very small droplet formation, due to the small interfacial tensions, which leads to flooding. Regarding reversed micelle systems, the presence of surfactants can cause emulsification of the mixture and lead to flooding.

The prime basis of scale-up of batch mixers is often the geometrical similarity, particularly equal power per unit volume. For columns scale-up, data on the following parameters are needed: protein mass transfer coefficients, dispersed phase hold-up, and extent of mixing in both phases. Usually, these parameters are dependent on the phase velocity, physical properties, and column geometry (Cunha and Aires-Barros 2000).

Liquid-liquid extraction techniques are easy to scale up, minimizing the problems encountered with other separation techniques when applied on a large scale, namely, loss of resolution capacity, complex equipment design, inability to operate in continuous mode, and economic limitations.

5.1 Aqueous Two-Phase Systems

Most large-scale applications of ATPSs are based on polymer-salt systems (Kula et al. 1982; Kroner et al. 1982; Hustedt et al. 1985, 1988; Standberg et al. 1991; Raghavarao et al. 1995). Examples of large-scale purification using ATPSs are presented in Table 7. These systems are attractive because of their low cost and rapid phase separation, although they are not particularly selective.

Table 7 Examples of Large-Scale Purification of Proteins Using ATPS

Protein	Organism	ATPS	No. steps	Yield (%)	Purifcation factor	Ref.
Formate dehydrogenase	*Candida boidini*	PEG/salt	3	78	4.4	Kroner et al. (1982a)
Leucine dehydrogenase	*Bacillus sphaericus*	PEG/crude dextran	1	98	2.4	Schütte et al. (1993)
β-Galactosidase	*E. coli*	PEG/salt	1	87	9.3	Veide et al. (1983)
Interferon	Human fibroblasts		1	75	> 350	Menge et al. (1983)
Aspartase	*E. coli*	PEG/salt	3	82	18	Hustedt et al. (1985a)
Human growth hormone	*E. coli*	PEG/salt	3	81	8.5	Datar et al. (1986)
β-Galactosidase	*E. coli*	PEG/salt	2	80	18	Szöke et al. (1988)
Fumarase	*Brevibacterium ammoniagenes*	PEG/salt	2	75	22	Hustedt et al. (1988)
Penicillin acylase	*E. coli*	PEG/salt	2	78	10	Hustedt et al. (1988)
Superoxide dismutase	Bovine liver	PEG/salt	2	83	4	Boland et al. (1991)
Fumarase	Baker's yeast	PEG/salt	2	85	2.7	Papamichael et al. (1992)

A general scheme for large-scale extraction of proteins, shown in Fig. 8, combines static mixers, for mixing the phase components, and centrifugal separators for phase separation (Hustedt et al. 1985). In the purification of intracellular microbial proteins with PEG/salt (Kula et al. 1982; Veide et al. 1984), the cell debris partitions to the lower salt phase while the target protein is mainly recovered in the PEG phase by choosing the suitable conditions. The proteins in the upper phase can then be extracted into a new salt phase by changing the variables of the system (pH, phase composition, PEG molecular weight). Most of the PEG can therefore be recycled (Papamichael et al. 1992; Hustedt et al. 1985).

Usually for ATPS, the experiments can be performed on a small scale and the results obtained can be directly used for large-scale extraction. It has been demonstrated by Kroner et al. (1982) that scale-up by a factor of

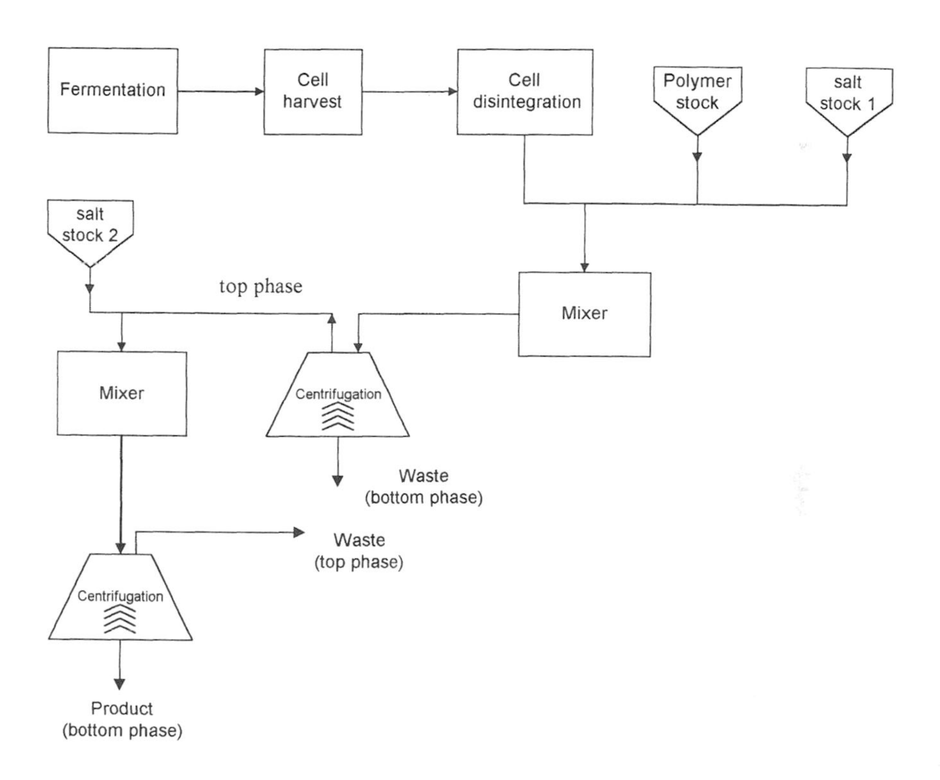

Figure 8 Process scheme for continuous purification of an intracellular recombinant protein, with aqueous two-phase systems. In the first step the protein is enriched in the top (polymer-rich) phase; in the second extraction step the protein is enriched in the bottom (salt-rich) phase.

25,000 could be easily accomplished, provided adequate mixing is maintained to achieve satisfactory mass transfer/equilibrium (Raghavarao et al. 1995). Furthermore, the scale-up is facilitated as the equipment used in ATPSs is of the same type as used for extraction in the chemical industry.

5.2 Reversed Micelle Systems

Most of the studies reported on the extraction of proteins with reversed micelles have been with pure proteins as model systems (Table 8). For the application of reversed micelles in large-scale there is a need to demonstrate the ability to recover partially purified proteins from biological complex media (Table 9).

The process of recovery of a given protein produced by fermentation using liquid-liquid extraction with reversed micelles is outlined in Fig. 9. Two strategies can be followed. In the usual situation (strategy 1), the method involves two consecutive steps: (1) the protein is extracted from the fermentation broth into the reversed micellar phase through contact of the extract with the organic solution and (2) the encapsulated protein is back-extracted to a fresh aqueous phase by contact of the loaded reversed micellar phase with a new aqueous solution. Alternatively (strategy 2), after the first step, the protein is recovered from the organic phase by inducing a phase change in the reversed micelle system by decreasing the temperature.

Studies conducted by Dekker et al. (1986) for the continuous extraction of α-amylase provide promising results for a continuous-mode operation using mixer-settler units. The extraction apparatus can be designed to provide maximal interface area and concentration of the bioproducts (ratio of volumes between aqueous and organic phases).

Scale-up studies were performed using a reversed micelle system of BDBAC/octane/hexanol for the extraction of α-amylase from 2 L clarified fermentation broth (Krei et al. 1995). The activity yield obtained for the forward extraction was 70% (approximately 15–20% lower than that observed in 10 ml scale). After back extraction, the overall yield drops further to about 60%. About 10–15% of activity was lost per extraction step in the stirred vessel, probably due to protein denaturation. The denaturation is probably due to enhanced shear forces generated by the stirrer and differences in the transfer rate between small scale and large scale (Krei et al. 1995.

A continuous perforated rotating disk contactor was used for the extraction of a recombinant cutinase from different complex media, namely, fermentation broth and supernatants after cell disruption by sonication and osmotic shock, to a reversed micellar phase of AOT in iso-octane (Carneiro-da-Cunha et al. 1994a,b, 1996). Cutinase was recovered with activity yields and purification factors ranging from about 5% to 50% and 1.2 and 10.2,

Table 8 Examples of Proteins Extracted into Reversed Micelles

Protein	pI	MW	Surfactant	Ref.
albumin	4.6	65,000	AOT	Wolbert et al. (1989), Armstrong and Li (1988)
alcohol dehydrogenase	5.7	68,000	TOMAC	Wolbert et al. (1989)
alkaline protease	9	33,000	TOMAC	Wolbert et al. (1989)
α-amylase	4.6	55,000	TOMAC; CTAB; BDBAC; CPB; CPC; AOT; ALIQUAT 336	Göklen and Hatton (1987), Wolbert et al. (1989), Krei and Hustedt (1992), Dekker et al. (1989), Hilhorst et al. (1995), Chang and Chen (1995), Chang et al. (1997)
carbonic anhydrase	4.4	31,000	DDAB	Göklen (1986), Wolbert et al. (1989)
α-chymotrypsin	6.0	25,000	AOT; TOMAC; NaDEHP	Wolbert et al. (1989), Marcozzi et al. (1991), Luisi et al. (1977), Luisi et al. (1979), Hentsch et al. (1992), Jolivalt (1990), Chang et al. (1994), Paradkar et al. (1994), Hentsch et al. (1996), Hu and Gulari (1996)
cytochrome b_5	4.4	13,600	CTAB; AOT	Pires and Cabral (1993), Pires et al. (1996)
cytochrome c	9.4	12,000	DDAB; Tween-85/ Span, AOT; NaDEHP; DOLPA	Göklen and Hatton (1987), Wolbert et al. (1989), Ishikawa et al. (1992), Ayala et al. (1992), Hentsch et al. (1996), Hu and Gulari (1996), Goto et al. (1997)
cytochrome c_3	5.2; 7.0; 10.2	13,000	AOT	Castro and Cabral (1988)
cutinase	7.8	22,000	AOT	Carneiro-da-Cunha et al. (1994a,b)
elastase	7.6	25,000	AOT	Göklen (1986)
flavodoxin	3.7	15,000	TOMAC	Wolbert et al. (1989)

Table 8 Continued

Protein	pI	MW	Surfactant	Ref.
hemoglobin	4.4–5.4	60,000–64,000	Tween 85/Span; TOMAC; DOLPA	Wolbert et al. (1989), Ayala et al. (1992), Goto et al. (1997)
hexokinsae	4.6–4.0	100,000	TOMAC	Wolbert et al. (1989)
trypsin inhibitor	4.5	24,500	TOMAC	Wolbert et al. (1989)
insulin	4.6	12,000	TOMAC	Wolbert et al. (1989)
lipase-a	3.6	120,000	AOT	Aires-Barros and Cabral (1991), Camarinha-Vicente et al. (1990)
lipase b	6.3	30,000	AOT	Aire-Barros and Cabral (1991), Camarinha-Vicente et al. 91990)
lysozyme	11.0	14,400	AOT; TOMAC	Göklen and Hatton (1987), Wolbert et al. (1989), Chang et al. (1994), Hentsch et al. (1996), Imai et al. (1997)
parvalbumin	4.3	12,000	TOMAC	Wolbert et al. (1989)
pepsin	< 1.0	34,000	TOMAC; AOT	Wolbert et al. (1989), Luisi et al. (1979), Chang et al. (1994)
peroxidase	5.6	40,000	TOMAC	Wolbert et al. (1989)
renin	4.6	43,000	AOT	Göklen (1986)
ribonuclease a	6.7	13,700	DDAB; TOMAC	Göklen and Hatton (1987), Wolbert et al. (1989)
rubredoxin	3.2	5,000	TOMAC	Wolbert et al. (1989)
superoxide dismutase	4.3	14,500	TOMAC	Wolbert et al. (1989)
trypsin	9.4	25,000	AOT; TOMAC	Göklen (1986), Wolbert et al. (1989), Luisi et al. (1979), Chang and Chen (1995), Hentsch et al. (1996)

Table 9 Examples of Protein Extraction from Complex Media Using Reversed Micelles

Protein	Organism	Surfactant	No. steps	Yield (%)	Purif. Factor	Ref.
Alkaline protease	*Bacillus* sp.	AOT	3	56	6	Rahaman et al. (1988)
α-Amylase		CTAB;	2	89	8.9	Krei and Hustedt (1992)
		BDBAC;	1	90	3.5	Krei et al. (1995)
		Aliquat 336	1	85	1.5	Chang and Chen (1995)
Peroxidase	*Bacillus licheniformis* *Bacillus amyloliquefaciens*	AOT	1	77	19	Huang and Lee (1994)
Cutinase	*Bacillus subtilis*	AOT	1	52.5	10.2	Carneiro-da-Cunha et al. (1996)
Inulinase	*Kluveromyces marxianus*	BDBAC	1	77	2.8	Pessoa and Vitolo (1997)
Lipase	*Penicillium citrinum*	AOT	1	68[a]	810[a]	Krieger et al. (1997)

After hydrophobic interaction chromatography.

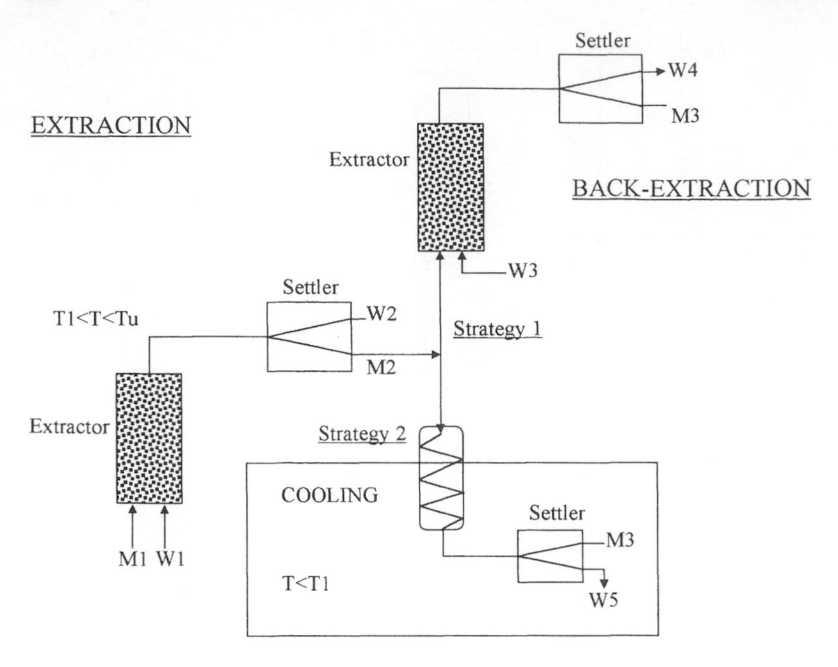

Figure 9 Liquid-liquid extraction of proteins with reversed micelles: M, reversed micellar phase; W, water phase, T_u and T_l upper and lower temperatures of phase transition. W1 is the aqueous feed, containing the protein to be extracted into M1. M2 is the collected reversed micellar phase containing the protein to be back-extracted into W4, through the aqueous feed W3 (strategy 1), or to be collected in its microaqueous phase W5, through a phase change, induced by a cooling step.

respectively, depending on the biological medium. The highest activity yield (52.5%) and purification factor (10.2) were obtained for the recovery of cutinase from fermentation medium diluted with 50 mM Tris-HCl buffer solution at pH 9.0 with 100 mM KCl. The cutinase was extracted to the organic phase with a protein yield of 78% after 70 min of operation (Carneiro-da-Cunha et al. 1994a,b). The results show that this type of equipment can be successfully used as continuous extraction equipment for protein and enzyme extraction using reversed micellar systems.

Extraction of a model protein, lysozyme, using reversed micellar phase was successfully carried out in a liquid-liquid spray column (Lye et al. 1996). When operated in a semibatch mode, the presence of surfactant in the organic phase did not lead to problems of emulsion formation within the column. This type of contactor provides one route by which reversed micelle extraction processes may be scaled up.

ABBREVIATIONS AND SYMBOLS

Aliquat 336	quaternary ammonium salt
AOT	sodium dioctylsulfosuccinate
ATPS	aqueous two-phase system
BDBAC	benzyldodecylbishydroxyethylammonium chloride
CF	concentration factor
CPB	cetylpyridinium bromide
CPC	cetylpyridinium chloride
CPT	cloud point temperature
CTAB	cetyltrimethylammonium bromide
DDAB	didodecyldimethylammonium bromide
Dex	dextran polymer
DOLPA	dioleyl phosphoric acid
EOPO	ethylene oxide–propylene oxide copolymer
Hps	hydroxypropyl starch polymer
pI	isoelectric point
K	partition coefficient
LCST	lower critical solution temperature
MW	molecular weight
NaDEHP	sodium bis(2-ethylhexyl)phosphate
PEG	polyethylene glycol
PF	purification factor
PVP	polyvinylpyrrolidone
Span	sorbitan
STL	slope of the tie line
TLL	tie-line length
TOMAC	trioctylmethylammonium chloride
Tween-85	polyoxyethylenesorbitan trioleate
VR	volume ratio
Y	yield

REFERENCES

Abbot N.L., Blankschtein D., Hatton T.A. (1990) On protein partitioning in two-phase aqueous polymer systems. Bioseparation 1, 191–225.

Adachi M., Harada M. (1993) Solubilization mechanism of cytochrome c in sodium bis(2-ethylhexyl) sulfosuccinate water/oil microemulsion. J. Phys. Chem. 97, 3631–3640.

Adachi M., Harada M., Shioi A., Takahashi H., Katoh S. (1996) Selective separation of concanavalin A using bioaffinity in reverse micellar system. ISEC '96, Melbourne, Australia, pp. 1393–1398.

Adachi M., Yamazaki M., Harada M., Shioi A., and Katch S. (1997) Bioaffinity separation of trypsin using trypsin inhibitor immobilized in reverse micelles composed of a nonionic surfactant. Biotechnol. Bioeng. 53, 406–408.

Aires-Barros M.R., Cabral J.M.S. (1991) Selective separation and purification of two lipases from *Chromobacterium viscosum* using AOT reversed micelles. Biotechnol. Bioeng. 38, 1302–1307.

Albery W.J., Choudhery R.A., Atay N.Z., Robinson B.H. (1987) Rotating diffusion cell studies of microemulsion kinetic. J. Chem. Soc. Faraday Trans. I 83, 2407–2419.

Albertsson P. (1986) Partition of Cell Particles and Macromolecules, 3rd ed., Wiley, New York.

Alred P.A., Kozlowski A., Harris J.M., Tjerneld F. (1994) Application of temperature-induced phase partitioning at ambient temperature for enzyme purification. J. Chromatogr. A 659, 289–298.

Alred P.A., Tjerneld F., Kozlowski A., Harris J.M. (1992) Synthesis of dye conjugates of ethylene oxide-propylene oxide copolymers and application in temperature-induced phase partitioning. Bioseparation 2, 363–373.

Ananthapadmanabhan K.P., Goddard E.D. (1986) A correlation between clouding and aqueous biphase formation in polethylene oxide/inorganic salt systems. J. Colloid Int. Sci. 113, 294–296.

Ananthapadmanabhan K.P., Goddard E.D. (1987) Aqueous biphase formation in polyethylene oxide-inorganic salt system. Langmuir 3, 25–31.

Andrews B.A., Head D.M., Dunthorne P., Asenjo J.A. (1990) PEG activation and ligand binding for the affinity partitioning of proteins in aqueous two-phase systems. Biotechnol. Techn. 4, 49–54.

Armstrong D.W., Li W. (1988) Highly selective protein separations with reversed micellar liquid membranes. Anal. Chem. 60, 86–88.

Ayala G.A., Kamat S., Beckmen E.J., Russel A.J. (1992) Protein extraction and activity in reverse micelles of a non-ionic detergent. Biotechnol. Bioeng. 39, 806–814.

Bailey F.E., Koleske J.V. (1976) Poly(ethylene oxide). Academic Press, New York.

Bamberger S., Seaman G.V.F., Brown J.A., Brooks D.E. (1984) The partition of sodium phosphate and sodium chloride in aqueous dextran poly(ethylene glycol) two-phase systems. J. Colloid Int. Sci. 99, 187–193.

Barker T.B. (1985) Quality by Experimental Design. Marcel Dekker, New York.

Barlow D.J., Thornton J.M. (1986) The distribution of charge groups in proteins. Biopolymers 25, 1717–1733.

Bausch T.E., Plucinski P.K., Nitsch W. (1992) Kinetics of the reextraction of hydrophilic solutes out of AOT-reversed micelles. J. Colloid Interface Sci. 150, 226–234.

Berggren K., Veide A., Nygren P.Å., Tjerneld F. (1999) Genetic engineering of protein-peptide fusions for control of protein partitioning in thermoseparating aqueous two-phase systems. Biotechnol. Bioeng. 62, 135–144.

Berggren K., Nilsson A., Johansson G., Bandmann N., Nygren P.Å., Tjerneld F. (2000a) Partitioning of peptides and recombinant protein-peptide fusions in

thermoseparating aqueous two-phase systems: effect of peptide primary structure. J. Chromatogr. A. 743, 295–306.
Berggren K., Egmond M.R., Tjerneld F. (2000b) Substitutions of surface amino acid residues of cutinase probed by aqueous two-phase partitioning (submitted).
Bergmann C.A., Landmeier B.J., Kaplan D.R. (1991) Phase separation analysis of interleukine 2. Mol. Immunol. 28, 99–105.
Boland M.J., Hesselink P.G.M., Papamichael N., Hustedt H. (1991) Extractive purification of enzymes from animal using aqueous two-phase systems: pilot scale studies. J. Biotechnol. 19, 19–34.
Bordier C. (1981) Phase separation of integral membrane proteins in Triton X-114 solution. J. Biol. Chem. 25, 1604–1607.
Box G.E.P., Hunter W.G., Hunter J.S. (1978) Statistics for Experimenters. An Introduction to Design, Data Analysis and Model Building. John Wiley, New York.
Bradford M.M. (1976) A rapid and sensitive method for the quantification of microgram quantities of protein utilizing the principle of protein-dye binding. Anal. Biochem. 72, 248–251.
Cabral J.M.S., Aires-Barros M.R. (1993) Reversed micelles in liquid-liquid extraction. In: Recovery Processes for Biological Materials (Kennedy J.F., Cabral J.M.S., eds.). John Wiley, Chichester, pp. 247–271.
Camarinha-Vicente M.L.C., Aires-Barros M.R., Cabral J.M.S. (1990) Purification of *Chromobacterium viscosum* lipases using reverse micelles. Biotechnol. Techn. 4, 137–142.
Carneiro-da-Cunha M.G., Cabral J.M.S., Aires-Barros M.R. (1994a) Studies on the extraction and back-extraction of a recombinant cutinase in a reversed micellar extraction process. Bioprocess Eng. 11, 203–208.
Carneiro-da-Cunha M.G., Aires-Barrow M.R., Tambourgi E., Cabral J.M.S. (1994b) Recovery of a recombinant cutinase with reversed micelles in a continuous perforated rotating disc contractor. Biotechnol. Techn. 8, 413–418.
Carneiro-da-Cunha M.G., Melo E.P., Cabral J.M.S., Aires-Barros M.R. (1996) Liquid-liquid extraction of a recombinant cutinase from fermentation media with AOT reversed micelles. Bioprocess Eng. 15, 151–157.
Carvalho C.M.L., Serralheiro M.L.M., Cabral J.M.S., Aires-Barros M.R. (1997) Application of factorial design to the study of tranesterification reactions using cutinase in AOT reversed micelles. Enzyme Microb. Technol. 21, 117–123.
Carvalho C.M.L., Cabral J.M.S., Aires-Barros M.R. (1999) Cutinase stability in AOT reversed micelles: system optimization using the factorial design methodology. Enzyme Microb. Technol. 24, 569–576.
Castro M.J.M., Cabral J.M.S. (1988) Reversed micelles in biotechnological processes. Biotechnol. Adv. 6, 151–167.
Chakhovskoy N. (1956) Critical saturation temperatures of polyethylene glycol monoether-water mixtures and the influence of a third constituent. Bull. Soc. Chim. Belg. 65, 474–493.

Chang Q., Liu H., Chen J. (1994) Extraction of lysozyme, α-chymotrypsin, and pepsin into reverse micelles formed using anionic surfactant, isooctane, and water. Enzyme Microb. Technol. 16, 970–973.

Chang Q-I., Chen J-Y. (1995) Reversed micelle extraction of trypsin: effect of solvent on the protein transfer and activity recovery, Biotechnol. Bioeng. 46, 172–174.

Chang Q-L., Chen J-Y., Zhang X-F., Zhao N-M. (1997) Effect of the cosolvent type on the extraction of α-amylase with reversed micelles: circular dichroism study. Enzyme Microb. Technol. 20, 87–92.

Chen J-P., Jen J-T. (1994) Extraction of concanavalin A with affinity reversed micelle system. Sep. Sci. Technol. 29, 1115–1132.

Cordes A., Kula M.R. (1986) Process design for large-scale purification of formate dehydrogenase from *Candida boidinii* by affinity partition. J. Chromatogr. 376, 375–384.

Coughlin R.W., Baclaski J.B. (1990) N-Laurylbiotinamide as affinity cosurfactant. Biotechnol. Prog. 6, 307–309.

Cunha M.T., Tjerneld F. (1995) Purification of proteins with temperature induced phase separation. Unpublished.

Cunha M.T., Aires-Barros M.R. (2000) Large-scale extraction of proteins. In: Aqueous Two-Phase Systems. Methods and Protocols (Hatti-Kaul R., ed.). Humana Press, Totowa, NJ, pp. 391–409.

Cunha M.T., Cabral J.M.S., Aires-Barros M.R. (1997) Quantification of phase composition in aqueous two-phase systems of Breox/phosphate and Breox/ Reppal PES 100 by isocratic HPLC. Biotechnol. Tech. 11, 351–353.

Cunha M.T., Tjerneld F., Cabral J.M.S., Aires-Barros M.R. (1998) Effect of electrolytes and surfactants on the thermoseparation of an ethylene oxide-propylene oxide random copolymer in aqueous solution. J. Chromatogr. B 711, 53–60.

Cunha M.T., Cabral J.M.S., Aires-Barros M.R., Tjerneld F. (2000) Effect of salts and surfactants on the partitioning of *Fusarium solani pisi* cutinase in aqueous two-phase systems of thermoseparating ethylene oxide/propylene oxide random copolymer and hydroxypropyl starch. Biosepacetion 9, 203–209.

Curtis R.A., Prausnitz J.M., Blanch H.W. (1998) Protein-protein and protein-salt interactions in aqueous protein solutions containing concentrated electrolytes. Biotechnol. Bioeng. 57, 11–21.

Datar R., Rosen C.G. (1986) Studies on the removal of *Escherichia coli* cell debris by aqueous two-phase polymer extraction. J. Biotechnol. 3, 207–219.

Dekker M., van't Riet K., Weijers S.R., Baltussen J.W.A., Laane C., Bijsterbosh B.H. (1986) Enzyme recovery by liquid-liquid extraction using reversed micelles. Chem. Eng. J. 33, B27–B33.

Dekker M., Hilhorst R., Laane C. (1989) Isolating enzymes by reversed micelles. Anal. Biochem. 178, 217–226.

Diamond A.D., Hsu J.T. (1989) Phase diagrams for Dextran-PEG aqueous two-phase systems at 22°C. Biotechnol. Techn. 3, 119–124.

Diamond A.D., Hsu J.T. (1990) Protein partitioning in PEG/dextran aqueous two-phase systems. AIChE J. 36, 1017–1024.

Diamond A.D., Hsu J.T. (1992) Aqueous two-phase systems for biomolecule separation. In: Advances in Biochemical Engineering/Biotechnology (Fiechter A., ed.). Springer-Verlag, Berlin, pp. 89–135.

Dungan S.R., Bausch T., Hatton T.A., Plucinski P., Nitsch W. (1991) Interfacial transport processes in the reversed micellar extraction of proteins. J. Colloid Interface Sci. 145, 33–50.

Elling L., Kula M-R., Hadas E., Katchalski-Katzir E. (1991) Partition of free and monoclonal-antibody-bound horseradish peroxidase in a two-phase aqueous polymer system. Novel procedure for the determination of the apparent binding constant of monoclonal antibody to horseradish peroxidase. Anal. Biochem. 192, 74–77.

Evans D.F. (1994) The Colloidal Domain: Where Physics, Chemistry, Biology and Technology Meet. VCH, New York.

Fletcher P.D.I., Howe A.M., Robinson B.H. (1987) The kinetics of solubilisate exchange between water droplets of water-in-oil microemulsion. J. Chem. Soc. Faraday Trans. 83, 985–1006.

Florin E., Kjellander R., Eriksson J.C. (1983) Salt effects on the cloud point of the poly(ethylene oxide) + water system. J. Chem. Soc. Faraday Trans. 80, 2889–2910.

Forciniti D., Hall C.K., Kula M.R. (1991a) Influence of polymer molecular weight and temperature on phase composition in aqueous two-phase systems. Fluid Phase Equilibria 61, 243–262.

Forciniti D., Hall C.K., Kula M.R. (1991b) Protein partition at the isoelectric point: influence of polymer molecular weight and concentration and protein size. Biotechnol. Bioeng. 38, 986–994.

Forciniti D., Hall C.K., Kula M.R. (1991c) Temperature dependence of the partition coefficient of proteins in aqueous two-phase systems. Bioseparation 2, 115–128.

Forland G.M., Samseth J., Gjerde M.I., Hoiland H., Jense A.O., Mortensen K. (1998) Influence of alcohol on the behaviour of sodium dodecylsulfate micelles. J. Colloid Interface Sci. 203, 328–334.

Fujimatsu H., Ogasawara S., Kuroiwa S. (1988) Lower critical solution temperature and theta temperature of aqueous solutions of nonionic surface active agents of various polyoxyethylene chain lengths. Colloid Polym. Sci 266, 594–600.

Galaev I.Y., Mattiasson B. (1993) Thermoreactive water-soluble polymers, nonionic surfactants, and hydrogels as reagents in biotechnology. Enzyme Microb. Technol. 15, 354–366.

Göklen K.E., Hatton T.A. (1985) Protein extraction using reverse micelles. Biotechnol. Prog. 1, 69–74.

Göklen K.E. (1986) Liquid-Liquid Extraction of Biopolymers: Selective Solubilization of Proteins in Reverse Micelles. Doctoral thesis, Massachusetts Institute of Technology.

Göklen K.E., Hatton T.A. (1987) Liquid-liquid extraction of low molecular proteins by selective solubilization in reversed micelles. Sep. Sci. Technol. 22, 831–841.

Goto M., Ono T., Nakashio F., Hatton T.A. (1997) Design of surfactants suitable for protein extraction by reversed micelles. Biotechnol. Bioeng. 54, 26–32.

Greve A., Kula M.R. (1991) Recycling of salts in partition protein extraction processes. J. Chem. Tech. Biotechnol. 50, 27–42.

Guan Y., Wu X., Treffry T.E., Lilley T.H. (1992) Studies on the isolation of penicillin acylase from *Escherichia coli* by aqueous two-phase partitioning. Biotechnol. Bioeng. 40, 517–524.

Guan, Y., Lilley, T.H., Treffry, T.E., Zhou, C., and Wilkinson, P.B. (1996) Use of aqueous two-phase systems in the purification of human interferon-alpha 1 from recombinant *Escherichia coli*. Enzyme Microb. Technol. 19:446–455.

Harris P.A., Karlström G., Tjerneld F. (1991) Enzyme purification using temperature-induced phase formation. Bioseparation 2, 237–246.

Hart R.A., Lester P.M., Reifsnyder D.H., Ogez J.R., Builder S.E. (1994) Large scale in situ isolation of periplasmic IGF-I from *E. coli*. Bio/Technology 12, 1113–1117.

Hart R.A., Ogez J.R., Builder S.E. (1995) Use of multifactorial analysis to develop aqueous two-phase systems for isolation of non-native IGF-I. Bioseparation 5, 113–121.

Hatton T.A. (1989) In Surfactant-Based Processes, Vol. 33 (Scamehorn J.F., Horwell J.H., eds.), Marcel Dekker, New York, pp. 55–90.

Hayes D.G., Gulari E. (1990) Esterification reactions in reverse micelles. Biotechnol. Bioeng. 35, 793–801.

Helenius A., Simons K. (1975) Solubilization of membranes by detergents. Biochim. Biophys. Acta. 415, 29–70.

Hentsch M., Menoud P., Steiner L., Flaschel E., Renken A. (1992) Optimization of the surfactant (AOT) concentration in a reverse micellar extraction process. Biotechnol. Techn. 6, 359–364.

Hentsch M., Warnery P., Renken A., Flaschel E. (1996) Study of the simultaneous partitioning of proteins and surfactant in reverse micelle two-phase systems: potential impact upon purification efficiency. Bioseparation 6, 67–76.

Hilhorst R., Sergeeva M., Heering D., Rietveld P., Fijneman P., Wolbert R.B.G., Dekker M., Bijsterbosch B.H. (1995) Protein extraction from an aqueous phase into a reversed micelle phase: effect of water content and reversed micelle composition. Biotechnol. Bioeng. 46, 375–387.

Hou M.J., Kim M., Shah D.O. (1988) A light scattering on the droplet size and interdroplet interactions in microemulsions of AOT-oil-water system, J. Colloid. Interface Sci. 123, 398–412.

Hu Z., Gulari E. (1996) Protein extraction using the sodium bis (2-ethylhexyl) phosphate (NaDEHP) reverse micelle system. Biotechnol. Bioeng. 50, 203–206.

Huang S-Y., Chang H-L. (1995) Enhancement of enzyme through temperature alternation in reverse micelle extraction. Bioseparation 5, 225–234.

Huang S-Y., Lee Y-C. (1994) Separation and purification of horseradish peroxidase from *Armoracia rusticana* root using reversed micelle extraction. Bioseparation 4, 1–5.

Huddleston J.G., Ottomar K.W., Ngonnyani D.M., Lyddiat A. (1991) Influence of system and molecular parameters upon fractionation of intracellular proteins from *Saccharomyces* by aqueous two-phase partition. Enzyme Microb. Technol. 13, 24–32.

Hustedt H., Kroner K.H., Menge U., Kula M.R. (1985) Protein recovery using two-phase systems. Trends Biotechnol. 3, 139–143.

Hustedt H. (1986) Extractive enzyme recovery with simple recycling of phase forming chemicals. Biotechnol. Lett. 8, 791–796.

Hustedt H., Kroner K.H., Papamichael N. (1988) Continuous cross-current aqueous two-phase extraction of enzymes from biomass; automated recovery in production scale. Process Biochem 23, 129–137.

Imai M., Natsume T., Naoe K., Shimizu M., Ichikawa S., Furusaki S. (1997) Hydrophilic surroundings for the solubilization of proteins related with their hydrophobicity in the AOT reversed micelle extraction. Bioseparation 6, 325–333.

Ishikawa H., Noda K., Oka T. (1990) Kinetic properties of enzymes in AOT-isooctane reversed micelles. J. Ferment. Bioeng. 6, 381–385.

Ishikawa S., Imai M., Shimizu M. (1992) Solubilizing water involved in protein extraction using reversed micelles. Biotechnol. Bioeng. 39, 20–26.

Johansson G., Hartman A., Albertsson P. (1973) Partition of proteins in two-phase systems containing charged poly(ethylene glycol). Eur. J. Biochem. 33, 379–386.

Johansson H.-O., Karlström G., Tjerneld F. (1993) Experimental and theoretical study of phase separation in aqueous solutions of clouding polymers and carboxylic acids. Macromolecules 26, 4478–4483.

Johansson H-O., Karlström G., Mattiasson B., Tjerneld F. (1995) Effects of hydrophobicity and counter ions on the partitioning of amino acids in thermoseparating Ucon-Water two-phase systems. Bioseparation 5, 269–279.

Johansson H-O., Lundh G., Karlström G., Tjerneld F. (1996) Effects of ions in partitioning of serum albumin and lysozyme in aqueous two-phase systems containing ethylene oxide/propylene oxide co-polymers. Biochim. Biophys. Acta 1290, 289–298.

Johansson H-O., Karlström G., Tjerneld F. (1997) Effect of solute hydrophobicity on phase behaviour in solutions of thermoseparating polymers. Colloid Polym. Sci. 257, 458–466.

Jolivalt C., Minier M., Reinon R. (1990) Extraction of α-chymotrypsin using reverse micelles. J. Colloid Interface Sci. 135, 86–96.

Kelley B.D., Wang D.I.C., Hatton T.A. (1993a) Affinity-based reversed micellar protein extraction: I. Principles and protein-ligand systems. Biotechnol. Bioeng. 42, 1199–1208.

Kelley B.D., Wang D.I.C., Hatton T.A. (1993b) Affinity-based reversed micellar protein extraction: II. Effect of cosurfactant tail length. Biotechnol. Bioeng. 42, 1209–1217.

Kenjo K. (1966) Colloid chemical studies of phase diagrams of binary polyoxyethylene nonylphenyl ether-water systems. Bull. Chem. Soc. Jpn 39, 685–694.

Köhler K., Veide A., Enfors S. (1991a) Partitioning of β-galactosidase fusion proteins in PEG/potassium phosphate aqueous two-phase systems. Enzyme Microb. Technol. 13, 204–209.

Köhler K., Ljungquist C., Kondo A., Veide A., Nilsson B. (1991b) Engineering proteins to enhance their partitioning coefficients in aqueous two-phase systems. Bio/Technol. 9, 642–646.

Krei G.A., Hustedt H. (1992) Extraction of enzymes by reverse micelles. Chem. Eng. Sci. 47, 99–111.

Krei G., Meyer U., Borner B., Hustedt H. (1995) Extraction of α-amylase using BDBAC-reversed micelles. Bioseparation 5, 175–183.

Krieger N., Taipa M.A., Aires-Barros M.R., Melo E.H.M., Lima-Filho J.L., Cabral J.M.S. (1997) Purification of the *Penicillium citrinum* lipase using AOT reversed micelles. J. Chem. Tech. Biotechnol. 69, 77–85.

Kroner K.H., Stach W., Schutte H., Kula M.R. (1982) Scale-up of formate dehydrogenase isolation by partition. J. Chem. Technol. Biotechnol. 32, 130–137.

Kula M.R. (1985) Liquid-liquid extraction of biopolymers. In: Comprehensive Biotechnology. The Principles of Biotechnology: Engineering Considerations (Cooney C.L., Humphrey A.E., eds.). Pergamon Press, New York, Vol. 28, pp. 451–471.

Kula M.R., Kroner K.H., Hustedt H. (1982) Purification of enzymes by liquid-liquid extraction. In: Advances in Biochemical Engineering (Fiechter A., ed.). Springer-Verlag, Berlin, Vol. 24, pp. 73–118.

Labrou N., Clonis Y.D. (1994) The affinity technology in downstream processing. J. Biotechnol. 36, 95–119.

Lang J.C., Morgan R.D. (1980) Nonionic surfactant mixtures. I Phase equilibriums in 3, 6, 9, 12-tetraoxadocosanol-water and closed loop coexistence. J. Chem. Phys. 73, 5849–5861.

Leser E.M., Wei G., Luisi P.L., Maestro M. (1986). Application of reversed micelles for the extraction of proteins. Biochem. Biophys. Res. Commun. 135, 629–635.

Leydet A., Boyer B., Lamaty G., Roque J.P. (1994) Nitrogen analogs of AOT. Synthesis and properties. Langmuir 10, 1000–1002.

Liu C.-L., Nikas Y.J., and Blankschtein D. (1996) Novel bioseparations using two-phase aqueous micellar systems. Biotechnol. Bioeng. 52, 185–192.

Louai A., Sarazin D., Pollet G., François J., Moreaux F. (1991) Effect of additives on solution properties of ethylene oxide-propylene oxide statistical copolymers. Polymer 32, 713–720.

Lu M., Albertsson P., Johansson G., Tjerneld F. (1994) Partitioning of proteins and thylakoid membrane vesiscles in aqueous two-phase system with hydrophobically modified dextran. J. Chromatogr. A 668, 215–228.

Luisi P.L., Henninger F., Joppich M., Dossena A., Casnati G. (1977) Solubilization and spectroscopic properties of α-chymotrypsin in cyclohexane. Biochem. Biophys. Res. Commun. 74, 1384–1389.

Luisi P.L., Bonner F.J., Pellegrini A., Wiget P.E., Wolf R. (1979) Micellar solubilization of protein in aprotic solvents and their spectroscopic characterization. Helv. Chim. Acta 62, 740–752.

Luisi P.L., Imre V.E., Kaeckle H., Pande H. (1983) TITLE In: Topics in Pharmaceutical Sciences (Breiner D.D., Speiser P., eds.). Elsevier, Amsterdam, pp. 243–254.

Luisi P.L., Magid L.J. (1986) Solubilization of enzymes and nucleic acids in hydrocarbon micellar solution. Crit. Rev. Biochem. 20, 409–474.

Luther J.R., Glatz C.E. (1994) Genetically engineered charge modifications to enhance protein separation in aqueous two-phase systems: electrochemical partitioning. Biotechnol. Bioeng. 44, 147–153.

Luther J.R., Glatz, C.E. (1995) Genetically engineered charge modifications to enhance protein separation in aqueous two-phase systems: charge directed partitioning. Biotechnol. Bioeng. 46, 62, 68.

Lye G.J., Asenjo J.A., Pyle D.L. (1996) Reverse micelle mass-transfer process: spray column extraction of lysozyme. AIChE J. 42, 713–726.

Maitra A. (1984) Determination of size parameters of water-Aerosol OT-oil reverse micelles from their nuclear magnetic resonance data. J. Phys. Chem. 8, 5122–5125.

Marcozzi G., Correa N., Luisi P.L., Caselli M. (1991) Protein extraction by reverse micelles: a study of the factors affecting the forward and backward transfer of α-chymotrypsin and its activity. Biotechnol. Bioeng. 38, 1239–1246.

Mat H.B., Stukey D.C. (1993) The effect of solvents on water solubilization and protein partitioning in reverse micelle system. In: Solvent Extraction in the Process Industry, Vol. 2 (Logsdail D.H., Slater M.J., eds.). Elsevier, London, pp. 848–855.

Meier, P., Imre E., Flesher M., Luisi P.L. (1984) Further investigations on the micelle solubilization of biopolymers in apolar solvents. In: Surfactants in Solution (Mital K.L., Lindam B., eds.). Plenum Press, New York, pp. 999–1012.

Melo E.P., Costa S.M.B., Cabral J.M.S. (1996) Denaturation of a recombinant cutinase from *Fusarium solani* in AOT-iso-octane reversed micelles. Appl. Biochem. Biotechnol 50, 45–56.

Melo E.P., Carvalho, C.M.L., Aires-Barros M.R., Costa S.M.B., Cabral J.M.S. (1998) Deactivation and conformational changes of cutinase reverse micelles. Biotechnol. Bioeng. 58, 380–386.

Menge U., Morr M., Mayr U., Kula M.R. (1983) Purification of human fibroblast interferon by extraction in aqueous two-phase systems. J. Appl. Biochem. 5, 75–90.

Minuth T., Thömmes J., Kula M.R. (1996) A closed concept for purification of the membrane bound cholesterol-oxidase from *Nocardia rhodochrous* by surfactant-based cloud-point extraction, organic solvent extraction and anion-exchange chromatography. Biotechnol. Appl. Biochem. 23, 107–116.

Minuth T., Gieren H., Pape U., Raths H.C., Thömmes J., Kula M.R. (1997) Pilot scale processing of detergent-based aqueous two-phase systems. Biotechnol. Bioeng. 55, 239–347.

Mottl H., Keck W. (1991) Purification of penicillin-binding protein 4 of *Escherichia coli* as a soluble protein by dye-affinity chromatography. Eur. J. Biochem. 200, 767–772.

Müller W. (1985) Partitioning of nucleic acids. In: Partitioning in Aqueous Two-Phase Systems: Theory, Methods, Uses, and Applications to Biotechnology (Walter H., Brooks D.E., Fisher D., eds.). Academic Press, Orlando, Florida, pp. 227–266.

Papamichael N., Börner B., Hustedt H. (1992) Continuous aqueous phase extraction of proteins: automated processing and recycling of process chemicals. J. Chem. Technol. Biotechnol. 51, 47–55.

Paradkar V.M., Dordick J.S. (1991) Purification of glycoprotein by selective transport using concanavalin-mediated reverse micellar extraction. Biotechnol. Prog. 7, 330–334.

Paradkar V.M., and Dordick J.S. (1994) Mechanism of extraction of chymotrypsin into isooctane at very low concentrations of aerosol OT in the absence of reversed micelles. Biotechnol. Bioeng. 43, 529–540.

Persson J., Nyström L., Ageland H., Tjerneld F. (1998) Purification of recombinant apolipoprotein A-1_{Milano} expressed in *Escherichia coli* using aqueous two-phase extraction followed by temperature-induced phase separation. J. Chromatogr. B 711, 97–109.

Pessoa Jr., A., Vitolo M. (1997) Separation of inulinase from *Kluyveromyces marxianus* using reversed micelle extraction. Biotechnol. Techn. 11, 421–422.

Piculell L., Lindman B. (1992) Association and segregation in aqueous polymer/polymer, polymer/surfactant, and surfactant/surfactant mixtures: similarities and differences. Adv. Colloid Interface Sci. 41, 149–178.

Pires M.J., Cabral J.M.S. (1993) Liquid-liquid extraction of a recombinant protein with a reverse micelle phase. Biotechnol. Prog. 9, 647–650.

Pires M.J., Martel P., Baptista A., Petersen S.B., Willson R., Cabral J.M.S. (1994) Improving protein extraction in reversed micellar systems through surface charge engineering. Biotechnol. Bioeng. 44, 773–780.

Pires M.J., Aires-Barros M.R., Cabral J.M.S. (1996) Liquid-liquid extraction of proteins with reversed micelles. Biotechnol. Prog. 12, 290–301.

Plunkett S.D., Arnold F.H. (1990) Metal affinity extraction of human hemoglobin in an aqueous polyethylene glycol–sodium sulfate two-phase system. Biotechnol. Techn. 4, 45–48.

Poppenborg L., Flashel E. (1994) Affinity extraction of proteins by means of reverse micellar phases containing metal-chelating surfactant. Biotechnol. Techn. 8, 307–312.

Raghavarao K.S.M.S., Rastogi N.K., Gowthaman M.K., Karanth N.G. (1995) Aqueous two-phase extraction for downstream processing of enzymes/proteins. Adv. Appl. Microbiol. 41, 97–171.

Rahaman R.S., Chee J.Y., Cabral J.M.S., Hatton T.A. (1988) Recovery of an extracellular alkaline protease from whole fermentation broth using micelles. Biotechnol. Prog. 4, 217–224.

Ramelmeier R.A., Terstappen G.C., Kula M.R. (1991) The partitioning of cholesterol oxidase in Triton X-114 based aqueous two-phase systems. Bioseparation 2, 315–324.

Saitoh T., Hinze W.L. (1991) Concentration of hydrophobic organic compounds and extraction of protein using alkylammoniosulfate zwitterionic surfactant mediated phase separations (cloud point extractions). Anal. Chem. 63, 2520–2525.

Sánchez-Ferrer A., Gilabert-Pérez M., Núñez E., Bru R., García-Carmona F. (1994) Triton X-114 phase partitioning in plant protein purification. J. Chromatogr. A 668, 75–83.

Schütte H., Kroner K.H., Hummel W., Kula M.R. (1983) Recent developments in separation and purification of biomolecules. Ann. N.Y. Acad. Sci. 413, 270–282.

Sebastião M.J., Cabral J.M.S., Aires-Barros M.R. (1993a) *Fusarium solani pisi* recombinant cutinase partitioning in PEG/potassium phosphate aqueous two-phase systems. Biotechnol. Techn. 7, 631–634.

Sebastião M.J., Cabral J.M.S., Aires-Barros M.R. (1993b) Synthesis of fatty acid esters by a recombinant cutinase in reversed micelles. Biotechnol. Bioeng. 42, 326–332.

Sebastião M.J., Cabral J.M.S., Aires-Barros M.R. (1996) Improved purification protocol of a *Fusarium solani pisi* recombinant cutinase by phase partitioning in aqueous two-phase systems of polyethylene glycol and phosphate. Enzyme Microb. Technol. 18, 251–261.

Sepulveda L., Macritchie F.J. (1968) Thermodynamics of nonionic-water systems. J. Colloid. Interface Sci. 28, 19–23.

Shanbhag V.P., Axelsson C. (1975) Hydrophobic interaction determined by partition in aqueous two-phase systems. Partition of proteins in systems containing fatty-acids esters of poly(ethylene glycol). Eur. J. Biochem. 60, 17–22.

Shioi A., Harada M., Takahashi H., Adachi M. (1996) Protein extraction in polyoxyethylene alkyl ether/AOT microemulsion systems. ISEC '96, Melbourne, Australia, pp. 1381–1386.

Sjöberg A., Karlström G. (1989) Temperature dependence of the phase equilibria for the system (polyethyleneglycol)/dextran/water. A theoretical and experimental study. Macromolecules 22, 1325–1330.

Sjöberg A., Karlström G., Tjerneld F. (1989) Effects on the cloud point of aqueous poly(ethylene glycol) solutions upon addition of low molecular weight saccharides. J. Am. Chem. Soc. 22, 4512–4516.

Strandberg L., Köhler K., Enfors, S-O. (1991) Large-scale fermentation and purification of a recombinant protein from *Escherichia coli*. Process. Biochem. 26, 225–234.

Suh S., Arnold F.H. (1990) A mathematical model for metal affinity protein partitioning. Biotechnol. Bioeng. 35, 682–690.

Sun Y., Ichikawa S., Sugiura S., Furusaki S. (1998) Affinity extraction of proteins with reversed micellar system composed of Cibacron blue-modified lecithin. Biotechnol. Bioeng. 58, 58–64.

Suzuki M., Kamihira M., Shiraishi T., Takeuchi H., Kobayshi T. (1995) Affinity partitioning of protein a using a magnetic aqueous two-phase system. J. Ferment. Bioeng. 80, 78–84.

Szöke A., Campagna R., Kroner K., Hustedt H. (1988) Improved extraction procedure for the isolation of β-galactosidase from *Escherichia coli*. Biotechnol. Techn. 2, 35–40.

Terstappen G.C., Geerts A.J., Kula M.R. (1992). The use of detergent-based aqueous two-phase systems for the isolation of extracellular proteins: purification of a lipase from *Pseudomonas cepacia*. Biotechnol. Appl. Biochem. 16, 228–235.

Tjerneld F. (1989) New polymers for aqueous two-phase systems. In: Separations Using Aqueous Two-Phase Systems: Applications in Cell Biology and Biotechnology (Fisher D., Sutherland I.A., eds.). Plenum Press, New York, pp. 429–438.

Tjerneld F., Berner S., Cajarville A., Johansson, G. (1986) New aqueous two-phase system based on hydroxypropyl starch useful in enzyme purification. Enzyme Microb. Technol. 8, 417–423.

Veide A., Smeds A.L., Enfors S. (1983) A process for large-scale isolation of β-galactosidase from *E. coli* in an aqueous two-phase system. Biotechnol. Bioeng. 25, 1789–1800.

Veide A., Lindback T., Enfors, S-O. (1984) Continuous extraction of β-galactosidase from *Escherichia coli* in an aqueous two-phase system: effects of biomass concentration on partitioning and mass transfer. Enzyme Microb. Technol. 6, 325–330.

Walter H., Johansson G., Brooks D.E. (1991) Partitioning in aqueous two-phase systems: recent results. Anal. Biochem. 197, 1–18.

Wise K.S., Kim M.F. (1987) Major membrane surface proteins of *Mycoplasma hyopneumoniae* selectively modified by covalently bound lipid. J. Bacteriol. 169, 5546–5555.

Wolbert R.B.G., Hilhorst R., Voskuilen G., Nachtegaal H., Dekker M., van't Riet K., Bijsterbosch B.H. (1989) Protein transfer from an aqueous phase into reversed micelles—the effect of protein size and charge distribution. Eur. J. Biochem. 184, 627–633.

Woll J.M., Hatton T.A. (1989) A simple phenomenological thermodynamic model for protein partitioning in reversed micelle systems Bioprocess Eng. 4, 193–199.

Woll J.M., Hatton T.A., Yarmush M.L. (1989) Bioaffinity separations using reversed micellar extraction. Biotechnol. Prog. 5, 57–62.

Zaslavsky, B.Y. (1995) Aqueous Two-Phase Partitioning. Physical Chemistry and Bioanalytical Applications. Marcel Dekker, New York.

Zaslavsky B.Y., Miheeva L.M., Aleschko-Ozhevski Y.P., Mahmudov A.U., Bagirov T.O., Garaev E.S. (1988) Distribution of inorganic salts between the coexisting phases of aqueous polymer two-phase systems. Interrelationship between the ionic and polymer composition of the phases. J. Chromatogr. 439, 267–281.

Zhao G., Xiao J. (1996) Aqueous two-phase system of the aqueous mixtures of cationic-anionic surfactants. J. Colloid Int. Sci. 177, 513–518.

Ziljstra G.M., Michielsen M.J.F., Gooijer C.D., Van der Pol L.A., Tramper J. (1998) IgG and hybridoma partitioning in aqueous two-phase systems containing a dye-ligand. Bioseparation 7, 117–126.

10

Adsorption as an Initial Step for the Capture of Proteins

Robert H. Clemmitt
BioProducts Laboratory, Elstree, Hertfordshire, England

Howard A. Chase
University of Cambridge, Cambridge, England

1. INTRODUCTION TO ADSORPTION AS A PRIMARY RECOVERY OPERATION

Many of the therapeutic and diagnostic products of the biotechnological industry are natural or recombinant proteins. These may be produced in a variety of systems such as bacterial, yeast, or mammalian cell cultures, and for most applications are required in final formulations of high purity. Techniques for the recovery of proteins have different requirements than those of traditional chemical processing. This is in part due to the sensitivity of the native protein structure and function to factors such as temperature, pressure, proteases, and interfacial contact (Kaufmann 1997). The other important difference in the separation of chemically produced compounds is the medium from which the material has to be isolated. Protein sources usually contain particulate material, together with a wide variety of compounds such as other proteins, lipids, and nucleic acids, each in a relatively low concentration, the desired protein being a minor component in the mixture.

Traditional downstream processing, as outlined in Fig. 1a, requires that the feedstream be clarified using centrifugation or filtration and possibly concentrated before application to some form of packed bed for purification. Problems with this strategy include the fact that clarification operations such as centrifugation or microfiltration are often less effective where small particles (Datar and Rosen 1996) or shear-sensitive organisms

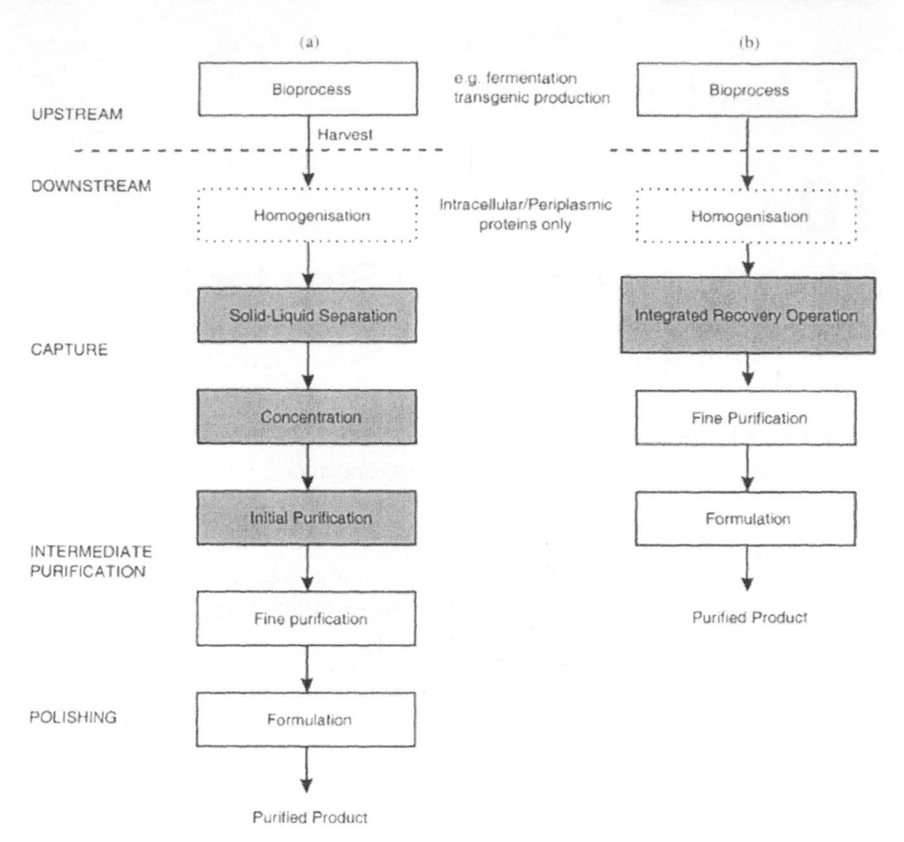

Figure 1 Comparison of downstream processing flowsheets: (a) a conventional scheme and (b) a simplified scheme utilizing an integrated recovery operation.

are involved (Kempken et al. 1995). Indeed, at each step in the fractionation protein is lost, which ultimately leads to a low process yield. Bonnerjea et al. (1986) reviewed 100 papers on protein purification for the year 1984 and found that, on average, four purification steps were required to purify a protein to homogeneity with an overall yield of 28%. The maximal number of steps was 9. At the industrial scale, high-stage yields are difficult to achieve due to further problems such as pipeline losses and time-dependent damage. The introduction of an integrative technique would ensure greater process yields by, for example, reducing the number of stages and removing proteolytic enzymes released during lysis (Barnfield-Frej et al. 1994).

Integrative techniques that on the one hand simplify solid-liquid separation and on the other combine originally independent steps to form new unit operations are substantial improvements on traditional downstream processing (as highlighted in Fig. 1). These integrative operations should therefore be capable of tolerating biological suspensions containing particulates as initial feedstock and deliver a clarified protein concentrate. Ideally a degree of initial purification of the target protein would be achieved, thus combining clarification, concentration, and purification in a single step.

Three industrial approaches to integration are available: aqueous two-phase extraction (Chap. 9), adsorption on derivatized microfiltration membranes (Chap. 6), and expanded bed adsorption (EBA) (Chap. 10). Adsorption operations have been very popular methods for the purification of proteins (Bonnerjea et al. 1988; Sofer and Hagel 1997), and in most cases, the required reduction in volume is obtained by including an adsorptive step in the recovery process. The conventional operating format has been the packed bed. However, the packed bed is not suitable for processing feedstocks containing particulate material. For such feedstreams, alternative contactor designs have been developed, based on batch or continuous loading onto adsorbents, in stirred tanks in fluidized or expanded beds, or into derivatized membranes. Factors that are critical to the success of such downstream processing operations include the careful design of the contactor, the adsorbent (or surface for adsorption), and the operating protocol. With this in mind, the following discussion covers the range of contactors available to achieve the separation and the desirable characteristics of adsorbents and ligands. The focus of the discussion and the majority of the examples concerns the expanded bed concept. This is the unit operation, of all those included, which has the greatest potential and indeed is now evolving into an invaluable tool for downstream processing engineers.

2. OPERATING FORMATS FOR ADSORPTION

A number of different reactor designs exist for the primary recovery of proteins from unclarified feedstocks. Those that operate in a truly continuous fashion, thereby producing a product stream continuously, usually require specialized equipment and operating protocols (Sec. 2.2). The more common format is the batch type of adsorption, in which a batch of feedstock is processed by the separation system at a time and the system runs in campaigns or cycles.

2.1 Batch Adsorption

The most common operating formats for batch adsorption—the stirred tank, the packed bed, the expanded bed, and membranes—are compared in Table 1. Stirred tanks may be used to contact adsorbent particles with

Table 1 Comparison of Operating Formats for Recovery of Proteins by Adsorption

Format	Operating principle	Ability to cope with particulates	Adsorption efficiency	Elution efficiency	Advantages/disadvantages	Ref.
Stirred tank	Stirred-tank contact of liquid and adsorbent	Yes	Poor	Poor	Handling of adsorbent and liquid; and their subsequent separation	Gordon and Cooney 1990; McCreath et al. 1993, 1994; Nigram et al. 1998
Packed bed	Plug flow of liquid through packed adsorbents	No	Good	Good	Widely used and well-understood method	—
Fluidized bed	Mixed flow of liquid through suspended adsorbent; gross mixing in the bed	Yes	Poor	Fair	Recirculating operation may be required to achieve a high degree of capture	Somers et al. 1989
Expanded bed	Near plug flow of liquid through suspended adsorbent; grading of adsorbent on fluidization reducing mixing	Yes	Good	Good	Still lack of adsorbents for operation in crude feedstreams	Barnfield-Frej et al. 1994; Chang et al. 1995; Hansson et al. 1994
Membrane	Cross-flow of liquid across membrane; adsorption through pores of membrane but rejection of debris	Yes	Good	Good	Low surface area and dependence on trans-membrane flux could limit application	Finger et al. 1995; Kroner et al. 1992; Thömmes and Kula 1995

Adapted from Chase 1994.

unclarified feedstocks for a period so as to obtain equilibrium, the adsorbent must then be separated from the depleted feed and the adsorbent washed and product eluted. This approach, although conceptually simple, is the least efficient as it constitutes at best one equilibrium stage, meaning that high partition coefficients are required to utilize the available capacity. Difficulties also arise in mechanical handling of the adsorbent, usually requiring novel and robust matrices. Packed beds cannot process feedstreams containing particulates due to void blinding and the formation of a plug of trapped solids. The resultant pressure drop can lead to adsorbent deformation and pipe failure. Considerable research has gone into the development of contactors based on the principle of the fluidized bed, which maintains plug flow characteristics. Several solutions exist, including compartmentalized fluidized beds (Buijs and Wesselingh 1980; van der Wiel and Wesselingh 1989), magnetically stabilized fluidized beds (Chetty and Burns 1991), and expanded beds. The focus here is on the expanded bed because it is the simplest solution to the problem of minimizing particle mixing and the resultant axial dispersion in the liquid (Chase 1994). In this section the theoretical basis, existing adsorbents and apparatus, applications, and limitations of the technique are discussed. Mention is also made of membrane chromatography, which is also capable of process integration through the recovery of proteins from unclarified feeds, although perhaps not as efficiently as EBA.

Fluidized and Expanded Bed Adsorption

A distinction is usually made between a fluidized bed and an expanded bed (Fig. 2). Liquid-solid fluidized beds with small density differences between the phases show particulate fluidization behavior rather than the aggregative, bubbling behavior seen in gas-solid systems. The mixing in liquid-solid fluidized beds, though less severe, can still be significant, leading to a reduction in the number of theoretical plates. Such columns are usually used in a recycle mode during adsorption in order to achieve significant capture (Thömmes et al. 1995a). The simplest solution to the problem of reducing mixing within the bed is the introduction of a distribution in both particle size and density such that a classification occurs upon fluidization, leading to a stratified bed with the lower buoyant weight particles at the top and the higher buoyant weight particles at the base. The reduction in solid and associated liquid dispersion results in adsorption performance closer to packed beds. The term "expanded bed" has been used extensively in the literature and by manufacturers to describe fluidized beds that have been stabilized by careful adsorbent and equipment design to give limited axial dispersion and higher plate numbers. A number of recent reviews of EBA

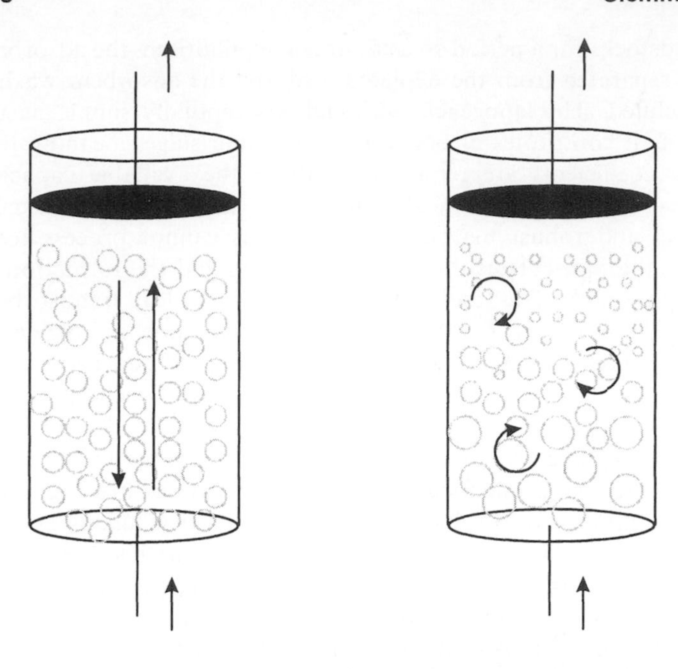

Figure 2 Schematic diagram of the difference between a fluidized and an expanded bed. The arrows within the bed indicate particle motion, while the overall flow is upward.

have been published, covering operational and design aspects, as well as more theoretical considerations (Chase 1994; Hjorth 1997; Thömmes 1997).

Fluidization. In order to achieve fluidization, the superficial velocity, u, of the fluid through the bed must exceed the minimum velocity, u_{mf}, for incipient fluidization but not exceed the terminal settling velocity of the particle, u_t, above which the particle is elutriated from the bed. This range of flow rates, which result in a stable fluidized bed, is dependent on the properties of the adsorbent particles and the fluidizing liquid. The terminal velocity of the particle can be estimated using the Stokes equation:

$$u_t = \frac{d_p^2 g(\rho_p - \rho)}{18\mu} \quad (1)$$

where d_p is the particle diameter, g the acceleration due to gravity, ρ_p the particle density, ρ the liquid density, and μ the liquid viscosity. This equation is valid only where the particle Reynolds number (Re_p) is less than 0.2, this quantity being defined as:

$$Re_p = \frac{u d_p \rho}{\epsilon \mu} \tag{2}$$

where u is the superficial velocity of the fluidizing liquid and ϵ is the overall bed voidage. The Re_p of the particles discussed here for use in EBA are of the order of 0.2 (Johansson 1994). The minimal fluidization velocity, u_{mf}, may be calculated using either the Ergun equation (e.g., in Bird et al. 1960) or velocity-voidage relationships (Riba et al. 1978). The expansion characteristics of a fluidized bed are most simply described using the correlation of Richardson and Zaki (1954).

$$\frac{u}{u_t} = \epsilon^n \tag{3}$$

where n is an index and u_t is the terminal velocity. These two parameters are usually determined experimentally by fitting a double log plot of ϵ and u to data measuring the bed height at various superficial velocities. This set of equations has been used successfully to characterize several matrices used in expanded bed adsorption (Chang et al. 1995; Chase and Draeger 1992; Finette et al. 1996; Thömmes et al. 1995a, 1996).

It should be noted that in reality most of the adsorbents used have a range of diameters and densities, and modified equations can be used to describe the system better (Al-Dibuoni and Garside 1979; Couderc 1985; Kennedy and Bretton 1966). Bed expansion can also be affected or controlled using features such as mechanical stabilizers, including static mixers (Fanquex et al. 1984), sieve plates (Buijs and Wesselingh 1980), or magnetic fields and magnetically susceptible particles (Burns and Graves 1985; Chetty and Burns 1991). Parameters such as column verticality (Bruce et al. 1999), temperature [through its effect on density and viscosity (Finette et al. 1998)], and distributor fouling can also affect bed expansion.

Variations in the degree of fluidization occur during start-up and shutdown and as a result of system disturbances, such as altered viscosity or density of the fluidizing liquid. These transients have been studied experimentally by a number of authors (Chang 1995; de Luca et al. 1994). More recently, a distributed parameter model has been developed to predict bed height transients under step changes in the fluidization velocity (Thelen and Ramirez 1997, 1997).

Solids Dispersion and Bed Stability. The principal difference between the overall mixing within the fluidized bed and the expanded bed is the mobility of the particles, which adds the dispersive element of solids mixing. And as has been outlined above, this dispersive element can be reduced by using particles that segregate upon fluidization giving an expanded bed. Particle movement is then reduced to a random motion about a reasonable consistent position, so that in the extreme, the bed can be considered to be composed of layers of noninteracting particles (Thömmes 1997). The degree of classification is dependent on the ratio of the largest to smallest particle present (in the simple case of monodensity beads). Al-Dibouni and Garside (1979) define classification as becoming important with a size ratio greater than 1.3, and a ratio of more than 2.2 as being required for the formation of a perfectly classified bed. The particle size ratio in commercial adsorbents for expanded beds is now effectuated deliberately. For example, the ratio for Streamline is about 3.0 [assuming a size range of 100–300, μm (Hjorth 1997)]. However, these adsorbents also have a distribution of particle densities and would therefore be expected to show complex behavior dependent on the linear flow rate.

Bed stability or axial dispersion in EBA is as important as, if not more important than that in conventional chromatography. The controlling mechanisms are the solid phase dispersion and the mass transport contributions as described by the van Deemter equation (van Deemter et al. 1956). The resulting axial dispersion is measured by generating a residence time distribution (RTD) using some form of tracer stimulus-response experiment. A suitable tracer is injected into the system as a pulse or a step and the concentration profile in the exit stream is monitored. The resulting RTD is fitted to either the axial dispersion or tanks in series model (Levenspiel 1972) to give a parameter that describes the mixing in the system. The axial dispersion model gives the Peclet number, defined as

$$\mathrm{Pe} = \frac{Hu}{\epsilon D_{\mathrm{axl}}} \tag{4}$$

where H is the height of the expanded bed and D_{axl} is the axial dispersion coefficient. If the system is in plug flow, $\mathrm{Pe} = \infty$ and if the system is perfectly mixed $\mathrm{Pe} = 0$. Several authors have indicated that a certain minimum Peclet number is required for successful protein adsorption (Chang 1995; Slater 1992; Thömmes 1997). Another dimensionless number sometimes used to characterize mixing is the Bodenstein number (Bo). The tanks in series model describes the degree of mixing using a number of tanks, N; the larger the number of tanks, the lower the level of mixing in the bed.

Straightforward measurement or detailed study of axial dispersion has been used to define when the bed is stable and ready for the application of the unclarified feedstock, to investigate the effects of operating variables on mixing (Bruce et al. 1999; Karau et al. 1997; Thömmes et al. 1995b), or, more recently, to evaluate the effects of the feedstock itself on the bed (Fernandez-Lahore et al. 1999).

Operational Procedure. In principle, expanded bed processes are operated in a similar fashion to packed bed processes, the main difference being the direction of liquid flow. The standard sequence of frontal chromatography is generally followed as in Fig. 3: equilibration, sample application (loading), washing, elution, and cleaning in place (CIP).

The bed is first expanded by the application of equilibration buffer in an upward direction. The flow rate is selected to give a bed expansion of 2–3 times the settled height. During the equilibration process the bed stabilizes as the particles find their equilibrium position along the length of the bed.

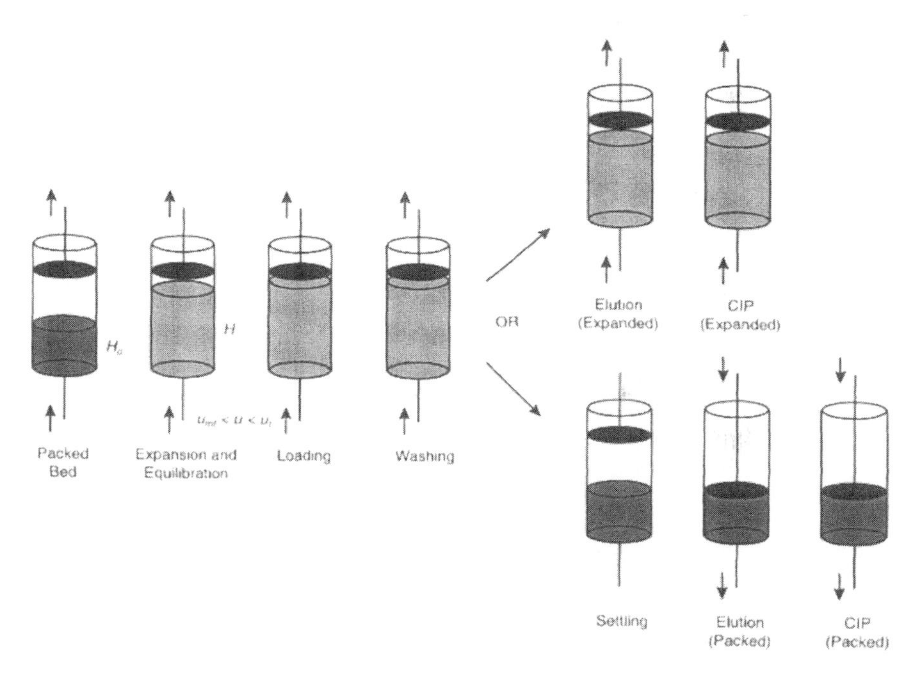

Figure 3 Schematic diagram of the stages in expanded bed adsorption. The elution stage may be performed either in an expanded (upper scheme) or packed (lower scheme) configuration. The arrows indicate the direction of liquid flow through the column (either upward or downward). H_0 is the settled height of adsorbent, H is the expanded height of adsorbent, u_{mf} is the minimal fluidization velocity, u is the superficial velocity, u_{t} is the terminal velocity, and CIP is the cleaning-in-place stage.

After approximately 30 min the bed should have stabilized and only small circulatory movements of the adsorbent beads are observed. Prior to sample application, bed stability may be assessed (in order of increasing accuracy) by the degree of bed expansion, visual inspection, or an RTD test.

After expansion, equilibration, and stabilization, the biomass-containing broth is applied to the bed. The properties of the feedstock are usually very different to those of the equilibration buffer, often being of higher viscosity and density and containing a large quantity of cells or cell debris. If the flow rate is not reduced the degree of bed expansion increases. The feedstock can have other effects on the quality of expansion, as described by Barnfield-Frej et al. (1994). The effects of cell debris adsorbent interactions are discussed in the section, "Limitations of Expanded Bed Adsorption." Two different loading methods have been used by practitioners during the loading phase of operation, the bed can either be operated at a constant flow rate (Barnfield-Frej et al. 1994; Hansson et al. 1994), in which case it will expand, or at a constant height (Chase and Draeger 1992; Chang et al. 1995; Chang and Chase 1996a), in which case the flow rate will need to be reduced progressively.

Following the load, residual biomass and unbound proteins are washed from the bed using the same buffer as was used in equilibration. Washing may also be carried out using wash buffers containing viscosity enhancers such as glycerol (Chase and Draeger 1992; Chang et al. 1995; Chang and Chase 1996a), with the aim of reducing the number of column volumes needed to remove particulate material from the bed. The reason for this is that when applying a negative step change in solution density or viscosity, such as occurs when beginning the wash, instabilities arise at the step interface due to "viscous fingering." This loss of plug flow reduces the effectiveness of the wash, due to increased buffer consumption and therefore overall wash time. In studies using the viscous buffers to correct this it was concluded that the presence of glycerol may affect the elution process, but no investigation was made into the amount of buffer required to remove the viscosity enhancer. This extra washing step and the use of the viscosity enhancer would be expected to affect the wash time and costs, respectively. Further investigation is required to fully quantify the advantages of using a viscous wash buffer.

Elution may be performed in either the packed or expanded configuration. As the washing procedure is supposed to remove all residual particles, the bed may be allowed to settle and elution performed with either upflow or downflow. For this purpose, the upper adapter needs to be lowered onto the top surface of the bed, requiring a movable adapter which is usually sealed to the environment using O rings. This is a drawback in pharmaceutical production where movable parts may be unwanted due to biological safety reasons, especially if production in a closed system is required. Elution can be performed in the expanded mode; the only disadvantage is the increased elution volume (measured at approximately 40% by Hjorth et al. 1995) and therefore decreased product concentration, while being able to dispense with the movable upper adapter (Hjorth 1998).

Application of crude cell homogenates to adsorbents used for protein adsorption increases the contact of the adsorbent with nucleic acids, lipids, and cellular compounds, which are removed in conventional primary recovery steps prior to standard column chromatography. Therefore, specific care has to be taken to achieve a thorough cleaning of the adsorbent. The exact CIP procedure used is dependent on the nature of the feedstream and the type of adsorbent used. Several different protocols have been used, involving irrigation of the bed with solutions containing reagents such as 1 M NaOH, 2 M NaCl, 70% ethanol, 6 M urea, or 30% isopropanol (Barnfield-Frej et al. 1994; Chang and Chase 1996a; Jagersten et al. 1996). In some situations, where interaction between cells and/or cell debris and the adsorbent have occurred extended CIP procedures, and possibly refluidization may be required.

Ligand Types. The purification achieved in an adsorption operation depends on the selectivity of the interaction between the immobilized ligand and the target protein. The dissociation constant, K_d, for the interaction should ideally be in the range 10^{-3} to 10^{-8} M (Boschetti 1994; Narayanan 1994) in order to attain significant retention and desorption within physiological constraints, respectively. The ligand (and coupling chemistry used) must also be suitable for operation in crude feedstocks, and in many cases the appropriate choice is a compromise between ligand stability and high selectivity. The stability of ligands is loosely related to their size; the smaller the ligand, the less sensitive it is to external factors. Larger groups such as antibodies, receptors, and lectins are particularly susceptible. Repeated association-dissociation steps, extended contract with crude material, treatment with drastic washing solutions, and the presence of protease traces in the feeds can modify or destroy such ligands (Boschetti 1994).

Suitable ligands for the primary recovery of proteins from crude feedstocks are outlined below, in order of ascending selectivity, and are compared and summarized in Table 2. Examples typical of each group are also illustrated in Fig. 4. The first group includes ion exchange and hydrophobic groups, which may be classified as general-purpose ligands. The charged groups are simple in nature and stable; however, the presence of large amounts of similarly charged molecules can reduce capacity and selectivity. In addition, the presence of charged cell debris and large macromolecules

Table 2 General Comparison of the Advantages and Disadvantages of the Different Ligand Types[a]

Ligand type	Example(s)	Advantages	Disadvantages
Ion exchange	DEAE, SP, or Q groups	High capacity Mild elution Predictable behavior	Poor selectivity Ionic strength dependence Cell or nucleic acid fouling
Hydrophobic	Butyl or phenyl groups	High capacity Ionic strength promoted Minimal cell or nucleic acid fouling	Poor selectivity Possibility of precipitation or denaturation
Dye-ligand	Procion red H-E7B, Procion yellow H-E3G or Cibacron blue 3GA	Group selective Mixed mode of adsorption	Dye leaching
Metal chelate	IDA or NTA groups	Group selective Mild elution Ionic strength independent	Empiricism in binding
Synthetic	Polyeptides and others based on triazine ring structure	High capacity Infinite range of binding (can engineer group specificity) Low risk of product contamination Eliminate animal sourced products from production	In majority of cases a new ligand is required for each new product Dependence on a ligand discovery company at a sensitive time in product development

[a]See Fig. 4 for structures typical of each group.

DEAE, diethylaminoethyl; IDA, iminodiacetic acid; SP, sulfopropyl; NTA, nitrolotriacetic acid; Q, quaternary amine.

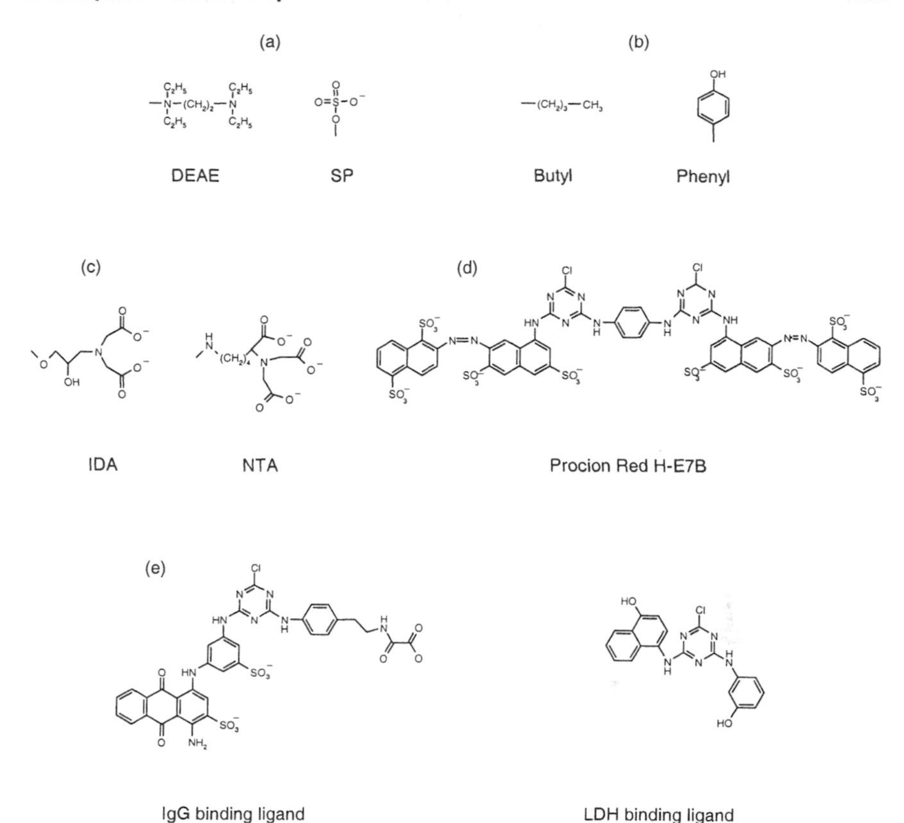

Figure 4 Example structures from each of the ligand groups: (a) DEAE and SP groups, (b) butyl and phenyl groups, (c) IDA and NTA groups, (d) Procion red H-E7B, and (e) synthetic IgG (Teng et al. 1999) and LDH (Labrou et al. 1999) binding groups.

such as nucleic acids can cause operational difficulties and contamination of eluates. Hydrophobic interaction chromatography is based on the interaction between hydrophobic regions on protein surfaces and hydrophobic or hydrophobic substituted adsorbents. This is usually driven by high concentrations of "salting-out" salts such as ammonium sulfate, an effect that parallels the ability to precipitate proteins from aqueous solutions according to the Hofmeister (lyotropic) series (Pahlman et al. 1977). This makes the technique extremely useful for the recovery of proteins directly from fermentation broths or following salt precipitation. Disadvantages of the technique are slow aggregation or precipitation of target protein, and denaturation on the column or during elution.

The second classification, including, for example, dye ligands and metal chelates, are group selective because they bind proteins of a functional class. Immobilized textile dyes may be used to isolate proteins whose function depends on nucleotide cofactors such as NAD^+, $NADP^+$, or ATP. The dyes have the advantage of being considerably cheaper and easier to immobilize than the biological cofactors and can be used repeatedly in acidic or alkaline solvents containing chaotropic agents. The disadvantages are that they are rarely totally selective and the leached ligand is potentially toxic. Immobilized metal affinity chromatography (IMAC) exploits the affinities for metal ions that are exhibited by functional groups on the surfaces of proteins (Porath 1992). The advantages of IMAC include (1) the high stability of the metal chelates over a wide range of solvent conditions and temperature; (2) the high metal loadings that result in high protein loading capacities; and (3) the ease of product elution and ligand regeneration (Arnold 1991). Over the past few years there has been considerable interest in the generation of small ligand molecules that mimic the chemical specificity of larger natural or recombinant molecules. These molecules can be used for the affinity purification of bioproducts with the major advantage of their intrinsic chemical and biological stability, and form the third and final group of ligands discussed here.

Ligand generation methods may be rational, as typified by structure-based approaches, or irrational, such as with combinatorial libraries or phage display methods. Rational approaches employ molecular modeling and X-ray crystallographic data to create ligands that fit specific areas of the target protein. The approach requires detailed structural information on the protein and the interaction site that is to be mimicked. Companies such as Affinity Chromatography Limited (ACL, Cambridge, UK) have created structures based on stable nuclei, such as the triazine ring. Trial compounds are then synthesized and tested in binding studies, after which the best candidate is isolated, characterized, and synthesized in bulk. This type of strategy has been successful in the generation of ligands for IgG (Teng et al. 1999), LDH (Labrou et al. 1999), an insulin precursor, and factor VIII. It is also now being extended to the high-resolution separation of different glycoforms (Lowe and Palanisamy 1999). This area of ligand discovery looks set to make an important contribution to selectivity in complicated separations such as in direct extraction.

The alternative, irrational methods are useful when detailed structural information is not available. Combinatorial chemistry methods are based on the generation of vast numbers of chemical structures that are subsequently screened for appropriate protein binding properties. Peptide libraries are now a proven method for the generation of ligands (Carbonell 1999; Clackson and Wells 1994; Fassina et al. 1996; Huang and Carbonell

1999). Phage display methods are based on the generation of genetic constructs for the production of stable microproteins and transfecting these into bacteriophage. Genetic variants of the microprotein are then induced and these are subsequently expressed by the bacteriophage. The phage displayed peptides are then screened for effectiveness as ligands for affinity purification. This method has been extended and developed most notably by the Dyax Corporation (Cambridge, MA, USA) (Kelley et al. 1999). Phage display has been used to select peptides with affinity for α-chymotrypsin (Krook et al. 1995) and single-stranded DNA (Krook et al. 1994), and has even found an application in the generation of antibodies (Winter et al. 1994). A recent review of protein and peptide libraries for in vitro selection highlights these and many other points (Clackson and Wells 1994). Although such methodology holds great promise, it remains largely untested in the expanded mode.

Adsorbents. Adsorbents for use in expanded beds must have basic characteristics common to all chromatographic media. Important features include particle size, chemical and physical resistance, porosity, surface area, ease of derivatization, and low nonspecific binding. They must also have fluidization properties so as to expand 2–3 times at suitable flow rates, while having a range of diameters and densities so as to generate a classified bed. Again, the immobilized ligand must be suitable for repeated use in crude feeds and have a high selectivity for the protein.

Adsorbent particles are characterized using Eqs. (1) to (3). Table 3 summarizes the properties of many of the adsorbents used to generate fluidized or expanded beds. Adsorption in stable expanded beds was originally carried out using Sepharose-type adsorbents, well-characterized packed bed media (Draeger 1991). However, due to the physical properties of the media, the required bed expansion was achieved at very low velocities (10–30 cm h^{-1}). Thus, denser materials with larger diameters were generated to give the required buoyant weight. Composite materials such as the Streamline range [based on a quartz (Hansson et al. 1994) core coated with agarose] have been developed by Amersham Pharmacia Biotech (Uppsala, Sweden). This material has been used in the majority of successful applications (as described later; see Tables 4 and 5).

Other commercial adsorbents include the PROSEP range, produced by Bioprocessing Ltd (Durham, UK), of highly porous glass beads with a selectable pore size between 70 and 1000 Å. Composite media are produced by UpFront Chromatography (Copenhagen, Denmark) using glass cores surrounded by porous agarose and Biosepra AS (Villeneuve la Garenne Cedex, France) with their HyperD range, which has a porous ceramic skeleton whose pores are filled with a stable derivative of polyacrylamide.

Table 3 Fluidization Characteristics of Matrices for EBA[a]

Matrix	Description	Size distribution, mean (μm)	Mean density ($g\,ml^{-1}$)	u_{mf} ($cm\,h^{-1}$)	u_t ($cm\,h^{-1}$)	n	Manufacturer	Ref.
Q Sepharose Fast Flow	6% agarose, highly cross-linked	44–180; 93.5	1.13	2.6	252	4.9	APB	Draeger 1991
Streamline	6% agarose	NG; 185	1.22	14.4–36	612–1296	3.1–4.8	APB	de Luca et al. 1994
Streamline (quartz weighted)	6% agarose, quartz core	100–300; NG	NG	NG	1404	5.4	APB	Hjorth et al. 1995
Streamline Procion red H-E7B	6% agarose, quartz core	100–300; 206	1.15	NG	886	4.8	MAR	Chang et al. 1995
Streamline rProtein A	4% agarose, metal alloy core	80–165; 130	1.40	114	990	3.4	APB	Thömmes et al. 1996
Bioran CPG	Controlled pore glass	130–250; NG	0.41	84	1724	4.5	Schott	Thömmes et al. 1995a
Perfluoropolymer	Perfluoropolymer, coated with PVA and then cross-linked	NG; 80	1.67	NG	612	4.26	MAR	Owen et al. 1997; Owen and Chase 1999

Zirconium oxide	Porous zirconia particles modified by fluoride adsorption	20–140; 54 25–104; 43	3.48 3.33	NG	972 1116	5.6	MAR	Griffith et al. 1997
Fractosil 1000	Porous silica	63–100; 81.5	1.35	6	773	4.9	E. Merck	Finette et al. 1996
DEAE Spherodex LS	Porous silica, covered with derivatized dextran	100–260; 180	NG	18.8	3540	4.9	Biosepra SA	Finette et al. 1998
PROSEP A	Highly porous, dense glass beads	63–150; NG	1.3	NG	534	5.9	Bioprocessing Ltd	Beyzhavi 1994
S Hyper D LS	Porous ceramic skeleton, the pores filled with polyacrylamide	NG; 203	1.47	NG	1685	4.7	Biosepara SA	Voute et al. 1996

[a]Physical properties on the range of materials used to date are included, and where possible the fluidization properties (u_{mf}, u_t, and n) were calculated using the raw data and Eqs. (1) to (3).

MAR, made according to reference (and the adsorbent is not commercially available in this form); APB, Amersham Pharmacia Biotech (Uppsala, Sweden) product; NG, not given in reference (and could not be calculated from raw data).

Table 4 Adsorption Characteristics of Adsorbents for EBA

Ligand	Matrix	Size distribution (μm)	Porosity (exclusion limit, Da)	Ligand density[a]	Breakthrough capacity (mg ml^{-1})[b]	Equilibrium binding capacity (mg ml^{-1})	Application notes and reference(s)
DEAE	Streamline	100–300	4×10^6	0.13–0.21	35	55	Recovery of G6PDH from unclarified *S. cerevisiae* homogenate (Chang and Chase 1996a)
SP	Streamline	100–300	4×10^6	0.17–0.24	65	75	Purification of MAb from whole hybridoma fermentation broth (Thömmes et al. 1995a)
SP XL	Streamline	100–300	4×10^6	0.18–0.24	140	243	Manufacturers information and Thömmes, 1999
rProtein A	Streamline	80–165	2×10^7	6	20	50	Recovery of MAb from mammalian cell culture (Thömmes et al. 1996)
Heparin	Streamline	100–300	4×10^6	4	NG	NG	Manufacturers information

Chelating (IDA)	Streamline	100–300	4×10^6	40.6	8	70	Facilitated recovery of GST-$(His)_6$ from unclarified *E. coli* homogenate (Clemmitt and Chase 2000)
Phenyl (low sub)	Streamline	100–300	4×10^6	NG(~ 20)	NG	NG	Recovery of ADH from unclarified *S. cerevisiae* homogenate (Smith 1997)

[a] Ligand density for ion exchange groups, DEAE, SP, and SP XL determined as mmol charged groups per ml adsorbent, for rProtein A as mg protein bound per ml, for heparin as mg heparin bound per ml adsorbent, for chelating as Cu^{2+} binding capacity in μmol per ml, for phenyl as μmol phenyl groups per ml.

[b] Breakthrough capacity evaluated at 1% breakthrough 300 cm h^{-1} in expanded mode (15 cm settled bed) with BSA (DEAE) and (chicken egg white) lysozyme (SP). For SP XL at 10% breakthrough at 300 cm h^{-1} in packed mode (10 cm settled bed, 4.4 ml) with lysozyme. For rProtein A breakthrough capacity was determined using IgG at 1% breakthrough at 300 cm h^{-1} in expanded mode (15 cm settled height). Chelating was evaluated in packed mode at 1% breakthrough at 300 cm h^{-1} (5 cm settled height) using BSA.

[c] Total binding capacity was determined for DEAE, SP, and rProtein A in packed mode at 50 cm h^{-1} and 50% breakthrough using the same conditions as in note b. Equilibrium adsorption isotherms were determined for SP XL (lysozyme (Thömmes 1999) and chelating (BSA), and the values displayed are the equilibrium binding capacities (q_m). NG indicates this information was not given by manufacturers or in reference.

Table 5 Literature Review of Some of the Recent Applications of Expanded Bed Adsorption

Feedstock	Protein	Ligand	Column diameter (cm)	Feed volume (L)	Yield (%)	Purification factor	Ref.
Renatured inclusion bodies and *E. coli* extract	IL-8	SP	5	16	97	4.35	Barnfield-Frej 1996
Homogenised *E. coli*	Annexin V	DEAE	20	26.5	95	2.22	Barnfield-Frej et al. 1994
Periplasmic *E. coli* extract	Exotoxin A	DEAE	20	180	79	1.98	Johansson et al. 1996
Whole *E. coli* broth	ZZ-M5	DEAE	5	16	90	NG†	Hansson et al. 1994
Alkaline lysed *E. coli* broth	Plasmid DNA	DEAE	20	NG	NG	NG	Varley et al. 1999
Homogenised *S. cerevisiae* suspension	G6PDH	DEAE	5	1.1	98	12	Chang and Chase 1996a
Homogenised *S. cerevisiae* suspension	G6PDH	Procion Red H-E7B	5	0.8	99	103	Chang et al. 1995
Homogenised *S. cerevisiae* suspension	ADH	Phenyl	20	21.3	80	8	Smith 1997
H. polymorpha fermentation broth	Aprotinin	SP	5	6.4	76	3.8	Zurek et al. 1996
CHO cell culture	rMAb	SP	5	26	70–85	7	Batt et al. 1995
Hybridoma fermentation broth	MAb	rProtein A	5	60	83	NG†	Thömmes et al. 1996
Milk	Lysozyme	SP	5	5	89	8.3	Noppe et al. 1996
Crude porcine muscle extract	LDH	Cibacron blue 3GA (polymer shielded)	1.6	NG	78	4.1	Garg et al. 1996

IL-8, interleukin-8; G6PDH, glucose-6-phosphate dehydrogenase; ADH, alcohol dehydrogenase; rMAb, recombinant monoclonal antibody; LDH, lactate dehydrogenase; NG, not given in reference; †indicates that the eluate was of a high purity (SDS-PAGE).

Another interesting candidate is a controlled pore glass produced by Merck (Poole, Dorset, UK), as modified by Thömmes et al. (1995a). Noncommercial composites such as dextran-silica (Morton and Lyddiatt 1994) or cellulose-titanium dioxide (Gilchrist et al. 1994) have also been successful. Chase and collaborators have developed adsorbents based on hydrophilized perfluoropolymers, and separate studies have shown how this material can be derivatized with a whole range of ligands (McCreath 1993; McCreath et al. 1995, 1997). The most recent addition has been very-high-density materials that allow a significant reduction in particle size and improved mass transport capabilities. Such matrices include fluoride-modified zirconia (Griffith et al. 1997). It is likely that such adsorbents will not be the answer to every expanded separation problem as it is possible to have too high an operating window of flow rates such that the residence times in the bed are too low for capture.

Studies have also been aimed at developing adsorbents with reduced mass transfer limitation due to porous diffusion, as has occurred in conventional packed bed chromatography through the use of convective protein transport through perfusive pores (Nash and Chase 1998; Whitney et al. 1998) or accelerated diffusion in gels of high charge density (Boschetti et al. 1995). The characteristics of the HyperD range from Biosepara AS give reduced diffusional lengths and increased surface area, and are claimed to have solid diffusion-controlled mass transport giving improved recovery from dilute feeds (Boschetti et al. 1995). Amersham Pharmacia Biotech recently introduced a new high-capacity adsorbent range, Streamline Q or SP XL. These possess the usual quartz core, coated with cross-linked agarose to which is grafted dextran molecules that are flexible enough to allow the passage of charged protein molecules through the pores and greatly increase the exposure of the Q or SP groups. The effect is to increase binding capacity, increase the rate of protein uptake, and improve resolution. A recent study found decreasing ratios of breakthrough capacities to equilibrium binding capacity with increasing feed concentration for a range of proteins, again suggesting significant contribution of solid diffusion to mass transport (Thömmes 1999).

Table 4 summarizes the adsorption characteristics of some of the resins available commercially and synthesized in house. The ligands available largely reflect the discussion above. For example, the Streamline range is available with SP, DEAE, IDA, heparin, or rProtein A. Other manufacturers offer additional types of ligands, such as UpFront Chromatography's mixed-mode ligands. The rProtein A Streamline with metal alloy core is a new development from Amersham Pharmacia Biotech. This has a high density (1.3 g ml^{-1}) enabling a reduction in particle size (80–165 μm). This, together with the use of the more open-porosity 4% agarose, gives significantly improved mass transport capabilities. The recombinant Protein A ligand is genetically modified by deleting non-IgG binding regions and introducing an extra cysteine residue at the C terminus. These modifications allow an oriented coupling to the base matrix and better binding capacity; there is claimed to be no significant chromatographic change after 100 cycles of normal use. The adsorbent was effective in the capture of monoclonal antibody (MAb) from cells containing hybridoma fermentation broth with a dynamic capacity of more than 14 mg MAb per ml and gave a clarified and highly concentrated eluate (Thömmes et al. 1996).

Column Design and Equipment. The development of a stable expanded bed is dependent on the design of the column used; the two important parameters are the distributor design and the verticality of the column. The distributor should ideally provide a flat velocity profile across the column cross-section and prevent the formation of stagnant zones and channels. Due to the low pressure drop across a fully expanded bed, the pressure drop across the distribution system must be high in order to generate a plug flow. A distributor design may be tested by measuring bed stability and bed expansion. Various types of liquid distribution systems have been described in the literature for the generation of expanded beds, including perforated glass plates (de Luca et al. 1994), sintered glass (Chase and Draeger 1992), and beds of glass ballotini (Thömmes et al. 1996). As expanded beds are employed for the processing of biomass-containing feeds, the distributor must also allow the passage of cell debris without becoming blocked and without damaging shear-sensitive cells. An alternative column design, produced by UpFront Chromatography, is based on a novel stirrer in the base of the column, which agitates the feed and lower part of the bed to provide distribution. This design is reported to facilitate processing of extremely crude feedstreams without blockage. A detailed study of the features of distributors for expanded beds, as well as the best combination for even flow distribution and good fouling resistance; has yet to be performed.

The other important consideration influencing bed stability is the vertical alignment of the column. Van der Meer et al. (1984) have shown that even very small deviations from the vertical can lead to significant liquid flow inhomogeneity and thus reduced efficiency. This was quantified further by Bruce et al. (1999), in terms of the effect on column mixing and adsorption performance in an EBA system and is discussed below with other factors affecting axial dispersion.

Commercial expanded bed equipment is provided by all the major manufacturers. The materials used in construction must be resistant to the process solutions used, such as the harsh CIP reagents. Amersham

Pharmacia Biotech produce a range of columns with a glass (2.5–20 cm diameter) or stainless steel (20 to 100 cm diameter) tube and a perforated plate and mesh cap distributor. The larger, opaque columns require level detection systems based on ultrasound transducers in the upper adapter. These provide automation for the positioning of this adapter. Such sensors have also been investigated as tools for process monitoring (Thelen et al. 1997).

Method Development and Scale-up. The rational development of a fully optimized production scale process generally involves up to four stages of experimentation (Hjorth 1997). Initial investigations are performed to identify optimal binding conditions for the target protein to the adsorbent in packed beds with clarified feedstreams. The influence of cells and cell debris on the adsorption process is then tested using small-scale expanded beds with unclarified feedstock. At this stage the sample preparation, protein and biomass concentration, load volume, and operating conditions (such as loading velocity) are all optimized. The third and fourth steps involve scaling up the separation to the pilot scale (typically 3–6 L of adsorbent and up to 20-cm-diameter columns) and then production scale (typically (50–300 L of adsorbent and up to 100-cm diameter columns) (Hjorth 1997).

Scale-up is achieved by increasing the column diameter while maintaining solutions, feedstock characteristics, adsorbent, linear velocity, and bed expansion. The critical part of the scaling operation is the design and performance of the inlet and outlet liquid distributor.

Scale up has been demonstrated using the purification of Annexin V from *E. coli* homogenates (Barnfield-Frej et al. 1994) and also whole *S. cerevisiae* suspensions spiked with BSA (Barnfield-Frej et al. 1997). There was only a slight increase in axial dispersion in moving from a 2.5-cm- to a 60-cm-diameter column, while breakthrough capacities were maintained. The smallest column commercially available for developmental expanded bed work is the Streamline 25 column produced by Amersham Pharmacia Biotech. However, the high cost of many biological products makes optimization studies very expensive. For this reason, scaled-down columns of 1 cm diameter have been developed (Ghose and Chase 2000a,b; Bruce et al. 1999). Nonchromatographic factors, such as feedstock composition, which can vary between different scales of fermentation or culture and batches of the same process, or cell disruption where changes in the type of homogenizer may be necessary can cause differences in performance.

Performance in Expanded Bed Adsorption. The overall performance of a protein purification operation may be characterized by its productivity and, additionally, in the case of a primary recovery operation, the clarification efficiency and volume reduction. In expanded bed adsorption, utilizing porous adsorbents, five main system parameters may be defined that govern the overall performance (Thömmes 1997): adsorption equilibrium, liquid phase dispersion, solid phase dispersion, particle side mass transport, and liquid side mass transport. By manipulating the operating variables the limitations of the system parameters can be minimized. The operating variables include adsorbent size and density distribution, adsorbent bed height, linear flow rate, and feedstock viscosity, density, and particle and protein concentration.

Several studies have been performed on these various aspects of performance. Batch binding studies of the effect of cells on adsorption to ion exchangers illustrated the dominance of electrostatic effects, with anion exchangers badly affected whereas cation exchangers and affinity adsorbents usually perform well (Draeger 1991). Liquid phase dispersion has been an area of particular interest due to the influence of particle design on solids dispersion, the interdependence of bed height and liquid velocity, and operating variables such as increased viscosity. The axial dispersion coefficient increases with increasing flowrate in expanded bed adsorption; however, the effect on the Bodenstein number is complicated and has been observed by Thömmes (1997) to either remain constant (Draeger 1991), decrease through a minimum and then increase (Thömmes et al. 1995b), or show no clear trend (Dasari et al. 1993). Beds of particles suitable for generation of stable expanded beds have axial dispersion coefficients in the range $2–9 \times 10^{-6}$ m^2 s^{-1} (Batt et al. 1995; Chang and Chase 1996b). There has been little research directly on solid phase dispersion within fluidized beds or on the effect of using particles suitable for the formation of classified beds. However, there is evidence to suggest that solids dispersion is just 10% of the level of liquid dispersion (van der Meer et al. 1984; Thömmes 1997) and therefore constitutes a minor part of the overall mixing. Column vertical alignment can effect separation efficiency through its effect on particle movement. A column misalignment of $0.15°$ resulted in a reduction of the Bodenstein number from 75 to 45 for a 5-cm-diameter expanded bed column (Bruce et al. 1999).

Protein adsorption in expanded beds is usually achieved by selective binding to the internal pore surfaces within highly porous particles. Mass transfer from the bulk liquid therefore occurs by film diffusion followed by porous diffusion. The porous diffusion mass transfer step is often the controlling step in packed bed adsorption and is also likely to dominate in expanded bed adsorption (Boschetti 1994; Thömmes 1997). Several studies have demonstrated the fall in dynamic capacity that occurs with increase in velocity (Hjorth et al. 1995). The increased liquid viscosity and biomass present can also reduce film and porous mass transfer efficiency (Draeger and Chase 1991; Chase and Draeger 1992).

Chang and Chase (1996b) provide an interesting study of how the height control philosophy applied during loading can affect utilization of available capacity and resultant productivity. The study involved the adsorption of lysozyme to Streamline SP in feedstocks of varying viscosity, either at a constant flow rate (allowing the bed to expand) or at a constant bed height (reducing the flow rate during the load). The productivity was found to be relatively invariant. But it was found that the dynamic capacity was higher when operated at constant bed height, although the process time was longer. This was thought to be the most efficient operating protocol, although the situation might change where labile products are to be processed.

This type of overall study illustrates the difficulty in optimizing the expanded bed process. It should be noted that in many cases increasing the selectivity of the chromatographic process will dramatically increase throughput, whereas an increase in column efficiency may have only a minor effect. This indicates the requirement for highly discriminating media for use in EBA to give processes of fewer steps and higher productivity (Chase 1998).

Advances in Separation Methodology. Different methods of generating novel ligands for separations include, as highlighted above (see section "Ligand Types") the new techniques of generating and screening phage display and combinatorial libraries. In addition, the slightly older method of fusing a sequence to the protein of interest followed by purification using existing methods is also gaining in popularity (see Chap. 4). Upstream genetic engineering has been used to aid the purification of ZZ-M5 by anion exchange EBA and GST using polyhistidine fusions and metal affinity expanded bed adsorption (see examples). The latter method has the advantages of being insensitive to ionic strength and tolerating the presence of denaturants or detergents. The combination of expanded bed adsorption and an affinity fusion strategy could result in a simple and efficient protein production and purification process. A potential drawback to this method is that in circumstances where the native product is required, some form of cleavage is needed followed by further chromatographic steps to recover the pure protein.

Feedstocks. Expanded bed adsorption may be used to separate proteins from a variety of particulate-containing feedstreams and a selection of the successful applications described in the literature are given in Table 5. The particulates may be whole cells, cell fragments, subcellular structures, or any other suspended particle that would normally be removed by clarification, and the protein may be a native enzyme or protein, or a recombinant one expressed in a host organism. The range of hosts for recombinant protein production has increased greatly. These include bacteria, yeast, mammalian cells, insect cells, and even transgenic organisms producing proteins in secretions such as milk. Expression may be intracellular, periplasmic, or extracellular.

Feedstocks can affect the hydrodynamic and chromatographic performance of the expanded bed in several ways. Therefore, when scouting, careful consideration must be made of upstream factors such as fermentation media composition and additives, as the link between fermentation and downstream processing is especially close where primary recovery operations are concerned. Other such factors include ionic strength, particle and protein concentration, viscosity and biomass density (homogenized feeds), temperature, pH adjustment (can cause precipitation or aggregation of proteins or cells), and stirring of the sample during application.

Limitations of Expanded Bed Adsorption. Expanded bed adsorption is not the answer to every problem. There are situations in which it will simplify the downstream processing and situations where more problems are created than solved. EBA cannot be used in some situations due to certain properties of the broth (Chase 1994). Where the particulates and either the adsorbent or the distributor porosity (usually mesh width about 60 μm) are of similar size, elutriation or blockage is possible. This situation may arise where cells are growing as large mycelia or in clumps. A simple solution to this problem might be the use of coarse-mesh screen filters prior to application to the expanded bed.

In other situations where broths of very high biomass content or cell disruptates of very high viscosity are applied, problems can occur. For instance, Barnfield-Frej et al. (1994) found that *E. coli* homogenates of up to 5% dry weight and viscosities of up to 10 mPa gave good fluidization characteristics with Streamline DEAE. At slightly higher values the bed expanded dramatically so that periodic flow reversal was required, and at still higher values the bed was seen to collapse. In situations where the biomass content is too high dilution is an option; however, the decreased feed concentration can reduce adsorption efficiency. Where the viscosity is too high as a result of nucleic acids, the disruption protocol can be improved by either using a greater number of cycles or greater pressure (high-pressure homogenization), or alternatively nucleases can be used to hydrolyze the nucleic acid instead of shearing it.

Problems can occur in EBA due to the interaction of either nucleic acids and/or cells with the adsorbent causing interparticle binding. Such events create aggregates, leading to deteriorating quality of fluidization and in bad cases complete bed collapse. The deteriorated quality of fluidization can create areas of the bed considered as dead zones with little flow as well as fast-moving channels that essentially bypass the adsorbent. This causes reduced capacity. A recent study has looked at the interaction of a hybridoma cell culture with expanded beds of Streamline SP and rProtein A Streamline (Feuser et al. 1999). The hybridoma cells were seen to interact significantly with the cation exchange resin. Cell breakthrough occurred after 17–20 sedimented bed volumes. This was in contrast to the affinity adsorbent, where no retardation of the cells was observed. In neither case was any cell damage found indicating the low shear nature of the expanded operation. This study indicated the role that nucleic acid and cell binding could play in reducing efficiency.

Fernandez-Lahore et al. (1999) took this study further, using carefully selected tracers to monitor the quality of fluidization during the application of whole *S. cerevisiae* cell broth and cell homogenates of *E. coli* and *S. cerevisiae* to the naked Streamline base matrix and the derivatized alternatives SP, DEAE, Chelating (Loaded with Cu^{2+}), and Phenyl. The resulting RTD traces were analyzed using the model of van Swaaij et al. (1969) giving three parameters to characterize flow: the number of transfer units for mass exchange between mobile and stagnant phase, the Peclet number for overall axial dispersion, and the mobile fraction of the liquid in axially dispersed plug flow. The method described could theoretically be used to evaluate certain adsorbent feedstock combinations to check straightforward compatibility or preparation conditions for the feedstock.

The conductivity of feedstocks can also be so high as to prevent binding to ion exchangers. In situations where cell disruptates are applied, feedstocks may be diluted directly with a low ionic strength buffer. On the other hand, mammalian cell cultures need dilution while maintaining the cells in an intact form requiring diluents such as 200 mM D-glucose so as to maintain osmolality but reduce the ionic strength.

Comparisons Between Expanded Bed Adsorption and Conventional Routes. Comparisons between expanded bed approaches and conventional processing strategies at either the laboratory, pilot, or full-process scale enable a proper assessment of the advantages of the technique. Several studies have been performed to assess the comparative benefits within overall processes.

Batt et al. (1995) compared expanded bed adsorption and filtration processes followed by affinity purification for the recovery of an antibody from whole mammalian cell culture broth. The expanded process was found to be less labor intensive and potentially more economical than the filtration process. Although the process time with the expanded bed was greater, it did not require the constant monitoring as did the filtration. Suding and Tomusiak (1993) compared their conventional approach, consisting of centrifugation and filtration prior to packed bed chromatography, with EBA for the recovery of an unspecified intracellular product from *E. coli*. Although the expanded route required more matrix, resulted in lower yield, and had higher production costs, the overall productivity was found to be twice that of the packed route. Johansson et al. (1996) presented packed and expanded bed processes for the recovery of a periplasmically expressed exotoxin A from *E. coli*. The expanded route was three times as fast as the packed route. Smith (1997) presented data comparing the recovery of ADH at laboratory scale using expanded bed recovery and conventional approaches based on polyethyleneimine flocculation, ammonium sulfate fractional precipitation, or centrifugation prior to packed bed adsorption. The expanded route was found to give higher process yields and reduced process time; however, cell debris was found to reduce the capacity of the matrix resulting in lower capacity and greater buffer consumption.

EBA is now becoming recognized as a realistic alternative to existing procedures for protein recovery. Indeed examples of full-scale industrial processes are beginning to appear. This may be attributed to the increasing number and diversity of successful applications at bench scale, the development of purpose designed adsorbents and equipment, and increasingly tighter process control and understanding. Such processes are proving efficient and compliant with the tough requirements of regulatory bodies such as the Food and Drug Administration (FDA). Validation considerations have been reviewed with regard to EBA applications (Sofer 1998), and the first fully commissioned processes are beginning to appear (Sumi et al. 1998).

Membrane Adsorption

Membrane chromatography is another technology that can be used for the integration of process steps (see Table 1). The main and fundamental advantage of the membrane over the conventional approach using columns of porous particles is the absence of pore diffusion. By attaching ligands to the inner surface of the throughpores of membranes, mass transport occurs by convective flow, thereby eliminating the main mass transport limitation in porous media and leaving film mass transfer and surface reaction. A large number of different types of membrane and modes of operation are used, as identified in a recent review (Thömmes and Kula 1995). Derivatization with a variety of different groups, ranging from ion exchange and hydrophobic groups to affinity ligands such as protein A, is possible. Processing of crude feedstocks has been attempted; indeed, dye ligand membranes have been used for the recovery of enzymes from crude extracts (Champluvier and Kula 1992) and thiophilic membranes for the purification of monoclonal antibodies from cell culture (Finger et al. 1995). Cross-flow operation of derivatized membranes has been shown to allow the processing of particulate-containing feedstreams (Kroner et al. 1992). However, performance of the process is dependent on the trans membrane flux, which can fall due to fouling with particulates.

2.2 Continuous Adsorption Operations

A problem with cyclic batchwise modes of operation is the fact that units must go through a load, wash, elution, and regeneration phase during each purification cycle. This means that recovery only occurs for a small fraction of the overall operating time. Continuous protein recovery operations would provide inherently smaller equipment and capital investment, reduced labor requirements, the possibility for greater throughputs, and more consistent product quality (Hudstedt et al. 1988; Kroner et al. 1987). In addition, despite problems with on-line monitoring associated with the purification of proteins, continuous processes are typically more amenable to control and optimization—two important features for large-scale operation.

Continuous adsorption operations have been used on a very large scale for the recovery of small molecules such as uranium or residues from mining (Slater 1992); however, biotechnological applications have been scarce. A possible reason for this is that the material requirement for proteins for therapeutic or diagnostic use is comparatively small, ranging from a few grams to a few kilograms per annum, whereas for other bulk products, such as enzymes destined for use in detergents, this figure is relatively small. Other reasons for the lack of examples include the failure of any continuous processes to be established in industry. This means that all newly developed processes would need to be designed from first principles and undergo a whole series of validation procedures prior to approval. Batchwise processes are also more amenable to monitoring and quality control and are therefore more likely to receive approval by the regulatory agencies. A final consideration is that the majority of proteins of interest are produced in batchwise fermentations, so that the hypothetical feedstream is produced in a discontinuous fashion.

In this section various approaches to continuous contacting for the primary recovery of proteins are identified. These include the stirred-tank or mixer-settler unit, magnetically stabilized fluidized beds (MSFBs), and a continuous countercurrent contactor. An example is also included of a process whereby a fluidized bed adsorber was integrated with a fluidized bed bioreactor for the production of monoclonal antibodies to illustrate how direct extraction can be integrated into continuous culture. It is important to distinguish at this point between truly continuous processes, which continuously process the input stream and produce a continuous level of product, and semicontinuous systems where the input stream is processed continuously but the product is eluted in periodic bursts.

Continuous Affinity Recycle Extraction

Continuous affinity recycle extraction (CARE) is a procedure for carrying out affinity separations on a continuous basis using two interconnected continuous stirred-tank reactors (CSTRs) for contacting adsorbent and process solutions (Gordon and Cooney 1990). The feedstream is supplied to the adsorption stage where it is contacted with an affinity adsorbent; the desired product binds while the contaminants are washed out with a wash buffer. Then the concentrated loaded adsorbent resin is separated from the effluent using a membrane and pumped to an eluting stage where product is released selectively, thus regenerating the adsorbent before it is recirculated to the adsorption stage. CARE has been used successfully to purify β-galactosidase from unclarified *E. coli* homogenates using *p*-aminobenzyl-1-thio-β,D-galactopyranoside agarose (Pungor et al. 1987). β-Galactosidase could be produced continuously at 70% yield and with a 35-fold purification from a crude lysate diluted 1 in 50 with wash buffer. Many subsequent improvements have been made to the initial unit, including multiple adsorption stages, extra wash stages, and elution stages, and attempts have been made to model the process (Gordon et al. 1990; Rodrigues et al. 1992). Although the system is of simple design and is easily scaled up, it utilizes standard adsorbent particles that would be expected to break down under the mechanical action of stirring and peristaltic circulation.

Perfluorocarbon Emulsion Reactor for Continuous Affinity Separations

The perfluorocarbon emulsion reactor for continuous affinity separations (PERCAS) unit developed by McCreath et al. (1993, 1994) is a similar device to CARE. It consists of a system of four mixer-settler units arranged in series and in a closed loop (Fig. 5). The unloaded adsorbent (a perfluorocarbon emulsion derivatized with either CI Reactive blue 2 or Procion red H-E7B) is continuously pumped into the first mixer-settler unit where it is contacted with a continuous supply of unclarified feedstream. The exiting stream of this CSTR-like chamber supplies the resulting mixture of loaded adsorbent and depleted feedstream into a settler tank where the two are separated by virtue of their density difference. The depleted feedstream goes to waste whereas the product-laden adsorbent is fed to three further mixer-settler units where it is submitted to wash, elution, and regeneration prior to recirculation to the adsorption stage. This system improves upon the CARE system by utilizing robust nonporous perfluorocarbon droplets that are resistant to the mechanical action of peristaltic pumps and fouling (McCreath 1993). The settler unit may also prove more resilient than the macroporous membrane used in CARE although it might be expected to limit the processing flow rate operational window.

Magnetically Stabilized Fluidized Beds

Continuous operation is possible in MSFBs in a countercurrent mode of operation. Burns and Graves (1985) used two such contactors in series for the affinity purification of human serum albumin (HSA) using Cibacron blue F3-GA derivatized adsorbents. The solids were supplied to the adsorption stage from a reservoir, where they were contacted countercurrently with the incoming HSA solution. Use of specially designed magnets allowed entrapment of the magnetically susceptible particles in a downward-moving magnetic field. This enabled low dispersion flow of adsorbent and effective countercurrent contacting of the two streams. Loaded particles were collected manually in a reservoir and delivered to the elution column before being recycled to carry out further HSA extraction. Chetty and Burns (1991) and Chetty et al. (1991) used a slight modification of the above apparatus whereby magnetically susceptible particles were only used to control the flow of more conventional nonmagnetic adsorbents for the fractionation of myoglobin and lysozyme. Continuous countercurrent MSFBs have also found applications in the separation of paraffin from olefins (Sikavitsas et al. 1995). This type of contactor may have limited application in industry for protein purification because of the complicated nature of the equipment involved and the need for specially designed and fabricated supports.

Semicontinuous Fluidized Bed Adsorption

Continuous culture may be an efficient way of producing proteins that are susceptible to secondary processing in the course of fermentations where substrate limitations and product inhibitions can occur. Short residence times within these systems support the production of correctly assembled proteins and minimize the contact of sensitive bioproducts with degrading enzymes. The downstream processing operation following continuous culture must be capable of handling large volumes of dilute feedstream. A recent application therefore coupled a 25-ml continuous bench-scale fluidized bed bioreactor containing hybridoma cells immobilized on porous glass microcarriers producing MAbs to an expanded bed adsorption column containing 300 ml of Streamline SP (Born et al. 1995). Although not operated in a truly continuous fashion, the configuration yielded up to 20 mg/day of MAb in a cell-free solution at fourfold concentration and a fivefold purification. The application demonstrated how, in principle, contactors such as

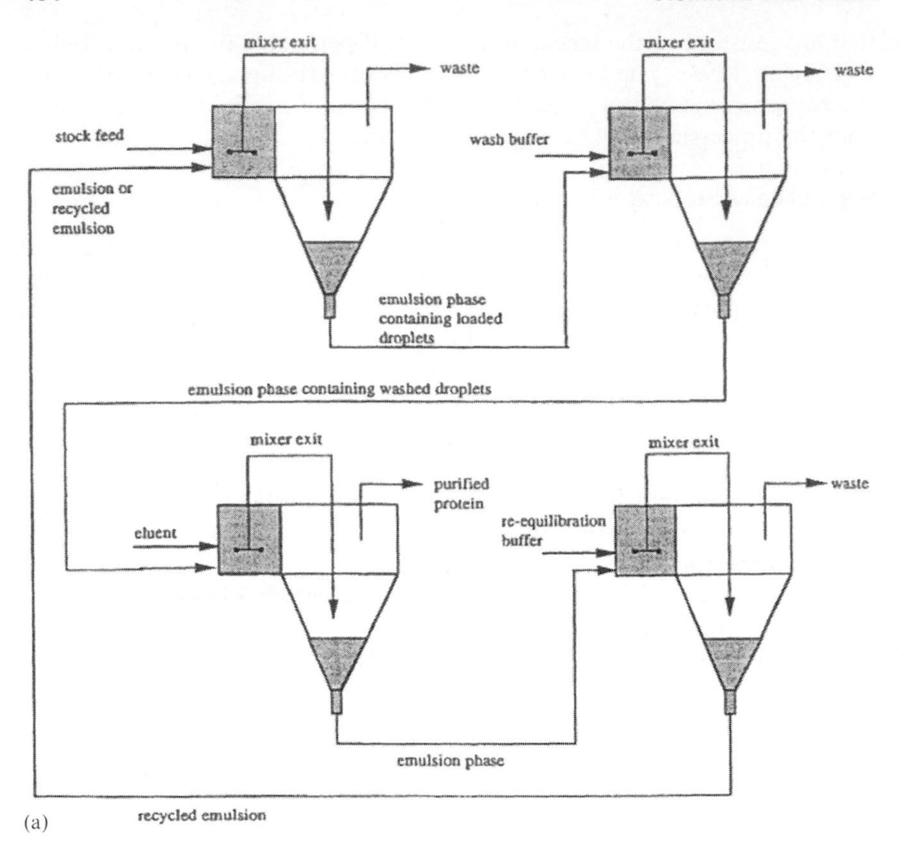

Figure 5a Principle of PERCAS operation: (a) the operating principle of the unit and (b) a photograph of it in use for the recovery of HSA from an unclarified *S. cerevisiae* suspension. (Reproduced with permission from McCreath 1993.)

those described here might be deployed alongside continuous culture for continuous primary recovery.

Studies have also been performed on fluidized beds as external loops to microbial batch fermentors (Hamilton et al. 1999). A comparison has also been made of the relative efficacy of conventional fluidized bed processing, interfaced fluidized bed adsorption (FBA) (where the bed is attached directly to the fermentor effluent), and integrated FBA or direct product sequestration (DPS) (where the bed is integrated and the effluent returned to the fermentor). The model studied was the recovery of an extracellular acid protease from a fermentation of *Yarrowia lipolytica*. A threefold increase in enzyme productivity and the elimination of product autolysis was achieved

(b)

Figure 5b

using direct product capture or DPS by uncoupling product drive feedback inhibition and avoiding product degradation during the long batch fermentation.

Continuous Countercurrent Contactor

Owen (1998) designed, tested, and modeled a continuous contactor based on the expanded bed principle. It consists of four fluidized bed units arranged in series and in a closed loop, in a similar way to the PERCAS system (Fig. 6). Each fluidized bed was of a design such that crude feedstream was applied at the base, flowed through a stream of falling adsorbent and was then collected in a depleted form at the bed exit. The adsorbent was added to the top of each column and allowed to sediment in the column before being collected just above the distributor and passed onto the next stage. Such an arrangement enabled countercurrent contact between the process streams and, as a result, the most efficient depletion of the feed due to the maintenance of a high concentration gradient as a driving force for adsorption. Four stages were used to adsorb the target protein, wash the adsorbent of residual solids and nonbound protein, selectively elute the bound protein, and then regenerate the

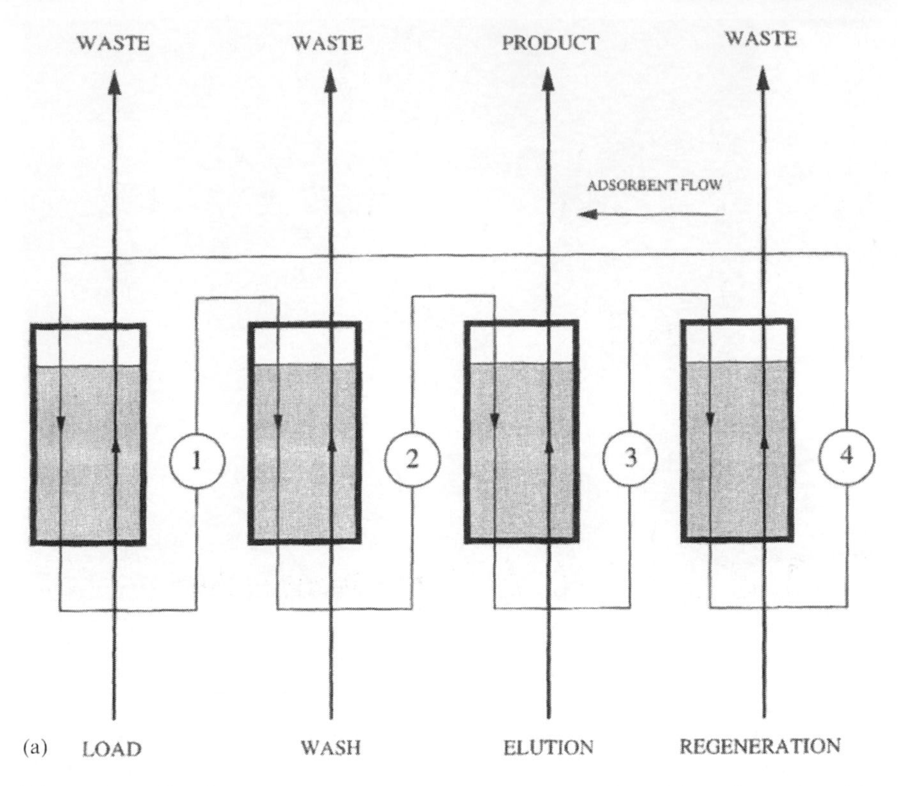

Figure 6a Continuous countercurrent contactor operation: (a) a schematic of the process and (b) a photograph of the four-stage unit just prior to homogenate application. (Reproduced with permission from Owen 1998.)

adsorbent. The adsorbent used was a solid perfluoropolymer particle (Table 3), coated with poly(vinyl alcohol) (PVA), cross-linked and derivatized with the dye-ligand Procion red HE-7B. The contactor was used successfully with a variety of unclarified feedstocks such as egg whites, yeast (Owen et al. 1997), and *E. coli* homogenates (Owen 1998) and milk. The adsorption stage was studied and modeled in depth using lysozyme as a test protein (Owen and Chase 1999).

3. FUTURE PROSPECTS

Primary recovery operations such as EBA, which is a relatively recent discovery, seem destined for an important role as the initial stage in

(b)

Figure 6b

downstream processing. Such operations represent a simple way to eliminate multiple processing steps such as clarification, concentration, and initial purification. This has been identified as a way of reducing equipment and labor costs and improving process yields by decreasing processing time and eliminating losses due to protease action and secondary modification.

Through judicial selection of adsorbents and associated ligands further purification is possible in this initial capture step so as to produce products of purity sufficient as to require minimal further processing. Such ligands must be an appropriate balance between the robustness required for repeated operation in crude feedstreams and subsequent CIP operations and the selectivity for the target protein. At present, general-purpose groups based on ion exchange or hydrophobic interaction chromatography dominate separations. However, group affinity ligands such as textile dyes and metal chelates could be utilized in order to achieve slightly higher purities. Techniques for the generation of synthetic ligands already applied to drug discovery by the pharmaceutical companies are likely to generate specific ligands for selected proteins with the required specificity and binding strength while being chemically resistant. Fusion strategies, based on the upstream modification of the protein with a tail or tag in order to facilitate downstream processing, may represent an extremely efficient and simple purification route. Recent developments in the cleavage of tails for release of the native protein using intein or aminopeptidases are likely to improve the success of such strategies. Each of the latter techniques has the advantage of being a way to tailor a purification to a protein rather than performing some sort of screening of existing chromatography media, thus simplifying the purification of a product about which little is known. The fusion technique has the added advantage of providing a route of identifying peaks and quantifying yield using assays and ELISA protocols based on the properties and binding of the tail.

The expanded bed unit operation seems to have the greatest potential of all the batchwise processes for primary adsorptive recovery and over recent years has been applied to an increasing range of separations. Further investigation into different protein ligand interactions is required to enable closer tailoring of the adsorbent to the feedstream and target protein. Synthetic ligand design should play a major role. In this contactor, as well as others, matrix design is moving toward particles with enhanced mass transport capability. Research is also needed into the fundamentals of the process such as solids dispersion and modeling to enable optimization and greater process control. Other adsorption operations such as those involving the continuous countercurrent contactor need more fundamental research into such issues as scale-up, further modeling, and applications data in order to be considered for commercial development.

4. EXAMPLES

4.1 Purification of Recombinant Annexin V and ZZ-M5 from Unclarified *E. coli* Homogenates Using Expanded Bed Anion Exchange Chromatography

In some of the first examples of EBA the anion exchange adsorbent Streamline DEAE was used to clarify, concentrate, and partially purify the two proteins annexin V (Barnfield-Frej et al. 1994) and ZZ-M5 (Hansson et al. 1994) from unclarified fermentation broths. Annexin V is a small (34,000 Da) anticoagulant protein found in the human placenta with a low pI of 4.9. It was cloned and expressed intracellularly in *E. coli* and the resulting fermentation broth was disrupted by high-pressure homogenization (three passages) and pumped directly onto the expanded bed. The separation was optimized in packed beds and scaled to pilot level using a Streamline 200 column (20-cm diameter). The yield of annexin V was more than 95% at all scales, and scanning SDS-PAGE gels demonstrated an increase in purity from 9% to 20%.

ZZ-M5 is a fusion protein consisting of two synthetic IgG binding domains (ZZ) derived from staphylococcal protein A and a repeat structure (M5) from the central region of the malaria antigen Pf155/RESA. This was of interest as a potential component of a malaria vaccine. A subtlety of the strategy was the design of the fusion protein to have a low pI of 4.5 to allow anion exchange at a pH of 5.5 such that the majority of the host proteins did not adsorb. The gene product was also secreted into the culture medium during fermentation. This made it possible for the broth, with only on-line dilution to control conductivity and viscosity, to be applied directly to the column. ZZ-M5 was eluted at 90% yield and a very high purity as judged by SDS-PAGE. An additional immunoaffinity step was included to reduce DNA and endotoxin to levels acceptable to the regulatory authorities. The two applications demonstrate how anion exchange principles can be applied very effectively in primary recovery operations, such as EBA, and how with proteins of low pI impressive purifications are possible.

4.2 Recovery of Alcohol Dehydrogenase from Unclarified *S. cerevisae* Homogenates Using Expanded Bed Hydrophobic Interaction Chromatography

Hydrophobic interaction chromatography was found to be superior to both anion exchange and metal chelate chromatography for the capture of native intracellular ADH from yeast homogenates (Smith 1997). Adsorption conditions were developed in packed beds in terms of ligand selection, pH, and loading volume. The yeast broth was homogenized in a high-pressure homogenizer and then diluted and brought to 0.78 M ammonium sulfate in potassium phosphate buffer at pH 7. The separation was then developed along two lines. Different clarification methods (centrifugation, batch and continuous polyethyleneimine flocculation, and fractional precipitation) were tested prior to packed bed chromatography on phenyl Sepharose FF (low sub) for comparison with optimized EBA on Streamline phenyl (low sub). The yield of ADH in the expanded bed was higher (95%) than the packed bed (85%) due mainly to the reduction in the number of stages. The expanded bed also maintained its yield through successive cycles, whereas the packed bed was seen to foul resulting in a reduction in ADH yield from 85% to 58% over 10 cycles. However, the dynamic capacity in the expanded bed was lower due to the presence of cell debris.

4.3 Expanded Bed Affinity Chromatography of Dehydrogenases from *S. cerevisiae* Using Dye Ligand Perfluoropolymer Supports

McCreath et al. (1995) used hydrophilized perfluoropolymer matrices derivatized with the dye ligands Procion yellow H-E3G and Procion red H-E7B to partially purify malate dehydrogenase (MDH) and glucose-6-phosphate dehydrogenase (G6PDH), respectively, from unclarified *S. cerevisiae* homogenates. Small packed bed chromatography experiments were used to develop elution strategies based on the use of sodium chloride and the selective eluents NADH or $NADP^+$. Expanded bed affinity chromatography of MDH resulted in an eluted fraction containing 89% of the applied activity with a purification factor of 113. The corresponding results for G6PDH were 84% of the applied activity and a purification factor of 172. This application demonstrates how dye-ligand affinity chromatography, with appropriate elution conditions, can give highly pure eluate fractions. A disadvantage of this type of chromatography is that the dyes have complex structures with multiple ionic and hydrophobic groups present, which can interact in a nonspecific manner with unwanted proteins and reduce the capacity for the target.

4.4 Polymer Shielded Dye Ligand Affinity Chromatography of Lactate Dehydrogenase

Garg et al. (1996) improved the selectivity of the dye-ligand support, Cibacron blue 3GA immobilized on Streamline, using polymer shielding with the water-soluble polymer poly(vinylpyrrolidone) (PVP). Such nonionic polymer molecules are thought to attach to the dye matrix via nonspecific interactions, and their presence restricts the nonspecific interactions between the dye and other proteins. The adsorbent was used in expanded mode for the recovery of LDH from a crude porcine muscle extract. The recovery of LDH improved from 17% to 78% with the shielded adsorbent and the associated purification factor improved from 1.8 to 4.1.

4.5 Facilitated Downstream Processing of Histidine-Tagged Glutathione *S*-transferase [GST-$(His)_6$] Using Metal Affinity Expanded Bed Adsorption

The successful marriage of the fusion approach and EBA was recently demonstrated for the facilitated downstream processing of GST-$(His)_6$ (Clemmitt and Chase 2000). The enzyme was expressed intracellularly in *E. coli* and processed using metal affinity expanded bed adsorption.

Dynamic capacities of Ni^{2+} Streamline Chelating in packed mode (loaded to 5% breakthrough) were around 357 U ml^{-1} (36 mg ml^{-1}); however, problems occurred when using immobilized Cu^{2+} or Zn^{2+} with metal ion transfer and subsequent metal affinity precipitation of $\text{-[-(GST-(His)}_6\text{)-Me}^{2+}\text{-]}_n$ complexes. Batch binding tests demonstrated the high affinity of GST-$(His)_6$ for immobilized Ni^{2+}, with a q_m of 695 U ml^{-1} (70 mg ml^{-1}) and a K_d of 0.089 U ml^{-1} (0.0089 mg ml^{-1}). Ni^{2+} loaded Streamline Chelating was then used to purify GST-$(His)_6$ from unclarified *E. coli* homogenate, resulting in an eluted yield of 80% and a 3.34-fold purification. The high dynamic capacity in expanded mode of 357 U ml^{-1} (36 mg ml^{-1}) demonstrated that this specific interaction was not affected by the presence of *E. coli* cell debris.

4.6 Comparison of Expanded Bed Adsorption and PERCAS for the Direct Extraction of G6PDH from Unclarified *S. cerevisiae* Homogenates

The PERCAS unit was applied to the separation of HSA from blood plasma (McCreath et al. 1993) and to the primary recovery of G6PDH from unclarified *S. cerevisiae* homogenates (McCreath et al. 1994). In each case the adsorbent was a flocculated perfluorocarbon affinity emulsion generated by the homogenization of perfluorodecalin with PVA and subsequent cross-linking and flocculation. This adsorbent exhibited limited fouling with this crude feedstream, was found to be stable to extended periods of peristaltic pumping, and has good settling properties. Interestingly, the continuous PERCAS unit was compared with the batchwise EBA for the G6PDH extraction (McCreath et al. 1994). In each case G6PDH was recovered with an average purification factor of 18; however, the productivity of PERCAS was some 2.25 times higher than EBA under similar process conditions.

4.7 Application of Continuous Countercurrent Contactor to the Direct Purification of Malate Dehydrogenase (MDH) from Unclarified *S. Cerevisiae* Homogenates

The continuous countercurrent contactor was applied to the continuous purification of MDH from unclarified *S. cerevisiae* homogenates using a perfluoropolymer derivatized with Procion red H-E7B (Owen et al. 1997). Although not optimized the system delivered a product stream with 78% of the applied MDH, at a rate of 70 U min^{-1} and 10-fold purification. The specific productivity of approximately 0.35 U ml^{-1} min^{-1} settled adsorbent was higher than could be achieved with the same adsorbent in an expanded bed.

5. MEDIA AND EQUIPMENT MANUFACTURERS

Affinity Chromatography Ltd (ACL), 307 Huntingdon Rd, Girton, Cambridge, CB3 0JX, UK.
Amersham Pharmacia Biotech, S-75182 Uppsala, Sweden.
Bioprocessing Ltd, Medomsley Road, Consett, Durham, DH8 6TJ, UK.
BioSepra S. A., 35 avenue Jean Jaures, F, 92395 Villeneuve la Garenne Cedex, France.
Dyax Corporation, 1 Kendall Square, Cambridge, Massachusetts 02139, USA.
Merck, Merck House, Poole, Dorset, BH15 1TD, UK.
Schott, Schott Glass Technologies Inc., 400 York Avenue, Duryea, PA 18642, USA.
UpFront Chromatography A/S, Lerso Parkalle 42, DK-2100 Copenhagen, Denmark.

NOMENCLATURE

Abbreviations

ADH	alcohol dehydrogenase
ATP	adenosine triphosphate
BSA	bovine serum albumin
CARE	continuous affinity recycle extraction
CIP	cleaning in place
CPG	controlled pore glass
CSTR	continuous stirred-tank reactor
DEAE	diethylaminoethyl
DPS	direct product sequestration
EBA	expanded bed adsorption
ELISA	enzyme-linked immunosorbent assay
FBA	fluidized bed adsorption
FDA	Food and Drug Administration
G6PDH	glucose-6-phosphate dehydrogenase
GST-$(His)_6$	glutathione *S*-transferase, hexahistidine tagged
HSA	human serum albumin
IDA	iminodiacetic acid

IL-8	interleukin-8
IMAC	immobilized metal affinity chromatography
LDH	lactate dehydrogenase
MAb	monoclonal antibody
MDH	malate dehydrogenase
MSFB	magnetically stabilized fluidized bed
NAD	nicotinamide adenine dinucleotide
NADP	nicotinamide adenine dinucleotide phosphate
NG	not given
NTA	nitrilotriacetic acid
PERCAS	perfluorocarbon emulsion reactor for continuous affinity separations
PVA	poly(vinyl alcohol)
PVP	poly(vinylpyrrolidone)
Q	quaternary amine group
RTD	residence time distribution
rtPA	recombinant tissue plasminogen activator
SDS-PAGE	sodium dodecyl sulfate–polyacrylamide gel electrophoresis
SP	sulfopropyl
ZZ-M5	affinity fusion created from a repeat of the A domain (ZZ) and a repeat structure (M5) of the malaria antigen (Hansson et al. 1994)

Symbols

Bo	Bodenstein number
D_{axl}	axial dispersion coefficient
d_{p}	particle diameter
g	acceleration due to gravity
H	height of the expanded bed
H_0	settled height of bed
K_{d}	dissociation constant
N	number of theoretical tanks
n	Richardson and Zaki coefficient
Pe	Peclet number
pI	isoelectric point
q_{m}	equilibrium binding capacity
$\mathrm{Re_p}$	particle Reynolds number
u	superficial velocity
u_{mf}	minimal fluidization velocity
u_{t}	terminal velocity

Greek Symbols

ϵ	bed voidage
ρ	liquid density
ρ_p	particle density
μ	liquid viscosity

REFERENCES

Al-Dibuoni M.R., Garside J. (1979) Particle mixing and classification in liquid solid fluidised beds. Trans. I Chem. E. 57, 95–103.

Arnold F.H. (1991) Metal affinity separations: A new dimension in protein processing. Bio. Technology 9, 151–156.

Barnfield-Frej, A.-K., Hjorth R., Hammarström A. (1994) Pilot scale recovery of recombinant annexin V from unclarified *Escherichia coli* homogenate using expanded bed adsorption. Biotechnol. Bioeng. 44, 922–929.

Barnfield-Frej A.K. (1996) Expanded bed adsorption for the recovery of renatured human recombinant interleukin-8 from *Escherichia coli* inclusion bodies. Bioseparation 6, 265–271.

Barnfield-Frej A.-K., Johansson S., Leijon P. (1997) Expanded bed adsorption at process scale: Scale-up verification, process example and sanitisation. Bioproc. Eng. 16, 57–63.

Batt B.C., Yabannavar V.M., Singh V. (1995) Expanded bed adsorption process for protein recovery from whole mammalian cell culture broth. Bioseparation 5, 41–52.

Beyzhavi K. (1994) Personal communication. Fluidised bed adsorption using conventional chromatography columns. Bioprocessing Ltd., Consett, County Durham, UK.

Bird R.B., Stewart W.E., Lightfoot N. (1960) Transport Phenomena, Wiley, New York.

Bonnerjea J., Oh S., Hoare M., Dunnill P. (1986) Protein purification. The right step at the right time. Bio/Technol. 4, 954–958.

Bonnerjea J., Jackson M., Hoare M., Dunnill P. (1988) Affinity flocculation of yeast cell debris by carbohydrate specific compounds. Enzyme Microb. Tech. 10, 357–360.

Born C., Thömmes J., Biselli M., Wandrey C., Kula M.-R. (1996) An approach to integrated antibody production. Coupling of fluidised bed cultivation and fluidised bed adsorption. Bioproc. Eng. 15, 21–29.

Boschetti E. (1994) Advanced sorbents for preparative protein separation purposes. J. Chromatogr. A 658, 207–236.

Boschetti E., Guerrier L., Girot P., Horvath J. (1995) Preparative high performance liquid chromatographic separation of proteins with HyperD ion exchange supports. J. Chromatogr. B 664, 225–231.

Bruce L.J., Ghose S., Chase H.A. (1999) The effect of column verticality on separation efficiency in expanded bed adsorption. Bioseparation 8, 69–75.

Buijs A., Wesselingh J.A. (1980) Batch fluidised ion exchange column for streams containing suspended particles. J. Chromatogr. 201, 319–327.

Burns M.A., Graves D.J. (1985) Continuous affinity chromatography using a magnetically stabilised fluidised bed. Biotechnol. Prog. 1, 95–103.

Carbonell R.G. (1999) Affinity purification of proteins using ligands derived from combinatorial peptide libraries. Oral presentation. Recovery of Biological Products IX, Whistler, Canada, May.

Champluvier B., Kula M.-R. (1992) Dye-ligand membranes as selective adsorbents for rapid purification of enzymes. A case study. Biotechnol. Bioeng. 40, 33–40.

Chang Y.K. (1995) Development of expanded bed techniques for the direct extraction of proteins from unclarified feedstocks. PhD thesis, University of Cambridge.

Chang Y.K., McCreath G.E., Chase H.A. (1995) Development of an expanded bed technique for an affinity purification of G6PDH from unclarified yeast cell homogenates. Biotechnol. Bioeng. 48, 355–366.

Chang Y.K., Chase H.A. (1996a) Ion exchange purification of G6PDH from unclarified yeast cell homogenates using expanded bed adsorption. Biotechnol. Bioeng. 49, 204–216.

Chang Y.K., Chase H.A. (1996b) Development of operating conditions for protein purification using expanded bed techniques: The effect of degree of bed expansion on adsorption performance. Biotechnol. Bioeng. 49, 512–526.

Chase H.A. (1994) Purification of proteins by adsorption chromatography in expanded beds. Trends Biotechnol. 12, 296–303.

Chase H.A. (1998) The use of affinity adsorbents in expanded bed adsorption. J. Mol. Recog. 11, 217–221.

Chase H.A., Draeger N.M. (1992) Affinity purification of protein using expanded beds. J. Chromatogr. 597, 129–145.

Chetty A.S., Burns M.A. (1991) Continuous protein separation in a magnetically stabilised fluidised bed using non-magnetic supports. Biotechnol. Bioeng. 38, 963–971.

Chetty A.S., Gabis D.H., Burns M.A. (1991) Overcoming support limitations in magnetically stabilised fluidised bed separators. Powder Technol. 64, 165–174.

Clackson T., Wells J.A. (1994) In vitro selection from protein and peptide libraries. Trends Biotechnol. 12, 173–184.

Clemmitt R.H., Chase H.A. (2000) Facilitated downstream processing of a histidine-tagged protein from unclarified *E. coli* homogenates using immobilised metal affinity expanded bed adsorption. Biotechnol. Bioeng. 67, 206–216.

Couderc J.P. (1985) Incipient fluidisation and particulate systems. In: Fluidisation (Davidson J.F., Clift R., Harrison D., eds.), Academic Press, London.

Dasari G., Prince I., Hearn M.T.W. (1993) High performance liquid chromatography of amino acids, peptides and proteins 124: Physical characterisation of fluidised bed behaviour of chromatographic packing materials. J. Chromatogr. A 631, 115–124.

Datar R.V., Rosén C.-G. (1996) Cell and cell debris removal: Centrifugation and cross flow filtration. In: Bioprocessing (Stephanopoulos G., ed.), VCH, Weinheim, pp. 469–503.

de Luca L., Hellenbroich D., Titchener-Hooker N.J., Chase H.A. (1994) A study of the expansion characteristics and transient behaviour of expanded beds of adsorbent particles suitable for bioseparations. Bioseparation 4, 311–318.

Draeger N.M. (1991) The use of liquid fluidised beds for the purification of proteins. PhD thesis, University of Cambridge.

Draeger N.M., Chase H.A. (1991) Liquid fluidised bed adsorption of proteins in the presence of cells. Bioseparation 2, 67–80.

Fanquex P.F., Flaschel E., Renken A. (1984) Development of an enzyme fluidised bed reactor equipped with motionless mixers: Application of the lactose hydrolysis in whey. Chimia 38, 262–269.

Fassina G., Verdoliva A., Odierna M.R., Ruvo M., Cassini G. (1996) Protein A mimetic peptide ligand for affinity purification of antibodies. J. Mol. Recog. 9, 564–569.

Fernandex-Lahore M., Kleef R., Kula M.-R., Thömmes J. (1999) The influence of complex biological feedstock on the fluidisation and bed stability in expanded bed adsorption. Biotechnol. Bioeng. 64, 484–496.

Feuser J., Halfar M., Lutkemeyer D., Ameskamp N., Kula M.-R., Thömmes J. (1999) Interaction of mammalian cell culture broth with adsorbents in expanded bed adsorption of monoclonal antibodies. Proc. Biochem. 34, 159–165.

Finette G.M.S., Mao Q.M., Hearn M.T.W. (1996) Studies on the expansion characteristics of fluidised beds with silica based adsorbents used in protein purification. J. Chromatogr. A 743, 57–73.

Finette G.M.S., Mao Q.M., Hearn M.T.W. (1998) Examination of protein adsorption in fluidised bed and packed bed columns at different temperatures using frontal chromatographic methods. Biotechnol. Bioeng. 58, 35–46.

Finger U.B., Thömmes J., Kinzelt D., Kula M.-R. (1995) Application of thiophilic membranes for the purification of monoclonal antibodies from cell culture media. J. Chromatogr. A 664, 69–78.

Garg N., Galaev I.Yu, Mattiasson B. (1996) Polymer-shielded dye-ligand chromatography of lactate dehydrogenase from porcine muscle in an expanded bed system. Bioseparation 6, 193–199.

Ghose S., Chase H.A. (2000a) Expanded bed chromatography of proteins in small diameter columns. I. Scale down and validation, Bioseparation 9, 21–28.

Ghose S., Chase H.A. (2000b) Expanded bed chromatography of proteins in small diameter columns. II. Methods development and scale up. Bioseparation 9, 29–36.

Gilchrist G.R., Burns M.T., Lyddiatt A. (1994) Solid phases for protein adsorption in liquid fluidised beds: Comparison of commercial and custom assembled particles. In: Separations for Biotechnology 3 (Pyle D.L. ed.), Royal Society of Chemistry, pp. 186–192.

Gordon N.F., Cooney C.L. (1990) In: Protein Purification: From Molecular Mechanisms to Large Scale Processes, ACS Symposium Series 240 (Ladisch M.R., Wilson R.C., Painton C.-D.C., Builder S.E., eds.), Washington DC, pp.

Gordon N.F., Moore C.M.V., Cooney C.L. (1990) An overview of continuous protein purification processes. Biotechnol. Adv. 8, 741–762.

Griffith C.M., Morris J., Ribichaud M., Annen M.J., McCormick A.V., Flickinger M.C. (1997) Fluidization characteristics of and protein adsorption on fluoride modified porous zirconium oxide particles. J. Chromatogr. A 776, 179–195.

Hamilton G.E., Morton P.M., Young T.W., Lyddiatt A. (1999) Process intensification by direct product sequestration from batch fermentations: Application of a fluidised bed, multibed external loop contactor. Biotechnol. Bioeng. 64, 310–321.

Hansson M., Ståhl S., Hjorth R., Uhlén M., Moks T. (1994) Single-step recovery of a secreted recombinant protein by expanded bed adsorption. BioTechnology 12, 285–288.

Hjorth R. (1997) Expanded bed adsorption in industrial bioprocessing. Recent developments. Trends Biotechnol. 15, 230–235.

Hjorth R. (1998) Expanded bed adsorption: Elution in expanded mode. Poster presentation. 2nd International Conference on Expanded Bed Adsorption EBA '98, Napa Valley, CA, June.

Hjorth R., Kampe S., Carlsson M. (1995) Analysis of some operating parameters of novel adsorbents for recovery of proteins in expanded beds. Bioseparation 5, 217–223.

Huang P.Y., Carbonell R.G. (1999) Affinity chromatographic screening of soluble combinatorial peptide libraries. Biotechnol. Bioeng. 63, 633–641.

Hustedt H., Kroner K.H., Papamichael N. (1988) Continuous cross-current aqueous two phase extraction of enzymes from biomass. Automated recovery in production scale. Proc. Biochem. 23, 129–137.

Jagersten C., Johansson S., Bonnerjea J., Pardon R. (1996) Capture of humanised IgG4 directly from the fermentor using Streamline rProtein A. Presentation. Recovery of Biological Products VIII, Tucson, AZ, October.

Johansson U. (1994) Hydrodynamics of liquid solid fluidised beds in the low Reynolds number region. Licentiate Thesis, Royal Institute of Technology, Stockholm, Sweden.

Johansson H.J., Jagersten C., Shiloach J. (1996) Large scale recovery and purification of periplasmic recombinant protein from *E. coli* using expanded bed adsorption chromatography followed by new ion exchange media. J. Biotechnol. 48, 9–14.

Karau A., Benken C., Thömmes J., Kula M.-R. (1997) The influence of particle size distribution and operating conditions on the adsorption performance in fluidised beds. Biotechnol. Bioeng. 55, 54–64.

Kaufmann M. (1997) Unstable proteins: How to subject them to chromatographic separations for purification procedures. J. Chromatogr. A 699, 347–369.

Kelley B.D., Booth J., Vunnum S., Tannatt M., Trevino R., Wright R., Jennings P., Yu J., Potter D., Ransohoff T., Deetz J. (1999) Development of an affinity ligand for purification of a recombinant protein using phage display.

Oral presentation. Recovery of biological products IX, Whistler, Canada, May

Kempken R., Preissmann A., Berthold W. (1995) Assessment of a disk stack centrifuge for use in mammalian cell separation. Biotechnol. Bioeng. 46, 132–138.

Kennedy S.C., Bretton R.H. (1966) Axial dispersion of spheres fluidised with liquids. AICHE J. 12, 24–30.

Kroner K.H., Krause S., Deckwer W.D. (1992) Cross-flow anwendung von affinitätsmembranen zur primarseparation von proteinen. BioForum 12, 455–458.

Kroner K.H., Hustedt H., Cordes A., Schütte H., Recktenwald A., Papamichael N., Kula M.-R. (1987) New trends in enzyme recovery. Ann. NY Acad. Sci. 501, 403–412.

Krook M., Lindbladh C., Birnbaum S., Naess H., Eriksen J.A., Mosbach K. (1995) Selection of peptides with surface affinity for α-chymotrypsin using page display library. J. Chromatogr. A, 711, 119–128.

Krook M., Mosbach K., Lindbladh C. (1994) Selection of peptides with affinity for single stranded DNA using a phage display library. Biochem. Biophys. Res. Commun. 204, 849–854.

Labrou N.E., Eliopoulos E., Clonis Y.D. (1999) Molecular modeling for the design of a biomimetic ligand. Application to the purification of bovine heart L-lactate dehydrogenase. Biotechnol. Bioeng. 63, 322–332.

Levenspiel O. (1972) Chemical Reaction Engineering, Wiley, New York.

Lowe C.R., Palanisamy U.D. (1999) New affinity adsorbents for the purification of glycosylated biopharmaceuticals. Oral presentation. Recovery of biological products IX, Whistler, Canada, May.

McCreath G.E. (1993) Development and applications of perfluorocarbon affinity emulsions. PhD thesis, University of Cambridge.

McCreath G.E., Chase H.A., Purvis D.R., Lowe C.R. (1993) Novel affinity separations based on perfluorocarbon affinity emulsions. Development of a perfluorocarbon emulsion reactor for continuous affinity separations and its application in the purification of human serum albumin from blood plasma. J. Chromatogr. A 629, 201–213.

McCreath G.E., Chase H.A., Lowe C.R. (1994) Novel affinity separations based on perfluorocarbon affinity emulsions. Use of a perfluorocarbon affinity emulsion for the direct extraction of G6PDH from homogenised bakers yeast. J. Chromatogr. A 659, 275–287.

McCreath G.E., Chase H.A., Owen R.O., Lowe C.R. (1995) Expanded bed affinity chromatography of dehydrogenase from bakers yeast using dye ligand supports. Biotechnol. Bioeng. 48, 341–354.

McCreath G.E., Owen R.O., Nash D.C., Chase H.A. (1997). Preparation and use of ion-exchange chromatographic supports based on perfluoropolymers. J. Chromatogr. A 773, 73–83.

Morton P.M., Lyddiatt A. (1994) Direct integration of protein recovery with productive fermentations. In: Separations for Biotechnology 3 (Pyle D.L., ed.), The Royal Society of Chemistry, London, pp. 329–335.

Narayanan S.R. (1994) Preparative affinity chromatography of proteins. J. Chromatogr. A, 658, 237–258.

Nash D.C., Chase H.A. (1998). A comparison of diffusion and diffusion-convection matrices for use in the ion exchange separation of proteins. J. Chromatogr. A 807, 185–207.

Nigram S.C., Sakoda A., Wang H.Y. (1988) Bioseparation recovery from unclarified broths and homogenates using immobilised adsorbents. Biotechnol. Prog. 4, 166–172.

Noppe, W., Hanssens I., de Cuyper M. (1996) Simple 2-step procedure for the preparation of highly-active pure equine milk lysozyme. J. Chromatogr. A 719, 327–331.

Owen R.O. (1998) Continuous counter-current contacting for the direct extraction of proteins. PhD thesis, University of Cambridge.

Owen R.O., Chase H.A. (1999) Modelling of the continuous counter-current expanded bed adsorber for the purification of proteins. Chem. Eng. Sci. 54, 3765–3781.

Owen R.O., McCreath G.E., Chase H.A. (1997) A new approach to continuous counter-current protein chromatography: Direct purification of malate dehydrogenase from a *Saccharomyces cerevisiae* homogenate as a model system. Biotechnol. Bioeng. 53, 427–441.

Påhlman S., Rosengren J., Hjerten S. (1977) Hydrophobic interaction chromatography on uncharged Sepharose derivatives—Effects of neutral salts on the adsorption of proteins. J. Chromatogr. 131, 99–108.

Porath J. (1992) Immobilised metal affinity chromatography. Protein Exp. Purif. 3, 263–281.

Pungor E., Afeyan N.B., Gordon N.F., Cooney C.L. (1987) Continuous affinity recycle extraction: A novel protein separation technique. Bio/Technology 5, 604.

Riba J.P., Routie R., Couderc J.P. (1978) Conditions minimales de mise en fluidisation par un liquide. Can. J. Chem. Eng. 46, 26–30.

Richardson J.F., Zaki W.N. (1954) Sedimentation and fluidisation: Part 1. Trans. I. Chem. E. 32, 35–53.

Rodrigues M.I., Zaror C.A., Maugeri F., Asenjo J.A. (1992) Dynamic modelling, simulation and control of continuous adsorption recycle reaction. Chem. Eng. Sci. 47, 263–269.

Sikavitsas V.I., Yang R.T., Burns M.A., Laugenmayr E.I. (1995) Magnetically stabilised fluidised bed for gas separations. Ind. Eng. Chem. Res. 34, 2873–2880.

Slater M.J. (1992) Principles of Ion Exchange Technology, Butterworth and Heinemann, Oxford, UK.

Smith M.P. (1997) An Evaluation of Expanded Bed Adsorption for the Recovery of Proteins from Crude Feedstocks. PhD Thesis, University College London, UK.

Sofer G. (1998) Streamline validation issues. Oral presentation. 2nd International Conference on Expanded Bed Adsorption EBA '98, Napa Valley, CA, June.

Softer G., Hagel L. (1997) Handbook of Process Chromatography: A Guide to Optimisation, Scale-up and Validation, Academic Press Ltd., London.

Somers W., van't Reit K., Rozie H., Rombouts F.M., Visser J. (1989) Isolation and purification of endo-polygalacturonase by affinity chromatography in a fluidised bed reactor. Chem. Eng. J. 40, B7–B19.

Suding A., Tomusiak M. (1993) Protein recovery from *E. coli* homogenate using expanded bed adsorption chromatography. Presentation 205th American Chemical Society National Meeting, Colorado.

Sumi A., Okuyama K., Kobayashi K., Ohtani W., Ohmura T., Yokoyama K. (1998). Purification of recombinant human serum albumin (efficient purification using Streamline system). Oral Presentation. 2nd International Conference on Expanded Bed Adsorption EBA '98 Napa Valley, CA, June.

Teng S.F., Sproule K., Hussain A., Lowe C.R. (1999) A strategy for the generation of biomimetic ligands for affinity chromatography. Combinatorial synthesis and biological evaluation of an IgG binding ligand. J. Mol. Recog. 12, 67–75.

Thelen T.V., Ramirez W.F. (1999) Modelling of solid liquid fluidisation in the Stokes flow regime using two phase theory. AICHE J. 45, 708–723.

Thelen T.V., Mairal A.P., Horsen C.S., Ramirez W.F. (1997) Application of ultrasonic backscattering for level measurement and process monitoring of expanded bed adsorption columns. Biotechnol. Prog., 13, 681–687.

Thelen T.V., Ramirez W.F. (1997) Bed height dynamics of expanded beds. Chem. Eng. Sci. 52, 3333–3344.

Thömmes J. (1997) Fluidised bed adsorption as a primary recovery step in protein purification. Adv. Biochem. Eng. Biotechnol. 58, 185–230.

Thömmes J. (1999) Investigations on protein adsorption to agarose-dextran composite media. Biotechnol. Bioeng. 62, 358–362.

Thömmes J., Kula M.-R. (1995) Membrane chromatography: An integrative concept in the downstream processing of proteins. Biotechnol. Prog. 11, 357–367.

Thömmes J., Halfar M., Lenz S., Kula M.-R. (1995a) Purification of monoclonal antibodies from whole hybridoma fermentation by fluidised bed adsorption. Biotechnol. Bioeng. 45, 205–211.

Thömmes J., Weiher M., Karau A., Kula M.-R. (1995b) Hydrodynamics and performance in fluidised bed adsorption. Biotechnol. Bioeng. 48, 367–374.

Thömmes J., Badar A., Halfar M., Karau A., Kula M.-R. (1996) Isolation of monoclonal antibodies from cell containing hybridoma broth using a protein A coated adsorbent in expanded beds. J. Chromatogr. A 752, 111–122.

van Deemter J.J., Zuiderweg F.J., Klinkenberg A. (1956) Longitudinal diffusion and resistance to mass transfer as causes of non-ideality in chromatography. Chem. Eng. Sci. 5, 271–289.

van der Meer A.P., Blanchard C.M.R.J.P., Wesselingh J.A. (1984) Mixing of particles in liquid fluidised beds. Chem. Eng. Res. Des. 62, 214–222.

van Swaaij W.P.M., Charpentier J.C., Villermaux J. (1969) Residence time distribution in the liquid phase of trickle flow in packed bed columns. Chem. Eng. Sci. 24, 1083–1095.

van der Wiel. P., Wesselingh J.A. (1989) Continuous adsorption in biotechnology. In: Adsorption: Science and Technology (Rodriguez M.D., LeVan M.D., Tondeur, D., eds.), Kluwer Academic, Dordrecht, The Netherlands.

Varley D.L., Hitchcock A.G., Weiss A.M.E., Horler W.A., Cowell R., Peddie L., Sharpe G.S., Thatcher D.R., Hanak J.A.J. (1999) Production of plasmid DNA for human gene therapy using modified alkaline cell lysis and expanded bed anion exchange chromatography. Bioseparation 8, 209–217.

Voute N.A., Girot P., Boschetti E. (1996) HyperD: A new ion exchanger for fluidised bed adsorption. Post presentation. 1st International Conference on Expanded Bed Adsorption EBA '96, Cambridge, UK, December.

Whitney D., McCoy M., Gordon N., Afeyan N. (1998) Characterisation of large pore polymeric supports for use in perfusion biochromatography. J. Chromatogr. A 807, 165–184.

Winter G., Griffiths A.D., Hawkins R.E., Hoogenboom H.R. (1994) Making antibodies by phage display technology. Annu. Rev. Immunol. 12, 433–455.

Zurek C., Kubis E., Keup P., Horlein D., Beunink J., Thömmes J., Kula M.-R., Hollenberg C.P., Gellissen G. (1996) Production of two aprotinin variants in *Hansenula polymorpha*. Proc. Biochem. 31, 679–689.

11
Fast Chromatography of Proteins

Per-Erik Gustavsson and Per-Olof Larsson
Lund University, Lund, Sweden

1. INTRODUCTION

Proteins are complex biomolecules and by nature unstable. During the isolation step they are exposed to changes in environment, enzymatic degradation by proteases, or even time itself, leading to protein losses and loss of biological activity. Purification of proteins is always a race against time and there is a need to accelerate the isolation procedure. In addition to saving the yield of the target protein, fast separations also help to improve the cost-effectiveness of the production process. Clearly, analytical protein separations also benefit from rapid techniques.

Steady development over the years has resulted in a number of reliable and efficient chromatography materials that are standard in today's laboratory scale as well as large-scale purification of proteins. Table 1 gives a few examples. During the last decade several new concepts also have emerged, resulting in separation media with exceptional properties and often particularly well suited for fast or very fast protein separations, analytical as well as preparative. Is there a common denominator for these new materials? In a sense there is: all are focusing on the reduction of slow diffusion. An immediate step toward fast chromatography is otherwise simply accomplished by increasing the flow rate through the column. However, here we encounter a problem, since proteins diffuse very slowly, in fact 10–30 times more slowly than small molecules. Molecules inside a normal chromatographic particle are only transported by diffusion and if the diffusion is slow it cannot match an increased flow rate outside the particle. A result of this imbalance is band broadening/decreased resolution or reduced dynamic capacity of the chromatographic column. Ideally, the time needed

Table 1 Examples of Workhorses in Protein Purification

Trade name	Material	Chromatography mode	Manufacturer/ supplier
Sepharose Fast Flow	Agarose	IEC/HIC/AC	1
Sephadex G-25	Dextran	Desalting	1
Trisacryl	Polymer	SEC/IEC/AC	2

1, Amersham Biosciences, Uppsala, Sweden; 2, BioSepra Inc. Marlborough, MA, USA. IEC, ion exchange chromatography; HIC, hydrophobic interaction chromatography; AC, affinity chromatography; SEC, size exclusion chromatography.

for a molecule to diffuse through a particle should not be too much longer than the time needed to pass the particle on the outside. Table 2 lists the mean diffusion times for a protein and for a small molecule in particles with different diameters and shows dramatic differences due to the squared relationship between diffusion time and diffusion distance. A standard particle for protein chromatography is usually around 100 μm. The table shows that a flow velocity of 1 mm s^{-1} is far too high for such a particle in combination with a slowly diffusing protein if the ideal matching between outside and inside transport should apply.

Thus, most approaches to accomplish fast or very fast separations of proteins have in one way or another focused on reducing this diffusion time for proteins in the chromatographic sorbent and thereby make it possible to

Table 2 Mean Diffusion Times for a Small and a Large Molecule, Calculated According to Einstein's Equation for Solute Diffusion: $t_D = d^2/2D$, where t_D is mean diffusion time, d is diffusion distance, and D is diffusion coefficient in a large-pore particle[a]

Particle diameter (μm)	Diffusion time (small molecule) (s)	Diffusion time (protein) (s)	Time for a molecule to pass outside the particle at a flow velocity of 1 mm s^{-1} (s)
100	1.25	21	0.1
10	0.0125	0.21	0.01
1	0.000125	0.0021	0.001

[a]Data are based on the diffusion coefficients in free solution for azide ($D = 10^{-5}$ cm^2 s^{-1}) and for bovine serum albumin ($D = 5.9 \times 10^{-7}$ cm^2 s^{-1}).

use increased flow rates with retained resolution/capacity, as illustrated in Fig. 1. The approaches are as follows: small particles (a), nonporous particles (b), flowthrough particles, (c) and solid diffusion particles (d). The figure also includes another principle for faster protein chromatography—turbulent chromatography (e) that concentrates on enhancing the mass transfer outside the particles. This chapter describes these approaches for fast chromatography but does not include techniques such as expanded bed adsorption, which is covered in Chap. 10. Expanded bed adsorption/ chromatography is certainly a rapid preparative technique, taking into account that pretreatment of the process stream (filtering/centrifugation) is obviated. However, the expanded bed technique itself does not qualify as a fast chromatography principle (unless small and very dense particles are used. Table 3 summarizes the properties of these new approaches to achieve faster protein separation.

2. CHROMATOGRAPHY WITH SMALL POROUS PARTICLES: HPLC/FPLC

As was pointed out in the introduction, high chromatographic flow in the column requires short diffusion distances if the chromatographed proteins

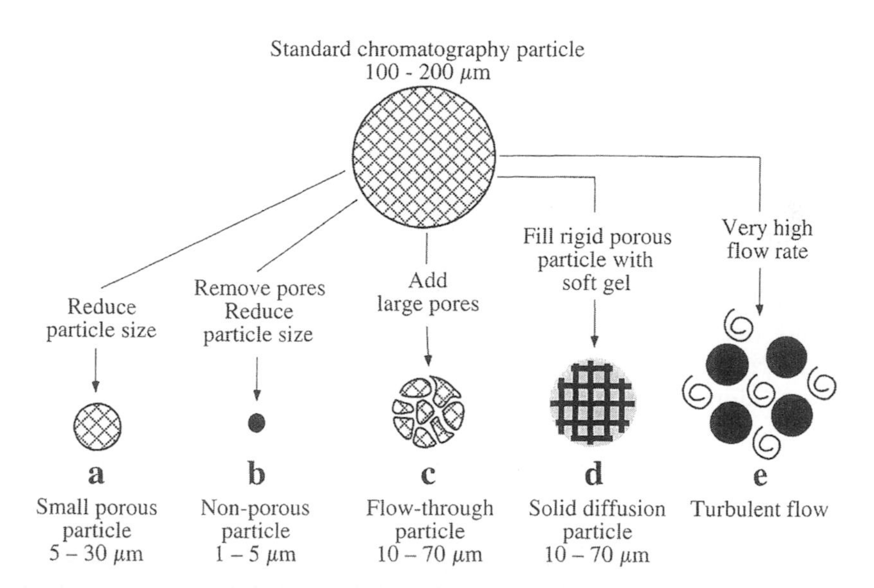

Figure 1 Approaches to fast chromatography of proteins.

Table 3 Different Chromatographic Approaches to Accomplish Fast Protein Separations[a]

Approach	Design/mechanism	Properties
Small porous particles	5–30 μm particle size	Fast separation, high capacity
Nonporous particles	1–5 μm particle size	Very fast separation, very low capacity
Flowthrough particles	Convective flow inside particle	Fast separation, medium–high capacity
Solid diffusion particles	Enhanced diffusion	Fast separation, very high capacity
Turbulent flow	Enhanced mass transfer outside the particles	Very fast separation, low capacity

[a]Capacity implies the static loading capacity (maximal binding capacity).

and the sorbent are to be reasonably well equilibrated with each other. A somewhat pragmatic approach to achieve this in adsorption/desorption chromatography is to utilize only an outer shell of a large chromatography particle, say a 30% layer. Under such conditions the particle will behave, to some extent, as three times as small particle, meaning that a much higher flow rate can be adopted. Obviously, this approach means a reduction in capacity, although not especially serious—the 30% outer layer has a binding capacity that is 66% of that of the whole particle (Janson and Pettersson 1992).

However, a more general approach is to reduce the size of the chromatography particle considerably, as was done with the introduction of high-performance liquid chromatography (HPLC) and later by fast protein liquid chromatography (FPLC). While HPLC is a general term relating to all kinds of molecules, FPLC caters for the special requirements associated with protein separations. Both techniques offer high-performance separations, not only as a result of small particle size but also as a result of a consistent adoption of materials, equipment, and procedures that diminish band spreading (pumps, valves, tubing, column construction, detectors, etc.). The term FPLC—besides being a trademark (Amersham Biosciences)—is often used as a generic term and is associated with properties such as protein compatibility, salt/buffer tolerance, and moderate pressure demands.

It is illustrative to discuss the influence of various chromatography parameters using the van Deemter equation as a starting point. The equation describes the plate height (HETP), which is a general measure of the chromatographic efficiency—the smaller the HETP value, the better efficiency—as the sum of three independent terms: multipath effect (*A* term), longitudinal diffusion (*B* term), and intraparticle diffusion (*C* term), which are schematically illustrated in Fig. 2.

The *A* term accounts for band spreading due to the different paths taken by the protein molecules between the particles in the packed bed. The *A* term is constant, independent of the flow rate, and is numerically usually given as $2d_p$ (particle diameter), i.e., 0.2 mm for a 100-μm particle. However, poorly packed columns can have substantially higher values.

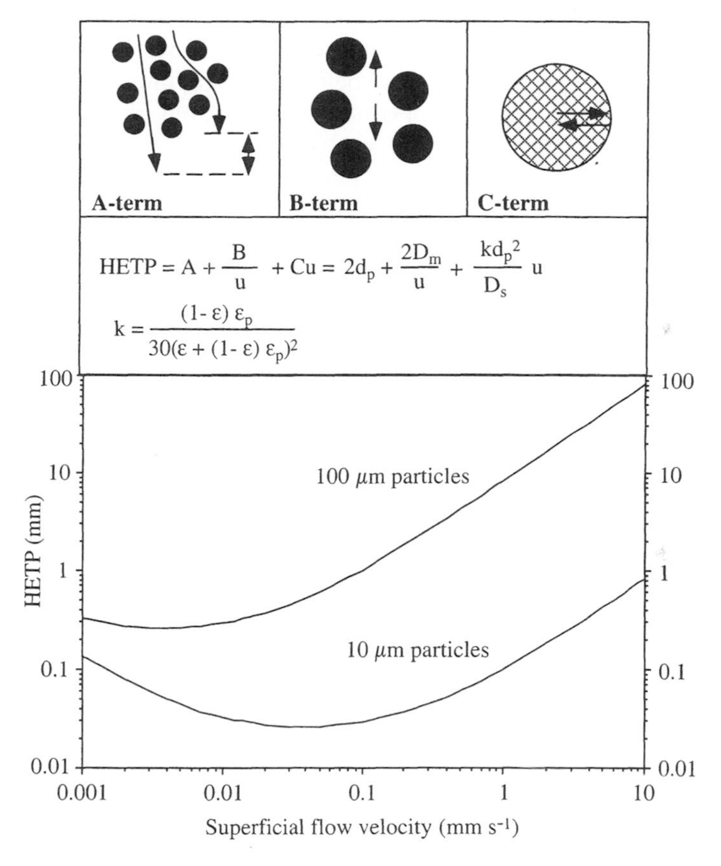

Figure 2 The van Deemter equation for nonretained solutes with the three main causes of band broadening. The resulting HETP curves for 100-μm and 10-μm particles are calculated using the following values, typical for bovine serum albumin in agarose particles: $D_m = 5.9 \times 10^{-7}\ \text{cm}^2\ \text{s}^{-1}$; $D_s = 2.5 \times 10^{-7}\ \text{cm}^2\ \text{s}^{-1}$; $\epsilon = 0.4$ (bed porosity); $\epsilon_p = 0.88$ (particle porosity).

The B term relates to longitudinal diffusion of proteins in the space between the particles in the column, and is usually taken to be $2D_m$, where D_m is the diffusion coefficient of the protein. The HETP contribution of longitudinal diffusion B/u decreases as a function of flow rate and can almost always be neglected in practical chromatography situations due to the low diffusion rate of proteins. For example, the value of B/u for BSA at a linear flow velocity of $0.1\,mm\,s^{-1}$ is approximately 0.001 mm.

The C term is usually the most important contributor to HETP and is here attributed to the slow diffusion inside the chromatography particle. Slow extraparticle diffusion also contributes to the C term, but is neglected here because its contribution is much smaller than the intraparticle contribution. The C term can be expanded as kd_p^2/D_s, where k is a constant related to the packing used and D_s is the diffusion coefficient inside the chromatography particle. Importantly, the C term contains the squared particle diameter. As can be seen in Fig. 2, particle size plays an enormous role in chromatographic efficiency. Small porous particles can be operated at high flow rates maintaining a high efficiency and capacity.

Table 4 lists some selected chromatography materials presently used for the high-performance separation of proteins, all based on small porous particles. In almost all cases the entries refer to prepacked columns. The list covers only a few suppliers of high-performance chromatography media and is by no means complete. Also, Table 4 does not cover variations with respect to coupled ligand and pore size; the reader is referred to the manufacturers' catalogues for details. As can be noted in Table 4, the particle sizes are in the range of 5–30 μm. For analytical or micropreparative separations all particle sizes may be used, including the smallest ($< 5\,\mu m$) so as to fully realize the efficiency of particles with short diffusion distances. For preparative applications the larger particle size are often preferred ($> 30\,\mu m$), since the high efficiency/small size constellation has its price, i.e., expensive packing materials and an increased pressure drop over the column. The Kozeny-Carman equation (Coulson et al. 1991) shows a troublesome relationship between particle size (d_p) and generated pressure drop in the column (ΔP):

$$\Delta P = L\eta uK/d_p^2 \tag{1}$$

where L is the length of the column, η the viscosity of the mobile phase, u the linear velocity of the mobile phase, and K a constant for a particular column relating to the geometry and the porosity of the bed. Thus, the pressure over the column increases by a factor of 100 when the particle diameter is reduced by a factor of 10. This may be perfectly acceptable with an analytical setup (HPLC equipment) but may be a major drawback in preparative applica-

Table 4 Small Conventional[a] Particles for Rapid Protein Separation (HPLC)

Trade name	Mean particle diameter (μm)	Base material	Manufacturer/ supplier
Size exclusion chromatography			
Superdex	13,34	Agarose/dextran	1
Superose	10,13,30	Agarose	1
SigmaChrom GFC	13.5	Polysaccharide[b]	2
Bio-Sil	5	Silica	3
TSK-gel SW	5,10	Silica	4
Fractogel EMD (bulk)	30	Methacrylate[b]	5
BioSep	5	Silica	6
Protein-Pak	10	Silica	7
Ion exchange chromatography			
Mono beads	10	PS/DVB[c]	1
Source	15,30	PS/DVB	1
Bio-Scale	10	Methacrylate	3
TSK gel 5PW	10	Methacrylate	4
Fractogel EMD (bulk)	30,65	Methacrylate[b]	5
BioSep	7	Polymer[b]	6
Protein Pak	8,15	Methacrylate	7
Bakerbond WP	5,15,40	Silica	8
Hydrocell	10	PS/DVB	9
Hydrophobic interaction chromatography			
Source	15	PS/DVB	1
Sepharose HP	34	Agarose	1
Macro-Prep (bulk)	50	Methacrylate	3
TSK gel 5PW	10	Methacrylate	4
Fractogel EMD (bulk)	30	Methacrylate[b]	5
Protein-Pak	10	Methacrylate	7
Bakerbond WP HI	5,15,40	Silica	8
Hydrocell	10	PS/DVB	9
Affinity chromatography			
HiTrap AC	34	Agarose	1
SigmaChrom AF	20	Methacrylate	2
TSK gel 5PW	10	Methacrylate	4
Fractogel EMD (bulk)	30,65	Methacrylate[b]	5
Protein-Pak	40	Not specified	7
Reversed phase chromatography			
Source	5,15	PS/DVB	1
Supelcosil LC	5	Silica	2
TSK gel 4PW	7	Methacrylate	4
Delta-Pak	5,15	Silica	7

Table 4 Continued

Trade name	Mean particle diameter (μm)	Base material	Manufacturer/ supplier
Hypersil, PEP	5	Silica	10
Kromasil-10	10	Silica	11

1, Amersham Biosciences, Uppsala, Sweden; 2, Supelco, Bellefonte, PA, USA; 3, Bio Rad Laboratories, Hercules, CA, USA; 4, TosoHaas, Montgomeryville, PA, USA; 5, Merck KGaA, Darmstadt, Germany; 6, Phenomenex, Torrance, CA, USA; 7, Waters, Milford, MA, USA; 8, J. T. Baker, Deventer, Holland; 9, BioChrom Labs Inc., Terre Haute, IN, USA; 10, Hypersil, Cheshire, UK; 11, EKA Chemicals AB, Separation Products, Bohus, Sweden.

[a] Flowthrough and solid diffusion media not included.

[b] Only information available from the manufacturer.

[c] PS/DVB, polystyrene/divinylbenzene.

tions (compatible equipment is expensive or unavailable, and the particles themselves may have problems handling the generated pressure drops). The pressure drop problem is to some extent relieved if monodispersed, spherical particles are chosen. Again, such particles are usually considerably more expensive than other qualities.

Figure 3 shows a gel filtration separation of a five-protein mixture plus a low molecular weight marker on different agarose-based media and illustrates the strong influence of particle size on the chromatographic efficiency of the column (Andersson et al., 1985). By decreasing the particle size from 110 μm to 13 μ, a 15-fold reduction in separation time could be achieved. Compared to the other chromatographic modes, gel filtration chromatography is noninteractive and is therefore heavily dependent on the chromatographic efficiency (low HETP value), which explains the rather moderate flow rates used in this example (2.5–38 cm h^{-1}).

3. CHROMATOGRAPHY WITH NONPOROUS MEDIA

Nonporous supports increase the speed of separation by completely eliminating the problem of slow diffusion of proteins in the interior of the chromatography support. The main drawback of nonporous supports is their low capacity, attributable to the fact that only the outer surface is available for protein binding. Nonporous supports are available in many formats where particles and various forms of membranes are the most common. The term nonporous is somewhat misleading in case of membranes since they are certainly transected by pores in the micrometer

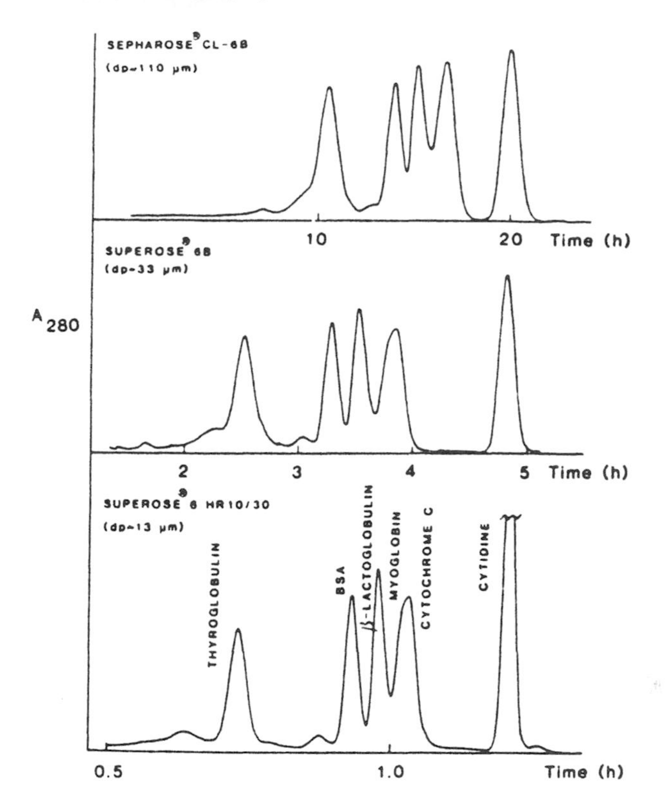

Figure 3 Comparison of three agarose-based gel filtration media. Mobile phase: 0.05 M phosphate, 0.15 M NaCl, pH 7. Sample: 0.5 mg ml^{-1} thyroglobulin, 0.8 mg ml^{-1} bovine serum albumin, 0.25 mg ml^{-1} β-lactoglobulin, 0.1 mg ml^{-1} myoglobin, 0.1 mg ml^{-1} cytochrome *c*, 0.01 mg ml^{-1} cytidine. Column dimensions, sample volume, and linear flow rate: 500 × 16 mm, 1000 μl, 2.5 cm h^{-1}; 500 × 16 mm, 100 μl, 10 cm h^{-1}; 300 × 10 mm, 50 μl 38 cm h^{-1} from top to bottom, respectively. (Reproduced with permission from Andersson et al. 1985.)

range, where the chromatographic flow is channeled. However, diffusion pores in the nanometer range, typical for ordinary, porous chromatography materials, are absent (with occasional exceptions; Tennikova et al. 1991). Some commercially available nonporous supports are listed in Table 5.

Analogous to membranes, some continuous beds may also be described as nonporous supports, including a commercially available one (Table 5). We are describing all continuous beds in connection with flow-through media.

Table 5 Some Selected Nonporous Chromatography Supports Commercially Available for the Fast Separation of Proteins

Trade name		Material	Chrom. mode[a]	Format	Company/supplier
Prepacked columns with nonporous beads					
	Part. diam. (μm)				
NPR II	1.5	Silica	RPC	Various columns	1
Kovasil MS	1.5	Silica	RPC	33 × 4.6 mm	2
Micra NPS	1.5	Silica	RPC	33 × 4.6 mm	3
TSKgel-NPR	2.5	Methacrylate	IEC/RPC/HIC	35 × 4.6 mm	4
Shodex-420N	3	Methacrylate	IEC	35 × 4.6 mm	5
MiniBeads	3	Polyether	IEC	30 × 3.2 mm	6
Hamilton PRP-∞	4	PS/DVB	RPC	30 × 4.1 mm	7
Hydrocell NP10	10	PS/DVB	IEC/HIC	Various columns	8
Continuous beds					
	Pore size (μm)				
UNO	—	Acrylate	IEC	Various columns	9
Membranes					
Sartobind	3–5	Cellulose	IEC/AC	Single sheet, stacked sheets, rolled sheet	10
Fractoflow	0.6	polyamide	IEC	Hollow fiber	11

1, YMC Co. Ltd., Kyoto, Japan; 2, Phenomenex, Torrance, CA, USA; 3, Eichrom Techn. Inc., Darien, IL, USA; 4, TosoHaas, Montgomeryville, PA, USA; 5, Showa Denko K. K., Tokyo, Japan; 6, Amersham Biosciences, Uppsala, Sweden; 7, Hamilton Co., Reno, NE, USA; 8, BioChrom Labs Inc., Terre Haute, IN, USA; 9, Bio Rad Laboratories, Hercules, CA, USA; 10, Sartorius AG, Göttingen, Germany; 11, Merck KGaA, Darmstadt, Germany; RPC, reversed phase chromatography; IEC, ion exchange chromatography; HIC, hydrophobic interaction chromatography; AC, affinity chromatography.

3.1 Particles

In the latter part of the 1960s, Horvath et al. (1967) investigated the properties of columns packed with nonporous glass beads derivatized with different ion exchange groups. The particle size ranged from 50 to 100 μm and they achieved fast separations of ribonucleotides, although the capacity was very low. Twenty years later, in order to increase the capacity, several research groups decreased the particle size to the range of 1–10 μm, and nonporous beads were coming into use for fast separation of proteins in all chromatographic modes (Burke et al. 1986; Unger et al. 1986; Kalghatgi and Horvath 1987; Kato et al. 1987; Anspach et al. 1988; Hjertén and Liao 1988; Liao and Hjertén 1988; Maa and Horvath 1988; Nimura et al. 1991). Despite the decrease in particle size, the available surface area was still quite low (Table 6), which has restricted the use of nonporous beads to analytical applications such as quality control, on-line monitoring, and purity check.

A wide range of nonporous beads are commercially available in particle sizes ranging from 1 to 5 μm for reversed phase chromatography (RPC), ion exchange chromatography (IEC), and hydrophobic interaction chromatography (HIC) separations of proteins. The nonporous beads are usually sold in prepacked columns because efficient packing of very small particles is difficult. Nonporous beads are mainly prepared from mechanically strong materials such as silica and polystyrene/divinylbenzene (PS/DVB) to cope with the high pressures generated in columns packed with very small particles.

Figure 4 gives an example of the very rapid separations that are possible with nonporous particles. Six proteins were separated on a column packed with 1.5-μm monodisperse silica particles in 7 s (Issaeva et al. 1999).

Table 6 Surface Area Related to Particle Size for Nonporous Beads

Particle diam. (μm)	Suface area ($m^2\ ml^{-1}$ packed bed)
0.5	7.2
1	3.6
2	1.8
5	0.72
10	0.36
100	0.036

Typical value for porous beads, regardless of particle size: silica 300-Å pores: 50 $m^2\ ml^{-1}$ packed bed.

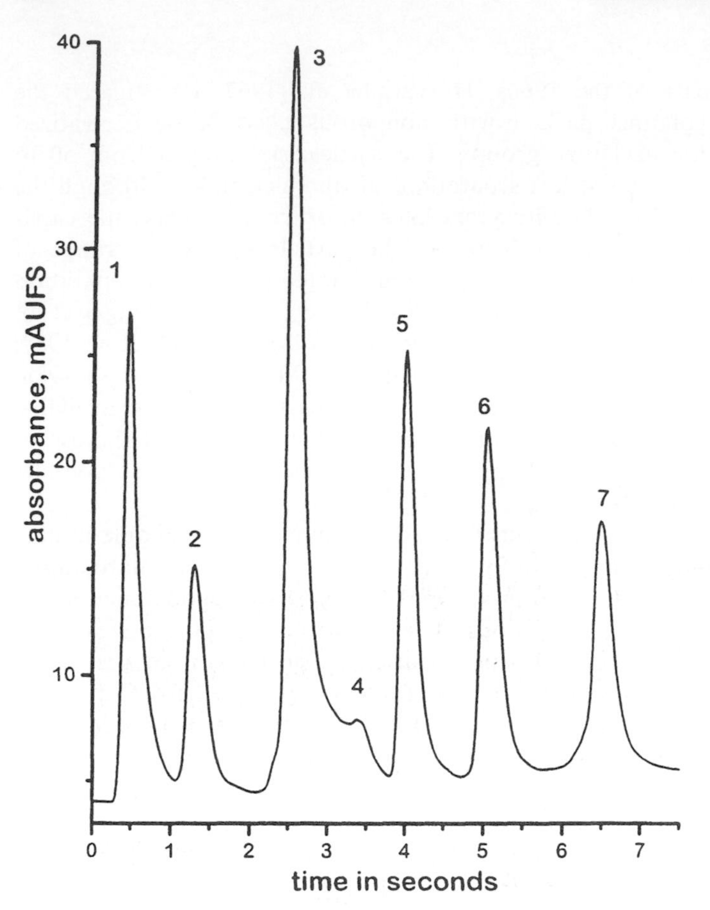

Figure 4 Fast reversed phase chromatography separation of six proteins on a nonporous particle column. Sample: 0.5 μl containing ribonuclease A (1), cytochrome *c* (2), lysozyme (3), unknown (4), bovine serum albumin (5), catalase (6) and ovalbumin (7). Column: 15 × 4.6 mm Micra NPS-RP. Flow rate: 4 ml min^{-1}. Pressure: 405 bar. Linear gradient in 5 s from 30% to 100% acetonitrile in water containing 0.1% TFA. Detection of proteins were made at 210 nm. (Reproduced with permission from Issaeva et al. 1999.)

The use of a shorter column than the commercially available 33 × 4.6 mm column was tried to allow a higher flow rate so as to reduce the separation time of the proteins. This can often be achieved since the column length does not significantly affect the resolution in gradient elution conditions (Issaeva et al. 1999). Apart from a shorter column length, a somewhat peculiar approach to reduce the separation time is to run the separation at an elevated temperature (Horvath et al. 1967; Kalghatgi and Horvath 1987; Maa and Horvath 1988; Nimura et al. 1991; Huber et al. 1997; Lee 1997; Issaeva et al. 1999). The elevated temperature lowers the eluent viscosity and thereby gives a lower pressure drop in the column [Eq. (1)]. For instance, Maa and Horvath (1988) were able to reduce the separation time by a factor of 2 when the separation of six proteins was carried out at 80°C compared with 25°C. This approach is suitable for special analytical applications where the proteins do not have to retain their structure or biological activity. A review of protein separations using nonporous particles is given by Lee (1997).

3.2 Membranes

Membranes for the separation of proteins started to appear in the literature at the end of the 1980s (Brandt et al. 1988; Huang et al. 1988). Various formats, such as single sheet (Roper and Lightfoot 1995), stacked sheets (Briefs and Kula 1992; Gerstner et al. 1992; Roper and Lightfoot 1995; Weiss and Henricksen 1995; Charcosset 1998; Josic and Strancar 1999), rolled sheet (Yang et al. 1992; Roper and Lightfoot 1995; Charcosset 1998; Josic and Strancar 1999), and hollow fibers (Roper and Lightfoot 1995; Kubota et al. 1997; Charcosset 1998; Josic and Strancar 1999), have been developed and are commercially available, as can be seen from Table 4. A common material for membranes is cellulose but several other materials are also used (Roper and Lightfoot 1995; Charcosset 1998; Josic and Strancar 1999). The membranes in analogy to nonporous beads lack diffusion pores, leading to a low surface area for protein binding and consequently a low capacity. Additional problems, which are claimed to have been solved, are the sample application and maldistribution of flowthrough membrane columns (Josic and Strancar 1999). Figure 5 shows an IgM purification from a cell culture supernatant using anion exchange membrane chromatography (Santarelli et al. 1998). After centrifugation of the cells, the supernatant was diluted with adsorption buffer and applied to the membrane at a flow rate of 10 ml min^{-1}. The IgM was detected by ELISA and corresponded to peak 4 in the diagram. The purification was completed in 8 min with an IgM purity of 95%.

The example demonstrates the attractive features of membranes: high flow rate and fast adsorption in combination with a low pressure drop. In the example (Santarelli et al. 1998), the flow was 10 ml min^{-1} through a 25-mm-diameter giving a pressure of around 0.1 bar. This makes membrane chromatography especially suitable for purification of proteins from dilute feedstreams with a low content of the target protein using ion exchange or affinity membranes (Charcosset 1998). Numerous examples of membrane

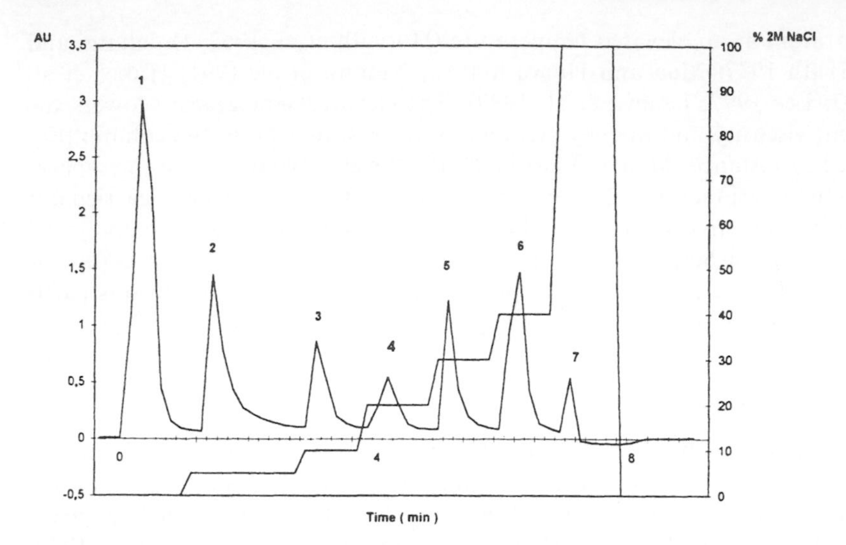

Figure 5 Purification of human monoclonal antibodies from a cell culture media using membrane chromatography. Membrane: A Sartorius single-sheet unit (Sartobind Q5) containing quaternary ammonium groups, 0.257×25 mm with a pore size of 3–5 μm. Adsorption buffer: 25 mM MES pH 7. Elution was accomplished with a multistep gradient of increasing concentration of NaCl. Sample volume: 9 ml. Detection at 280 nm. (Reproduced with permission from Santarelli et al. 1998.)

use for the isolation of proteins are documented. However, their survival on the market seems rather cumbersome since many earlier available products have been discontinued a few years after their introduction (e.g., MemSep and ActiDisc).

4. CHROMATOGRAPHY WITH FLOW-THROUGH PARTICLES

4.1 Intraparticle Convection

The nonporous particles described above allowed very fast chromatographic separations but had two serious drawbacks, at least for preparative use: low capacity and high pressure drop. The recently introduced flowthrough particles have a unique construction that takes care of these problems while still allowing very rapid separations. The flowthrough particles are comparatively large and speed up the mass transfer by an intraparticle convective flow (Fig. 6). This is achieved by a bimodal pore configuration, large transport pores, and considerably smaller diffusion pores. The intraparticle flow transports the proteins into the interior of each particle leaving only short distances to be covered by the slow diffusion, as opposed to a normal porous chromatography particle, where the chromatographed proteins must rely solely on diffusion (Fig. 6). This means that large flowthrough particles inherit advantageous properties from several types of particles, without suffering from their drawbacks: the chromatographic efficiency characteristic of smaller particles/nonporous particles, the high capacity of porous particles, and the low pressure drop of large particles (Fig. 6). However, the high-capacity statement needs some modification. The flow pores in effect constitute voids inside the particle, where no binding can occur. Thus, the total (static) binding capacity of a flowthrough particle could be expected to be somewhat reduced (25–50%) by the flow pores.

Although the concept of intraparticle convective flow started to appear in the literature in the beginning of the 1970s [size exclusion chromatography (Guttman and DiMarzio 1970; DiMarzio and Guttman 1971; Van Kreveld and Van Den Hoed 1978; Grüneberg and Klein 1981) and large-pore catalysts particles (Neale et al. 1973; Komiyama and Inoue 1974; Nir

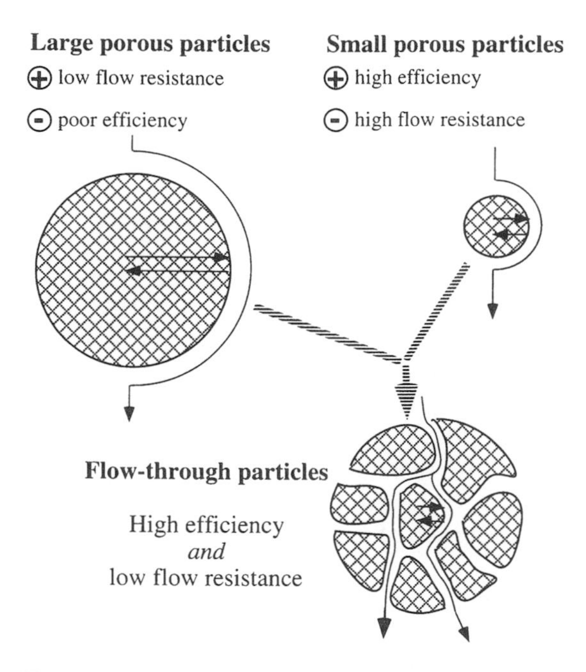

Figure 6 The concept of intraparticle convective flow combines the advantages of conventional small and large particles.

and Pismen 1977; Rodrigues et al. 1982], it was not until the late 1980s that materials suitable for protein chromatography were described by Lloyd and Warner (1990) and by Afeyan et al. (1990) who also coined the term "perfusion chromatography."

The effects of intraparticle convective flow can be modeled, e.g., via a modified van Deemter equation. A number of reports have appeared from several groups (Afeyan et al. 1990; Rodrigues et al. 1991; Carta et al. 1992; Liapis and McCoy 1992; Rodrigues et al. 1992a,b, 1993; Carta and Rodrigues 1993; Frey et al. 1993; Freitag et al. 1994; Liapis and McCoy 1994; Gustavsson and Larsson 1996; Heeter and Liapsis 1996; McCoy et al. 1996; Gustavsson et al. 1998, 1999), including a number of reviews (Potschka 1993; Li et al. 1995; Hamaker and Ladisch 1996; Collins 1997; Rodrigues 1997). Carta and Rodrigues (1993) give the following equation applicable to the chromatographic behavior of nonretained solutes:

$$\mathrm{HETP} = 4R_\mathrm{p} + 2D_\mathrm{m}/u + \left[2(1-\epsilon)\epsilon' b^2 R_\mathrm{p}^2\right] / \left\{15 D_\mathrm{p}\left[\epsilon + (1-\epsilon)\epsilon' b\right]^2\right\} \left[f(\lambda) + (b-1)/(b^2 T)\right] u \tag{2}$$

which may look complicated, but most of these parameters are constants relating to the column packing and the chromatographed substance. By choosing appropriate values for these parameters, see Table 7 [as was done by Nash and Chase (1998)] and inserting these in Eq. (2), it will appear more attractive:

$$\mathrm{HETP} = 0.04 + 8 \times 10^{-5}/u + 0.389[f(\lambda) + 0.00086]u\ (\mathrm{mm}) \tag{3}$$

HETP is now a function of the flow rate (u) and $f(\lambda)$, the convective enhancement factor. However $f(\lambda)$ in turn is also a function solely of the flow rate and varies between 1 and 0, being 1 at very low flow and 0 at very high flow. When $f(\lambda)$ has a value of 1, the equation becomes essentially the same as the equation for standard porous particles. The reason is that at such low flow rates the intraparticle flow will also be too low to contribute to an improved mass transport. A theoretical model for predicting the behavior of these particles under preparative conditions (breakthrough curves) has been developed by the group of Liapis, mainly for affinity and ion exchange chromatography systems (Liapis and McCoy 1992, 1994; Heeter and Liapis 1996).

Figure 7 shows an HETP comparison between flowthrough particles and standard particles accomplished by plotting Eq. (3) and the van Deemter equation in Fig. 2 as a function of the superficial flow velocity of the column. Looking at Fig. 7, some characteristic features can be noticed.

Table 7 Constants and Functions Used for Eq. (2) and for the van Deemter Equation in Fig. 2*

Equation	Constants and functions
2	$D_p = 4 \times 10^{-5}\ mm^2\ s^{-1}$ (diffusion coefficient in through pore)
	$D_c = 1 \times 10^{-5}\ mm^2\ s^{-1}$ (diffusion coefficient in microparticle)
	$R_m = 0.3\ \mu m$ (microparticle radius)
	$\epsilon' = 0.4$ (throughpore voidage)
	$\epsilon'' = 0.4$ (microparticle voidage)
	$K' = \epsilon''$ (distribution between the through-pore fluid and the microparticle)
	$\alpha = 0.008$ (split ratio/volumetric fraction of flow through the particle
	$f(\lambda) = 3/\lambda[(1/\tanh\lambda) - (1/\lambda)]$ (convective enhancement factor)
	$\lambda = (uR_p\alpha)/(3D_p(1-\epsilon)\epsilon')$ (intraparticle Peclet number)
	$b = 1 + (1-\epsilon)K'/\epsilon' = 1.65 T = (D_c R_p^2)/(D_p R_m^2) = 278$
van Deemter	$\epsilon_p = 0.64$ (particle porosity)
	$D_s = 1 \times 10^{-5}\ mm^2\ s^{-1}$ (diffusion coefficient in pore)
Both equations	$R_p = 10\ \mu m$ (particle radius)
	$D_m = 4 \times 10^{-5}\ mm^2\ s^{-1}$ (diffusion coefficient in mobile phase for IgG)
	$\epsilon = 0.35$ (column voidage)

*Constants and functions taken from Nash and Chase (1998) and Whitney et al. (1998)

At low mobile phase velocities in the column, the diffusional transport of proteins in the bead will dominate over the convective transport, which means that diffusion from the outside of the bead to its center is faster than the convective transport in the flow pores. This is easy to understand considering that the intraparticle flow velocity is usually around 1% of the flow velocity outside the beads. Accordingly, flowthrough particles will behave like normal particles. However, when the flow in the column is increased, the flow velocity inside the beads will increase correspondingly and the mass transport inside the beads will rely more and more on the intraparticle flow than on diffusion from the outside of the bead. This is clearly seen in Fig. 7, where the HETP curve for the flowthrough particles gradually loses its steepness and finally continues with a much lower directional coefficient.

4.2 Pore Flow Measurements

The most important factor that determines the degree of pore flow is the relation of pore size to particle size or, more precisely, the relation of the

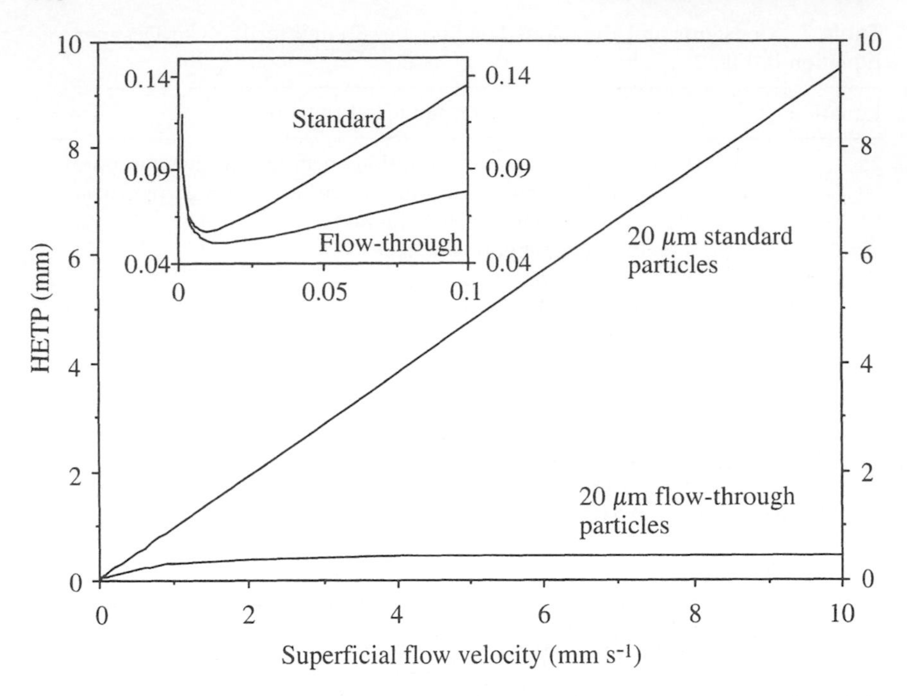

Figure 7 HETP comparison between flowthrough particles and standard particles (calculated for IgG). The curve for flowthrough particles is described by Eq. (3) and the curve for standard particles by the equation given in Fig. 2, using the parameter values in Table 7. The curves were constructed using the MS Excel spread sheet program. The insert shows the curves at very low flow velocities.

particle permeability to bed permeability (Afeyan et al. 1990; Rodrigues et al. 1991, 1992a; Carta et al. 1992; Frey et al. 1993; Heeter and Liapis 1996; McCoy et al. 1996). Direct experimental measurements of pore flow have been carried out. For instance, Pfeiffer et al. (1996) measured the volumetric flow through single particles as a function of applied pressure. The permeability of these single particles could then be determined from the pressure drop–flow rate relationship and were compared to the theoretical values based on the Kozeny-Carman equation. Direct measurements of pore flow in packed columns have also been made (Gustavsson et al. 1998) and can be quite illustrative. Figure 8 shows a microparticle (dyed yeast cell) passing through the interior of an agarose-based flowthrough particle (150 μm diameter) having very large flow pores (30 μm diameter). The figure gives the position at 1-s intervals.

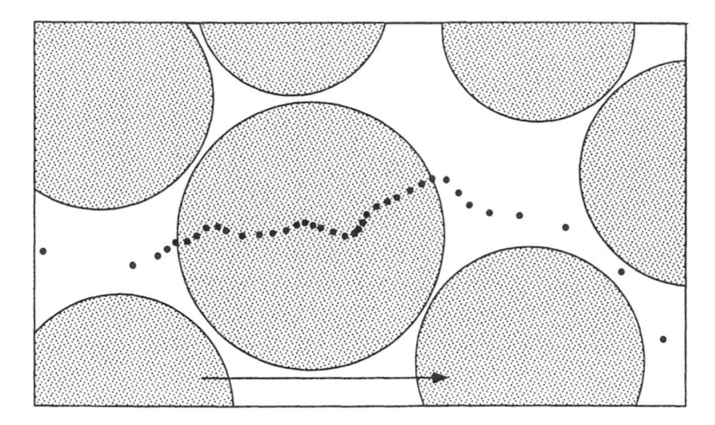

Figure 8 The position of a single microparticle (dyed yeast cell) at 1-s intervals during its passage through a superporous agarose bead situated in a packed bed. The arrow indicates the direction of flow. (Adapted from Gustavsson et al. 1998 with permission.)

The figure shows that the velocity of the yeast cell (and indirectly the velocity of the liquid) is much lower on the inside of the particle. As can also be noticed, the yeast cell slows down before entering the superpores and accelerates after leaving the superpores. The figure was constructed from video recordings through a microscope that made it possible to measure the velocities of the yeast cells both inside and outside the superporous particles. Table 8 summarizes some of the results of these measurements and emphasizes the strong influence the ratio of pore size to particle size has on the magnitude of the pore flow. Some of the split ratios achieved in these mea-

Table 8 Split Ratio[a] as a Function of the Pore Size/Particle Size Ratio for Superporous Agarose Beads (Gustavsson et al. 1998)

Pore diameter as % of particle diameter	Split ratio (%),[a] observed value	Split ratio (%),[a] calculated value
7.5	2	1.7
12.5	4.4	4.7
21	10.4	13.3

[a]The split ratio is defined as the volumetric fraction of the flow rate that goes through the particles in a packed bed and can be expressed as a percentage of the flow rate through the column.

surements are exceptionally high compared with the reported values for commercial flowthrough particles, which are around 1%.

4.3 Commercial Products/Applications

Table 9 lists commercially available flowthrough particles with bimodal pore structure, i.e., particles dedicated for protein separation and provided with both flow pores and diffusion pores. Not listed are very-large-pore (gigapores) particles, e.g., 4000-Å silica particles. Such particles have been on the market for a considerable time and will certainly have pore flow. However, they have only very large pores, meaning that the total pore surface is small, e.g., around $5\,m^2\,ml^{-1}$ packed bed, which is much less than particles with bimodal pore structure.

Flowthrough particles are preferably used in adsorption chromatography modes, such as ion exchange chromatography and affinity chromatography. An example of an industrially important affinity chromatography step is isolation of the recombinant factor VIII (r-VIII SQ) developed by Pharmacia & Upjohn (Stockholm, Sweden). The established procedure utilizes Sepharose Fast Flow derivatized with monoclonal antibodies directed against this protein.

Figure 9 shows a comparison of the breakthrough profiles for Sepharose Fast Flow and a new, agarose-based, flowthrough support (superporous agarose beads), both derivatized with anti-factor VIII antibodies (Pålsson et al. 1999). The purpose of the comparison was to find faster methods for isolating r-VIII from a large-scale animal cell culture. As can be seen from the figure, the superporous agarose matrix allowed a four-

Table 9 Some Selected Flowthrough Particles Commercially Available for High-Speed Protein Purification

Trade name	Particle diameter (μm)	Material	Chrom. mode	Manufacturer/ supplier
POROS (type 1,2)	10, 20, 50	PS/DVB	IEC/HIC/AC/RPC	1
PL-4000	12.5, 20, 60	PS/DVB	RPC/IEC	2
Unisphere	8	Alumina	RPC	3

1, Appled Biosystems, Foster City, CA, USA; 2, Polymer Laboratories, Shropshire, UK; 3, Biotage Inc., Charlottesville, VA, USA. IEC, ion exchange chromatography; HIC, hydrophobic interaction chromatography; AC, affinity chromatography; RPC, reversed phase chromatography.

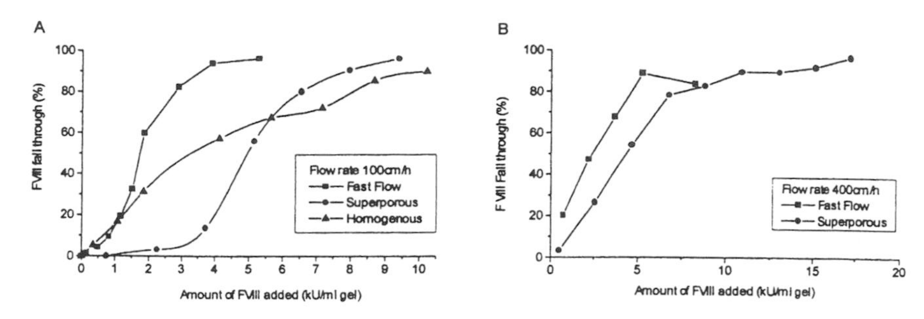

Figure 9 Comparison of breakthrough curves for normal and flow-through agarose particles. Column materials: Sepharose Fast Flow and superporous agarose beads, both derivatized with anti-factor VIII antibodies and packed in 25 × 10 mm columns. Sample: A cell culture broth containing approximately 0.03% r-VIII SQ of total protein. The detection was done by measuring the coagulant activity of r-VIII SQ. (Reproduced with permission from Pålsson et al. 1999.)

fold increase in flow rate, resulting in a much higher throughput at a 1000-fold purification in a single step. This was interpreted as a result of the better mass transfer properties of the superporous gel matrix. The homogeneous agarose matrix used in Fig. 9A was included as a reference since it had the same basic properties and was treated in the same way as the superporous agarose matrix. The investigation also pointed out the benefit of using an agarose-based material for this purification, since a commercially available polystyrene-based flowthrough support, earlier used for the same kind of purification, gave too much nonspecific adsorption (Stefansson 1993).

Ion exchange chromatography is probably the most commonly used chromatography technique in the lab as well as in industrial applications. Figure 10 illustrates the rapid preparative purification of β-galactosidase from a crude *E. coli* lysate with 20-μm flowthrough particles (Afeyan et al. 1990). The figure illustrates a scale-up, from 3.8 mg to 76 mg of protein, using normal scale-up rules, i.e., keeping the column length constant and increasing column diameter and pump speed matching to give the same linear flow velocity (1500 cm h^{-1}). The slope of the gradient of increasing NaCl concentration was kept constant.

Figure 11 shows a very-high-speed analytical separation of a protein mixture, using a pore flow material in reversed phase chromatography mode (Fulton et al. 1991). The linear velocity in the column was 8700 cm h^{-1}, i.e., 2.4 cm s^{-1} allowing the separation to occur in slightly more than 10 s.

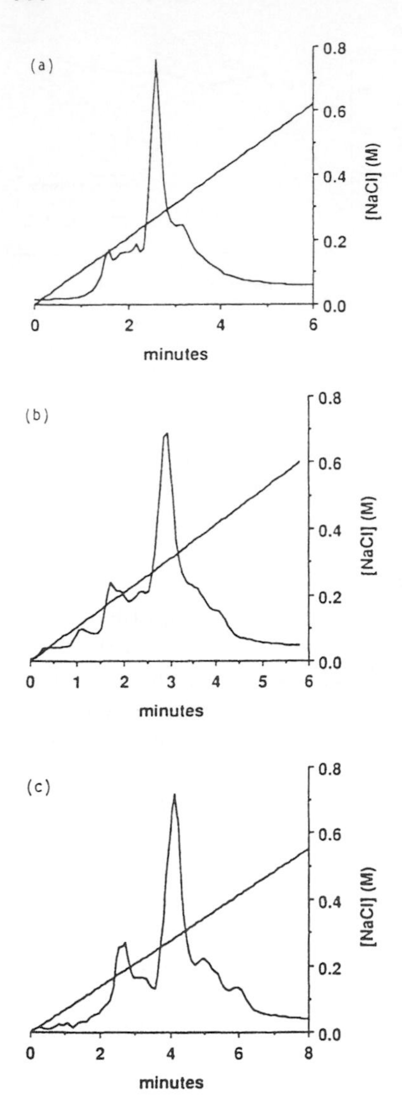

Figure 10 Scale-up of a β-galactosidase purification with perfusion particles. A crude *E. coli* lysate was used as starting material. Column material: POROS Q/M anion exchanger. Adsorption buffer: 0.02 M Tris-HCl, pH 8. Elution was carried out by a linear gradient of NaCl to a concentration of 0.5 M. Detection was made at 280 nm. Column dimensions, total protein load, flow rate, and gradient time: 100×4.6 mm, 3.8 mg, 4.2 ml min^{-1}, 4.8 min. (a); 100×10 mm, 18 mg, 20 ml min^{-1}, 4.8 min. (b); 100×25.4 mm, 76 mg, 84 ml min^{-1}, 7.2 min. (c). (Reproduced with permission from Afeyan et al. 1990.)

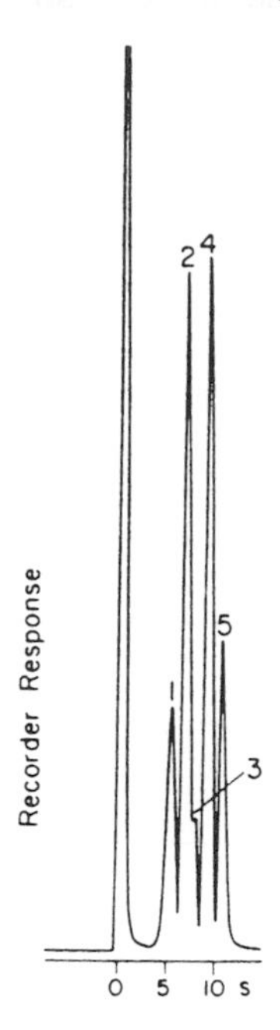

Figure 11 High-speed reversed phase chromatography separation of five proteins. Column: 30 × 2.1 mm POROS R/M (20 μm particle size). Sample: 2 μl containing 1 2 μg ribonucleasc A (1), 8 μg cytochrome *c* (2), 6 μg lysozyme (3), 8 μg *β*-lactoglobulin (4), 12 μg ovalbumin (5). Flow rate: 5 ml min^{-1} (8700 cm h^{-1}). A lincar gradient in 24 s from 20% to 50% acetonitrile in water containing 0.1% TFA was used for elution. Detection at 220 nm. (Reproduced with permission from Fulton et al. 1991.)

4.4 Continuous Beds

An interesting alternative to columns packed with beads is the continuous bed format wherein the chromatography column is made of a single piece of material (a monolith) transected by channels large enough to permit a chromatographic flow through the column. These continuous beds/rods/plugs are potentially easier and cheaper to manufacture than packed columns—at least in some configurations—and often exhibit a favorable flow rate/pressure drop relationship. Most of these continuous beds display two classes of pores: the flow pores (channels) and diffusion pores, which provide a large surface area/high binding capacity (Svec and Fréchet 1992; Minakuchi et al. 1996; Gustavsson and Larsson 1999). The continuous rods developed by Svec and Fréchet (1992) based on methacrylate or polystyrene, were prepared in the presence of a porogenic solvent to create a bimodal pore structure and have been used for the chromatographic separation of proteins. Continuous silica rods (Minakuchi et al. 1996) have been developed and also contain a bimodal pore structure. The silica rods are commercialized by Merck for the separation of small molecules. Gustavsson and Larsson (1996, 1999) have described several types of continuous beds (columns, thick membranes, fibers), all based on so-called superporous agarose. The beds were used for protein isolation, as supports for antibodies and for enzymes in rapid on-line ELISA applications (Nandakumar et al. 2000). and as an electrophoretic medium (internal cooling via circulation of a water-immiscible liquid in the flow pores (Gustavsson and Larsson 1999).

One type of continuous bed developed by Hjertén and coworkers (Hjertén et al. 1989, 1993; Hjertén 1999) could be described as nonporous in the sense that they do not contain any diffusion pores (at last not for proteins), only flow pores. This was achieved by polymerizing different acrylamide-acrylate monomers in the presence of a high salt concentration, directly in the column. The length of the bed was then decreased 10- to 15-fold by compression to produce a continuous network of nonporous bundles surrounded by channels large enough to allow a chromatographic flow through the column. The bed showed a resolution practically independent of flow rate when applied in protein separations and is now available on the market in the ion exchange mode (Liao et al. 1998; Table 5). The small nonporous bundles (1 μm) are claimed to create an enhanced capacity in comparison with nonporous beads and have replaced the nonporous bead column (Burke et al. 1986) in the chromatography media product range of Bio-Rad Laboratories.

5. CHROMATOGRAPHY WITH SOLID DIFFUSION/SURFACE DIFFUSION MEDIA

Diffusion is normally the only transport mechanism inside a particle (the flowthrough particles described in the previous section providing an exception). Usually, the diffusion can be described as pore diffusion, meaning that the solute diffuse more or less freely in the fluid inside the pores, and once the solute is adsorbed it becomes stationary and fixed to the pore wall. However, with specially designed adsorbents, an enhanced diffusion mechanism is operative, called *solid diffusion*. Solid diffusion has been well known for many years in the context of ion exchange of small molecules, but only recently for proteins (Holton et al. 1993; Boschetti 1994; Horvath et al. 1994; Boschetti et al. 1995; Boschetti and Coffman 1998). Solid diffusion particles do not contain any real pores but instead a very flexible gel matrix.

The solute is immediately sorbed by the gel and is then transported further into the particle in a sorbed state (solid diffusion). If an ion exchange particle is considered, the charges on the flexible gel matrix are partially overlapping. When the solute is transported through such a matrix it experiences a near homogeneous electric field without any local potential wells where it would be trapped. A consequence of this transport mechanism is that it is the binding capacity of the sorbent that provides the driving force for diffusion and not the concentration outside the particle, which is the case for conventional diffusion particles (and also flowthrough particles). This circumstance makes solid diffusion adsorbents especially suited to capture proteins from very dilute solutions, such as in the purification of monoclonal antibodies (Holton et al. 1993).

A peculiar aspect of solid diffusion adsorbents is that unretained proteins are excluded from the interior of the particle altogether—there are simply no pores for normal diffusion. Again, if the conditions are changed (e.g., the pH of the buffer), the protein will adsorb and gain access to the whole particle via solid diffusion through the gel matrix. Since the gel matrix usually harbor an abundance of charges, the total capacity of solid diffusion adsorbents may be very high.

A commercially available solid diffusion particle, trade name HyperD (BioSepra), is a composite consisting of a rigid porous material whose pores are filled with a soft, flexible, charge-carrying acrylamide gel (ion exchanger), as shown in Fig. 12. The material for the rigid porous material can be varied as seen in the figure. The commercial ion exchange product is available in a number of particle sizes (10–70 μm), has a very high protein binding capacity (125 mg ml^{-1} for BSA with an anion exchanger), and is made of

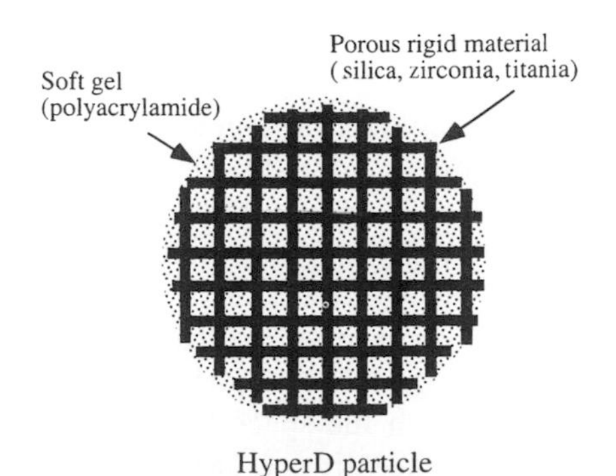

Figure 12 Schematic illustration of a HyperD particle.

a ceramic material that resists solutions of high pH, commonly used to regenerate/sanitize chromatography adsorbents.

The HyperD particle takes advantage of the mechanical strength of the ceramic material, which prevents the swelling of the soft gel matrix. Such swelling may be difficult to avoid in other particles with less rigid structure, when the repelling effect of overlapping charges causes the gel to swell. The resulting charge separation diminishes the chances for solid diffusion.

Also other materials on the market have been shown to be at least partially operating according to the solid diffusion principle, e.g., the XL series of adsorbents from Amersham Biosciences (Thömmes 1999). Here long flexible chains of dextran with ion exchange groups fill the pores of highly cross-linked agarose particles. These particles show very high binding capacities, e.g., 160 mg ml^{-1} for lysozyme with a cation exchanger. Possibly also the so-called tentacle media from Merck (Fractogel EMD, Table 4) could have a solid diffusion element (or surface diffusion element as described below).

Closely related to the solid diffusion concept is surface diffusion. Here the overlapping charges are confined to a surface, preferably on flexible spacers, and the diffusing proteins are moving over the surface in a sorbed state. The membranes from Sartorius (Sartobind, Table 5; Chap. 6) have been shown to operate according to this principle (Gebauer et al. 1997).

Several papers describing the theoretical modeling of solid diffusion/surface diffusion mechanism in chromatography are available (Yoshida et al. 1994; Fernandez and Carta 1996; Fernandez et al. 1996; Weaver and Carta 1996; Gebauer et al. 1997; Boschetti and Coffman 1998; Hansen and Mollerup 1998; Thömmes 1999).

6. TURBULENT FLOW CHROMATOGRAPHY

An interesting but seldom used approach to achieve fast protein separations is to use flow rates so high that a turbulent regime is entered in the column. The basic concept has been well known for a long time, but practical attempts have been scarce until Cohesive Technologies introduced "turbulent chromatography." Turbulent flow chromatography is performed on columns packed with 50-μm conventional porous chromatography particles at very high flow rates (e.g., 14,000 cm h^{-1} or 4 cm s^{-1}). The turbulent flow give a more efficient distribution of the proteins in the space between the particles in the bed. The resulting higher solute concentration at the particle surface also improves the mass transfer in the pores (Cohesive Technologies 1999).

However, as was pointed out by Knox (1999), the slow intraparticle diffusion will still be the major cause of band broadening when turbulent chromatography is run together with porous supports. Clearly the turbulent chromatography concept can only be efficient if nonporous particles are used. So far, turbulent flow chromatography has been demonstrated in separations of a few proteins in gradient elution mode (Knox 1999).

ACKNOWLEDGMENT

Advice from Dr. Jörg Thömmes at the Institut for Enzyme Technology, Heinrich-Heine University, Düsseldorf, Jülich, Germany and the economic support from the Swedish Center for Bioseparation is gratefully acknowledged.

REFERENCES

Afeyan N.B., Gordon N.F., Mazsaroff I., Varady L., Fulton S.P., Yang Y.B., Regnier F.E. (1990) Flow-through particles for the high-performance liquid chromatographic separation of biomolecules: perfusion chromatography. J. Chromatogr. 519, 1–29.

Andersson T., Carlsson M., Hagel L., Pernemalm P.-Å., Janson J.-C. (1985) Agarose-based media for high-resolution gel filtration of biopolymers. J. Chromatogr. 326, 33–44.

Anspach B., Unger K.K., Davies J., Hearn M.T.W. (1988) Affinity chromatography with triazine dyes immobilized onto activated non-porous monodisperse silicas. J. Chromatogr. 457, 195–204.

Boschetti E. (1994) Advanced sorbents for preparative protein separation purposes. J. Chromatogr. A 658, 207–236.

Boschetti E., Coffman J.L. (1998) Enhanced diffusion chromatography and related sorbents for biopurification. In: Bioseparation and Bioprocessing, Vol. 1 (G. Subramanian, ed.), Wiley-VCH, Weinheim, pp. 157–198.

Boschetti E., Guerrier L., Girot P., Horvath J. (1995) Preparative high-performance liquid chromatographic separation of proteins with HyperD ion-exchange supports. J. Chromatogr. B 664, 225–231.

Brandt S., Goffe R.A., Kessler S.B., O'Connor J.L., Zale S.E. (1988) Membrane-based affinity technology for commercial scale purifications. Bio/Technology 6, 779–782.

Briefs K.-G., Kula M.-R. (1992) Fast protein chromatography on analytical and preparative scale using modified microporous membranes. Chem. Eng. Sci. 47, 141–149.

Burke D.J., Duncan J.K., Dunn L.C., Cummings L., Siebert C.J., Ott G.S. (1986) Rapid protein profiling with a novel anion-exchange material. J. Chromatogr. 353, 425–437.

Carta G., Gregory M.E., Kirwan D.J., Massaldi H.A. (1992) Chromatography with permeable supports: theory and comparison with experiments. Sep. Technol. 2, 62–72.

Carta G., Rodrigues A.E. (1993) Diffusion and convection in chromatographic processes using permeable supports with a biodisperse pore structure. Chem. Eng. Sci. 48, 3927–3935.

Charcosset C. (1998) Purification of proteins by membrane chromatography. J. Chem. Technol. Biotechnol. 71, 95–110.

Cohesive Technologies, Acton, MA, USA (1999) Turbulent flow chromatography, Information via Website www.cohesivetech.com.

Collins W.E. (1997) Protein separation with flow-through chromatography. Sep. Purif. Meth. 26, 215–253.

Coulson J.M., Richardson J.F., Backhurst J.R., Harker J.H. (1991) Chemical Engineering, Vol. 2, 4th ed., Pergamon Press, Oxford.

DiMarzio E.A., Guttman C.M. (1971) Separation by flow and its application to gel permeation chromatography. J. Chromatogr. 55, 83–97.

Fernandez M.A., Carta G. (1996) Characterization of protein adsorption by composite silica–polyacrylamide gel anion exchangers. I. Equilibrium and mass transfer in agitated contactors. J. Chromatogr. A 746, 169–183.

Fernandez M.A., Laughinghouse W.S., Carta G. (1996) Characterization of protein adsorption by composite silica–polyacrylamide gel anion exchangers. I. Mass transfer in packed columns and predictability of breakthough behavior. J. Chromatogr. A 746, 185–198.

Freitag R., Frey D., Horváth C. (1994) Effect of bed compression on high-performance liquid chromatography columns with gigaporous polymeric packings. J. Chromatogr. A 686, 165–177.

Frey D.D., Schweinheim E., Horváth C. (1993) Effect of intraparticle convection on the chromatography of biomacromolecules. Biotechnol. Prog. 9, 273–284.

Fulton S.P., Afeyan N.B., Gordon N.F., Regnier F.E. (1991) Very high speed separation of proteins with a 20-μm reversed-phase sorbent. J. Chromatogr., 547, 452–456.

Gebauer K.H., Thömmes J., Kula M.R. (1997) Breakthrough performance of high-capacity membrane adsorbers in protein chromatography. Chem. Eng. Sci. 52, 405–419.

Gerstner J.A., Hamilton R., Cramer S.M. (1992) Membrane chromatographic systems for high-throughput protein separations. J. Chromatogr. 596, 173–180.

Grüneberg M., Klein J. (1981) Mass transfer of macromolecules in steric exclusion chromatography. 2. Convective transport in internal pores (hydrodynamic chromatography). Macromolecules 14, 1415–1419.

Gustavsson P.-E., Axelsson A., Larsson P.-O. (1998) Direct measurements of convective fluid velocities in superporous agarose beads. J. Chromatogr. A 795, 199–210.

Gustavsson P.E., Axelsson A., Larsson P.-O. (1999) Superporous agarose beads as a hydrophobic interaction chromatography support. J. Chromatogr. A 830, 275–284.

Gustavsson P.-E., Larsson P.-O. (1996) Superporous agarose, a new material for chromatography. J. Chromatogr. A 734, 231–240.

Gustavsson P.-E., Larsson P.-O. (1999) Continuous superporous agarose beds for chromatography and electrophoresis. J. Chromatogr. A 832, 29–39.

Guttman C.M., DiMarzio E.A. (1970) Separation by flow. II. Application to gel permeation chromatography. Macromolecules 3, 681–691.

Hamaker K.H., Ladisch M.R. (1996) Intraparticle flow and plate height effects in liquid chromatography stationary phases. Sep. Purif. Meth. 25, 47–83.

Hansen E., Mollerup J. (1998) Application of the two-film theory to the determination of mass transfer coefficients for bovine serum albumin on anion-exchange columns. J. Chromatogr. A 827, 259–267.

Heeter G.A., Liapis A.I. (1996) Effects of structural and kinetic parameters on the performance of chromatographic columns packed with perfusive and purely diffusive adsorbent particles. J. Chromatogr. A 743, 3–14.

Hjertén S. (1999) Standard and capillary chromatography, including electrochromatography, on continuous polymer beds (monoliths), based on water-soluble monomers. Ind. Eng. Chem. Res. 38, 1205–1214.

Hjertén S., Liao J.-L. (1988) High-performance liquid chromatography of proteins on compressed, non-porous agarose beads. I. Hydrophobic-interaction chromatography. J. Chromatogr. 457, 165–174.

Hjertén S., Liao J.-L., Zhang R. (1989) High-performance liquid chromatography on continuous polymer beds. J. Chromatogr. 473, 273–275.

Hjertén S., Nakazato K., Mohammad J., Eaker D. (1993) Reversed-phase chromatography of proteins and peptides on compressed continuous beds. Chromatographia 37, 287–294.

Holton III, O.D., Blizzard C.D., Vicalvi Jr. J.J., Pearson A.J. (1993) Hyperdiffusion chromatography for hi-speed, hi-capacity antibody purification. Gen. Eng. News, Jan, p. 10.

Horvath C., Preiss B.A., Lipsky S.R. (1967) Fast liquid chromatography: an investigation of operating parameters and the separation of nucleotides on pellicular ion exchangers. Anal. Chem. 39, 1422–1428.

Horvath J., Boschetti E., Guerrier L., Cooke N. (1994) High-performance protein separations with novel strong ion exchangers. J. Chromatogr. A 679, 11–22.

Huang S.H., Roy S., Hou K.C., Tsao G.T. (1988) Scaling-up of affinity chromatography by radial-flow cartridges. Biotechnol. Prog. 4, 159–165.

Huber C.G., Kleindienst G., Bonn G.K. (1997) Application of micropellicular polystyrene/divinylbenzene stationary phases for high-performance reversed-phase liquid chromatography electrospray-mass spetrometry of proteins and peptides. Chromatographia 44, 438–448.

Issaeva T., Kourganov A., Unger K.K. (1999) Super-high-speed liquid chromatography of proteins and peptides on non-porous Micra NPS-RP packings. J. Chromatogr. A 846, 13–23.

Janson J.-C., Pettersson T. (1992) Large-scale chromatography of proteins. In: Preparative and Production Scale Chromatography (G. Ganetsos, P.E. Barker, eds.), Marcel Dekker, New York, pp. 559–590.

Josic D., Strancar A. (1999) Application of membranes and compact, porous units for the separation of biopolymers. Ind. Eng. Chem. Res. 38, 333–342.

Kalghatgi K., Horváth C. (1987) Rapid analysis of proteins and peptides by reversed-phase chromatography. J. Chromatogr. 398, 335–339.

Kato Y., Kitamura T., Mitsui A., Hashimoto T. (1987) High-performance ion-exchange chromatography of proteins on non-porous ion exchangers. J. Chromatogr. 398, 327–334.

Knox J.H. (1999) Band dispersion in chromatography—a new view of A-term dispersion. J. Chromatogr. A 831, 3–15.

Komiyama H., Inoue H. (1974) Effects of intraparticle flow on catalytic reactions. J. Chem. Eng. Japan 7, 281–286.

Kubota N., Konno Y., Saito K., Sugita K., Watanabe K., Sugo T. (1997) Module performance of anion-exchange porous hollow-fiber membranes for high-speed protein recovery. J. Chromatogr. A 782, 159–165.

Lee W.-C. (1997) Protein separation using non-porous sorbents. J. Chromatogr. B, 699, 29–45.

Li Q., Grandmaison E.W., Hsu C.C., Taylor D., Goosen M.F.A. (1995) Interparticle and intraparticle mass transfer in chromatographic separation. Bioseparation 5, 189–202.

Liao J.-L., Hjertén S. (1988) High-performance liquid chromatography of proteins on compressed, non-porous agarose beads. II. Anion-exchange chromatography. J. Chromatogr. 457, 175–182.

Liao J.-L., Lam W.-K., Tisch T.L., Franklin S.G. (1998) Continuous bed chromatography. Gen. Eng. News, Jan 1, p. 17.

Liapis A.I., McCoy M.A. (1992) Theory of perfusion chromatography J. Chromatogr. 599, 87–104.

Liapis A.I., McCoy M.A. (1994) Perfusion chromatography, effect of micropore diffusion on column performance in systems utilizing perfusive adsorbent particles with a biodisperse porous structure. J. Chromatogr. A 660, 85–96.

Lloyd L.L., Warner F.P. (1990) Preparative high-performance liquid chromatography on a unique high-speed macroporous resin. J. Chromatogr. 512, 365–376.

Maa Y.-F., Horvath C. (1988) Rapid analysis of proteins and peptides by reversed-phase chromatography with polymeric micropellicular sorbents. J. Chromatogr. 445, 71–86.

McCoy M.A., Kalghatgi K., Regnier F.E., Afeyan N.B. (1996) Perfusion chromatography—characterization of column packings for chromatography of proteins. J. Chromatogr. A 743, 221–229.

Minakuchi H., Nakanishi K., Soga N., Ishizuka N., Tanaka N. (1996) Octadecylsilylated porous silica rods as separation media for reversed-phase liquid chromatography. Anal. Chem. 68, 3498–3501.

Nandakumar M.P., Pålsson E., Gustavsson P.-E., Larsson P.-O., Mattiasson B. (2000) Superporous agarose monoliths as mini-reactors in flow injection systems, on-line monitoring of metabolites and intracellular enzymes in microbial cultivation processes. Bioseparation 9, 193–202.

Nash D.C., Chase H.A. (1998) Comparison of diffusion and diffusion–convection matrices for use in ion-exchange separations of proteins. J. Chromatogr. A 807, 185–207.

Neale G., Epstein N., Nader W. (1973) Creeping flow relative to permeable spheres. Chem. Eng. Sci. 28, 1865–1875.

Nimura N., Itoh H., Kinoshita T., Nagae N., Nomura M. (1991) Fast protein separation by revered-phase high-performance liquid chromatography on octadecylsilyl-bonded non-porous silica gel. Effect of particle size of column packing on column efficiency. J. Chromatogr. 585, 207–211.

Nir A., Pismen L.M. (1977) Simultaneous intraparticle forced convection, diffusion and reaction in a porous catalyst. Chem. Eng. Sci. 32, 35–41.

Pålsson E., Smeds A.-L., Petersson A., Larsson P.-O. (1999) Faster isolation of recombinant factor VIII SQ, with a superporous agarose matrix. J. Chromatogr. A 840, 39–50.

Pfeiffer J.F., Chen J.C., Hsu J.T. (1996) Permeability of gigaporous particles. AIChE J. 42, 932–939.

Potschka M. (1993) Mechanism of size-exclusion chromatography I. Role of convection and obstructed diffusion in size-exclusion chromatography. J. Chromatogr. 648, 41–69.

Rodrigues A.E. (1997) Permeable packings and perfusion chromatography in protein separation. J. Chromatogr. B 699, 47–61.

Rodrigues A.E., Ahn B.J., Zoulalian A. (1982) Intraparticle-forced convection effect in catalyst diffusivity measurements and reactor design. AIChE J. 28, 541–546.

Rodrigues A.E., Lu Z.P., Loureiro J.M. (1991) Residence time distribution of inert and linearly adsorbed species in a fixed bed containing "large-pore" supports: applications in separation engineering. Chem. Eng. Sci. 46, 2765–2773.

Rodrigues A.E., Lopes J.C., Lu Z.P., Loureiro J.M., Dias M.M. (1992a) Importance of intraparticle convection in the performance of chromatographic processes. J. Chromatogr. 590, 93–100.

Rodrigues A.E., Ramos A.M.D., Loureiro J.M., Diaz M., Lu Z.P. (1992b) Influence of adsorption-desorption kinetics on the performance of chromatographic processes using large-pore supports. Chem. Eng. Sci. 47, 4405–4413.

Rodrigues A.E., Lu Z.P., Loureiro J.M., Carta G. (1993) Peak resolution in linear chromatography, effects of intraparticle convection. J. Chromatogr. A 653, 189–198.

Roper D.K., Lightfoot E.N. (1995) Separation of biomolecules using adsorptive membranes. J. Chromatogr. A 702, 3–26.

Santarelli X., Domergue F., Clofent-Sanchez G., Dabadie M., Grissely R., Casagne C. (1998) Characterization and application of new macroporous membrane ion exchangers. J. Chromatogr. B 706, 13–22.

Stefansson K. (1993) Diploma work, Royal Institute of Technology, Stockholm.

Svec F., Fréchet J.M.J. (1992) Continuous rods of macroporous polymer as high-performance liquid chromatography separation media. Anal. Chem. 64, 820–822.

Tennikova T.B., Bleha M., Svec F., Almazova T.V., Belenkii B.G. (1991) High-performance membrane chromatography of proteins, a novel method of protein separation. J. Chromatogr. 555, 97–107.

Thömmes J. (1999) Investigations on protein adsorption to agarose-dextran composite media. Biotechnol. Bioeng. 62, 358–362.

Unger K.K., Jilge G., Kinkel J.N., Hearn M.T.W. (1986) Evaluation of advanced silica packings for the separation of biopolymers by high-performance liquid chromatography. II. Performance of non-porous monodisperse 1.5 μm silica beads in the separation of proteins by reversed-phase gradient elution high performance liquid chromatography. J. Chromatogr. 359, 61–72.

Van Kreveld M.E., Van Den Hoed N. (1978) Mass transfer phenomena in gel permeation chromatography. J. Chromatogr. 149, 71–91.

Weaver L.E., Carta G. (1996) Protein adsorption on cation exchangers: comparison of macroporous and gel-composite media. Biotechnol. Prog., 12, 342–355.

Weiss A.R., Henricksen G. (1995) Membrane adsorbers for rapid and scaleable protein separations. Gen. Eng. News, May 1, p. 22.

Whitney D., McCoy M., Gordon N., Afeyan N. (1998) Characterization of large-pore polymeric supports for use in perfusion biochromatography. J. Chromatogr. A 807, 165–184.

Yang Y., Velayudhan A., Ladisch C.M., Ladisch M.R. (1992) Protein chromatography using a continuous stationary phase. J. Chromatogr. 598, 169–180.

Yoshida H., Yoshikawa M., Kataoka T. (1994) Parallel transport of BSA by surface and pore diffusion in strongly basic chitosan. AIChE J. 40, 2034–2044.

12

Novel Approaches to the Chromatography of Proteins

Ruth Freitag
Swiss Federal Institute of Technology, Lausanne, Switzerland

1. INTRODUCTION

No book on the isolation and purification of proteins would be complete without a chapter on protein chromatography. The reason for this is that many proteins of interest occur in very complex environments and their isolation requires a high-resolution method that can exploit even minor differences in structure, such as one amino acid or glycosylation residue, or configuration. Currently, chromatography in its various forms is unsurpassed in this regard. Moreover, since any separation involves at some point the creation of two phases, it is at present difficult to even imagine some other method to replace chromatography with its comparatively "physiological" conditions of temperature and buffer composition and pH in protein separation. Especially in the life sciences, chromatographic separations have become the workhorse for protein separation. In the preparative range there is no other technique of similar performance and in the analytical range only electrophoresis comes close. Moreover, as biotechnology develops as an industry, preparative scale has progressed from the milligram to the gram and kilogram scale. This trend is likely to continue, and other branches of industry may profit from the concomitant progress in preparative chromatographic methods. Examples are plasma fractionation or the dairy industry.

At present, protein chromatography suffers from a well-known "dilemma," i.e., that of combining resolution with speed and capacity.

Chromatographic separations typically involve a stationary phase, placed inside a column, and a mobile phase, percolating through this column. If a column volume is to be used to advantage, a high adsorption capacity (surface area) is required. In order to achieve this, the surface of the stationary phase is maximized by the use of porous particles. However, the inner surface of these particles can only be reached by molecular diffusion, a time-consuming process especially for large biologicals such as proteins, which typically have diffusivities of two to three orders of magnitude less than, for example, small salt ions. If the mobile phase flow through the column is too fast, the slow transport in and out of the particles will increasingly have a broadening effect (dispersion) on the peak. Especially in protein chromatography, the mobile phase flow rate and concomitantly the speed of a given separation are therefore restricted. This interrelation between mobile phase flow and peak width is usually depicted in form of the so-called van Deemter curve, which gives the theoretical plate height, H (a measure of the peak width/column efficiency), as a function of the mobile phase flow rate, u. In its simplest form, the following mathematical description can be used to describe the van Deemter curve:

$$H = B/u + A + Cu \tag{1}$$

where A, B, and C are constants.

For conventional columns packed with porous particles, the curve passes through an optimum (minimum). For a well packed column the smallest H value should be in the range of 2–3 times the particle diameter. For flow rates above the optimum one, H increases more or less linearly with u (C-term region). A similar increase in plate height is observed for very low flow rates (B-term region), this time due to band broadening by molecular diffusion. The B-term region of the curve is seldom a cause for concern in protein chromatography, whereas the rapid decrease of the column efficiency with higher flow rates is.

For a long time, the "solution" to the dilemma was to achieve high resolution at the price of average capacity and fairly low speeds. Gradient separations that took 1 h to separate the amino acids contained in one 10-μl sample were accepted. However, this has changed since high-throughput screening became a major challenge to analytical chemistry and preparative chromatography entered the truly industrial scale. The powerful chromatographic principle is most likely to stay; however, its current realization and application needs to change drastically. Changes have to occur at all levels: the chromatographic mode, the instrumentation, and even the stationary phase morphology. In the following sections, three "nonconventional" approaches to chromatography are discussed, which may help to transform chromatography into the unit operation needed in the upcoming era of bioindustrial production. With displacement chromatography, a more productive mode of chromatography is discussed. Monolith chromatography offers an escape from the dilemma of biochromatography, as it promises to combine capacity, speed, and resolution in a new way. Finally, continuous annular chromatography is the first system for the continuous chromatographic separation of multicomponent mixtures. This selection is somewhat arbitrary, and certain areas of bioseparation that have recently received much interest will not be touched upon. These include a number of important preparative methods, such as the simulated moving bed (Nicoud 1998), but also principles that are more suited to adsorption-desorption type of separations, such as expanded bed (Anspach et al. 1999; Chap. 10) and radial flow chromatography (Wallworth 1998).

2. DISPLACEMENT CHROMATOGRAPHY

People tend to associate a separation as shown in Fig. 1 with chromatography, i.e., an approach called elution chromatography. The components of the sample mixture are separated into individual "peaks" separated by volumes of "empty" mobile phase. This facilitates detection because most biopolymers, such as proteins and DNA, can be followed by a UV detec-

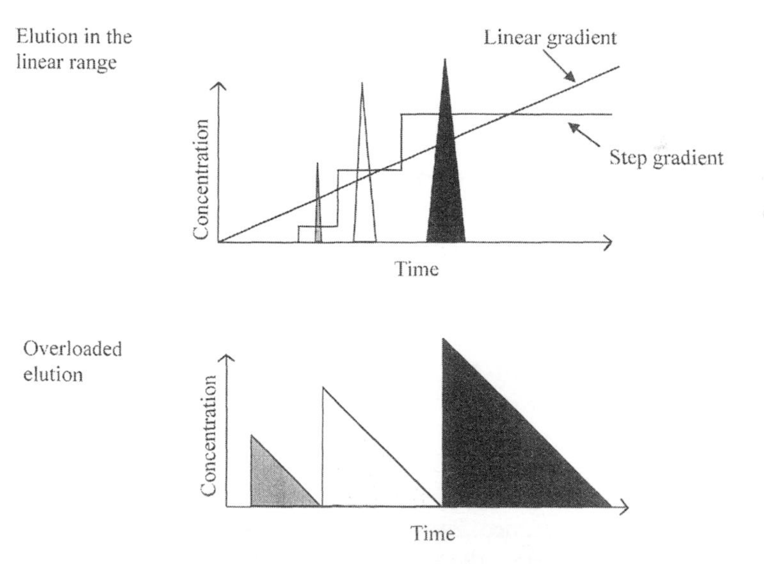

Figure 1 Schematic presentation of a separation by elution chromatography.

tor. Given that each peak contains only one substance, quantification is easy. Even "identification" becomes possible, since elution (retention) times should not change much for a given set of (analytical) conditions. If the mobile phase does not change during the separation, we speak of isocratic elution, a term coined by Csaba Horvath in the late 1960s, meaning elution "at constant strength" (from the Greek *iso* meaning "the same" and *cratos* meaning "strength"). If the elutive power of the mobile phase is increased during the separation, we speak of gradient elution. This approach is almost always necessary in the case of protein mixtures, since proteins tend to show a wide range of stationary phase affinities. It is therefore almost impossible to define a single mobile phase composition that allows all proteins of a given mixture to pass through the column in a reasonable time.

In gradient elution the proteins bind strongly to the stationary phase until the elutive strength of the mobile phase enforces desorption. Once this is the case, the protein ceases to interact and moves at mobile phase speed through the remainder of the column. An important consequence of this model is that the gradient is necessarily superimposed on the separation. Gradient elution is a powerful principle for analytical chromatography. Scale-up is possible, however, there are disadvantages to using the elution approach for preparative chromatography. One is the fact that either the stationary phase capacity is not fully exploited or the column has to be overloaded. In overloaded elution chromatography, the separation (resolution) suffers; otherwise the throughput becomes less than optimal. A second problem is the gradient. The creation of a smooth, reproducible linear gradient is no problem in a small analytical column but is very difficult for the typical large-diameter preparative columns. Thus, step gradients are more popular in preparative elution chromatography, which renders the system either more complex (one step per substance) or less powerful (one step for the low-affinity impurities/contaminants, one step for the product, one step for the high-affinity impurities/contaminants).

Displacement chromatography is a viable alternative to elution chromatography in the preparative mode. The course of a typical displacement separation is depicted in Fig. 2. First the substance mixture to be separated is loaded on the column until the stationary phase capacity is almost exhausted. Decidedly nonlinear conditions, i.e., heavy column "overloading," are necessary for a separation in the displacement mode. Afterward, the column is flushed with a solution containing the so-called displacer. This displacer step is responsible for the separation of the sample components into consecutive pure substance zones. Since the substance mixture is resolved into consecutive zones, the column's capacity is used much more efficiently in displacement then in elution chromatography.

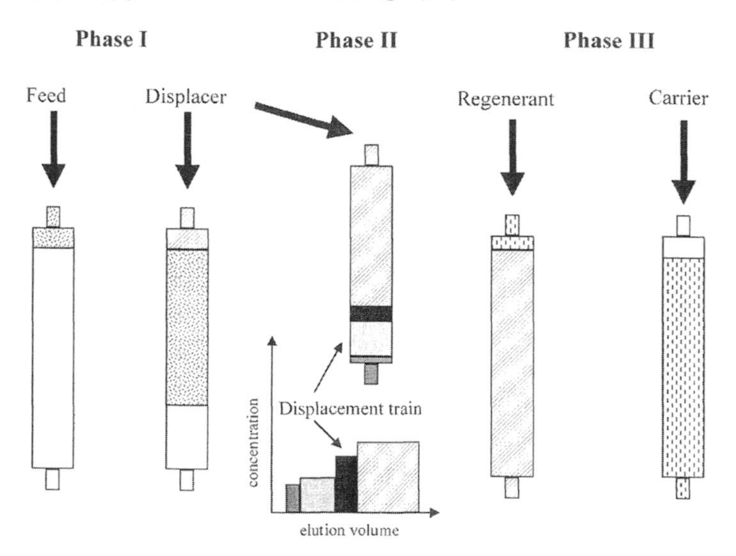

Figure 2 Stages of a typical displacement separation.

2.1 Theoretical Background of Displacement Chromatography

The separation in displacement chromatography is caused by the competition for the stationary phase binding sites. The displacer has a higher stationary phase affinity than the sample components. Under nonlinear conditions, the binding of any of the substances cannot be considered independently; instead, there is competition for the binding sites and under these conditions the isotherms become suppressed. The easiest way to demonstrate this effect is the competitive Langmuir isotherms.

For any individual substance, this formalism links the amount of a given substance i adsorbed to the stationary phase (q_i) to the concentration of i in the mobile phase (c_i).

$$q_i = \frac{b_i c_i}{1 + a_i c_i} \tag{2}$$

with a and b as constants.

If two substances compete for the binding sites, the equation becomes:

$$q_i = \frac{b_i c_i}{1 + a_i c_i + a_j c_j} \tag{3}$$

Hence for a given c_i the amount q_i becomes smaller if a substance j is also present.

As the displacer front advances through the column, the number of binding sites becomes smaller and smaller and competition between the proteins increases. As a result, the proteins with the lowest stationary phase affinity are pushed forward by those with medium and high affinity, and finally even the protein with the highest affinity leaves the column immediately before the displacer front. In the final phase of the displacement separation, the stationary phase is regenerated and reequilibrated.

This separation mechanism renders displacement chromatography somewhat attractive for preparative applications. A single displacer step is sufficient to separate a multicomponent protein mixture. Concomitantly, the displacer stays behind the protein zones rather than intermingling with them, as the gradient in elution chromatography does. In addition, the concentration in the protein zones can be controlled to some extent. In elution chromatography, on the other hand, it is by no means easy to influence the concentration of a given protein as it elutes from the column. In many cases a dilution of the original concentrations has to be accepted, whereas in others high viscosities or aggregation may occur at the peak maximum, where the concentration of the biological macromolecule is highest.

In displacement chromatography the concentration in the substance zones depends—for a given chromatographic situation (set of mobile and stationary phase)—only on the displacer concentration. Once this is fixed, an operating line can be drawn in the manner indicated in Fig. 3 in order to determine the concentrations in the sample zones. The procedure is as follows: For a given displacer concentration, the speed of the advancing displacer front, u_i, is given by the following equation (DeVauldt 1943):

$$u_i = \frac{u_0}{(1 + \phi\ \partial q_i/\partial c_i)} \tag{4}$$

Since mobile phase flow rate, u_0, and the phase ratio, ϕ, are constants for a given system, and all substances in the displacement train move at the speed of the displacer front (for this reason the displacement train is also called isotachic train), the q/c ratio is necessarily also identical for all substances. Consequently, the concentration in each substance zone can be taken from the intercept of the operating line—drawn to connect the origin of the isotherm plot with the point on the displacer isotherm defined by the adjusted displacer concentration, c_D—with the single-component isotherm of the substance of interest. An adjustment of the concentration of a certain zone, albeit only one at a time, is easily possible by adjusting the displacer concentration. From Eq. (4) it follows also that high displacer concentrations result in fast separations and high protein concentrations in the individual zones.

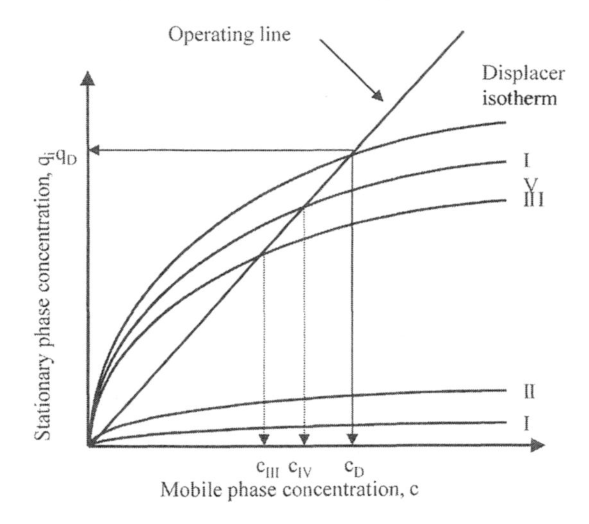

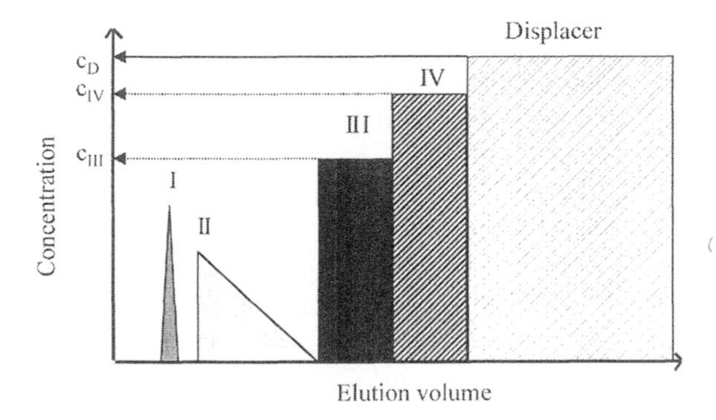

Figure 3 Determination of the protein concentrations in the displacement train by the "operating line" plot.

This treatment has the merit of allowing some access to the principle of displacement chromatography; however, it is also the result of many oversimplifications and does not consider some important aspects. One of them concerns the isotherm model. In Fig. 3 a typical Langmuir shape is used. However, for many biopolymers this is not necessarily the case. Crossing isotherms, S-shaped isotherms, and many other non-Langmurian isotherm shapes and systems have been described (Freitag 1999). The ideal adsorbed solution approach has been suggested as an improvement; however, even with this approach many of the peculiarities of multicomponent protein isotherm systems remain elusive (Antia and Horvath 1991; Frey 1987). A second caveat concerns the relevance of single-substance isotherms recorded for the bulk mobile phase. It has been shown that the local microenvironment of a protein may be much more important for adsorption and competition. In ion exchange displacement chromatography, currently by far the most common mode for protein displacement separations, it has, for example, been shown that the displacer through its very interaction with the stationary phase also displaces small salt counterions from the surface (Brooks and Cramer 1992). This induces a salt gradient, which in turn creates a local salt environment for the proteins. In some cases—especially for the higher displacer concentrations recommended for fast separations—the local salt concentration becomes high enough to prevent protein adsorption altogether and cause the protein to elute in front of the displacement train (Brooks and Cramer 1992; Gallant and Cramer 1997).

Another common simplification is the assumption of ideal chromatographic conditions. Only under ideal conditions will highly concentrated, rectangular zones be able to exist adjacent to each other. Under realistic conditions some mixing is unavoidable (for an excellent treatment of the theory of displacement chromatography, see, for example, Guiochon et al. 1994). Real displacement separations show abrupt but continuous changes in the concentration profile and some zone overlap always occurs. Narrow shock layers require high-performance stationary phases, displacers of sufficient affinity (otherwise the final protein zone will tail into the displacer zone), and sufficiently fast adsorption kinetics. In fact, the adsorption kinetics determine the maximal adsorption energies that can still be considered to be of practical use. If the binding is too strong and hence the adsorption-desorption equilibrium is reached only slowly, increased zone overlap may cause a problem (Kundu et al. 1995).

2.2 Typical Areas of Application of Displacement Chromatography

The concentration effect associated with displacement chromatography renders this method especially attractive for certain applications. One area is the processing of expensive materials and feeds. In this context a recent case study has yielded some impressive results (Gerstner 1996). In this particular case, the production costs for synthetic phosphorothioate oligonucleotides were considered. The standard technique for this purification problem is gradient elution. When the costs for the preparation of 5 kg of the material were calculated for both elution and displacement chromatography, the result was a reduction of the production costs by 27% if displacement chromatography was used. This was largely due to the higher yield of product with the required purity (95–97% plus the complete removal of phosphodiester containing oligonucleotides), if displacement was used. Gradient elution results in a "peak" shaped zone, with the desired product appearing toward the "tail" and at the concomitant low concentration. In displacement chromatography, the failure sequences were concentrated in the early fractions and the desired product oligos could be collected at high purity and high concentration in the later fractions. As a consequence, product recoveries were 50–70% in the case of displacement and only 30–50% in case of elution chromatography. Solvent consumption was reduced to a third (170 versus 525 L) and the number of production days from 39 to 26 days. A similar improvement as for the oligonucleotides discussed above is to be expected for the downstream process efficiency of precious recombinant proteins. In each case the higher recovery yield will be a major advantage, while the superior column utilization will reduce the number of runs for a given column.

A second promising area of application for displacement chromatography is the final polishing step, which is part of many downstream processes for recombinant proteins. In these steps, some minor impurities (aggregates, minor product variants, failure sequences) are removed. In displacement chromatography, these impurities are focused into very narrow and highly concentrated bands adjacent to the product zone. Their removal becomes possible at very low product loss. For example, when displacement chromatography was used for the polishing of a degraded recombinant growth hormone preparation, the impurities were found in the final fractions, while over 98% of the product was collected in pure form (Hancock et al. 1992).

2.3 Developing Areas of Displacement Chromatography

Given its many advantages and clear potential, it is not surprising that several areas of displacement chromatography are currently subject to intense research. A very important area is the development of tools for modeling/simulation of the separation which is prerequisite for less labor-intensive process development. Another very active area is displacer design. Unfortunately, the latter is usually done with little regard to the stationary phase chemistry, perhaps with the aim of creating the universal protein displacer, e.g., for anion exchange chromatography. Without wanting to go into the peculiarities of rational displacer design, this having been the topic of recent review (Freitag and Wandrey 2002), it is highly unlikely that such a universal displacer exists. The physicochemical character of the displacer, on the other hand, is crucial for the success of a displacement separation. As it was already pointed out above, high displacer concentrations are required for fast and efficient separations; thus, the displacer should dissolve well in the mobile phase. For obvious reasons it should be homogeneous, nontoxic, biocompatible, cheap, and detectable, the latter being very important for pharmaceutical applications. Removal from the product zone should also be possible. On top of all these requirements, the relationship between the stationary phase and the displacer must be optimal. The following points must be taken into consideration:

The stationary phase affinity of the displacer. This will often increase with the number of interaction points per molecule. Consequently, a polyion may be a displacer candidate for ion exchange chromatography.

The possibility of induced gradients (salt steps in case of ion exchange displacement chromatography). All other things being equal, this effect will be more pronounced for a small oligomeric molecule of a given chemistry than the corresponding large polymer (Gadam et al. 1993).

The possibility of secondary interaction, which may be used to fine-tune the displacer's affinity. A recent series of publications has, for example, demonstrated that the possibility of hydrophilic interactions on top of the electrostatic ones can considerably increase the affinity of a given displacer for polystyrene-divinylbenzene stationary phases as opposed to an agarose based one (Shukla et al. 1998).

Small molecules have certain advantages over true polymers as protein displacers. They are usually more homogeneous and thus less likely to contain low-affinity fractions that may pollute the displacement train (Zhu et al. 1991). Concomitantly, their separation from the protein zones is easier, since the difference in size can be exploited, e.g., in gel filtration or dialysis. A problem is their inherently lower affinity, which can be ameliorated by either increasing the charge density (Kundu et al. 1995) or by introducing secondary nonspecific forms of interaction (Shukla et al. 1998) as pointed out above. Since the affinity of small molecules depends more strongly on the chromatographic condition, column regeneration is also facilitated when small displacers are applied. The use of very-high-affinity polymers as protein displacer is not advisable, on the other hand, since these molecules are extremely difficult to remove from the column.

A beginner may experience some apprehension in regard to how to begin the development of the separation procedure. The database in elution chromatography is, of course, impressive and most analytical, and protein chemists have some experience with that technique. Experience with displacement chromatography is much less ubiquitous. In elution chromatography, analytical and linear data can be scaled up and a number of good expert systems exist for process development. In displacement chromatography some insight can be gained from a simulation according to the steric mass action model, first introduced in 1992 (Brooks and Cramer 1992, 1996; Kundu et al. 1997). This is a very simple model that has been developed for ion exchange chromatography of proteins under nonlinear conditions. While originally developed for displacement chromatography, it may equally well be applied to overload elution chromatography (Gallant et al. 1995; Iyer et al. 1999). The model takes the influence of the induced salt gradients into account as well as the fact that most macromolecules not only interact with a number of sites on the stationary phase surface but also block a certain number of others (Fig. 4). These two characteristic features of each molecule are called the *characteristic charge* and the *steric factor* of the molecule. In addition, the affinity (desorption) constants are needed. All of these can usually be calculated from linear elution data. The results can then be used to simulate actual—if idealized—displacement separations. In addition, they allow predictions concerning the suitability of a certain substance as displacer (displacer grading plots) or the suitability of certain process parameters (salt concentration in the carrier, displacer concentration) for displacement rather than, for example, elution separations. Once the parameters have been determined, it is easy to construct the

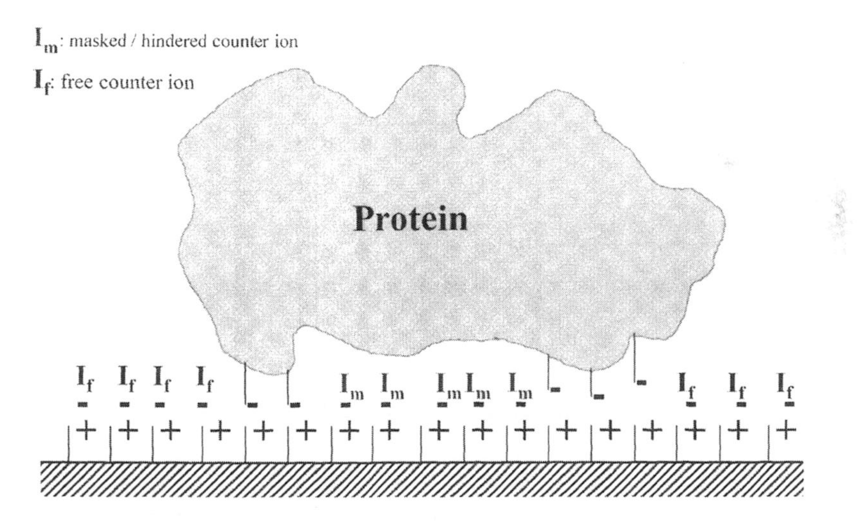

Figure 4 Schematic presentation of the binding of a protein molecule to an ion exchanger taking both active interaction and passive shielding of certain binding sites into account.

dynamic affinity plot (affinity constant as a function of the characteristic charge). For this a Δ point (q_D/c_D as defined by the displacer isotherm) is located on the y axis and connected with the points defined by the affinity constants and characteristic charges of the individual substances. The sum of the affinity lines of a given system represents the affinity plot (Fig. 5). The affinity plot allows a prediction of the possible effect of a change in displacer concentration as well as of the order of the substances in the displacement train. To date the model has mostly been used to simulate and predict displacement separations of small proteins (<100,000 g/mol) using low molecular mass (<1000 g/mol) displacers. Recent publications seem to indicate that especially in the case of monolithic stationary phases the model also applies to larger biologicals such as plasmid DNA (Freitag and Vogt 1999).

The monitoring of the displacement train has also been identified as a problem specific to displacement as opposed to elution chromatography. Since the substances leave the column in highly concentrated and consecutive zones, a UV monitor is incapable of differentiating between the different compounds. While this problem may also be present in overloaded elution chromatography, it is clearly worse in displacement. In the past the solution was to collect fractions at regular and tightly spaced intervals for subsequent analysis. It was not unusual to spend several hours with fraction analysis, when the actual separation had only taken 30 min. Together with the low

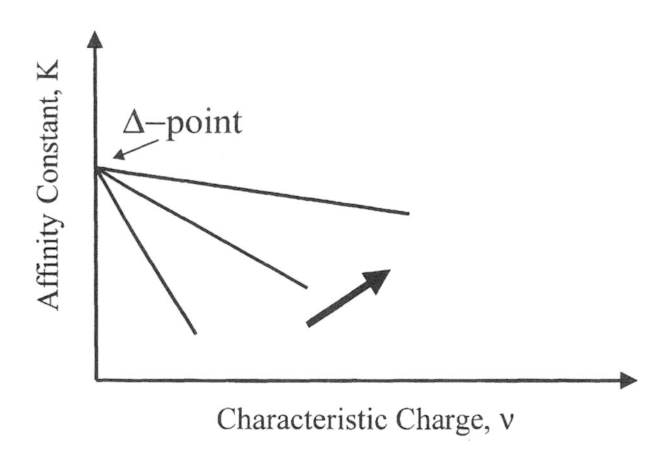

Figure 5 Example for an affinity plot created according to the SMA formalism. The order of the substances in the displacement train will be as indicated by the arrow. Any substances A can only displace another substance B, when the affinity line of A lies above that of B.

mobile phase flow rates typical of displacement chromatography (see below), this made displacement a slow and awkward technique to use. The arrival of high-speed chromatographic systems and correspondingly fast chromatographic software has changed this drastically. It is now possible to analyze the effluent from the preparative displacement column every 30 s and to use the data for fraction collection and pooling according to preset criteria (Freitag and Breier 1995).

Another recent advance in analytical chemistry has rendered the detection/monitoring problem even less critical. Modern high molecular mass spectrometry combined with electrospray ionization (ESI-MS) has become comparatively cheap and its use for "universal" detection in liquid chromatography is no longer unthinkable. Linked to the displacement column the mass spectrometer can be used as a detector—and this has yielded some very promising results already—or the displacement column can be used in a sample preparation (separation plus concentration) step for the MS. The resolution power of the mass spectrometer makes perfect separation unnecessary. Concomitantly, the high concentration factors of displacement chromatography become a distinct advantage, since they considerably improve the detection limit of the mass spectrometer. In the past such a setup has been used to find trace peptides in a peptide digest of a recombinant protein, which are telltales of malformation/degeneration of the product (Frenz et al. 1990) or for the identification of trace compounds in biological matrices in general (Hancock et al. 1992).

2.4 Displacement Approach in Relation to Other Aspects of Optimizing Chromatographic Separations

In future, perhaps no chromatographic mode will ever dominate preparative chromatography as exclusively as elution chromatography has in the past. Displacement chromatography, which combines the resolution of true gradient chromatography with the simplicity of single-step elution, holds much promise in regard to robust large-scale separation processes. However, below it will be shown that the optimization of a chromatographic protein separation is hardly ever a one-parameter problem. Superior results are achieved only in combination with stationary phase and instrument design. As discussed in Sec. 3, an optimized stationary phase morphology can make displacement more convenient and faster. In Sec. 4 it will be demonstrated how chromatographic mode and stationary phase morphology can come together to revolutionize the chromatographic separation of multicomponent mixtures by changing chromatography from a predominantly batch operation to a truly continuous one.

3. MONOLITHIC STATIONARY PHASES

The column morphology is clearly a very important aspect of any chromatographic separation. The conventional approach of using (small) porous particles to pack chromatographic columns is a direct cause of the "dilemma of biochromatography" outlined above. Over the past years a number of solutions have been suggested to overcome the "*C*-term restriction" of biochromatography, in the hope to produce columns where the peak width does not increase with increasing mobile phase flow rate. Among them are gigaporous "perfusion" and gel-filled "hyperdiffusion" particles. Columns packed with such particles are said to combine high resolution with high speed/throughput and—ideally—constant capacity. However, the essential problem of the particle-based stationary phase morphology remains and, according to personal experience, the plate height of such columns tends to deteriorate with increasing mobile phase flow rate in very much the same manner as observed for conventional porous particles.

Another solution to the problem of intraparticulate mass transfer are nonporous particles, where for obvious reasons no pore diffusion is possible. However, the corresponding columns have fairly low capacities, even when small particles are used and thus are better suited to analytical than to preparative applications. Other ways to reduce, if not circumvent, the problem of intraparticle mass transfer are the use of very small particles (1–2 μm) or the use of high temperatures. With small particles the distances to be covered by diffusion become smaller, but only at the price of a dramatic increase in back pressure. The use of elevated temperatures improves the diffusivities, but at least in the case of preparative protein chromatography a temperature-induced denaturation of the product is always possible.

Currently, the most elegant way to circumvent the problem of mass transfer resistance inside the particle is the use of a monolithic stationary phase. Such stationary phases consist of a porous polymer block. To create the polymer, monomer and cross-linker molecules as well as molecules bearing the interactive groups (e.g., an ionomer in the case of ion exchange chromatography) are polymerized in the presence of a porogen directly inside the column. The porogen is a good solvent for the molecules of the start mixture but a nonsolvent for the polymer. As the polymer forms it "precipitates" into small coils, which subsequently aggregate into "nodules" (Fig. 6). Due to the high cross-linking, the nodules can be considered "nonporous" from the viewpoint of biochromatography. The channels, which are formed between the aggregates, have average channel diameters between 1 and 5 μm. Their exact size can be influenced by the polymerization conditions. Such columns are still porous enough to permit injection and washing/regeneration by pressure. The fissured surface structure of the polymer

Figure 6 Scanning electron microscopy (SEM) pictures of the stationary phase structure of a monolithic stationary phase (UNO, courtesy of Bio-Rad, USA).

provides a large surface area per unit volume (up to $10\,m^2/g$ of polymer), and the capacity of such columns is usually comparable to that found for the conventional particle based ones (see Section 3.2). In some cases, for example, for very large molecules such as plasmid DNA, considerably higher capacities are found for the monolithic columns compared to the particle-based ones (Freitag and Vogt 2000). This can be explained by a steric effect present in porous beads. If a large molecule adsorbs close to the mouth of a pore (a likely event if the entire pore space is available), it may in fact block the entrance for all other biomolecules of similar charge.

In monolith chromatography the mobile phase can be considered to flow through the single "particle." A monolith has more similarities to a bundle of capillaries than to a packed particle bed, and a recent treatment of the theory of monolith disk chromatography (see Sec. 3.1) uses this approach to model the separations (Tennikova and Freitag 1999). A suitable average pore size and an optimized pore structure is most important for the performance of a monolithic column. The pore diameter needs to be optimized also in regard to the intended chromatographic interaction mechanism. Affinity interactions need larger pores than the comparatively fast ion exchange interactions. In monolithic stationary phases the adsorption-desorption process takes place on the walls of the flowthrough pores under conditions (pressure, shear stress) defined by the flow rate of the mobile phase. Pore diffusion is not an issue, and of all the diffusion processes typically considered in chromatography only a single one remains, i.e., the diffusion through the stagnant liquid film covering the stationary phase surface. A monolithic column may hence be expected to combine high capacity with fast separation and high resolution.

Chromatographic monoliths have become commercially available in various forms. One example is the CIM (Convective Interaction Media) disks produced by BIA d.o.o., Slowenia. These are flat (diameter in the centimeter range, thickness in the millimeter range) disks on the basis of a poly(glycidyl methacrylate-*co*-ethylene dimethacrylate) copolymer. Currently, the disks are available as strong and weak cation and anion exchangers. They can also be used directly to immobilize proteins and peptides in a very efficient and straightforward way, since the base polymer in the underivatized state contains epoxy groups (surface concentration between 3 and 5 mmoles per gram of polymer). The amount of epoxy groups on the disk surface can be controlled via the polymerization conditions. Ligands that carry free $-NH_2$ groups, such as proteins, can be covalently linked to the disks by a reaction with the epoxy groups. The CIM material has recently also become available as a tube for cross-flow chromatography (Tennikova and Freitag 2000). The second commercially available monolithic stationary phase is the UNO column commercialized by Bio-Rad (USA). These monoliths have the typical column dimensions, with a length in the centimeter and a diameter in the millimeter range. UNO-type columns are available as strong cation and strong anion exchangers (UNO Q and UNO S respectively). In the following sections, both types of monolithic stationary phase will be discussed in more detail.

3.1 Monolithic Disk Chromatography

Monolithic disks and monolithic columns have many similarities. A major difference is the length of the separation distance. For monolithic disks, this distance is only in the millimeter range, which has in the beginning caused some doubts in regard to the ability of these monoliths to separate complex protein mixtures. However, it is now accepted that the minimal length ("operative thickness of adsorption layer," OTAL) required for a protein separation by gradient elution is rather short and well within the thickness of, for example, a BIA d.o.o., CIM disk. A useful experiment to demonstrate this has been pointed out before (Tennikova and Freitag 1999). If a protein mixture is injected into a column and the column is inverted afterward, the usual gradient will often achieve a separation of similar quality as in the standard experiment, where the entire column length "contributes" to the separation. The situation is different for small molecules regardless of whether isocratic or gradient elution is used.

The phenomenon can be explained by considering the pronounced differences in the stationary phase affinity of proteins (Tennikova and Freitag 1999). The differences in the binding constants are largely due to differences in the desorption rate constants. In columns, adsorption and desorption takes place repeatedly; in a disk, a single interaction event can be postulated. With large molecules differing widely in binding strength, such a one-step process is nevertheless sufficient for a separation. Protein separation by monolithic disk chromatography has been said to be based on "the selective separation of adsorbed substances as a consequence of the desorption process" (Tennikova and Freitag 1999). Small molecules, which differ less in binding energy, are less well separated by monolithic disks, even though some examples exist (Podgornik et al. 1999).

An investigation of the mobile phase flow rate dependency of the chromatographic performance of the monolithic disks will demonstrate that indeed "plate height" (peak width), resolution, and capacity show little change up to 10 ml/min, a limit imposed by many of today's high-performance liquid chromatography (HPLC) pumps. The pressure versus linear flow rate relationship remains linear even at high flow. Since the disks are too thin to cause much back pressure (in our hands a 3-mm-thick BIA d.o.o., CIM disk created a back pressure of 12 bar when phosphate buffer was pumped through at a flow rate of 5 ml/min), such high flow rates can be obtained at relatively moderate pressure, an aspect that may make disk chromatography an attractive alternative for the production environment.

A second advantage of using the disk rather than the column geometry concerns the possibility of using mixed-mode chromatography. Many protein purification schemes use more than one column and interaction mode. Since a number of disks can be usually inserted into a cartridge, e.g., the one provided by BIA d.o.o., it is possible to create a multi-interaction system with a single unit. It has been shown for the related membrane chromatography that mixed-mode affinity/ion exchange chromatography can resolve mixtures, which could otherwise only be resolved in a two-column approach (Freitag et al. 1996). An additional advantage of this approach is the fact that whenever no single gradient can be found for the elution, the two disks can be easily disconnected and eluted separately. At present, the many advantages of monolithic disk (and membrane) chromatography are not fully realized by most applicants. Many tend to use these units more like affinity filters than as high-performance stationary phases. Below a number of applications are discussed in detail to demonstrate the potential of the disks.

Fast Semipreparative Isolation of a Recombinant Protein from Crude Mixture

Bacterial surface receptors such as protein A and protein G are important agents in affinity chromatography, because they are groups and subtype specific ligands for antibodies. The demand and also the price for these molecules is high. Their production as recombinant molecules has advantages, since the risk of contaminating the final ligand with bacterial toxins is not nearly as high under these conditions.

During process development, product samples need to be regularly isolated and evaluated. At this stage the process scale tends to be on the small side, and a robust, fast system for the purification of semiquantitative amounts (milligram range) from the complex production environment is required. This is an excellent application for monolithic affinity disks. While in conventional affinity chromatography the speed of the separation is limited by either the low molecular diffusivity of the target molecules or the column's back pressure, the same separation can be performed on monolithic disks at the maximum speed still compatible with the affinity interaction. Often this will be orders of magnitude higher (Nachmann 1992).

Semipreparative applications require slightly larger disks than those provided currently by BIA d.o.o. In the case discussed here, disks of 250 mm diameter and 2 mm thickness were synthesized according to standard procedures and activated by coupling h-IgG molecules to the epoxy groups in one-step reaction (Kasper et al. 1998). The h-IgG was dissolved (concentration at least 7 mg/ml) in a 20 mM sodium carbonate buffer, pH 9.3, and added to the disk. After 16 h at 30°C the maximal quantity (up to 9 mg/ml of support) has been bound. The ligand density of the disks was thus considerably higher than that of the previously investigated chromatographic membranes (Kasper et al. 1997).

While in the case of the affinity membranes either a spacer or the expensive tresyl/tosyl coupling chemistry was necessary to couple biologically active ligands (Kasper et al. 1997), no spacer was required in the case of the monolithic disks. Instead the epoxy groups could be used directly to produce affinity disks while retaining the full biological activity of the ligand. The disks were subsequently used for the recovery of IgG-binding fragments of protein G variants of human strain G148 produced by recombinant, nonpathogenic *E. coli* from the corresponding cell lysates. Due to the specific morphology of the monolithic stationary phase, the ligands were presumably freely accessible to the target molecules. Affinity constants were between 10^{-7} and 10^{-8}. The effect of the mobile phase flow rate was negligible (Fig. 7). When the disks were used to isolate the recombinant protein G from a cell lysate, the recovered amount corresponded to amount determined by enzyme-linked immunosorbent assay (ELISA). About 0.4 mg was recovered per run (~ 5 min). According to the analysis by sodium dodecyl sulphate–polyacrylamide gel electrophoresis (SDS-PAGE) (silver stain) the product was highly pure.

Use of Monolithic Disks in Flow Injection Analysis

Bioprocess monitoring is becoming more and more involved as the life science–based industries mature and the time span allowed for process development becomes shorter. Flow injection analysis (FIA) is the most common method to automate analytical procedures from simple pH measurements to protein and enzyme assays (Freitag 1996). The instrumental setup is simple and in the minimal version consists of a tube through which a buffer/regent solution flows, an injection for sample injection into the flow channel, and a detector. Whenever assays involving reusable expensive reagents (e.g., antibodies, enzymes) are automated, the trend goes toward the heterogeneous system, and these agents are immobilized on solid and often porous particles and subsequently packed into a "cartridge." As a consequence, the majority of the ligands are immobilized inside the particles where they can only be

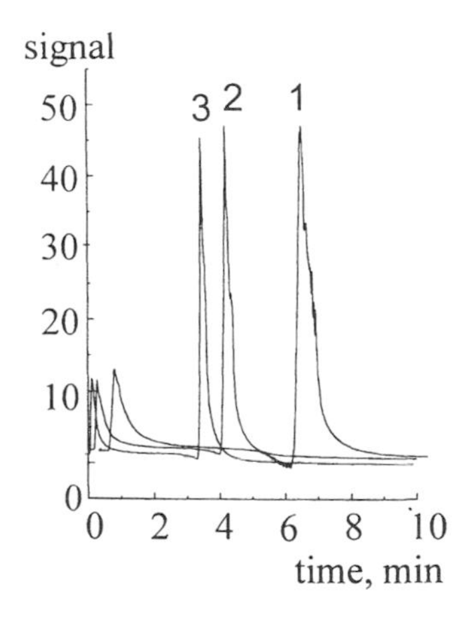

Figure 7 Influence of the flow rate on the isolation of recombinant protein G using a monolithic disk. Flow rates were 1 ml/min (peak 1), 2 ml/min (peak 2), and 3 ml/min (peak 3). Adsorption buffer was PBS pH 7.0 and desorption buffer B 0.01 M HCl. (Reproduced with permission from Kasper et al. 1998.)

reached by molecular diffusion. Whenever large product molecules are to be quantified it becomes a challenge to combine, capacity, speed, and resolution. In many aspects a heterogeneous FIA systems resembles a chromatographic apparatus (Fig. 8). Ideally, the cartridge should

Be compatible to high flow rates (in a production environment preferably at low to medium pressure)
Be readily modifiable by stable immobilization of biologically active ligands
Be cheap
Show good long-term stability
Show little to no dependence of the "performance" on the flow rate
Show little mass transfer resistance

Unfortunately, currently used FIA cartridges tend to be prone to bed instabilities, unfit for elevated flow rates, and generally not reliable enough for automatic process monitoring in a production environment. Substituting a monolithic disk for the hitherto used particle-based cartridge thus has the

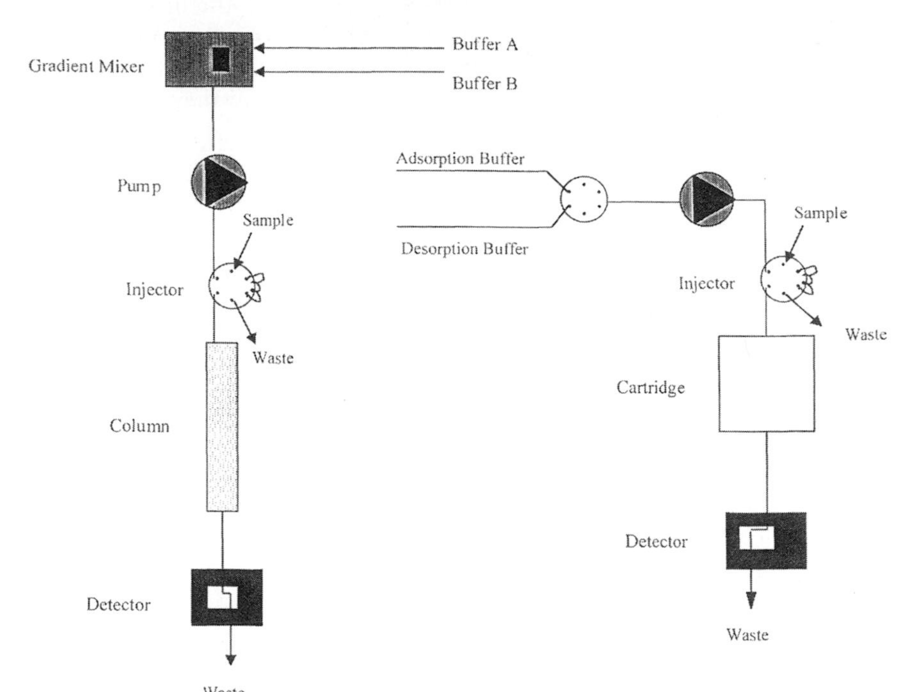

Figure 8 Schematic presentation of a chromatographic (left) and a heterogeneous FIA system (right).

potential of fulfilling the majority if not all of the above-mentioned requirements without increasing the costs to a noticeable degree (the ligand itself will in all likelihood remain the most expensive part of the system).

Since a FIA system is usually intended for analytical purposes, the commercially available small disks (10 mm diameter and 3 mm thickness) can be used. In a first application, the production of recombinant protein G by genetically modified *E. coli* was monitored (Hagedorn et al. 1999). Up to 8.5 mg/ ml of h-IgG ($\pm$1 mg/ml) could be immobilized on the disk. The ligand density of the disks was in the same range as that previously reported for the conventional heterogeneous FIA cartridges (Freitag 1990). In addition, the epoxy-mediated linkage was found to be stable under elution conditions such as a pH of 2 necessary to detach the captured protein G from the affinity disk.

For actual process monitoring the disks were inserted into a conventional FIA system (flow rate 2 ml/min). Fifty microliters of sample was injected and after a stable baseline was again observed (usually after 2.5 min, depending on the exact sample composition) the bound protein G was eluted with elution buffer and quantified either by peak height or area (Fig. 9). The calibration curve remained linear between 0.01 and 1.0 mg/ ml ($R = 0.998$ in case of the peak area). The correlation between the FIA quantification and that by ELISA or the previously used conventional protein analysis was excellent ($R > 0.997$). By using a flow rate of 2 ml/min, an entire analytical cycle including regeneration of the disk took 10 min. The analysis could be accelerated without problem by increasing the flow rate to 3 or even 5 ml/min. Neither the normalized peak height nor the peak area showed any change.

Antibody Binding Group Specific Affinity Ligands in HPMC

While the isolation and quantification of recombinant protein G may still be considered a special case, the opposite, i.e., the isolation of antibodies using columns with immobilized group-specific ligands (e.g., protein G), is one of the most common applications of affinity chromatography. In the third application example, three group-specific ligands (protein A, protein G, and protein L) were coupled to the disks and subsequently compared in regard to their interaction with a variety of antibodies (bovine, polyclonal human, and a recombinant human IgG_1 antibody with a kappa light chain, IgG_1-κ) (Berruex et al. 2000). Protein A and protein G are known to bind to the Fc part of antibodies albeit with some species and subtype specificity, whereas protein L binds specifically to the kappa light chain (Åkerstrom and Björck 1989).

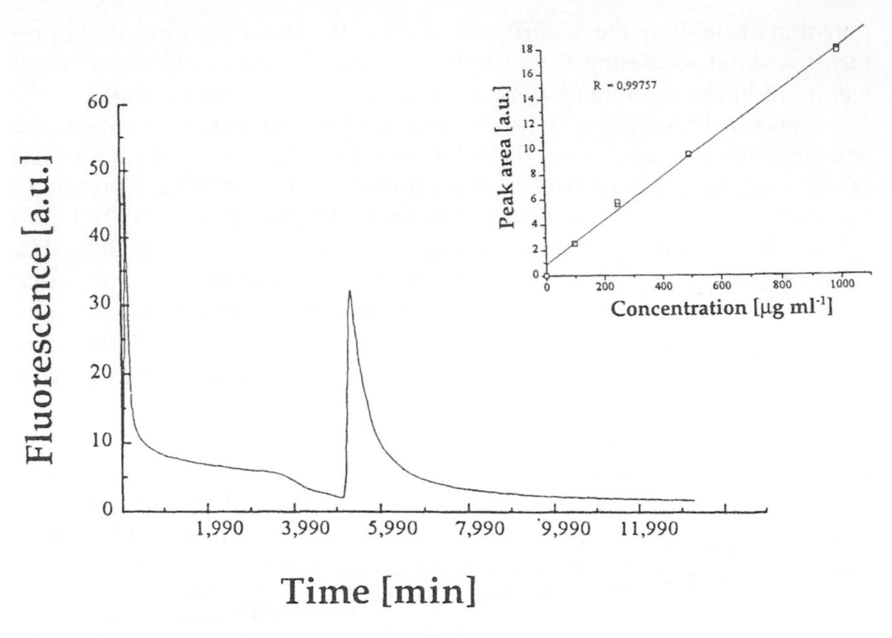

Figure 9 Signal obtained during quantification of recombinant protein G from cell lysate using an FIA system with monolithic affinity disk. The sample volume was 5 ml, the sample concentration 0.56 mg/ml. The calibration obtained using a standard (see insert) was linear over three orders of magnitude. (Reproduced with permission from Hagedorn et al. 1999.)

Between 0.02 and 0.03 μmol of ligand could be immobilized per disk (10 × 3 mm). The ligands were biologically active and showed the expected species- and subtype-dependent differences in their binding behavior. The quantification of bovine IgG with all three disk types took less than 1 min (Fig. 10) and could presumably be accelerated even further, since the limiting part was the maximal flow rate supported by the pump of the chromatographic system. In all cases quantitative recovery was achieved, even of small amounts of product (μg/ml range) in the presence of some "background" proteins in the supernatant. In all cases the purity of the isolated antibody was high and fully comparable to that of the corresponding standard. In the case of the antibody purified using protein A and protein G disks, a small band corresponding to the heavy chain (50 kDa) was also present in the gel, while a small band corresponding to the mass of the light chain was seen in the antibody sample prepared by the protein L disk.

An investigation of the flow rate dependence of the binding constants showed that binding was slightly improved at high flow rate (6 versus 4 ml/

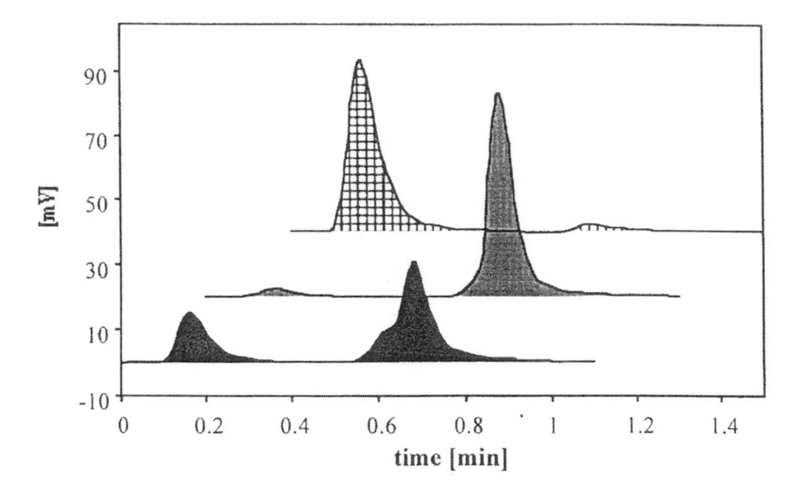

Figure 10 Chromatogram of bovine IgG with affinity monolithic disks. Top: Protein A disk, Middle: Protein G disk. Bottom: Protein L disk. Since protein L binds only to antibodies with κ light chains but not to those with λ light chains, two fractions are observed. Flow rate: 4 ml/min; adsorption buffer: PBS pH 7.4; desorption buffer: 0.01 M HCl; sample: 25 µg bovine IgG in 0.1 ml adsorption buffer. (Reproduced with permission from Berruex et al. 2000.)

min). In the case of protein L 7.34×10^{-9} M was calculated at 6 ml/min in contrast to 1.46×10^{-8} for 4 ml/min. The maximal binding capacity was almost independent of the flow rate. The desorption constants measured for the ligands (Table 1) are orders of magnitude lower (i.e., stronger binding) than those reported previously for conventional affinity chromatography (Labrou and Clonis 1994) and are in fact very close to those measured in free solution. It can be assumed that the easy accessibility of the affinity ligands contributes to this improvement. The adsorption isotherms for all systems were almost rectangular (Fig. 11). This has important consequences

Table 1 Affinity Parameters Determined for Protein A, Protein G, and Protein L Activated Disks

	Protein A	Protein G	Protein L	
			4 ml/min	6 ml/min
K_d [M]	1.82×10^{-8}	1.34×10^{-8}	1.46×10^{-8}	7.34×10^{-9}

Data was obtained from the linearized adsorption isotherms. (Reproduced with permission from Berruex et al. 2000).

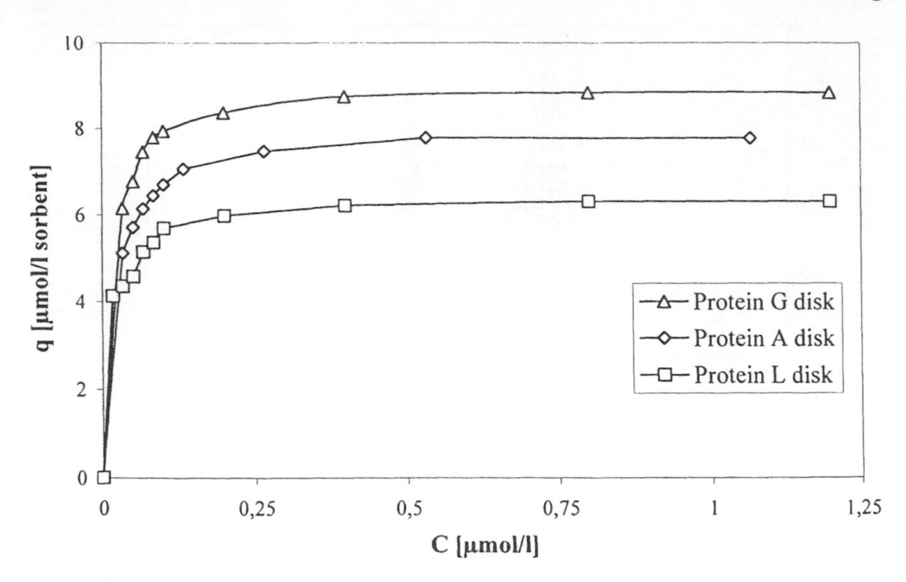

Figure 11 Adsorption isotherms measured for recombinant human IgG_1-κ antibody using affinity monolithic disks. Flow rate: 4 ml/min; buffer: PBS pH 7.4; antibody concentrations were varied from 2.5 ml/L to 180 mg/L. (Reproduced with permission from Berruex et al. 2000.)

in terms of the use of affinity disks for product quantification. In the case of a rectangular isotherm, the adsorbed amount, q_i, becomes quickly independent of the concentration in solution, c_i. Consequently, it becomes possible to simply increase the sample volume in order to adapt the analytical system to changing product concentrations. If, for instance, the minimal quantity for reliable quantification (LOQ) of substance x corresponds to 1 mg of said substance, it doesn't matter whether a 10-ml sample containing 0.1 mg/ml of x or 1-ml sample containing 1 mg/ml of x or a 100-μl sample containing 10 mg/ml of x is used (Table 2).

Chromatography of Plasmid DNA on Monolithic Disks

DNA is the second common biological polymer molecule besides proteins. Increasing amounts of pure plasmid DNA are, for example, needed in gene therapy, while fast and reliable analysis of various DNA samples remains a major challenge in molecular biotechnology. Chromatography is increasingly discussed as a powerful technique for DNA preparation. From a chemical point of view, DNA molecules are polyelectrolytes and, more specifically, polyanions. Other than in proteins the (negative) charges are

Table 2 Recovery of IgG (Peak Areas) After Loading Solutions of Different Concentrations and Volumes Using Protein A, Protein G, and Protein L Activated Disks

	IgG [μG][a]	Area	Dilution factor	Loop (μl)
Protein A HPMAC	12	177.9	10	25
	12	170.8	20	50
	24	331.8	10	50
	24	341.8	20	100
Protein G HPMAC	12	183.5	10	25
	12	167.7	20	50
	24	331.7	10	50
	24	342.7	20	100
Protein L HPMAC	12	170.1	10	25
	12	164.0	20	50
	24	324.8	10	50
	24	321.7	20	100

[a] Based on OD_{280} measurement result of 4.82 mg/ml initial recombinant IgG_1-κ antibody (purified solution). (Reproduced with permission from Berruex et al. 2000.)

distributed evenly over the molecule. Plasmid DNA molecules diffuse slowly and show a pronounced steric effect in chromatography columns packed with porous particles. Unlike many synthetic polyelectrolytes that share with DNA the homogeneous charge structure, all (plasmid) DNA molecules of a given type are supposed to be identical in their molecular mass.

Very little has been done on DNA separation using monoliths. Currently, one paper describes the behavior of polynucleotides on monolithic disks, taking a 7.2-kb predominantly supercoiled plasmid as example (Giovannini et al. 1998). This paper shows some important differences in the behavior of plasmid DNA as opposed to proteins. First of all, isocratic elution is possible in DNA chromatography on monolithic disk. Under the respective optimized conditions the plasmid sample is split into three peaks, by isocratic or gradient elution respectively (Figs. 12a and 12b). Thesc fractions correspond to three bands in the agarose gel. In the agarose gel these were aligned to the supercoiled (strongest band), nicked, and open circular form of the plasmid DNA. After linearization of the plasmid a more homogeneous peak was observed in the monolith chromatogram, concomitantly the binding strength was reduced and the DNA eluted earlier.

The current theory of monolithic disk chromatography interprets the separation, e.g., of proteins, as a single desorption step. Given the simple chemistry of the DNA molecule, the presence of three peaks under opti-

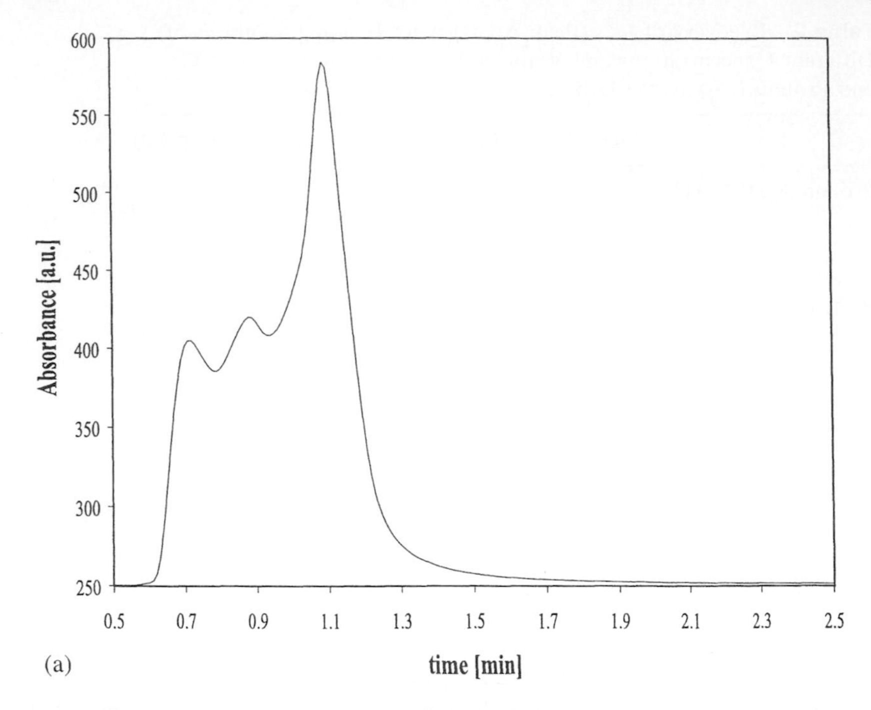

Figure 12a Separation of plasmid DNA on monolithic anion exchanger disk using isocratic elution. Stationary phase: Q CIM disk, flow rate: 1 ml/min; buffer A: 20 mM Tris HCl pH 7.4; Buffer B: buffer A containing an additional 1 M NaCl; mobile phase: 25% buffer A and 75% buffer B; sample: 5 μg plasmid DNA (pCMVβ) in 1 0 μl buffer A. (Reproduced with permission from Giovannini et al. 1998.)

mized isocratic but also under gradient conditions needs to be explained. From the agarose gel it can be deduced that three forms of the same plasmid are present in the sample. These forms can be expected to differ in binding strength due to differences in flexibility. A more flexible molecule should have a lower characteristic charge density. A similar change in adsorption strength can be observed for native and denatured protein molecules of otherwise identical primary structure. It is also possible that shear effects are operative. An adsorbed plasmid DNA molecule is large enough to extend beyond the stagnant film layer on the surface into the region, where there is a noticeable mobile phase flow. Again, the flexibility of the molecule may render this effect more or less significant.

While the finding that isocratic separation of plasmid DNA was possible at all was already surprising, the observed effect of the flow rate was even more so. In protein monolithic disk chromatography, but also for

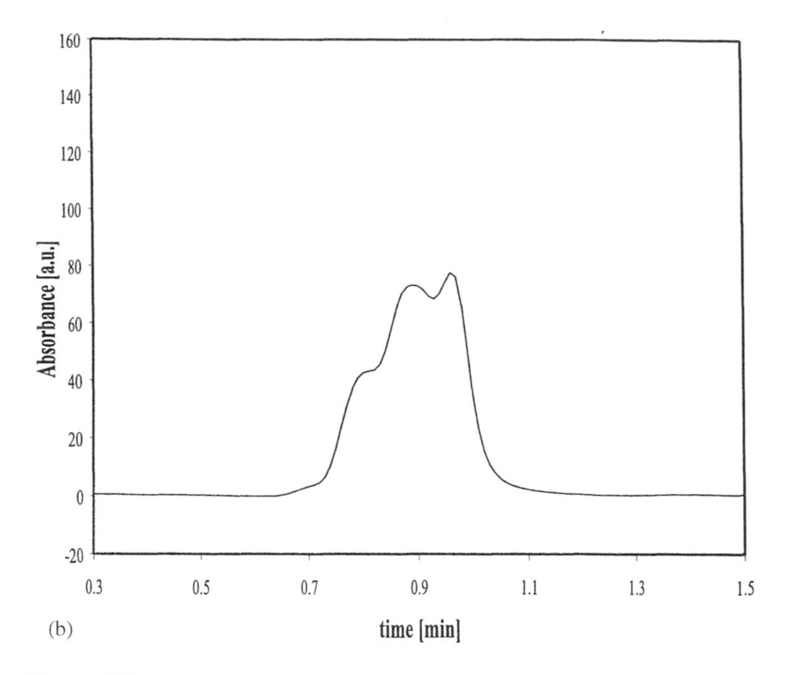

Figure 12b Separation of plasmid DNA on monolithic anion exchanger disk using gradient elution. Stationary phase: DEAE CIM disk; flow rate: 4 ml/min; buffer A: 20 mM Tris HCl pH 7.4; buffer B: buffer A containing an additional 1 M NaCl; gradient: linear from 75% to 100% B; gradient time: 2 min; sample: 5 μg plasmid DNA (pCMVβ) in 10 μl buffer A. (Reproduced with permission from Giovannini et al. 1998.)

gradient elution DNA chromatography on the disks, higher flow rates improved the separation. In isocratic elution instead, an optimal flow rate of 1 ml/min was observed. The use of higher flow rates, i.e., 2 and 3 ml/min, led to a loss in resolution. The application of lower flow rates, such as 0.2 ml/min, reduced the smoothness of the chromatogram without improving the overall separation.

When the disks were compared to standard (particle packed) and monolithic columns (UNO), the monolithic column also allowed a separation of the plasmid preparation, albeit only under isocratic and not under gradient conditions. The longer column length did improve the separation in terms of distance between the peak maxima (Fig. 13). However, since the peaks also became broader, to be expected under isocratic conditions, the resolution was not really improved. The particle-based column did not resolve the plasmid at all.

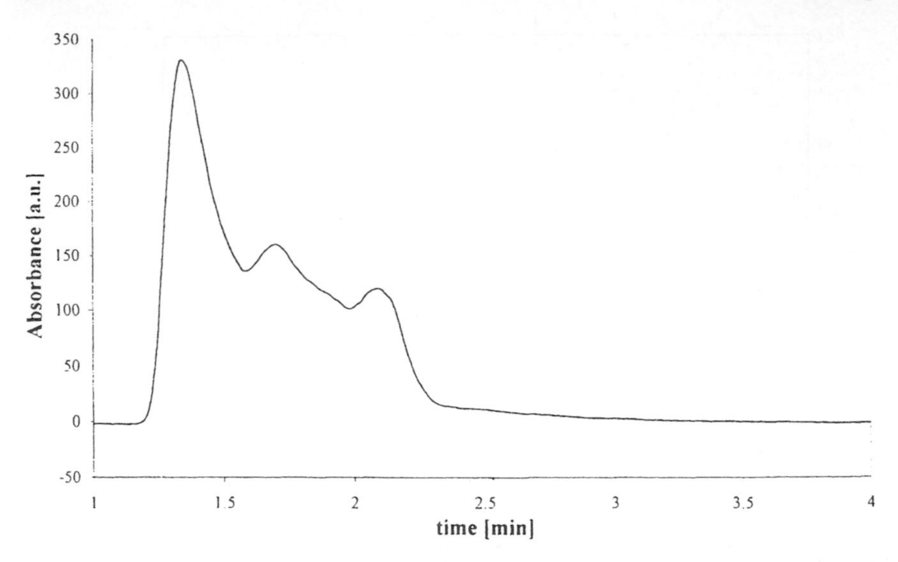

Figure 13 Separation of plasmid DNA on monolithic anion exchanger column using isocratic elution. Stationary phase: UNO Q1; flow rate: 0.5 ml/min; buffer A: 20 mM Tris HCl pH 7.4; buffer B: buffer A containing an additional 1 M NaCl; mobile phase: 20% buffer A and 80% buffer B; sample 2: 2 μg plasmid DNA (pCMVβ) in 10 μl buffer A. (Reproduced with permission from Giovannini et al. 1998.)

3.2 Monolithic Column Chromatography

Monolithic column, such as the UNO column, have a similar structure as the monolithic disks discussed in the previous section. Their van Deemter curve also shows a very flat *C*-term region. In Fig. 14 the van Deemter curve of the UNO Q1 column (column dimensions 35 × 7 mm, 1.3 ml) is shown in comparison with that of a particle-based high-performance anion exchanger column, the BioScale Q2 column (also from Bio-Rad, column dimensions: 52 × 7 mm, 2 ml), packed with porous 10-μm particles. A low molecular mass tracer (tyrosine, M_w 181.2 g/mol) was used under nonretaining conditions for the experiment (Freitag and Vogt 2000). The latter was done in order to avoid effects caused by the surface reaction and restrict the investigation to band-broadening effects caused by, for example, mass transfer resistances and extracolumn effects. The curve of the BioScale Q2 column is typical for columns packed with porous particles. The plate height increases with increasing flow rate. With approximately 50 μm the minimum is barely in the region of the "well-packed" column. The UNO has a similar plate height minimum; however, there is no increase with increasing flow rate.

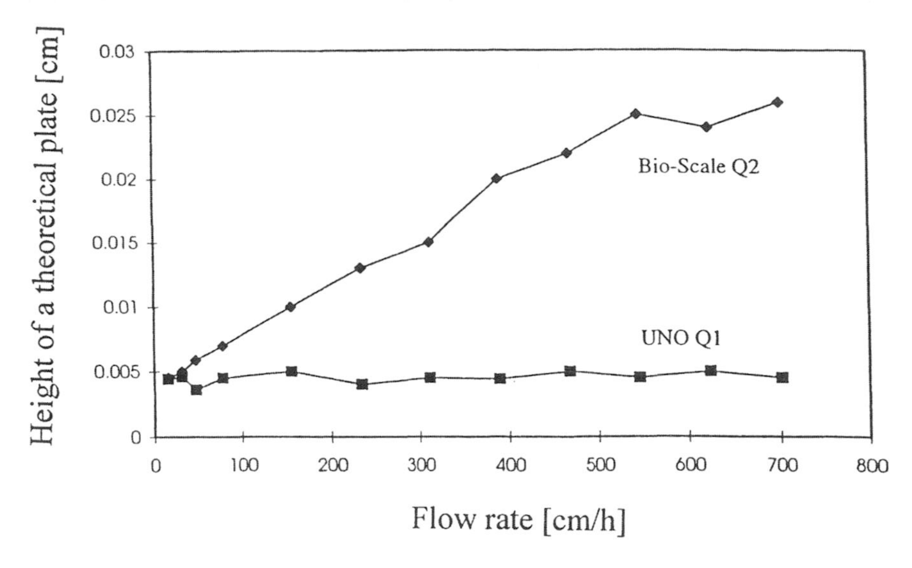

Figure 14 Van Deemter curves measured for the BioScale Q2 and the UNO Q1 column using 1 mg/ml tyrosine as tracer. Mobile phase: 0.02 M Tris HCl pH 8.0 containing 0.2 M NaCl. (Reproduced with permission from Freitag and Vogt 2000.)

Within the investigated range (0.01–4.5 ml/min), the *C*-term region is completely flat. According to the manufacturer, the column also has an almost flow rate–independent capacity as well as a low and linear dependence of the pressure drop on the mobile phase flow.

One area that should profit tremendously from the use of monolithic rather than particle based columns is displacement chromatography. Other than in gradient elution, in displacement chromatography a certain distance is needed for the development of the displacement train. Concomitantly, mass transfer phenomena have been taken to be responsible for the fact that mobile phase flow rates are inherently one order of magnitude lower in displacement than in elution chromatography. Below some applications of the UNO column in displacement chromatography are discussed. Unless mentioned otherwise, the UNO Q1 column is usually compared with the BioScale Q2 column from the same manufacturer. In spite of the difference in volume (1.3 for the UNO versus 2.0 ml for the BioScale column), the absolute number of interactive groups of the two columns seems comparable (Table 3). In spite of the considerably smaller total column volume, the stationary phase volume of the UNO Q1 column is almost equal to that of the BioScale Q2 column. The capacity for small ions, i.e., the value for λ_0 in Table 3, and for proteins (Fig. 15) is considerably higher for the UNO Q1 column than for the BioScale Q2 one.

Table 3 Column Parameters of the Particle-Based Column, BioScale Q2, and the Monolithic Column, UNO Q1

Parameter	BioScale Q2	UNO Q1
Column dimensions	7 × 52 mm	7 × 35 mm
Empty column volume, V_c	2.00 ml	1.35 ml
Column dead volume, V_0	1.60 ml	1.01 ml
Stationary phase volume, V_{st}	0.40 ml	0.34 ml
Stationary phase porosity, ϵ	0.802[a]	0.747[a]
	0.640[b]	0.642[b]
Absolute number of interactive groups	434 μmol	538 μmol
Low molecular weight ion capacity (nitrate), λ_0	0.89 M	1.64 M

[a] Measured for Acetone.
[b] Measured for β-lactoglobulin. (Reproduced with permission from Freitag and Vogt 2000.)

Table 4 compiles the parameters calculated according to the steric mass action model (Brooks and Cramer 1992; Freitag and Vogt 2000) for two proteins, i.e., α-lactalbumin and β-lactoglobulin, for both column types. As expected, the characteristic charge is a molecular property, i.e., similar values are measured for each protein on both anion exchangers. The equilibrium constant differs considerably (factor 2) between the two columns in case of the proteins but not for small salt ions. Apparently, the proteins interact much stronger with the monolithic than with the particle-based column. The steric factor of the proteins is also considerably larger in case of the particle-based column. Since the steric factor represents a passive blockage of surface sites (see also Fig. 4), this could mean that proteins block entire pores in the case of porous particles.

Speeding Up Displacement Chromatography

Although exceptions do exist (Subramanian and Cramer 1989), most authors agree that resolution in protein displacement chromatography already suffers at flow rates that are perfectly compatible with gradient elution chromatography. Even under these circumstances the throughput of the displacement column may still be higher, so the low flow rates are a nuisance rather than a strong handicap. However, any improvement of this characteristic should make displacement chromatography even more attractive for preparative separations. The rapid loss in resolution has been attributed to mass transfer phenomena inside the porous particle, so that a monolithic column should help to improve matters considerably.

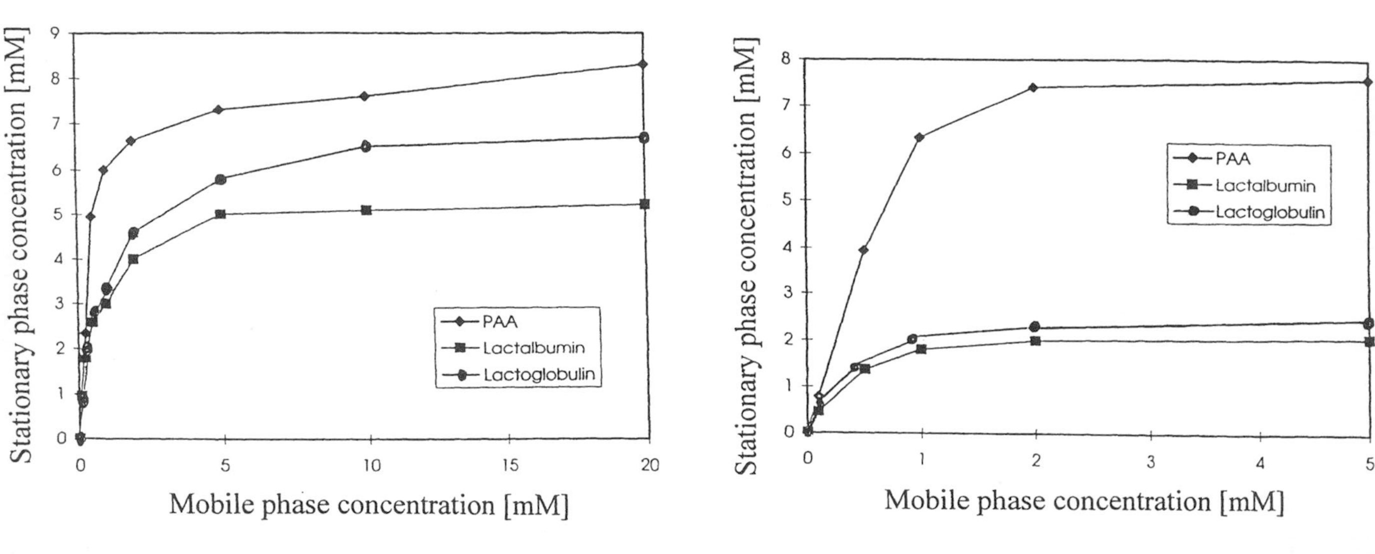

Figure 15 Adsorption isotherms measured for the whey proteins α-lactalbumin and β-lactoglobulin as well as for the displacer polyacrylic acid (mass 5100 g/mol) for (left) the UNO Q1 column and (right) the BioScale Q2 column. Mobile phase: 0.02 M Tris HCl pH 8.0 containing 0.2 M NaCl. (Reproduced with permission from Freitag and Vogt 2000.)

Table 4 SMA Parameters Determined for the Particle-Based Column, BioScale Q2, and the Monolithic Column, UNO Q1

Parameter	BioScale Q2	UNO Q1
Characteristic charge, ν (α-lactalbumin)	5.81	5.62
Characteristic charge, ν, (β-lactoglobulin)	7.68	7.38
Characteristic charge, ν (PAA)	40.2	41.2
Steric factor, σ (α-lactalbumin)	412	318
Steric factor, σ (β-lactoglobulin)	356	236
Steric factor, σ (PAA)	78	165
Equilibrium constant, K (α-lactalbumin)	0.265	0.447
Equilibrium constant, K (β-lactoglobulin)	0.26	0.63
Equilibrium constant, K (PAA)	0.63	0.60

Reproduced with permission from Freitag and Vogt 2000.

A mixture of 10 mg of the whey proteins α-lactalbumin and β-lactoglobulin were separated using a UNO Q1 column and a BioScale Q2 column, respectively, and polyacrylic acid (PAA, M_w 5100 g/mol) as displacer (Vogt and Freitag 1998). The displacer concentration was optimized using the affinity plot routine from the steric mass action model. At a flow rate of 0.1 ml/min, the separation between the two proteins is excellent for both column types. It is possible, however, that the displacement train was not fully developed for the UNO column, since the β-lactoglobulin tails into the displacer zone (Fig. 16a). Up to this point the UNO column showed no superiority to the BioScale Q2 column, which can be used to separate a similar amount of proteins with much the same resolution (Vogt and Freitag 1997) (Fig. 16b). However, while any increase in the flow rate beyond 0.1 ml/min reduced the separation efficiency of the particle-based column drastically, the separation on the UNO column could be performed at 1 ml/min and more (Fig. 17). For the evaluation of Fig. 17 it should be noted that for flow rates of 0.1 and 0.2 ml/min the fraction size was 50 µl. This value increased to 100 µl and 150 µl for flow rates of 0.5 and 1.0 ml/min, respectively. Flow rates of more than 1.0 ml/min could not be investigated due to the increasing difficulties to collect small enough fractions for a meaningful representation of the separation.

Separating Molecules of Different Size by Displacement Chromatography

The preparation of plasmid DNA at large scale may well be the coming major challenge to preparative bioseparation. The plasmids are produced

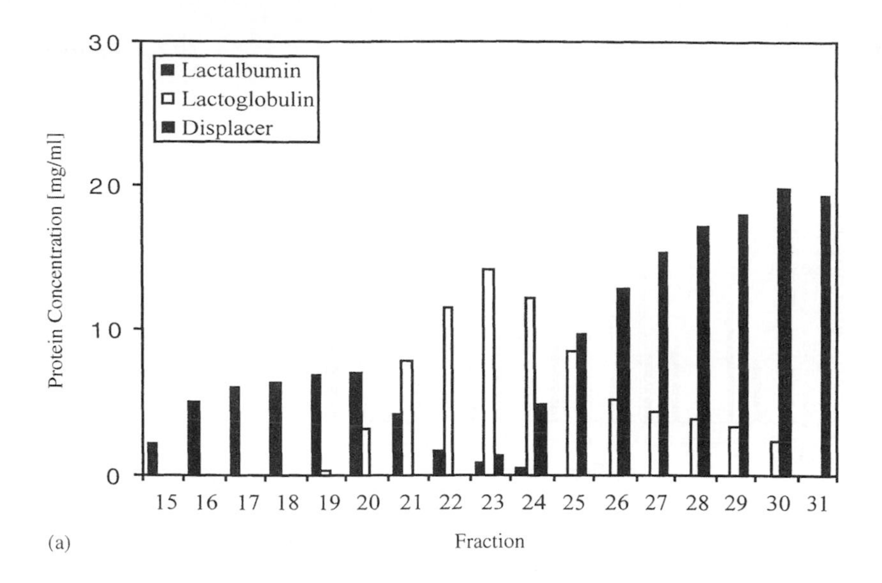

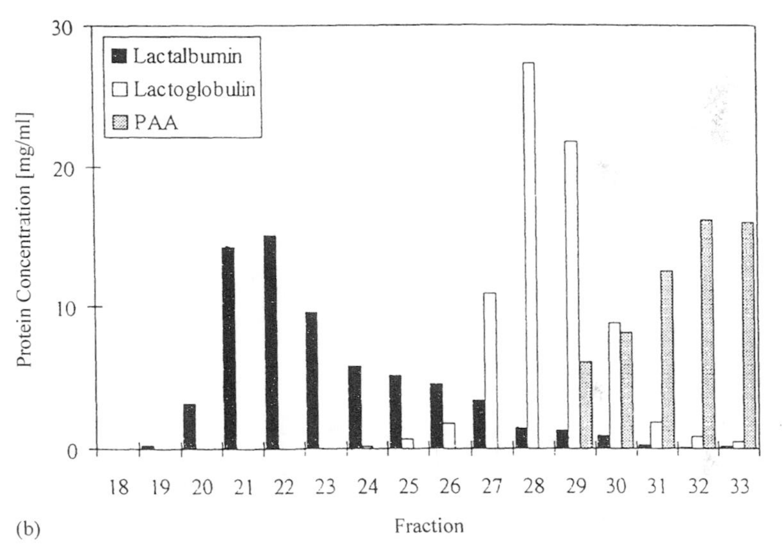

Figure 16 (a) Displacement of the whey proteins α-lactalbumin and β-lactoglobulin using polyacrylic acid (mass 5100 g/mol) as displacer. Stationary phase: UNO Q1 column; carrier 0.02 M Tris HCl pH 8.0; flow rate 0.1 ml/min; fraction size: 50 μl; sample: 1 ml containing 3.4 mg α-lactalbumin and 4.3 mg β-lactoglobulin. (Reproduced with permission from Vogt and Freitag 1998.) (b) Displacement of the whey proteins α-lactalbumin and β-lactoglobulin using polyacrylic acid (mass 5100 g/mol) as displacer. Stationary phase: BioScale Q2 column; carrier: 0.02 M Tris HCl pH 8.0; flow rate: 0.1 ml/min; fraction size: 50 μl; sample: 1 ml containing 4.1 mg α-lactalbumin and 4.0 mg β-lactoglobulin. (Reproduced with permission from Vogt and Freitag 1997.)

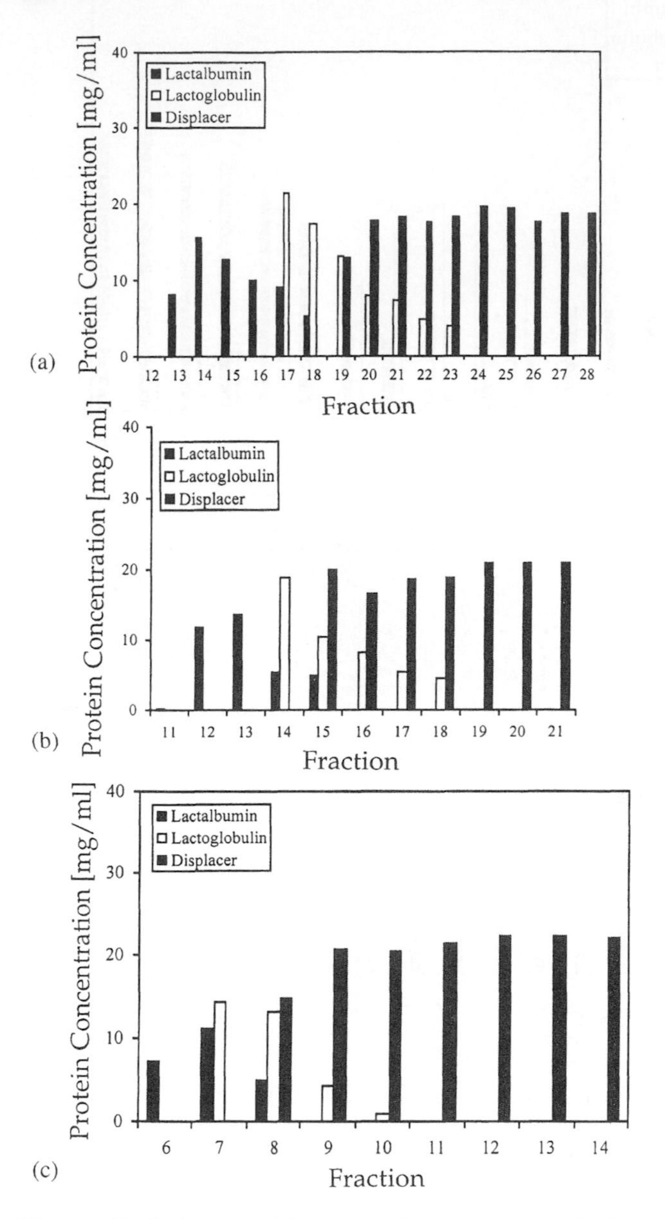

Figure 17 Influence of the carrier flow rate on the displacement of the whey proteins α-lactalbumin and β-lactoglobulin with polyacrylic acid (mass 5100 g/mol) as displacer on a monolithic column. Stationary phase: UNO Q1 column, carrier: 0.02 M Tris HCl pH 8.0; flow rate: (a) 0.2 ml/min; (b) 0.5 ml/min; (c) 1.0 ml/min. Fraction size: (a) 50 µl; (b) 100 µl; (c) 150 µl. Sample: 1 ml containing 5.0 mg α-lactalbumin and 5.2 mg β-lactoglobulin. (Reproduced with permission from Vogt and Freitag 1998.)

inside bacteria like *E. coli* and subsequently must be recovered from the cell lysates. Plasmid purification thus involves the removal of proteins, other polynucleotides (such as genomic DNA and RNA), and lipopolysaccharides (LPSs). Especially the latter is a major problem. LPS molecules form a major and necessary part of the cell membrane of gram-negative bacteria. If even a small amount of LPS enters the bloodstream, a violent reaction of the immune system occurs, which may even lead to death by septic shock.

Concentrated DNA solutions are very viscous. Displacement chromatography, where the concentration in the substance zone can be controlled, has therefore certain advantages over elution chromatography for plasmid preparation. Both DNA and LPS molecules are polyanions (Fig. 18) and hence bind strongly to anion exchangers. Since both molecules contain

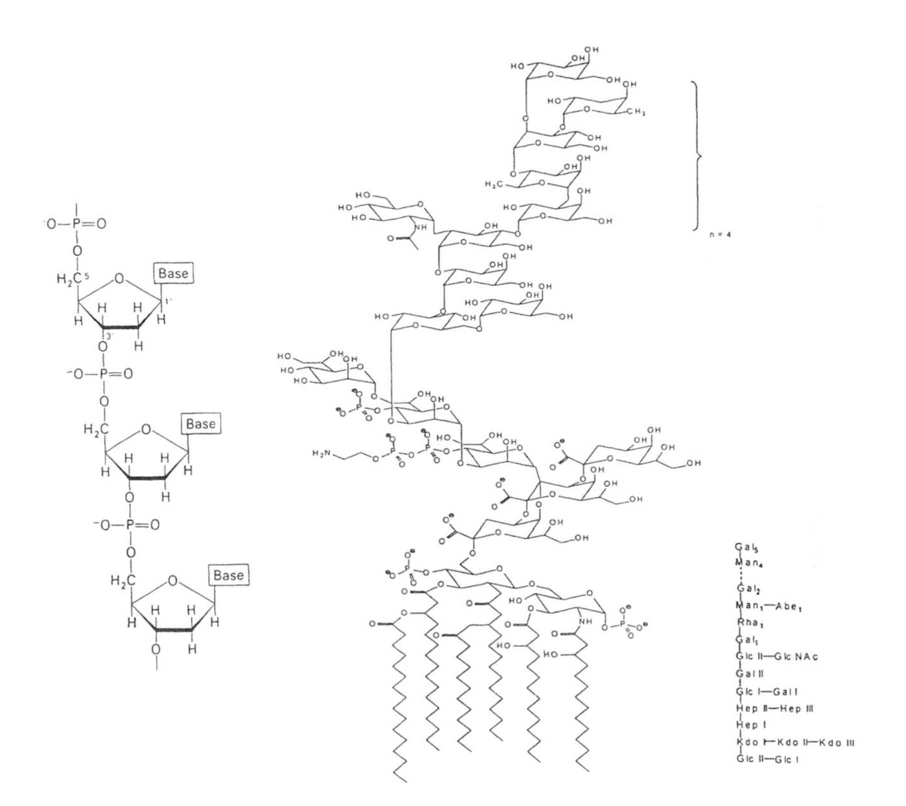

Figure 18 Chemical structure of (left) DNA and (right) LPS (model substance consisting of lipoid A structure from *E. coli*, oligosaccharide region of *S. typhimurium*, and O-specific side chain from *Salmonella* serogroup B. (According to Kastowsky et al. 1992.)

phosphate groups, interaction with hydroxyapatite (HA) is also possible. HA is a calcium phosphate modification with known affinity to DNA and certain basic proteins (Bernardi 1971; Hjertén and Liao 1988). The DNA molecules are assumed to bind to the calcium-containing "*C* sites" on the HA surface, Bio-Rad provides a special HA-type II material, which is presumably rich in such *C* sites and therefore suited for plasmid preparation EGTA (ethylenglycol-bis(β-aminoethylether)-*N*,*N*,*N*′,*N*′-tetraacetic acid) has given good results as displacer of *C*-site interacting substances (Kasper et al. 1996). However, the attempt to separate LPS complexes (800,000 g/mol) from a protein (holo-transferrin) using EGTA as displacer demonstrated a basic problem of displacement chromatography (Fig. 19). While the large LPS complexes were displaced and appeared in a roughly rectangular zone in front of the displacer, the protein appeared in a peak-shaped zone superimposed on the LPS zone (Freitag and Vogt 1999). This was interpreted as the result of the considerable size difference. While the inner particle surfaces is almost inaccessible to the LPS, the protein molecules can enter the pores and access the adsorptive surface within. As a consequence, the direct competition for the binding site, a necessity for

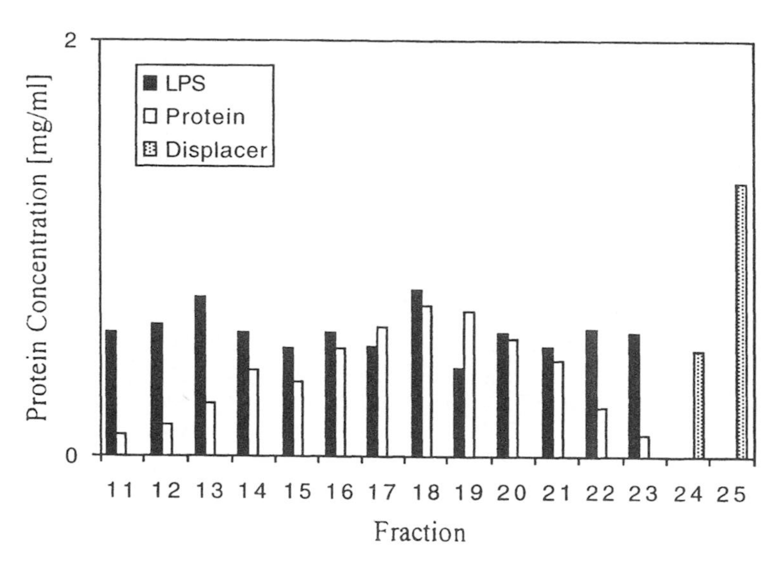

Figure 19 Attempt to separate two molecules of different size by displacement chromatography. Stationary phase: porous HA type II (10 µm particles); column dimensions: 250 × 4 mm; carrier: 0.005 M phosphate pH 7.2; flow rate: 0.1 ml/min; fraction size: 50 µl; sample: 1 ml containing LPS and model protein; displacer: EGTA. (Reproduced with permission from Freitag and Vogt 1999.)

the development of well-defined zones in displacement chromatography, is not taking place and separation is not possible. The same problem was observed for a conventional anion exchanger column (BioScale Q2). The separation of DNA and LPS molecules was possible with particle-based columns, however, in this case molecules of more or less identical size can be assumed, since the LPS molecules form large aggregates in aqueous solution (> 80,000 g/mol) (Freitag et al. 1997). However, even this separation was unsatisfactory due to the broad overlap between the zones. This overlap was assumed to be due to pronounced mass transfer limitations.

If these assumptions are correct, the utilization of monolithic column should improve both the plasmid DNA/protein and the plasmid DNA/LPS separation. The steric mass action mode was used to define the displacer for the three substance classes from the UNO column (PAA, M_w 5100 g/mol) and to fine-tune the chromatographic conditions. The predictions of the model corresponded to the experimental results. The displacement of all three substance classes by PAA was possible. The shape of the substance zones was rectangular and the concentration increased in the expected order. Experiments with two component mixtures (DNA/protein, DNA/LPS, protein/LPS) pointed to the desired direction. Molecules of different size behaved as expected, i.e., competed nevertheless with each other for the binding sites, and compared to the result obtained with the particle-based anion exchanger column, the overlap between the DNA and the LPS zones was less pronounced (Fig. 20).

3.3 Role of Monolithic Stationary Phases in Protein Chromatography

By combining speed, capacity, and resolution, monolithic stationary phases solve a major problem in protein chromatography. Since both strong and weak ion exchanger types are provided, one of the major interaction mode of protein chromatography can be performed using monoliths. The range of activated phases for affinity monolith chromatography is still somewhat small, but the epoxy-activated disks, to which most proteineous ligands can easily be coupled, mark a beginning. A present, it is perhaps the lack of hydrophobic stationary phases, such as hydrophobic interaction for preparative applications and reversed phase for analytical ones, which is limiting protein monolith chromatography in general. A second limitation concerns preparative monolith chromatography. Currently, the maximal volume of monolithic "columns"—the term is here used to include the disks—seems to be around 20 ml. Even when displacement chromatography is used, this is much too small for an industrial scale protein preparation.

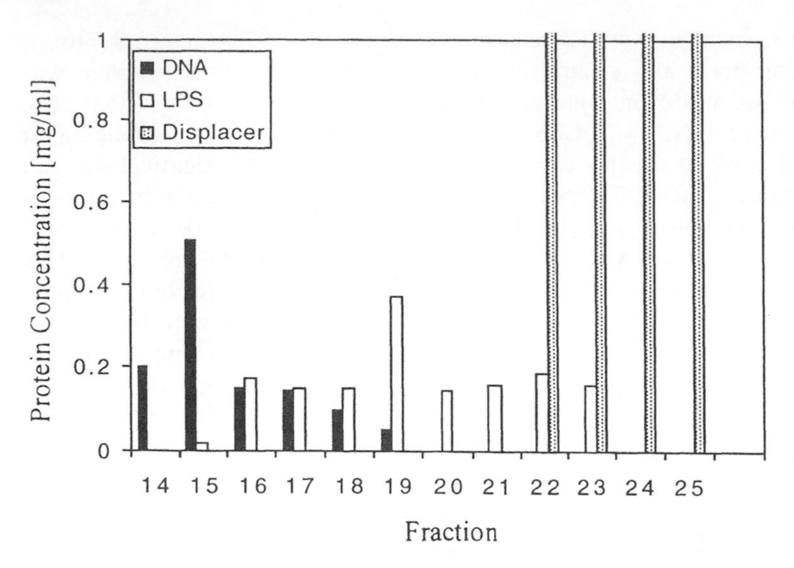

Figure 20 Separation of plasmid DNA and LPS by monolith displacement chromatography. Stationary phase: UNO Q1; carrier: 0.02 M Tris HCl pH 8.0; flow rate: 0.1 ml/min; displacer polyacrylic acid (mass 5100 g/mol); fraction size: 50 μl. (Reproduced with permission from Freitag and Vogt 1999.)

4. CONTINUOUS ANNULAR CHROMATOGRAPHY

Chromatographic separations are traditionally batch procedures, which means that in the day and age when even mammalian cell culture–based processes routinely reach the 1000- to 10,000-L scale, the columns used for downstream processing by the biotech industry can reach impressive dimensions. In spite of the theoretical advantages, few attempts to render chromatography a continuous operation were successful. An important exception is the so-called simulated moving bed (SMB) (Nicoud 1998). In an SMB the two chromatographic phases, the solid and the liquid one, are in "motion." For the latter, true movement is simulated by judicious switching of normally up to eight columns. An SMB allows separating two components or fractions in a continuous countercurrent mode. The high-binding compound (fraction) is enriched in the solid phase, the low binding one in the liquid phase. In the pharmaceutical industry, SMBs have become increasingly popular for the separation of enantiomers, a task that has become routine, since most new drugs will only be allowed to enter the market in pure enantiomeric form. For application in the biotech industry, SMB have some drawbacks. While they are truly compatible with very large-scale operations, they adapt less well to the kilogram scale, which is more prominent in bioproduction. The restriction to two-component separations is also a problem. With the exception of affinity chromatography, it is usually difficult to find conditions under which only the product binds, while all other impurities/contaminants do not (or vice versa). Most biochromatographic separations require at least a three-compound/fraction separation: low-binding contaminants/impurities, product, high-binding contaminants/impurities. Given the price and the limited lifetime of affinity supports, even affinity chromatography may be difficult to combine with the SMB approach for simple economic reasons. A minor drawback of the SMB as the standard downstream process system for the bioindustry may also be the fact that some software backup is necessary for method development and setup. The popular trial-and-error approach or the simple transfer and scale-up of an analytical batch elution chromatography is usually not possible.

It is precisely the area of industrial bioseparation, where another type of continuous chromatography may become most successful—so-called continuous annular chromatography (CAC). This principle has already established itself in industry for the separation of precious metals in the isocratic mode. It is highly compatible with the usual gram to kilogram scale of bioproducts and contrarily to the situation with the SMB, multicomponent mixtures can be separated by CAC. The transfer of separation protocols developed for batch columns to the CAC is usually straightforward, and little computer work is necessary to achieve good separations.

4.1 Principle of Preparative Continuous Annular (Bio-)Chromatography (P-CAC)

The principle of a multicomponent separation by continuous annular chromatography is illustrated in Fig. 21. In a P-CAC the solid phase is realized in form of a hollow cylinder, which rotates slowly around its axis. In the simplest version, the mobile phase percolates continuously through the rotating bed and the feed is continuously introduced from a fixed inlet nozzle. The sample components are separated along the axis of the column, in a manner very similar to conventional batch chromatography. Since they reach the lower end of the rotating cylinder after different migration times, their outlet position will be shifted in a varied degree in relation to the fixed inlet. High-binding components leave the column later, hence their elution point will be farther away from the inlet, while less well-binding substances elute earlier and their elution point will therefore be close to the inlet. Once equilibrium is reached, the substance mixture can be continuously fed at the inlet point and the purified substances can be continuously collected from the respective outlets.

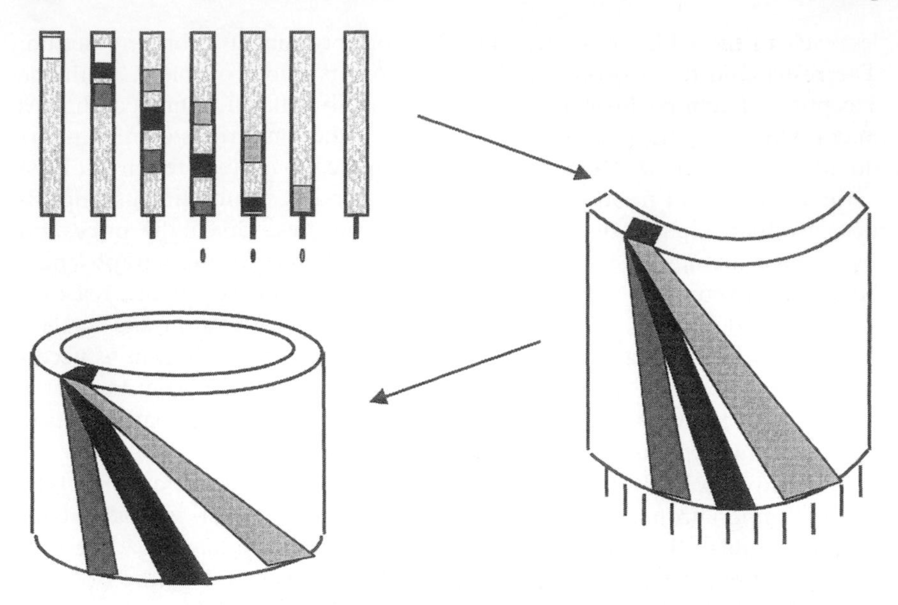

Figure 21 Principle of the preparative continuous annular chromatography.

If the substance zones are made visible by the use of colored substances, the impression of helical bands is given (Fig. 21). In spite of this appearance, however, the migration of the substances occurs exclusively in an axial direction and no flow in radial direction takes place. An investigation of radial mass transfer and rotation speed on the van Deemter curve clearly showed that these were negligible (Wolfgang et al. 1998). Therefore, it can be said that all substances are resolved along the column's axis much as in the usual batch column. As a result, it is easy to transfer a separation from the batch to the continuous mode as long as isocratic elution is possible.

Currently, the only commercially available P-CAC is produced by Prior Engineering (Götzis, Austria). The system is predominantly used for the separation of precious platinum group metals. However, according to the manufacturer, much effort is currently put into the development of a P-CAC for bioseparation purposes. The stationary phase cylinder in the Prior system has a wall thickness in the millimeter range (5–20 mm). The inner diameter of the cylinder is 14 cm and the cylinder's height can be adjusted between 10 and 20 cm. During the separation, the cylinder rotates around its axis with an adjustable speed between 200° and 500°/h. The packing of the stationary phase is easy, since the instrument is capable of a self-packing routine involving a complex set of shaking and rotating motions. The results are excellent and the plate heights, which are reproducibly obtained by this method, are at least comparable and often superior to those found in commercially available batch columns containing the same stationary phase (Wolfgang et al. 1998). Most applicants pack on top of the actual stationary phase several centimeters of glass beads to avoid any disturbance of the packed bed by the introduction of the liquids.

4.2 Application of the P-CAC for Bioseparation

The standard P-CAC system uses exclusively isocratic elution. Among the standard methods in protein chromatography, it is gel filtration that is most easily adapted to the continuous system. Since gel filtration, which is usually applied to the end of the downstream process for polishing, is also a time-consuming and cumbersome method, which usually has low throughputs, continuous annular gel filtration may already be an attractive option. A first successful use of continuous gel filtration on an industrial scale for plasma fractionation, has been reported by the Austrian company Octapharma Pharmazeutika (Josic D., Octapharma Pharmazeutika Produktionsges. m.b.H., Vienna, Austria, personal communication). Compared with the batch column, the throughput was reported to be higher and the product more uniform. Problems with column fouling were not observed in spite of prolonged continuous use of the system.

A second approach that should adapt well to P-CAC is affinity chromatography. Modern P-CAC systems, like the one supplied by Prior Engineering, provide up to five buffer inlets. Thus it is possible to partition the CAC in a feed, an elution, and a washing/regeneration zone (Fig. 22). A first use of the P-CAC in the affinity mode using hydroxyapatite as a pseudo-affinity stationary phase was recently reported (Giovannini and Freitag 2001). In this case, the system was packed with porous hydroxyapatite type I beads (Bio-Rad), a material that is known to interact strongly with basic proteins such as antibodies. Since the majority of all known proteins and hence most proteineous contaminants/impurities are acidic, hydroxyapatite chromatography is usually well suited for antibody purification since it is easy to define conditions under which exclusively the basic molecules bind to the stationary phase.

In Fig. 23a the isolation of antibody (human IgG_1-κ) from a serum-free CHO cell culture supernatant is shown. A 2-ml batch column packed with hydroxyapatite type I (10-μm porous particles) was used. Elution took place in the indicated linear phosphate gradient. The separation was subsequently transferred to the P-CAC system, with the only exception of using a phosphate step for the elution rather than a continuous gradient (Fig. 23b). The components of the original feed were well separated into three fractions: first the low binding one containing the majority of the cell culture proteins;

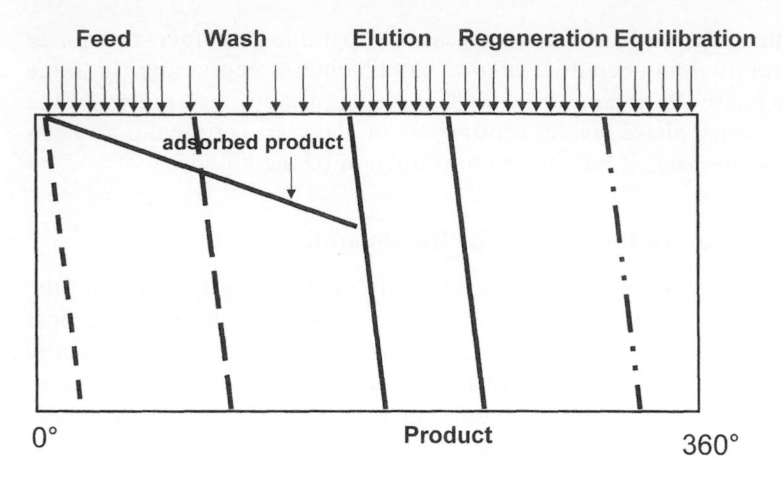

Figure 22 If necessary, the cylinder of the chromatograph can be divided into feed, elution, and regeneration zones.

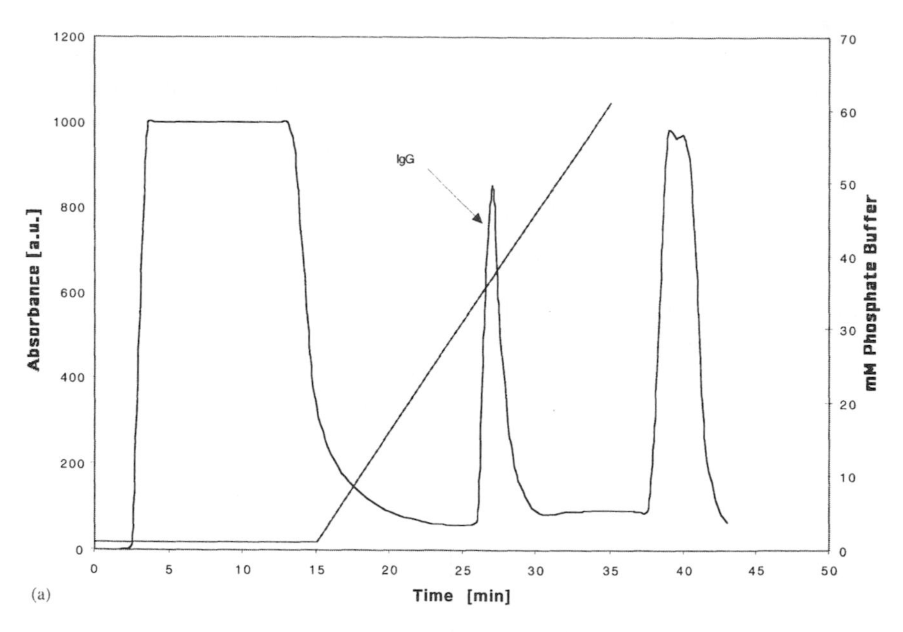

Figure 23a Isolation of a recombinant human antibody from a serum-free CHO cell culture supernatant on a conventional batch column. For chromatographic conditions, see Table 5. (Reproduced with permission from Giovannini and Freitag 2001.)

then the product fraction containing the nearly pure antibody, and finally the high-affinity fraction containing mostly DNA. A similar separation was also attempted using a fluidized bed (for all chromatographic conditions, see Table 5).

The purity of the recovered antibody fractions was determined by SDS-PAGE and compared with that of a standard IgG_1-κ purified by protein A affinity chromatography. While the quality of the hydroxyapatite-purified material was not quite in the same range as that of the protein A–purified material, there was little difference between the antibody obtained with the batch compared with the antibody obtained with the P-CAC hydroxyapatite column. By comparison, the fluidized bed yielded in this case an inferior product. A comparison of the performance in general (Table 6) showed that better results in terms of yield and final purity are obtained with the P-CAC and the conventional batch column, whereas the expanded bed is superior in terms of throughput. The buffer consumption of the P-CAC is between that of the fluidized bed and the conventional col-

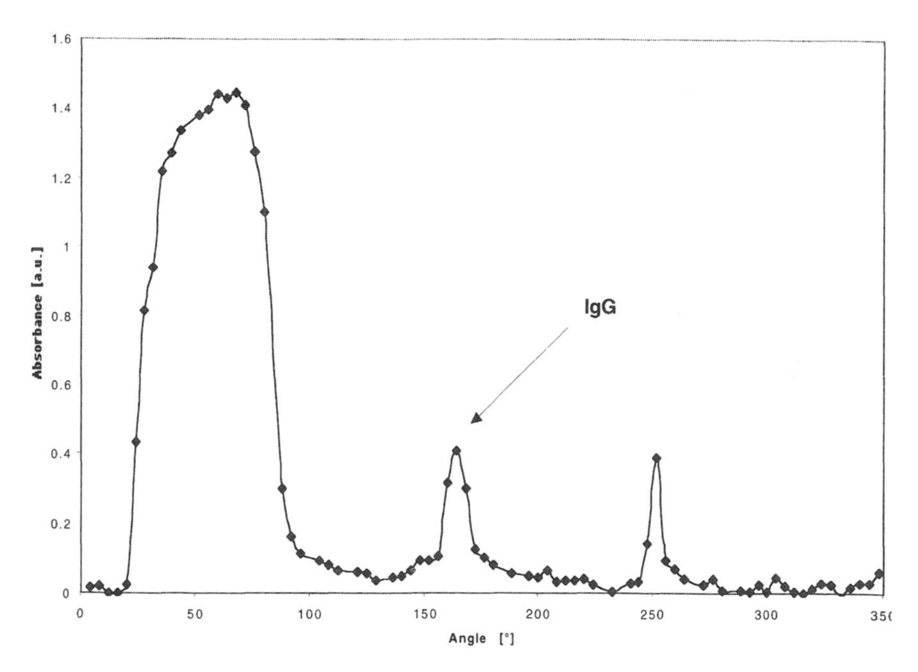

Figure 23b Isolation of a recombinant human antibody from a serum-free CHO cell culture supernatant using the P-CAC. For chromatographic conditions, see Table 5. (Reproduced with permission from Giovannini and Freitag 2001.)

Table 5 Conditions for the Chromatographic Isolation of a Recombinant Human IgG_1-κ Antibody Using a Conventional Batch Column, a Preparative Continuous Annular Chromatograph (P-CAC), and a Fluidized Bed (FB)

Chromatographic mode	Chromatographic conditions
Batch	Sample: 2.5 ml of supernatant diluted 1:1 with buffer A Column: 250 × 4 mm Stationary phase: porous HA 20 μm Buffer A: 1 mM phosphate + 1 M NaCl, pH 6.0 Buffer B: 400 mM phosphate, pH 6.0 Gradient: 0% B to 15% B in 20 min Flow rate: 8 cm/min
P-CAC	Sample: Cell culture supernatant dilute 1:3 with buffer A Column: bed height 18.5 cm, cylinder wall thickness 5 mm, cylinder inner diameter 14 cm Stationary phase: porous 20 μm HA Buffer A: 1 mM phosphate + 1 M NaCl, pH 6.0 Buffer B: 400 mM phosphate, pH 6.0 Step product elution: 85% buffer A/15% buffer B Flow rates: feed 0.18 cm/min, buffer A (binding) 0.7 cm/min, product elution 0.35 cm/min, buffer B (washing) 0.15 cm/min Rotation speed: 180°/h
Fluidized bed	Sample: 80 ml of supernatant, dilute 1:1 with buffer A Column: Streamline 25 (Pharmacia Biotech) height: 12 cm, extended bed height 25 cm flow rate: 5.1 cm/min Stationary phase: HA 120 μm Buffer A: 1 mM phosphate + 1 M NaCl, pH 6.0 Buffer B: 400 mM phosphate, pH 6.0 Elution: 0% B to 15% B in 48 min Flow rate: 1.6 cm/min

Reproduced with permission from Giovannini and Freitag 2001.

umn. While DNA removal is nearly quantitative with either the batch or the P-CAC column, this is less well achieved with the fluidized bed.

4.3 Future Trends for Continuous Biochromatography

The application of the P-CAC system for biochromatography is still under development. Gel filtration and affinity chromatography are exceptions and, as stated earlier, most protein separations currently require gradient elution.

Table 6 Comparison of the Results of Chromatographic Isolation of a Recombinant Human IgG_1-κ Antibody Using a Conventional Batch Column (column), a Preparative Continuous Annular Chromatograph (P-CAC), and a Fluidized Bed (FB)

	Column	P-CAC	FB
Yield	92%	88%	77%
DNA removal	99%	99%	91%
Throughput (ml/min)	0.05	3.0	3.5
Buffer (L/L)	17	10.5	3.5
Purity	High	High	Low

Reproduced with permission from Giovanni and Freitag 2001.

While modern P-CAC systems are compatible with a simple three-step gradient, in general the method would benefit from the use in the displacement mode, since this would allow a multicomponent separation using a single displacer step. The possibility of doing so has been demonstrated already for small molecules (De Carli et al. 1990). A second intriguing possibility would be the insertion of a monolithic stationary phase in the P-CAC. On one side, this would reduce the problem of scaling up monolith chromatography. Since such a cylinder could be prepared with nearly any diameters, the problem of polymerizing a large solid column would be circumvented. On the other side, this approach would allow acceleration of the P-CAC in general due to the improvement of the mass transfer effects as discussed above for monolithic disks and columns. Displacement P-CAC would also benefit from such a development.

5. CONCLUSIONS

It is not easy and certainly not necessary to invent a completely new bioseparation methodology to eventually replace preparative chromatography in order to arrive at high-performance, high-throughput but cost-efficient options for protein isolation. What is necessary is a clear understanding of how proteins interact and behave within the chromatographic system in order to use this information for the engineering of suitable separation units. Some of the recent approaches to this task have been outlined above. In all likelihood, no isolated solution is possible and instead all aspects of biochromatography should be considered together. Currently, the best approach to a high-resolution preparative protein separation would perhaps consists of using a P-CAC system in the displacement mode with a monolithic column as stationary phase. However, this will hopefully change in the near future.

REFERENCES

Åkerstrom B, Björck L (1989) Protein L: an immunoglobulin light chain-binding bacterial protein. Characterization of binding and physicochemical properties. J. Biol. Chem. 264, 19740–19746.

Anspach FB, Curbelo D, Hartmann R, Garke G, Deckwer W-D (1999) Expanded bed chromatography in primary protein purification. J. Chromatogr. A 865, 129–144.

Antia FD, Horvath CG (1991) Analysis of isotachic patterns in displacement chromatography. J. Chromatogr. 556, 119–143.

Bernardi G (1971) Chromatography of nucleic acids on hydroxyapatite columns. Meth. Enzymol 21, 95–139.

Berruex L, Freitag R, Tennikova T (2000) High performance monolith affinity chromatography. J. Pharm. Anal. 24, 95–104.

Brooks CA, Cramer SM (1992) Steric mass-action ion exchange: displacement profiles and induced salt gradients. AIChE J. 38, 1969–1978.

Brooks CA, Cramer SM (1996) Solute affinity in ion-exchange displacement chromatography. Chem. Eng. Sci. 51(15), 3847–3860.

De Carli JP, Carta G, Byers CH (1990) Displacement separations by continuous annular chromatography. AIChE Journal 36(8), 1220–1228.

DeVauldt D (1943) The theory of chromatography. J. Am. Chem. Soc. 65, 532–540.

Freitag R (1990) Entwicklung und Automatisierung von Immunochemischen Nachweisverfahren zur On-Line-Detektion Hochmolekulrer Medienkomponenten in Fermentationsprozessen. PhD thesis. University of Hannover, Germany.

Freitag R (1996) Biosensors in Analytical Biotechnology. RG Landes, Academic Press, Austin, Texas.

Freitag R (1999) Displacement chromatography of biomolecules. In: Analytical and Preparative Separation Methods of Biomacromolecules (Hassan Y Aboul-Enein ed.). Marcel Dekker, New York, pp. 203–253.

Freitag R, Breier J (1995) Displacement chromatography in biotechnological downstream processing. J. Chromatogr. 691, 101–112.

Freitag R, Fix M, Brüggemann O (1997) Analysis of endotoxins by capillary electrophoresis. Electrophoresis 18(10), 1899–1905.

Freitag R, Splitt H, Reif OW (1996) Controlled mixed-mode interaction chromatography on membrane adsorbers. J. Chromatogr. 728, 129–137.

Freitag R, Vogt S (1999) Preparation of plasmid DNA using displacement chromatography. Cytotechnol. 30, 159–167.

Freitag R, Vogt S (2000) Comparison of particulate and continuous bed columns for protein displacement chromatography. J. Biotechnol. 78, 69–82.

Freitag R, Wandrey Ch (2002) Synthetic displacers for preparative biochromatography. In: Synthetic Polymers for Biotechnology and Medicine (R Freitag, ed.). RG Landes, Academic Press, Austin, Texas (submitted).

Frenz J, Bourell J, Hancock WS (1990) High-performance displacement chromatography-mass spectrometry of tryptic peptides of recombinant human growth hormone. J. Chromatogr. 512, 299–314.

Frey DD (1987) Free energy consumption at self-sharpening concentration fronts in isothermal fixed-bed adsorption. J. Chromatogr. 409, 1–13.

Gadam SD, Jayaraman G, Cramer SM (1993) Characterization of non-linear adsorption properties of dextran-based polyelectrolyte displacers in ion-exchange systems. J. Chromatogr. 630, 37–52.

Gallant SR, Cramer SM (1997) Productivity and operating regimes in protein chromatography using low-molecular-mass displacers. J. Chromatogr. 771, 9–22.

Gallant SR, Kundu A, Cramer SM (1995) Modeling non-linear elution of proteins in ion-exchange chromatography. J. Chromatogr. 702, 125–142.

Gerstner JA (1996) Economics of displacement chromatography—a case study: purification of oligonucleotides. BioPharm. 9, 30–35.

Giovannini R, Freitag R (2001) Isolation of a recombinant antibody from cell culture supernatant—Continuous annular versus batch and expanded bed chromatography. Biotechnol. Bioeng. 73, 522–529.

Giovannini R, Freitag R, Tennikova T (1998) High-performance membrane chromatography of supercoiled plasmid DNA. Anal. Chem. 70(16), 3348–3354.

Guiochon G, Golshan-Shirazi S, Katti A (1994) Fundamentals of Preparative and Nonlinear Chromatography. Academic Press, San Diego.

Hagedorn J, Kasper C, Freitag R, Tennikova T (1999) High performance flow injection analysis of recombinant protein G. J. Biotechnol. 69, 1–7.

Hancock WS, Wu S, Frenz J (1992) Frontiers of biopolymer purification: displacement chromatography. Bio Separations Int. 5(12), 18–21.

Hjertén S, Liao JL (1988) High-performance liquid chromatography of proteins on compressed, non-porous agarose based I. Hydrophobic interaction chromatography. J. Chromatogr. 457, 165–174.

Iyer H, Tapper S, Lester P, Wolk B, Van Reis R (1999) Use of the steric mass action model in ion-exchange chromatographic process development. J. Chromatogr. 832, 1–9.

Josic D, Octapharma Pharmazeutika Produktionsges. m.b.H., Vienna, Austria, personal communication.

Kasper C, Meringova L, Freitag R, Tennikova T (1998) Fast isolation of protein receptors from streptococci G by means of macroporous affinity discs. J. Chromatogr. 798, 65–72.

Kasper C, Reif OW, Freitag R (1997) Evaluation of affinity filters for protein isolation. Bioseparation 6(6), 373–382.

Kasper C, Vogt S, Breier J, Freitag R (1996) Protein displacement chromatography in hydroxy- and fluoroapatite columns. Bioseparation 6, 247–262.

Kastowsky M, Gutberlet Th, Bradaszek H (1992) Molecular modelling of the three-dimensional structure and conformational flexibility of bacterial lipopolysaccharide. J. Bacteriol. 174, 4798–4806.

Kundu A, Barnthouse K, Cramer SM (1997) Selective displacement chromatography of proteins. Biotech. Bioeng. 56(2), 119–129.

Kundu A, Suresh V, Cramer SM (1995) Antibiotics as low-molecular-mass displacers in ion-exchange displacement chromatography. J. Chromatogr. 707, 57–67.

Labrou N, Clonis YD (1994) The affinity technology in downstream processing. J. Biotechnol. 36, 95–119.

Nachmann M (1992) Kinetic aspects of membrane-based immunoaffinity chromatography. J. Chromatogr. 597, 167–172.

Nicoud RM (1998) Simulated moving bed (SMB): some possible application for biotechnology. In: Bioseparation and Bioprocessing: A Handbook, Vol. 1 (Subramanian G, ed.). Wiley-VCH, Weinhelm, pp. 3–39.

Podgornik A, Barut M, Jancar J, Strancar A, Tennikova T (1999) High-performance membrane chromatography of small molecules. Anal. Chem. 71, 2986–2991.

Shukla AA, Bae S, Moore JA, Cramer SM (1998) Structural characteristics of low-molecular-mass displacers for cation-exchange chromatography. II. Role of the stationary phase. J. Chromatogr. 827, 295–310.

Subramanian G, Cramer SM (1989) Displacement chromatography of proteins under elevated flow rate and crossing isotherm conditions. Biotechnol. Progr. 5, 92–97.

Tennikova T, Freitag R (1999) High-performance membrane chromatography of proteins. In: Analytical and Preparative Separation Methods of Biomacromolecules (Hassan Y Aboul-Enein, ed.). Marcel Dekker, New York, pp. 255–300.

Tennikova T, Freitag R (2000) An introduction to monolithic disks as stationary phase for high performance biochromatography. J. High Resol. Chromatogr. 23(1), 27–38.

Vogt S, Freitag R (1997) Comparison of anion-exchange and hydroxyapatite displacement chromatography for the isolation of whey proteins. J. Chromatogr. 760, 125–137.

Vogt S, Freitag R (1998) Displacement chromatography using the UNO continuous bed column as stationary phase. Biotechnol. Progr. 14(5), 742–748.

Wallworth DM (1998) Radial flow chromatography: developments and application in bioseparations. In: Bioseparation and Bioprocessing: A Handbook, Vol. 1. (Subramanian G, ed.). Wiley-VCH, Weinheim, pp. 145–156.

Wolfgang J, Herttig A, Prior A, Byers Ch (1998) Theoretical plate numbers in a batch column and in a P-CAC. Oral presentation at the HPLC '98 meeting, May 2–8, St Louis.

Zhu J, Katti AM, Guiochon G (1991) Effect of displacer impurities on chromatographic profiles obtained in displacement chromatography. Anal. Chem. 63, 2183–2188.

13
Scale-up and Commercial Production of Recombinant Proteins

Richard Carrillo and Gene Burton
Bayer Corporation, Berkeley, California, U.S.A.

1. INTRODUCTION

The successful scale-up of a protein purification process to commercial production is the culmination of all research and development activities. We will discuss the elements of a successful scale-up strategy as well as process control and monitoring during production. The main emphasis will be on chromatographic separations, since they are at the heart of virtually all recombinant protein purification schemes. The scale-up strategy for a well-developed process can be relatively straightforward. However, consideration of commercial processing constraints during the design phase is critical to ensure that a viable, efficient, and validatable process is delivered to the production environment.

2. PROCESS DESIGN FOR PROTEIN PRODUCTION

It is advantageous to have an understanding of the ultimate throughput requirements for a given process/product during the design phase as well as an appreciation of the capabilities (layout, utilities, equipment) of the eventual production suite. Obviously, process design can influence plant design and vice versa. Because of this, communication and collaboration between development scientists, process engineers, and manufacturing personnel is important to ensure that a developed process is a "good fit" for the production facility.

The design of an integrated purification scheme is a topic covered extensively elsewhere (Wheelright 1991), but for the purpose of this work, there are some general considerations that should be taken into account when formulating a process that will ultimately be conducted at a commercial scale.

In general, upstream, initial capture steps have their greatest utility as concentration/dewatering operations. Therefore, there is a premium for high throughput. Chromatography resins employed in these operations will generally be of large particle diameter with column efficiency (measured by number of theoretical plates) being lower, relative to the smaller particle diameter resins used downstream. However, the benefits of lower pressure drop and resultant equipment and running costs usually outweigh increased column efficiency gains (Jagschies 1988; Wheelright 1991).

In selecting the logical sequence of steps in an integrated process scheme, considerations that will affect equipment and plant design should be taken into account. If the overall process design is of an orthogonal nature (all different chemistries, targeted purification functions), then the order of steps should impart the highest level of efficiency to process flow. For example, ion exchange steps can follow hydrophobic interaction chromatography (HIC) steps and vice versa. Chromatographies that are relatively insensitive to salt concentration, such as immobilized metal ion affinity chromatography (IMAC), are excellent steps to transition eluates from one ion exchange step to another. The relative capacity of individual steps should also be considered. There is little logistic advantage to having a very-high-capacity step following a low-capacity one, leading to inordinately long load times on the second column. Minimizing transition times between steps by cutting down on operator manipulation (titrations, dilutions, etc.) ultimately increases throughput and decreases the number of process control points.

Finally, an additional consideration for groups that will historically develop a series of purification process designs is that of working toward a "generic" process in conjunction with the facility.

In general, there are the physical and chemical aspects of each process to take into consideration. Using the average mammalian cell culture process as an example, the primary physical consideration is volume reduction in the first one- or two-unit operations. Even with limitations imposed by media surfactants commonly found in most protein-free cell culture processes today, 20-fold volume reduction utilizing membrane systems works adequately to move to the "thousand liter" working volume range for downstream steps. Similarly, affinity steps, if employable, also contribute to the volume reduction scenario, along with a greatly improved purification impact. Once having reached a reduced volumetric level, most downstream unit operations will be similarly sized and equipment readily interchangeable from process to process. In addition, and with few exceptions, subsequent downstream chemistries can usually be accommodated within the reduced volume levels at or below 1000–2000 L.

The results of the above approach suggest that unit operation space, sizing, and equipment are more readily predicted along with associated costs. These factors fit well into multiuse facilities and support the interests of process economics through multiple processes, rather than just the current effort.

3. SCALE-UP OF CHROMATOGRAPHY

A scale-up from the bench to pilot scale can be on the order of 100-fold. The subsequent scale-up from pilot to production scale is generally smaller, usually no more than 25-fold. The defined (developed) process is scaled using the following general guidelines outlined in Table 1. These parameters reflect implementing a desired increase in throughput by increasing the column volume.

In general, the volume of a column is increased by making it wider and maintaining a constant bed height. By keeping the linear velocity constant, the volumetric flow rate increases and mobile phase volumes are directly proportional to the increased column volume.

There are practical caveats to following the above guidelines to the letter. A linear flow rate that can be achieved with laboratory columns may not translate to production scale columns due to lack of mechanical support from the column wall. Efforts to maintain the same flow rate may result in a prohibitive pressure drop. For this reason, it is recommended, where possible, that final optimization of flow rate be done at a scale employing columns that are $\geq$ 10 cm in diameter.

Table 1 Guidelines for Scale-up of Chromatographic Process

Kept constant	Increased
Resin bed height	Column diameter
Linear velocity of mobile phase	Volumetric flow rate
Sample concentration	Sample and mobile phase volumes
Gradient slope	Gradient volume

Columns readily available from manufacturers may not be of the exact diameter that will give both the desired volume and bed height. In cases such as these a column with a diameter closest to ideal is packed to the designated volume, resulting in a change in bed height at the new scale. Provided no pressure drop issues (for the packed bed and/or system) manifest as a result of an increased bed height, maintaining the same residence (contact) time should provide comparable chromatographic performance. This does assume that plate height (HETP), essentially a measure of the consistency of the packed bed, remains relatively unchanged during scale-up.

Maintaining comparable resin bed efficiency is an important consideration during scale-up. However, loss of efficiency must always be evaluated in the context of what real impact it has on the separation process. Since this is sometimes difficult to assess generally HETP/N as well as bed asymmetry are tested for comparability. Extra column (dilution) effects not directly related to the consistency of the packed bed will be discussed in more detail below in the section on equipment. The procedure for packing representative columns during scale-up may change to ensure a homogeneous bed. Development of the appropriate procedure can be a significant subproject in itself, but the results often justify the effort. Table 2 outlines testing results derived from a bind, step gradient wash and elute column packed representing development, pilot, and commercial scales—a scale-up of approximately 5000-fold. The stationary phase was a polymethacrylate-based ion exchanger. The packing protocol was modified at each scale-up point to accommodate the larger volumes and hardware. The change in plate count observed at the final scale was found to not affect the separation in any way.

4. PROCESS CONTROL

A commercial scale process must be implemented and maintained under a high degree of control. The rigorous treatment of the subject of validation

Table 2 Number of Theoretical Plates (N) and Asymmetry (A_f) Measurements for Lab, Pilot, and Production Anion Exchange Columns

Column dimensions (cm)	Resin bed volume (L)	N	A_f
1.6 × 15	0.03	554	1.2
30 × 15	10.6	528	1.1
115 × 15	156	333	1.4

can be found elsewhere (Jagschies 1988; Martin-Moe et al. 2000b; Sofer and Hagel 1997), but in essence the equipment and processes utilized during commercial manufacturing must be in a validated state, i.e., they must demonstrate through documentation that they perform in accordance with predetermined acceptance criteria during a qualification stage. Thereafter, monitoring ensures that they remain so.

Validation strategy is best formulated during the process development stage. In summary, the following guidelines can be applied (Martin-Moe et al. 2000a):

1. Process variables are identified and categorized into operational (input) and tested (output) parameters.
   Input parameters are further examined to identify those that are critical from a technical success standpoint.
   Finally, critical process input parameters are evaluated to identify those that are difficult to control in a production environment.
2. Critical process input parameters that are difficult to control (e.g., protein/product loading on downstream chromatography columns) should have studies performed to establish ranges under a validation protocol with predetermined acceptance criteria. Range finding studies done at production scale can put large amounts of expensive material at risk due to potential failure to meet acceptance criteria. Because of this, processes that are scalable can have operational limits established at a much smaller scale. Of course, comparability must be demonstrated among the different scales.
3. Critical process input parameters that are easily controlled (e.g., flow rate/contact time and pH) can be validated as set points with a narrow range during the qualification of the commercial scale.

Equipment and process monitoring is necessary to maintain a validated process. A discussion of general equipment issues follows below, but from a control standpoint "qualifying" a system before each use/batch is looked on favorably by regulatory agencies. This may involve such things as visual inspection, where possible, of a chromatographic gel bed, or performing an efficiency/asymmetry test on a column. Filtration systems can be subjected to an integrity test.

The input and output parameters of a production process should be formally monitored and reviewed to identify any changes or deviations indicating that the process may not be in control. Parameters are generally judged against established action/alert levels. It is important, however, to note that a statistically significant number of runs employing a representative feedstream must be used to establish levels that reflect meaningful process variability. Periodic reviews of the process performance may result in action/alert levels being adjusted or in cases where recurring deviations are found to have a root cause; the process may be changed. For an established process, this must be carried out under a formal change control system.

5. EQUIPMENT

Purification equipment employed in the production of recombinant proteins has evolved into hardware and software that impart a high level of control over the process. These automated systems not only marginally decrease labor costs, but perhaps more importantly contribute to the reproducibility of operation and recording of important data, resulting in a more robust process. Custom and off-the-shelf columns and pumping systems are available from manufacturers such as Amersham-Pharmacia Biotech and Millipore. The purchase of this hardware often involves long lead times (> 6 months) and therefore must be timed appropriately. We will outline design and selection considerations for such systems.

Industrial chromatography columns are commonly run by software-controlled automated pumping devices that include output measuring (e.g., pressure, flow rate, pH, and conductivity) instruments. The design of these "skids" must have input from the development group to ensure that the appropriate number of inlet and outlet valves, flow rate requirements, and instruments that measure critical parameters are all incorporated. There must be input from the engineering and technical operations/manufacturing group regarding acceptable materials of construction, proper design of control systems, and sanitary design.

Specifically to the criteria for selection of equipment, materials of construction should be nonreactive with the feedstream and leachates should be minimal and/or benign. The components and system should be of a sanitary design, i.e., they should be able to pass a rigorous cleaning qualification/validation test. These criteria apply to both chromatography and filtration systems.

Chromatography column and skid hardware must address all the "materials clearance" issues raised above and in addition must meet the mechanical requirements (e.g., must meet pressure drop requirements) of the developed process. This again stresses the importance of the development group to understand the limitations of existing production scale equipment and to either accommodate this or influence the replacement with equipment that is suitable.

As described in Sec. 3, packing procedures for chromatography columns may have to be reoptimized as part of scale-up to ensure that the quality of packing or plate height, HETP, is comparable between scales. In addition, because extra column effects resulting from dilution caused by the skid hardware and decreases in flow distributor efficiency can potentially affect chromatographic performance, it is important to design efficiency/asymmetry testing such that it in essence tests the system, as opposed to the packed bed alone, i.e., potential effects on chromatographic performance resulting from scale-up of the hardware are illuminated. It has been our experience that extra column dilution effects actually become less of a factor during scale-up. This makes sense when you consider that the lab scale typically involves employing columns that have a volume of approximately 1–10 ml on pumping/monitoring systems with a comparable extra column volume. The extra column volume in a well-designed industrial skid is generally significantly less than the packed bed volume. There are, however, cases such as in gel filtration where dilution as the result of the system should be minimized in delivering the sample to the column.

In order to keep equipment (and therefore the process) in a controlled or validated state, regular preventive maintenance and or calibration procedures must be in place. In addition, requalification may be scheduled at discrete time points for key pieces of equipment and instruments. As with process changes described in Sec. 4, any modifications or replacement of existing equipment in a production environment must be managed by a formal change control system.

6. DISPOSAL/TREATMENT OF WASTE STREAMS

The disposal/treatment of wastestreams is an issue that must be considered as early as possible in process development. A promising purification technique may not be feasible at production scale because a waste stream requiring treatment may exceed the limits of the plant's and or municipality's treatment system. Furthermore, in situations where organic solvents are used, an acceptable facility must either be in existence or be built. In general, when a waste stream from a potential commercial process is suspected of containing a toxic or hazardous substance, it is prudent to consult the firm's health and environmental safety department. This will not only clarify possible safety issues associated with operator handling but will also help define facility considerations and disposal requirements.

Since commercial scale processes commonly utilize equipment (e.g., tanks and transfer lines) that is cleaned in place with acid and/or caustic solutions, the resulting waste streams are generally neutralized before they are ultimately released. Because this operation may require holding tanks and other components, the capacity of such a neutralization system bears consideration when determining overall process throughput.

7. ECONOMIC CONSIDERATIONS AND IMPACT ON PROCESS DESIGN

The economics of downstream processing and its importance and impact vary according to project and within the industry. Most discussions on process economics focus on the details of costs and relative currency amounts related to major contributing factors to cost of goods sold (COGS). For this type of information, the readers are referred to works that address this topic in some detail (Jagschies 1988; Kennedy and Thompson 1987; Peskin and Rudge 1992; Sofer and Hagel 1997). We will focus our discussion in this chapter on some of the global trends or approaches that organizations are using to address the issue of cost of goods in addition to highlighting factors that the process designer will need to consider to minimize cost of goods, especially with regard to mammalian cell culture processes.

As most readers are aware, the development of recombinant protein pharmaceuticals includes clinical and production phases that encompass the evolution of a process bound for the market. The production phase may need to respond to changing conditions presented by clinical studies and market estimates of need (both amount and configuration). Furthermore, the scale of production supporting early clinical work will not necessarily conform to market scale or needs. The effort to balance the commitment of resources (people and currency), raw materials and effort in process design with the regulation aspects of drug development has in recent years prompted some organizations to adopt a biphasic type of development approach. The rush of information from the genome search activities with regard to candidate recombinant drug molecules has also contributed to searching for ways to combine balancing cost with the chance of technical success while speeding up development. Finally, consideration is also given to the effect of drug development success rate for recombinant pharmaceuticals, which has been stated to be as low as about 20%.

Although this biphasic approach can have several labels, the tenet of such an approach is generally the same, i.e., to limit the commitment of time and resources until the drug candidate has proven the principle of its development. An overview of such an approach can be summarized in Table 3.

The early-phase processes will need to meet clinical and regulatory standards regarding product quality, without necessarily meeting the man-

Table 3 Considerations and Features of Biphasic Development Strategy

Early phase	Late phase
To support clinical phase I and II	To support clinical phase III and beyond to market
Speed	COGS
Adequacy	Scalable
Basis for the late-phase process	Transferable to manufacturing
Simple	Market sustaining

ufacturing and market standards for optimization and efficiency. The main goal in this phase is to develop a safe and effective product. The late-phase process will be initiated, once the product has demonstrated its worth and will focus on the manufacturing and market criteria while maintaining the previously established safety and efficacy. The late-stage development, for example, may center on process improvements in COGS and scalability, as well as final containment. The degree of process improvements between the two phases may be governed by quality of the initial expression cell line, whether affinity chromatography could be employed in downstream processing, etc.

Another way to look at the costs of the development of a process relates to examining issues driven by the selection of the post production system. In Table 4, we have listed for consideration some general process characteristics that could have a significant impact on costs, especially as they relate to equipment, raw materials, facility use, and testing requirements.

Comparisons are frequently made in the industry between bacterial and mammalian cell line hosts. The advantage goes to bacterial hosts for absolute expression yields and minimal to no issues surrounding viral clearance/killing. However due to the structurally complex nature of most recombinant proteins (secondary structure—disulfide bridging and carbohydrates), the impact of the need for protein folding greatly limits these advantages to a few cases. The resultant physical parameters necessary for the successful protein folding process, along with the ability to assess and assure the quality of the final product, exert the strongest influence in this case.

In comparison, mammalian cell systems afford lower absolute yields (although this is being challenged constantly via expression improvements) and longer production phases, but generally more widely accepted answers to the aforementioned molecular complexities. Both host systems, after isolation and/or folding, are downstream processed in similar manners. We

Table 4 Host-Related Factors

	Host		
	Bacteria	Mammalian cells	Transgenics
Stable expression:			
Time to establish	Short (< 6 mo)	Median (6–12 mo)	Long (> 6 mo)
Run length	2–3 days	1–2 weeks (batch)	Continuous, variable
Relative absolute yields	High	Median	Median to high
Isolation:			
Extraction	Yes	No	Variable
Relative operation volume	Low to high volume	High volume	Low to high volume
Key issue(s)	Solids separation DNA	Volume reduction	Commercial handling systems available
Purification:			
Folding	Heterogeneity questions Volume impact, cost	Not applicable	Not applicable
Specific pyrogen removal step(s)	May be necessary	Generally not applicable	Source related?
Host cell DNA removal	May require additional steps to ensure clearance	May require additional steps to ensure clearance	May require additional steps to ensure clearance
Viral clearance/killing	Not considered an issue	May require additional steps to ensure clearance	Ongoing regulatory discussions
Quality/analytics:			
Folding impact	Simple or complex (folding)	Not applicable	Not applicable
Carbohydrate	Not applicable	Currently accepted	Plants questioned

have also indicated transgenics as a third host class, for comparison, and as a group that is under active consideration by the industry. Many of the factors of the host class are similar to those for mammalian cells. However, discussions continue on the issues confronting this host class, especially with regard to virus clearance needs and the impact(s) of carbohydrate differences.

Some individuals process components that generally impact costs of goods are listed in the following table (Table 5). This table illustrates a fairly standard list of items for the process designer to consider Although each cost grouping can vary from product to product, several stand out routinely as major contributors: water, labor, and quality control/release.

To improve COGS for purification, productivity in cell culture and the streamlining of purification have, perhaps, the greatest effect. With higher cell culture yields comes higher downstream purification yield and enhanced product-to-host contaminant ratios. With purification streamlining comes potential consistency of and reduction in the number of unit operations. Creating homogeneous unit operations from process to process or minimizing the number of events can result in appropriate facilities sizing and

Table 5 Components Contributing to Cost of Goods

Item
Raw materials—commercial:
membranes
filters
chemicals
chromatographic media
Raw materials—manufacturer prepared:
specialty chemicals
resins
Equipment
Facilities
Facilities overhead
Utilities
water
Labor
Formulated bulk storage
Quality control testing
in-process testing
lot release
Validation/maintenance

character, which help to enable a multiuse facility approach. Uniform operations can then minimize labor, operational complexity, as well as create the ability to update operations or facilities from the technological perspective. The combination of streamlining and higher culture yields can then lead to increased economy of scale (i.e., minimizing the number of downstream trains or lots), which provides opportunity for reduction of quality costs.

With the constant development of new technologies and increased experience with the complexities of recombinant drug development, process designers have a constantly evolving experience at their disposal.

REFERENCES

Jagschies G (1988) Process-scale chromatography. In: Ullmann's Encyclopedia of Industrial Chemistry. VCH, New York.

Kennedy RN, Thompson RG (1987) R&D facility budgets and costs that drive design. Pharm. Eng. 7, 13–17.

Martin-Moe S, Ellis J, Coan M, Victor R, Savage J, Bogren N, Leng B, Lee C, Burnett M (2000a) Validation of critical process input parameters in the production of protein pharmaceutical products: a strategy for validating new processes or revalidating existing processes. PDA J. Pharm. Sci. Technol. 54(4), 314–319.

Martin-Moe S, Kelsey WH, Ellis J, Kamarck ME (2000b) Process validation in biopharmaceutical manufacturing. In: Biopharmaceutical Process Validation. Sofer G, Zabriskie DW, eds.: New York, Dekker, pp. 287–298.

Peskin AP, Rudge SR (1992) Optimization of large-scale chromatography for biotechnological applications. Appl. Biochem. Biotechnol. 34,35, 49–59.

Sofer G, Hagel L (1997) Handbook of Process Chromatography. Academic Press, San Diego.

Wheelwright SM (1991) Protein Purification, Design and Scale up of Downstream Processing. Oxford University Press, New York, pp. 10–23.

14

Process Monitoring During Protein Purification

Bo Mattiasson and Rajni Hatti-Kaul
Lund University, Lund, Sweden

1. WHY MONITORING OF DSP?

Downstream processing constitutes a significant part of a protein production process, usually involving several steps and a major share of the total processing cost. In view of this, monitoring of downstream processing becomes an important consideration so as to ensure that the purification sequence is performed optimally (Kenney and Chase 1987). Following are some of the important reasons for monitoring downstream processing of proteins:

- It gives a very good documentation on the processing conditions. This is especially important in the pharmaceutical industry.
- It provides a good overview of the separation process and forms a fundament on which the process control can be built.
- Monitoring and control will increase the reliability concerning reproducibility between different batches of a process.

2. HOW IS MONITORING PERFORMED TODAY?

A fairly primitive mode of operation with regard to monitoring and control of downstream processing has prevailed so far. The normal procedure is to withdraw a sample and analyze it off-line the process. This concerns a broad spectrum of analyses, ranging from purity to biological activity. However, on-line monitoring of a few parameters, mainly in the effluent stream of the chromatography columns, has been in operation for many years.

2.1 UV Absorbance

Registration of absorbance at wavelengths where the target products have a specific absorbance is a convenient mode of monitoring their presence in the process fluids. Most proteins do not have specific absorbance patterns; thus, the general protein absorbance is registered at 280 nm or, when more sensitive analyses are to be carried out, at 220 nm. If the content of nucleic acids is to be quantified, absorbance at 260 nm is also registered and then the value read at 280 nm is corrected for the contribution from the nucleic acids at 260 nm. Proteins with specific prosthetic groups that have typical absorbance patterns are registered at a specific wavelength. For example, the enzyme peroxidase is characterized by its heme content and of course the total protein absorbance. Thus, a specific ratio of absorbance at 403 and 280 nm is used to characterize a peroxidase fraction.

UV absorbance provides a simple mode for obtaining a trace of where the protein can be found. Chromatography operations are normally followed by continuous recording of the absorbance at 280 nm of the elution flow. This is done by pumping the eluate through a flowthrough cell placed in a UV monitor. The signal is registered either by a recorder or a computer. In recent years, it has also become possible to utilize diode array detectors for monitoring absorbance at a range of wavelengths simultaneously, and then specific ratios between certain absorbance bands can be read. When production processes are set up, it is often regarded as safe to analyze one chromatogram very carefully and if the others are similar in shape, then the protein content being represented by the different peaks in the chromatogram are taken to be the same from time to time. However, the absorbance measurements are not very high resolving, giving only an indication of the presence of proteins but not of the biological activity or the homogeneity.

2.2 Ionic Strength/Conductivity and pH

Elution from a chromatography adsorbent often involves a change in ionic strength or pH. By registering changes in conductivity and pH, one gets good information on whether chromatography is run under desired conditions and when to expect the eluted material. This allows one to modify the elution conditions, if necessary. Registration of these parameters is normally carried out using a flowthrough conductivity sensor and pH electrode, respectively, that are integrated components of the commercial fast protein liquid chromatography (FPLC) setups.

Typical profiles of the three parameters discussed above are presented in Fig. 1.

2.3 Separation Systems with Integrated Computer Control

Computers with high-speed data processing capacity, high storage capacity, and user-friendly interfaces have become available in recent years. This has made it possible to store chromatograms from different separations and to compare them directly on-line. Thus, when one successful separation has been done and all off-line analyses have been made, one can compare the behavior of the chromatographic procedure with respect to the few parameters that are monitored routinely.

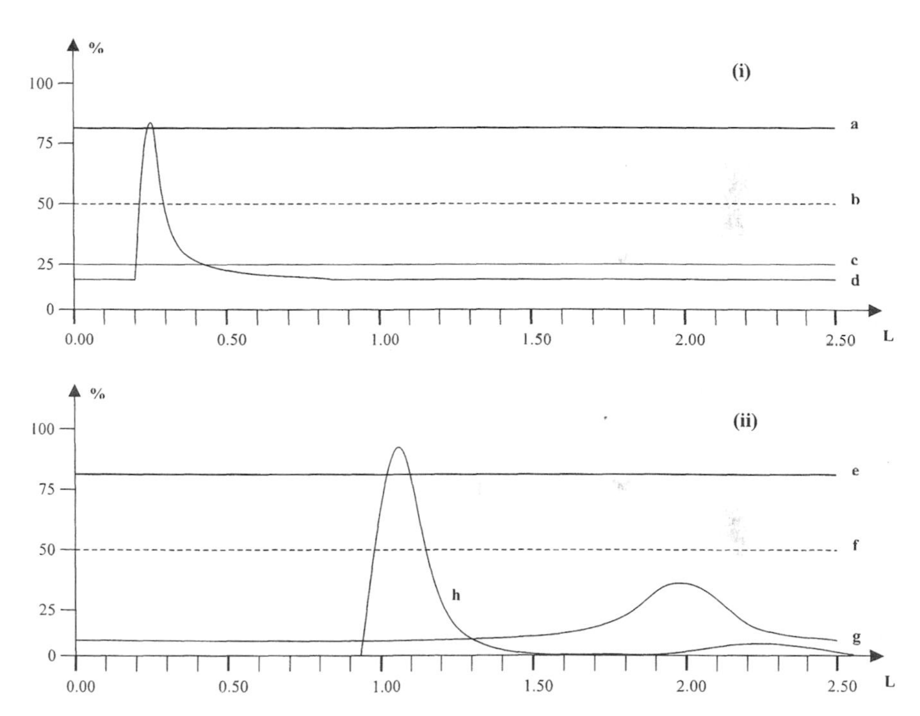

Figure 1 Registration of gel chromatographic desalting of hemoglobin on Sephadex G-25. Registration of parameters was done (i) prior to the column and (ii) after passage through the column. The different parameters monitored represented in the figure are: (a) pH (interval 7.5–8.8); (b) temperature (°C); (c) pressure (0.0–1.0 bar); (d) conductivity (0.1–58.2 mS/cm); (e) pH (7.5–8.8); (f) temperature (°C); (g) conductivity (0.1–58.2 mS/cm); (h) UV absorbance at 280 nm (0.0–6.0). (Reprinted with permission of Mattiasson and Håkanson 1991.)

It is also possible to preselect certain conditions when a fraction shall be recovered simply by programming the computer to switch a valve or turn on a pump, and so forth. Such possibilities are available in a few chromatography setups supplied by Amersham Biosciences (Uppsala, Sweden), Biorad (Richmond, CA), and Perseptive Biosystems (Cambridge, MA).

3. REQUIREMENTS IN DSP MONITORING

The high proportion of the process cost attributed to downstream processing stage for many proteins suggests the importance of performing the purification sequence within specified limits at high yields and knowing rapidly when such limits are crossed so as to make alterations without delay. For example, adsorption to a chromatography matrix is traditionally monitored by assaying the levels of key components in the effluent stream. However, in order to control the loading, it is important to obtain information of the progress of information within the bed (Chase et al. 1999).

As downstream processing aims at separating the target molecule from any impurities, it becomes necessary also to keep track of those. The situation is especially important when low levels of an impurity may cause severe problems for the intended application of the target molecule. Such cases are often found when dealing with pharmaceutical compounds where the presence of trace amounts of nucleic acids or endo-/enterotoxins will restrict the use of the preparation. Likewise, when isolating enzymes for diagnostics, it is important to eliminate interfering activities. In addition, it is mandatory to detect the presence of product variants that may arise from primary structure alterations as a result of error in gene expression, faulty posttranslational processing, improper folding, oxidation, and so forth. Quantitation of target molecule on-line would involve monitoring of fairly high concentrations of analyte and would require defined limits with regard to the concentration range in which the analytical procedure should operate. On the other hand, one needs ultrasensitive assays when tracing the fate of impurities. The levels of allowed impurities in pharmaceutical preparations are often defined based on the analytical techniques; thus, it is imperative that the highest possible sensitivities be used. This feature must be balanced with the desire to make quick readings in order to make an on-line registration of the concentration level.

The desirable features in monitoring devices are high specificity to cope with complex biological materials accompanied by low interference or the ability to correct for interferences, simple operation and maintenance, and reliability over a period of time without operator involvement. If, in addition, prior process knowledge or a mathematical model allows feedback control using the on-line information, then costly run stoppage times or disposal of nonspecification material can be avoided.

4. DEVELOPMENTS IN MONITORING OF TARGET PROTEINS

For many years there has been very little development in the area of process monitoring in downstream processing (Mattiasson 1984). However, it has started attracting attention as technology becomes available as a result of developments in other areas, such as monitoring of fermentation processes and, to some extent, the on-line clinical monitoring of patients. Concomitantly, the need for on-line monitoring has become more pronounced. Sophisticated Kalman filtering techniques have been introduced to enable control of fermentation processes, which basically use available measurements such as pH, temperature, gas analysis to a continuously updated state model to predict factors such as growth rate and product formation. However, in downstream processing the focus has been on developing rapid on-line analytical techniques for directly measuring product concentrations.

A somewhat nonconventional approach to process monitoring was by the use of green fluorescent protein of the jellyfish *Aequorea victoria* as the fusion tag of the target protein (Poppenborg et al. 1997). In general, however, many of the conventional off-line methods based on immunoassays and enzyme assays for measuring the amount of active protein have been adapted to on-line monitoring. A primary consideration for on-line analysis is the means of sampling the analyte from the process stream and sample handling. Perhaps the most advanced tool in this respect is flow injection analysis (FIA), which has been successfully applied to downstream processing monitoring and its potential in process control identified.

During the last decade, analytical technologies have also undergone tremendous advances that have the potential to be integrated with the protein production process for on-line monitoring. For example, rapid high-performance liquid chromatography (HPLC), capillary zone electrophoresis, and mass spectrometry have the potential to offer detailed information about the process stream and to detect changes in the primary structure of proteins (Paliwal et al. 1993a). Developments in microtechnology have now made possible enzyme assays and immunoassays on microchips, resulting in reduced analysis times and minimal reagent consumption (Chiem and Harrison 1997; Hadd et al. 1997; Zugel et al. 2000). Although currently used for discrete assays, they should have potential for on-line analysis. Also, biosensors based on a variety of physical parameters are now available. Less readily available however are well-developed algorithms for feedback control, system troubleshooting and the like (Paliwal et al. 1993a). Limited studies reported in this direction include modelling of protein precipitation process to enable process control in the face of variable influent stream (Foster et al. 1986; Holwill et al. 1997).

4.1 Flow Injection Analysis: Adapting Protein Assays to On-line Monitoring

On-line monitoring of protein activity has been made possible using FIA, which has become a dominating modality in modern analytical instruments (Ruzicka and Hansen 1975, 1991; Schmid and Künnecke 1990). In FIA, the sample is injected into a nonsegmented continuous carrier stream of a suitable liquid and carried as a peak-shaped zone through a reaction coil to the detector, which continuously records the absorbance, electrode potential, or other physical parameters (Fig. 2). In its simplest form, the equipment consists of a pump for propelling the reagent stream though a thin tube, an injection valve by which a defined sample volume is injected reproducibly, and a reaction coil in which the sample disperses and reacts with the reagent producing a signal peak to be sensed by a flowthrough detector. The peak is related to the concentration of the species to be analyzed.

A laboratory assay can be easily adapted to the flow injection analysis by controlling the sample dispersion, which is achieved by variation of the injected sample volume, or also by changing the length of mixing and reagent coils, flow velocities, and dilution. Although FIA does not generate results continuously, it provides a possibility of rational sample handling with high sampling frequency (more than 100 per hour), small sample volumes, low reagent consumption, and high flexibility in adaptation to different reaction procedures and detection methods. For monitoring of unclarified samples, a provision for removing the particulate matter, normally a 0.45 μm membrane filter, is incorporated into the analysis system. A miniature centrifugal separator has also been developed as an alternative means of on-line sample clarification, which is designed to automatically sample, spin, and deliver supernatant to an analyzer within 30–60 s, and wash out solids from the bowl (Richardson et al. 1996a,b).

Enzymes have been regarded as the simplest to monitor using FIA, since they are catalytically active. By mixing a side stream from a separation process with a stream of substrate and appropriate buffer, a stream of liquid is obtained that can be read continuously for registering the enzymatic activity. An example of analysis of enzyme content in the effluent from a chromatographic column is shown in Fig. 3. Here the detection is done using a calorimetric principle, thereby registering the reaction heat from the enzyme-catalyzed process (Danielsson and Larsson 1990). Automated

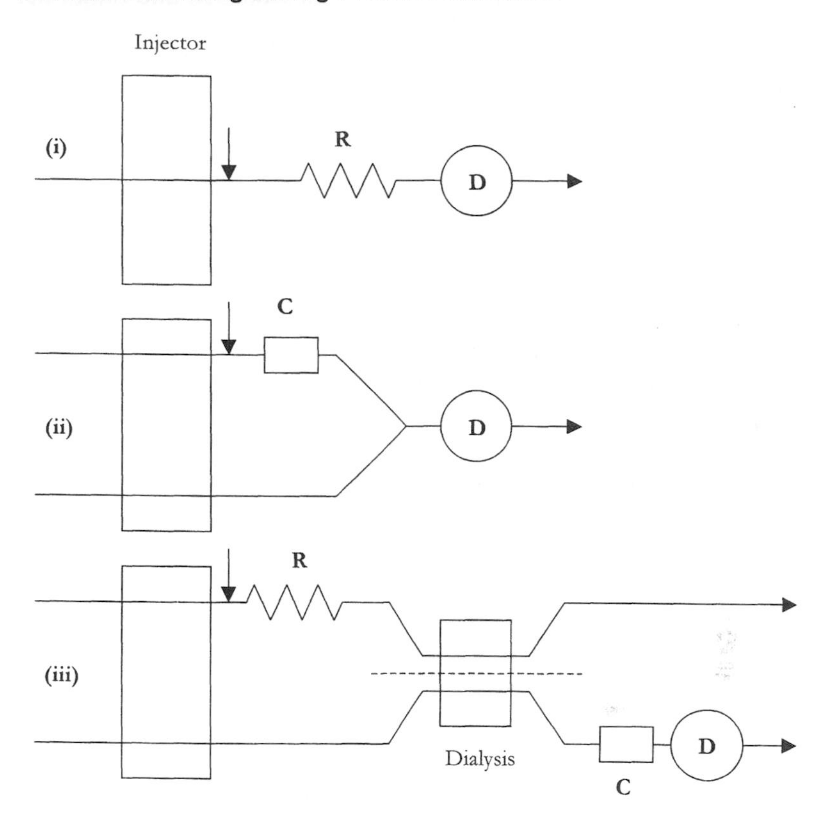

Figure 2 Schematic presentation of three different arrangements for flow injection analysis systems. (i) A simple system where the sample is introduced and passed through a reaction coil (R) and then the detector (D), (ii) contains two streams, one with a column reactor (C) and one for supplying medium after passage through the reactor and before entering the detector, and (iii) a more complex arrangement with two liquid streams—one sample stream and the other containing receiving buffer, a reaction coil, and a dialysis unit. After the dialysis step, the sample passes through a column reactor before entering the detector zone.

FIA assays have been used for on-line monitoring of various oxidoreductases and other enzymes during different downstream operations (Recktenwald et al. 1985; Stamm and Kula 1990; Richardson et al. 1996b) (Table 1). Frequencies of up to 180 samples per hour and response times of 10–30 s using FIA with fluorometric detection have been reported

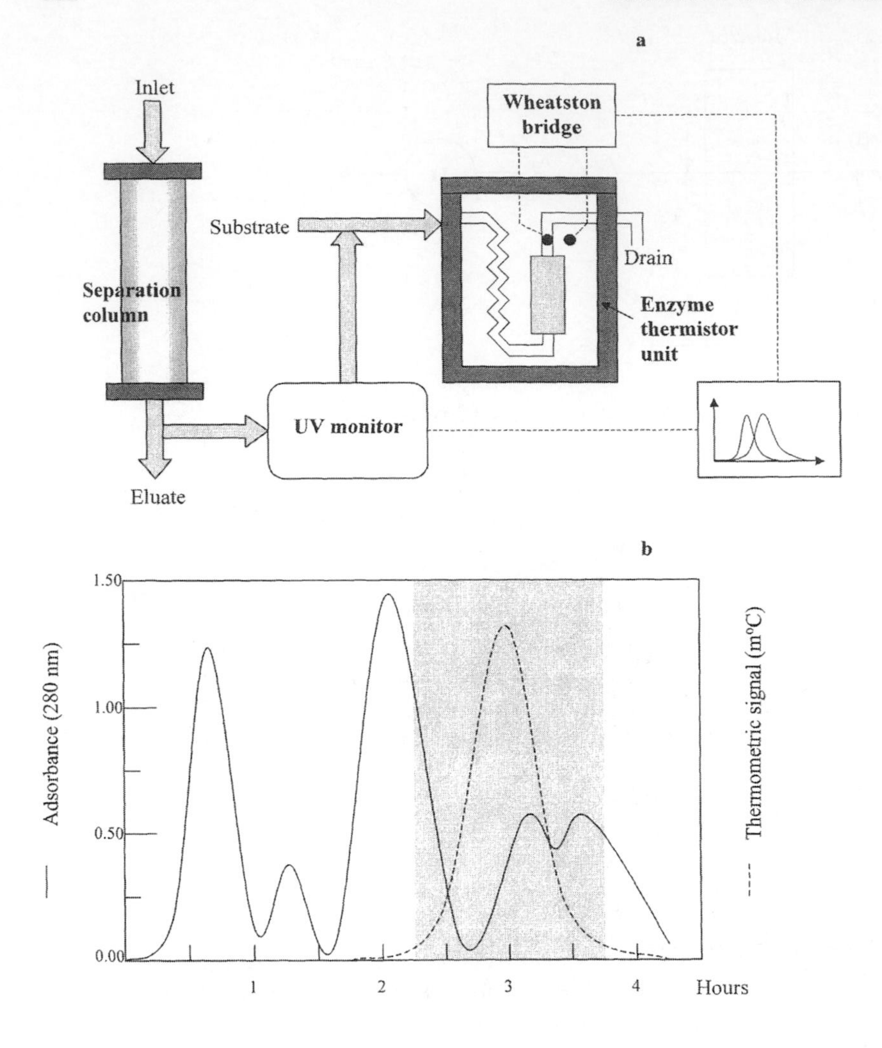

Figure 3 (a) Flow scheme for on-line monitoring of enzyme activity using a side stream mixed with substrate before passing through a flow calorimeter. The flow calorimeter ("enzyme thermistor") registers the reaction heat from the enzymatic process, thereby indicating the presence of enzyme. (After Danielsson and Larsson 1990.) (b) Elution profiles from a chromatographic column where the effluent is monitored using A_{280} and a side stream mixed with substrate and passed through the flow calorimeter unit. The reaction heat generated was registered as the dashed line, indicating the presence of the specific enzyme in the chromatogram. (Modified from Danielsson and Larsson 1990.)

Table 1 Examples of On-Line Enzyme Assays Using Flow Injection Analysis

Enzyme	Downstream processing step	Ref.
Formate dehydrogenase from *Candida boidinii*	Cell disintegration and diafiltration of cell homogenate; chromatography	Recktenwald et al. 1985; Stamm and Kula 1990
L-Leucine dehydrogenase from *Bacillus cereus*	Cell disintegration of cell suspension	Recktenwald et al. 1985
Alanine dehydrogenase from *Bacillus cereus* Phenylalanine dehydrogenase from *Rhodococcus* sp.	Chromatography	Stamm and Kula 1990
Alcohol dehydrogenase	Continuous precipitation using ammonium sulfate	Richardson et al. 1996a,b
Dimethylformamidase	Chromatography	Lüdi et al. 1990
Calf bone alkaline phosphatase	Chromatography	Takahasi et al. 1989

(Stamm and Kula 1990). The FIA concept was also used for combined determination of enzymatic activity and protein concentration during chromatographic purification of dimethylformamidase (DMFase), allowing on-line calculation of the specific activity (Lüdi et al. 1990).

Immunoassays have been performed by integrating the FIA concept with a binding reaction on an insoluble matrix (see following).

4.2 Chromatographic Monitoring

Rapid chromatographic monitoring provides a general method for measuring the protein levels and purity (Chase 1986; Frej et al. 1986). Different components in a process fluid may interact to different extents with a chromatography matrix; thus, they can be eluted as separate peaks from a chromatography column after a certain liquid volume has passed through or by varying the mobile phase. For on-line monitoring, a provision for continuous or semicontinuous withdrawal of process fluid and in-line filtration for removal of particulate matter is made. Detection of the compound leaving the column is generally done using a nonspecific technique such as flow spectrophotometry either at a single wavelength or with photodiode array detectors. Fluorometic detection is more sensitive and can also impart selectivity necessary to discriminate between proteins (Riggin et al. 1991; O'Keefe et al. 1992). Detection and integration of the peaks of interest on the chromatogram are performed on-line by a microprocessor connected to the detector output to obtain information on the level of the product in the process stream. The same microprocessors will also be responsible for control of sampling and the chromatographic cycle. After comparison of the integrated values with stored calibration data, quantitative information on the level of the compounds of interest can be sent in a digital or analogue form to the main process controller. The entire cycle of sampling, chromatographic separation, and data collection can be performed repeatedly.

For the chromatography to be useful for on-line analysis, a prerequisite of course is that the cycle time is short compared with the time scale over which the levels of the analyzed compounds are changing. Chase and coworkers (Chase et al. 1999; Clemmitt et al. 1999) have used chromatographic analysis for monitoring the development of adsorbate profiles during loading, washing, and elution phases of protein separation by expanded bed adsorption. This was achieved by modifying the column to allow automated abstraction of samples at various heights along the bed.

The use of matrices allowing fast adsorption kinetics has greatly increased the sensitivity, accuracy, and applicability of chromatography-based analysis (Chase 1986). Nonporous chromatography matrices allow rapid separation of proteins because (1) the interactions are localized on the surface, (2) equilibrium is reached rapidly, and (3) high mobile phase velocities may be used. However, the limitations are low surface area leading to easy overloading of the columns, and high operating pressures that stress both the column and the equipment. Matrices with open-pore structures, e.g., POROS, allow very-high-speed separations and may further obviate the need for sample preparation by filtration (Paliwal et al. 1993b). Accuracy and speed of analyses is further determined by optimization of the chromatographic separation in terms of flow rate, bed size, sample size, elution strategy and so forth. Reduction in analysis time of proteins by 50% was reported by running parallel-column, gradient elution chromatography in which two gradient elution trains were operated 180° out of phase. (Thévenon-Emeric and Regnier 1991). Sample volume can also influence the sensitivity of detection and can be automatically adjusted by considering the size of the peak obtained in the preceding cycle. However, it is essential to avoid the protein load exceeding the adsorption capacity of the bed (Chase 1986).

Adsorption chromatography has been a method of choice for analysis because sufficiently short cycle times can be achieved. Nonspecific techniques, such as ion exchange chromatography, have the advantage of being rather robust and for allowing simultaneous detection of levels of more than one protein. Monitoring the downstream processing of β-galactosidase using ion exchange chromatography was more convenient over the conventional assay method in that the purity of the sample could be directly observed on the chromatogram (Frej et al. 1986). The ability of ion exchange and hydrophobic chromatography to discriminate between protein variants differing by a single amino acid has also been demonstrated (Chicz and Regnier 1989, 1990). However, at times the resolution may not be sufficient and the chromatograms produced may be difficult to analyze in a truly on-line manner. Also, the relatively high dissociation constant of the interactions in crude samples may lower the sensitivity of detection.

Affinity chromatographic separations permit good resolution of the protein of interest due to high selectivity of interaction making peak identification easy. Furthermore, the lower dissociation constants make adsorption efficient at low concentrations, and larger volumes can be applied greatly increasing the sensitivity of the assay. The affinity columns based on protein G and A, respectively, have been used for rapid quantification of antibodies (Riggin et al. 1991; Paliwal et al. 1993b). Chromatographic analysis systems based on immunoadsorbents can potentially be used to monitor the level of almost any protein. But the protein ligands in general are invariably susceptible to denaturation as a result of fouling and particularly when the binding constants between the ligand and protein product are high. Less specific affinity adsorbents may be used successfully provided the sample contains only one protein having affinity for the adsorbent. Examples are the measurement of lysozyme using an immobilized dye column (Chase 1986), and, more recently, monitoring separation of histidine-tagged glutathione *S*-transferase from an unclarified *Escherichia coli* homogenate by expanded bed adsorption using Zn^{2+}-loaded NTA-silica as the affinity chromatographic sensor (Clemmit et al. 1999).

Chromatographic Immunoassay

The specificity of antibodies makes them useful for differentiating between product polypeptide and the other proteins present in the process fluid. The traditional immunoassays are complex, labor-intensive procedures involving a number of manual operations. Normally the protein analyte bound to an immobilized antibody in a microtitre plate is indirectly quantified using a labeled antibody (sandwich enzyme-linked immunosorbent assay (ELISA) or antigen (competitive ELISA). Although several samples can be analyzed simultaneously, the time required for an assay (a couple of hours) is incompatible with the rapid time frame required for process monitoring.

Immunochromatographic methods have been developed in a flow-through mode allowing detection of very low concentrations of antigen (Ohlson et al. 1978). The target molecule in the flow stream is rapidly brought into contact with excess immobilized antibody in a flowthrough reactor. Once the nonbinding components are washed off, the captured analyte may be eluted and quantified directly or may be measured in the bound form by using common signal amplification techniques as in the ELISA format (Mattiasson et al. 1977). When used in the chromatographic format, the immunosorbent is recycled, reagents and washing buffers are continuously pumped into the analytical reactor, reaction time is controlled through the flow rate, and the reaction product formed is transported to an external detector for quantitation. Using such heterogeneous immunoassays, it has been possible to reduce the experimental errors substantially and thereby also to operate the system far from equilibrium.

Antibodies have been immobilized to various matrices, including chromatography gels, chromatography disks, magnetic particles, membranes, fused silica capillaries, and so forth (Stöcklein and Schmid 1990; de Frutos et al. 1993; Nilsson et al. 1993, 1994; Hagedorn et al. 1999; Klockewitz et al. 2000). In some studies, antibody immobilization has been achieved through binding to protein A coupled to the matrix (Cassidy et al. 1992; Nadler et al. 1994a).

Direct Analysis. Direct measurement of the analyte is typically done by UV absorbance (Afeyan et al. 1992), or more sensitive fluorometric detection (Riggin et al. 1991; Klockewitz et al. 2000). The latter mode could even provide information on purity of the protein fractions as demonstrated for on-line detection of recombinant antithrombin III and tissue plasminogen activator (Klockewitz et al. 2000). Based on the matrices used, very-high-speed immunosassays allowing 200 assays per hour are possible, and the analytical data obtained can be used to control fraction collection (Paliwal et al. 1993a).

Regnier and coworkers have attempted different formats of immunochromatographic analysis to determine product concentration and quality (de Frutos and Regnier 1993; Riggin et al. 1993; Nadler et al. 1994a,b). Dual column immunochromatographic analysis has been proposed for analysis of protein variants (Riggin et al. 1993; Nadler et al. 1994a,b). In one form of the technique, a precolumn immunosorbent cartridge is used to subtract antigens from the chromatographic stream. The antibody is immobilized by biospecific adsorption to the protein A or protein G column. The product antigen is harvested by capture on the column and subsequently reconcentrated on another chromatography column. Comparing the chromatograms to those in which antibody was not used (subtractive immunoassay) allows peak identification in a chromatogram without using protein standards. Monitoring a γ-interferon production using an immunosorbent in tandem with an ion exchange column could distinguish glycosylated and nonglycosylated forms of the protein (Nadler et al. 1994a).

Another strategy is to separate variants and interfering substances before immunochromatography. For determination of IgG multimers, analysis by size exclusion chromatography (SEC) in the first dimension was done to allow determination of molecular size, and fractions from the SEC column were then automatically transferred to the protein A column for capture of immunoglobulins (Nadler et al. 1994b). The presence of contaminating proteins did not affect the analysis by protein A affinity chromatography in the second dimension.

Chromatography-Based ELISA. Both competitive and sandwich ELISA have been performed in the chromatographic format (Stöcklein and Schmid 1990; Cassidy et al. 1992; de Frutos and Regnier 1993; Nilsson et al. 1992, 1993, 1994). Sample is passed through a cartridge containing an immobilized primary antibody where the target molecule is captured. Subsequently, in sandwich ELISA an enzyme-tagged secondary antibody is injected that forms a sandwich with the primary antibody and the target molecule, whereas in competitive mode a known amount of pure labeled antigen is added (these can also be added along with the sample). Substrate solution is then pumped over the cartridge, and formation of a product from the enzymatic turnover of the substrate is monitored using a visible light detector. A typical assay is schematically presented in Fig. 4. The heterogeneous ELISAs have also been adapted to the flow injection mode (de Alwis and Wilson 1985; Stöcklein and Schmid 1990). Table 2 lists some examples of the protein monitored by flow ELISA. The total assay times including washing and reequilibration of the chromatography cartridge have been in the range of a few minutes. Figure 5 shows a chromatographic separation of three model proteins recorded by both conventional modes and by flow ELISA. Flow ELISA also turned out to be an attractive analytical system for monitoring the product concentration during an integrated process for α-amylase production by *Bacillus amyloliquefaciens* (Nilsson et al. 1994) (Fig. 6).

Sensitivity of the ELISA is strongly dependent on the method used for detecting the enzyme reaction product, which could be based on spectrophotometric, fluorometric, electrochemical, calorimetric analyses, and so forth. Immunoglobulin G was detected at the femtomole level with an on-line electrochemical detector (de Alwis and Wilson 1985). As quantitation in ELISA is based on measuring the concentration of the product accumulating as a result of enzyme amplification reaction, significant increase in assay sensitivity and reduction in detection time was achieved by reducing the reaction volumes by performing the assay in fused silica capillaries of 100 μm internal diameter (providing a volume of 79 nl/cm) (de Frutos et al. 1993). Additional modifications of ELISA obtained by complexing the analyte with the

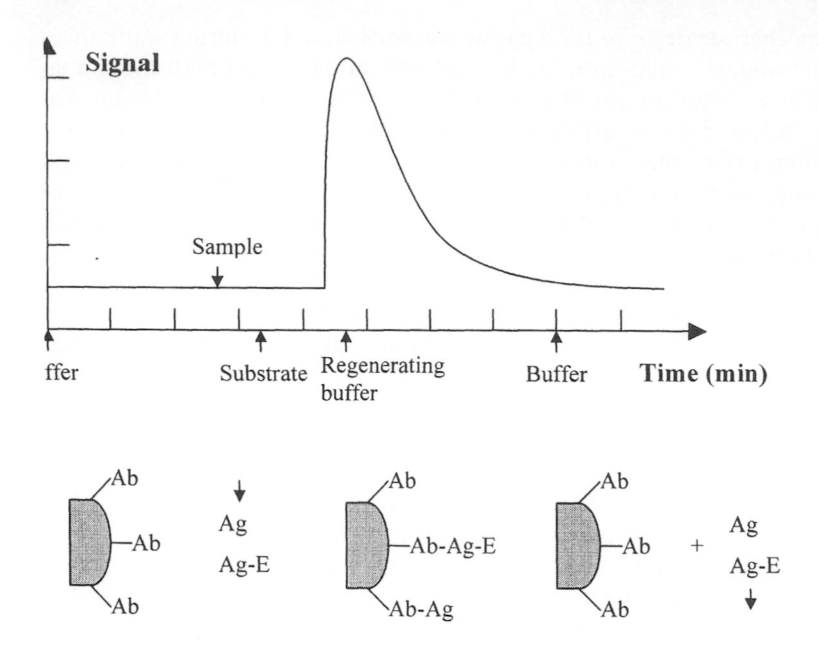

Figure 4 Schematic presentation of a flow-competitive ELISA assay cycle. The different reagents are introduced into the continuous buffer stream at the arrows indicated. Schematically shown at the bottom of the figure is the immunosorbent during the different stages of the assay cycle. The cycle starts with the native immunosorbent in a packed bed. A sample containing native antigen is mixed with a fixed amount of enzyme-labeled antigen and this mixture is then introduced into the flow. On passage, binding will take place in a competitive manner between native and labeled antigen. After a short washing period, substrate for the marker enzyme is introduced. Soon thereafter, a dissociating buffer is introduced to rinse the immunosorbent. In the final step, running buffer is introduced and the equipment is ready for another cycle.

enzyme-labeled second antibody external to the column and subsequently capturing on the immunosorbent column where the enzyme amplification and detection portions of the assay are executed improved the detection limit further (de Frutos et al. 1993; Johns et al. 1996). Using such an approach, an analyte concentration as low as 4.3×10^{-13} M could be determine. Moreover, lower concentrations of the second antibody were required.

Antibody Stability. Stability properties of the antibodies becomes increasingly important in immunoassays as the demand on life expectancy

Table 2 Examples of Analyses Using Flow Injection Binding Assay

Target compound	Application	Ref.
Human serum albumin	Clinical analysis	Mattiasson et al. 1977
Creatine kinase	Clinical analysis	Luque de Castro and Fernandez-Romero 1992
α-Fetoprotein	Clinical analysis	Maeda and Tsuji 1985
Monoclonal antibodies	Monitoring during hybridoma cell cultivation	Stöcklein and Schmid 1990
Immunoglobulin G	Model system	de Alwis and Wilson 1985
α-Amylase	Process monitoring	Nilsson et al. 1994
β-Galactosidase	Monitoring during cultivation of *E. coli*	Tocaj et al. 1999
L-Lactate dehydrogenase	Monitoring during fermentation of *Lactococcus lactis*	Nandakumar et al. 2000
Plasmid DNA	Monitoring during cultivation of recombinant *E. coli*	Nandakumar et al. 2001
17-α-Hydroxyprogesterone	Serum analysis	Maeda and Tsuji 1985
Carbohydrates	Model studies of weak affinity	Ohlson et al. 2000

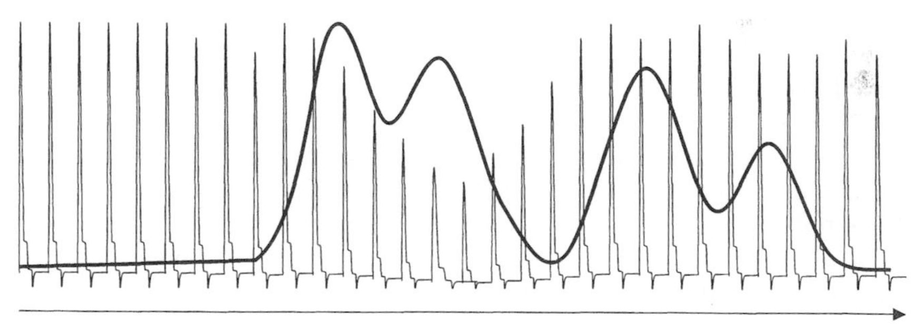

Figure 5 Registration of the absorbance at 280 nm of the effluent after separating IgG, HSA, lysozyme, and *N*-acetyltryptophan (in the same order as the absorbance peaks) on a gel permeation column. The content of HSA was monitored using the sequential competitive flow ELISA (seen as spikes) (From Nilsson et al. 1992).

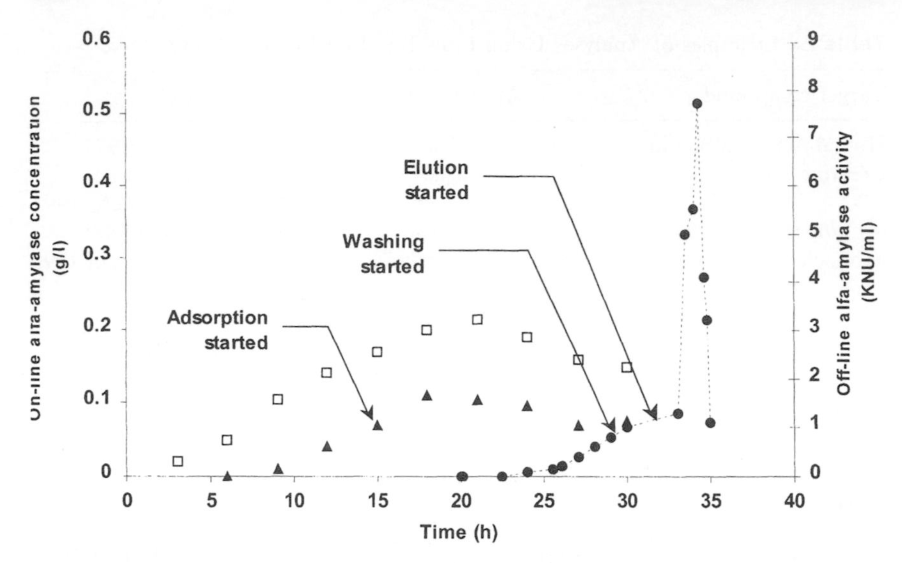

Figure 6 Flow ELISA for monitoring α-amylase production. The bioreactor is integrated with cell recycling and an affinity adsorption step for harvesting the enzyme. The concentration of the target protein was low in the medium as long as the adsorbent was capturing what was being produced. The enzyme level in the medium started to rise when the capacity of the adsorbent was saturated. The release of the bound α-amylase could also be followed using the flow ELISA. (□) Off-line analysis of sample from inside the fermenter; (●) on-line analysis after passage over an affinity column; (▲) on-line assay before the affinity column. (From Nilsson et al. 1994.)

and reliability is increased (Nilsson et al. 1993). In some cases, excellent stability of the immunoadsorbent has been reported, allowing reuse of at least 1000 times (Johns et al. 1996).

The most critical step in the immunoassay that compromises antibody stability is the one where the antibody-antigen complex is dissociated, usually done by applying low pH for high-affinity immune complexes. The antibody is partially denatured under these conditions but recovers most of its former conformation and thus the ability to bind antigen in the reconditioning step. The low degree of denaturation ultimately results in a major loss of binding activity during recycling of the antibody for analyses. This could be regarded as detrimental to the concept of using flow injection ELISA for process control since the assay will change its performance with subsequent runs. However, in a competitive binding situation when a fixed amount of labeled antigen has to compete with a native antigen, the conditions will be the same with regard to the relative probability for the labeled antigen to bind, irrespective of the number of binding sites available. Thus, if one expresses the binding in terms of percentage of the original binding value with only labeled antigen molecules, then a constant situation prevails in spite of the fact that the number of antibody binding sites may change dramatically during the process. This concept of analyses is illustrated in Fig. 7 (Nilsson et al. 1993).

However, as a general rule of thumb, it is more appropriate for continuous assays or repeated discrete assay to utilize antibodies that are binding the antigen less strongly and also have better dissociation kinetics than those antibodies selected for discrete immunoassays.

4.3 Capillary Electrophoresis

Capillary electrophoresis (CE) in different formats has emerged as a complementary technique for rapid protein separations. The technique combines the quantification and handling benefits of HPLC with the separating power of conventional electrophoretic techniques. The advantage over the latter is that spectrophotometric detectors are built into the equipment. Capillary zone electrophoresis (CZE) is the simplest form of CE and the most widely used. Separation in CZE occurs because solutes migrate in discrete zones and at different velocities. Separation of cationic and anionic solutes by CZE occurs due to electro-osmotic flow that results from the effect of the applied electrical field on the solution double layer at the interior of the capillary wall. In the area of protein analysis, CZE has been used for purity validation, screening protein variants, and conformational studies (Paliwal et al. 1993a). It has also been applied for monitoring separation of antibody isoforms (Wenisch et al. 1990), recombinant proteins (Reif and Freitag 1994), and vaccines (Hurni and Miller 1991), and holds great promise for on-line analysis. The technique allows real-time analysis providing the possibility to ensure that purification parameters are met before proceeding to the next step. Besides the product, the pattern of associated contaminants is also displayed (Hurni and Miller 1991). The unique fingerprint provided would be invaluable in confirming the lot-to-lot consistency in the purification process.

Interaction between the solutes and the capillary wall has been one of the problems faced during protein separations, leading to peak tailing or even total adsorption. A variety of strategies, such as inclusion of zwitterionic buffers, surface-active components, extremes of pH, or coating the capillary walls, have been employed to minimize these effects (Chiem and Harrison 1997; Schultz and Kennedy 1993; Schmalzing et al. 1995).

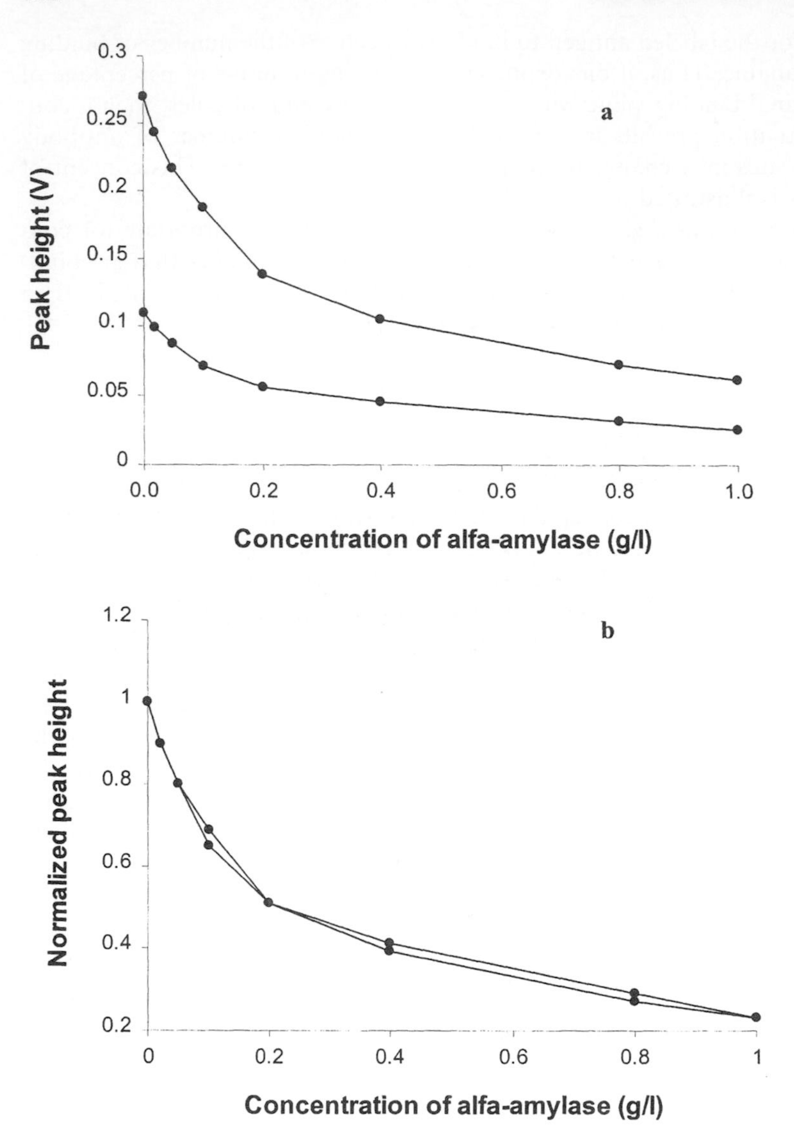

Figure 7 Two calibration graphs for determination of α-amylase using the same antibody column repeatedly. (a) The upper curve represents the values from a fresh column starting curve and the lower one after a few days use and more than 200 assays. (b) The same curves after correction for denaturation based on the use of capacity value of the adsorbent (From Nilsson et al. 1993).

CE-Based Enzyme Microassays

Small amounts of enzyme can be detected by the performance of assays in capillary electrophoretic systems (Bao and Regnier 1992; Wu and Regnier 1993; Wu et al. 1994; Harmon et al. 1996). The electrophoretically mediated microassay (EMMA) is based on the fact that mobility of the enzyme, reagents, and product is different under applied potential and may be used to mix the reactants and separate the enzyme from the product after catalysis. At constant potential, the enzyme and the product are continuously separated, and only when the operating potential of the system is interrupted does the separation cease and the product accumulate. The potential is then reapplied to transport the product to the detector for quantitation. The detection limit using EMMA was reported to be about three orders of magnitude lower than by conventional methods (Bao and Regnier 1992). The microassay of alkaline phosphatase using electrochemical and spectrophotometric detection showed the former to provide better detection limits (Wu et al. 1994). Wu and Regnier (1993) have subsequently demonstrated that the use of gel-filled capillaries instead of open capillaries reduces the product diffusion due to the high viscosity of the gel matrix and in turn increases the sensitivity of detection. The EMMA format has recently been adapted to a microchip-based assay (Zugel et al. 2000).

CE Immunoassays

Separation of antigen, antibody, and antigen-antibody complex is rapidly achieved by CZE (Nielsen et al. 1991). Separation is due to the fact that the complex will have a different charge-to-mass ratio than either free antigen or antibody. Capillary electrophoresis is also useful in the performance of rapid, automated immunoassays (Schultz and Kennedy 1993; Chiem and Harrison 1997). Fluorescence-based detection increases the sensitivity of the assays as also stated earlier for chromatographic monitoring. By labeling either the antigen or the antibody with a fluorescent tag, the complex produced is also labeled. Homogeneous assays can be run in different modes for determination of either the antibody or the antigen.

The electric field strength used influences the separation achieved in CZE; at higher strengths the separation time is shorter and the complex is stable during the course of the run. Separation times of 1–3 min have been achieved at electric field strength of 1000–500 V/cm (Schultz and Kennedy 1993). Using a microchip CE device, separations within about 40 s have been reported (Chiem and Harrison 1997).

4.4 Affinity Sensors

Electrode-based enzyme immunoassays are among the earliest tools applied for measurement of biomolecules (Robinson et al. 1986; Locher et al. 1992). The immunoassays are in intimate contact with a physical transducer, e.g., potentiometric, amperometric, conductometric, etc., which converts the chemical to an electrical signal. Subsequently, development of a number of sensitive sensors has taken place that record changes in a physical parameter directly as a result of binding of a protein to an antibody or another binding molecule. This concept of measuring is in principle more attractive than indirect monitoring because no additional reagents have to be added. In Table 3 are listed some different monitoring principles that have been used. At present these devices generally require sample to be delivered ex situ, however their ability to produce data in real time is of considerable importance for monitoring and control purposes. Of a variety of physical parameters that change during antigen-antibody interaction, the optical property changes seem to provide the most readily detectable system (Place et al. 1985).

Surface Plasmon Resonance

The technology that has attracted much attention in recent years for following biomolecular interactions in real time is the one based on the phenomenon of surface plasmon resonance (SPR), which occurs when surface plasmon waves are excited at a metal-liquid interface (Flanagan and Pantell 1984). The principle is illustrated in Fig. 8. SPR has been utilized in commercialized automated instruments like BIAcore (Biacore, Uppsala, Sweden) and Iasys (Fisons Applied Sensor Technology, Cambridge, UK). The sensor surface is normally a glass plate covered with a thin layer of gold. When laser light is directed at, and reflected from, the inside of the glass at the gold surface, SPR causes a reduction in the reflected light intensity at a specific combination of angle and wavelength. Changes in the refractive index at the surface layer as a

Table 3 Principles of Some Affinity Sensors

Phenomenon measured	Measuring principle	Ref.
Refractive index	Surface plasmon	Jönsson et al. 1985
Refractive index	Ellipsometry	Mandenius and Mosbach 1988
Charge distribution	Capacitance	Berggren and Johansson 1997
Charge distribution	Streaming potential	Glad et al. 1986
Change in weight	Piezoelectric crystal	Bastiaans 1988

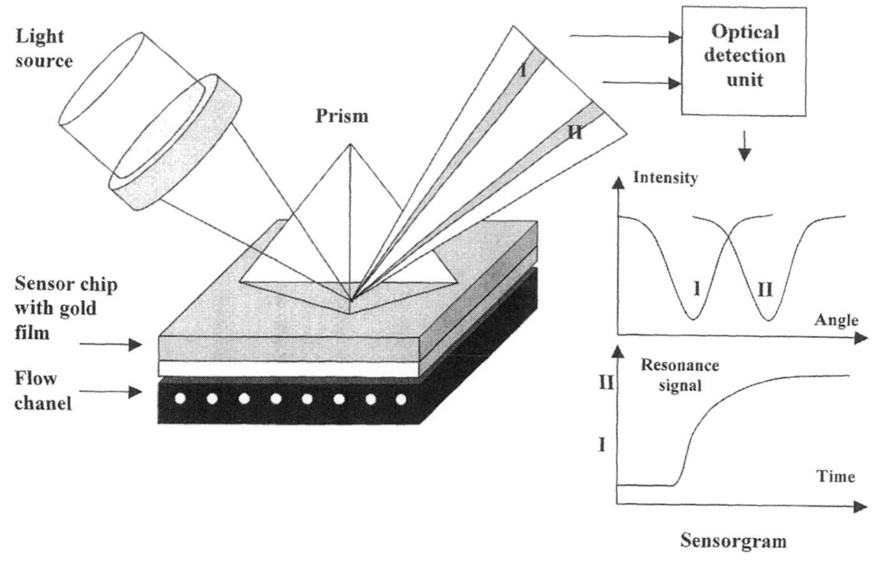

Figure 8 Schematic presentation of the surface plasmon resonance measurement. In the Biacore instrument a plane-polarized light is focused into a wedge-shaped beam. An increased concentration of sample molecules in the surface coating of the sensor chip causes a corresponding increase in refractive index. This results in a change in the angle of the incidence required to create a surface plasmon resonance phenomenon (the SPR angle). This SPR angle is monitored as a change in the detector position for the reflected intensity dip (from I to II). By registering the SPR angle as a function of time, it is possible to monitor kinetic events when binding to the sensor chip takes place.

result of biomolecular binding events are detected as changes in the SPR signal. To perform an analysis, one interactant is bound to the sensor surface while the sample containing the other interactant is injected over it in a precisely controlled flow. Fixed-wavelength light is directed at the sensor surface, and binding is detected as changes in the particular angle where SPR creates extinction of light. This change is measured continuously to form a sensorgram, which provides a complete record of the progress of association and dissociation of the interactants (Fig. 9).

This technology has made possible rapid evaluation of binding properties of potential ligands for protein purification. Even weak affinity interactions with binding constants in the millimolar range, and thus having fast on/off rates, could be monitored using the SPR principle (Ohlson et al. 2000). This has been demonstrated in the case of a monoclonal antibody with weak binding for its antigen (a blood group substance). The antibodies

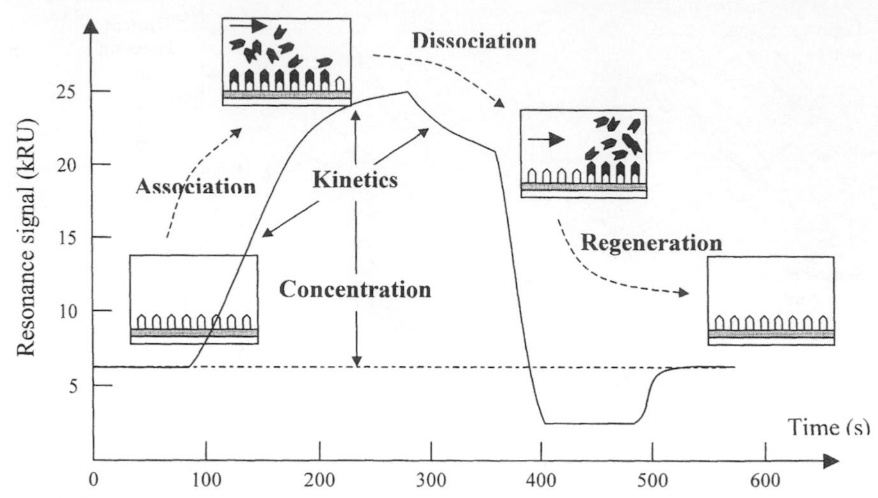

Figure 9 Sensorgram from a surface plasmon resonance measurement. Binding of an analyte to the surface of the sensor chip is registered as a change in the resonance signal. Since the registration is continuous, it is possible to calculate the association kinetics from the shape of the increasing part of the graph. The amplitude reflects the concentration of the analyte, when no more analyte is added to the incoming buffer, a decrease in signal is registered. The decrease is related to the spontaneous dissociation of the analyte from the sensor surface, and this part of the curve can be used to calculate the dissociation constant. Upon regeneration when all interactions are destroyed, the signal from the sensor chip goes back to the start level and the chip is ready for a new assay.

are immobilized with a high density such that the antigen interacts with many potential bindings sites, often resulting in only a retardation of the antigen. The attractive feature is that the reading directly reflects the actual concentration in the medium. When the concentration is lowered, there is a direct influence on the signal (Fig. 10).

The same technology is potentially very useful for process monitoring. A simple two-site BIAcore assay that can measure the quality of recombinant antibody throughout the purification process has been reported (Lawson et al. 1997).

Ellipsometry

Ellipsometry is used to measure the thickness (down to less than 0.1 nm) and the refractive index of very thin films on solid surfaces utilizing the alteration of polarized light from planar toward a more or less circular polariza-

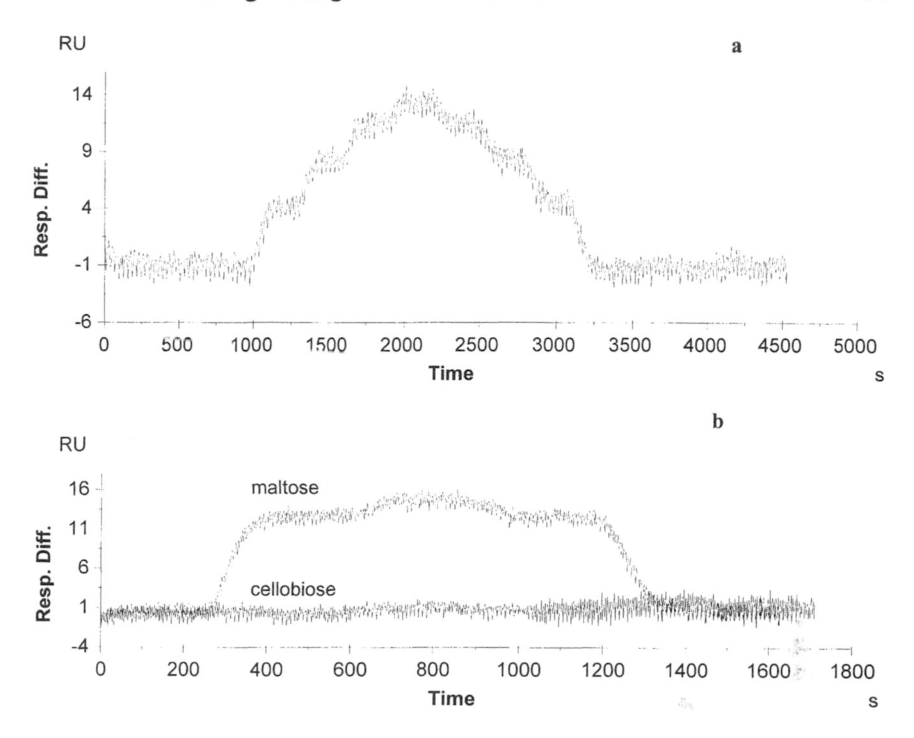

Figure 10 Continuous immonosensing in a surface plasmon resonance biosensor with immobilized monoclonal antibody specific against maltose. (a) The concentration of maltose was gradually increased in steps to 0.05, 0.12, 0.25, and 0.50 mM and subsequently decreased back to zero. Buffer: 10 mM Hepes pH 7.4, 150 mM NaCl, 3.4 mM EDTA 0.005% (v/v) P20 (surfactant). (b) Specificity of the binding was verified by injecting cellobiose in the absence or presence of maltose (altering the concentration in steps to 0.25, 0.50, and back to 0.25 mM, for each disaccharide. (Reproduced with permission from Ohlson et al. 2000.)

tion (i.e., elliptical polarization) when it is reflected from the surface (Place et al. 1985). It is a sensitive method that can detect the interaction between antigen in a solution and antibodies immobilized on a surface or vice versa (Horisberger 1980; Mandenius and Mosbach 1988). The method has been used for determination of protein A and IgG at concentrations down to the 10^{-8} M range (Jönsson et al. 1985).

Turbidimetry and Nephelometry

Light scattering properties of antibody-antigen aggregates can be used in turbidimetric or nephelometric detection. In the former, the intensity of light after passage of the solution is measured, while in the latter the intensity of the scattered light is measured. New antibodies have to be added to every sample, but keeping the sample volume small can reduce their consumption. This technique has been demonstrated in automated instruments with both turbidimetric and laser nephelometric devices (Worsfold et al. 1985; Worsfold 1986; Freitag et al. 1991). The interference due to background light absorption is solved by the use of a reference flow cell in the instrument. Contributions from larger particles and bubbles can be filtered from the signal. Both turbidimetric and nephelometric measurements on antigen-antibody aggregates are accelerated by making kinetic measurements of aggregate formation rates instead for waiting for the equilibrium to be established.

Potentiometric Sensor

Potentiometric immunosensors have been known since 1975 (Janata 1975). These measure a potential between the sensing element and a reference element. Binding of a protein to a ligand/antibody immobilized to a polymer-coated metallic conductor results in a change in the surface charge of polymer-solution interface that is measured potentiometrically against a reference electrode immersed in the same solution (Janata 1975; Lowe 1979; Taylor et al. 1991). The potential difference between the reference and the affinity electrode is related to the concentration of the protein in solution The exact mechanism underlying the electrical changes is not entirely clear but is attributed to the conformational changes of the antibodies or ligands during the binding process. Using an interdigitated electrode transducer, the biochip is seen to act as a capacitor, storing and releasing charge as binding to the analyte occurs (Taylor et al. 1991). Integration of such a system with flow injection analysis, e.g., by positioning the affinity and reference chip opposite each other in a flow cell, allows real-time detection in a process stream (Taylor et al. 1991).

Internal Reflection Sensing

Another technological development relevant to bioprocess monitoring is based on a sensor that works on the principle of optical waveguiding within a resonant mirror to follow analyte interaction (Holwill et al. 1996). An antibody is immobilized on the surface of an optically transparent waveguide, a glass plate or an optical fiber. Interaction with its complementary antigen is monitored by absorption of light derived from multiple internal reflections within the waveguide. The sensitivity of the system is enhanced as the light beam has repeated opportunities to interact with the antigen-antibody complexes between directing the light into the guide and its ultimate recovery (Lowe 1985). Sensor response has been demonstrated over a broad range of protein concentration. This system was used to monitor the performance of a continuous precipitation process on-line involving measurement of a rate of change of absorbance at single wavelength after addition of reagent to a representative sample stream (Holwill et al. 1997). The information was fed to a control algorithm programmed to maintain predefined set points by feedback control through adjustments to the overall feed saturation.

Capacitance

Capacitance measurements are used for registering binding to analytes to immobilized binders on a gold surface that is modified with alkyl thiols in order to form a self-assembling monolayer. A small fraction of the alkyl thiols have a carboxyl group in the terminal end of the alkyl chain that are used for immobilization of a protein, whereas the alkyl layer functions as an insulator. When the analyte binds to the immobilized protein, a conformational change takes place and a capacity signal is registered. This assay can be made very sensitive because minute changes in capacitance can be measured. Provided weak affinity binders are not used, this assay configuration is useful for on-line monitoring.

There are reports on very high sensitivity in assays of DNA, interleukins, and heavy metals. In these cases different affinity groups, including antibodies (Berggren and Johansson 1997), nucleic acids (Berggren et al. 1999), repressor proteins (Bontidean et al. 2001), and proteins with high selectivity for heavy metal ions (Bontidean et al. 1998), were immobilized on the sensor surface. It has furthermore been shown possible to use the sensor system for continuous monitoring as long as the capacity on the electrode surface is sufficient (Hedström and Mattiasson, unpublished results). After saturation, cleaning of the surface is required in order to release the bound material.

Microgravimetric Detection

This method is based on the shift in resonance frequency of a piezoelectric crystal. Addition of a layer to the crystal reduces the frequency that is linearly correlated to the mass of the layer (Saubrey 1959). The relation can be presented in a simplified form as

$$\Delta F = -k\,\Delta M$$

where ΔF is the change in fundamental frequency, ΔM the adsorbed or coated mass, and k a constant that refers to the basic frequency of the crystal and the area coated.

Provided an antibody is immobilized on the crystal, it should be possible to register binding of the antigen (Roeder and Bastiaans 1983). Continuous addition of the antigen may also be possible, and then increments in change of frequency will be correlated to the concentrations of antigen in the sample. The crystal may be used until the surface is saturated after which it needs to be regenerated. Piezoelectric crystals may be used online as long as there is a reasonable balance between on and off rates, i.e., binding and dissociation.

Streaming Potential

Streaming potential is a method based on the principle that if a solid surface comes in contact with a solution, there is an electrochemical potential built up between the bulk solution and the surface. This potential is dependent on the charge distribution on the surface, the flow rate, and the buffer used. The charge distribution over an affinity column can be monitored, and one can thereby register binding of target molecules to the affinity ligand. The system can therefore be used to study biospecific binding to an affinity surface (Glad et al. 1986). When measuring the streaming potential over a column with native affinity support, a baseline value is obtained. Exposure to solutions containing the target product leads to changes in streaming potential that can be correlated to the amount of product that is bound. The streaming potential measurements are very sensitive to the flow. It is actually a characteristic of the measurements that on reversing the flow a negative potential of the same amplitude is registered.

Streaming potential measurements have been used to follow production of monoclonal antibodies from a cell culture (Fig. 11). Good agreement between the streaming potential readings and those from discrete assays using traditional radioimmunoassay was observed (Miyabayashi and Mattiasson 1990). On-line monitoring can be performed as long as the capacity of the affinity column is not saturated.

4.5 On-Line Molecular Size Detection

Registration of molecular weight gives good possibilities for tracing molecules in a downstream processing step. It should be possible to detect any heterogeneity in the product, especially the degraded forms or other structural variants of the target protein. Mass spectrometry has become the most powerful technique for protein analysis and has turned out to be very useful for proteomics research. Monitoring of chromatographic separations using liquid chromatography–electrospray ionization mass spectrometry has been attempted (Linnemayer et al. 1998; Canarelli et al. 2002). There is a risk of

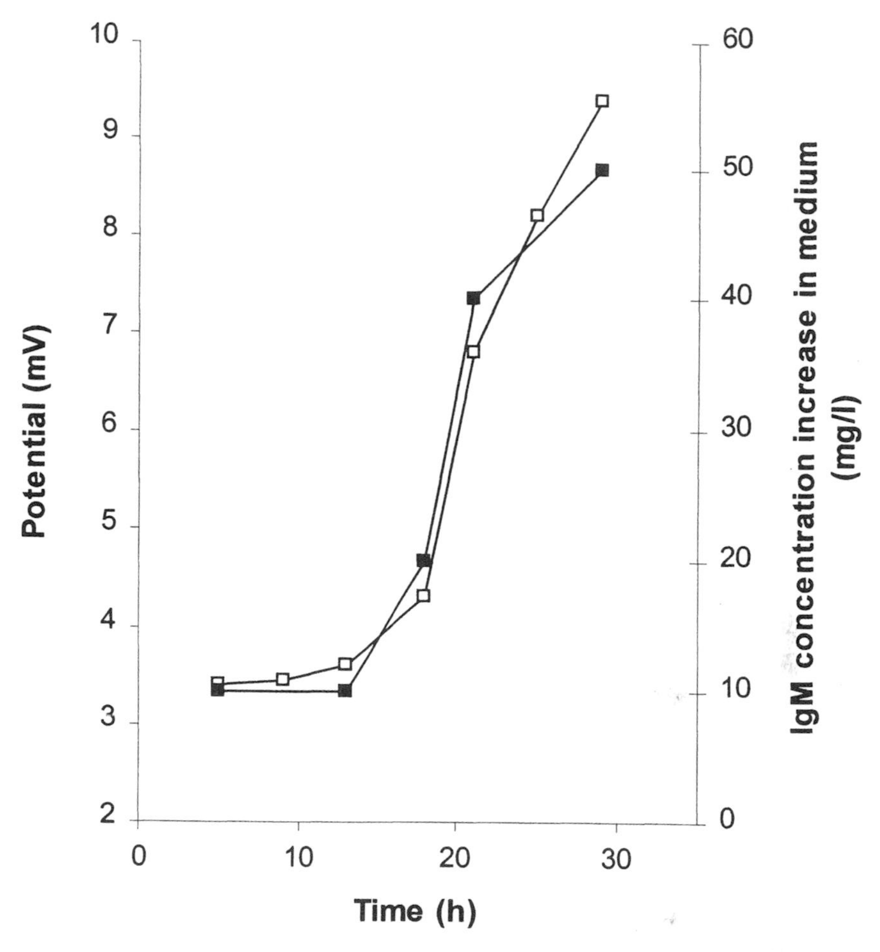

Figure 11 Registration of IgM produced from a hybridoma cultivation by streaming potential measurements. Assays using radioimmunoassay (■) and streaming potential (□). (From Miyabayashi and Mattiasson 1990, with permission.)

lowered sensitivity due to the presence of salts and impurities that leads to suppression of ionization of the target product. This could be avoided by coupling the MS detection via an on-line microdialyzer (Canarelli et al. 2002). Both chromatography and capillary electrophoresis coupled through electrospray interface with time-of-flight or ion trap mass spectrometer are likely to have an impact on rapid process monitoring.

Photon correlation spectroscopy, a method based on measurement of diffusion coefficients, may also be used to monitor the molecular size and polydispersity of proteins eluting from a chromatography column. Short sampling time (10 s) makes it applicable for in-line continuous monitoring (Clark et al. 1987).

5. REMARKS

When choosing the monitoring technique for a DSP process, the choice must be governed by the molecular characteristics of the system studied as well as by the need for data points. If it is sufficient with intermittent readings—a few up to ten times an hour—then flow injection binding assays may be a good choice. If, however, more intense monitoring is required, then a continuous monitoring principle is needed.

Measurement of molecules based on their biological activities offers good chances for selectivity. However, it should be borne in mind that truncated molecules and other slightly modified molecules may not be distinguished from the native molecules, which otherwise would give a separate signal in the mass detection step.

The need for better documentation may be the driving force behind development of better sensor systems to monitor DSP processes. It will of course also positively influence the chances to carry out the separation processes under better control, both with regard to the target molecule and to key impurities. The ability to perform process monitoring rapidly will allow enhanced product quality through the control of system variables, and process validation would be carried out continuously, thus reducing the time.

Miniaturization and integration of all aspects of measurement systems, such as chromatography and electrophoresis, will also have an impact on process monitoring as very small amounts of reagents will be needed, and single samples may be easily split for multiple analyses (Regnier et al. 1999).

ACKNOWLEDGMENTS

Dr. Benoit Guieysse is gratefully acknowledged for help with the figures.

REFERENCES

Afeyan NB, Gordon NF, Regnier FE (1992) Automated real-time immunoassay of biomolecules. Nature 358, 603–604

Bastiaans GJ (1988) Sensors having piezoelectric crystal for microgravimetric immunoassays. US Patent No. 4,735,906.

Bao J, Regnier FE (1992) Ultramicro enzyme assays in capillary electrophoretic system. J. Chromatogr. 608, 217–224.

Berggren C, Johansson G (1997) Capacitance measurements of antibody-antigen interactions in a flow system. Anal. Chem. 69, 3651–3657.

Berggren C, Stålhandske P, Brundell J, Johansson G (1999) A feasibility study of a capacitive biosensor for direct detection of DNA hybridization. Electroanalysis 11, 156–160.

Bontidean I, Berggren C, Johansson G, Csöregi E, Mattiasson B, Lloyd JR, Jakerman K, Brown NL (1998) Detection of heavy metal ions at femtomolar levels using protein-based biosensors. Anal. Chem. 70, 4162–4169.

Bontidean I, Kumar A, Csöregi E, Galaev IY, Mattiasson B (2001) Highly sensitive novel biosensor based on immobilized *lac* repressor. Angewandte Chem. 113, 2748–2750.

Canarelli S, Fisch I, Freitag R (2002) On-line microdialysis of proteins with high-salt buffers for direct coupling of electrospray ionization mass spectrometry and liquid chromatography. J Chromatogr. A 948, 139–149.

Cassidy SA, Janis LJ, Regnier FE (1992) Kinetic chromatographic sequential addition immunoassays using protein A affinity chromatography. Anal. Chem. 64, 1973–1977.

Chase HA (1986) Rapid chromatographic monitoring of bioproceses. Biosensors 2, 269–286.

Chase HA, Nash DC, Bruce LJ (1999) On-line monitoring of breakthrough curves within an expanded bed adsorber. Bioprocess. Eng. 20, 223–229.

Chicz RM, Regnier FE (1989) Single amino acid contributions to protein retention in cation-exchange chromatography: resolution of genetically engineered subtilisin variants. Anal. Chem. 61, 2059–2066.

Chicz RM, Regnier FE (1990) Microenvironmental contributions to the chromatographic behavior of subtilisin in hydrophobic-interaction and reversed-phase chromatography. J. Chromatogr. 500, 503–518.

Chiem N, Harrison DJ (1997) Microchip-based capillary electrophoresis for immunoassays: analysis of monoclonal antibodies and theophylline. Anal. Chem. 69, 373–378.

Clark DJ, Stansfield AG, Jepras RI, Collinge TA, Holding FP, Atkinson T (1987) Some approaches to downstream processing monitoring and the development of new separation techniques. In: Sep. Biotechnol. (Pap. Int. Conf.), Verrall MS, Hudson MJ, eds, Harwood, Chichester, UK, pp. 419–429.

Clemmitt RH, Bruce LJ, Chase HA (1999) On-line monitoring of the purification of GST-$(His)_6$ from an unclarified *Escherichia coli* homogenate within an immobilised metal affinity expanded bed. Bioseparation 8, 53–67.

Danielsson B, Larsson P-O (1990) Specific monitoring of chromatographic procedures. Trends Anal. Chem. 9, 223–227.

de Alwis WU, Wilson GS (1985) Rapid sub-picomole electrochemical enzyme immunoassay for immunoglobulin G. Anal. Chem. 57, 2754–2756.

de Frutos M, Regnier FE (1993) Tandem chromatographic immunological analyses. Anal. Chem. 65, 17A–25A.

de Frutos M, Paliwal SK, Regnier FE (1993) Liquid chromatography based enzyme-amplified immunological assays in fused-silica capillaries at the zeptomole level. Anal. Chem. 65, 2159–2163.

Flanagan MT, Pantell RH (1984) Surface plasmon resonance and immunosensors. Electr. Lett 20, 968–970.

Foster PR, Dickson AJ, Stenhouse A, Walker EP (1986) A process control system for the fractional precipitation of human plasma proteins. J. Chem. Tech. Biotechnol. 36, 461–466.

Freitag R, Scheper T, Schügerl K (1991) Development of a turbidimetric immunoassay for on-line monitoring of proteins in cultivation processes. Enzyme Microb. Technol. 13, 969–975.

Frej A-K, Gustafsson J-G, Hedman P (1986) Recovery of β-galactosidase by adsorption from unclarified *Escherichia coli* homogenate. Biotechnol. Bioeng. 28, 133–137.

Glad C, Sjödin K, Mattiasson B (1986) Streaming potential—a general affinity sensor. Biosensors 2, 89–100.

Hadd AG, Raymond DE, Halliwell JW, Jacobson SC, Ramsey JM (1997) Microchip device for performing enzyme assays. Anal. Chem. 69, 3407–3412.

Hagedorn J, Kasper C, Freitag R, Tennikova T. (1999) High performance flow injection analysis of recombinant protein G. J. Biotechnol. 69, 1–7.

Harmon BJ, Leesong I, Regnier FE (1996) Moving boundary electrophoretically mediated microanalysis. J. Chromatogr. A 726, 193–204.

Holwill I, Gill A, Harrison J, Hoare M, Lowe PA (1996) Rapid analysis of biosensor data using initial rate determination and its application to bioprocess monitoring. Process Control Qual. 8, 133–145.

Holwill IJ, Chard SJ, Flanagan MT, Hoare M (1997) A Kalman filter algorithm and monitoring apparatus for at-line control of fractional protein precipitation. Biotechnol. Bioeng. 53, 58–70.

Horisberger M (1980) An application of ellipsometry. Assessment of polysaccharide and glycoprotein interaction with lectin at liquid/solid interface. Biochim. Biophys Acta 632, 298–309.

Hurni WM, Miller WJ (1991) Analysis of a vaccine purification process by capillary electrophoresis. J. Chromatogr. 559, 337–343.

Janata J (1975) An immunoelectrode. J. Am. Chem. Soc. 97, 2914–2916.

Jönsson U, Malmquist M, Rönnberg I (1985) Adsorption of immunoglobulin G, protein A, and fibronectin in the submonolayer region evaluated by a combined study of ellipsometry and radiotracer techniques. J. Colloid Interface Sci. 103, 360–372.

Johns MA, Rosengarten LK, Jackson M, Regnier FE (1996) Enzyme-linked immunosorbent assay in a chromatographic format. J. Chromatogr. A 743, 195–206.

Kenney AC, Chase HA (1987) Automated production scale affinity purification of monoclonal antibodies. J. Chem. Tech. Biotechnol. 39, 173–182.

Klockewitz K, Riechel P, Hagedorn J, Scheper T, Noe W, Howaldt M, Vorlop J (2000) Fast FIA-immunoanalysis systems for the monitoring of downstream processes. Chem. Biochem. Eng. Q 14, 43–46.

Lawson ADG, Chaplin LC, Lang V, Sehdev M, Spitali M, Popplewell A, Weir N, King DJ (1997) Two-site assays for measuring recombinant antibody quality. BIA J. 1, 23.

Linnemayer K, Rizz A, Josic D, Allmaier G (1998) Comparison of microscale cleaning procedures for (Glyco) proteins prior to positive ion matrix-assisted laser desorption ionization mass spectrometry. Anal. Chim. Acta 372, 187–199.

Locher G, Sonnleitner B, Fiechter A (1992) On-line measurement in biotechnology: techniques. J. Biotechnol. 25, 23–53.

Lowe C (1979) The affinity electrode. Application to the assay of human serum albumin. FEBS Lett. 106, 405–408.

Lowe C (1985) An introduction to the concepts and technology of biosensors. Biosensors 1, 3–16.

Luque de Castro MD, Fernandez-Romero JM (1992) Total individual determination of creatine kinase isoenzyme activity by flow injection and liquid chromatography. Anal. Chim. Acta 263, 43–52.

Lüdi H, Garn MB, Bataillard P, Widmer HM (1990) Flow injection analysis and biosensors: Applications for biotechnology and environmental control. J. Biotechnol. 14, 71–79.

Maeda M, Tsuji A (1985) Enzymatic immunoassay of alpha-fetoprotein, insulin and 17-alpha-hydroxprogesterone based on chemiluminescence in a flow-injection system. Anal. Chim. Acta 167, 241–248.

Mandenius CF, Mosbach K (1988) Detection of biospecific interactions using amplified ellipsometry. Anal. Biochem. 70, 68–72.

Mattiasson B (1984) Immunochemical assays for process control: potentials and limitations. Trends Anal. Chem. 3, 245–250.

Mattiasson B, Borrebaeck C, Sanfridsson B, Mosbach K (1977) Thermometric enzyme lined immunosorbent assay: TELISA. Biochim. Biophys. Acta 483, 221–227.

Mattiasson B, Hakanson H (1991) Measurement and control in downstream processing. In: Measurement and Control in Bioprocessing, Carr-Brian K, ed., Elsevier Science, London, pp. 221–250.

Miyabayashi A, Mattiasson B (1990) A dual streaming potential device used as an affinity sensor for monitoring hybridoma cell cultivations. Anal. Biochem. 184, 165–171.

Nadler TK, Paliwal SK, Regnier FE, Singhvi R, Wang DIC (1994a) Process monitoring of the production of γ-interferon in recombinant Chinese hamster ovary cells. J. Chromatogr. 659, 317–320.

Nadler TK, Paliwal SK, Regnier FE (1994b) Rapid, automated, two-dimensional high-performance liquid chromatographic analysis of immunoglobulin G and its multimers. J. Chromatogr. 676, 331–335.

Nandakumar MP, Nandakumar R, Mattiasson B (2000) Fluorimetric quantification of intracellular lactate dehydrogenase during fermentation using flow injection analysis. Biotechnology Lett. 22, 1453–1457.

Nandakumar MP, Nordberg-Karlsson E, Mattiasson B (2001) Integrated flow-injection processing for on-line quantification of plasmid DNA during fermentation. Biotechnol. Bioeng. 73, 406–411.

Nielsen RG, Rickard EC, Santa PF, Sharknas DA, Sittampalam GS (1991) Separation of antibody-antigen complexes by capillary zone electrophoresis, isoelectric focusing and high-performance size-exclusion chromatography. J. Chromatogr. 539, 177–185.

Nilsson M, Håkanson H, Mattiasson B (1992) Process monitoring by flow-injection immunoassay. Evaluation of a sequential competitive binding assay. J. Chromatogr. 597, 383–389.

Nilsson M, Mattiasson G, Mattiasson B (1993) Automated immunochemical binding assay (flow-ELISA) based on repeated use of an antibody column placed in a flow-injection system. J. Biotechnol. 31, 381–394.

Nilsson M, Vijayakumar AR, Holst O, Schornack C, Håkanson H, Mattiasson B (1994) On-line monitoring of product concentration by flow-ELISA in an integrated fermentation and purification process. J. Ferm. Bioeng. 78, 356–360.

Ohlson S, Hansson L, Larsson P-O, Mosbach K (1978) High performance liquid affinity chromatography (HPLAC) and it applications to the separation of enzymes and antigens. FEBS Lett. 93, 5–9.

Ohlson S, Jungar C, Strandh M, Mandenius C-F (2000) Continuous weak-affinity immunosensing. Trends Biotechnol. 18, 49–52.

O'Keefe DO, Lee AL, Yamazaki S (1992) Use of monobromobrimane to resolve two recombinant proteins by reversed-phase high-performance liquid chromatography based on their cysteine content. J. Chromatogr. 627, 137–143.

Paliwal SK, Nadler TK, Regnier FE (1993a) Rapid process monitoring in biotechnology. Trends Biotechnol. 11, 95–101.

Paliwal SK, Nadler TK, Wang DIC, Regnier FE (1993b) Automated process monitoring of monoclonal antibody production. Anal. Chem. 65, 3363–3367.

Place JF, Sutherland RM, Dähne C (1985) Opto-electronic immunosensors: a review of optical immunoassay at continuous surfaces. Biosensors 1, 321–353.

Poppenborg L, Friehs K, Flaschel E (1997) The green fluorescent protein is a versatile reporter for bioprocess monitoring. J. Biotechnol. 58, 79–88.

Recktenwald A, Kroner K-H, Kula M-R (1985) On-line monitoring of enzymes in downstream processing by flow injection analysis. Enzyme Microb. Technol. 7, 607–612.

Regnier FE, He B, Linn S, Busse J (1999) Chromatography and electrophoresis on chips: critical elements of future integrated, microfluidic analytical systems for life science. Trends Biotechnol. 17, 101–106.

Reif O-W, Freitag R (1994) Control of the cultivation process of antithrombin III and its characterization by capillary electrophoresis. J. Chromatogr. 680, 383–394.

Richardson P, Molloy J, Ravenhall R, Holwill I, Hoare M, Dunnill P (1996a) High speed centrifugal separator for rapid on-line sample clarification. J. Biotechnol. 49, 111–118.

Richardson P, Ravenhall R, Flanagan MT, Holwill I, Molloy J, Hoare M, Dunnill P (1996b) Monitoring and optimisation of fractional protein precipitation by flow-injection analysis. Process Control Qual. 8, 91–101.

Riggin A, Regnier FE, Sportsman JR (1991) Quantification of antibodies of human growth hormone by high-performance protein G affinity chromatography with fluorescence detection. Anal. Chem. 63, 468–474.

Riggin A, Sportsman JR, Regnier FE (1993) Immunochromatographic analysis of proteins. Identification, characterization and purity determination. J. Chromatogr. 632, 37–44.

Robinson GA, Cole VM, Rattle SJ, Forrest GC (1986) Bioelectrochemical immunoassay for human chorionic gonadotrophin in serum using an electrode-immobilised capture antibody. Biosensors 2, 45–57.

Roeder JE, Bastiaans GJ (1983) Microgravimetric immunoassay with piezoelectric crystals. Anal. Chem. 55, 2333–2336.

Ruzicka J, Hansen EH (1975) Flow injection analyses. Part I. A new concept of fast continuous flow analysis. Anal. Chim. Acta 78, 145–157.

Ruzicka J, Hansen EH (1991) Flow injection is coming of age. Lab. Rob. Autom. 3, 113.

Saubrey GZ (1959) The use of oscillators for weighing thin layers and for microweighing. Z. Phys. 155, 206–212.

Schmalzing D, Nashabeh W, Yao X-W, Mhatre R, Regnier FE, Afeyan NB, Fuchs M (1995) Capillary electrophoresis-based immunoassay for cortisol in serum. Anal. Chem. 67, 606–612.

Schmid RD, Künnecke W (1990) Flow injection analysis (FIA) based on enzymes or antibodies—applications in the life sciences. J. Biotechnol. 14, 3–31.

Schultz NM, Kennedy RT (1993) Rapid immunoassays using capillary electrophoresis with fluorescence detection. Anal. Chem. 65, 3161–3165.

Stamm WW, Kula M-R (1990) Monitoring of enzymes during chromatographic separations. J. Biotechnol. 14, 99–114.

Stöcklein W, Schmid RD (1990) Flow-injection immunoanalysis for the on-line monitoring of monoclonal antibodies. Anal. Chim. Acta 234, 83–88.

Takahashi K, Taniguchi S, Kuroishi T, Yusuda K, Sano T (1989) Automated stopped flow/continuous flow apparatus for serial measurement of enzyme reactions and its application as real-time analyzer for column chromatography. Anal. Chim. Acta 220, 13–21.

Taylor RF, Marenchic IG, Spencer RH (1991) Antibody- and receptor-based biosensors for detection and process control. Anal. Chim. Acta 249, 67–70.

Thévenon-Emeric G, Regnier FE (1991) Process monitoring by parallel column gradient elution chromatography. Anal. Chem. 63, 1114–1118.

Tocaj A, Nandakumar MP, Holst O, Mattiasson B (1999) Flow injection analysis of intracellular β-galactosidase in *Escherichia coli* cultivations, using an on-line system including cell disruption, debris separation and immunochemical quantification. Bioseparation 8, 255–276.

Wenisch E, Tauer C, Jungbauer A, Katinger H, Faupel M, Righetti PG (1990) Capillary zone electrophoresis for monitoring r-DNA protein purification in multicompartment electrolysers with immobiline membranes. J. Chromatogr. 516, 133–146.

Worsfold PJ, Hughes A, Mowthorpe DJ (1985) Determination of human serum immunoglobulin G using flow injection analysis with rate turbidimetric detection. Analyst 110, 1303–1305.

Worsfold P (1986) Flow injection techniques for monitoring biochemically selective interactions. Anal. Chim. Acta 180, 56–58.

Wu D, Regnier FE (1993) Native protein separations and enzyme microassays by capillary zone and gel electrophoresis. Anal. Chem. 65, 2029–2035.

Wu D, Regnier FE, Linhares MC (1994) Electrophoretically mediated micro-assay of alkaline phosphatase using electrochemical and spectrophotometric detection in capillary electrophoresis. J. Chromatogr. B 657, 357–363.

Zugel SA, Burke BJ, Regnier FE, Lytle FE (2000) Electrophoretically mediated microanalysis of leucine aminopeptidase using two-photon excited fluorescence detection on a microchip. Anal. Chem. 72, 5731–5735.

15
Protein Formulation

Drew N. Kelner, Wei Wang, and D. Q. Wang
Bayer Corporation, Berkeley, California, U.S.A.

1. INTRODUCTION

Protein formulation is emerging as a distinguishable branch of pharmaceutical sciences, with a growing and sizable body of literature (Manning et al. 1989; Arakawa et al. 1993; Cleland et al. 1993; Carpenter et al. 1997; Wang 1999, 2000). The science of protein formulation is, however, still in its infancy such that no general and easily accessible solution exists for protein stabilization. While the two major dosage forms for protein pharmaceuticals, liquid and lyophilized solid, both present unique challenges, they share a common goal: to overcome the inherent physical and chemical instability of proteins such that a stable product with sufficient shelf life for commercialization can be achieved. The major challenge in protein formulation is therefore to identify the major pathway(s) of instability of the protein by investigating the biochemical and biophysical properties of the protein that impact its stability. Once these properties are understood, the formulation effort can progress to the identification of excipients and conditions that provide optimal stability of the formulated biopharmaceutical.

The purpose of this chapter is to update the reader on four topics: (1) the basic properties of proteins that relate to their mechanisms of instability; (2) the challenges and strategies inherent in the development of protein formulations; (3) the design of solid lyophilized protein formulations; and (4) analytical support of formulation development.

2. BASIC PROTEIN PROPERTIES

2.1 Protein Folding and Conformational Stability

A protein usually folds into its active three-dimensional structure during biosynthesis into a conformation that is defined by its amino acid sequence (Ellis et al. 1998). While proper protein folding is assisted by molecular chaperones (Hendrick and Hartl 1995), thermodynamic forces drive the protein folding process. The conformational stability of a protein is measured by the free energy change between the folded (f) and unfolded (u) states ($\Delta G_{f \to u}$) for the protein denaturation reaction. The larger the $\Delta G_{f \to u}$, the more stable the protein. While $\Delta G_{f \to u}$ values for proteins have been reported in the 5- to 20-kcal/mol range (Volkin and Klibanov 1989; Jaenicke 1990; Pace et al. 1996), a single hydrogen bond can impact the protein's free energy by 0.5–2 kcal/mol, and an ion pair by 0.4–1.0 kcal/mol (Vogt and Argos 1997). The $\Delta G_{f \to u}$ is therefore relatively small, such that the native state is only slightly more stable than the unfolded state. This thermodynamic consideration frames the difficulty of the protein formulation challenge whereby the inherent physical and chemical instability of proteins must be overcome to maintain a stable, active product.

2.2 Mechanisms of Protein Instability

The most commonly observed mechanisms of protein instability, which can be grouped into the major categories of chemical and physical degradation, have been extensively reviewed in the literature (Manning et al. 1989; Cleland et al. 1993; Volkin et al. 1997; Reubsaet et al. 1998).

Many chemical degradation pathways are responsible for the direct inactivation of proteins. The most commonly observed chemical degradation mechanisms include deamidation, oxidation, disulfide bond formation/breakage/exchange, hydrolysis, isomerization, succinimidation, and racemization. These degradation pathways and representative protein examples are listed in Table 1. In many cases, a number of reactions can happen simultaneously in proteins, making separation and identification of the degradation products very difficult, such as in basic fibroblast growth factor (bFGF) (Wang et al. 1996) and insulin (Strickley and Anderson 1997).

Certain amino acids are particularly sensitive to certain types of degradation under certain conditions (Table 2). Both Asn and Gln are susceptible to deamidation; Asn is far more labile (Powell 1994; Li et al. 1995b; Daniel et al. 1996). The most easily oxidizable residues are Met and Cys, although other residues, such as His, Trp, and Tyr, are also potential sites of oxidation (Manning et al. 1989; Li et al. 1995b,c; Daniel et al. 1996). Met residues in proteins can be oxidized easily under atmospheric conditions due to the high flexibility of methionine in proteins (Richards 1997). Cys residues can lead to disulfide bond formation or thiodisulfide exchange, a common cause of chemically induced protein aggregation, as observed in bFGF (Shahrokh et al. 1994; Wang et al. 1996). The -X-Asp-Y- sequence is subject to acid hydrolysis (Manning et al. 1989; Li et al. 1995b; Vieille and Zeikus 1996). Hydrolytic cleavage is particularly rapid at Asp-Gly and Asp-Pro (Powell 1994, 1996), such as for recombinant human macrophage colony-stimulating factor (rhM-CSF) in an acid solution (Schrier et al. 1993). Asp residues can also undergo succinimidation and racemization.

Although the aforementioned amino acids are relative labile, their actual reactivity in proteins depends strongly on their location in a protein and is also affected by neighboring groups (Paranandi et al. 1994). Usually the most mobile residues degrade first. Examples include deamidation of Asn^{1} in human epidermal growth factor (hEGF) (Son and Kwon 1995), oxidation of Met^{1} in recombinant human leptin (Liu et al. 1998), and succinimidation at Asp^{53} and Asp^{33} in recombinant Zn-hirudin (Gietz et al. 1999). In the determination of the effect of a neighboring group on the reactivity of a labile amino acid, Cross and Schirch (1991) demonstrated that increasing the size and branching of the residue on the carboxyl side of Asn in a pentapeptide decreases the rate of deamidation by as much as 70-fold (compared with a Gly residue).

Formulation pH is another important stability-influencing factor. The pH affects the physical stability of proteins by altering the electrostatic interactions in proteins and also strongly impacts the chemical stability of proteins, since the rate of chemical reactions is pH dependent (Table 2).

The rate of degradative reactions in proteins is also strongly influenced by the tertiary structure of proteins. Since a very limited number of water molecules exist inside a protein, the presence of buried labile amino acid(s) may not lead to any detectable degradation. This explains why the disulfide exchange of the sulfhydryl group in β-lactoglobulin A is very slow in its native state but increases significantly when the protein gradually unfolds with increasing urea concentrations (Apenten 1998). The native protein conformation, therefore, needs to be protected to prevent potential chemical degradation during storage. In addition, secondary structures in proteins may affect chemical degradation. For example, the deamidation rate of Asn^{8} in several growth hormone–releasing factor (GRF) analogues decreases with increasing helical content in aqueous methanol solution (Stevenson et al. 1993).

Changing formulation conditions may change not only the extent of chemical degradation but also the degradation pathways in proteins. Qi and Heller (1995) demonstrated that the recovery of intact insulinotropin in

Table 1 Common Chemical Degradation in Proteins

Chemical degradation	Proteins	Formulation	Storage temperature	Refs
Deamidation	Insulin	Aqueous solution containing methylparaben and NaCl	4–45°C	Brange et al. 1992a
	Il-1RA	Lyophilized in 4% mannitol, 2% glycine and 10 mM citrate, pH 6.5	−10–10°C	Chang et al. 1996b
Disulfide bond formation	bFGF	Aqueous solution containing EDTA and citrate, pH 5	25°C	Shahrokh et al. 1994
	Il-1RA	Aqueous solution containing 140 mM NaCl, and 10 mM citrate, pH 6.5	30°C	Chang et al. 1996a
Disulfide bond exchange	Insulin	Lyophilized at pH 7.3	50°C	Costantino et al. 1994
Glycation	rhDNase I	Spray-dried or lyophilized protein containing lactose	4, 25, and 40°C	Quan et al. 1999
Hydrolysis	Insulin	Aqueous solution containing methylparaben and NaCl	4–25°C	Brange et al. 1992a

Isomerization	bFGF	Aqueous solution containing EDTA and citrate, pH 5	25°C	Shahrokh et al. 1994
	tPA	Aqueous solution containing 0.2 M arginine, 5% glycerol, 0.02% sodium azide, and 0.1 M phosphate, pH 7.3	37°C	Paranandi et al. 1994
Maillard reaction	Alkaline phosphatase	Lyophilized formulation containing lactose	56°C	Ford and Dawson 1993
	Invertase	Lyophilized formulation containing raffinose, lactose, or maltose	95°C	Schebor et al. 1997
Oxidation	IGF-I	Aqueous solution containing 0.1 M phosphate, pH 6.3	25°C with or without light	Fransson and Hagman 1996
	Growth hormone	Lyophilized in phosphate buffer, pH 7.4	25°C under nitrogen	Pikal et al. 1991
Succinimidation	bFGF	Aqueous solution containing EDTA and citrate, pH 5	25°C	Shahrokh et al. 1994
Transamidation	Insulin	Aqueous solution containing methylparaben and NaCl		Brange et al. 1992a

Table 2 Favored pH Conditions for Protein/Peptide Degradation

Degradation reactions	Proteins/peptides	Reaction sites	Favored pH conditions	Ref.
Cleavage	bFGF	Asp-X, esp. X = Pro	Very acidic	Shahrokh et al. 1994
Deamidation	hEGF	Asn^{1}	Neutral to alkaline	Son and Kwon 1995
Deamidation	bFGF	Asn-X	Neutral to alkaline	Shahrokh et al. 1994
Deamidation	Insulin	Asn^{a-21}	pH < 5	Strickley and Anderson 1997
Deamidation	RNase A Lysozyme	Asn-X Gln-X	High pHs	Kristjansson and Kinsella 1991
Oxidation (by H_2O_2)	rhPTH	Met^{8} and met^{18}	10	Nabuchi et al. 1995
Oxidation (by ascorbic acid/$CuCl_2$/O_2)	Relaxin	His^{A12}, Met^{B4}, Met^{B25}, etc.	5 > 6 > 7 > 8	Li et al. 1995a
Oxidation (by ascorbic acid/$FeCl_3$/O_2)	His-Met	Met	7–8	Li et al. 1993
Succinimidation	bFGF	Asx-Gly	4–5	Shahrokh et al. 1994

water upon incubation at 50°C was higher than that in the presence of 20 mM phosphate or 23% dextran at pH 8.0, and that the degradation pathways in the three types of media were different as determined by reversed phase–high-performance liquid chromatography (RP-HPLC).

It should be noted that many chemical degradations do not affect the protein activity. Oxidized recombinant human relaxin at Met^{B4} and Met^{B25} has the same bioactivity as the wild type (Nguyen et al. 1993). Mono-oxidized recombinant human leptin at Met^{1} does not show any detectable change in tertiary structure or potency (Liu et al. 1998). Several other degraded proteins retain their original activity, including oxidized and deamidated human growth hormone (hGH) (Cholewinski et al. 1996), deamidated hEGF, deamidated insulin (Brange et al. 1992b), and deamidated recombinant interleukin-2 (rIL-2) (Sasaoki et al. 1992).

The most common physical instability pathway is protein unfolding, which often leads to aggregation and/or precipitation. Protein unfolding and subsequent aggregation may be induced by a variety of physical factors, such as temperature, pH, and ionic strength (Table 3). It may also result from chemical degradation or modification and the subsequent exposure of the hydrophobic surface(s), such as for human relaxin after oxidation of His and Met residues (Li et al. 1995a). Both unfolding and chemically induced aggregation may occur simultaneously, such as for insulin (Brange et al. 1992a), bFGF (Shahrokh et al. 1994), and β-galactosidase (Yoshioka et al. 1993).

Proteins can be unfolded easily in the presence of a denaturant (chemical unfolding) or by increasing the temperature (thermal unfolding). During the thermal unfolding process, the temperature at which 50% of the molecules are unfolded ($\Delta G = 0$ at this time) is defined as the unfolding (or melting/denaturation/transition) temperature (T_m). A protein may have two or more melting temperatures, depending on the experimental conditions and the analytical techniques used for T_m determination. Multimeric, chimeric, or modular proteins routinely have more than one melting temperature, such as recombinant human placental factor XIII (rFXIII) (Kurochkin et al. 1995), dimeric V55C mutant of Cro repressor protein (Fabian et al. 1999), and chimeric protein toxin sCD4(178)-PE40 (sCD4-PE40) (Davio et al. 1995).

Protein unfolding temperatures are usually in the range of 40–80°C (Wang 1999). Some, like thermophilic proteins, may have T_m values exceeding 100°C (Jaenicke 1996). Although generally a protein with a higher unfolding temperature is more stable, there is no clear relationship between T_m and $\Delta G_{f \rightarrow u}$ (Dill et al. 1989). It should be noted that the melting temperature determined by different analytical techniques may be significantly different and may be strongly affected by the experimental conditions (Vermeer and Norde 2000).

Table 3 Factors Affecting Stability of Proteins in Liquid State

Factor	Proteins	Effect(s)	Ref.
Temperature	RNase	Protein may denature at both low (−22°C) and high (40°C) temperatures	Zhang et al. 1995a
	Serum albumin	Protein may denature at < 0°C and > 50°C	Kosa et al. 1998
pH	Recombinant factor VIII SQ	Decreased aqueous stability at pHs outside 6.5–7.0	Fatouros et al. 1997a
	Bovine insulin	Faster fibril formation in solution of pH 2.5 than that of pH 7	Brange et al. 1997
	Il-1ra	More aggregation during storage of lyophilized formulation from pH 6 than from pH 7	Chang et al. 1996d
	hIGF-I	Increased oxidation in solution from pH 6 to 7.5	Fransson and Hagman 1996
Ionic strength	rhDNase	T_m was increased by 5°C when the NaCl concentration was increased from 75 to 200 mg/ml of NaCl	Chan et al. 1996
	Insulin	Increased fibrillation when ionic strength was increased from 0.04 to 0.11	Brange et al. 1997
	rhMGDF	Decreased thermal unfolding reversibility from 70% to 0% and decreased T_m from 52 to 40°C when NaCl concentration was increased from 50 to 200 mM	Narhi et al. 1999

Formulation excipients	Hemoglobin	Ascorbate (5mM) increased autoxidation and induced modification	Kerwin et al. 1999
	Human relaxin	Metal ions accelerated oxidation	Li et al. 1995a
	hGH	Metal ions accelerated oxidation	Zhao et al. 1997
Surface adsorption	aFGF	Loss of proteins to the surfaces of various containers	Volkin and Middaugh 1996
	IL-2	Adsorption-induced protein denaturation through silicone rubber tubing	Tzannis et al. 1997
Shaking	hGH	Increased protein aggregation by shaking	Katakam et al. 1995; Katakam and Banga, 1997; Bam et al. 1998
	IL-2	Reduced protein activity by shaking	Wang and Johnston 1993
Shearing	Fibrinogen	Reduced protein activity by shearing	Charm and Wong 1970
	rhGH	Increased aggregation by shearing	Maa and Hsu 1997
Freeze-thaw	L-Asparaginase	60% loss of protein activity at 10 μg/ml in 50 mM sodium phosphate buffer (pH 7.4)	Izutsu et al. 1994a
	Hemoglobin	Protein aggregation	Kerwin et al. 1998
Protein concentration	IL-1β	Increased aggregation with increasing concentration from 0.1 to 0.5 mg/ml	Gu et al. 1991
	Insulin	Increased aggregation with increasing concentration from 0.75% to 5%	Brange et al. 1997
Protein source	Insulin	Different aggregation rates for insulin from different manufacturers	Brange et al. 1992a
Pressure	RNase A	Protein unfolds completely under 4000 bar	Prehoda et al. 1998; Zhang et al. 1995a
	H^+-ATPase	Increased aggregation when the pressure was increased from 0 to 2000 bar	Tsai et al. 1998

Unlike chemical unfolding, thermal unfolding is often irreversible, such as for interferon-β-1a (IFN-β-1a) (Runkel et al. 1998) and β-galactosidase (Yoshioka et al. 1994a). However, the reversibility of thermal unfolding can be changed by adjusting the solution conditions. While thermal unfolding of recombinant human megakaryocyte growth and development factor (rhMGDF) is irreversible with the formation of precipitate in phosphate or citrate buffer (pH 4–7), the unfolding is about 50% reversible after cooling, with no visible precipitates in 10 mM imidazole, histidine, or Tris buffer from pH 6 to 8 (Narhi et al. 1999).

3. DEVELOPMENT OF A LIQUID PROTEIN FORMULATION

There are two major dosage forms for protein pharmaceuticals: liquid and freeze-dried solid. A liquid formulation has the advantages of simpler processing, more efficient production, and more convenient application, whereas a freeze-dried protein formulation generally offers more stability. The freeze-drying process, however, often causes significant protein damage. Therefore, it is usually the stability of a protein that dictates the selection of the final dosage form. In this section, we discuss general considerations in formulating a liquid protein product as well as stabilization of proteins by excipients and structural modifications.

3.1 General Considerations in Formulating a Liquid Protein Product

Formulating a liquid protein product generally requires at least three phases of experimentation: (1) preformulation, (2) formulation, and (3) stability studies.

Preformulation

Protein preformulation is the first and prerequisite step in formulating a protein product. In this stage, the physicochemical properties of a protein are determined, such as protein purity, pI, and solubility at different pHs. Generally, the purity of a protein should be 95% or better since impurities may, under some conditions, play a role in product stability.

Protein preformulation also explores the potential protein decomposition pathways at different pHs and ionic strengths, multiple freeze/thaw stability, and the effect of buffering agents.

Formulation

Protein formulation is the critical phase in the development of a liquid protein product. In this phase, formulation scientists need to screen tonicity-adjusting agents, protein stabilizers, preservatives, and container/closure compatibility. While NaCl is the most commonly used excipient for tonicity adjustment, the impact of NaCl on protein stability must be evaluated. The most commonly used buffering agents are phosphate for pHs between 6 and 8 and citrate for pHs below 6. A protein can sometimes be stabilized by a particular buffering agent, such as acetate for hEGF (Son and Kwon 1995) and bFGF (Wang et al. 1996), and citrate for recombinant human keratinocyte growth factor (rhKGF) (Chen et al. 1994) and maize leaf phosphoenol pyruvate carboxylase (*ml*-PEPC) (Jensen et al. 1996). The buffer concentration also needs to be chosen, as it may affect both the physical and chemical stability of a protein, such as in the case of hGH (Pikal et al. 1991).

In formulating a protein product, a minimal number of components is always desirable, as more components will increase not only the processing time and cost, but also the likelihood of contamination and container/closure delivery device incompatibility. Furthermore, the presence of additional formulation components increases the burden of analytical testing and quality control. Therefore, each component requires clear justification.

In addition to the selection of proper formulation components, formulation pH is another important parameter to consider, with particular emphasis on the stability of the protein and the route of drug administration. A protein usually requires a particular pH (or a pH range) for stability, which may or may not be ideal for patient application. Selection of a formulation pH should balance both aspects. For subcutaneous injection, an isotonic solution with a neutral pH is highly desirable to minimize injection irritation. For an intravenous preparation, both tonicity and pH may not be as critical to ensure patient comfort.

Finally, compatible containers need to be identified. Type 1 borosilicate glass is usually the material of choice for containers due to its strong chemical resistance and low level of leachables. However, glass containers may adsorb protein, especially at low concentrations. In that case, surface-treated containers may be tested, a surfactant may be included in the formulation to mitigate adsorption, and/or the ionic strength may be adjusted to minimize electrostatic interactions with the container surface. At the same time, suitable closures need to be carefully chosen depending on their compatibility with the protein formulation, moisture/vapor transfer properties, and seal integrity.

Rapid Screening of Protein Formulations

Due to both the ever-increasing competition in the biotechnology industry and the urgent medical need for efficacious new products, a protein drug candidate needs to be brought to the market as rapidly as possible. This undoubtedly increases the pressure of shortening each stage in the drug development process, including protein formulation. To meet this demand, a method for rapidly screening formulation excipients may be needed to expedite excipient selection without resorting to time- and labor-intensive real-time stability studies.

Several screening methods have been summarized in a recent review (Wang 1999). These include (1) comparison of the protein unfolding temperature (T_m) in different formulations, (2) comparison of half denaturation concentration, C_{half}, in the presence of a denaturant, and (3) comparison of IR spectra of different protein formulations with a reference spectrum. In many cases, protein thermal stability has been shown to correlate positively with protein stability against denaturing agents, including detergents and organic solvents (Cowan 1997). In other cases, such a correlation may not exist, such as for albumin (Farruggia and Pico 1999).

Among these three screening methods, determination of the protein unfolding temperature (T_m) seems to be the most informative. If the difference in T_m among different formulations is not detectable, the reversibility of thermal unfolding may also be a measure of protein storage stability (Remmele and Gombotz 2000). Narhi et al. (1999) demonstrated that when thermal unfolding of rhMGDF was reversible, protein monomers were recovered after storage, but soluble aggregates of tetramers to 14-mers were formed under irreversible conditions. The reversibility can be changed by adjusting the type and concentration of formulation excipients. Since the stability of a protein determined by these screening methods does not always correlate with real-time stability, these methods need to be used with caution.

Stability Studies

Stability studies are indispensable during the entire formulation development process. Early stability studies may help to identify potential degradation pathways and to screen formulation candidates. Stability studies on the final product are conducted to define the optimal storage conditions and expiration date. The current stability requirement by the FDA for a pharmaceutical product is less than 10% deterioration after storage for 2 years under the specified storage conditions.

To accelerate formulation development, stability studies are often conducted under accelerated (stressed) conditions. These stressed conditions include high temperature, high humidity, intensive lighting, extreme pHs, and increased air-water interfaces induced by vortexing or shaking. The key issue in interpreting the results of accelerated stability studies is whether the data from accelerated stability studies can be extrapolated to those under real-time conditions.

The stability results obtained at high temperatures do not necessarily reflect or predict what happens under real-time conditions. This is due to the multiple and changing protein degradation pathways at different temperatures (Pikal et al. 1991; Gu et al. 1991). If the multiple degradation processes in proteins can be described separately, of if the rate-limiting degradation step does not change with temperature, prediction of protein stability based on accelerated stability studies is very useful (Vemuri et al. 1993; Yoshioka et al. 1994b; Jensen et al. 1997; Volkin and Middaugh 1996; Fransson et al. 1996).

While accelerated stability studies are often used during formulation development, real-time stability studies are required to define the expiration date for the final protein product.

3.2 Protein Stabilization by Excipients

Most proteins need a stabilizer in the liquid state for long-term storage. These stabilizers are formulation excipients, which are also referred to as chemical additives (Li et al. 1995c), cosolutes (Arakawa et al. 1993), or cosolvents (Timasheff 1993, 1998; Lin and Timasheff 1996). A formulation excipient may affect both the protein drug stability and drug efficacy (Hancock et al. 2000). Therefore, selection of suitable formulation excipients is critical in protein formulation development.

Proteins are stabilized by excipients mainly through preferential interaction, a major stabilization mechanism that has been described elsewhere (Timasheff 1993, 1998; Lin and Timasheff 1996). Other stabilization mechanisms have also been proposed. One of these is the increased solution viscosity in the presence of excipients. Jacob and Schmid (1999) found that both refolding and unfolding of the cold-shock protein CspB from *Bacillus subtilis* were decelerated in the presence of sucrose or ethylene glycol. This observation was postulated to be due to an increase in solution viscosity, as refolding and unfolding were considered as diffusional processes, at least for certain proteins. Since the partially unfolded state has a higher diffusion coefficient, increasing the solution viscosity would have a greater effect on the partially unfolded state than on the folded state (Damodaran and Song 1988). This mechanism may explain why the population of unfolded RNase A was decreased in the presence of sucrose (Wang et al. 1995) and the folding rate of staphylococcal nuclease was increased in the presence of xylose (Frye and Royer 1997). Excipients may also stabilize proteins by reducing the solvent accessibility and conformational mobility (Kendrick et al. 1997).

Protein-stabilizing excipients can be broadly divided into the following types: sugars and polyols, amino acids, amines, salts, polymers, and surfactants. Many of these excipients have been long recognized as compatible osmolytes in many different organisms (Yancey et al. 1982). The excipients and examples of their use are listed in Table 4. It should be noted that the degree of stabilization by different types of excipients can be significantly different for a particular protein, such as acidic fibroblast growth factor (aFGF) in the presence of sodium sulfate, His, Gly, trehalose, dextrose, or sorbitol (Tsai et al. 1993). Excipients that stabilize one protein may actually destabilize another.

Sugars and polyols are the most commonly used nonspecific protein stabilizers. A concentration of 0.3 M (or 5%) sugar or polyol has been suggested as the minimal level required to achieve a significant protein stabilization (Arakawa et al. 1993; Wang 1999). During selection of sugars for protein stabilization, reducing sugars should be avoided whenever possible because of their potential to react with amino groups in proteins (Maillard reaction), leading to the formation of carbohydrate adducts.

Amino acids are another important class of stabilizers and osmolytes. Amino acids that stabilize one protein may destabilize another. For example, arginine and histidine have been shown to stabilize interleukin-1 receptor (IL-1R) (Remmele et al. 1998) and rhKGF (Zhang et al. 1995b), respectively, but both amino acids destabilized cytochrome *c* (Taneja and Ahmad 1994). A recent study showed that arginine and histidine destabilized several proteins in terms of thermal transition, including RNase A, holo-α-lactalbumin, apo-α-lactalbumin, lysozyme, and metmyoglobin (Rishi et al. 1998). It was argued that arginine and histidine might behave like denaturants, since their side chains are structurally similar to urea and guanidinium ion. This hypothesis does not explain the differential stabilizing effect of these amino acids on different proteins.

Among polymers, hydroxypropyl-β-cyclodextrin (HP-β-CD) is probably the most commonly used. HP-β-CD has a donut structure whereby the inner hydrophobic core can interact with the hydrophobic part of a protein, thereby preventing hydrophobic interaction–induced protein aggregation. HP-β-CD has been shown to stabilize several proteins, including porcine growth hormone (pGH) (Charman et al. 1993), IL-2 and insulin (Brewster et al. 1991), rhKGF (Zhang et al. 1995b), and lysozyme and ovalbumin (Sah 1999). However, it reduced the T_m and T_0 (melting onset) of hen egg lysozyme by about -0.2°C per 1% increase in cyclodextrin concentration (w/w) (Branchu et al. 1999).

Table 4 Various Excipients and Their Stabilization of Proteins in Liquid State

Category	Excipients	Stabilized proteins	Ref.
Sugars/polyols	Dextrose	Relaxin; thrombin	Li et al. 1996; Boctor and Mehta 1992
	Ethylene glycol	Relaxin	Li et al. 1996
	Glycerol	Relaxin; thrombin	Li et al. 1996; Boxtor and Mehta 1992
	Lactose	rhDNase; elastase; IL-1R	Chan et al. 1996; Chang et al. 1993; Remmele et al. 1998
	Mannitol	rhDNase; elastase; FVIII SQ; rhG-CSF; relaxin	Chan et al. 1996; Chang et al. 1993; Fatouros et al. 1997b; Herman et al. 1996; Li et al. 1996
	Sorbitol	Elastase; FVIII SQ; aFGF; IgG; RNase	Chang et al. 1993; Fatouros et al. 1997b; Tsai et al. 1993; Gonzalez et al. 1995; McIntosh et al. 1998
	Sucrose	rhDNase; elastase; bFGF; rhIFN-γ; IL-1ra; RNase A; whey proteins	Chan et al. 1996; Chang et al. 1993; Wang et al. 1996; Kendrick et al. 1998; Chang et al. 1996a; Liu and Sturtevant 1996, McIntosh et al. 1998; Kulmyrzaev et al. 2000
	Trehalose	rhDNase; baker's yeast ADH, GDH, LDH; aFGF; RNase A	Chan et al. 1996; Ramos et al. 1997; Tsai et al. 1993; Lin and Timasheff 1996
Buffers	Acetate	HEGF; bFGF	Son and Kwon 1995; Wang et al. 1996
	Citrate	KGF; α_1-antitrypsin	Chen et al. 1994; Vemuri et al. 1993
	Phosphate	aFGF	Won et al. 1998
	Imidazole/Tris	rhMGDF	Narhi et al. 1999
Amino acids	Alanine	Cytochrome *c*	Taneja and Ahmad 1994
	Arginine	IL-1R	Remmele et al. 1998
	Aspartate	rhKGF	Zhang et al. 1995b
	Cysteine	IL-1R	Remmele et al. 1998
	Glutamate	rhKGF	Zhang et al. 1995b

Table 4 Continued

Category	Excipients	Stabilized proteins	Ref.
	Glycine	bovine α-lactalbumin; RNase A	Sabulal and Kishore 1997; Liu and Sturtevant 1996
	Histidine	rhKGF; rhMGDF	Zhang et al. 1995b; Narhi et al. 1999
	Lysine	Bovine α-lactalbumin; cytochrome *c*; IL-1R; lysozyme	Sabulal and Kishore 1997; Taneja and Ahmad 1994; Remmele et al. 1998; Rishi et al. 1998
	Proline	Glutamine synthetase	Paleg et al. 1984
	Serine	Cytochrome *c*	Taneja and Ahmad 1994
	Threonine	Cytochrome *c*	Taneja and Ahmad 1994
	Tryptophan	Insulinotropin	Qi and Heller 1995
Polymers	Dextrans	Elastase	Chang et al. 1993
	Gelatin	LMW-UK	Vrkljan et al. 1994; Manning et al. 1995
	Heparin	aFGF; rhKGF	Volkin et al. 1993; Chen et al. 1994
	HP-β-CD	pGH; rhKGF	Charman et al. 1993; Zhang et al. 1995b
	Maltosyl-β-cyclodextrin	Insulin	Tokihiro et al. 1997
	PEGs	BSA; IL-1R; LMW-UK	Kita et al. 1994; Remmele et al. 1998; Vrkljan et al. 1994
Surfactants	Tween-20	rConIFN; EGF; rhFXIII; rhGH	Ip et al. 1995; Son and Kwon 1995; Kreilgaard et al. 1998b; Maa and Hsu 1997
	Tween-40	rhGH	Bam et al. 1996

	Tween-80	Bovine serum albumin; hemoglobin; TGF-β_1	Arakawa and Kita 2000; Kerwin et al. 1998; Gombotz et al. 1996
	Pluronic F68	rhG-CSF; hGH	Johnston 1996; Katakam et al. 1995
	Pluronic F88	rhGH	Maa and Hsu 1997
	Pluronic F 127	IL-2, urease	Wang and Johnston 1993
	SDS	BSA; aFGF, RNase	Giancola et al. 1997; Won et al. 1998
	Triton X-100	EGF	Son and Kwon 1995
Salts	NaCl	FVIII SQ; IL-1R	Fatouros et al. 1997a; Remmele et al. 1998
	KCl	Baker's yeast ADH, GDH, LDH	Ramos et al. 1997
Metal ions	Ca^{2+}	rhDNase; rFVIII SQ	Chan et al. 1996, 1999; Fatouros et al. 1997a
	Mn^{2+}	RNase H	Goedken and Marqusee 1998
	Zn^{2+}	hEGF	Son and Kwon 1995
Chelating agents	EDTA	Insulinotropin	Qi and Heller 1995
Miscellaneous	Betaine	Glutamine synthetase; RNase	Paleg et al. 1984; Yancey et al. 1982
	TMAO	RNase	Yancey et al. 1982
	Mannosylglycerate	Baker's yeast ADH, GDH, LDH	Ramos et al. 1997
	DPPG	RNase	Lo and Rahman 1998
	HSPC	IFN-γ	Kanaoka et al. 1999
	Glyeryl monooleate	Insulin	Sadhale and Shah 1999

Since proteins are polymers, their concentration may increase their own stability, although increasing the protein concentration often increases the rate of protein aggregation. Estrada-Mondaca and Fournier (1998) demonstrated that increasing the concentration of *Drosophila* acetylcholinesterase from 1.3 nM to 13 μM gradually decreased the rate of protein inactivation during incubation at 20°C. Although protein-protein interaction was proposed for the stabilization of the protein, saturation of surface-induced protein inactivation at high concentrations cannot be ruled out as an explanation for the increased stability.

Surfactants are often used to mitigate protein aggregation and/or protein surface adsorption (Table 4). Due to possible contamination with alkyl peroxides, surfactants should be carefully used with oxidation-sensitive proteins. For example, recombinant human ciliary neurotrophic factor (CNTF) has been found to dimerize by alkyl peroxides present in Tween-80 (Knepp et al. 1996). A correlation was found between the peroxide level in Tween-80 and the degree of oxidation in rhG-CSF, and the peroxide-induced oxidation appeared to be more serious than that induced by the atmospheric oxygen present in the vial head space (Herman et al. 1996). To inhibit this side reaction, a lower concentration of surfactants may be used, or an antioxidant, such as Cys, Met, or glutathione, may be included in the protein formulation.

The major effect of salts as protein stabilizers is the inhibition of intra- and intermolecular electrostatic interactions in proteins. This inhibition often leads to increased hydrophobic interaction and, thus, increased protein stability. Since electrostatic interactions in proteins may contribute to protein stability negatively (Honig and Yang 1995) or positively (Dill 1990; Imoto 1997), screening of these electrostatic interactions may lead to protein stabilization or destabilization. Within a protein, both stabilizing and nonstabilizing electrostatic interactions may coexist (Ooi 1994; Strop and Mayo 2000). The net effect of salt depends on the type and location of the electrostatic interactions, and, frequently, the type and concentration of salt.

Many excipients can also inhibit chemical degradations in proteins. Examples include inhibition of succinimidation of recombinant hirudin by zinc (Gietz et al. 1999), deamidation of rhDNase I by calcium (Chen et al. 1999), oxidation of insulinotropin by histidine or tryptophan (Qi and Heller 1995), and metal-catalyzed oxidation of several proteins, including human relaxin by sugars, polyols, and dextran (Li et al. 1996), aFGF by polyanions (Volkin et al. 1993), and papain by histidine (Kanazawa et al. 1994). This inhibition has been attributed to a variety of mechanisms, such as the chelating effect of amino acids, sugars, and polyols (Kanazawa et al. 1994; Li et al. 1996) and direct interaction with the protein (Volkin et al. 1993).

3.3 Stabilization of Proteins by Structural Modifications

Proteins may be stabilized by structural modifications (Table 5). Among the different chemical modifications, conjugation of polyethylene glycol molecules (PEGs) to proteins (PEGylation) is probably the most commonly used. PEGs are usually attached to lysine residues in proteins. PEG-modified proteins often exhibit increased stability and improved solubility (Li et al. 1995b).

Changing the protein structure may potentially stabilize a protein thermodynamically by increasing the protein's unfolding free energy change and/or kinetically by slowing down degradation processes. In most cases, modifications are targeted to increase the thermodynamic stability, which can be achieved either by stabilization of the native state (improvement of the stabilizing forces) or by destabilization of the unfolded state (such as introduction of a cross-link) (Imoto 1997).

Since chemical reactions often require harsh conditions, the structure of a protein can alternatively be modified by genetic manipulation. Site-specific mutations can replace labile or destabilizing amino acids with stabilizing ones (Spector et al. 2000). Due to the current lack of protein structural details and stabilization mechanisms, genetic manipulation for protein stabilization has to be done empirically. Even with clear-cut guidelines, such as those proposed by Querol et al. (1996), efforts to increase protein stability by amino acid substitution have not always been successful.

Modifying protein structure for increased stability has the potential to change the immunogenicity of the protein. For this reason, a mutant protein drug has to go through extensive toxicity studies. The solubility of a protein may also change after replacement of hydrophilic amino acids with hydrophobic ones, thereby presenting potential formulation issues. Therefore, any chemical or genetic modification of proteins should be carefully evaluated.

4. LYOPHILIZATION AND DESIGN OF A SOLID PROTEIN FORMULATION

Although a liquid formulation is preferable, a solid protein formulation may be required if the desired shelf life cannot be achieved with a liquid formulation. The most commonly used method to prepare a solid protein formulation is lyophilization (Cleland et al. 1993; Fox 1995). In this section, the lyophilization process and its associated stresses on proteins are briefly described, followed by discussions on stabilization of proteins during lyophilization and long-term storage, factors affecting the stability of solid proteins, and, finally, the design of a stable solid protein product.

Table 5 Stabilization of Proteins by Structural Modifications

Type	Protein(s)	Site of modification	Ref.
Mutation/permutation	aFGF	Cys-117 to Ser	Volkin and Middaugh 1996
	S6 ribosomal protein	Addition of Met to N terminus	Uversky et al. 1999
Glycosylation	Insulin	LysB29	Baudys et al. 1995
Disulfide formation	T4 Lysozyme	3-97, 9-164, 21-142	Matsumura et al. 1989
Other cross-linking	Hemoglobin	α-chain Lys99	Lundblad and Bradshaw 1997
PEGylation	Insulin	PheB1 or LysB29	Hinds et al. 2000
	rhMGDF	N-terminal amine	Guerra et al. 1998

4.1 Protein Stress During Lyophilization

The lyophilization (freeze-drying) process consists of three major steps: freezing of a protein solution, primary drying of the frozen solid, and secondary drying of the nonfrozen water. An optional thermal treatment step, such as annealing, is sometimes included before primary drying. Detailed reviews on lyophilization have been published (Pikal 1990a,b; Franks 1991; Skrabanja et al. 1994; Carpenter et al. 1997; Jennings 1999; Wang 2000).

Although some proteins can completely survive the lyophilization process, such as α_1-antitrypsin (Vemuri et al. 1994), porcine pancreatic elastase (Chang et al. 1993), and bovine pancreatic ribonuclease A (RNase A) (Townsend and DeLuca 1990), many lose varying levels of activity or structure during the process. Examples include bilirubin oxidase (BO) (Nakai et al. 1998), β-galactosidase (Izutsu et al. 1993, 1994a), phosphofructokinase (PFK), and lactate dehydrogenase (LDH) (Carpenter et al. 1986, 1990; Prestrelski et al. 1993; Anchordoquy and Carpenter 1996), *Erwinia* L-asparaginase (Adams and Ramsay 1996), and monoclonal antibodies (Ressing et al. 1992). Loss of structure and/or function across lyophilization is due to a variety of stresses that tend to destabilize and/or denature an unprotected protein.

Low-Temperature Stress

Proteins may denature at low temperatures, such as ox liver catalase (Shikama and Yamazaki 1961), ovalbumin (Koseki et al. 1990), and RNase A (Zhang et al. 1995a). Low-temperature-induced protein denaturation has been proposed as a universal feature of proteins (Franks 1995). Although the nature of cold denaturation has not been satisfactorily delineated, it is likely due to the increased solubility of nonpolar groups in water with decreasing temperatures (Dill et al. 1989; Graziano et al. 1997). The decreasing solvophobic interaction at low temperatures in proteins can reach a point where protein stability reaches zero, causing cold denaturation (Jaenicke 1990).

Concentration Effect

Freezing a protein solution rapidly increases the concentration of all solutes due to ice formation. Freezing at 0.9% NaCl solution to its eutectic temperature of −21°C can cause a 24-fold increase in its concentration (Franks 1990). The calculated concentration of small carbohydrates in the maximally freeze-concentrated matrices (MFCS) is as high as 80% (Roos 1993). Thus, all physical properties related to concentration may change, such as ionic strength and the relative composition of solutes due to selective crystallization. These changes may potentially destabilize a protein. Due to solute concentration at low temperatures, chemical reactions may actually accelerate in a partially frozen aqueous solution (Liu and Orgel 1997).

Formation of Ice-Water Interface

Freezing a protein solution generates an ice-water interface. Proteins can be adsorbed to the interface, resulting in surface-induced denaturation (Strambini and Gabellieri 1996). The area of the interface is determined by a variety of factors, such as solute composition, solute concentration, and cooling rate (Sarciaux et al. 1999). Rapid (quench) cooling generates a large ice-water interface, whereas a smaller interface is induced by slow cooling. Denaturation of several proteins has been shown to be due to the formation of ice crystals, including aldolase, bfGf, CNTF, glutamate dehydrogenase (GDH), IL-1ra, LDH, maleate dehydrogenase (MDH), PFK, and tumor necrosis factor–binding protein (TNFbp) (Kendrick et al. 1995; Chang et al. 1996d).

pH Changes During Freezing

Freezing a buffered protein solution may selectively crystallize one buffering species, causing pH changes. Na_2HPO_4 crystallizes more readily than NaH_2PO because the solubility of the disodium form is considerably lower than that of the monosodium form. Because of this, the molar $[NaH_2PO_4]/[Na_2HPO_4]$ ratio of phosphate buffer at pH 7 can change from 0.72 to 0.57 at the ternary eutectic temperature during freezing (Franks 1990, 1993). This can lead to a significant pH drop during freezing, which can denature pH-sensitive proteins, such as LDH (Nema and Avis 1992; Anchordoquy and Carpenter 1996). The pH drop during freezing can also potentially affect the storage stability of lyophilized proteins, such as lyophilized IFN-γ (Lam et al. 1996) and IL-1ra (Chang et al. 1996d).

Annealing

Annealing is sometimes included before primary drying to promote recrystallization of an excipient for more efficient drying. Annealing may have an adverse effect on protein stability, as has been shown for β-galactosidase (2 μg/ml) in a mannitol formulation (Izutsu et al. 1993), IL-6 in a sucrose/glycine (1:1 at 20 mg/ml) formulation (Lueckel et al. 1998b), LDH in a mannitol formulation (Izutsu et al. 1994b), and trypsin in a sucrose formulation (Millqvist-Fureby et al. 1999). Annealing at $-7°C$ for 1 or 12 h was also found to destabilize hemoglobin in a PEG/dextran (1:1 weight ratio) system during lyophilization, as monitored by IR spectroscopy (Heller et al. 1999).

Dehydration Stresses

Generally, the water content of a lyophilized protein product is less than 10%. Therefore, lyophilization removes part of the original protein hydration shell. Removal of the hydration layer may disrupt the native state of a protein and cause denaturation. A hydrated protein, when exposed to a water-poor environment during dehydration, tends to transfer protons to ionized carboxyl groups, thereby reducing the charge of the protein (Rupley and Careri 1991). The decreased charge density may facilitate protein-protein hydrophobic interactions, potentially causing protein aggregation. Water molecules can also be an integral part of the active site(s) in proteins. Removal of these functional water molecules during dehydration can cause loss of bioactivity (Nagendra et al. 1998).

4.2 Stabilization of Proteins During Lyophilization and Long-Term Storage

Protein pharmaceuticals usually need to be stabilized during lyophilization. Although stabilization can be achieved through chemical modifications or genetic engineering, as discussed above, inclusion of a stabilizing excipient(s) in the formulation during lyophilization usually achieves satisfactory results. Theoretically, all of the stabilizing excipients used in liquid formulation (see Table 4) can be used to inhibit freezing-induced protein denaturation. Some of these excipients, such as sugars, may be used to stabilize proteins during both freezing and drying.

Since freezing and drying stresses imposed on proteins during lyophilization are different, the mechanisms of protein stabilization by stabilizers are not the same in the two stages of lyophilization. The preferential interaction mechanism applies equally well in freeze-thaw processes (Arakawa et al. 1993; Carpenter et al. 1991; Crowe et al. 1993b). Other stabilization mechanisms also apply, such as viscosity increase and suppression of pH changes during freezing (Anchordoquy and Carpenter 1996; Heller et al. 1997).

During lyophilization, the preferential interaction mechanism is no longer applicable. As the hydration shell of proteins is removed, many excipients, such as salts, that stabilize proteins in solution do not offer the same effect during lyophilization (Ramos et al. 1997). Two major hypotheses were proposed to explain the stabilization of proteins by excipients during drying: (1) formation of an amorphous glass and (2) formation of hydrogen bonds between the excipient(s) and the protein as the hydration shell is removed. These two mechanisms have been discussed elsewhere (Carpenter et al. 1990, 1993; Arakawa et al. 1991; Roser 1991; Crowe et al. 1993a; Franks 1994; Fox 1995; Allison et al. 1996, 1998, 1999; Wang 2000). Additional mechanisms were also proposed, including inhibition of protein-protein interaction by physical dilution and separation of protein molecules (Liu et al. 1990; Costantino et al. 1995a; Chang et al. 1996d; Strickley and Anderson 1997) and excipient-water interactions to facilitate protein refolding (Costantino et al. 1995a,b).

Usually a higher protein concentration requires higher levels of stabilizers for adequate protection. The amount of saccharides needed to achieve maximum protection of catalase during lyophilization was found to be equivalent to that forming a monomolecular layer on the protein surface (Tanaka et al. 1991). This is also the case for the stabilization of L-asparaginase by saccharides during lyophilization (Ward et al. 1999). On the other hand, excessive use of a stabilizer(s) in a protein formulation may eventually reach a limit of stabilization or even destabilize a protein during lyophilization (Carpenter and Crowe 1989; Allison et al. 1998). One of the possible explanations for this stabilization limit is the crystallization of the stabilizer, which diminishes the interaction of the stabilizer with the protein (Carpenter and Crowe 1989; Izutsu et al. 1994b).

Polymers may stabilize proteins by increasing the T_g of a solid protein formulation (te Booy et al. 1992; Prestrelski et al. 1995). In addition, polymers (proteins or other stabilizing agents) may indirectly stabilize proteins by inhibiting the crystallization of other stabilizing excipients in a solid formulation (Izutsu et al. 1995). For these reasons, polymers seem to be used more often in formulating solid proteins than with liquid protein products. Commonly used polymers include dextran (Liu et al. 1990; Chang et al. 1993; Kreilgaard et al. 1998a, 1999), gelatin (Takeshita et al. 2000), HP-β-CD (Hora et al. 1992a,b; Ressing et al. 1992; Katakam and Banga 1995), maltodextrin (Schebor et al. 1996; Cardona et al. 1997), and polyethyleneimine (Andersson and Hatti-Kaul 1999).

One of the frequently used polymeric stabilizers in protein drug development is serum albumin (Dawson 1992; Nema and Avis 1992; Anchordoquy and Carpenter 1996; Page et al. 2000). Many marketed protein products contain albumin, such as Betaseron, Epogen, Kogenate, and Recombinate (Physicians' Desk Reference 1999). However, the ever-increasing concern about the potential contamination of serum albumin with blood-borne pathogens limits its future application in protein products. The current trend is to develop albumin-free formulations for protein pharmaceuticals.

Proteins at higher concentrations are often more resistant against both freezing- and lyophilization-induced protein denaturation/aggregation. The activity recovery or stability of many labile proteins after freeze-thawing correlates directly with the initial protein concentration, such as in the case of rhFXIII (Kreilgaard et al. 1998b), LDH (Carpenter et al. 1990; Anchordoquy and Carpenter 1996), and ovalbumin (Koseki et al. 1990). The correlation of increased protein concentration with improved stability across lyophilization has also been demonstrated, such as for bovine and human IgG species (Sarciaux et al. 1998, 1999), LDH (Anchordoquy and Carpenter 1996), and PFK (Carpenter et al. 1987). The stabilization of proteins with increased concentration during freezing and/or lyophilization has been attributed to (1) a general polymeric stabilization effect, (2) the saturation of surface-induced protein denaturation at high concentrations, and (3) favorable protein-protein interactions (Mozhaev and Martinek 1984; Allison et al. 1996).

4.3 Factors Affecting Stability of Solid Proteins

A variety of instability mechanisms have been reported for lyophilized proteins during storage. These instability pathways are similar to those observed in the liquid state (Table 1), although chemical degradation occurs to a significantly lower extent than in the liquid state. A variety of factors can contribute to one or more of these instability pathways during storage of solid protein pharmaceuticals. These factors include storage temperature, glass transition temperature, formulation pH, residual moisture content, type and concentration of formulation excipients, crystallization of amorphous excipients, and reconstitution medium (Table 6).

The glass transition temperature (T_g) of a protein formulation is considered as one of the major determinants of protein stability (Hatley and Franks 1991), since positive correlations have been observed between T_g and the storage stability of certain proteins (Table 6). There are, however, many examples where formulations of a lower T_g are more stable than those of a higher T_g, especially when sugar formulations are compared with polymer formulations (Cardona et al. 1997; Rossi et al. 1997). This is partly due to inefficient hydrogen bonding between polymeric stabilizers and proteins. Recent studies indicate that the molecular mobility-changing temperature (T_{mc}) appears more closely related to protein storage stability than T_g (Yoshioka et al. 1997, 1998, 1999).

The pH of a solid protein formulation can directly affect storage stability (Table 6). It may also indirectly influence protein stability by changing the hygroscopic properties of a protein (Strickley and Anderson 1996) or the

Table 6 Factors Affecting Stability of Proteins in Solid State

Factors	Protein(s)	Effect(s)	Ref.
Temperature	Lactase	Increased loss of activity with increasing storage temperatures from 37°C to 70°C	Mazzobre et al. 1997
	rhIL-1ra	Increased aggregation and deamidation of lyophilized rhIL-1ra between 8°C and 50°C	Chang et al. 1996b,d
Glass transition temperature (T_g)	Il-2	Storage stability correlated with T_g of the protein formulation	Prestrelski et al. 1995
	Invertase	Storage stability correlated with T_g of the protein formulation	Schebor et al. 1996
Formulation pH	RNase	Increasing pH accelerates protein aggregation during storage (pH 10.0 > pH 4.0 > pH 6.4)	Townsend and DeLuca 1990
	Il-1ra	protein deamidation decreases with decreasing pH (pH 7.0 > pH 6.5 > pH 6.0 > pH 5.5)	Chang et al. 1996d
Residual moisture	tPA	Increased loss of protein activity with increasing moisture content from 5% to 18%	Hsu et al. 1991
	Insulin	Increased aggregation/degradation with increasing water content or storage relative humidity	Costantino et al. 1994; Strickley and Anderson 1996, 1997

Excipient type	Elastase	Ascorbic acid at 5 mM reduced storage stability at 40°C and 79% relative humidity	Chang et al. 1993
	rFIX	Protein storage stability in buffers: histidine > phosphate or Tris	Bush et al. 1998
Excipient concentration	RNase	Loss of activity increased with increasing phosphate concentration from 0 to 0.2 M	Townsend and DeLuca 1990
	IL-2	Reduced protein activity by shaking	Wang and Johnston 1993
Crystallization of excipients	bFGF	Crystallization of sucrose caused increased formation of degradants	Wu et al. 1998
	Anti-IgE	Decreased storage stability due to mannitol crystallization	Costantino et al. 1998
Reconstitution medium	Interferon-γ	Inclusion of Tween-20 in the medium decreased protein aggregation	Webb et al. 1998
	Bovine IgG	Inclusion of Tween-80 in the medium decreased protein aggregation	Sarciaux et al. 1999
	IL-2	Decreased aggregation with changing medium pH from 7 to 4	Zhang et al. 1996

rate of chemical degradation of a stabilizer(s) such as sucrose (te Booy et al. 1992; Skrabanja et al. 1994).

The residual moisture content after lyophilization often controls long-term protein stability, both physically and chemically (Franks 1990; Hatley and Franks 1991). Water can affect protein stability both indirectly as a plasticizer or reaction medium and directly as a reactant or a product (Shalaev and Zografi 1996). Water dramatically decreases the glass transition temperature of proteins, polymers, or other formulation excipients (Slade et al. 1989; Roos and Karel 1991; Buera et al. 1992; te Booy et al. 1992; Roos 1993; Wolkers et al. 1998). Therefore, a lyophilized protein may adsorb sufficient amounts of moisture during storage to reduce its T_g below the storage temperature, accelerating its instability and sometimes causing product collapse (Oksanen and Zografi 1990; Strickley and Anderson 1997). High moisture content also facilitates crystallization of formulation excipients, such as various sugars, indirectly affecting protein stability (Table 6).

Usually, a lower moisture content leads to a more stable protein product. As a general rule, the moisture content of a lyophilized protein formulation should not exceed 2% for maximal stability (Daukas and Trappler 1998). This general rule was corroborated with several examples, including lyophilized bFGF (Wu et al. 1998) and monoclonal antibody cA2 IgG (Katakam et al. 1998). Exceptions exist, such as lyophilized BSA and bovine γ-globulin (BGG) (Yoshioka et al. 1997). In many cases, a bell-shaped relationship has been demonstrated between protein stability and moisture content. Examples include lyophilized tetanus toxoid (Schwendeman et al. 1995), recombinant human albumin (rHA) (Costantino et al. 1995b) and BSA (Liu et al. 1990), ovalbumin, glucose oxidase, and bovine β-lactoglobulin (Liu et al. 1990), and insulin (Katakam and Banga 1995; Separovic et al. 1998). Therefore, to find a moisture content that confers maximal stability for a lyophilized protein product, long-term stability studies should be conducted on formulations with different moisture contents. Only through these real-time stability studies can the optimal moisture content be determined for the final protein product.

The reconstitution step may potentially affect protein stability. Several mechanisms have been proposed. Rapid reconstitution with water may not allow a dried protein to rehydrate as slowly as it was dehydrated during the dehydration step, causing denaturation and/or aggregation (Cleland et al. 1993). In addition, the pH of water-reconstituted protein formulations may be different from the pH before lyophilization, possibly due to loss of some formulation components, such as HCl (Strickley and Anderson 1996) and acetic acid (Hatley et al. 1996). Finally, many proteins are temperature sensitive and the temperature of the reconstitution medium can make a significant difference in protein stability (Sampedro et al. 1998).

4.4 Design of a Stable Lyophilized Protein Product

Generally speaking, three types of lyophilized protein formulations can be designed: crystalline, amorphous, and mixed type. A crystalline protein formulation contains a crystalline bulking agent, such as glycine, which provides strong cake structure and acceptable dissolution upon reconstitution. Since the crystalline agent alone rarely provides any stabilization to proteins, a crystalline formulation is only suitable for stable proteins that do not need an amorphous stabilizer(s). An amorphous protein formulation does not have any crystalline structure after lyophilization and may allow maximal interaction between the amorphous agent(s) and the protein. Such formulations may or may not form a strong cake after lyophilization, depending on the type and concentration of the bulking agent. The third type is a mixed formulation, where crystalline and amorphous components coexist. In these formulations, the amorphous component stabilizes the protein during processing and storage while the crystalline component provides the cake structure.

Lyophilization starts with preparation of a protein solution. Therefore, the first step in designing a solid formulation is to select a stable pH and an appropriate buffering agent for the protein in the liquid state. This pH should also be optimal to minimize protein denaturation during lyophilization and to allow maximal long-term stability for the lyophilized protein. To meet all of these requirements the formulation pH must be carefully chosen. Commonly used buffering agents include acetate, citrate, glycine, histidine, and phosphate, A preferable buffering agent should also stabilize a protein. For pH-sensitive proteins, sodium phosphate should be avoided because of a possible pH change during freezing. In addition, selection of a proper buffer concentration is also important, as the buffer concentration not only affects the storage stability of lyophilized proteins but also plays a critical role in stabilizing proteins during lyophilization (Izutsu et al. 1993).

The most commonly used crystalline agents for lyophilized proteins are mannitol and glycine. Glycine has several advantages, including low toxicity, high solubility, and high eutectic temperature (Akers et al. 1995). Mannitol may be used both as a bulking and stabilizing agent (Gombotz et al. 1996). In addition, most amino acids are potential bulking agents as they easily crystallize out (Mattern et al. 1999). Another crystalline agent, NaCl, is not a preferable bulking agent due to its low eutectic and glass transition temperature, making lyophilization inefficient (Franks 1990; Carpenter et al. 1997).

A solid protein product should be stable during storage minimally above 0°C, preferably under room temperature conditions. To achieve this goal, an amorphous stabilizer(s) is usually needed to protect a protein during lyophilization and/or long-term storage. The most widely used stabilizers in solid protein products are sugars. Reducing sugars are generally not preferable due to their tendency to undergo the Maillard reaction. Among sugars, sucrose seems to be the most commonly used. Recently, trehalose has been approved by the FDA as an injectable excipient. The relative effect of these two sugars in stabilizing solid proteins for long-term storage is protein dependent (Chang et al. 1996d; Lueckel et al. 1998b). Some polymers can be used as protein stabilizers in the solid state, as they can increase the T_g of protein formulations. On the other hand, polymers may not be as effective as sugars in stabilizing proteins due to inefficient hydrogen bonding with proteins (Schebor et al. 1996; Cardona et al. 1997; Kreilgaard et al. 1998a). An alternative is to combine polymers and sugars in a formulation to achieve enhanced protein stabilization. However, this strategy has not been very effective, at least for certain proteins (Lueckel et al. 1998b).

In a protein formulation, the physical properties of all components are mutually affected. While the amorphous stabilizers(s) may inhibit crystallization of the bulking agent(s), the bulking agent may reduce the T_g' of an amorphous excipient (Bush et al. 1998; Kasraian et al. 1998; Kim et al. 1998; Lueckel et al. 1998a). Therefore, the proper selection of their relative amounts is critical. In addition, the total quantity of solid in protein formulations should be kept between 2% and 10%. While a solid content lower than 2% may not form a strong cake, higher amounts (> 10%) may be difficult to process and reconstitute (Hatley et al. 1996; Carpenter et al. 1997; Breen et al. 1998; Jennings 1999; Willemer 1999).

To expedite the selection of formulation excipients, two screening methods have been used: determination of T_g by differential scanning calorimetry (DSC) or IR spectroscopy of a lyophilized formulation in the presence of different excipients. Both methods should be used with caution, as these parameters may not always reflect changes in protein activity or stability.

Selection of the final protein formulation requires stability studies. A variety of formulation parameters in stability studies can be used to compare different formulations, including protein activity, physical attributes of the cake (shape, color, and texture), particulate formation, moisture content, and ease of reconstitution. To expedite selection of the final protein formulation, accelerated stability studies are frequently conducted. As in the case for liquid protein pharmaceuticals, the key issue is whether the data obtained at high temperatures can be extrapolated to reflect those under real-time conditions. Only through real-time stability testing can the final protein formulation be determined.

5. ANALYTICS IN FORMULATION DEVELOPMENT

5.1 Strategy for Analytical Support of Formulation Development

One of the major objectives of the preformulation work should be to define the mechanism(s) of protein degradation observed under stressed conditions. The data obtained in this early phase of the formulation project can be used to devise an effective analytical strategy for support of formulation development.

In order to correlate the observed instability mechanism(s) with loss of bioactivity, it is crucial to develop a reproducible activity assay during this early phase of formulation development. Among the in vitro activity assays available, cell-based bioassays, in vitro activity assays, and binding assays are the most commonly used methodologies. Cell-based bioassays provide a direct readout of a biological response, such as T-cell proliferation for a cytokine (Takashima et al. 1999), and therefore can be considered as true potency assays. Cell-based potency assays are, however, labor intensive, time consuming, and relatively imprecise due to the dependence on biological systems, which contribute significant variability.

In vitro bioassay, when available, can provide an excellent means for evaluating the impact of formulation excipients and storage conditions on the biological activity of the protein. Examples of in vitro bioassays include coagulation assays for clotting factors such as factor VIII and protease inhibition assays for protease inhibitors such as antithrombin III. In vitro bioassays generally have the distinct advantages of high throughput and improved precision relative to cell-based bioassays. However, since in vitro bioassays provide only an indirect assessment of the biological activity of the active ingredient, they may not provide a true measurement of the potency of the protein under consideration. Binding assays have often been used with antibodies and antagonists, since binding is the major biological activity of interest for these types of products. It should be noted that binding in and of itself is a necessary but insufficient measure of biological activity because the actual biological effect desired results from the binding activity, which is presumably blocking a metabolic pathway that leads, in turn, to a biological response. In the case of drugs that act as agonists, binding assays can be informative, but the data cannot directly reflect the bioactivity of the protein because the signaling event resulting from agonist binding is a necessary component of the desired biological response.

While cell-based bioassays are labor intensive and difficult to automate, both in vitro activity assays and binding assays are amenable to high-throughput, automated formats. Since in vitro activity assays and binding assays can often be carried out in a format similar to that of enzyme-linked immunosorbent assays (ELISA), automated dilutors and fully integrated immunoassay systems can be used for these assays. An alternative system for binding assays that has recently gained popularity is the biosensor technology, such as the Biacore (Ehnebom et al. 1997; Worn and Puckthun 1998) and IAsys (Huber et al. 1999; Shuck 1997), which provide kinetic data on binding constants not available from ELISA-like systems, which provide only a "snapshot" of the extent of binding at equilibrium. The data provided by the biosensor technology can provide useful information about the active ingredient, particularly for detailed characterization, lot comparisons, and evaluation of the impact of process changes on the product. However, the kinetic parameters evaluated by this technology likely provide more information than is required during formulation development.

5.2 Analytical Methods for Monitoring Chemical Degradation

High-performance liquid chromatography (HPLC) is the most commonly employed technique for monitoring chemical degradation of proteins (Snyder et al. 1997; Weston and Brown 1997; Cunico et al. 1998). HPLC is a relatively mature technology that has broad application to the evaluation of protein integrity. Modern chromatographic instruments provide high-resolution, excellent precision and high throughput due to the availability of automated systems. Although methods development is required on a case-by-case basis, it is feasible to monitor deamidation for many proteins by ion exchange and/or reversed-phase chromatography, since deamidation alters the charge and hydrophobicity of the protein. The most commonly used chromatographic method for monitoring methionine oxidation is reversed-phase HPLC, while disulfide interchange can be monitored by reversed-phase and/or hydrophobic interaction chromatography. While HPLC can also be used to monitor proteolysis, peptide bond cleavage is usually easier to monitor and quantitate using electrophoretic techniques.

During chromatographic methods development, it is important to demonstrate the impact of chemical degradation on the chromatographic profile of the analyte. Chemical treatment of the protein can be carried out to demonstrate the alteration of the chromatographic pattern resulting from deamidation (Patel and Borchardt 1990), methionine oxidation (Keck 1996; Gao et al. 1998; Liu et al. 1998) and/or disulfide interchange (Browning et al. 1986). Verification of the identity of the component(s) of the peak(s) resulting from "forced" chemical degradation can be accomplished by collecting the material in the peak(s) and subjecting the degradation product(s) to structural analysis. Modern methods in protein mass spectrometry (MS) provide an array of high precision methods for verification of structural variants and degradation products. Methionine oxidation is relatively easy to confirm by electrospray ionization–mass spectrometry (ESI-MS), since oxidation results in mass increases in multiples of 16 amu. Confirmation of deamidation is more difficult using MS because deamidation of a single residue results in a mass change of only 1 amu. While it is not feasible to precisely measure protein molecular weights to the required level of accuracy to confirm deamidation using ESI-MS, the availability of a new, higher precision technology, quadrupole–time-of-flight mass spectrometry (QTOF), brings verification of deamidation to the realm of feasibility. Alternatively, more laborious peptide mapping with N-terminal sequencing of isolated peptides can be used.

Capillary electrophoresis (CE) is a relatively new separation technique that is gaining widespread acceptance in bioanalytical laboratories (Polanski and Shintani 1996; Landers 1997). CE is a rapid and quantitative method that can be used to monitor chemical degradation of proteins such as deamidation (Frenz et al. 1989) and oxidation (Landers et al. 1993). In addition, methods have been developed for comparing the stability of glycoforms of a protein. An example of the use of CE for evaluation of glycoform stability is shown in Fig. 1, which demonstrates the use of neutral CE for analysis of the glycoform of a cytokine produced in mammalian cell culture. Recent advances in CE have greatly expanded the utility of the technique. Examples include the development of laser-induced fluorescence detection, which has significantly increased the sensitivity of CE (Hughes and Richberg 1993), as well as tandem CE-MS methods (Carbeck et al. 1998).

SDS-polyacrylamide gel electrophoresis (SDS-PAGE) is used throughout the formulation development process, as it is a simple and informative method that provides data on protein purity and the presence of proteolytic and/or hydrolytic degradation products. Monitoring the product by SDS-PAGE throughout development of the protein drug offers the possibility of detecting the presence of contaminating protease(s) in the purification process streams and/or purified protein bulk such that the proteolytic activity can be removed during development of the purification process. The potential presence of a contaminating protease activity is generally indicated by the appearance of lower molecular weight bands upon storage, particularly at ambient and higher temperatures. While removal of the protease from the process stream is the preferred approach to drug development, in some instances this may not be feasible. In that case, it is important to characterize the class of protease present because such knowledge can be used to select an effective protease inhibitor as a formulation component and/or determine the storage conditions that will minimize proteolytic degradation.

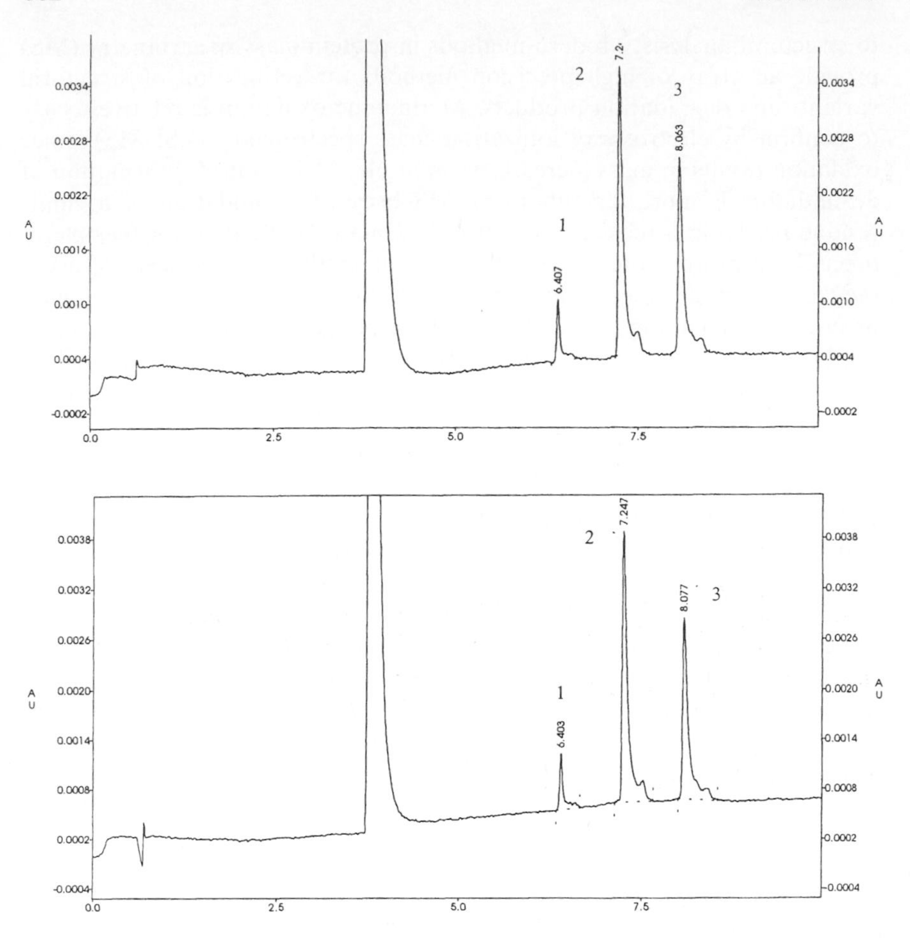

Figure 1 Neutral CE analysis of a lyophilized glycoprotein containing one *O*-linked glycan. Top panel: initial results; bottom panel: sample held at 40°C for 17 weeks. Peak 1: non-glycosylated; peak 2: mono-sialylated; peak 3: di-sialylated.

Identification of the class of protease activity present can be achieved by incubating the samples under conditions known to cause proteolytic degradation in the presence of various classes of protease inhibitors.

While the addition of reducing agents to samples prior to electrophoretic analysis can be used to determine if disulfide-bonded aggregates are present, SDS-PAGE may not be capable of detecting non-covalent protein aggregates, which are usually dissociated under denaturing conditions. To overcome this limitation, native PAGE may be needed (Arakawa and

Kita 2000) if an alternative method for analysis of aggregation is not feasible.

5.3. Analytical Methods for Monitoring Physical Instability

Since aggregation is a major degradative pathway for many protein pharmaceuticals, analytical methods are often chosen to monitor possible protein aggregation/precipitation during formulation development. The simplest method available is UV/visible spectroscopy, which can be used to measure optical density resulting from light scattering of protein aggregates/precipitates at or around 340 nm. However, this method provides only a relative assessment of aggregation and does not provide absolute quantitation of protein aggregates/precipitates. Similarly, static or dynamic light scattering methods only measure the relative degree of protein aggregation/precipitation, although they do provide information on the size distribution of protein aggregate/precipitates. Analytical ultracentrifugation can be used to determine protein aggregation as well as conformation. This technique, which can provide quantitative data on the extent of aggregation, is highly specialized and complex in terms of sample handling and data interpretation. For this reason, it is likely not to be useful for routine analytical purposes. In comparison, SEC-HPLC is a commonly used technique for quantitative assessment of aggregate content. This method only monitors soluble aggregates and has the potential limitation of all chromatographic methods: interaction of aggregate species with the column or the inlet frit. Such interaction will retard aggregate species on the column, thereby rendering interpretation of the data difficult. Nevertheless, this method is relatively simple and very useful. The power of SEC-HPLC can be significantly enhanced when it is coupled with a multiangle light scattering (MALS) detector for determination of the absolute molecular weight of soluble aggregates in protein solutions (Wyatt, 1997a,b).

The conformational stability of proteins is another parameter for evaluation during formulation development. The most commonly used methods for evaluating changes in protein structure include circular dichroism (CD), Fourier transform infrared spectroscopy (FTIR), fluorescence spectroscopy, and analytical ultracentrifugation. Among these techniques, CD is likely the most widely used method, since it is relatively straightforward and requires less expertise for data interpretation than the other methods, particularly when compared to analytical ultracentrifugation. The availability of modern CD instruments with advanced software makes it possible to obtain data on the relative content of α-helix, β sheet, and random structure in proteins as well as to monitor changes in structure during stability assessment (Fasman 1996; Pelton and McLean 2000). In addition, CD can be used to obtain thermodynamic data such as melting temperature and enthalphy change when temperature ramping methods are used (Yadav and Ahmed 2000). FTIR is probably the most widely used method to monitor protein conformation in the solid state (Dong et al. 1995; Carpenter et al. 1998). While FTIR can be used to obtain data on secondary structure of proteins in solution, interpretation of data is complex due to the requirements for removing the contribution that water makes to the spectrum in the region of the amide bond. While the significance of the data obtained from CD and FTIR in absolute terms may be difficult to interpret with respect to the structural integrity of a protein, the relative changes observed can be used to evaluate the impact of formulation and storage conditions on the structure of a protein in solution or the solid state.

The formulation development strategy must take into account the potential for proteins to adsorb to surfaces by means of both hydrophobic and electrostatic interactions. Since electrostatic interactions can be reduced by increasing the ionic strength, whereas hydrophobic interactions are enhanced by increased ionic strength, the formulation development plan should target an ionic strength that minimizes protein adsorption via these two mechanisms while providing maximal product stability. Development studies focused on minimizing product adsorption to the vials used for product packaging, as well as to syringes, infusion bags, infusion sets, and so forth, used for delivery of the drug to the patient, require precise methods for product quantitation (see Section 5.4). In addition, adsorption studies require a reliable potency assay (Sec. 5.1) to determine the impact, if any, of surface adsorption on the activity of the protein. Since adsorption can induce protein aggregation as a result of protein unfolding followed by protein-protein interactions, an analytical method for monitoring aggregation, such as SEC-HPLC and/or UV spectroscopy, should be used during adsorption studies.

5.4 Other Methods Used in Formulation Development

Evaluation of stability samples generally requires a means for measuring the protein quantity. The simplest method for determination of protein quantity is to measure the UV absorbance at 280 nm (A_{280}) of a protein solution (or a reconstituted protein solution). Due to possible protein aggregation, A_{280} needs to be corrected for the contribution of light scattering by making a measurement of the optical density in a non-UV-absorbing region, such as at 340 nm (Mach et al. 1992). However, the light scattering contribution can only be roughly estimated as the size, shape, and distribution of the protein aggregates/precipitates will impact the light scattering intensity. The extinction coefficient of a protein can be experimentally determined by carrying out amino acid analysis of the purified protein, calculated based on the amino acid sequence (Mach et al. 1992; Pace et al. 1995; Edelhoch 1967), or obtained through available databases, such as Swiss Prot (on the National Library of Medicine web site). As a rule of thumb, the determined extinction coefficients generally fall within about 5% of the theoretical values (Gill and von Hippel 1989).

The classical dye binding methods for protein content determination, such as Bradford (Bradford 1976) and BCA assays (Smith et al. 1985), are relatively rapid and precise, but subject to interference by a number of substances commonly used in formulations, such as surfactants. A method for protein content determination with good precision and high throughput is HPLC, such as reversed-phase HPLC (RP-HPLC). The use of chromatographic methods offers the advantage that they are excellent tools for monitoring protein structural integrity during protein processing or storage. Many chemical degradation products can be easily detected, quantitated, and identified when used with other analytical tools such as MS (Sec. 5.2).

During development of a solid protein product, additional analytical methods are needed. The most commonly used methods include DSC, lyo-microscopy, and a moisture assay. DSC is probably the most commonly used method for determination of the glass transition temperature (T_g') (Her and Nail 1994). For an amorphous formulation T_g' is equivalent to the collapse temperature. Proper determination of T_g' is critical as it not only dictates the maximum temperature for primary drying but also affects the stability of the dried protein product. Lyo-microscopy provides real-time images of freezing, melting, crystallization, collapse, and melt-back during lyophilization processes (Adams and Ramsay 1996; Overcashier et al. 1997). Lyo-microscopy is an excellent technique to confirm DSC results. Using both methods, the necessity of including an annealing step before primary drying and the maximal allowable temperature for primary drying can be determined. Knowledge of this maximal temperature allows optimization of primary drying by selecting a proper combination of shelf temperature and pressure, as both of them affect product temperature.

5.5 Choice of Analytical Methods for Support of Formulation Development

Due to the complexity of protein structure, the choice of analytical methods for support of formulation development is case specific. Among the factors for consideration, knowledge about the mechanism(s) of protein instability is perhaps the most critical one. Such knowledge provides the basis for choosing appropriate analytical methods for monitoring product stability. In the absence of such information, it is difficult to devise a coherent strategy, although use of a potency assay, in conjunction with chromatographic approaches to evaluate changes in the chromatographic profile, may be adequate for monitoring stability studies. In such instances, the appearance of new peaks or alterations in the shape of the main peak(s) will generate questions about the reason(s) for the observed change(s). Experimental investigations along these lines will in some instances lead to discovery of instability mechanisms for the protein. An approach to probing instability mechanisms of a completely new protein is to examine the protein unfolding temperature (T_m) in saline at a physiological pH and then to incubate the protein solution for a short time at a temperature just below T_m. Analysis of the resulting stability samples by established methods may provide some clue about the instability mechanism(s). Analytical methods that detect differences in stability samples can, in turn, be used throughout the drug development process to monitor product stability.

ACKNOWLEDGMENTS

We appreciate the support of Dr. Sheryl Martin-Moe and Dr. Robert Kuhn in writing this manuscript. We thank Michael Dumas for providing the capillary electrophoresis data shown in Fig. 1.

REFERENCES

Adams GDJ, Ramsay JR (1996) Optimizing the lyophilization cycle and the consequences of collapse on the pharmaceutical acceptability of *Erwinia* L-asparaginase. J. Pharm. Sci. 85, 1301–1305.

Akers MJ, Milton N, Byrn SR, Nail SL (1995) Glycine crystallization during freezing: the effect of salt form, pH, and ionic strength. Pharm. Res. 12, 1457–1461.

Allison SD, Chang B, Randolph TW, Carpenter JF (1999) Hydrogen bonding between sugar and protein is responsible for inhibition of dehydration-induced protein unfolding. Arch. Biochem. Biophys. 365, 289–298.

Allison SD, Dong A, Carpenter JF (1996) Counteracting effects of thiocyanate and sucrose on chymotrypsinogen secondary structure and aggregation during freezing, drying and rehydration. Biophys. J. 71, 2022–2032.

Allison SD, Randolph TW, Manning MC, Middleton K, Davis A, Carpenter JF (1998). Effects of drying methods and additives on structures and function of actin: Mechanisms of dehydration-induced damage and its inhibition. Arch. Biochem. Biophys. 358, 171–181.

Apenten RKO (1998) Protein stability function relations: β-lactoglobulin-A sulphydryl group reactivity and its relationship to protein unfolding stability. Int. J. Biol. Macromol. 23, 19–25.

Anchordoquy TJ, Carpenter JF (1996) Polymers protect lkactate dehydrogenase during freeze-drying by inhibiting dissociation in the frozen state. Arch. Biochem. Biophys 332, 231–238.

Andersson MM, Hatti-Kaul R (1999) Protein stabilizing effect of polyethyleneimine. J. Biotechnol. 72, 21–31.

Arakawa T, Kita Y (2000) Protection of bovine serum albumin from aggregation by Tween 80. J. Pharm. Sci. 89, 646–651.

Arakawa T, Prestrelski SJ, Kenney WC, Carpenter JF (1993) Factors affecting short-term and long term stabilities of proteins. Adv. Drug Deliv. Rev. 10, 1–28.

Arakawa T, Yoshiko K, Carpenter JF (1991) Protein-solvent interactions in pharmaceutical formulations. Pharm. Res. 8, 285–291.

Bam L, Cleland JL, Yang J, Manning MC, Carpenter JF, Kelley RF, Randolph TW (1998) Tween protects recombinant human growth hormone against agitation-induced damage via hydrophobic interactions. J. Pharm. Sci. 87, 1554–1559.

Bam NB, Cleland JL, Randolph TW (1996) Molten globule intermediate of recombinant human growth hormone: stabilization with surfactants. Biotechnol. Prog. 12, 801–809.

Baudys M, Uchio T, Mix D, Wilson D, Kim SW (1995) Physical stabilization of insulin by glycosylation. J. Pharm. Sci. 84, 28–33.

Boctor AM, Mehta SC (1992) Enhancement of the stability of thrombin by polyols: microcalorimetric studies. J. Pharm. Pharmacol. 44, 600–603.

Bradford MM (1976) A rapid and sensitive method for quantitation of microgram quantities of protein utilizing the principle of protein-dye binding. Anal. Biochem. 72, 248–254.

Branchu S, Forbes RT, York P, Nyqvist H (1999) A central composite design to investigate the thermal stabilization of lysozyme. Pharm. Res. 16(5), 702–708.

Brange J, Havelund S, Hougaard P (1992a) Chemical stability of insulin. 2. Formation of higher molecular weight transformation products during storage of pharmaceutical preparations. Pharm. Res. 9, 727–734.

Brange J, Langkjaer L, Havelund S, Volund, A (1992b) Chemical stability of insulin. 1. Hydrolytic degradation during storage of pharmaceutical preparations. Pharm. Res. 9, 715–726.

Brange J, Andersen L, Laursen ED, Meyn G, Rasmussen E (1997) Toward understanding insulin fibrillation. J. Pharm. Sci. 86, 517–525.

Breen ED, Costantino HR, Hsu CC, Shire SJ (1998) Effect of bulk concentration on reconstitution of lyophilized protein formulations. Pharm. Sci. (Suppl) 1, S-540.

Brewster ME, Hora MS, Simpkins JW, Bodor N (1991) Use of 2-hydroxypropyl-β-cyclodextrin as a solubilizing and stabilizing excipient for protein drugs. Pharm. Res. 8, 792–795.

Browning JL, Mattaliano RJ, Chow EP, Liang SM, Allet B, Rosa J, Smart JE (1986) Disulfide scrambling of interleukin-2: HPLC resolution of the three possible isomers. Anal. Biochem. 155, 123–128.

Buera MP, Levi G, Karel M (1992) Glass transition in polyvinylpyrrolidone: effect of molecular weight and diluents. Biotechnol. Prog. 8, 144–148.

Bush L, Webb C, Bartlett L, Burnett B (1998) The formulation of recombinant factor IX: stability, robustness, and convenience. Semin. Hematol. 35, 18–21.

Carbeck JD, Severs JC, Gao J, Wu Q, Smith RD, Whitesiobs GM (1998) Correlation between the charge of proteins in solution and in the gas phase investigated by protein charge ladders, capillary electrophoresis and electrospray ionization mass spectrometry. J. Phys. Chem. B. 102, 10596–10601.

Cardona S, Schebor C, Buera MP, Karel M, Chirife J (1997) Thermal stability of invertase in reduced-moisture amorphous matrices in relation to glassy state and trehalose crystallization. J. Food Sci. 62, 105–112.

Carpenter JF, Arakawa T, Crowe JH (1991) Interactions of stabilizing additives with proteins during freeze-thawing and freeze-drying. Develop. Biol. Standard. 74, 225–239.

Carpenter JF, Crowe JH (1989) An infrared spectroscopic study of the interactions of carbohydrates with dried proteins. Biochemistry 28, 3916–3922.

Carpenter JF, Crowe LM, Crowe JH (1987) Stabilization of phosphofructokinase with sugars during freeze-drying: characterization of enhanced protection in the presence of divalent cations. Biochim. Biophys. Acta 923, 109–115.

Carpenter JF, Crowe JH, Arakawa T (1990) Comparison of solute-induced protein stabilization in aqueous solution and in frozen and dried state. J. Dairy Sci. 73, 3627–3636.

Carpenter JF, Hand SC, Crowe LM, Crowe JH (1986) Cryoprotection of phosphofructokinase with organic solutes: characterization of enhanced protection in the presence of divalent cations. Arch. Biochem. Biophys. 250, 505–512.

Carpenter JF, Pikal MJ, Chang BS, Randolph TW (1997) Rational design of stable lyophilized protein formulations: some practical advice. Pharm. Res. 14, 969–975.

Carpenter JF, Prestrelski SJ, Arakawa T (1993) Separation of freezing and drying-induced denaturation of lyophilized proteins using stress-specific stabilization. I. Enzymatic activity and calorimetric studies. Arch. Biochem. Biophys. 303, 456–464.

Carpenter JF, Prestrelski SJ, Dong A (1998) Application of infrared spectroscopy to development of stable lyophilized protein formulations. Eur. J. Pharm. Biopharm. 45, 231–238.

Chan H-K, Au-Yeung K-L, Gonda I (1996) Effects of additives on heat denaturation of rhDNase in solutions. Pharm. Res. 13, 756–761.

Chan HK, Au-Yeung KL, Gonda I (1999) Development of a mathematical model for the water distribution in freeze-dried solids. Pharm. Res. 16, 660–665.

Chang BS, Beauvais RM, Arakawa T, Narhi LO, Dong A, Aparisio DI, Carpenter JF (1996a) Formation of an active dimer during storage of interleukin-1 receptor antagonist in aqueous solution. Biophys. J. 71, 3399–3406.

Chang BS, Beauvais RM, Dong A, Carpenter JF (1996b) Physical factors affecting the storage stability of freeze-dried interleukin-1 receptor antagonist: glass transition and protein conformation. Arch. Biochem. Biophys. 331, 249–258.

Chang BS, Reeder G, Carpenter JF (1996d) Development of a stable freeze-dried formulation of recombinant human interleukin-1 receptor antagonist. Pharm. Res. 13, 243–249.

Chang BS, Randall CS, Lee YS (1993) Stabilization of lyophilized porcine pancreatic elastase. Pharm. Res. 10, 1478–1483.

Charm SE, Wong BL (1970) Shear degradation of fibrinogen in the circulation. Science 170, 466–468.

Charman SA, Mason ML, Charman WN (1993) Techniques for assessing the effects of pharmaceutical excipients on the aggregation of porcine growth hormone. Pharm. Res. 10, 954–962.

Chen B-L, Arakawa T, Hsu E, Narhi LO, Tressel TJ, Chien SL (1994) Strategies to suppress aggregation of recombinant keratinocyte growth factor during liquid formulation development. J. Pharm. Sci. 83, 1657–1661.

Chen B, Costantino HR, Liu J, Hsu CC, Shire SJ (1999) Influence of calcium ions on the structure and stability of recombinant human deoxyribonuclease I in the aqueous and lyophilized states. J. Pharm. Sci. 88, 477–482.

Cholewinski M, Luckel B, Horn H (1996) Degradation pathways, analytical characterization and formulation strategies of a peptide and a protein. Calcitonin and human growth hormone in comparison. Pharm. Acta Helv. 71(6), 405–419.

Cleland JL, Powell MF, Shire SJ (1993) The development of stable protein formulations: a close look at protein aggregation, deamidation, and oxidation. Crit. Rev. Ther. Drug Carrier Syst. 10, 307–377.

Costantino HR, Andya JD, Nguyen P-A, Dasovich N, Sweeney TD, Shire SJ, Hsu, CC, Maa, Y-F (1998) Effect of mannitol crystallization on the stability and aerosol performance of a spray-dried pharmaceutical protein, recombinant humanized anti-IgE monoclonal antibody. J. Pharm. Sci. 87, 1406–1411.

Costantino HR, Griebenow K, Mishra P, Langer R, Klibanov AM (1995a) Fourier transform infrared spectroscopic investigation of protein stability in the lyophilized form. Biochim. Biophy. Acta 1253, 69–74.

Costantino HR, Langer R, Klibanov AM (1994) Moisture-induced aggregation of lyophilized insulin. Pharm. Res. 11, 21–29.

Costantino HR, Langer R, Klibanov AM (1995b) Aggregation of a lyophilized pharmaceutical protein, recombinant human albumin: effect of moisture and stabilization by excipients. Biotechnology 13, 493–496.

Cowan DA (1997) Thermophilic proteins: stability and function in aqueous and organic solvents. Comp. Biochem. Physiol. A Physiol. 118(3), 429–438.

Cross RT, Schirch V (1991) Effect of amino acid sequence, buffers, and ionic strength on the rate and mechanism of deamidation of asparagine residues in small peptides. J. Biol. Chem. 266, 22549–22556.

Crowe JH, Crowe LM, Carpenter JF (1993a) Preserving dry biomaterials: the water replacement hypothesis, part 1. Biopharm. 6, 28–37.

Crowe JH, Crowe LM, Carpenter JF (1993b) Preserving dry biomaterials: the water replacement hypothesis, part 2. Biopharm. 6, 40–43.

Cunico RL, Gooding KM, Wehr J (1998) Basic HPLC and CE of Biomolecules. Bay Bioanalytical Laboratory, Richmond, CA.

Damodaran S, Song KB (1988) Kinetics of adsorption of proteins at interfaces: role of protein conformation in diffusional adsorption. Biochim. Biophys. Acta 954, 253–264.

Daniel RM, Dines M, Petach HH (1996) The denaturation and degradation of stable enzymes at high temperature. Biochem. J. 317, 1–11.

Daukas LA, Trappler EH (1998) Assessing the quality of lyophilized parenterals. Pharm. Cosmetic Quality 2, 21–25.

Davio SR, Kienle KM, Collins BE (1995) Interchain interactions in the chimeric protein toxin sCD4(178)-PE40: a differential scanning calorimetry (DSC) study. Pharm. Res. 12, 642–648.

Dawson PJ (1992) Effect of formulation and freeze-drying on the long-term stability of rDNA-derived cytokines. Dev. Biol. Stand. 74, 273–282.

Dill KA (1990) Dominant forces in protein folding. Biochemistry 29, 7133–7155.

Dill KA, Alonso DOV, Hutchinson K (1989) Thermal stabilities of globular proteins. Biochemistry 28, 5439–5449.

Dong A, Prestrelski SJ, Allison SD, Carpenter JF (1995) Infrared spectroscopic studies of lyophilization- and temperature-induced protein aggregation. J. Pharm. Sci. 84, 415–423.

Edelhoch H (1967) Spectroscopic determination of tryptophan and tyrosine in proteins. Biochem. 6, 1948–1955.

Ehnebom J, Pula S, Bjorquist P, Deinum J (1997) Comparison of chromogenic substrates for tissue plasminogen activator and the effects on the stability of plasminogen activator inhibitor type-1. Fibrinol. Proteol. 11, 207–293.

Ellis RJ, Dobson C, Hartl U (1998) Sequence does specify protein conformation. Trends Biochem. Sci. 23(12), 468.

Estrada-Mondaca S, Fournier D (1998) Stabilization of recombinant *Drosophila* acetylcholinesterase. Protein Expr. Purif. 12(2), 166–172.

Fabian H, Falber K, Gast K, Reinstadler D, Rogov VV, Naumann D, Zamyatkin DF, Filimonov VV (1999) Secondary structure and oligomerization behavior of equilibrium unfolding intermediates of the lambda cro repressor. Biochemistry 38(17), 5633–5642.

Farruggia B, Pico GA (1999) Thermodynamic features of the chemical and thermal denaturations of human serum albumin. Int. J. Biol. Macromol. 26(5), 317–323.

Fasman GD (1996) Circular Dichroism and the Conformational Analysis of Biomolecules. Kluwer Academic, Norwell, MA.

Fatouros A, Osterberg T, Mikaelsson M (1997a) Recombinant factor VIII SQ-influence of oxygen, metal ions, pH and ionic strength on its stability in aqueous solution. Int. J. Pharm. 155, 121–131.

Fatouros A, Osterberg T, Mikaelsson M (1997b) Recombinant factor VIII SQ-inactivation kinetics in aqueous solution and the influence of disaccharides and sugar alcohols. Pharm. Res. 14, 1679–1684.

Ford AW, Dawson PJ (1993) The effect of carbohydrate additives in the freeze-drying of alkaline phosphatase. J. Pharm. Pharmacol. 45, 86–93.

Fox KC (1995) Putting proteins under glass. Science 267, 1922–1923.

Franks F (1990) Freeze-drying: from empiricism to predictability. Cryoletters 11, 93–110.

Franks F (1991) Freeze-drying: from empiricism to predictability. The significance of glass transitions. Dev. Biol. Stand. 74, 9–19.

Franks F (1993) Solid aqueous solutions. Pure Appl. Chem. 65, 2527–2537.

Franks F (1994) Long-term stabilization of biologicals. Biotechnology 12, 253–256.

Franks F (1995) Protein destabilization at low temperature. Adv. Protein Chem. 46, 105–139.

Fransson J, Florin-Robertsson E, Axelsson K, Nyhlén C (1996) Oxidation of human insulin-like growth factor I in formulation studies: kinetics of methionine oxidation in aqueous solution in solid state. Pharm. Res. 13, 1252–1257.

Fransson J, Hagman A (1996) Oxidation of human insulin-like growth factor I in formulation studies. II. Effect of oxygen, visible light, and phosphate on methionine oxidation in aqueous solution and evaluation of possible mechanisms. Pharm. Res. 13, 1476–1481.

Frenz J, Wu SL, Hancock WJ (1989) Characterization of human growth hormone by capillary electrophoresis. J. Chromatogr. 480, 379–391.

Frye KJ, Royer CA (1997) The kinetic basis for the stabilization of staphylococcal nuclease by xylose. Protein Sci. 16, 789–793.

Gao J, Yin DH, Yao Y, Sun H, Qin Z, Schoneich C, Williams JD, Squier TC (1998) Loss of conformational stability in calmodulin upon methionine oxidation. Biophys. J. 74(3), 1115–1134.

Giancola C, De Sena C, Fessas D, Graziano G, Barone G (1997) DSC studies on bovine serum albumin denaturation: effects of ionic strength and SDS concentration. Int. J. Biol. Macromol. 20, 193–204.

Gietz U, Arvinte T, Alder R, Merkle HP (1999) Inhibition of succinimide formation in aqueous Zn-r Hirudin suspensions. Pharm. Res. 16, 1626–1632.

Gill SC, von Hippel PH (1989) Calculation of protein extinction coefficients from amino acid sequence data. Anal. Biochem. 182(2), 319–326.

Goedken ER, Marqusee S (1998) Folding the ribonuclease H domain of Moloney murine leukemia virus reverse transcriptase requires metal binding or a short N-terminal extension. Proteins Struct. Funct. Genet. 33, 135–143.

Gombotz WR, Pankey SC, Bouchard LS, Phan DH, MacKenzie AP (1996) Stability, characterization, formulation and delivery system development for transforming growth factor-beta. In: Formulation, Characterization, and Stability of Protein Drugs. Pearlman R, Wang YJ (eds). Plenum Press, New York, pp. 219–245.

Gonzalez M, Murature DA, Fidelio GD (1995) Thermal stability of human immunoglobulins with sorbitol. Vox Sang 68, 1–4.

Graziano G, Catanzano F, Riccio A, Barone G (1997) A reassessment of the molecular origin of cold denaturation. J. Biochem. 122, 395–401.

Gu LC, Erdos EA, Chiang H-S, Calderwood T, Tsai K, Visor GC, Duffy J, Hsu W-C, Foster LC (1991) Stability of interleukin 1β(IL-1β) in aqueous solution: analytical methods, kinetics, products, and solution formulation implications. Pharm. Res. 8, 485–490.

Guerra P, Acklin C, Kosky AA, Davis JM, Treuheit MJ, Brems DN (1998) PEGylation prevents the N-terminal degradation of megakaryocyte growth and development factor. Pharm. Res. 15, 1822–1827.

Hancock GE, Smith JD, Heers KM (2000) The immunogenicity of subunit vaccines for respiratory syncytial virus after co-formulation with aluminum hydroxide adjuvant and recombinant interleukin-12. Viral Immunol. 13(1), 57–72.

Hatley RHM, Franks F (1991) Applications of DSC in the development of improved freeze-drying processes for labile biologicals. J. Therm. Anal. 37, 1905–1914.

Hatley RHM, Franks F, Brown S, Sandhu G, Gray M (1996) Stabilization of a pharmaceutical drug substance by freeze-drying: a case study. Drug Stability 1, 2–12.

Heller MC, Carpenter JF, Randolph TW (1997) Manipulation of lyophilization-induced phase separation: implications for pharmaceutical proteins. Biotechnol. Prog. 13, 590–596.

Heller MC, Carpenter JF, Randolph TW (1999) Application of a thermodynamic model to the prediction of phase separations in freeze-concentrated formulations for protein lyophilization. Arch. Biochem. Biophys. 363, 191–201.

Hendrick JP, Hartl F-U (1995) The role of molecular chaperones in protein folding. FASEB J. 9, 1559–1569.

Her, L-M, Nail SL (1994) Measurement of glass transition temperatures of freeze-concentrated solutes by differential scanning calorimetry. Pharm. Res. 11, 54–59.

Herman AC, Boone TC, Lu HS (1996) Characterization, formulation, and stability of Neupogen (Filgrastim), a recombinant human granulocyte-colony stimulating factor. In: Formulation, Characterization, and Stability of Protein Drugs. Pearlman R, Wang YJ (eds). Plenum Press, New York, pp. 303–328.

Hinds K, Koh JJ, Joss L, Liu F, Baudys M, Kim SW (2000) Synthesis and characterization of poly(ethylene glycol)-insulin conjugates. Bioconj. Chem. 11(2), 195–201.

Honig B, Yang A-S (1995) Free energy balance in protein folding. Adv. Protein Chem. 46, 27–58.

Hora MS, Rana RK, Smith FW (1992a) Lyophilized formulations of recombinant tumor necrosis factor. Pharm. Res. 9, 33–36.

Hora MS, Rana RK, Wilcox CL, Katre NV, Hirtzer P, Wolfe SN, Thomson JW (1992b) Development of a lyophilized formulation of interleukin-2. Dev. Biol. Stand. 74, 295–303.

Hsu CC, Ward CA, Pearlman R, Nguyen HM, Yeung DA, Curley JG (1991) Determining the optimum residual moisture in lyophilized protein pharmaceuticals. Dev. Biol. Stand. 74, 255–271.

Huber A, Demartis S, Aoyagi M (1999) The use of biosensor technology for the engineering of antibodies and enzymes. J. Mol. Recog. 12, 198–216.

Hughes DE, Richberg P (1993) Capillary micellar electrokinetic, sequential multi-wavelength chromatographic characterization of a chimeric monoclonal antibody-cytotoxin conjugate. J. Chromatogr. 635, 313–318.

Imoto T (1997) Stabilization of protein. Cell Mol. Life Sci. 53(3), 215–223.

Ip AY, Arakawa T, Silvers H, Ransone CM, Niven RW (1995) Stability of recombinant consensus interferon to air-jet and ultrasonic nebulization. J. Pharm. Sci. 84, 1210–1214.

Izutsu K, Yoshioka S, Kojima S (1995) Increased stabilizing effects of amphiphilic excipients on freeze-drying of lactate dehydrogenase (LDH) by dispersion into sugar matrices. Pharm. Res. 12, 838–843.

Izutsu K, Yoshioka S, Kojima S (1994a) Physical stability and protein stability of freeze-dried cakes during storage at elevated temperatures. Pharm. Res. 11, 995–999.

Izutsu K, Yoshioka S, Terao T (1993) Decreased protein-stabilizing effects of cryoprotectants due to crystallization. Pharm. Res. 10, 1232–1237.

Izutsu K, Yoshioka S, Terao T (1994b) Effect of mannitol crystallinity on the stabilization of enzymes during freeze-drying. Chem. Pharm. Bull. (Tokyo) 42, 5–8.

Jacob M, Schmid FX (1999) Protein folding as a diffusional process. Biochemistry 38(42), 13773–13779.

Jaenicke R (1990) Protein structure and function at low temperatures. Phil. Trans. R. Soc. Lond. B Biol. Sci. 326, 535–551.

Jaenicke R (1996) Glyceraldehyde-3-phosphate dehydrogenase from *Thermotoga maritima*: strategies of protein stabilization. FEMS Microbiol. Rev. 18, 215–224.

Jennings TA (1999) Lyophilization: Introduction and Basic Principles. Interpharm Press, Englewood, Colorado.

Jensen WA, Armstrong JM, De Giorgio J, Hearn MT (1996) Stability studies on pig heart mitochondrial malate dehydrogenase: the effect of salts and amino acids. Biochim. Biophys. Acta 1296, 23–34.

Johnston TP (1996) Adsorption of recombinant human granulocyte colony stimulating factor (rhG-CSF) to polyvinyl chloride, polypropylene, and glass: effect of solvent additives. PDA J. Pharm. Sci. Technol. 50, 238–245.

Kanaoka E, Nagata S, Hirano K (1999) Stabilization of aerosolized IFN-gamma by liposomes. Int. J. Pharm. 188(2), 165–172.

Kanazawa H, Fujimoto S, Ohara A (1994) Effect of radical scavengers on the inactivation of papain by ascorbic acid in the presence of cupric ions. Biol. Pharm. Bull. 17, 476–481.

Kasraian K, Spitznagel TM, Juneau JA, Yim K (1998) Characterization of the sucrose/glycine/water system by differential scanning calorimetry and freeze-drying microscopy. Pharm. Dev. Technol. 3, 233–239.

Katakam M, Banga AK (1995) Aggregation of insulin and its prevention by carbohydrate excipients. PDA J. Pharm. Sci. Technol. 49, 160–165.

Katakam M, Banga AK (1997) Use of poloxamer polymers to stabilize recombinant human growth hormone against various processing stresses. Pharm. Dev. Technol. 2, 143–149.

Katakam M, Bell LN, Banga AK (1995) Effect of surfactants on the physical stability of recombinant human growth hormone. J. Pharm. Sci. 84, 713–716.

Katakam M, Robillard P, Brown J, Tolman G (1998) Effects of moisture and lyophilization cycle on aggregation of a monoclonal antibody product. Pharm. Sci. (Suppl.) 1, S-543.

Keck RG (1996) The use of t-butyl hydroperoxide as a probe for methionine oxidation in proteins. Anal. Biochem. 236, 56–62.

Kendrick BS, Chang BS, Arakawa T, Peterson B, Randolph TW, Manning MC, Carpenter JF (1997) Preferential exclusion of sucrose from recombinant interleukin-1 receptor antagonist: role in restricted conformational mobility and compaction of native state. Proc. Natl. Acad. Sci. USA 94, 11917–11922.

Kendrick BS, Carpenter JF, Cleland JL, Randolph TW (1998) A transient expansion of the native state precedes aggregation of recombinant human interferon-gamma. Proc. Natl. Acad. Sci. USA 95, 14142–14146.

Kendrick BS, Chang BY, Carpenter JF (1995) Detergent stabilization of proteins against surface and freezing denaturation. Pharm. Res. 12 (Suppl.), S-85.

Kerwin BA, Alkers MJ, Apostol I, Moore-Einsel C, Etter JE, Hess E, Lippincott J, Levine J, Mathews AJ, Revilla-Sharp P, Schubert R, Looker DL (1999) Acute and long-term stability studies of deoxy hemoglobin and characterization of ascorbate-induced modifications. J. Pharm. Sci. 88, 79–88.

Kerwin BA, Heller MC, Levin SH, Randolph TW (1998) Effects of Tween 80 and sucrose on acute short-term stability and long-term storage at −20°C of a recombinant hemoglobin. J. Pharm. Sci. 87, 1062–1068.

Kim AI, Akers MJ, Nail SL (1998) The physical state of mannitol after freeze-drying: effect of mannitol concentration, freezing rate, and a noncrystallizing cosolute. J. Pharm. Sci. 87, 931–935.

Kita Y, Arakawa T, Lin T-Y, Timasheff SN (1994) Contribution of the surface energy perturbation to protein-solvent interactions. Biochemistry 33, 15178–15189.

Knepp VM, Whatley JL, Muchnik A, Calderwood TS (1996) Identification of antioxidants for prevention of peroxide-mediated oxidation of recombinant human ciliary neurotrophic factor and recombinant human nerve growth factor. J. Pharm. Sci. Technol. 50, 163–171.

Kosa T, Maruyama T, Otagiri M (1998) Species differences of serum albumins. II. Chemical and thermal stability. Pharm. Res. 15, 449–454.

Koseki T, Kitabatake N, Doi E (1990) Freezing denaturation of ovalbumin at acid pH. J. Biochem. 107, 389–394.

Kreilgaard L, Frokjaer S, Flink, JM, Randolph TW, Carpenter JF (1998a) Effects of additives on the stability of recombinant human factor XIII during freeze-drying and storage in the dried solid. Arch. Biochem. Biophys. 360, 121–134.

Kreilgaard L, Frokjaer S, Flink JM, Randolph TW, Carpenter JF (1999) Effects of additives on the stability of *Humicola lanuginosa* lipase during freeze-drying and storage in the dried solid. J. Pharm. Sci. 88, 281–290.

Kreilgaard L, Jones LS, Randolph TW, Frokjaer S, Flink JM, Manning MC, Carpenter JF (1998b) Effect of Tween 20 on freeze-thawing- and agitation-induced aggregation of recombinant human factor XIII. J. Pharm. Sci. 87, 1597–1603.

Kristjansson MM, Kinsella JE (1991) Protein and enzyme stability: structural, thermodynamic, and experimental aspects. Adv. Food Nutr. Res. 35, 237–316.

Kulmyrzaev A, Bryant C, McClements DJ (2000) Influence of sucrose on the thermal denaturation, gelation, and emulsion stabilization of whey proteins. J. Agric. Food Chem. 48(5), 1593–1597.

Kurochkin IV, Procyk R, Bishop PD, Yee VC, Teller DC, Ingham KC, Medved LV (1995) Domain structure, stability and domain-domain interactions in recombinant factor XIII. J. Mol. Biol. 248, 414–430.

Lam XM, Costantino HR, Overcashier DE, Nguyen TH, Hsu CC (1996) Replacing succinate with glycocholate buffer improves the stability of lyophilized interferon-gamma. Int. J. Pharm. 142, 85–95.

Landers JP (ed.) (1997) Handbook of Capillary Electrophoresis. CRC Press, Boca Raton, FL.

Landers JP, Oda RP, Liebnow JA, Spelsberg TC (1993) Utility of high resolution capillary electrophoresis for monitoring peptide homo- and hetero-dimer formation. J. Chromatogr. 652(1), 109–117.

Li S, Patapoff TW, Nguyen TH, Borchardt RT (1996) Inhibitory effect of sugars and polyols on the metal-catalyzed oxidation of human relaxin. J. Pharm. Sci. 85, 868–872.

Li, S., Nguyen TH, Schoneich C, Borchardt RT, (1995a) Aggregation and precipitation of human relaxin induced by metal-catalyzed oxidation. Biochemistry 34, 5762–5772.

Li S, Schoneich C, Borchardt RT (1995b) Chemical instability of proteins. Pharm. News 2, 12–16.

Li S, Schoneich C, Borchardt RT (1995c) Chemical instability of protein pharmaceuticals: mechanisms of oxidation and strategies for stabilization. Biotechnol. Bioeng. 48, 491–500.

Li S, Schoneich C, Wilson GS, Borchardt RT (1993) Chemical pathways of peptide degradation. V. Ascorbic acid promotes rather than inhibits the oxidation of methionine to methionine sulfoxide in small model peptides. Pharm. Res. 10, 1572–1579.

Lin T-Y, Timasheff SN (1996) On the role of surface tension in the stabilization of globular proteins. Protein Sci. 5, 372–381.

Liu JL, Lu KV, Eris T, Katta V, Westcott KR, Narhi LO, Lu HS (1998). *In vitro* methionine oxidation of recombinant human leptin. Pharm. Res. 16, 632–640.

Liu R, Orgel L (1997) Efficient oligomerization of negatively charged beta-amino acids at −20°C. J. Am. Chem. Soc. 119, 4791–4792.

Liu Y, Bolen DW (1995) The peptide backbone plays a dominant role in protein stabilization by naturally occurring osmolytes. Biochemistry 34, 12884–12891.

Liu Y, Sturtevant JM (1996) The observed change in heat capacity accompanying the thermal unfolding of proteins depends on the composition of the solution and on the method employed to change the temperature of unfolding. Biochemistry 35, 3059–3062.

Liu WR, Langer R, Klibanov AM (1990) Moisture-induced aggregation of lyophilized proteins in the solid state. Biotechnol. Bioeng. 37, 177–184.

Lo YL, Rahman YE (1998) Effect of lipids on the thermal stability and conformational changes of proteins: ribonuclease A and cytochrome c. Int. J. Pharm. 161, 137–148.

Lueckel B, Bodmer D, Helk B, Leuenberger H (1998a) Formulations of sugars with amino acids or mannitol—influence of concentration ratio on the properties of the freeze-concentrate and the lyophilizate. Pharm. Dev. Technol. 3, 325–336.

Lueckel B, Helk B, Bodmer D, Leuenberger H (1998b) Effects of formulation and process variables on the aggregation of freeze-dried interleukin-6 (IL-6) after lyophilization and on storage. Pharm. Dev. Technol. 3, 337–346.

Lundblad RL, Bradshaw RA (1997) Applications of site-specific chemical modification in the manufacture of biopharmaceuticals. I. An overview. Biotechnol. Appl. Biochem. 26 (Pt 3), 143–151.

Maa Y-F, Hsu CC (1997) Protein denaturation by combined effect of shear and air-liquid interface. Biotechnol. Bioeng. 54, 503–512.

Mach H, Middaugh CR, Lewis RV (1992) Statistical determination of the average values of the extinction coefficients of tryptophan and tyrosine in native proteins. Anal. Biochem. 200(1), 74–80.

Manning MC, Matsuura JE, Kendrick BS, Meyer JD, Dormish JJ, Vrkljian M, Ruth JR, Carpenter JF, Shefter E (1995) Approaches for increasing the solution stability of proteins. Biotechnol. Bioeng. 48, 506–512.

Manning MC, Patel K, Borchardt RT (1989) Stability of protein pharmaceuticals. Pharm. Res. 6, 903–917.

Matsumura M, Signor G, Matthews BW (1989) Substantial increase of protein stability by multiple disulphide bonds. Nature 342, 291–293.

Mattern M, Winter G, Kohnert U, Lee G (1999) Formulation of proteins in vacuum-dried glasses. II. Process and storage stability in sugar-free amino acid systems. Pharm. Dev. Technol. 4, 199–208.

Millqvist-Fureby A, Malmsten M, Bergenstahl B (1999) Surface characterisation of freeze-dried protein/carbohydrate mixtures. Int. J. Pharm. 191(2), 103–114.

Mozhaev VV, Martinek K (1984) Structure-stability relationships in proteins: new approaches to stabilizing enzymes. Enzyme Microb. Technol. 6, 50–59.

Mazzobre MF, Buera MP, Chirife J (1997) Glass transition and thermal stability of lactase in low-moisture amorphous polymeric matrices. Biotechnol. Prog. 13, 195–199.

McIntosh KA, Charman WN, Charman SA (1998) The application of capillary electrophoresis for monitoring effects of excipients on protein conformation. J. Pharm. Biomed. Anal. 16, 1097–1105.

Nabuchi Y, Fujiwara E, Ueno K, Kuboniwa H, Asoh Y, Ushio H (1995) Oxidation of recombinant human parathyroid hormone: effect of oxidized position on the biological activity. Pharm. Res. 12, 2049–2052.

Nagendra HG, Sukumar N, Vijayan M (1998) Role of water in plasticity, stability, and action of proteins: The crystal structures of lysozyme at very low levels of hydration. Proteins: Struct. Funct. Genet. 32, 229–240.

Nakai Y, Yoshioka S, Aso Y, Kojima S (1998) Solid-state rehydration-induced recovery of bilirubin oxidase activity in lyophilized formulations reduced during freeze-drying. Chem. Pharm. Bull. 46, 1031–1033.

Narhi LO, Philo JS, Sun B, Chang BS, Arakawa T (1999) Reversibility of heat-induced denaturation of the recombinant human megakaryocyte growth and development factor. Pharm. Res. 16, 799–807.

Nema S, Avis KE (1992) Freeze-thaw studies of a model protein, lactate dehydrogenase, in the presence of cryoprotectants. J. Parenter. Sci. Technol. 4, 76–83.

Nguyen TH, Burnier J, Meng W (1993) The kinetics of relaxin oxidation by hydrogen peroxide. Pharm. Res. 10, 1563–1571.

Oksanen CA, Zografi G (1990) The relationship between the glass transition temperature and water vapor absorption by poly(vinylpyrrolidone). Pharm. Res. 7, 654–657.

Ooi T (1994) Thermodynamics of protein folding: effects of hydration and electrostatic interactions. Adv. Biophys. 30, 105–154.

Overcashier DE, Brooks DA, Costantino HR, Hsu CC (1997) Preparation of excipient-free recombinant human tissue-type plasminogen activator by lyophilization from ammonium bicarbonate solution: an investigation of the two-stage sublimation phenomenon. J. Pharm. Sci. 86, 455–459.

Pace CN, Shirley BA, McNutt M, Gajiwala K (1996) Forces contributing to the conformational stability of proteins. FASEB J. 10, 75–83.

Pace CN, Vajdos F, Fee L, Grimsley G, Gray T (1995) How to measure and predict the molar absorption coefficient of a protein. Prot. Sci. 4(11), 2411–2423.

Page C, Dawson P, Woollacott D, Thorpe R, Mire-Sluis A (2000) Development of a lyophilization formulation that preserves the biological activity of the platelet-inducing cytokine interleukin-11 at low concentrations. J. Pharm. Pharmacol. 52(1), 19–26.

Paleg LG, Stewart GR, Bradbeer JW (1984) Proline and glycine betaine influence protein solvation. Plant Physiol. 75, 974–978.

Paranandi MV, Guzzetta AW, Hancock WS, Aswad DW (1994) Deamidation and isoaspartate formation during *in vitro* aging of recombinant tissue plasminogen activator. J. Biol. Chem. 269, 243–253.

Patel K, Borchardt RT (1990) Chemical pathways of peptide degradation. II: Kinetics of deamidation of an asparaginyl residue in a model hexapeptide. Pharm. Res. 7(7), 703–711.

Pelton JT, McLean LR (2000) Spectroscopic methods for analysis of protein secondary structure. Anal. Biochem. 277(2), 167–176.

Physicians' Desk Reference 53rd ed. (1999) Medical Economics Company, Inc. Montvale, NJ.

Pikal MJ (1990a) Freeze-drying of proteins. I: Process design. Biopharm. 3(8), 18–27.

Pikal MJ (1990b) Freeze-drying of proteins. II: Formulation selection. Biopharm. 3(9), 26–30.

Pikal MJ, Dellerman KM, Roy ML, Riggin RM (1991) The effects of formulation variables on the stability of freeze-dried human growth hormone. Pharm. Res. 8, 427–436.

Polanski J, Shintani H (1996) Handbook of Capillary Electrophoresis Applications. Kluwer Academic, Norwell, MA.

Powell MF (1994) Peptide stability in aqueous parenteral formulations. In: Formulation and Delivery of Proteins and Peptides. Cleland JL and Langer R (eds.). American Chemical Society, Washington DC, pp. 101–117.

Powell MF (1996) A compendium and hydropathy/flexibility analysis of common reactive sites in proteins: reactivity at Asn, Asp, Gln, and Met motifs in neutral pH solution. In: Formulation, Characterization, and Stability of Protein Drugs. Pearlman R, Wang YJ (eds). Plenum Press, New York, pp. 1–140.

Prehoda KE, Mooberry ES, Markley JL (1998) Pressure denaturation of proteins: evaluation of compressibility effects. Biochemistry 37, 5785–5790.

Prestrelski SJ, Arakawa T, Carpenter JF (1993) Separation of freezing and drying-induced denaturation of lyophilized proteins using stress-specific stabilization. II. Structural studies using infrared spectroscopy. Arch. Biochem. Biophys. 303, 465–473.

Prestrelski SJ, Pikal KA, Arakawa T (1995) Optimization of lyophilization conditions for recombinant human interleukin-2 by dried-state conformational analysis using Fourier-transform infrared spectroscopy. Pharm. Res. 12, 1250–1259.

Qi H, Heller DL (1995) Stability and stabilization of insulinotropin in a dextran formulation. PDA J. Pharm. Sci. Technol. 49(6), 289–293.

Quan CP, Wu S, Dasovich N, Hus C, Patapoff T, Canova-Davis E (1999) Susceptibility of rhDNase I to glycation in the dry-powder state. Anal. Chem. 71, 4445–4454.

Querol E, Pervez-Pons JA, Mozo-Villarias A (1996) Analysis of protein conformational characteristics related to thermostability. Protein Eng. 9, 265–271.

Ramos A, Raven NDH, Sharp RJ, Bartolucci S, Roswsi M, Cannio R, Lebbink J, vanderOost J, deVos WM, Santos H (1997) Stabilization of enzymes against thermal stress and freeze-drying by mannosylglycerate. Appl. Environ. Microbiol. 63, 4020–4025.

Remmele Jr, RL, Gombotz WR (2000) Differential scanning calorimetry, a practical tool for elucidating stability of liquid pharmaceuticals. BioPharm 13, 36–46.

Remmele Jr, RL, Nightlinger NS, Srinivasan S, Gombotz WR (1998) Interleukin-1 receptor (IL-1R) liquid formulation development using differential scanning calorimetry. Pharm. Res. 15, 200–208.

Ressing ME, Jiskoot W, Talsma H, van Ingen CW, Beuvery EC, Crommelin DJ (1992) The influence of sucrose, dextran, and hydroxypropyl-beta-cyclodextrin as lyoprotectants for a freeze-dried mouse IgG2a monoclonal antibody (MN12). Pharm. Res. 9, 266–270.

Reubsaet JLF, Beijnen JH, Bult A, van Maanen RJ, Marchel JAD, Underberg, WJM (1998) Analytical techniques used to study the degradation of proteins and peptides chemical instability. J. Pharm. Biomed. Anal. 17, 955–978.

Richards FM (1997) Protein stability: still an unresolved problem. Cell. Mol. Life Sci. 53, 790–802.

Rishi V, Anjum F, Ahmad F, Pfeil W (1998) Role of non-compatible osmolytes in the stabilization of proteins during heat stress. Biochem. J. 329 (Pt 1), 137–143.

Roser B (1991) Trehalose drying: a novel replacement for freeze-drying. Biopharm. 4(9), 47–53.

Rossi S, Buera MP, Moreno S, Chirife J (1997) Stabilization of the restriction enzyme EcoRI dried with trehalose and other selected glass-forming solutes. Biotechnol. Prog. 13, 609–616.

Roos Y (1993) Melting and glass transitions of low molecular weight carbohydrates. Carbohydr. Res. 238, 39–48.

Roos Y, Karel M (1991) Phase transitions of mixtures of amorphous polysaccharides and sugars. Biotechnol. Prog. 7, 49–53.

Runkel L, Meier W, Pepinsky RB, Karpusas M, Whitty A, Kimball K, Brickelmaier M, Muldowney C, Jones W, Goelz SE (1998) Structural and functional differences between glycosylated and non-glycosylated forms of human interferon-beta (IFN-β). Pharm. Res. 15, 641–649.

Rupley JA, Careri G (1991) Protein hydration and function. Adv. Protein Chem. 41, 37–173.

Sabulal B, Kishore N (1997) Amino acids and short peptides do not always stabilize globular proteins: a differential scanning calorimetric study on their interactions with bovine α-lactalbumin. J. Chem. Soc., Faraday Trans. 93, 433–436.

Sadhale Y, Shah JC (1999) Stabilization of insulin against agitation-induced aggregation by the GMO cubic phase gel. Int. J. Pharm. 191(1), 51–64.

Sah H (1999) Stabilization of proteins against methylene chloride/water interface-induced denaturation and aggregation. J. Controlled Release 58(2), 143–151.

Sampedro JG, Guerra G, Pardo JP, Uribe S (1998) Trehalose-mediated protection of the plasma membrane H^+-ATPase from *Kluyveromyces lactis* during freeze-drying and rehydration. Cryobiology 37, 131–138.

Sarciaux J-M, Mansour S, Hageman MJ, Nail SL (1998) Effect of species, processing conditions and phosphate buffer composition on IgG aggregation during lyophilization. Pharm. Sci. (Suppl.) 1, S-545.

Sarciaux J-M, Mansour S, Hageman MJ, Nail SL (1999) Effects of buffer composition and processing conditions on aggregation of bovine IgG during freeze-drying. J. Pharm. Sci. 88, 1345–1361.

Sasaoki K, Hiroshima T, Kusumoto S, Nishi K (1992) Deamidation at asparagine-88 in recombinant human interleukin 2. Chem. Pharm. Bull. 40, 976–980.

Schebor C, Buera MP, Chirife J (1996) Glassy state in relation to the thermal inactivation of enzyme invertase in amorphous dried matrices of trehalose, maltodextrin and PVP. J. Food Eng. 30, 269–282.

Schebor C, Burin L, Buera MP, Aguilera JM, Chirife J (1997) Glassy state and thermal inactivation of invertase and lactase in dried amorphous matrices. Biotechnol. Prog. 13, 857–863.

Schrier JA, Kenley RA, Williams R, Corcoran RJ, Kim Y, Northey Jr RP, D'Augusta D, Huberty M (1993) Degradation pathways for recombinant human macrophage colony-stimulating factor in aqueous solution. Pharm. Res. 10, 933–944.

Schwendeman SP, Costantino HR, Gupta RK, Siber GR, Klibanov AM, Langer R (1995) Stabilization of tetanus and diphtheria toxoids against moisture-induced aggregation. Proc. Natl. Acad. Sci. USA 92, 11234–11238.

Separovic F, Lam YH, Ke X, Chan HK (1998) A solid-state NMR study of protein hydration and stability. Pharm. Res. 15, 1816–1821.

Shahrokh Z, Eberlein G, Buckley D, Paranandi MV, Aswad DW, Strantton P, Mischak R, Wang YJ (1994) Major degradation products of basic fibroblast growth factor: detection of succinimide and iso-aspartate in place of aspartate. Pharm. Res. 11, 936–944.

Shalaev EY, Zografi G (1996) How does residual water affect the solid-state degradation of drugs in the amorphous state? J. Pharm. Sci. 85, 1137–1141.

Shikama K, Yamazaki I (1961) Denaturation of catalase by freezing and thawing. Nature 190, 83–84.

Skrabanja AT, de Meere AL, de Ruiter RA, van den Oetelaar PJ (1994) Lyophilization of biotechnology products. PDA J. Pharm. Sci. Technol. 48, 311–317.

Shuck P (1997) Use of surface resonance to probe the equilibrium and dynamic aspects of interactions between biological macromolecules. Annu. Rev. Biomol. Struct. 26, 541–566.

Slade L, Levine H, Finley JW (1989) Protein-water interactions: water as a plasticizer of gluten and other protein polymers. In: Protein Quality and the Effects of Processing. Philips RD, Finley JW (eds). Marcel Dekker, New York, pp. 9–124.

Smith PK, Krohn RI, Hermansson, Mallia AK, Gartner F (1985) Measurement of protein using bicinchoninic acid. Anal. Biochem. 150, 76–85.

Snyder LR, Glajch JL, Kirkland J (1997) Practical HPLC Method Development, 2nd ed. John Wiley and Sons, New York.

Son K, Kwon C (1995) Stabilization of human epidermal growth factor (hEGF) in aqueous formulation. Pharm. Res. 12, 451–454.

Spector S, Wang M, Carp SA, Robblee J, Hendsch ZS, Fairman R, Tidor B, Raleigh DP (2000) Rational modification of protein stability by the mutation of charged surface residues. Biochemistry 39(5), 872–879.

Stevenson CL, Freidman AR, Kubiak TM, Donlan ME, Borchardt RT (1993) Effect of secondary structure on the rate of deamidation of several growth hormone releasing factor analogs. Int. J. Pept. Protein Res. 2, 497–503.

Strambini GB, Gabellieri E. (1996) Proteins in frozen solutions: evidence of ice-induced partial unfolding. Biophys. J. 70, 971–976.

Strickley RG, Anderson BD (1996) Solid-state stability of human insulin I. Mechanism and the effect of water on the kinetics of degradation in lyophiles from pH 2-5 solutions. Pharm. Res. 13, 1142–1153.

Strickley RG, Anderson BD (1997) Solid-state stability of human insulin. 2. Effect of water on reactive intermediate partitioning in lyophiles from pH 2-5 solutions: stabilization against covalent dimer formation. J. Pharm. Sci. 86, 645–653.

Strop P, Mayo SL (2000) Contribution of surface salt bridges to protein stability. Biochemistry 39(6), 1251–1255.

Takashima I, Yamaguchi Y, Yorishima J, Ohta K, Manami K, Hihara J, Toge T (1999) Immunological properties of tumor cells genetically modified to secrete interleukin-2. Anticancer Res. 19(6B), 5299–5306.

Takeshita A, Saito H, Toyama K, Horiuchi A, Kuriya S, Furusawa S, Tsuruoka N, Takiguchi T, Matsuda T, Utsumi M, Shiku H, Matsui T, Egami K, Tamura K, Ohno R (2000) Efficacy of a new formulation of lenograstim (recombinant glycosylated human granulocyte colony-stimulating factor) containing gelatin for the treatment of neutropenia after consolidation chemotherapy in patients with acute myeloid leukemia. Int. J. Hematol. 71(2), 136–143.

Tanaka K, Takeda T, Miyajima K (1991) Cryoprotective effect of saccharides on denaturation of catalase by freeze-drying. Chem. Pharm. Bull. 39, 1091–1094.

Taneja S, Ahmad F (1994) Increased thermal stability of proteins in the presence of amino acids. Biochem. J. 303, 147–153.

te Booy MP, de Ruiter RA, de Meere AL (1992) Evaluation of the physical stability of freeze-dried sucrose-containing formulations by differential scanning calorimetry. Pharm. Res. 9, 109–114.

Timasheff SN (1993) The control of protein stability and association by weak interactions with water: how do solvents affect these processes? Ann. Rev. Biophys. Biomol. Struct. 22, 67–97.

Timasheff SN (1998) Control of protein stability and reactions by weakly interacting cosolvents: The simplicity of the complicated. Adv. Protein Chem. 51, 355–432.

Tokihiro K, Irie T, Uekama K (1997) Varying effects of cyclodextrin derivatives on aggregation and thermal behavior of insulin in aqueous solution. Chem. Pharm. Bull. 45, 525–531.

Townsend MW, DeLuca PP (1990) Stability of ribonuclease A in solution and the freeze-dried state. J. Pharm. Sci. 79, 1083–1086.

Tsai YR, Yang SJ, Jiang SS, Ko SJ, Hung SH, Kuo SY, Pan RL (1998) High-pressure effects on vacuolar H^+-ATPase from etiolated mung bean seedlings. J. Protein Chem. 17, 161–172.

Tsai PK, Volkin DB, Dabora JM, Thompson KC, Bruner MW, Gress JO, Matuszewska B, Keogan M, Bondi JV, Middaugh CR (1993) Formulation design of acidic fibroblast growth factor. Pharm. Res. 10, 649–659.

Tzannis ST, Hrushesky WJM, Wood PA, Przbycien TM (1997) Adsorption of a formulated protein on a drug delivery device surface. J. Colloid. Interface Sci. 189, 216–228.

Uversky VN, Abdullaev ZK, Arseniev AS, Bocharov EV, Dolgikh DA, Latypov RF, Melnik TN, Vassilenko KS, Kirpichnikov MP (1999) Structure and stability of recombinant protein depend on the extra N-terminal methionine residue: S6 permutein from direct and fusion expression systems. Biochim. Biophys. Acta 1432, 324–332.

Vemuri S, Yu CT, Roosdorp N (1993) Formulation and stability of recombinant alpha 1-antitrypsin. Pharm. Biotechnol. 5, 263–286.

Vemuri S, Yu CD, Roosdorp N (1994) Effect of cryoprotectants on freezing, lyophilization, and storage of lyophilized recombinant alpha-l-antitrypsin formulations. PDA J. Pharm. Sci. Technol. 48, 241–246.

Vermeer AW, Norde W (2000) The thermal stability of immunoglobulin: unfolding and aggregation of a multi-domain protein. Biophys. J. 78(1), 394–404.

Vielle C, Zeikus JG (1996) Thermoenzymes: identifying molecular determinants of protein structural and functional stability. Trends Biotechnol. 14, 183–191.

Vogt G, Argos P (1997) Protein thermal stability: hydrogen bonds or internal packing? Fold. Des. 2, S40–46.

Volkin DB, Burke CJ, Sanyal G, Middaugh CR (1996) Analysis of vaccine stability. Dev. Biol. Stand. 87, 135–142.

Volkin DB, Klibanov AM (1989) Minimizing protein inactivation. In: Protein Function: A Practical Approach. Creighton TE (ed.). Information Press, Oxford, pp. 1–24.

Volkin DB, Mach H, Middaugh CR (1997) Degradative covalent relations important to protein stability. Mol. Biotechnol. 8, 105–122.

Volkin DB, Middaugh CR (1996) The characterization, stabilization, and formulation of acidic fibroblast growth factor. In: Formulation, Characterization, and Stability of Protein Drugs. Pearlman R, Wang YJ (eds). Plenum Press, New York, pp. 181–217.

Volkin DB, Tsai PK, Dabora JM, Gress JO, Burke CJ, Linhardt RJ, Middaugh CR (1993) Physical stabilization of acidic fibroblast growth factor by polyanions. Arch. Biochem. Biophys. 300, 30–41.

Vrkljan M, Foster TM, Powers ME, Henkin J, Porter WR, Staack H, Carpenter JF, Manning MC (1994) Thermal stability of low molecular weight urokinase during heat treatment. 2. Effect of polymeric additives. Pharm. Res. 11, 1004–1008.

Wang A, Robertson AD, Bolen DW (1995) Effects of a naturally occurring compatible osmolyte on the internal dynamics of ribonuclease A. Biochemistry 34, 15069–15104.

Wang P-L, Johnston TP (1993) Enhanced stability of two model proteins in an agitated solution environment using poloxamer 407. J. Parent. Sci. Technol. 47, 183–189.

Wang W (1999) Instability, stabilization, and formulation of liquid protein pharmaceuticals. Int. J. Pharm. 185, 129–188.

Wang W (2000) Lyophilization and development of solid protein pharmaceuticals. Int. J. Pharm. 203, 1–60.

Wang, YJ, Shahrokh Z, Vermuri S, Eberlein G, Beylin I, Busch M (1996) Characterization, stability, and formulation of basic fibroblast growth factor. In: Formulation, Characterization, and Stability of Protein Drugs. Pearlman R, Wang YJ (eds). Plenum Press, New York, pp. 141–180.

Ward KR, Adams GD, Alpar HO, Irwin WJ (1999) Protection of the enzyme L-asparaginase during lyophilisation—a molecular modelling approach to predict required level of lyoprotectant. Int. J. Pharm. 187, 153–162.

Webb S, Randolph T, Carpenter J, Cleland J (1998) An investigation into reconstitution of lyophilized recombinant human interferon-gamma. Pharm. Sci. (Suppl.) 1, S-542.

Weston A, Brown PR (1997) HPLC and CE: Principles and Practice. Academic Press, San Diego.
Willemer H (1999) Experimental freeze-drying: procedures and equipment. In: Rey L, May JC (eds). Freeze-Drying/Lyophilization of Pharmaceutical and Biological Products, Vol. 96. Marcel Dekker, New York, pp. 79–121.
Wolkers WF, Oldenhof H, Alberda M, Hoekstra FA (1998) A Fourier transform infrared microspectroscopy study of sugar glasses: application to anhydrobiotic higher plant cells. Biochim. Biophys. Acta 1379, 83–96.
Won CM, Molnar TE, McKean RE, Spenlehauer GA (1998) Stabilizers against heat-induced aggregation of RPR 114849, an acidic fibroblast growth factor (AFGF). Int. J. Pharm. 167, 25–36.
Worn A, Pluckthun A (1998) An intrinsically stable antibody scfv fragment can tolerate the loss of both disulfide bonds and fold correctly. FEBS Lett. 427, 357–361.
Wu S, Leung D, Vemuri S, Shah D, Yang B, Wang J (1998) Degradation mechanism of lyophilized basic fibroblast growth factor (BFGF) protein in sugar formulation. Pharm. Sci. (Suppl.) 1, S-539.
Wyatt PJ (1997a) Multiangle light scattering combined with HPLC. LC-GC 15, 160–168.
Wyatt PJ (1997b) Multiangle light scattering: the basic tool for macromolecular characterization. Instrumentation Sci. Technol. 25, 1–18.
Yadav S, Ahmed F (2000) A new method for the determination of stability parameters of proteins from their heat-induced denaturation curves. Anal. Biochem. 283, 207–213.
Yancey PH, Clark EE, Hand SC, Bowlus RD, Somero GN (1982) Living with water stress: evolution of osmolyte systems. Science 217, 1214–1222.
Yoshioka S, Aso Y, Izutsu K, Kojima S (1994a) Is stability prediction possible for protein drugs? Denaturation kinetics of β-galactosidase in solution. Pharm. Res. 11, 1721–1725.
Yoshioka S, Aso Y, Izutsu K, Terao T (1993) Aggregates formed during storage of beta-galactosidase in solution and in the freeze-dried state. Pharm. Res. 10, 103–108.
Yoshioka S, Aso Y, Izutsu K-I, Terao T (1994b) Application of accelerated testing for shelf-life prediction of commercial protein preparations. J. Pharm. Sci. 83, 454–456.
Yoshioka S, Aso Y, Kojima S (1997) Dependence of the molecular mobility and protein stability of freeze-dried γ-globulin formulations on the molecular weight of dextran. Pharm. Res. 14, 736–741.
Yoshioka S, Aso Y, Kojima S (1999) The effect of excipients on the molecular mobility of lyophilized formulations, as measured by glass transition temperature and NMR relaxation–based critical mobility temperature. Pharm. Res. 16, 135–140.
Yoshioka S, Aso Y, Nakai Y, Kojima S (1998) Effect of high molecular mobility of poly(vinyl alcohol) on protein stability of lyophilized gamma-globulin formulations. J. Pharm. Sci. 87, 147–151.

Zhang J, Peng X, Jonas A, Jonas J (1995a) NMR study of the cold, heat, and pressure unfolding of ribonuclease A. Biochemistry 34, 8631–8641.

Zhang MZ, Pikal K, Nguyen T, Arakawa T, Prestrelski SJ (1996) The effect of the reconstitution medium on aggregation of lyophilized recombinant interleukin-2 and ribonuclease A. Pharm. Res. 13, 643–646.

Zhang MZ, Wen J, Arakawa T, Prestrelski SJ (1995b) A new strategy for enhancing the stability of lyophilized protein: the effect of the reconstitution medium on keratinocyte growth factor. Pharm. Res. 12, 1447–1452.

Zhao F, Ghezzo-Schoneich E, Aced GI, Hong J, Milby T, Schoneich C (1997) Metal-catalyzed oxidation of histidine in human growth hormone. J. Biol. Chem. 272, 9019–9029.

16

Regulatory Requirements, Cleaning, and Sanitization Issues in Protein Purification

Gail Sofer
Bioreliance Corporation, Rockville, Maryland, U.S.A.

1. INTRODUCTION

Protein products used as therapeutics or diagnostics for health care as well as proteins used for many research projects must be highly purified. Consistently achieving high levels of purity requires that attention be paid to sanitization and cleaning of purification equipment, chromatography media, and filters. For health care products, regulatory agencies issue guidelines, points to consider, guidances, and other documents that delineate the requirements. This chapter addresses some of the regulatory documents that apply to protein purification with an emphasis on cleaning and sanitization. It also presents current issues related to cleaning and sanitization of protein recovery and purification operations.

2. REGULATORY REQUIREMENTS THAT APPLY TO PROTEIN PURIFICATION

High purity of protein therapeutics and diagnostics enhances the quality of health care. Worldwide regulatory agencies oversee pharmaceutical and biopharmaceutical firms and institutions to ensure they meet their commitments to provide products meeting predetermined specifications for purity and safety. The regulations issued by the regulatory agencies often describe only broad concepts, and it is up to each manufacturer to decide on the specifics that will enable them to comply with the regulations. There are some regulations that are the law and, as such, are enforceable. There are also documents that are guidances, points to consider, concept letters, and the like, which are not enforceable by law. In addition, there are reports, letters, approval notices, warning letters, and other documents that provide insights into the latest regulatory thinking on specific issues related to protein purification. Regulatory documents are produced by most countries as well as international organizations such as the World Health Organization (WHO). The European Committee for Proprietary Medicinal Products (CPMP) also produces numerous documents, most of which can be accessed on the internet [1]. The documents that are discussed in this chapter include those from the U.S. Food and Drug Administration (FDA) and the International Conference on Harmonization (ICH).

2.1 ICH

The International Conference on Harmonization resulted from a cooperative effort by Japan, the United States, and Europe. The documents issued by ICH are produced and agreed on by the three regions, and firms can use the guidelines to comply with worldwide requirements. The guidelines are broadly written. The lack of specificity has the advantage of enabling improvements to be made as technology improves without revising license applications for therapeutic proteins. On the other hand, the lack of specificity often leads to different interpretations. Many other nations follow the ICH guidelines, which can be accessed on the internet [2]. Several ICH documents provide guidance on protein purification issues, some of which are listed in Table 1.

The *ICH guideline on viral safety* describes the use of scaled-down purification processes for demonstrating virus removal. It also describes

Table 1 Some Key Points from ICH Documents

Stability
Analysis of purification process
Raw materials
Reference materials
Consistency
Scaled-down purification processes

the need to ensure that chromatography columns and other devices used in the purification scheme perform consistently over time. This document notes that assurance should be provided that any virus potentially retained by the production system would be adequately destroyed or removed prior to reuse of the system [3].

Purification processes are also used to analyze genetic stability for recombinant DNA–derived protein products [4]. Protein analytical techniques are employed to demonstrate that the purified product is consistently produced from production cells expanded under pilot plant or full scale to the proposed in vitro cell age or beyond. The validation of these and other analytical methods is described in two ICH documents that present a description of the types of analytical procedures to be validated, validation characteristics that need to be validated, and methodology [5,6].

Analytical methods are also used to measure residual solvents. Particularly applicable for high-performance liquid chromatography (HPLC) is an ICH guideline that classifies residual solvents based on risk assessments. For other types of chromatography in which solvents that should be limited are employed, such as ethylene glycol, the guidelines should be consulted [7].

In addition to genetic stability, protein stability issues should be addressed. *The ICH guideline on stability testing of new drug substances and products* applies in general to biotechnology/biological products [8]. And the document *Stability Testing of Biotechnological/Biological Products* provides more information specifically for proteins, which are usually sensitive to temperature, oxidation, light, ionic content, and shear [9]. The type of information required to assess the stability of intermediates, drug substance (bulk material), and drug product is presented in the latter document.

The *ICH document on specifications of biotechnological/biological products* provides information on process controls, such as in-process acceptance criteria, action limits, and raw materials [10]. It is noted that the quality of the raw materials used in production should meet standards appropriate for their intended use. All raw materials should be evaluated for the absence of deleterious endogenous or adventitious agents. Procedures that make use of affinity chromatography should be accompanied by appropriate measures to ensure that such process-related impurities or potential contaminants arising from their production and use do not compromise the quality and safety of the drug substance or drug product. Reference materials for product-related substances, product-related impurities, and process-related impurities are often needed, and a description of their purification should be provided in an application to market the product.

2.2 FDA

The FDA publishes points to consider, guidance for industry, and letters, as well as regulations such as those in the Code of Federal Regulation (CFR). Most of the current regulatory documents are available on the FDA's web page [11]. The Freedom of Information Act (FOI) in the United States provides for the dissemination of information that can be quite useful in assessing the current regulatory thinking. FOI also provides insights into the problems that occur in the industry related to sanitization and cleaning for purification processes. Many of these problems are listed in FDA form 483s. Table 2 lists some of the key issues with regard to protein purification dealt with in the FDA documents.

Points to Consider

Two of the points-to-consider documents that describe protein purification are the *1995 Points to Consider in the Manufacture and Testing of Therapeutic Products Derived from Transgenic Animals* and the *1997 Points to Consider in the Manufacture and Testing of Monoclonal Antibody Products for Human Use* [12,13]. The document on transgenics advises that "it is possible that adventitious agents or chemical contaminants could enter the animals and be concentrated in the product by the purification procedure." It is also noted that the quality of the manufacturing processes requires product purity to be determined at multiple points using tests found to be sensitive to product structure and potency. The *1997 Points to Consider in the Manufacture and Testing of Monoclonal Antibody Products for Human Use* describes the need to employ production techniques that will prevent introduction of and eliminate impurities. These impurities include animal proteins and materials, DNA, endotoxin, other pyrogens, culture media constituents, components that may leach from columns, and viruses. This document notes that limits should be set prospectively on the number of times a purification component (e.g., a chromatography column) can be reused. There is also a recommendation that firms save retention samples

Table 2 Some Key Points from FDA Documents

Potential for concentration of impurities
Analysis of purification processes
Elimination of impurities
Lifetime determination for chromatography columns
Prevention of cross-contamination
Cleaning and sanitization

from each production under appropriate conditions so that side-to-side comparisons may be made to determine product comparability.

Guidance for Industry

A series of documents entitled *Guidance for Industry* addresses the content and format of the Chemistry, Manufacturing, and Controls (CMC) section of *Biologics License Applications* (BLA) [14]. These documents describe the necessity of providing a detailed description of the purification and downstream processing, including a rationale for the chosen methods and the precautions taken to ensure containment and prevention of contamination or cross-contamination. Any equipment that is used for more than one product should also be described. In-process controls are also described in the CMC section, and in-process bioburden and endotoxin limits should be specified where appropriate. The validation of the purification process to demonstrate removal of chemicals used for purification, column contaminants, endotoxin, antibiotics, residual host cell proteins, DNA, and viruses, where appropriate, should be provided. This document also notes that raw materials and reagents should be listed. For special reagents (e.g., a monoclonal antibody used in affinity chromatography), a detailed description of the preparation and characterization of the reagent should be submitted.

One of the more confusing issues is what to do when. A preliminary draft, *Guidance on the CMC content and format of INDs (Investigational New Drugs) for Phases 2 and 3*, provided an outline of the FDA's expectations for clinical trials. Although this document has been updated, the older version provides more specifics [15,16]. For example, it was noted that for Phase 1 a brief description of the extraction/purification process would be sufficient. This should be updated in Phase 2. But by Phase 3 there should be a step-by-step description of the process and validation of the extraction and purification process to remove viruses and other impurities.

Another draft, *Guidance for Industry*, addresses the manufacturing, processing, or holding of active pharmaceutical ingredients (APIs) [17]. This guidance addresses equipment construction and notes that surfaces that contact raw materials, intermediates, or APIs should not be reactive, additive, or absorptive so as to alter the quality and purity of the API. Equipment should be designed, constructed, and installed to allow for ease of cleaning and, as applicable, for sanitization. Written procedures should be established and followed for cleaning and maintaining equipment, including storage vessels. The document also notes that clean equipment should be protected and inspected prior to use. A maximal time that may elapse between the completion of processing and equipment cleaning, the choice of cleaning methods, cleaning agents, and levels of cleaning should be established and justified. Residue limits should be practical, achievable, and verifiable. Compatibility with equipment construction materials must also be considered, as well as cleaning agent residues and the ability of the cleaning agent to remove residues. Cleaning validation should employ validated analytical methods sufficiently sensitive to detect residuals. Microbial and endotoxin contamination should be addressed in the equipment cleaning and sanitization studies.

FDA 483 and Other FOI Documents

Lifetime of chromatography media is addressed in several documents. In one BLA review letter, a firm was requested to provide supporting data demonstrating the performance of the column over its proposed lifetime by including an evaluation of the removal of contaminants from the harvest, the cleaning validation and sanitization effectiveness, and bacteriostatic performance. It was also noted that actual processing conditions should be used, including protein load and flow rates.

Another firm was asked to develop and use a host cell protein assay to complete their validation study to extend chromatography column usage. The assay was to be employed to determine the relative level of host cell proteins in the in-process product pool batches from two chromatography steps. These batches were then to be compared to earlier batches to justify the extension of the column lifetime. In one company, lifetime studies included an evaluation of protein load capacity; elution profiles, host cell proteins, media components, DNA, and viral reduction.

During an establishment inspection, one firm was cited for using different conditions in the lifetime studies and in manufacturing. Buffer conditions for elution were modified, and the frequency of cleaning and sanitization was increased during manufacturing, thus invalidating the lifetime study.

In-process analysis is also an issue addressed in approval letters and during inspections by regulatory authorities. A firm agreed to submit the results of hold period studies for in-process intermediates, including biochemical, bioburden, and endotoxin studies that were to be performed on a periodic basis to demonstrate stability of the intermediates. In addition to product-intermediate analysis, firms have been asked to show that storage conditions and buffers routinely maintain a bacteriostatic effect for purification columns by monitoring bioburden for every column run. Firms have been told to provide in-process bioburden specifications based on manufacturing history, and to establish routine bioburden monitoring alert and action limits for production. One firm was cited for not having provided validation of hold periods of buffers used for purification.

Cleaning of equipment and purification devices is addressed in several FDA 483 documents. In one document it was noted that there was discoloration of a column, the discoloration was not removed during regeneration, and the firm had not identified the nature of the discoloration. Another firm was cited for neither characterizing nor evaluating the impact of carry-over proteins. One FDA inspector observed that there was no periodic monitoring of chromatography columns following cleaning. The lack of cleaning validation for multiuse equipment has also been noted.

3. ACHIEVING COMPLIANCE WITH THE REGULATIONS

Based on the topics of the regulatory documents, it is clear that in purification of protein therapeutics and diagnostics, there are multiple factors to consider. Five of these factors are discussed below. It is necessary to start out with suitable raw materials, remove impurities to the necessary level warranted by the product's intended use and dose, keep retention samples, avoid contamination, and use good cleaning and sanitization protocols.

3.1 What Are Suitable Raw Materials?

As pointed out in the regulatory documents, raw materials should meet standards appropriate for their intended use. If purification processes ultimately intended for manufacturing are designed in research laboratories, it is important to understand the quality of the raw materials. How can one design a purification process if the detergent contains unknown impurities, buffer salts have been contaminated with endotoxins, or the chromatography media contain microorganisms with proteases? Even if the resulting product is not for human use, the purity of raw materials is important. If, for example, the product is to be used for safety testing, how will a firm know the cause if the animals lose weight? Is the weight loss due to the product or to impurities that have been copurified? Even if a protein product is not a diagnostic or a therapeutic, if it is not produced consistently due to unsuitable raw materials or any other cause, the results of the analytical testing can be difficult to interpret. Certainly, for applications such as crystallography, the purity is critical.

Raw materials from animal sources pose a potential risk from viruses and transmissible spongiform encephalopathies (TSEs) that have the potential to infect humans. While most of the known raw materials with these risks are used in cell culture (e.g., bovine serum, serum replacements of animal origin, growth factors in buffer containing bovine serum albumin) and cell dissociation (e.g., trypsin, elastase), others are used in purification (e.g., gelatin). It is essential to ensure that when these products are used they are derived from sources shown to be free from viruses and TSE-causing agents.

3.2 What Degree of Purity Does One Need to Achieve?

Claims of 99.99% purity are often misleading, since the claim is only as good as the analytical method employed. Using sodium dodecyl sulfate–polyacrylamide gel electrophoresis (SDS-PAGE) with Coomassie blue stain, it is quite possible to achieve such a claim, but the claim is not very meaningful. Today's analytical techniques include highly sensitive methods such as mass spectrometry, carbohydrate analysis, and biosensors. These methods allow us to perform extensive purity testing during processing and on the final product. The product indication, dose, and patient population determine the final product purity requirements. It is therefore quite difficult when starting out to state how pure the final product must be. But we can determine what impurities must be removed based on the product source.

Impurities are derived from the process and are dependent on the feedstream source and the processing materials. Aggregates formed during processing will have to be removed during the final purification steps. Inconsistent purification may lead to different glycoform patterns that can affect product potency. Leachables from chromatography media and filters are considered impurities. Contaminants, on the other hand, are those substances that enter the process due to a breakdown in good manufacturing practices (GMPs) and may include microorganisms, adventitious viruses, and endotoxins.

Impurities are sometimes unknown. For example, if we use a natural source as starting material (e.g., blood or plasma), we may not know exactly what it contains. While we increasingly gather information about previously unknown impurities, such as newly found viruses, it is the unknown that poses the greatest risk. Therefore, for products such as those from blood, we need to ensure that the purification process will provide us with assurance that any unknown agents will be removed and/or inactivated. On the other hand, for products derived from sources such as *Escherichia coli* or yeast, there is quite a bit of knowledge, the source is well controlled, and the risk of harmful effects to the patient is well understood. The production processes for products from such sources usually do not require inactivation steps, and fewer purification steps may be necessary to achieve the requisite purity. Regardless of the source, it is common to see three or more purification steps for production of therapeutic proteins. Risks associated with sources such as *E. coli* include host cell proteins and endotoxins. Host cell proteins may cause immunogenic reactions. Endotoxins can cause toxic shock. In terms of risk, Chinese hamster ovary (CHO) cells fall between blood and *E. coli*, and have the potential to contaminate a product with both known endogenous retroviruses and adventitious viruses. Table 3 illustrates some of the potential impurities associated with various protein product sources.

3.3 Why Keep Retention Samples?

In order to achieve consistency in protein purification, it is necessary to analyze the impact of process changes. Raw materials, environmental factors, and processing conditions may affect protein purity. Retention samples should be kept from each step of the process for future comparisons. These samples must be properly stored and their stability determined. When a change is made, analysis of the change can be performed by using orthogonal analytical techniques that facilitate evaluation of the impact of the change. For example, we might use carbohydrate analysis and mass spectrometry to evaluate the effect of a new cell culture media on an ion exchange step used to purify a monoclonal antibody.

When changes are made to a purification process, the removal of specific impurities may be reduced. Of course, it is also possible to make changes that provide a higher degree of purity. But purer is different. For example, if we remove one glycoform from a glycoslyated protein, the biological activity can be affected. Therefore, as changes are made to a purification process, it is essential to asses the impact of the changes on safety and potency as well as purity.

Table 3 Some Impurities Associated with Commonly Used Protein Product Sources

	Source				
Potential impurities	Mammalian cell culture	*E. coli*	Yeast	Insect cells	Blood
Virus	√			√	√
Host cell proteins	√	√	√	√	
Endotoxins		√			
DNA[a]	√	√	√	√	
Protein product variants	√	√	√	√	√

[a]DNA is considered less of a risk than previously thought with up to 10 ng/dose accepted by WHO and Europe for previously used continuous cell lines.

In 1996, comparability protocols were introduced in the United States for licensed products [18]. A firm may write a comparability protocol for a specified change and agree to adhere to a series of planned activities with specified results. If the FDA approves the protocol, the firm can make the changes and reduce their reporting level to the FDA, thereby saving time for implementation of the change—sometimes by as much as 6 months. It is essential that the firm follows the protocol and obtain the specific results. As of June 1999, the FDA approved more than a dozen protocols. Improved changes have included changes in load, scale, and optimization of chromatography conditions [19]. It is noteworthy that changes in glycosylation are considered beyond the scope of comparability protocols. Although comparability protocols apply only to licensed products, comparability must also be demonstrated as a firm progresses from safety testing to the clinic and licensure. Retention sample are essential for evaluating changes and demonstrating comparability.

3.4 How Can We Avoid Contamination?

Prevention of contamination during purification requires attention to good process hygiene; control of air, water, and raw material quality; adherence to standard operating procedures (SOPs); and maintenance of change control. Good hygiene includes the use of suitable cleaning and sanitization protocols. While some firms have stated that they can tolerate a certain amount of bioburden during early purification steps, others have found a deleterious effect from the presence of microorganisms. The potential impact of bioburden should be assessed for each process and each in-process hold step [20]. Microorganisms can release proteases and toxins. While the presence of some bioburden might not affect a protein diagnostic, it has the potential to result in inaccurate diagnoses. Figure 1 illustrates aglycosylation that can occur when microorganisms are present. Above is a chromatogram from a noncontaminated glycosylated protein, while below is the chromatogram from a low-level contamination (less than 10 cfu/ml). The sample containing microbial contamination showed a significant increase in aglycosylated species.

3.5 How Do We Achieve Suitable Cleaning and Sanitization?

Prior to designating cleaning and sanitization protocols, it is important to determine what components of the purification system we are trying to clean and sanitize; what we are trying to remove; if the items being treated are compatible with the cleaning/sanitizing agents; what level of cleanliness is required; what assays are available to determine the level of cleanliness;

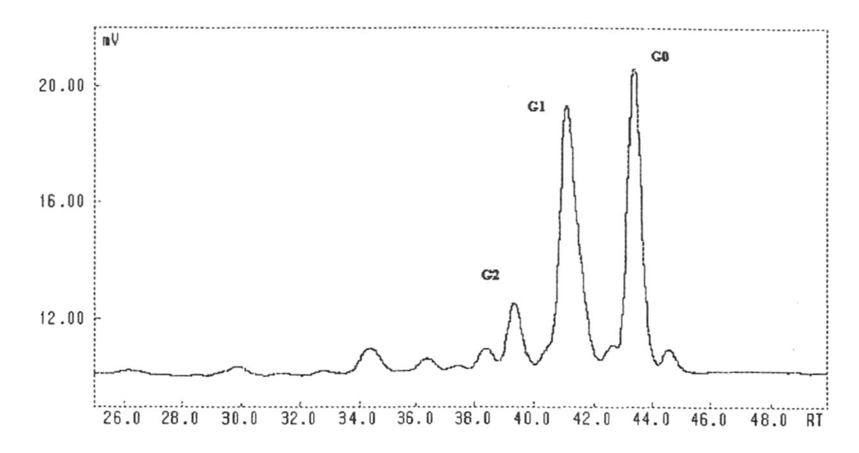

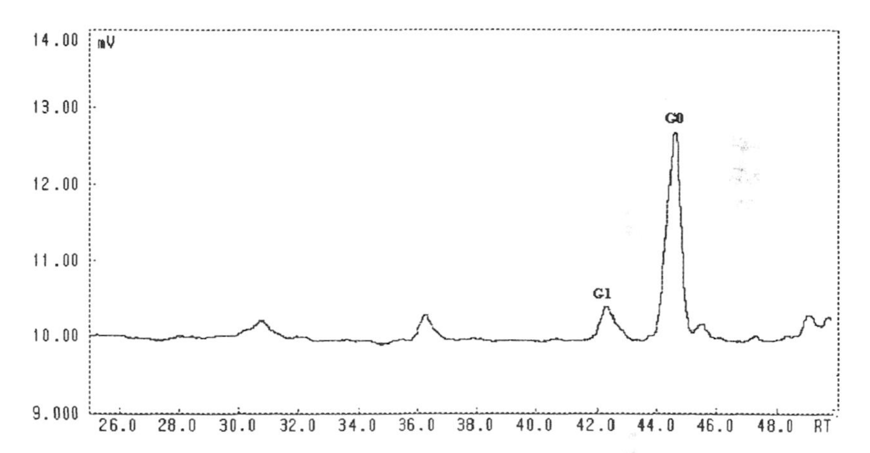

Figure 1 Normal glycoform pattern (above). Agalactosylated species (below) resulting from low-level bioburden. (Courtesy of R. Francis.)

and how we can validate removal of any cleaning and/or sanitizing agents. Cleaning validation must also be addressed.

Cleaning

For purification and recovery operations, we have to ensure cleanliness of storage and buffer tanks, chromatography and filtration systems, chromatography columns, chromatography media, and filters.

Removal of Specific Impurities. The cleaning protocol should be designed to remove specific impurities such as proteins, nucleic acids, endotoxin, and virus. Whereas high salt might be suitable for cleaning hydrophilic proteins from hydrophilic chromatography media, hydrophobic proteins will only bind more firmly, making additional cleaning steps more difficult. Even hydrophilic molecules might bind through hydrophobic interactions to a surface. It is, therefore, necessary to evaluate not only the cleaning agents but also the sequence in which they are used.

Nucleic acids often bind tightly to positively charged anion exchangers [21]. In fact, it appears that nucleic acids might be bound irreversibly in some cases. While this does not seem to present a problem for consistent purification of proteins, it is an issue to be considered when determining capacity of a chromatography column. If all the available capacity is used, a decline in product or intermediate purity may be seen over time due to small amounts of irreversibly bound nucleic acids.

Endotoxins are inactivated by sodium hydroxide, but it is useful to know if they are also removed. In one study, a high load of endotoxin was used to challenge a chromatography system and packed anion exchanger. Results from both the *Limulus amoebocyte* lysate (LAL) and total organic carbon (TOC) assays showed that there was neither biological activity of the endotoxin, nor were the carbon fragments detected [22]. Typically, removal of endotoxins from chromatography columns is monitored prior to each use with the LAL assay. One firm was specifically asked by FDA to "provide data on endotoxin level and contaminant protein level on product lots made using regenerated resin."

Although viral inactivation by sodium hydroxide and other virucidal agents has been demonstrated for chromatography columns, removal of inactivated viral particles is an issue that is poorly understood. The use of quantitative polymerase chain reaction may enhance our ability to understand if virus particles are removed as well as inactivated during cleaning and sanitization [23].

Sufficiently sensitive assays for proteins, nucleic acids, endotoxins, and viruses, as well as other impurities related to the product source and processing agents, are essential for determining and validating cleaning efficiency. Some useful assays include TOC, total protein assays, SDS-PAGE, UV, and product-specific assays. For each piece of equipment, filter, or chromatography resin used in purification, appropriate assays should be selected. Suitable assays enable us to determine the operating conditions that ensure consistent cleaning. Conditions such as contact time, flow rate, temperature, and cleaning agent concentration must be defined within reasonable ranges that can be used in manufacturing. It is essential to define the maximal time that will elapse after use and prior to cleaning, and removal of cleaning agents must be demonstrated. All equipment must be labeled to define whether it is clean or dirty. An acceptable residual limit for any remaining impurities or cleaning agents must be established by performing a risk assessment that considers dose and patient safety or accuracy of a diagnostic protein product.

Sampling of a chromatography resin is illustrated in Fig. 2. The cleaning efficacy for this anion exchanger was evaluated by testing for bioburden, endotoxin, small-ion capacity, and residual product. In some cases, it can be valuable to extend the cleaning contact time or the concentration of the cleaning agent to see if additional soil is removed. This should be done with small-scale models, not production columns. Compatibility with the components to be cleaned should be evaluated in development.

After the cleaning protocol is established, it must be validated, typically during manufacture of Phase III clinical trials. For packed chromatography columns, cleaning validation can be performed by assaying rinse water. For equipment, swabbing and the use of coupons are common practices. Coupons are cut-out pieces of equipment, which can be treated with a heavy soil, even baked onto the surface. The cleaning SOP is then applied to ensure its effectiveness. Visual inspection is also required for equipment. Where feasible, the use of disposable equipment can eliminate the need for cleaning validation in early clinical trials.

Chromatography columns often become discolored during use. This has become an issue during some GMP inspections. It is up to the firm to demonstrate that the discoloration does not have an impact on the product quality. This includes evaluating, to the best of one's ability, the source of discoloration and its composition. Furthermore, one should assess if the discoloration can be removed by a more stringent cleaning protocol, if it is carried over into the next production run, and its potential to have a detrimental effect on the product.

Although chromatography columns and filters are dedicated to one product, equipment is often campaigned, i.e., used for one product, cleaned, and used again for another product. Multiuse equipment must be cleaned and cleaning validation performed every time there is a product changeover. It must be demonstrated that there is no carry-over from one product to the next. Demonstrating the absence of carry-over requires highly sensitive assays. Each situation must be evaluated for risk associated with the previously run product along with its associated impurities and processing agents. Additional information on cleaning validation for multiuse facilities is presented by Sherwood [24]. Additional information on cleaning and cleaning validation is provided in the book, *Cleaning and Cleaning Validation: A Biotechnology Perspective* [25].

PURPOSE: To demonstrate chromatography resin through direct sampling and testing.

SCOPE: This protocol applies to the sampling and testing of resin only.

PROCEDURE:

1. When chromatography column xxx is opened for planned or unplanned reslurry, before resin is slurried, take the following 15 mL samples:
   1.1. Four samples, one each separated by 90 degrees from each other (samples 1-4).
   1.2. Two samples from the center (samples 5 & 6).
   1.3. One sample from any region that appears to have a darker or different color from the rest of the column. Indicate the location of these samples on attachment I (samples 7a-7x).
2. After the resin is slurried, take six 15 mL samples. Take three samples (samples 8-10) in polypropylene tubes, and three in polystyrene tubes (samples 11-13).
3. Testing
   3.1. Test samples according to the table.

Sample No.	Bioburden	Endotoxin	Small ion capacity	Product sensitive
1-4			X	X
5	X			
6			X	X
7	X		X	X
8	X			
9		X		
10			X	X
11	X			
12		X		
13			X	X

Figure 2 Sampling of a chromatography resin. (Courtesy of S. Rudge.)

Sanitization

As noted earlier in this chapter, the presence of microbial organisms can result in contamination with proteases, toxins, and other substances that may have an adverse effect on a patient or a protein product. However, with few exceptions, protein purification operations are classified as aseptic, not sterile. The few exceptions include immobilized monoclonal antibody columns that are time consuming and expensive to develop and may not tolerate the harsh sanitization conditions commonly used in purification steps. Bioburden specifications should be based on process experience. As noted by one FDA investigator, firms often try to set their bioburden specifications higher to avoid a failure. If a firm is operating in a range and then sets the specifications 25% higher, it will not be apparent how the high level of bioburden will affect product quality. Gram-negative microorganisms, for example, can contaminate a product with endotoxins; whereas gram-positive microorganisms may lead to enterotoxin contamination. Filters used in purification operations can retain microorganisms, but the toxins and other potentially harmful substances have the potential to pass through the commonly used 0.2- and 0.45-μm filters. The products of the microorganisms could be concentrated along with a protein product.

It is generally unnecessary to perform challenge studies with microorganisms for purification systems. Vendors sometimes provide data that can support the justification of the use of a particular sanitizing agent. But each firm must perform routine analysis of bioburden, employing their own personnel, environment, and SOPs. In a 1998 approval letter, the FDA asked a firm to "institute for every column run bioburden monitoring of the ion exchange column storage solution to ensure that storage conditions and storage buffer routinely maintain a bacteriostatic effect." In another comment available from FOI, it was noted that "an in-process bioburden and endotoxin specification should be established base on manufacturing history, not a failure in GMPs."

4. CONCLUSIONS

Protein purification processes are relied on to provide safe therapeutic products, accurate diagnostics, and reliable research tools. For regulated products such as therapeutics and diagnostics, it is important to determine which regulations are applicable and stay current with the latest regulatory opinions and developing technologies for processes and their analysis. Some of the important issues that should be considered are the quality of raw materials, level of purity required, retention samples, cleaning, and sanitization. Suitable cleaning and sanitization protocols should be developed very early in a purification process and refined with process experience if necessary. Robust cleaning and sanitization methods aid in providing consistently pure proteins.

REFERENCES

1. www.eudra.org/en_home.htm.
2. www.ifpma.org/ich.
3. ICH. Quality of Biotechnological Products: Viral Safety Evaluation of Biotechnology Products Derived from Cell Lines of Human or Animal Origin, 1997.
4. ICH. Quality of Biotechnological Products: Analysis of the Expression Construct in Cells Used for Production of r-DNA Derived Protein Products, 1995.
5. ICH. Text on Validation of Analytical Procedures, 1994.
6. ICH. Validation of Analytical Procedures: Methodology, 1996.
7. ICH. Impurities: Guidelines for Residual Solvents, 1997.
8. ICH. Stability Testing of New Drugs and Products Q1A and Q1A (R)., 1993 and 1999.
9. ICH. Stability Testing of Biotechnological/Biological Products, 1995.
10. ICH. Specifications: Test Procedures and Acceptance Criteria for Biotechnological/Biological Products, 1999.
11. www.fda.gov.
12. US Department of Health and Human Services; Food and Drug Administration; Center for Biologics Evaluation and Research. Points to Consider in the Manufacture and Testing of Therapeutic Products for Human Use Derived from Transgenic Animals, 1995.
13. US Department of Health and Human Services; Food and Drug Administration; Center for Biologics Evaluation and Research. Points to Consider in the Manufacture and Testing of Monoclonal Antibody Products for Human Use, 1997.
14. US Department of Health and Human Services; Food and Drug Administration; Center for Biologics Evaluation and Research; Center for Drugs Evaluation and Research. Guidance for Industry for the Submission of Chemistry, Manufacturing, and Controls Information for a Therapeutic Recombinant DNA-Derived Product or a Monoclonal Antibody Product for In Vivo Use, 1996.
15. US Department of Health and Human Services; Food and Drug Administration; Center for Biologics Evaluation and Research; Center for Drugs Evaluation and Research. Preliminary Draft Guidance for Industry CMC Content and Format of INDs for Phases 2 and 3 Studies of Drugs, Including Specified Therapeutic Biotechnology-Derived Products, 1997.
16. US Department of Health and Human Services; Food and Drug Administration; Center for Biologics Evaluation and Research; Center for Drugs Evaluation

and Research. Guidance for Industry INDs for Phases 2 and 3 Studies of Drugs, Including Specified Therapeutic Biotechnology-Derived Products, CMC Content and Format Draft, 1999.
17. US Department of Health and Human Services; Food and Drug Administration; Center for Biologics Evaluation and Research; Center for Drugs Evaluation and Research and Center for Veterinary Medicine. Guidance for Industry Manufacturing, Processing, or Holding Active Ingredients Draft, 1998.
18. US Department of Health and Human Services; Food and Drug Administration; Center for Biologics Evaluation and Research and Center for Drugs Evaluation and Research. Guidance Concerning Demonstration of Comparability of Human Biological Products Including Therapeutic Biotechnology-Derived Products, April 1996.
19. S Moore, C Joneckis, Oral Presentation. Strategic Use of Comparability Studies and Assays for Well-Characterized Biologicals. IBC Second International Conference, Arlington, VA, June 1999.
20. ME Winkler. Purification issues. In: Biopharmaceutical Process Validation. G Sofer, D Zabriskie, eds. New York: Marcel Dekker, 2000, pp. 143–155.
21. Y Dasarathy. A validatable cleaning in-place protocol for total DNA clearance from an anion exchange resin. BioPharm. 9:41–44, 1996.
22. G Sofer, L Hagel. Handbook of Process Chromatography. London: Academic Press, 1997, pp. 160–161.
23. M Wiebe. The application of quantitative PCR assays to evaluate the ability of product purification steps to clear process impurities and potential contaminants. Oral Presentation. Second Annual Validation of Biologics Conference, September 1999, San Francisco, Institute for International Research.
24. D Sherwood. Cleaning: multiuse facility issues. In: Biopharmaceutical Process Validation. G Sofer, D Zabriskie, eds. New York: Marcel Dekker, 2000, pp. 235–249.
25. PDA Biotechnology Cleaning Validation Subcommittee. Cleaning and Cleaning Validation. Bethesda, MD: PDA, 1996.

Index